Finite Elements in Water Resources

Finite Elements in Water Resources

Proceedings of the Second International Conference on Finite Elements in Water Resources held at Imperial College, London, in July, 1978

Edited by
C. A. Brebbia, *Southampton University*
W. G. Gray and G. F. Pinder, *Princeton University*

PENTECH PRESS
London : Plymouth

First published, 1978
by Pentech Press Limited
Estover Road, Plymouth, Devon

ISBN 0 7273 0602 2

British Library Cataloguing in Publication Data

International Conference on Finite Elements
in Water Resources, *2nd, London, 1978*
Finite elements in water resources.
1. Water resources development – Mathematics – Congresses
I. Title II. Brebbia, Carlos Alberto
III. Gray, William G IV. Pinder, George F
628.1'61'01515353 TC409

ISBN 0–7273–0602–2

Printed in Great Britain by
Billing & Sons Limited, Guildford, London and Worcester

Contents

1. GROUND WATER FLOW

Application of the Finite Element Method to the Numerical Analysis of a Leaky Aquifer 1.3
L. Rodarte

Finite Element Models for Simultaneous Heat and Moisture Transport in Unsaturated Soils 1.19
R.G. Baca, I.P. King and W.R. Norton

Finite Elements Solution of Unsaturated Porous Media Flow 1.37
C. Tzimopoulos

A Galerkin-Finite Element Simulation of Solute Transport in Subsurface Drainage Systems 1.51
M.A. Marino and G.B. Matanga

Simulation of One-Dimensional Unsteady Drainage in Porous Media 1.69
F. Stauffer

Flow to Large Diameter Bottom Entry Wells 1.87
C.R. Dudgeon and R.J. Cox

Free Surface Seepage Flow in Earth Dams 1.103
I. Kazda

Unsteady Groundwater Regime During Dry-Dock Building 1.121
I. Seteanu and R. Popa

A Quantitative Comparison between Finite Element Solution and Experimental Result of Transient Unconfined Groundwater Flow 1.129
G. Ohashi

GROW1, A Program for Anisotropic Soil Fluid Flow with Free or Artesian Surface. Application to Excavations at Lock of Zeebrugge (Belgium) 1.143
J.F. Rammant and E. Backx

Automatic Computing of a Transmissivity Distribution Using Only Piezometric Heads 1.157
A. Yziquel and J.C. Bernard

Groundwater Flow Simulation Using Collocation Finite Elements 1.171
G.F. Pinder, E.O. Frind and M.A. Celia

Unsteady Flow in Porous Media Solved by Combined Finite Element Method of Characteristics Model 1.187
A.A. Hannoura and J.A. McCorquodale

A Finite Element 'Discrete Kernel Generator' for Efficient Groundwater Management 1.195
T. Illangasekare and H.J. Morel-Seytoux

Finite Element Description of Flow Field in Groundwater Management Models 1.213
K. Elango and H. Suresh Rao

Two Techniques Associated with the Galerkin Method for Solving Groundwater Flow Problems 1.227
S.P. Kjaran and S.T. Sigurdsson

Finite Element Analysis of Ground Water Flow and Settlements in Aquifers confined by Clay 1.249
K. Runesson, H. Tägnfors and N.E. Wiberg

2. SURFACE WATER FLOW

A Two-Dimensional Hydrodynamic Model of a Tidal Estuary 2.3
R.A. Walters and R.T. Cheng

Finite Element Simulation of Shallow Water Problems with Moving Boundaries 2.23
D.R. Lynch and W.G. Gray

Behaviour of a Hydrodynamic Finite Element Model 2.43
R.A. Harrington, N. Kouwen and G.J. Farquhar

Development of Generalized Free Surface Flow Models Using Finite Element Techniques 2.61
D.M. Gee and R.C. MacArthur

Recent Application of RMA's Finite Element Models for Two Dimensional Hydrodynamics and Water Quality 2.81
I.P. King and W.R. Norton

A Water Prediction System by the Finite Element Method 2.101
Y. Matsuda

Rayleigh-Ritz Formulation of Problems in Sedimentology 2.119
A. Prakash

Tsunami Wave Propagation Analysis by Finite Element Method 2.131
M. Kawahara, S. Nakazawa and S. Ohmori

Waves Generated by Landslide in Lakes or Bays 2.151
C.H.G. Koutitas and Th. S. Xanthopoulos

Long Wave Simulation Using a Finite Element Model 2.159
J.R. Houston and D.G. Outlaw

Computation of Stationary Water Waves Downstream of a Two-Dimensional Contraction 2.177
P.L. Betts

3. FLUID MECHANICS

A Finite Element Model of Turbulent Flow in Primary Sedimentation Basins 3.3
D.R. Schamber and B.E. Larock

Simulation of Stratified Turbulent Flows in Closed-Water Bodies Using the Finite Element Method 3.23
A.N. Findikakis, J.B. Franzini and R.L. Street

Solution of the Time-Dependent Navier-Stokes Equations Via F.E.M. 3.45
P.M. Gresho, R.L. Lee, T.W. Stullich and R.L. Sani

Optimal Control of the Boundary Layer Equations by the Outer Velocity Using Finite Element Method 3.65
G. Assassa, C.M. Brauner and B. Gay

Stresses in Oscillatory Converging or Diverging Flow by Finite Element Simulation 3.85
M. Durin and J. Ganoulis

The Numerical Treatment of Free Surface Flows by Finite Elements 3.97
H.J. Diersch and H. Martin

4. NUMERICAL TECHNIQUES

A Finite Element Method for the Diffusion-Convection Equation 4.3
E. Varoglu and W.D.L. Finn

Solution of the Convection-Diffusion Equation Using a Moving Coordinate System 4.21
O.K. Jensen and B.A. Finlayson

Variable Domain Finite Element Analysis of Unsteady Compressible Fluid Flow Problems 4.33
G. Van Goethem

A Time-Split Finite Element Algorithm for Environmental Release Prediction 4.53
A.J. Baker, M.O. Soliman and D.W. Pepper

Applications of Boundary Elements in Fluid Flow 4.67
C.A. Brebbia and L. Wrobel

Integral Equation Solutions to Non-Linear Free Surface Flows 4.87
P.L-F. Liu

Finite Element Formulations of Flows with Singularities 4.99
J.E. Akin

Comparison of Finite Element and Finite Difference Methods in Thermal Discharge Investigations 4.113
L.D. Spraggs

Numerical Smoothing Techniques Applied to Some Finite Element Solutions of the Navier-Stokes Equations 4.127
R.L. Lee, P.M. Gresho and R.L. Sani

A Finite Element Study of Large Amplitude Water Waves 4.147
P.L. Betts and B.L. Hall

A Digital Simulation Approach for a Tracer Case in Hydrological Systems (Multi-Compartmental Mathematical Model) 4.165
Y. Yurtsever and B.R. Payne

PREFACE

The First International Conference on Finite Elements in Water Resources was held at Princeton University, U.S.A. in July 1976, and attracted participants from many countries. This was the first meeting specifically designed to bring together the many engineers and scientists who are carrying out research or applying finite element techniques to water resources problems. The Proceedings of this Conference now published in book form, presented the participants with an exceptional 'state of the art'. Because of the success of this meeting it was decided to hold the Second International Conference on Finite Elements in Water Resources in London, England in July 1978.

Like the first, the Second International Conference brought together engineers and scientists who are applying finite elements to water resources problems. The Conference Proceedings includes papers dealing with Ground Water and Surface Water Flow, Fluid Mechanics and Numerical Techniques. Several papers deal with problems related to Marine and Coastal Waters and the topic of flow through porous media. Problems related to numerical simulation were treated in a special session, during which the application of boundary elements, the possibility of using moving coordinate grids and problems related to accuracy of the solutions were discussed. The meeting provided a forum for the discussion of recent advances and the applicability of the technique.

As a result of the success of the meeting the Organizing Committee plans to hold a third International Conference in July 1980.

The Editors

S E S S I O N 1

GROUND WATER FLOW

APPLICATION OF THE FINITE ELEMENT METHOD TO THE NUMERICAL ANALYSIS OF A LEAKY AQUIFER

Leopoldo Rodarte

Universidad Autónoma Metropolitana, Unidad Iztapalapa, México 13, D. F.

Introduction

Some problemas related to non-stationary flow in porous media can be analyzed by analytical methods only in those cases where the caracteristics of the media and the boundary conditions are simple enough. In other cases, where analytical mathematics are not applicable, analogous or digital models are used to solve these problems. In practical geohidrological problems it is frequent to find complex conditions such as non-homogeneous and anisotropic media, irregular geometry of the studied region, etc. To solve these types of problems, appropriate numerical methods are required.

The finite element method is a variational numerical method well suited for the analysis of geohidrological problems. This method has been used to solve non-stationary heat flow problems (Wilson and Nickell, 1966) as well as in Geohidrology (Taylor and Brown, 1967, Javandel and Witherspoon, 1968, 1968b, 1969).

The following are some of the most important advantages of this method:

1. It is not necessary to develop special expressions to define the boundary conditions. However, it is essential when the finite differences method is used (Remson, Hornebeger and Molz, 1971). This is important since most of the geohidrological regions subject to study have complex boundary conditions.

2. The dimensions of the elements can be fixed arbitrarily. Small elements are chosen in regions, such as in the vecinity of a well, where a function changes rapidly. On the other hand, where the function changes gradually, larger elements

may be used.

3. Compared with finite differences, this method permits a closer agreement with complex real boundaries.

4. The finite element method makes it possible to set up equations that produce symetric positive-definite matrices which can be reduced to band form and solved with a minimum of storage and computation time. (Javandel and Witherspoon, 1968a).

5. The cases of anisotropic and nonhomogeneous media, can be easily considered in the calculations.

In the following analysis, the numeric solution is obtained by the finite element method in accordance with Zienkiewicz's theory (1967) for non-stationary heat flow problems.

Theory

Let us consider a leaky aquifer, like the one shown in figure 1, formed by a main aquifer and two low permeability layers with a significant capacity of elastic storage, which can not be disregarded. In the main aquifer, a non-stationary yield

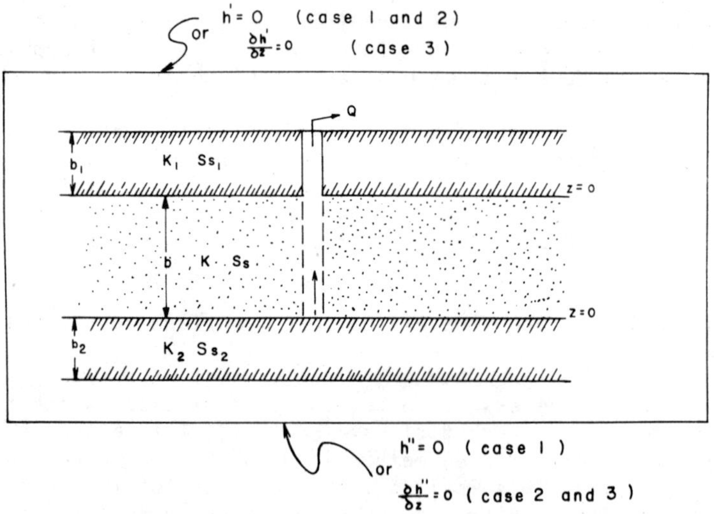

Figure 1. Sketch of the leaky system

Q is extracted through a system of wells. According to Herrera and Figueroa (1969), the following differential equation de fines the behaviours of a leaky aquifer for large values of time in anisotropic or nonhomogeneous porous media:

$$\frac{\partial}{\partial x}\left(Txx\frac{\partial s}{\partial x}\right) + \frac{\partial}{\partial y}\left(Tyy\frac{\partial s}{\partial y}\right) - ds + Q = \frac{1}{Vc}\frac{\partial s}{\partial t} \qquad (1)$$

In Equation (1), Txx and Tyy represent the transmissibilities of the main aquifer in the x and y directions respectively d and Vc are parameters that depend on the boundary conditions in the vertical plane as well as on the geohidrological characteristics of the low permeability layers.

For the boundary conditions shown in figure 1 these parameters are defined as follows:

Case 1 $\quad Vc = \dfrac{3}{3b\,Ss + b_1\,Ss_1 + b_2\,Ss_2}; \quad d = \dfrac{K_1}{b_1} + \dfrac{K_2}{b_2}$

Case 2 $\quad Vc = \dfrac{1}{b\,Ss + b_1 Ss_1 + b_2 Ss_2}; \quad d = 0$

Case 3 $\quad Vc = \dfrac{3}{b\,Ss + b_1\,Ss_1 + b_2\,Ss_2}; \quad d = \dfrac{K_1}{b_1}$

As previously indicated, Equation (1) is valid for large values of time. A detailed analysis of the elastic theory for a leaky aquifer can be found in Hantush (1960, 1964), Herrera and Figueroa (1969).

The equations for small and intermediate values of time can be found in Herrera and Rodarte (1973), Rodarte (1976, 1978).

Numerical Analysis

Application of the Finite Element Method.

For any given time $t=t_\circ$ let us define the functional

$$\chi\left(s, \frac{\partial s}{\partial x}, \frac{\partial s}{\partial y}\right) = \iint_e \left\{ \frac{Txx}{2}\left(\frac{\partial s}{\partial x}\right)^2 + \frac{Tyy}{2}\left(\frac{\partial s}{\partial y}\right)^2 - \left[Q - \frac{1}{Vc}\frac{\partial s}{\partial t} - \frac{1}{2}\,ds\right] s \right\} dx\,dy \qquad (2)$$

That is, $\frac{\partial s}{\partial t}$ is considered as an invariant. It can be shown that the functional (2) is equivalent to the differential equation (1) for those cases in which the horizontal boundaries are either impermeable or formed by contours with constant piezometric head (Kantorovich and Krilov, 1949)

Region R, where the problem will be analyzed, is divided in triangular elements, as shown in figure 2. The contribution of each element to the functional χ will be represented by χ^e. Then:

$$X = \Sigma X^e$$

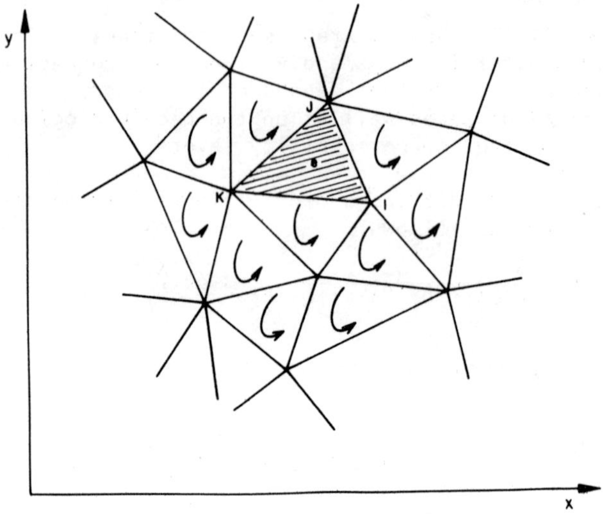

Figure 2. Division of a two-dimensional region into triangular elements

Let us assume that the nodes of the integration region are consecutively numbered from 1 to n'. If the elements are small enough the drawdown in each element can be expressed by means of the lineal function.

$$s = a + bx + cy \tag{3}$$

For the typical triangle with nodes j, k, 1, taken in a defined order (fig. 2) the drawdown in each node is calculated by the expressions

$$\begin{aligned} s_j &= a + bx_j + cy_j \\ s_k &= a + bx_k + cy_k \\ s_l &= a + bx_l + cy_l \end{aligned} \tag{4}$$

If we solve the system (4) with respect to the unknowns a, b and c, it is possible to evaluate the drawdown in each element by replacing (4) in (3)

$$s = \frac{1}{2\Delta} \Big[(a_j + b_j x + c_j y) \, s_j + (a_k + b_k x + c_k y) \, s_k + \\ + (a_l + b_l x + c_l y) \, s_l \Big] \tag{5}$$

where

$$\Delta = \begin{vmatrix} 1 & x_j & y_j \\ 1 & x_k & y_k \\ 1 & x_\ell & y_\ell \end{vmatrix} = \text{Area of the triangle with nodes } j,k,1 \qquad (6)$$

$$\begin{aligned} a_j &= x_k y_\ell - x_\ell y_k & b_j &= y_k - y_\ell & c_j &= x_\ell - x_k \\ a_k &= x_\ell y_j - x_j y_\ell & b_k &= y_\ell - y_j & c_k &= x_j - x_\ell \\ a_\ell &= x_j y_k - x_k y_j & b_\ell &= y_j - y_k & c_\ell &= x_k - x_\ell \end{aligned} \qquad (7)$$

$$\begin{aligned} N_j &= (a_j + b_j x + c_j y)/2\Delta \\ N_k &= (a_k + b_k x + c_k y)/2\Delta \\ N_\ell &= (a_\ell + b_\ell x + c_\ell y)/2\Delta \end{aligned} \qquad (8)$$

With the aid of (6) - (8) the equation (5) can also be represented as follows:

$$s = \langle N_j, N_k, N_\ell \rangle \begin{Bmatrix} s_j \\ s_k \\ s_\ell \end{Bmatrix} = \langle N \rangle \{s\}^e \qquad (9)$$

Similarly, $\frac{\partial s}{\partial t}$ can be expressed by means of the Equation

$$\frac{\partial s}{\partial t} = \langle N \rangle^e \left\{\frac{\partial s}{\partial t}\right\}^e \qquad (10)$$

From (9) it is possible to obtain the equations

$$\frac{\partial s}{\partial x} = \frac{1}{2\Delta} \langle b_j, b_k, b_\ell \rangle \begin{Bmatrix} s_j \\ s_k \\ s_\ell \end{Bmatrix} = \frac{1}{2\Delta} \langle B \rangle \{s\}^e$$

$$\frac{\partial s}{\partial y} = \frac{1}{2\Delta} \langle c_j, c_k, c_\ell \rangle \begin{Bmatrix} s_j \\ s_k \\ s_\ell \end{Bmatrix} = \frac{1}{2\Delta} \langle C \rangle \{s\}^e \qquad (11)$$

Also from (11)

$$\frac{\partial}{\partial s_m}\left(\frac{\partial s}{\partial x}\right) = \frac{1}{2\Delta} b_m \quad ; \qquad m = j, k, \ell$$

(12)

$$\frac{\partial}{\partial s_m}\left(\frac{\partial s}{\partial y}\right) = \frac{1}{2\Delta} c_m \quad ; \qquad m = j, k, \ell$$

Replacing (9) - (11) in equation (2), for each element we get

$$\chi^e = \iint_e \left\{ \frac{Txx}{8\Delta^2} \left(<B> \left\{ s \right\} \right)^2 + \frac{Tyy}{8\Delta^2} \left(<C> \left\{ s \right\} \right)^2 - \right.$$

$$\left. -\left(Q - \frac{1}{Vc} <N> \left\{ \frac{\partial s}{\partial t} \right\}^e - \frac{d}{2} < N> \left\{ s \right\}^e \right) <N> \left\{ s \right\}^e \right\} dx\, dy \qquad (13)$$

Derivating (13) with respect to s_j and considering (12) when grouping similar terms we have

$$\frac{\partial \chi^e}{\partial s_j} = \iint_e \left(< \frac{Txx}{4\Delta^2} <B> b_j + \frac{Tyy}{4\Delta^2} <C> c_j + \right.$$

$$\left. + d <N> N_j > \left\{ s \right\}^e \right) dx\, dy + \qquad (14)$$

$$+ \iint_e \left(\frac{1}{Vc} <N> \left\{ \frac{\partial s}{\partial t} \right\}^e N_j \right) dx\, dy -$$

$$- \iint_e Q N_j \, dx\, dy$$

The former equation can be expressed in a centroidal reference system (the origin of the reference system coincide with the centroid of each element). For this let us define

$$\bar{x} = (x_j + x_k + x_\ell)/3$$

$$\bar{y} = (y_j + y_k + y_\ell)/3$$

$$x = x - \bar{x}$$

$$y = y - \bar{y} \qquad (15)$$

$$\iint_e dx\, dy = \Delta \iint_e x\, dxdy = \iint y\, dxdy = 0$$

$$\iint_e x^2 dx\, dy = \frac{\Delta}{12} (x_j^2 + x_k^2 + x_\ell^2) = \overline{x^2}\Delta$$

$$\iint_e y^2 dx\, dy = \frac{\Delta}{12}(y_j^2 + y_k^2 + y_\ell^2) = \overline{y^2}\Delta \qquad (15)$$

$$\iint_e xy\, dx\, dy = \frac{\Delta}{12}(x_j y_j + x_k y_k + x_\ell y_\ell) = \overline{xy}\Delta$$

The translation of the reference system does not alter the mathematical solution of the problem.

Besides, let us define

$$[U] = \qquad \langle B\rangle^T \quad \langle B\rangle$$

$$[V] = \qquad \langle C\rangle^T \quad \langle C\rangle \qquad (16)$$

$$[W] = \iint_e \langle N\rangle^T \quad \langle N\rangle \quad dx\, dy$$

where the symbol T means the transposition of the matrix. In the Equations (16) U, V and W are matrices of 3 x 3 elements.

Replacing (15) and (16) in (14) it is obtained

$$\frac{\partial x^e}{\partial s_j} = \frac{Txx}{4\Delta} U_{jm}\, s_m + \frac{Tyy}{4\Delta} V_{jm}\, s_m + \frac{d}{4\Delta} W_{jm}\, s_m + \qquad (17)$$

$$+ \frac{1}{4\Delta Vc} W_{jm} \frac{\partial sm}{\partial t} - \frac{Q\Delta}{3}$$

In a similar manner it is possible to obtain $\frac{\partial x^e}{\partial s_k}$, $\frac{\partial x^e}{\partial s_\ell}$ and define

$$\left\{\frac{\partial x}{\partial s}\right\}^e = \left\{\begin{matrix} \frac{\partial x^e}{\partial s_j} \\ \frac{\partial x^e}{\partial s_k} \\ \frac{\partial x^e}{\partial s_\ell} \end{matrix}\right\} \qquad (18)$$

from where

$$\left\{\frac{\partial x}{\partial s}\right\}^e = [h]\{s\} + [p]\left\{\frac{\partial s}{\partial t}\right\} - \{F\} \qquad (19)$$

In these equations

$$h\, m\, n = \frac{Txx}{4\Delta} U\, m\, n + \frac{Tyy}{4\Delta} V\, m\, n + \frac{d}{4\Delta} W\, m\, n \qquad (20)$$

$$p_{mn} = \frac{1}{4\Delta Vc} \; W m n$$

$$F_m = \frac{Q\Delta}{3} \qquad \text{for } m, n = j, k, l \qquad (20)$$

The minimization equation is

$$\frac{\partial x}{\partial s_i} = \Sigma \frac{\partial x^e}{\partial s_i} = 0 \qquad (21)$$

The summation being taken over all the elements

Considering equation (18), (21) can be written as:

$$\frac{\partial x}{\partial s_j} = \Sigma\Sigma\, h_{jm}\, s_m + \Sigma\Sigma\, p_{jm} \frac{\partial s_m}{\partial t} - \Sigma F_j = 0 \qquad (22)$$

The summation being taken over all the elements and nodes.

The former equation can be expressed in the following manner:

$$\left[H\right]\left\{s\right\} + \left[P\right]\left\{\frac{\partial s}{\partial t}\right\} = \left\{F\right\} \qquad (23)$$

Equation (23) can be solved by the use of a finite differen-ce scheme.

<u>Application of the finite differences method.</u>
In accordance with Wilson and Clough (1962), a stable solution of finite differences is given by the expression

$$\left\{s\right\}_t - \left\{s\right\}_{t-\Delta t} = \frac{\Delta t}{2}\left(\left\{\frac{\partial s}{\partial t}\right\}_t + \left\{\frac{\partial s}{\partial t}\right\}_{t-\Delta t}\right) \qquad (24)$$

Replacing (24) in (23) and defining

$$\left[C\right] = \left[H\right] + \frac{2}{\Delta t}\left[P\right]$$

$$\left\{D\right\} = \left[P\right]\left\{\frac{\partial s}{\partial t}\right\} \qquad (25)$$

$$\left\{E\right\} = \frac{2}{\Delta t}\left[P\right]\left\{s\right\} + \left\{D\right\}$$

It is obtained

$$\left\{s\right\}_t = \left[C\right]^{-1}\left\{\left\{F\right\}_t + \left\{E\right\}_{t-\Delta t}\right\} \qquad (26)$$

$[C]^{-1}$ is the inverse of the matrix $[C]$

For the calculation of the initial conditions we have:

$$\{D\}_{t=0} = \{F\}_{t=0} - [H]\{s\}_{t=0} \qquad (27)$$

The sequence of calculation can be defined in the following way:

$$\{E\}_{t-\Delta t} = \frac{2}{\Delta t}[P]\{s\}_{t-\Delta t} + \{D\}_{t-\Delta t} \qquad (28)$$

$$\{s\}_{t} = [C]^{-1}\left(\{E\}_{t-\Delta t} + \{F\}_{t}\right) \qquad (29)$$

$$\{D\}_{t} = -\{D\}_{t-\Delta t} + \frac{2}{\Delta t}[P]\left(\{s\}_{t} - \{s\}_{t-\Delta t}\right) \qquad (30)$$

Equation (27) is used for the evaluation of initial conditions, and when these are known, Equation (28), (29) and (30) are successively applied.

Some Remarks

The former equations allow us to analyze numerically the differential equation (1) which defines the behaviour of a leaky aquifer that is initially in hydrostatic equilibrium (Herrera and Figueroa, 1969).

A more general case of anisotropy and nonhomogenity in the main aquifer is given by the equation

$$\frac{\partial}{\partial x_i}\left(T_{ij}\frac{\partial s}{\partial x_j}\right) - ds + Q = \frac{1}{Vc}\frac{\partial s}{\partial t} \qquad (31)$$

where $x_1 = x, x_2 = y$ and T_{ij} are the components of the permeability tensor

Equation (31) can be simplified by selection of a new system of cartesian coordinates, in such a way that the main components T_{xx} and T_{yy} match with the new reference system and the terms T_{xy} and T_{yx} became zero. In this case, Equation (31) can be written in the form:

$$\frac{\partial}{\partial x}\left(T_{xx}\frac{\partial s}{\partial x}\right) + \frac{\partial}{\partial y}\left(T_{yy}\frac{\partial s}{\partial y}\right) - ds + Q = \frac{1}{Vc}\frac{\partial s}{\partial t} \qquad (32)$$

That is, in the main reference system (32) is identical to (1), his transformation of the reference system is included in the elaborated program (Rodarte, 1974). Using this program, other cases which correspond mathematically to simplifications of the differential equation (1), can be also analyzed. A non-leaky and a water - table aquifer (lineal approximation) are examples of such cases.

Analytical Verification

Considering that the program is formed by a series of logical routines by which it is possible to solve different problems of application it is convenient to analyze schemes that can be described by known functions.

If in the solution of these schemes the program is accurate enough then it can be utilized in the analysis of complex problems (non-homogeneous cases, anisotropy, different combinations of the boundary conditions, etc.), which can not be solved by means of strictly analytical procedures. In both cases the computing routines are the same.

Flow to a non-leaky aquifer of infinite radial extension

In this case, there is always a main aquifer, horizontal, isotropic, homogeneous and with constant thickness. A completely penetrating well acts in the aquifer, from which a constant yield Q is extracted. If the aquifer has an infinite radial extension, the mathematical problem is defined by the equation

$$T \frac{\partial^2 s}{r^2} + \frac{T}{r}\frac{\partial s}{\partial r} = S \frac{\partial s}{\partial t} \tag{33}$$

$$s(r,0) = 0 \tag{34.1}$$

$$s(\infty,t) = 0 \tag{34.2}$$

$$\lim_{r \to 0} r \frac{\partial s}{\partial r} = - \frac{Q}{4\pi T} \tag{34.3}$$

The analytical solution for this equation was developed by Theis (1935)

$$s = \frac{Q}{4\pi T} \int_{\frac{r^2 S}{4Tt}}^{\infty} e^{-y} \frac{dy}{y} \tag{35}$$

If we express (1) in cilindrical coordinates and consider that $d=0$, $Vc = \frac{T}{S}$ (33) is obtained. Equation (23) must be solved numerically

$$\left[H\right]\left\{s\right\} + \left[P\right]\left\{\frac{\partial s}{\partial t}\right\} = \left\{F\right\}$$

with simplifications of the matrix H and P since d = 0, $Vc = \frac{1}{S}$

In the numerical solution, the integration mesh shown in figure 3 was used. When comparing the analytical and numerical solutions it is easy to observe (fig. 3), that the drawdowns in the aquifer are four times greater than the ones obtained by Equations (35)

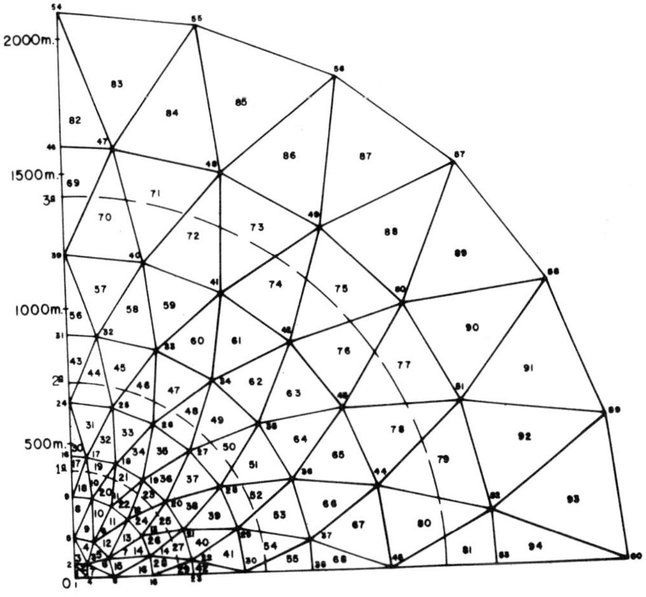

Figure 3. The integration mesh

In figure 4, we compare the analytical and numerical solutions for points 9 an 16.

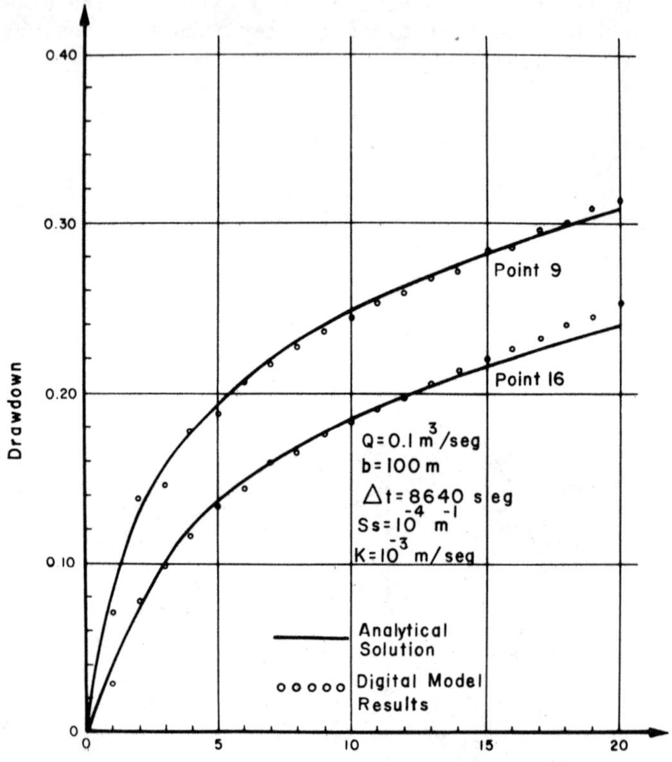

Figure 4. Non-leaky aquifer. Comparation of analytical and Numerical Results for Points 9 and 16.

<u>Analysis of a leaky aquifer</u>

Let us consider a two aquifer system, separated by a semipervious layer (figure 5). In the lower aquifer a well acts -- from which a constant yield is extracted. In addition, assume that the piezometric head in the upper aquifer remains constant and that the aquifers and the semipervious layers are homogeneous, isotropic, horizontal and of constant thickness, and of infinite radial extension. The solution to the problem for large values of time is given by Hantush (1960)

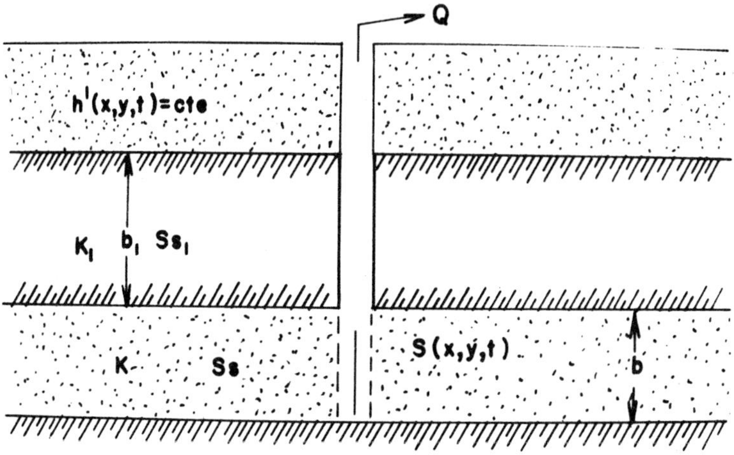

Figure 5. Schematic Diagram of a Leaky System

$$s = \frac{Q}{4\pi T} \int_{\frac{r^2}{4Vct}}^{\infty} exp(-y - \frac{dr^2}{4y}) \frac{dy}{y} \qquad (37)$$

In our Equation (1) this scheme corresponds to case 3 where the semipervious layer does not exist. The same integration mesh as the one used before (fig. 3) is used to solve this problem. Just as in the case of a non-leaky aquifer, the numerical results are four times greater than the ones given by equation (37). In figure 6, we compare the analytical and numerical solutions for points 9 and 16.

The numerical results obtained in both analyses as well as the detailed computer program can be found in Rodarte (1974)

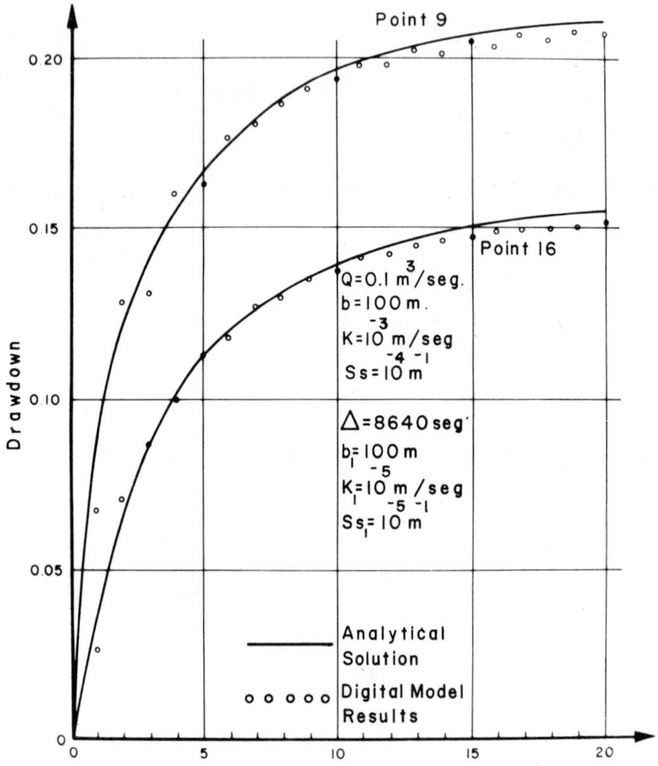

Figure 6. Leaky Aquifer. Comparation of Analytical and Numerical Results for points 9 and 16.

Conclusions

From the comparison of the numerical and the analytical results it is possible to conclude that the program gives accurate enough results for this type of problems. In further papers we will report the application of the analysis to regional problems.

NOTATION

b	Thickness of the main aquifer, L;
b_1, b_2	Thickness of the low permeability layers, L;
k	permeability of the main aquifer, L/T;
k_1, k_2	permeability of the low permeability layers, L/T;
Q	pumping rate from main aquifer, L^3/T;
r	radial distance to the pumping well, L;
s	drawdown in the main aquifer, L;
S_s	specific Storage of the main aquifer, L^{-1};
S_{s1}, S_{s2}	specific storage of the low permeability layers, L^{-1};
S	storage coefficient of the main aquifer, equal to $S_s b$
t	time, T;
T	Transmissibility of the main aquifer, equal to kb, L^2/T;
T_{xx}, T_{yy}	Transmissibilities of the main aquifer in the x and y directions, respectively, L^2/T;
x,y,z	coordinates, L;

REFERENCES

Hantush M. S., Modification of the Theory of Leaky Aquifers, J. Geophys, Res. 65, 3713-3725, 1960

Hantush M. S., Hydraulics of Wells, Vol. 1 in Advances in Hydroscience, edited by Ven Te Chow, Academic Press, New York, 281-432, 1964.

Herrera I., and L. Rodarte, Integrodifferential equations for systems of leaky aquifers and applications, 1, The nature of approximate theories, Water Resour Res., 9(4), 995-1005, 1973.

Javandel I., and P. A. Witherspoon, Analysis of Transient Fluid Flow in Multilayered System, Contribution R 4, Water Resources Center, University of California, Berkeley, 1968a.

Javandel I., and P. A. Witherspoon, Applications of the Finite Element Method to Transient Flow in Porous Media, Soc. Petrol. Engrs. J., 8(3), 241-252, 1968b.

Javandel I., and P. A. Witherspoon, A Method of Analyzing Transient Fluid Flow in Multilayered Aquifers, Water Resour. Res., 5(4), 856-869, 1969.

Kantorovick L. B., and V. I. Krilov, Priblizhennye Metody Bysshego Analiza, Giztexizdat, 1949.

Remson I., G. H. Hornebeger, and F. J. Molz, Numerical Methods in Subsurface Hydrology with an Introduction to the Finite Element Method. Wiley-Interscience, New York, 1971.

Rodarte L., Izzlevodanie Dvizhenia Podzemnix vod v Mnogoplastovyx Sistemax, Ph. D. Dissertation Institute for Civil Engineers V. V. Kkuibuishev, Moscou, 1974.

Rodarte L., Theory of Multiple Leaky Aquifer, 1, The Integrodifferential and Differential Equations for Short and Large Values of Time, Water Resources Res., Vol. 12, No. 2, 163-170, 1976.

Rodarte L., An Approximate Differential Equation to Describe Leaky Aquifer Behaviour During Intermediate and Large Values of Time, Water Resources Research, 1978

Taylor R. L., and C. V. Brown, Darcy Flow Solutions with a Free Surface, J. Hyd. Div. ASCE, Vol. 93, Hy2, 25., 1967.

Theis C. V., The relation between the lowering of the Piezometric Surface and the Rate and Duration of Discharge of a Well Using Ground Water Storage, Trans. Am. Geophys. Union, 16,519-524, 1935.

Wilson E. L., and R. W. Clough, Dynamic Response by a Step Analysis Proc. Symposium on The Use of Computers in Civil Engineering, Lisbon, Oct. 1962.

Wilson E. L., and R. E. Nickell, Application of Finite Element Method to Heat Conduction Analysis, Nuclear Engineering and Design, North Holland Pub. Co., Amsterdam (1966) 276

Zienkiewicz O. C., The Finite Element Method in Structural and Continuum Mechanics, London, Mc.Graw-Hill, 1967

FINITE ELEMENT MODELS FOR SIMULTANEOUS HEAT AND MOISTURE TRANSPORT IN UNSATURATED SOILS

R. G. Baca
Rockwell Hanford Operations, Richland, Washington 99352

I. P. King, W. R. Norton
Resource Management Associates, Lafayette, California 94549

ABSTRACT

A finite element model for use in the analysis of transport processes in arid site vadose zones is described and demonstrated. The numerical model is designed to provide detailed simulations of the interactions between climatic conditions and natural moisture movement in a dry soil system. The theoretical framework of the models considers processes of coupled heat and moisture transport, soil-atmosphere interactions and spatially varying soil properties.

The mathematical formulation of the non-isothermal model consists of the coupled form of the pressure based flow equation and the conduction equation. The flow equation models moisture flow as a function of the capillary, gravity and thermal driving forces. The heat equation considers conduction and surface fluxes arising from atmospheric heating and cooling.

A Galerkin finite element approach is used to solve the set of governing equations. The nonlinear, coupled equations are solved simultaneously using a Newton-Raphson iteration scheme. Quadratic basis functions are used to form the subdomain approximations for the principal variables and the spatially varying parameters. The soil hydraulic relationships are represented by cubic spline functions. A log transform approach is introduced to alter the form of the pressure based flow equation.

Results of numerical experiments are presented consisting of comparisons with classical analytical solutions for horizontal and vertical infiltration, linearized-coupled heat and moisture flow and actual field data. These comparisons demonstrate that the combined use of the cubic splines and the log transform approach significantly improve the solution convergences at low saturations.

1.20

INTRODUCTION

Disposal of low-level radioactive wastes in shallow land burial sites has been widely practiced for many years. Two important features of this disposal method are (1) its basic engineering simplicity and (2) its relatively low cost. Because of these attractive features and accumulating waste inventories, land burial of radioactive solid wastes may increase significantly in the next decade. To support a safe nuclear industry, the U.S. Department of Energy is currently sponsoring comprehensive land burial research and technology programs.

In certain arid sites, the unsaturated soils or vadose zone is an excellent medium for disposal of contaminated waste materials. Nominal rainfall in these areas limits the penetration of the vadose zone by meteoric waters. Consequently, the potential migration of contaminants to the ground surface or ground water will be minimal. Soil-waste interactions also reduce the rate of contaminant movement. Winograd (1974) gives an excellent overview of radioactive waste storage in arid zones.

A comprehensive predictive capability is a principal requirement for development of an advanced land burial technology which can be used to (1) assess the long-term impact of the existing land burial sites and practices, (2) evaluate new or alternative disposal site designs and (3) develop a general site criteria for shallow land burial grounds. Mathematical modeling is a logical way to develop such a predictive capability. Presently there are a number of general simulation models available for analysis of flow and contaminant transport in unsaturated soils (Reeves and Duguid, 1975 and Segol, 1977). The recent work of Frind, et al (1977) provides a good example of the versatility and usefulness of numerical models in the design of geologic environments for shallow waste storage facilities.

There are two major shortcomings in the existing modeling capability. First, most existing unsaturated flow models are based on the assumption of isothermal flow; consequently, temperature effects and thermal driving forces are neglected. In arid site vadose zones the thermal driving forces are relatively important to the soil-water cycle, and soil temperature significantly influences the rate and stage of evaporation. These aspects become even more important in the analysis of land burial grounds where the waste generates a thermal field. Secondly, nearly all existing unsaturated flow models have been developed using numerical techniques which perform well in so-called "wet" soil systems. When applied to flow into relatively dry soils, the governing equations behave so strongly nonlinear that the models produce unstable numerical solutions or are restricted to impractical solution constraints. Finlayson (1977) and Finlayson, et al (1978) have performed an in-depth study of these numerical difficulties which are related to the property of matrix stiffness.

The objective of this paper is to present the results of a recent research effort on the development of finite element models for arid site vadose zones.

MODELING APPROACH

The processes of heat and moisture transport in an arid site vadose zone are very diverse in terms of their physical nature, time and space scales. Soil-water transport processes are coupled with the processes of heat transport. Climatic and hydrologic factors influence these interacting cycles and determine the rate and direction of water and energy transport.

The modeling approach developed here is based on the consideration of (1) moisture flow as a function of capillary, gravity and thermal driving forces, rainfall and evaporation and (2) heat flow as a function of conduction, atmospheric heating and cooling at the air-soil interface. Figure 1 illustrates the basic conceptual model.

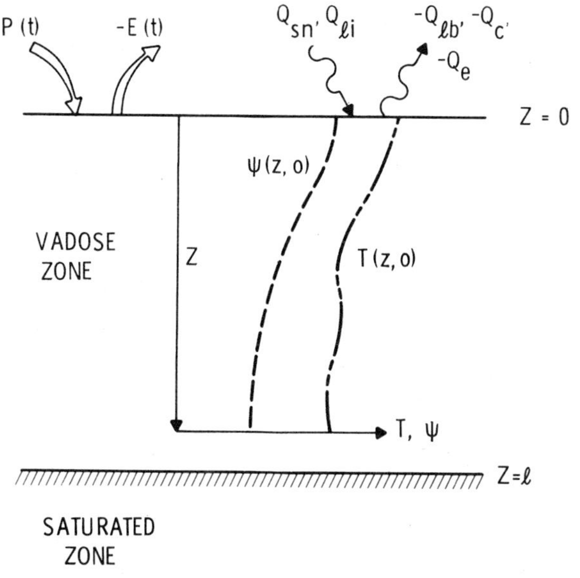

Figure 1 Definition Sketch for Soil System

Mathematical Models

The governing equations for non-isothermal flow in unsaturated soils can be derived using the basic conservation laws and Darcian flow concepts. In these equations, the coupling components which relate to the thermal driving forces are described by the theories of Philip and de Vries (1957). A detailed description of the theoretical basis of the mathematical models may be found in Baca, et al (1978). The governing equations for simultaneous heat and moisture transport are:

$$\frac{\partial\Theta(\Psi)}{\partial\Psi}\frac{\partial\Psi}{\partial t} = \nabla\cdot\{K(\Psi)[\nabla\Psi + \sigma\Psi\nabla T - 1] + D_{Tv}(\Psi)\ \nabla T \qquad (1)$$

$$C_h\frac{\partial T}{\partial t} = \nabla\cdot[\lambda(\Psi)\ \nabla T] \qquad (2)$$

with

$\frac{\partial\Theta(\Psi)}{\partial\Psi}$, specific moisture capacity (cm^{-1})
$K(\Psi)$, hydraulic conductivity ($cm\ sec^{-1}$)
σ, temperature coefficient of surface tension ($°C^{-1}$)
$D_{Tv}(\Psi)$, vapor phase diffusion coefficient ($cm^2\ sec^{-1}\ °C^{-1}$)
C_h, total specific heat capacity ($cal\ cm^{-3}$)
$\lambda(\Psi)$, thermal conductivity including both static and latent heat of vaporization effects ($cal\ cm^{-1}\ sec^{-1}\ °C^{-1}$)
$\Theta(\Psi)$, moisture content
Ψ, soil-water matric potential (cm)

The two thermal gradient terms in Equation (1) account for the liquid and vapor diffusion fluxes (Jury, 1973).

Boundary conditions at the surface are

$$-\ \{K(\Psi)\ \frac{\partial\Psi}{\partial z} + \sigma K(\Psi)\Psi\ \frac{\partial T}{\partial z} - K(\Psi) + D_{Tv}(\Psi)\ \frac{\partial T}{\partial z}\} = P(t) - E(t) \qquad (3)$$

$$-\ \{\lambda(\Psi)\ \frac{\partial T}{\partial z}\}_{z=0} = Q_{sn} + Q_{\ell i} - Q_{\ell b} + Q_c - Q_e \qquad (4)$$

where P(t) is the moisture flux downward due to precipitation while E(t) is the moisture flux due to evaporation. The heat exchange terms are defined as: Q_{sn} the net short wave radiation heat flux, $Q_{\ell i}$ the long wave incident radiation heat flux (primarily back radiation from clouds to earth), $Q_{\ell b}$ the long wave back radiation heat flux, Q_c the net heat flux due to convection and conduction and Q_e the heat flux due to evaporation. These quantities are calculated using semi-theoretical relationships, empirical equations and basic meteorologic data. In general, the bottom boundary condition is of the specified type.

In the lower range of a characteristic curve, i.e. $\Theta(\Psi)$ vs Ψ, the specific moisture capacity $\partial\Theta(\Psi)/\partial\Psi$ may vary from 5 to 25 orders of magnitude for certain soils. Similarly, the hydraulic conductivity $K(\Psi)$ can change very rapidly in the dry to intermediate moisture range. These characteristics are the principal causes of numerical difficulties associated with solving the pressure based flow equations (Braester, et al (1971). Two techniques can be used to ameliorate these nonlinear properties.

First, the functional relationships for the characteristic curve and hydraulic conductivity can be developed using cubic splines. These fits produce accurate evaluations of these parameters and generate their continuous derivatives. Secondly, the large range of Ψ typical of a dry condition can be reduced by introducing a change of variable:

$$\Psi = -e^{\beta} \tag{5}$$

or solving for β

$$\beta = \ln(-\Psi) \tag{6}$$

Substituting these relationships into Equations (1) and (2) yields the final governing equations:

$$\frac{\partial\Theta}{\partial\beta}\frac{\partial\beta}{\partial t} = \nabla \cdot \{K'(\beta)\,[-\nabla\beta - \sigma\nabla T - e^{-\beta}] + D_{Tv}(\beta)\,\nabla T\} \tag{7}$$

$$C_h\frac{\partial T}{\partial t} = \nabla \cdot \{\lambda(\beta)\,\nabla T\} \tag{8}$$

where $K'(\beta) = -\Psi K(\beta)$. In the log-transform space, the nonlinear coefficients are gradually varying and thus ameliorate the nature of nonlinearity. To test the effectiveness of these techniques, a one-dimensional finite element computer model was developed.

FINITE ELEMENT TECHNIQUE

Three major features of the general finite element method are (1) it can accommodate complex computational networks of arbitrary shape and size, (2) it can be formulated to yield high order approximations and (3) it can be extended to solve strongly nonlinear equations. The first two versatile features can be exploited to "optimize" the solution effort for a given simulation problem. For example, a simulation problem with a complex solution field can be solved using a refined mesh with high order accuracy elements in zones of rapid variations and with a coarse mesh and low order elements in zones where gradients are small. In the case of nonlinear problems, iterative algorithms with rapid convergence properties can be directly incorporated into the finite element integrals which yield efficient and stable solution techniques.

These three features of the finite element method are particularly applicable to the problem of modeling non-isothermal moisture flow in arid site vadose zones. In such soil systems distinct zones develop consisting of (1) diurnal zone (upper 500 cm), (2) seasonal zone (upper 5 m) and (3) a quasi-equilibrium zone. In addition, temperature and moisture profiles in dry soils develop steep gradients and sharp fronts; these characteristics translate into a strongly nonlinear behavior in the governing model equations which describe these processes. Consequently, the finite element method was specifically selected for this application.

Galerkin Method

By applying the Galerkin method of weighted residuals (Schechter 1967), the solution of the original partial differential equations can be recast as an integral problem. Fundamentally, the Galerkin method is based on the integral of the product of the residual error ε and some set of weighting functions ω_j. The result is

$$\chi = \int_R \varepsilon \, \omega_j \, dz \tag{9}$$

where the residual ε is defined in terms of the specific differential operators and the approximating functions for the dependent variables. By imposing the orthogonality constraint $\chi = 0$, the Galerkin functional becomes an error distribution principle (Collatz 1960) by which the residual ε is minimized over the domain R.

The integral over the entire domain can be partitioned into subdomains so that

$$\chi = \chi_1 + \chi_2 + \chi_3 + \ldots + \chi_n = \sum_{i=1}^{n} \chi_1 \tag{10}$$

where n is the number of subdomains. This may be expanded to give

$$\chi = \int_{R_1} \varepsilon \, \omega_j \, dz + \int_{R_2} \varepsilon \, \omega_j \, dz + \ldots + \int_{R_n} \varepsilon \, \omega_j \, dz \tag{11}$$

Since the governing equations for simultaneous heat and moisture transport are nonlinear, the expansion of the integral equations will give a set of nonlinear ordinary differential equations. At this point, however, we can incorporate a Newton-Raphson formulation into the integral equations which yields an iterative solution algorithm. More importantly, each iteration represents a problem of solving a system of linear equations.

Newton-Raphson Approach

The classical Newton-Raphson approach is based on a Taylor's series expansion and possesses second order convergence properties. If we define a two-component residual vector $\underline{\varepsilon} = [f,g]$, then the algorithm for a coupled set of equations is

$$\underline{\varepsilon}^{m+1} = \begin{bmatrix} \frac{\partial f}{\partial T} & \frac{\partial f}{\partial \beta} \\ \frac{\partial g}{\partial T} & \frac{\partial g}{\partial \beta} \end{bmatrix}^m \begin{Bmatrix} \Delta T \\ \Delta \beta \end{Bmatrix} + \begin{Bmatrix} f \\ g \end{Bmatrix}^m \qquad (12)$$

where m is the iteration index. Substituting this expression in the Galerkin functional for a typical element, one obtains:

$$\chi_i = \begin{bmatrix} \int \omega_j \frac{\partial f}{\partial T}\, dz & \int \omega_j \frac{\partial f}{\partial \beta}\, dz \\ \int \omega_j \frac{\partial g}{\partial T}\, dz & \int \omega_j \frac{\partial g}{\partial \beta}\, dz \end{bmatrix} \begin{Bmatrix} \Delta T \\ \Delta \beta \end{Bmatrix} + \begin{Bmatrix} \int \omega_j f\, dz \\ \int \omega_j g\, dz \end{Bmatrix} \qquad (13)$$

The iteration index m is omitted for simplicity. The quantities f, g and their partial derivatives are k component vectors where k is the number of collocation points on the subdomain R_i.

The residuals f and g are defined by the governing equations as

$$f = C_h \frac{\partial T}{\partial t} - \frac{\partial}{\partial z} \{\lambda(\beta) \frac{\partial T}{\partial z}\} \qquad (14)$$

and

$$g = \frac{\partial \Theta}{\partial \beta} \frac{\partial \beta}{\partial t} - \frac{\partial}{\partial z} \{K'(\beta) [- \frac{\partial \beta}{\partial z} - \sigma \frac{\partial T}{\partial z} - e^{-\beta}] + D_{Tv}(\beta) \frac{\partial T}{\partial z}\} \qquad (15)$$

The quantity $K'(\beta) = - \Psi K(\beta)$. Substituting these expressions into Equation (13) and integrating by parts yields the major components:

$$\int \omega_j f\, dz = \int \{\omega_j C_h \frac{\partial T}{\partial t} + \frac{\partial \omega_j}{\partial z} \lambda(\beta) \frac{\partial T}{\partial z}\}\, dz \qquad (16)$$

$$\int \omega_j \frac{\partial f}{\partial T}\, dz = \int \{\omega_j \frac{\partial}{\partial T} (C_h \frac{\partial T}{\partial t}) + \frac{\partial \omega_j}{\partial z} \frac{\partial}{\partial T} (\lambda(\beta) \frac{\partial T}{\partial z})\}\, dz \qquad (17)$$

$$\int \omega_j \frac{\partial f}{\partial \beta}\, dz = \int \frac{\partial \omega_j}{\partial z} \frac{\partial}{\partial \beta} (\lambda(\beta) \frac{\partial T}{\partial z})\, dz \qquad (18)$$

$$\int \omega_j g\, dz = \int \{\omega_j \frac{\partial \Theta}{\partial \beta} \frac{\partial \beta}{\partial t} + \frac{\partial \omega_j}{\partial z} [K'(\beta) (- \frac{\partial \beta}{\partial z} - \sigma \frac{\partial T}{\partial z} - e^{-\beta}) + D_{Tv}(\beta) \frac{\partial T}{\partial z}]\}\, dz \qquad (19)$$

$$\int \omega_j \frac{\partial g}{\partial \beta} dz = \int \{\omega_j \left[\frac{\partial \Theta}{\partial \beta} \frac{\partial}{\partial \beta} \left(\frac{\partial \beta}{\partial t}\right) + \frac{\partial^2 \Theta}{\partial \beta^2} \frac{\partial \beta}{\partial t}\right]$$

$$+ \frac{\partial \omega_j}{\partial z} \left[- K'(\beta) \frac{\partial}{\partial \beta} \left(\frac{\partial \beta}{\partial z} - e^{-\beta}\right)\right. \tag{20}$$

$$+ \frac{\partial}{\partial \beta} (K'(\beta)) \left(- \frac{\partial \beta}{\partial z} - \sigma \frac{\partial T}{\partial z} - e^{-\beta}\right) + \frac{\partial}{\partial \beta} (D_{Tv}(\beta)) \left.\frac{\partial T}{\partial z}\right]\} dz$$

$$\int \omega_j \frac{\partial g}{\partial T} dz = \int \{\frac{\partial \omega_j}{\partial z} \left[K'(\beta) \frac{\partial}{\partial T} \left(\sigma \frac{\partial T}{\partial z}\right)\right. \tag{21}$$

$$+ D_{Tv}(\beta) \frac{\partial}{\partial T} \left.\frac{\partial T}{\partial z}\right]\} dz$$

Additional terms are introduced through the integration by parts which vanish between subdomains (Baca, et al, 1978). The time derivatives in the above integrals are approximated using the general relation

$$\dot{\phi}_1 = \frac{\alpha}{\Delta t} (\phi_1 - \phi_0) + (1 - \alpha)\dot{\phi}_0 \tag{22}$$

The quantity ϕ represents β or T and the subscripts apply to the beginning and end of the time interval, Δt. The parameter α determines the order of approximation and generally ranges from 1.0 to 2.0.

The dependent variables β and T are approximated by quadratic basis functions of the general form:

$$\beta(z, t) \simeq \sum_{1}^{m} \beta_i(t)\ \omega_i(z) \tag{23}$$

$$T(z, t) \simeq \sum_{1}^{m} T_i(t)\ \omega_i(z) \tag{24}$$

The final integral equations are evaluated numerically. The nonlinear coefficients $\Theta(\beta)$, $K(\beta)$ and $\lambda(\beta)$ are generated by spline function evaluations.

EXAMPLES AND APPLICATIONS

Results from three numerical experiments are presented which demonstrate the general capability of the finite element simulation model. The test cases consist of the following: (1) solution of a linearized boundary value problem for non-isothermal flow, (2) solution of a nonlinear boundary value problem for imbibition in a vertical soil column and (3) simulation of the natural moisture patterns in an arid site vadose zone. Comparisons for other test problems may be found in Baca, et al (1978) and Gibbs and Baca (1978).

Linearized Non-isothermal Flow

A linearized non-isothermal flow problem was selected as a first test case to demonstrate the basic veracity and accuracy of the numerical model. The linearized non-isothermal flow model was originally developed by Hauk (1971) to study the coupling effects between temperature and moisture fields in soils. The mathematical model, derived from the concepts of irreversible thermodynamics, is expressed in non-dimensional form as:

$$\frac{1}{\kappa}\frac{\partial T'}{\partial t} = \frac{\partial^2 T'}{\partial \xi^2} + \eta \frac{\partial^2 \phi}{\partial \xi^2} \qquad (25)$$

$$\frac{1}{\lambda}\frac{\partial \phi}{\partial t} = \delta \frac{\partial^2 T'}{\partial \xi^2} + \frac{\partial^2 \phi}{\partial \xi^2} \qquad (26)$$

with the initial and boundary conditions

$$\phi(\xi, 0) = 0, \quad T'(\xi, 0) = 0 \qquad (27)$$

$$\frac{\partial \phi(0, t)}{\partial \xi} = H(\phi - 1), \quad \frac{\partial T'(0, t)}{\partial \xi} = -R_a \qquad (28)$$

$$\phi(\infty, t) = 0, \quad T'(\infty, t) = 0 \qquad (29)$$

The associated coefficients are defined in Table 1. The analytical solution derived by Hauk is compared with the numerical solution in Figure 2.

The finite element solution is based on the use of a variably spaced network consisting of 10 elements and 21 nodes. For convenience, the lower boundary condition was assigned a zero flux. As shown in the comparisons, the agreement between the analytical and numerical solutions is excellent.

Table 1 Constants for Non-isothermal Flow Problem

Constant	Numerical Value
Heat conduction factor, κ	3.10×10^{-7}
Moisture coupling factor, η	0.01
Hydraulic conductivity factor, λ	1.66×10^{-6}
Temperature coupling factor, δ	0.20
Evaporation-convection factor, H	2.00
Surface heat flux factor, R_a	28.00

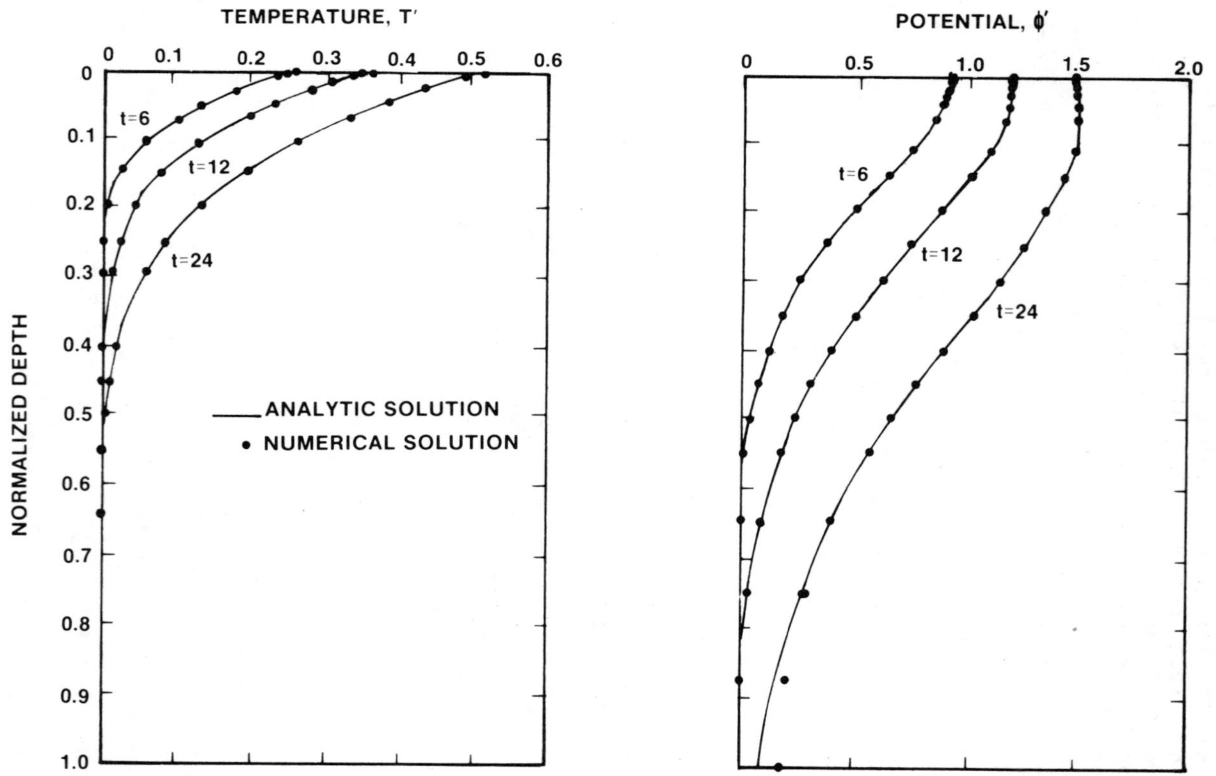

Figure 2 Comparison of Analytical and Numerical Solutions for Non-isothermal Flow Problem

Isothermal Imbibition Problem

To demonstrate the capability of the finite element approach to handle flow problems with sharp fronts, a one-dimensional isothermal imbibition problem was selected as a test case. Many examples for this class of problems exist in the literature which are based on approximate solutions of the moisture based flow equations. van Genuchten (1978) recently presented a numerical solution for such a problem which is used here as a basis for a relative comparison.

The Θ-based flow equation is solved by van Genuchten for a vertical soil column with an initial moisture content of 0.275 and a specified surface boundary condition of 0.52. The soil-water hydraulic properties for a Glendale clay loam were used which are characterized by the relationships:

$$K(\Psi) = \begin{cases} 75 \quad \left(\frac{5.4}{|\Psi|}\right)^{2.6}, \ \text{cm/day} & \Psi < -5.4 \text{ cm} \\ 75 \quad , \ \text{cm/day} & \Psi > -5.4 \text{ cm} \end{cases} \tag{30}$$

$$\Theta(\Psi) = \begin{cases} 0.52 \left(\frac{5.4}{|\Psi|}\right)^{0.2} & \Psi < -5.4 \text{ cm} \\ 0.52 & \Psi > -5.4 \text{ cm} \end{cases} \tag{31}$$

The numerical solution of the Θ-based equations was obtained by van Genuchten using a Galerkin finite element technique and Hermitian basis functions.

For this test case, the isothermal form of the pressure based flow equations were solved using a finite element network consistent with that used by van Genuchten. A total of 20 elements and 41 nodes was used. Constant time steps of 0.1 hours, with about three to five iterations per time step, were needed to obtain the numerical results shown in Figure 3.

The comparison of the Θ-based solution and the Ψ-based solution shows very good agreement. Both solutions show excellent tracking of the sharp moisture front, however, the present solution shows little or no undershoot-overshoot problems.

Natural Moisture Patterns

A final example is presented here from a recent application of the computer model to an arid site vadose zone. The application is an attempt to simulate the dynamic temperature and moisture fields which respond to climatic variations. Meteorologic observations were used to estimate the heat and moisture fluxes at the soil-atmosphere interface.

The computer simulation was performed from November 1974 through March 1975. This time was selected since it coincides with the wettest and most dynamic period. Data for this period provide a good basis for testing the over-all capability of the simulation model.

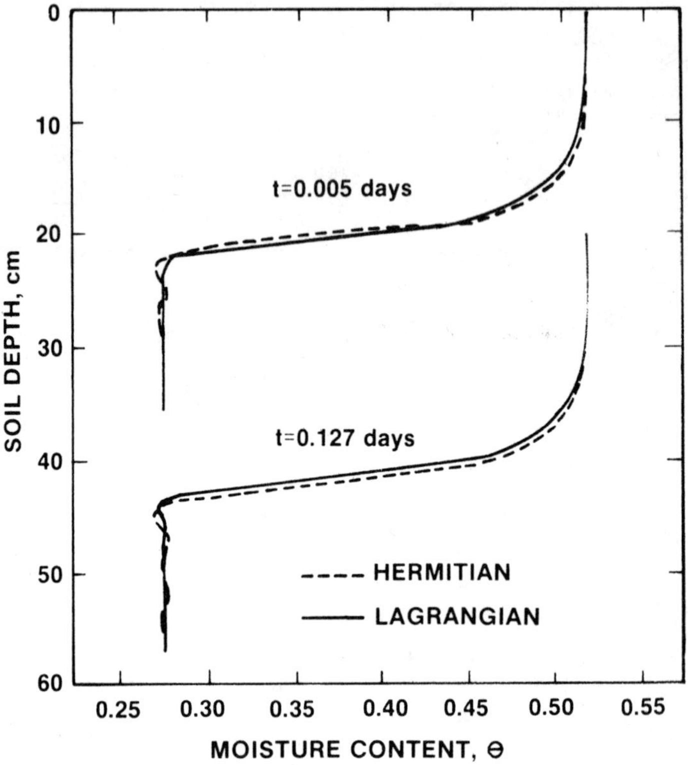

Figure 3 Comparison of Finite Element Solutions for Isothermal Imbibition Problem

The finite element representation of the soil system consisted of a variably spaced network with 30 elements and 61 nodes. Element lengths varied from 3 cm near the soil surface to 100 cm near the bottom. Fixed time steps of six hours were used throughout the simulation. The total number of iterations per time step varied from two to five depending on the nature of the surface boundary conditions, i.e., wetting or drying; time step size and iteration requirements were generally limited by the rate of change of the temperature field.

Soil moisture data for an open bottom lysimeter was selected for this application. The lysimeter, located on the Hanford Reservation, is about 18 meters deep and 3 meters in diameter.

The lysimeter soil consists of a homogeneous mixture of medium to coarse sands and is instrumented for moisture content measurements. Soil-water hydraulic properties for the lysimeter are shown in Figure 4.

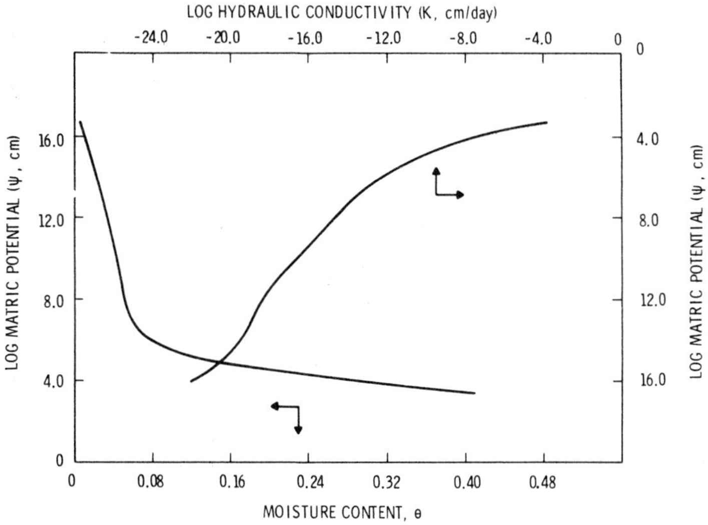

Figure 4 Hydraulic Properties for Lysimeter Soil

The comparison of numerical simulation and the field measurements is shown in Figure 5. The observed data, shown as the dots, represent an average of three spatially separated neutron probe readings. The simulated temperature profiles are shown in these comparisons, however, no temperature data were available for comparison. It should be kept in mind that the comparisons here are relative ones because the field measurements represent instantaneous values whereas the simulation results are daily average values.

By and large, the comparisons indicate that the model is reasonably tracking the principal features of the wetting and drying cycle. The significant discrepancies occur in the upper meter of the soil profile. In this surface zone, the simulation model appears to over-estimate the rate of moisture gain and under-estimate the evaporative loss rate. Several explanations can be offered for these effects.

Since the general behavior of the simulation model appeared consistent and reasonable, no particular effort was made to

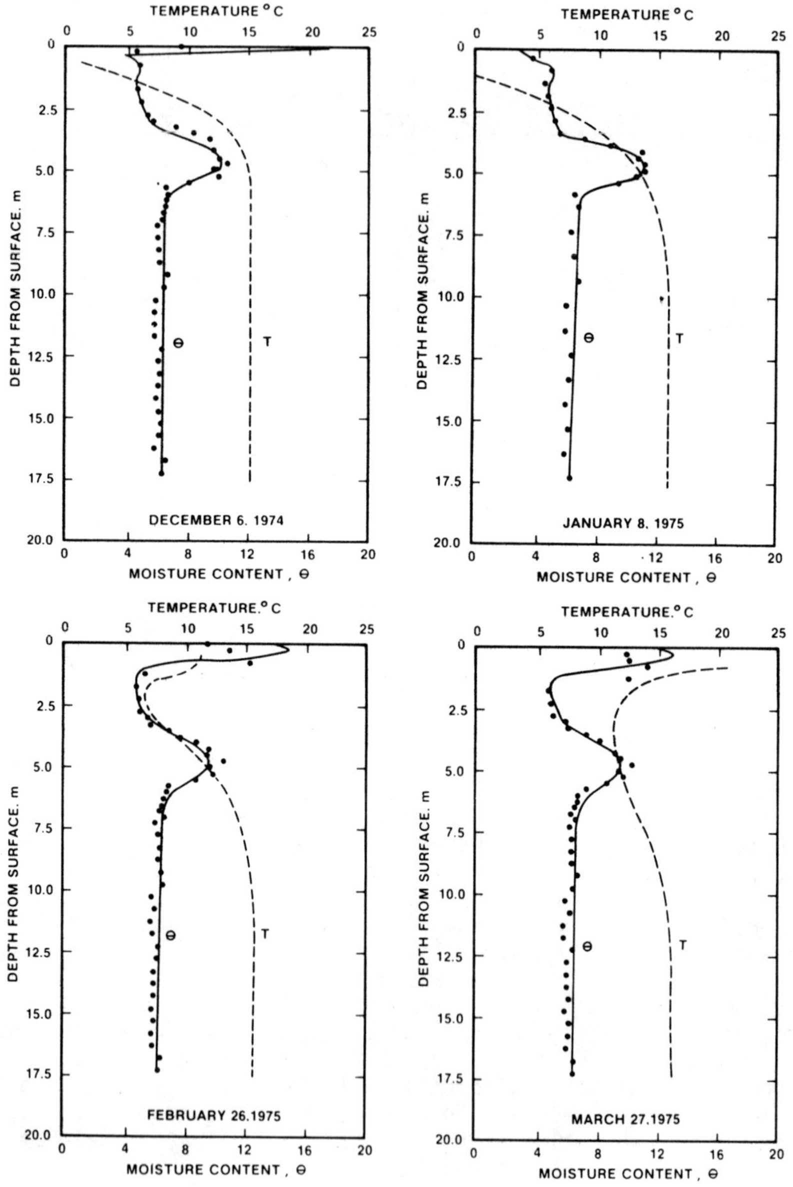

Figure 5 Simulated and Measured Moisture Profiles for Hanford Lysimeter

calibrate the model coefficients. A complete evaluation of the model's predictive capability will be performed in the near term with more complete field data.

CONCLUSIONS

A numerical model for the simulation of heat and moisture movement in arid site vadose zones is developed and applied. The model employs a new finite element approach which is based on the combined use of a log transform, cubic spline fitting and the Newton-Raphson iteration algorithm. Results of numerical experiments clearly illustrate that the new finite element approach can handle steep fronts and strong nonlinearities associated with flow into dry soils.

By this new approach, many existing isothermal flow models can be extended to solve a wider class of unsaturated flow problems. The only restriction is that the log transform is not applicable to saturated conditions. However, this restriction is seldom significant in arid site vadose zones.

ACKNOWLEDGMENTS

The authors wish to thank Mr. C. F. Manry of Rockwell Hanford Operations for his active interest during the conduct of research summarized in this paper. The contributions of Mr. D. B. McLaughlin of Resource Management Associates in the initial formulation of the mathematical models are also gratefully acknowledged. Special thanks are expressed to Mr. C. R. Schult for his assistance in computer applications and the preparation of the graphical comparisons. The work summarized in this paper was conducted under the Long-Term Low-Level Waste Technology Program for the U.S. Department of Energy by Rockwell Hanford Operations.

REFERENCES

Baca, R. G., I. P. King and W. R. Norton (1978) "Models for Simultaneous Heat, Moisture and Solute Transport in Arid Site Vadose Zones." Rockwell Hanford Operations, Richland, WA.

Braester, C., G. Dagan, S. P. Neuman and D. Zaslavsky (1971) "A Survey of the Equations and Solutions of Unsaturated Flow in Porous Media." Technion Report, Institute of Technology, Haifa, Israel.

Collatz, L. (1960) The Numerical Treatment of Differential Equations, Third Edition. Springer-Verlag OHG, Berlin, Germany.

Finlayson, B. A. (1977) "Water Movement in Dessicated Soils." in Finite Elements in Water Resources. Pentech Press, London.

Finlayson, B. A., R. W. Nelson and R. G. Baca (1978) "A Preliminary Investigation into the Theory and Techniques of Modeling the Natural Moisture Movement in Unsaturated Sediments." Rockwell Hanford Operations, Richland, WA.

Frind, E. O., R. W. Gillham and J. F. Pickens (1977) "Application of Unsaturated Flow Properties in the Design of Geologic Environments for Radioactive Waste Storage Facilities." in Finite Elements in Water Resources. Pentech Press, London.

Gibbs, A. G. and R. G. Baca (1978) "Analytical Models for Simultaneous Heat and Moisture Transport in Soils." Rockwell Hanford Operations, Richland, WA.

Hauk, R. W. (1971) "Soil Temperature and Moisture Fields Predicted by a System of Coupled Equations." Ph.D. Thesis, Department of Agricultural Engineering, Cornell University, Cornell, NJ.

Jury, W. A. (1973) "Simultaneous Transport of Heat and Moisture through a Medium Sand." Ph.D. Thesis, University of Wisconsin, Madison, WI.

Philip, J. R. and D. A. de Vries (1957) "Moisture Movement in Porous Materials under Temperature Gradients." Transcripts of the American Geophysical Union, 38, 222-232.

Reeves, M. and J. O. Duguid (1975) "Water Movement through Saturated-Unsaturated Porous Media: A Finite-Element Galerkin Model." ORNL-4927, Oak Ridge National Laboratory, Oak Ridge, TN.

Schechter, R. S. (1967) The Variational Method in Engineering. McGraw-Hill Book Company, New York, NY.

Segol, G. A. (1977) "A Three-Dimensional Galerkin-Finite Element Model for the Analysis of Contaminant Transport in Saturated-Unsaturated Porous Media." in Finite Elements in Water Resources. Pentech Press, London.

van Genuchten, M. T. (1978) "Mass Transport in Saturated-Unsaturated Systems, One-Dimensional Solutions." Research Report 78-WR-06, Water Resources Program, Department of Civil Engineering, Princeton University, Princeton, NJ.

Winograd, I. J. (1974) "Radioactive Waste Storage in the Arid Zone." EOS, American Geophysical Union, 55:884-894.

FINITE ELEMENTS SOLUTION OF UNSATURATED POROUS MEDIA FLOW

C. Tzimopoulos

Aristotle University of Thessaloniki - School of Technology - Thessaloniki - GREECE

INTRODUCTION

The flow of water in porous media in many practical situations is unsaturated. This is a special case of similtaneous flow of two immiscible fluids, where the nonwetting fluid (air) is assumed to be stagnant. Most of the work in the subject of immiscible fluids has been extensively treated by reservoirs engineering, whereas unsaturated flow has been done by soil physicists as a separate subject with agronomic or ecological aspects of hydrology (irrigation, drainage of agricultural lands, infiltration etc.). In recent years engineering hydrologists have exhibited an increasing interest in this field.

The flow equation describing the movement of water in porous media is

$$\frac{\partial C}{\partial t} = \nabla(K\nabla\Psi) + \frac{\partial K}{\partial z} . \tag{1}$$

Equation (1) is the result of the continuity equation and Darcy law, where C = the volumetric water content (or moisture content, cm^3/cm^3), $K = K(C)$ = the hydraulic conductivity (or capillary conductivity, cm/s) and Ψ = the capillary potential (cm).

For a homogeneous soil, we may express eq. (1) in more tractable form by introducing the diffusivity coefficient

$D = K\cdot(d\Psi/dC)$, assuming K and Ψ are single-valued functions of C and we take:

$$\frac{\partial C}{\partial t} = \nabla(D\cdot\nabla C) + \frac{\partial K}{\partial z} \tag{2}$$

The above equation is non linear and cannot be solved by analytical solutions in general. So, numerical solutions have been developed, such that the Philip's method (Philip, 1957, 1969), the finite differences method (Aschroft et all,1962) and the finite elements method (Bruch-Zyvoloski, 1973, Tzimopoulos, 1977).

In this paper we deal with the vertical infiltration into a homogeneous isotrope and isothermal soil column. In this case eq. (2) takes the form:

$$\frac{\partial C}{\partial t} = \frac{\partial}{\partial z}\left(D\cdot\frac{\partial C}{\partial z}\right) - \frac{\partial K}{\partial z}\ , \tag{3}$$

(with the z-axis taken positive downard)
with the following conditions:

Initial conditions

$$C = C_o \qquad t = 0, \qquad z \geq 0 \tag{3a}$$

Boundary condition on the surface

$$C = C_s, \qquad z = 0 \qquad t > 0 \tag{3b}$$

Boundary condition at the lower end

$$C = C_o, \qquad z = d \qquad t \geq 0 \tag{3c}$$

The finite elements method is used for the solution of the problem. Rectangular elements are used and the function C is approximated over the respective finite element by a linear combination of local basis functions. The Galerkin's method, which is a special case of the method of weighted residuals and a generalization of the Ritz method (Oden, 1972) was used for generating the finite element model, and the differential equation (3) was transformed in a system of non linear algebraic equations. After linearization the system is solved by the Thomas algorithm. Also the Crank-Nicolson finite differences method and Philip's method were used and compared with the above method.

FINITE ELEMENT APPROXIMATION

For the solution of eq. (3) the finite elements theory according Oden (Oden, 1972) is used. The domain of integration (z,t) is divided into rectangular elements re in space and time (fig.

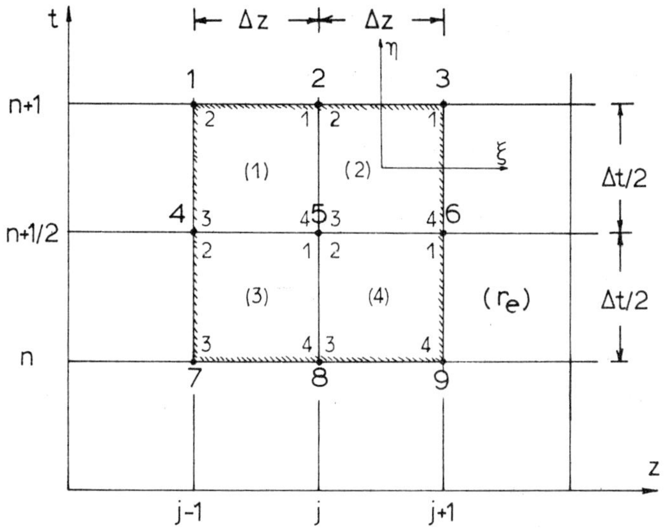

Figure 1. Solution domain divided in finite elements re in space and time.

1) with dimensions Δz, $\Delta t/2$. The global nodal points are numbered with large numbers $\Delta = 1,2,3,\ldots,G$, the local ones inside each element with small numbers $N = 1,2,3,4$. The rectangular elements are numbered with numbers enclosed in parentheses (e) = =(1), (2),...,(E).

The moisture content is approximated inside each element by the relation

$$\bar{c}^{(e)}(\vec{X}) = c^{N}_{(e)} \cdot \Psi^{(e)}_{N}(\vec{X}). \quad (N = 1,2,3,4) \qquad (4)$$

The index N means summation.

In (4) $c^{N}_{(e)}$ are the values which the function $\bar{c}^{(e)}(\vec{X})$ takes in the nodal points of elements re,

$$\bar{c}^{(e)}(\vec{X}^{M}) = c^{M}_{(e)} \quad \text{at node } \vec{X}^{M} \in re \tag{5}$$

The functions $\Psi^{(e)}_{N}(\vec{X})$ constitute a local base and have been chosen as follows (element (2) fig. 1)

$$\begin{aligned}
\Psi^{(2)}_{1}(\vec{X}) &= \tfrac{1}{4}(1+\xi)(1+\eta), \\
\Psi^{(2)}_{2}(\vec{X}) &= \tfrac{1}{4}(1-\xi)(1+\eta), \\
\Psi^{(2)}_{3}(\vec{X}) &= \tfrac{1}{4}(1-\xi)(1-\eta), \\
\Psi^{(2)}_{4}(\vec{X}) &= \tfrac{1}{4}(1+\xi)(1-\eta) .
\end{aligned} \tag{6}$$

The variables (ξ,η) are connected with (z,t) by means of

$$\xi = \frac{2(z-\bar{z})}{\Delta z} \quad , \quad \text{and} \quad \eta = \frac{4(t-\bar{t})}{\Delta t} \quad ,$$

$\bar{z},\bar{t}$ beeing the coordinates of the centroid of each element re. The global values C^{Δ} are connected with the local values $c^{N}_{(e)}$, by the relation

$$c^{N}_{(e)} = \Omega^{(e)N}_{\Delta} \cdot C^{\Delta} \qquad (\Delta = 1,2,3,\ldots G) \tag{7}$$

where $\Omega^{(e)N}_{\Delta}$ is a transformation with the property $\Omega^{(e)N}_{\Delta} = 1$,

when the global nodal point Δ coincides with node N of the element re, and in all other cases $\Omega^{(e)N}_{\Delta} = 0$. For the case of figure 1 the transformation $\Omega^{(e)N}_{\Delta}$ is given by the matrices:

$$\overset{(1)}{\Omega} = \begin{bmatrix} 0 & 1 & 0 & 0 & 0 & 0 & 0 & 0 & 0 \\ 1 & 0 & 0 & 0 & 0 & 0 & 0 & 0 & 0 \\ 0 & 0 & 0 & 1 & 0 & 0 & 0 & 0 & 0 \\ 0 & 0 & 0 & 0 & 1 & 0 & 0 & 0 & 0 \end{bmatrix} ,$$

$$\overset{(2)}{\Omega} = \begin{bmatrix} 0 & 0 & 1 & 0 & 0 & 0 & 0 & 0 & 0 \\ 0 & 1 & 0 & 0 & 0 & 0 & 0 & 0 & 0 \\ 0 & 0 & 0 & 0 & 1 & 0 & 0 & 0 & 0 \\ 0 & 0 & 0 & 0 & 0 & 1 & 0 & 0 & 0 \end{bmatrix},$$

$$\overset{(3)}{\Omega} = \begin{bmatrix} 0 & 0 & 0 & 0 & 1 & 0 & 0 & 0 & 0 \\ 0 & 0 & 0 & 1 & 0 & 0 & 0 & 0 & 0 \\ 0 & 0 & 0 & 0 & 0 & 0 & 1 & 0 & 0 \\ 0 & 0 & 0 & 0 & 0 & 0 & 0 & 1 & 0 \end{bmatrix},$$

$$\overset{(4)}{\Omega} = \begin{bmatrix} 0 & 0 & 0 & 0 & 0 & 1 & 0 & 0 & 0 \\ 0 & 0 & 0 & 0 & 1 & 0 & 0 & 0 & 0 \\ 0 & 0 & 0 & 0 & 0 & 0 & 0 & 1 & 0 \\ 0 & 0 & 0 & 0 & 0 & 0 & 0 & 0 & 1 \end{bmatrix},$$

The discrete model of $C(\vec{X})$ is

$$C(\vec{X}) \simeq \bar{C}(\vec{X}) = \sum_{e=1}^{E} \bar{c}^{(e)}(\vec{X}) \tag{8}$$

or in vieu of (4) and (7)

$$\bar{C}(\vec{X}) = \sum_{e=1}^{E} \Psi_N^{(e)}(\vec{X}) \cdot \overset{(e)N}{\Omega_\Delta} \cdot C^\Delta = \Phi_\Delta(\vec{X}) \cdot C^\Delta , \tag{9}$$

where $\Phi_\Delta(\vec{X})$ is the global approximation function shown in figure 2

$$\Phi_\Delta(\vec{X}) = \sum_{e=1}^{E} \overset{(e)N}{\Omega_\Delta} \cdot \Psi_N^{(e)}(\vec{X}) \tag{10}$$

Because of the nonlinearity of eq. (3), the most powerful technique for generating aceptable finite element models is the averaging method of Galerkin.

Using this method equation (3) is written

$$L(C) = \frac{\partial C}{\partial t} - \frac{\partial}{\partial z}(D \cdot \frac{\partial C}{\partial z}) + \frac{\partial K}{\partial z} = 0 , \tag{11}$$

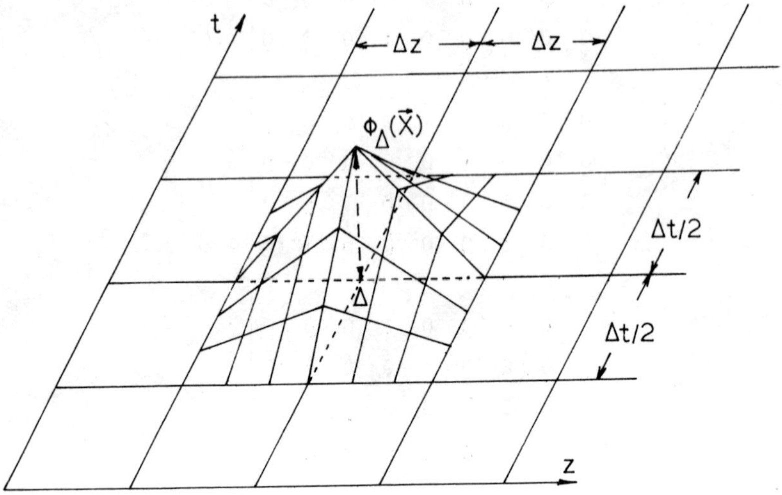

Figure 2. Global approximation function $\Phi_\Delta(\vec{X})$.

and the function $\Phi_\Delta(\vec{X})$ is orthogonal with respect to operator L(C), that is

$$\int_R L(\bar{C})\Phi_\Delta(\vec{X})\,dR = 0 \qquad (\Delta = 1,2,3,\ldots G) \tag{12}$$

or combining (4), (9), (11)

$$\sum_{e=1}^{E} {}^{(e)}\Omega_\Delta^M \cdot P_M^{(e)}(c_e^N) = 0 \quad . \tag{13}$$

In (13) $P_M^{(e)}(c_{(e)}^N)$ means

$$P_M^{(e)}(c_{(e)}^N) = \iint_{re} \Psi_M^{(e)}\left[\frac{\partial c}{\partial t} - \frac{\partial}{\partial z}\left(D\,\frac{\partial c}{\partial z}\right) + \frac{\partial K}{\partial z}\right]dzdt$$

which by using Green's theorem and (4) gives

$$P_M^{(e)}(c_{(e)}^N) = K^N \cdot b_N + D^L \cdot c^N \cdot a_{LN} + c^N \cdot d_N \quad (N = 1,2,3,4,\ L=1,2,3,4) \tag{14}$$

where

$$b_N = \iint_{re} \Psi_M^{(e)} \frac{\partial \Psi_N^{(e)}}{\partial z} dzdt, \quad a_{LN} = \iint_{re} \Psi_L^{(e)} \frac{\partial \Psi_N^{(e)}}{\partial z} \cdot \frac{\partial \Psi_M^{(e)}}{\partial z} dzdt .$$

$$d_N = \iint_{re} \Psi_M^{(e)} \frac{\partial \Psi_N^{(e)}}{\partial t} dzdt .$$

The resulting system (13) consists of nonlinear algebraic equations. The technique for solving this system is as follows:

1. We linearize the system (13), using the same procedure with Aschroft (Aschroft et all, 1962).
2. The functions at the time $(t+\Delta t/2)$ are approximated by

$$F_{t+\Delta t/2} = \frac{1}{2}(F_t + F_{t+\Delta t})$$

3. The solution of the linearized equations is obtained by an algorithm, which is reffered to Thomas (Tzimopoulos, 1977) and is a special adaptation of Gauss elimination method.

NUMERICAL APPLICATIONS

For the application of the given theory two soils are considered with the following data.

1st. The Columbia silt loam (from Kirkham and Powers, 1972)

Initial Condition

$C_o = 0,05 \qquad t = 0 , \qquad z \geq 0$

Boundary condition on the surface

$C_s = 0,34 \qquad z = 0 , \qquad t > 0$

Boundary condition at the lower end

$C_o = 0,05 \qquad z = 20 \text{ cm} \qquad t \geq 0$

Values of diffusivity D(C) and capillary conductivity K(C) for various moisture contents C are given in the following table.

TABLE 1

$C(cm^3/cm^3)$	K(cm/min)	$D(cm^2/min)$	$C(cm^3/cm^3)$	K(cm/min)	$D(cm^2/min$
0.34	$1,90\cdot10^{-2}$	2,83	0.19	$3,10\cdot10^{-6}$	0,0660
0.33	$1,58\cdot10^{-2}$	2,47	0.18	$1,60\cdot10^{-6}$	0,0561
0.32	$1,18\cdot10^{-2}$	2,200	0.17	$8,18\cdot10^{-7}$	0,0427
0.31	$6,82\cdot10^{-3}$	1,870	0.16	$4,21\cdot10^{-7}$	0,0355
0.30	$3,50\cdot10^{-3}$	1,600	0.15	$2,17\cdot10^{-7}$	0,0300
0.29	$1,87\cdot10^{-3}$	1,300	0.14	$1,11\cdot10^{-7}$	0,0227
0.28	$1,00\cdot10^{-3}$	0,980	0.13	$5,72\cdot10^{-8}$	0,0200
0.27	$5,30\cdot10^{-4}$	0,710	0.12	$2,95\cdot10^{-8}$	0,0178
0.26	$2,83\cdot10^{-4}$	0,530	0.11	$1,52\cdot10^{-8}$	0,0160
0.25	$1,49\cdot10^{-4}$	0,425	0.10	$7,90\cdot10^{-9}$	0,0146
0.24	$7,92\cdot10^{-5}$	0,355	0.09	$4,08\cdot10^{-9}$	0,0133
0.23	$4,23\cdot10^{-5}$	0,260	0.08	$2,12\cdot10^{-9}$	0,0121
0.22	$2,23\cdot10^{-5}$	0,118	0.07	$1,10\cdot10^{-9}$	0,0105
0.21	$1,15\cdot10^{-5}$	0,112	0.06	$5,40\cdot10^{-10}$	0,0091
0.20	$5,95\cdot10^{-6}$	0,086	0.05	$2,78\cdot10^{-10}$	0,0075

2nd. The Columbia silt loam (from Davidson et all, 1963)

Initial condition

$$C_o = 0,031 \qquad t = 0 \qquad z \geq 0$$

Boundary condition on the surface

$$C_s = 0,45 \qquad z = 0 \qquad t > 0$$

Boundary condition at the lower end

$$C_o = 0,031 \qquad z = 40 \text{ cm} \qquad t \geq 0 .$$

Experimental values of capillary conductivity and soil water diffusivity D are given in the figure (3).

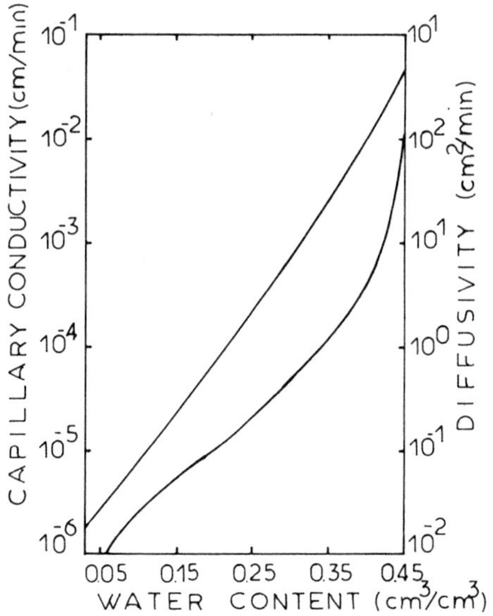

Figure 3. Capillary conductivity K and water diffisivity D versus water content C.

Time and length steps are correspondingly

1st case: $\Delta z = 1$ cm, $\Delta t = 1$ min

2nd case: $\Delta z = 2$ cm, $\Delta t = 0{,}5$ min.

To compare the numerical results with the finite element method, Crank-Nicolson computational sheme has also been used. The Crank-Nicolson sheme is written as:

$$\frac{C_j^{n+1}-C_j^n}{\Delta t} = \frac{D_{j+\frac{1}{2}}^{n+1}(C_{j+1}^{n+1}-C_j^{n+1})-D_{j-\frac{1}{2}}^{n+\frac{1}{2}}(C_j^{n+1}-C_{j-1}^{n+1})+D_{j+\frac{1}{2}}^{n+\frac{1}{2}}(C_{j+1}^{n}-C_j^{n})-D_{j-\frac{1}{2}}^{n+\frac{1}{2}}(C_j^{n}-C_{j-}^{n}}{2\Delta z^2}$$

$$-\frac{K_{j+1}^{n+1}-K_{j-1}^{n+1}+K_{j+1}^{n}-K_{j-1}^{n}}{4\Delta z} \qquad (15)$$

For the solution of (15) we linearize the nonlinear terms and then we solve the system by the Thomas algorithm. The same time and length steps have been used, as for the finite elements model. Water profiles developed during different time periods are presented in figures 4 and 5.

Philip (1962) solved eq. (3) by a lengthy process of successive approximations. His solution has the form

$$Z(c,t) = \lambda(c)t^{1/2} + \chi(c)\cdot t + \psi(c)\cdot t^{3/2} + \omega(c)\cdot t^{2} + \ldots \qquad (16)$$

For the case of figure (4) Warrick (Don Kirkham et all, 1972) calculated the parameters λ, χ, ψ as they are presented in figure 6.

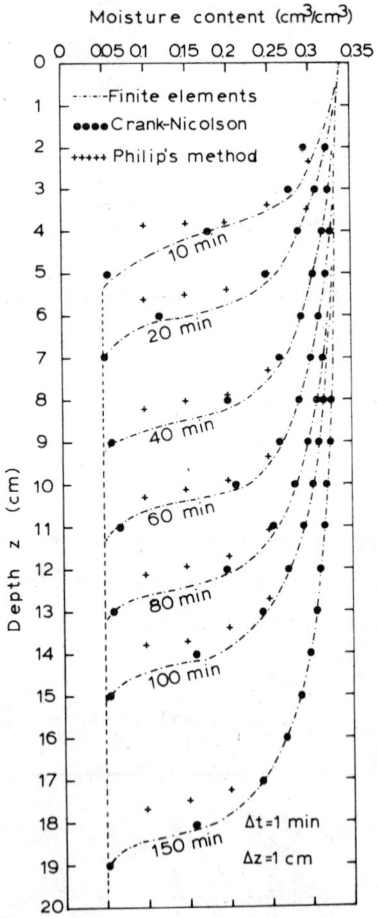

Figure 4. Computed moisture distribution curves (1st case).

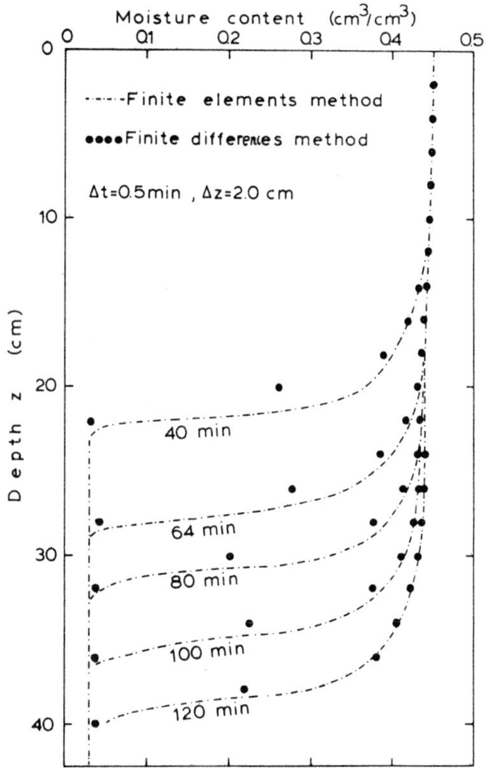

Figure 5. Computed moisture distribution curves (2nd case).

DISCUSSION

Figures 4 and 5 show the results of using finite elements method as compared to the Crank-Nicolson method. Each curve represents the moisture profile at a particulary time during the infiltration. The finite element technique used herein is an implicit sheme and is stable. For the elements size used, the results agree satisfactory with the Crank-Nicolson finite differences method.

The numerical results were computed with Univac 1106 computer installed at the University of Thessaloniki. The actual computer execution time is:

1st case	F.E.	C.P.U.	15,09	sec
	C-N	C.P.U.	11,00	sec
2nd case	F.E.	C.P.U.	11,04	sec
	C-N	C.P.U.	14,64	sec.

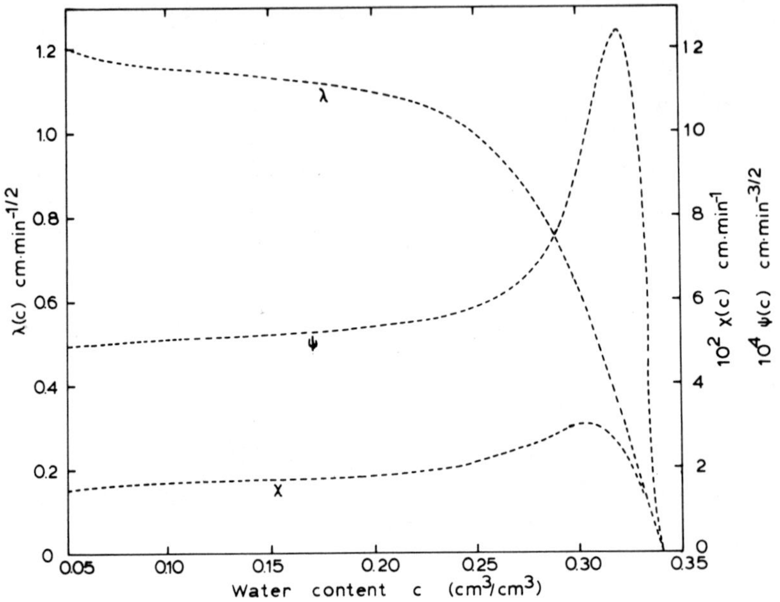

Figure 6. Values of λ,χ and ψ versus water content C (Kirham et all, 1972).

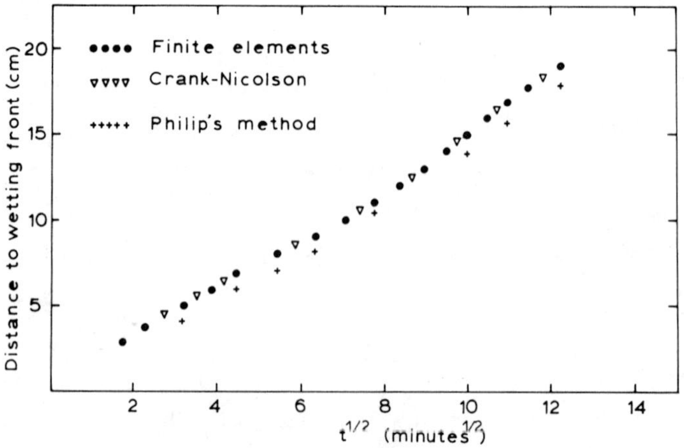

Figure 7. Distance to the wetting front versus square root of time (1st case).

In all cases the maximum number of time steps is K_{max} =250. Philip's method shows a small delay in all times (fig. 4), which is due, perhaps, to the truncation of the termes of the series (16). The same is presented also for the distance to the wetting front (Fig. 7).

REFERENCES

1. Aschroft G. - March D. - Evans D. - Boersmal (1962):"Numerical Method for solving the Diffusion Equation". Soil Sci. Soc. Am. Proc. pp. 522 - 525.
2. Bruch J.C. - Zyvoloski G. (1973): "Finite element Solution of Unsteady and Unsaturated Flow in Porous Media". The Mathematics of Finite Elements and Applications. pp. 201-211.
3. Davidson J.M. - Nielsen D.R. - Biggar J.W. (1963): "The measurement and Description of Water flow through Columbia silt loam and Hesperia Sandy loam" Hilgardia Vol. 34 No 15 pp. 601-617.
4. Don Kirkham - Powers L.S. (1971). Advanced Soil Physics. Wiley-Interscience.
5. Oden J.J. (1972): "Finite Elements of Nonlinear continua". McGraw-Hill Book Co.
6. Philip J.R. (1957)." The theory of infiltration: 1 The infiltration equation and its Solution". Soil Sci. 83, 345 - 357.
7. Philip J.R. (1969). "Theory of infiltration" Advances in Hydrosciences. Volume 5. Academic Press.
8. Tzimopoulos C. (1977).:"Un modèle aux elements finis pour l'étude du mouvement de l' humidité dans un milieu poreux isotherme". Symposium on Hydrodynamic Diffusion and Dispersion in Porous Media. I.A.H.R. Committee on Flow Through Porous Media April 20 - 22 Pavia - Italy.

A GALERKIN-FINITE ELEMENT SIMULATION OF SOLUTE TRANSPORT IN SUBSURFACE DRAINAGE SYSTEMS

Miguel A. Marino, George B. Matarga

Department of Land, Air and Water Resources and Department of Civil Engineering, University of California, Davis, Ca 95616

ABSTRACT

A Galerkin-type finite element method is employed to investigate the movement and distribution of a solute in a subsurface drainage system. The entire subsurface domain is treated as one region having saturated and unsaturated zones whose extent is determined as part of the analysis. To obtain continuous flow across elements and at the nodes, a Galerkin formulation of Darcy's law is constructed and seepage velocity vectors are calculated simultaneously at the nodes. Examples that demonstrate the capability of the technique are presented.

INTRODUCTION

Analysis of drainage problems, consisting mainly of the determination of drain-spacing requirements, has received the attention of many investigators (Luthin, 1957; Van Schilfgaarde, 1970). At present, estimation of drain-spacing requirements is based on steady state (Wesseling, 1964; Sakkas, 1975) and transient state concepts (Dumm, 1964; Renner and Mueller, 1974). Based on transient state concepts, analytical formulas have been derived for designing subsurface drainage systems for irrigated areas (Dumm and Winger, 1964; Dumm, 1968; Renner and Mueller, 1974). The derived formulas are based on one-dimensional flow in the saturated zone and neglect unsaturated flow. Unsaturated flow, however, plays an important role in the drainage of agricultural lands.

The purpose of this paper is to investigate the transient movement and distribution of a dissolved chemical substance in a saturated-unsaturated subsurface drainage system by means of the finite element method. In recent years, the finite element method has been used extensively to numerically simulate solute transport in porous media. The emphasis, however, has been in solute transport through a saturated medium with two-dimensional confined flow (Pinder, 1973; Cheng, 1974; Segol et al, 1975) and unconfined flow (Cabrera and Marino, 1976; Haji-Djafari and Wiggert, 1976; Marino, 1976). Little attention has been given to two-dimensional solute transport in unconfined flows where the entire subsurface domain is treated as one region having saturated and unsaturated zones whose extent is determined as part of the analysis (Marino, 1977).

The simulation of solute transport in a subsurface drainage system involves the solution of two partial differential equations. One differential equation is the transient flow equation that describes the flow of water in a saturated-unsaturated porous medium. If the pressure head distribution in the medium is given, the flow can be calculated by means of Darcy's law. The other differential equation is the convective-dispersion (transport) equation that describes the solute concentration in the system.

The flow equation is solved by a Galerkin-type finite element method described in detail by Neuman (1973). To obtain continuous flow across elements and at the nodes, a Galerkin formulation of Darcy's law is constructed (Haji-Djafari and Wiggert, 1976) and seepage velocity vectors are calculated simultaneously at the nodes. These transient velocities are subsequently used in computing dispersion coefficients and convective terms of the transport equation. A Galerkin-based finite element procedure is also employed to solve the convective-dispersion equation yielding the concentration of solute in the flow domain.

GOVERNING EQUATIONS

The equation describing transient flow in a saturated-unsaturated porous medium may be written (Neuman, 1973) as

$$\frac{\partial}{\partial x_i}\left(K_r K^s_{ij} \frac{\partial \psi}{\partial x_j} + K_r K^s_{i3}\right) = \left(\zeta + \frac{\theta}{\phi} S_s\right) \frac{\partial \psi}{\partial t} \qquad (1)$$

in which: ψ = pressure head (assumed to be a single-valued function of θ); θ = volumetric soil moisture content; K_r = relative hydraulic conductivity ($0 \leq K_r \leq 1$); K^s_{ij} = saturated hydraulic conductivity tensor; x_i ($i = 1,2,3$) = spatial coordinates (x_1 the horizontal and x_3 the vertical); t = time; $\zeta = \partial\theta/\partial\psi$ = specific moisture capacity; ϕ = porosity; and S_s = specific storage.

The appropriate initial and boundary conditions are: initial distribution of the pressure head:

$$\psi(x_i,0) = \psi_0(x_i) \qquad (2)$$

prescribed pressure head at each point along the boundary:

$$\psi(x_i,t) = \Psi(x_i,t) \qquad (3)$$

and prescribed flux normal to the boundary:

$$K_r\left[K^s_{ij}\ (\partial\psi/\partial x_j) + K^s_{i3}\right] n_i = -\ V(x_i,t) \qquad (4)$$

in which: ψ_0 = known function of x_i; Ψ and V = prescribed functions of x_i and t; and n_i = unit outer normal vector on the boundary of the flow region, S. It is not uncommon to have mixed boundary conditions, Equations (3) and (4), prescribed along the boundary of the flow region.

Using Darcy's law, the components of the seepage velocity may be written as

$$v_i = -\frac{K_r K^s_{ij}}{\theta}\frac{\partial h}{\partial x_j} \qquad i,j = 1,2,3 \qquad (5)$$

where h = $\psi + x_3$ = hydraulic head.

The equation describing the mass transport and dispersion of dissolved chemical constituents in a porous medium in which there are no chemical reactions may be expressed as

$$\frac{\partial}{\partial x_i}\left(D_{ij}\frac{\partial C}{\partial x_j}\right) - \frac{\partial}{\partial x_i}(v_i C) + J_i = \frac{\partial C}{\partial t} \qquad i,j = 1,2,3 \qquad (6)$$

in which: C = mass concentration of dissolved solids; D_{ij} = hydrodynamic dispersion coefficient, a second rank tensor; v_i = seepage velocity in direction i; J_i = mass flux of a source or sink; x_i = spatial coordinates; and t = time.

As an initial condition, the concentration distribution in the flow domain at some initial time t = 0 needs to be specified, i.e.,

$$C(x_i,0) = C_0(x_i) \qquad (7)$$

where C_0 = known function of x_i. The general equation of the boundary conditions for Equation (6) may be written as

$$f_1(x_i,t) + f_2(x_i,t)\ C + f_3(x_i,t)\ D_{ij}(\partial C/\partial x_j)n_i = 0 \qquad (8)$$

where f_1, f_2 and f_3 = prescribed functions of x_i and t. Three different conditions are applicable:
prescribed concentration on the boundary:

$$C = -f_1/f_2 \qquad (9)$$

prescribed flux normal to the boundary:

$$D_{ij}(\partial C/\partial x_j)n_i = - f_1/f_3 \tag{10}$$

and prescribed concentration and its normal derivative on the boundary:

$$(f_2/f_3)\ C + D_{ij}(\partial C/\partial x_j)n_i = - f_1/f_3 \tag{11}$$

Usually along the boundary one has mixed boundary conditions in which Equation (9) is specified over a part of the boundary and Equation (10) is specified for the remaining portion.

The coefficient of hydrodynamic dispersion may be expressed (Bear, 1972) as

$$D_{ij} = D'_{ij} + D''_{ij} \tag{12}$$

where D'_{ij} = coefficient of convective diffusion (mechanical dispersion) and D''_{ij} = coefficient of molecular diffusion. For most situations, the contribution of D''_{ij} to hydrodynamic dispersion is negligible when compared to D'_{ij}. Hence, for all practical purposes we may assume that $D_{ij} \simeq D'_{ij}$.

A comprehensive discussion of the factors affecting the dispersion coefficient is given by Bear (1972). For an isotropic porous medium, the coefficient of mechanical (or convective) dispersion can be written as

$$D'_{ij} = a_{II}\ \delta_{ij}\ v + (a_I - a_{II})\ v_i v_j/v, \qquad i,j = 1,2,3 \tag{13}$$

in which: a_I, a_{II} = longitudinal and transversal dispersivities of the porous medium, respectively; v_i, v_j = components of the seepage velocity in the i and j directions, respectively; v = magnitude of the velocity; and δ_{ij} = Kronecker delta (i.e., δ_{ij} = 1 if i = j and δ_{ij} = 0 if i ≠ j). Equation (13) is used in the present analysis to calculate $D'_{ij} \simeq D_{ij}$.

GALERKIN-FINITE ELEMENT METHOD

Galerkin's approach is a means of obtaining an approximate solution to a differential equation by requiring that the residual or error of the approximation be orthogonal to prescribed functions of coordinates or shape functions (Zienkiewicz, 1971; Pinder and Gray, 1977).

Equation (1) can be rewritten as

$$L(\psi) = \frac{\partial}{\partial x_i}\left(K_r K^s_{ij}\ \frac{\partial \psi}{\partial x_j} + K_r K^s_{i3}\right) - \left(\zeta + \frac{\theta}{\phi}\ S_s\right)\frac{\partial \psi}{\partial t} = 0 \tag{14}$$

in which L is a quasilinear differential operator and ψ denotes the exact solution. The approximate solution can be represented by

$$\overline{\psi} = \sum_{n=1}^{N} \beta_n(x_i)\ \psi_n(x_i,t) \tag{15}$$

where $\{\beta_n\ (x_i)\}_{n=1}^{N}$ is a set of linearly independent shape functions and $\{\psi_n\ (x_i,t)\}_{n=1}^{N}$ is a set of values of pressure heads to be determined at time t. Substitution of Equation (15) into Equation (14) results in a residual ε, i.e.,

$$L(\overline{\psi}) = \varepsilon \neq 0 \tag{16}$$

The objective is to make the residual as small as possible. This can be accomplished by requiring that

$$\int_R \beta_n(x_i)\ \varepsilon\ dR = 0 \tag{17}$$

for each $\beta_n(x_i)$ over a region of interest R. That is,

$$\int_R \beta_n(x_i)\ L(\overline{\psi})\ dR = 0, \qquad n = 1,2,\ldots,\ N \tag{18}$$

Equation (18), coupled with the finite element method, enables the determination of pressure heads in the flow region at a given instant of time.

The finite element discretization scheme involves the subdivision of the flow region into a network of elements. For a two-dimensional region it is convenient to use a network of linear triangular elements (Neuman, 1973) as shown in Fig. 1a. The corners of these elements are referred to as nodal points and x_i^n are the space coordinates of the nth node. Node n is associated with a region R^n (Fig. 1a) which is a subset of the total flow region R. For each node n there exists a shape function $\beta_n(x_i)$ which is linear in x_i inside each element and piecewise linear over R (Neuman, 1973) such that

$$\beta_n(x_i^m) = \delta_{nm} \quad \text{for all } x_i^m \text{ in } R \tag{19}$$

$$\beta_n(x_i) = 0 \quad \text{for all } x_i \text{ not in } R \tag{20}$$

Determination of the equations for $\psi_n\ (x_i,t)$, n = 1,2,..., N, involves the integration or derivation of the shape functions $\beta_n(x_i)$ over each element. These operations can be carried out most conveniently when the shape functions are written in terms of a local or elemental coordinate system. Following the approach of Neuman (1973), let $\{\beta_n^e(x_i)\}$ be a set of local shape functions for a single element R^e (Fig. 1b) that are linear in x_i and satisfy the requirements

$$\beta_n^e(x_i^m) = \delta_{nm} \quad \text{for all } x_i^m \text{ in } R^e \tag{21}$$

$$\beta_n^e(x_i) = 0 \quad \text{for all } x_i \text{ not in } R^e \tag{22}$$

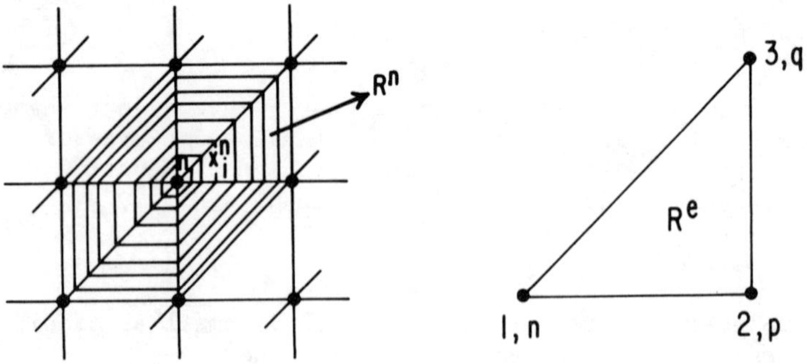

Figure 1 Schematic representation of (a) network of triangular elements and (b) single triangular element.

Equations (19) through (22) imply that the global shape functions $\beta_n(x_i)$ are the union of the local shape functions $\beta_n^e(x_i)$. Thus, Equation (18) can be rewritten in terms of local shape functions so that

$$\int_R \beta_n(x_i)\, L(\bar{\psi})\, dR = \sum_e \int_{R^e} L(\psi_m \beta_m^e)\, \beta_n^e\, dR = 0 \tag{23}$$

For unsaturated flow, Neuman (1973) recommends that the nodal values of the time derivatives be defined as weighted averages of $\partial\psi/\partial t$ over the entire flow region R, i.e.,

$$\frac{\partial \psi_n}{\partial t} = \frac{\int_R \left(\zeta + \frac{\theta}{\phi} S_s\right) \frac{\partial \psi}{\partial t} \beta_n\, dR}{\int_R \left(\zeta + \frac{\theta}{\phi} S_s\right) \beta_n\, dR} \tag{24}$$

In addition, K_{ij}^s, ϕ, and S_s are assumed to be constant in each element, while K_r, ζ, and θ vary linearly according to

$$K_r = K_{r\iota}\, \beta_\iota^e \tag{25}$$

$$\zeta = \zeta_\iota\, \beta_\iota^e \tag{26}$$

$$\theta = \theta_\iota\, \beta_\iota^e \tag{27}$$

in which ι stands for the corners of the triangle.

By combining Equations (23) through (27), and using Green's first identity, Neuman (1973) arrives at a system of first-order differential equations

$$A_{nm}\psi_m + P_{nm}\frac{\partial \psi_m}{\partial t} = Q_n - E_n, \qquad n,m = 1,2,\ldots, N \tag{28}$$

in which, for a vertical cross section

$$A_{nm} = \sum_e K_{r\iota} K^s_{ij} \int_{R^e} \beta^e_\iota \frac{\partial \beta^e_n}{\partial x_i} \frac{\partial \beta^e_m}{\partial x_j} \, dR \tag{29}$$

$$P_{nm} = \sum_e \int_{R^e} \beta^e_\iota \, \zeta_\iota + \frac{S_s}{\phi} \, \theta_\iota \, \beta^e_n \, dR \qquad \text{if } n = m \tag{30}$$

$$P_{nm} = 0 \text{ if } n \neq m \tag{31}$$

$$Q_n = - \sum_e \int_{S^e} V \beta^e_n \, dS \tag{32}$$

$$E_n = \sum_e K_{r\iota} K^s_{i3} \int_{R^e} \beta^e_\iota \frac{\partial \beta^e_n}{\partial x_i} \, dR \tag{33}$$

To provide continuous velocity functions across elements and at the nodes, the Galerkin-type finite element formulation of Equation (5) is employed. The concept has been presented by Haji-Djafari and Wiggert (1976) and used by Marino (1977). It is similar to that given by Zienkiewicz (1971) and used by Segol, et al. (1975). The approximate solutions of the hydraulic head and velocity vectors, respectively, can be written as

$$\bar{h} = \sum_{n=1}^{N} \beta_n(x_i) \, h_n(x_i, t) \tag{34}$$

$$\bar{v}_i = \sum_{n=1}^{N} \beta_n(x_i) \, v_{ni}(x_i, t) \tag{35}$$

The residual resulting from the substitution of Equation (35) into Equation (5) is

$$L(\bar{v}_i) = \bar{v}_i + K_r \frac{K^s_{ij}}{\theta} \frac{\partial \bar{h}}{\partial x_i} = \varepsilon_1 \neq 0 \tag{36}$$

It follows that

$$\int_R \beta_n(x_i) \, L(\bar{v}_i) \, dR = \sum_e \int_{R^e} L\left(v_{mi} \beta^e_m\right) \beta^e_n \, dR = 0 \tag{37}$$

By using Equations (34) through (37) one obtains a set of equations

$$[F] \, \{v_i\} = \{H_i\} \tag{38}$$

in which

$$[F] = \sum_e \int_{R^e} \beta_n \beta_m \, dR \tag{39}$$

$$\{H_i\} = - \sum_e \int_{R^e} \frac{K_r K^s_{ij}}{\theta} \beta_n \frac{\partial \beta_m}{\partial x_i} h_n \, dR \tag{40}$$

where [F] is a banded symmetrix matrix and $\{H_i\}$ is a known column vector. The solution of Equation (38) gives the ith component of the seepage velocity vectors at each node.

As with the flow equation, Equation (6) can be rewritten as

$$L(C) = \frac{\partial}{\partial x_i}\left(D_{ij} \frac{\partial C}{\partial x_j}\right) - \frac{\partial}{\partial x_i}(v_i C) + J_i - \frac{\partial C}{\partial t} = 0 \tag{41}$$

in which L is a differential operator and C is the exact solution. The approximate solution can be expressed as

$$\overline{C} \sum_{n=1}^{N} \beta_n(x_i) \, C_n(x_i, t) \tag{42}$$

in which $\{C_n(x_i, t)\}_{n=1}^{N}$ is a set of N values of nodal solute concentrations yet to be determined. Substitution of Equation (42) into Equation (41) results in a residual

$$L(\overline{C}) = \varepsilon_2 \neq 0 \tag{43}$$

To minimize the residual, the orthogonality condition requires that

$$\int_R \beta_n(x_i) \, L(\overline{C}) \, dR = 0, \qquad n = 1,2,\ldots, N \tag{44}$$

By applying Equations (19) through (22) one arrives at a functional similar to Equation (23), i.e.,

$$\int_R \beta_n(x_i) \, L(\overline{C}) \, dR = \sum_e \int_{R^e} L\left(C_m \beta^e_m\right) \beta^e_n \, dR = 0 \tag{45}$$

By combining Equations (41) and (44), and using Green's theorem, one obtains a system of differential equations which can be expressed as

$$[B]\,\{C\} + [F]\left\{\frac{\partial C}{\partial t}\right\} + [G]^T\,[I]\,\{C\} = [F]\,\{J\} + \{K\} \tag{46}$$

in which [F] is defined by Equation (39) and

$$[B] = \sum_e \int_{R^e} \left(D_{ij} \frac{\partial \beta_n}{\partial x_i} \frac{\partial \beta_m}{\partial x_j} + \beta_m \beta_n \frac{\partial v_i}{\partial x_i} + v_i \beta_n \frac{\partial \beta_m}{\partial x_i}\right) dR \tag{47}$$

$$\{G\} = \sum_e \int_{R^e} \beta_n (f_2/f_3)\, dR \tag{48}$$

$$= 0 \text{ if } f_3 = 0 \tag{49}$$

$$\{K\} = \sum_e \int_{R^e} \beta_n (f_1/f_3)\, dR \tag{50}$$

$$= 0 \text{ if } f_3 = 0 \tag{51}$$

where $[G]^T$ is the transpose of $\{G\}$ and $[I]$ is the identity matrix.

SOLUTION OF SYSTEM EQUATIONS

A fully implicit backward difference scheme (Neuman, 1973) is employed to approximate $\partial\psi_m/\partial t$ in Equation (28). Thus, Equation (28) becomes

$$\left[A_{nm}^{k+1/2} + \frac{1}{\Delta t_k} P_{nm}^{k+1/2}\right] \psi_m^{k+1} = Q_n^{k+1/2} - E_n^{k+1/2}$$

$$+ \frac{1}{\Delta t_k} P_{nm}^{k+1/2} \psi_m^k, \qquad n,m = 1,2,\ldots, N \tag{52}$$

in which k represents the time; $t = t_k$; $\Delta t_k = t_{k+1} - t_k$; $t_{k+1/2} = t_k + \Delta t_k/2$; and N = total number of nodes in the network of elements. Evaluation of Equation (52) leads to N simultaneous equations. The matrix resulting from $A_{nm}^{k+1/2}$ and $P_{nm}^{k+1/2}$ is banded and symmetric. An efficient Gaussian elimination technique is incorporated into a computer program to solve the simultaneous equations. The resulting nodal pressure heads allow the determination of the ψ=0 surface which separates between the saturated and the unsaturated portions of the porous medium, the hydraulic head distribution in the medium, and the seepage velocities. The latter are used in computing dispersion coefficients and convective terms of the transport equation.

The integration of the global relations for the transport equation, Equation (46), is carried out by means of the Crank-Nicholson Implicit Method. Thus, Equation (46) can be rewritten as

$$\left(\frac{1}{\Delta t_k}[F]^{k+1/2} + \frac{1}{2}\left([B]^{k+1/2} + [G]^{Tk+1/2}[I]\right)\right) \{C\}^{k+1/2}$$

$$= \left(\frac{1}{\Delta t_k}[F]^{k+1/2} - \frac{1}{2}\left([B]^{k+1/2} + [G]^{Tk+1/2}[I]\right)\right) \{C\}^{k}$$

$$+ [F]^{k+1/2}\{J\} + \{K\}^{1+1/2} \tag{53}$$

in which $\Delta t_k = t_{k+1} - t_k$ and $t_{k+1/2} = t_k + \Delta t_k/2$. Evaluation of Equation (53) leads to N simultaneous equations which upon solution yield the nodal solute concentrations.

APPLICATION

The Galerkin-finite element models are employed to determine the flow pattern and the distribution of a solute within subsurface drainage systems. A schematic representation of a saturated-unsaturated subsurface drainage system is presented in Fig. 2. Because of symmetry, only one half of the flow system needs to be considered. The initial flow conditions are those of equilibrium for zero infiltration. In the flow and dispersion problems to be investigated, the following conditions hold: the dimensions of the porous medium are 3.2 m high and 10 m long; a circular drain with a radius of 0.1 m is located 2.0 m below the soil surface; during the period of flow, the drain always runs half-full; the initial pressure head distribution is -2.0 m at the surface, 1.2 m at the impermeable stratum, and 0 m at the surface which separates between the saturated and the unsaturated portions of the porous medium; infiltration occurs at a specified uniform rate into the porous medium; functional relationships between pressure head, relative hydraulic conductivity, and moisture content are known (Fig. 3); the initial solute concentration in the porous medium is 1 mg/ℓ; the infiltration supplies a continuous source of solute at a known concentration of 10 mg/ℓ; and the longitudinal and transversal dispersivities of the medium are assumed to be constant over the entire domain and taken to be 20 and 5 m, respectively.

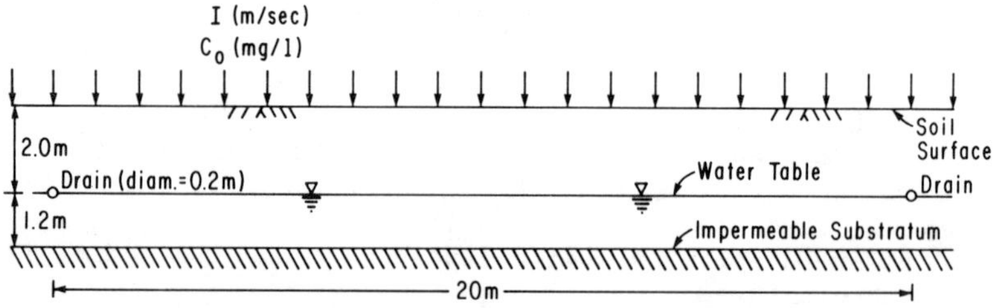

Figure 2 Schematic representation of subsurface drainage system.

First, we consider a subsurface drainage situation in which infiltration occurs at a rate I = 0.0033 m/min into a soil with moisture content at saturation $\theta = 0.342$ and saturated hydraulic conductivities $K_x = K_y = 0.0576$ m/min (Fig. 3). The Galerkin-finite element model was run for 86 min of simulated flow time with a variable time step ranging from 0.1 to 10 min. Some of the results obtained, indicating the position

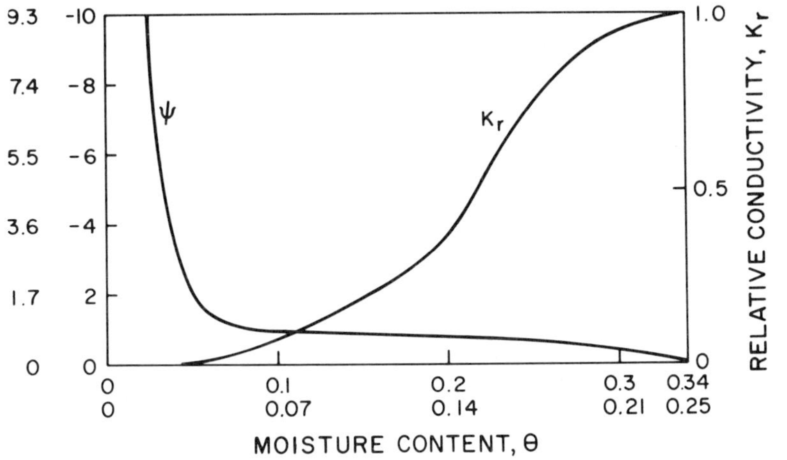

Figure 3 Relationships between pressure head, relative conductivity, and moisture content.

of the ψ=0 surface which separates between the saturated and the unsaturated portions of the porous medium, are presented in Fig. 4. As expected, the height of the ψ=0 surface increases with time. During the 86 min flow period the ψ=0

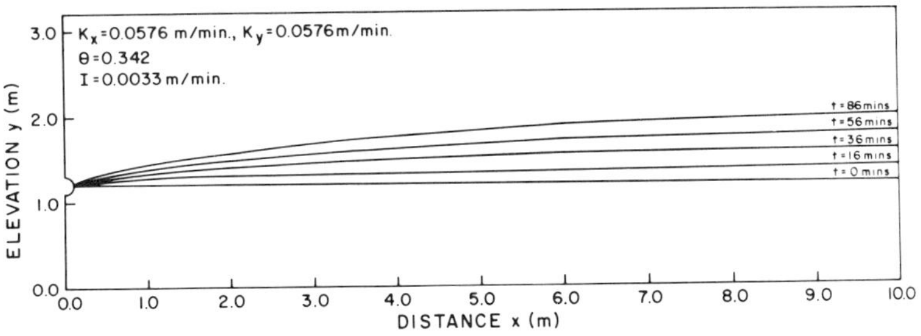

Figure 4 Position of the ψ=0 surface at the given times.

surface attains a maximum value of 1.98 m at x = 10 m. Lines of equal solute concentration at the end of 16, 36, 56, and 86 min after initiation of recharge are illustrated in Fig. 5. At t = 16 min (Fig. 5a), a solute concentration of 10 mg/ℓ has penetrated to approximately 1.1 m below the soil surface. The solute concentration near the drain is 8 mg/ℓ at t = 16 min (Fig. 5a) and 10 mg/ℓ at t = 36 min (Fig. 5b). At t = 56 min (Fig. 5c), a substantial portion of the porous medium has a concentration of 10 mg/ℓ. Nearly all of the medium has a concentration of 10 mg/ℓ at t = 86 min (Fig. 5d).

Second, the model was run for 86 min of simulated flow time with an infiltration rate I = 0.0006 m/min occurring into

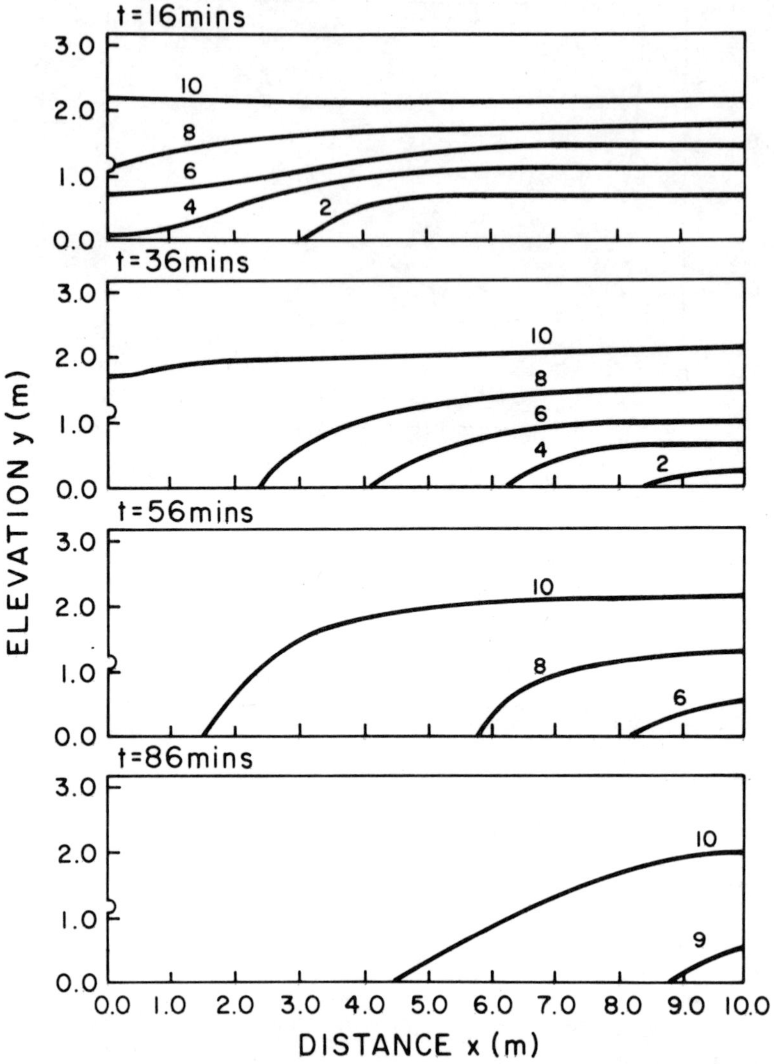

Figure 5 Lines of equal concentration at the given times (first problem).

a soil with θ = 0.250 and $K_x = K_y$ = 0.288 m/min at saturation (Fig. 3). When compared to the first problem, these parameter values reflect a decrease in I by a factor of 5, a decrease in θ by a factor of 1/4, and an increase in K by a factor of 5. Although not shown in an illustration, the position of the ψ=0 surface is almost stationary during the period of flow. This is due to the large value of K and small value of I used in the model. The ψ=0 surface attains a steady configuration at t = 16 min, with a maximum value of 1.24 m at x = 10 m. The distribution of the solute at various flow times is shown in Fig. 6. Because of the relatively high value of the hydraulic conductivity, the accumulation of solute in the porous medium is not as pronounced as in the previous problem. At t = 16 min (Fig. 6a) the solute concentration near the drain is 5 mg/ℓ, whereas at t = 36 min (Fig. 6b) it is 8 mg/ℓ. Concentrations of 9 mg/ℓ or more occur in approximately half of the porous medium at t = 56 min (Fig. 6c). At t = 86 min (Fig. 6d), concentrations of 8 mg/ℓ or more occur in the entire medium.

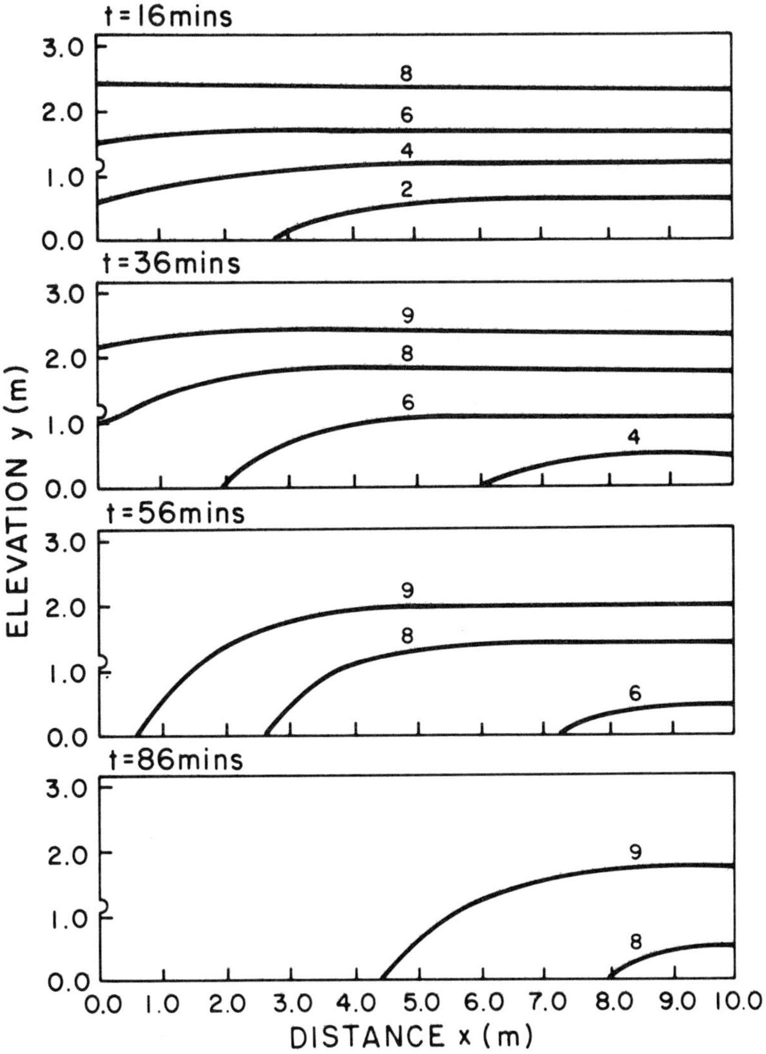

Figure 6 Lines of equal concentration at the given times (second problem).

Third, parameter values of I = 0.0006 m/min, θ = 0.025, K_x = 0.0576 m/min, and K_y = 0.0115 m/min are used in the model. The values of I and θ are the same as those used in the second problem but the conductivities reflect a decrease in K_x and K_y by a factor of 5 and 25, respectively. Inasmuch as the hydraulic conductivity in the horizontal direction is now five times larger than that in the vertical direction, the ψ=0 surface gradually increases during the period of flow. The resulting solute distribution in the drain system at various times after initiation of recharge is shown in Fig. 7. The concentration of solute in the porous medium is not as pronounced as in the previous problem. Figure 7a shows that at t = 16 min, the solute concentration gradually diminishes from 10 mg/ℓ at the soil surface to 2 mg/ℓ at approximately 2.5 m below the soil surface. At t = 36 min (Fig. 7b) the concentration near the drain is 6 mg/ℓ, whereas at t = 56 min (Fig. 7c) it is 8 mg/ℓ. At t = 86 min (Fig. 7d), concentrations of 5 mg/ℓ or more occur in the entire medium.

CONCLUSIONS

The Galerkin-type finite element model presented herein is useful in analyzing seepage through the subsurface. It is particularly well suited for the analysis of subsurface drainage systems, including the determination of drain-spacing requirements and salt distribution in the subsurface.

Inasmuch as seepage velocity components appear in the convective terms of the transport equation and are parameters in the dispersion coefficients, they play a major role in predicting solute transport in a saturated-unsaturated porous medium. Hence, accurate evaluation of the velocity components is very important. The Galerkin formulation of Darcy's law yields a reasonable estimate of seepage velocity vectors.

This study is a preliminary step in predicting the transient movement and distribution of a solute in a saturated-unsaturated subsurface drainage system. Future work should include a porous medium composed of various strata, chemical reactions taking place in the medium, and verification using field and laboratory data. In addition, particular attention should be given to proper determination in the field of the functional relationships between pressure head, relative conductivity, and moisture content and the variation of longitudinal and transversal dispersivities.

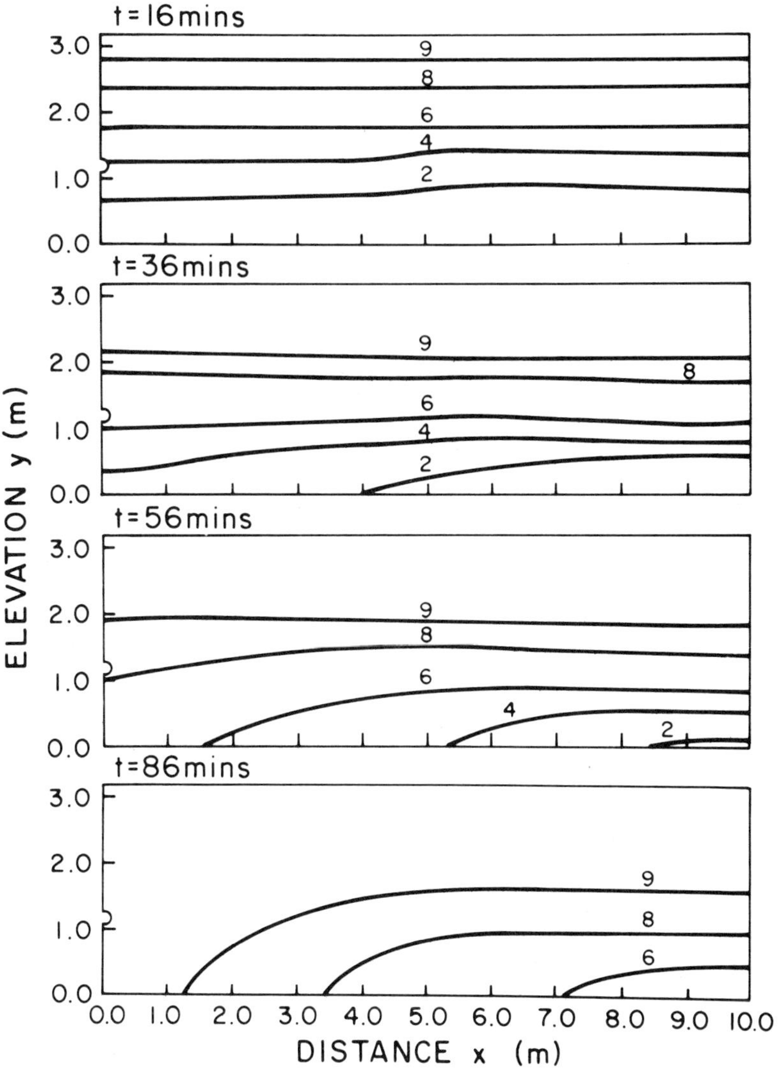

Figure 7 Lines of equal concentration at the given times (third problem).

ACKNOWLEDGMENT

The research leading to this report was supported by the University of California, Water Resources Center, as part of Water Resources Center Project UCAL-WRC-W-521. In addition, computational work was partially supported by CSRS Projects CA-D*-WSE-3081-H and CA-D*-WSE-3479-H.

REFERENCES

Bear, J. (1972) Dynamics of Fluids in Porous Media. American Elsevier Publishing Co., New York.

Cabrera, G. and M. A. Marino (1976) A Finite Element Model of Contaminant Movement in Groundwater. Water Resources Bulletin AWRA, 12, 2:317-335.

Cheng, R. T. (1974) On the Study of Convective Dispersion Equation, in Finite Element Methods in Flow Problems, edited by J. T. Oden et al., The University of Alabama Press, 29.

Dumm, L. D. (1964) Transient Flow Concept in Subsurface Drainage. Trans. Amer. Soc. Agric. Engrs., 7, 2:142-146.

Dumm, L. D. (1968) Subsurface Drainage by Transient-Flow Theory. J. Irr. Drain. Div. ASCE, 94, IR4:505-519.

Dumm, L. D. and R. J. Winger (1964) Subsurface Drainage System Design for Irrigated Areas Using the Transient-Flow Concept. Trans. Amer. Soc. Agric. Engrs., 7, 2:147-151.

Haji-Djafari, S. and D. C. Wiggert (1976) Two-dimensional Analysis of Tracer Movement and Transient Flow in Phreatic Aquifer. Paper presented at Second Int. Symp. on Finite Elem. Meth. in Flow Problems, Parallo, Italy.

Luthin, J. N., editor (1957) Drainage of Agricultural Lands. Amer. Soc. of Agron., Madison, Wisconsin.

Marino, M. A. (1976) Numerical Simulation of Contanimant Transport in Subsurface Systems. Proc. Symp. on Advances in Groundwater Hydrology, Amer. Water Resour. Assoc., Minnesapolis, Minn.

Marino, M. A. (1977) Solute Transport in a Saturated-Unsaturated Porous Medium. Paper presented at IFIP Working Conf. on Modeling and Simulation of Land, Air and Water Resources Systems, Ghent, Belgium.

Neuman, S. P. (1973) Saturated-Unsaturated Seepage by Finite Elements. Jour. Hydr. Div. ASCE, 99, HY12:2233-2250.

Pinder, G. F. (1973) A Galerkin-Finite Element Simulation of Groundwater Contamination on Long Island, New York. Water Resour. Res. 9, 6:1657-1669.

Pinder, G. F. and W. G. Gray (1977) Finite Element Simulation in Surface and Subsurface Hydrology. Academic Press, New York.

Renner, D. M. and C. C. Mueller (1974) Drainage System Design and Analysis by Computer. Jour. Irr. Drain. Div. ASCE, 100, IR3:255-265.

Sakkas, J. G. (1975) Generalized Nomographic Solution of Hooghoudt Equations. Jour. Irr. Drain. Div. ASCE, 101, IR1: 21-39.

Segol, G., G. F. Pinder, and W. G. Gray (1975) A Galerkin-Finite Element Technique for Calculating the Transient Position of the Saltwater Front. Water Resour. Res., 11, 2: 343-347.

Van Schilfgaarde, J. (1970) Theory of Flow to Drains, in Advances in Hydroscience, Vol. 6, edited by V. T. Chow, Academic Press, 43-106.

Wesseling, J. (1964) A Comparison of the Steady-State Drain-Spacing Formulas of Hooghoudt and Kirkham in Connection with Design Practice. Jour. of Hydrology, 2, 25-32.

Zienkiewicz, O. C. (1971) The Finite Element Method in Engineering Science. McGraw-Hill, New York.

SIMULATION OF ONE-DIMENSIONAL UNSTEADY DRAINAGE IN POROUS MEDIA

F. Stauffer

Federal Institute of Technology Zurich

INTRODUCTION

The important role of the unsaturated zone during unsteady drainage processes in porous media has often been shown. Compared to saturated flow, two phase flow involves much more complicated problems. One of these problems is the significance of experimentally found dynamic effects in the relation between capillary pressure and water content (Topp et al., 1967) and between water content and conductivity (Stauffer, 1977). These dynamic effects are approximatively taken into account in the mathematical simulation model presented here.

BASIC EQUATIONS

Drainage processes in one-dimensional vertical homogeneous and isotropic porous media are investigated for columns of fine sand. Consider the following situation: The sand column has a given initial water content distribution. The medium is drained through the lower end of the column by an outflow system (figure 1). The outflow system consists of a fine layer, filter layers and a tube connection with a fixed outflow level. The grain size distribution of the fine layer is such that air does not enter into the outflow system; the outflow system thus remains saturated. Medium and fluid are incompressible and isothermal. Flow condition, medium and fluid quality are so that the flow can macroscopically be described by Darcy's law

$$q = - k_w \cdot \frac{\partial}{\partial z} \left(- z + \frac{p_w}{\rho_w \cdot g} \right) \tag{1}$$

Here, q is the specific flux, k_w is the fluid conductivity, p_w is the fluid pressure, ρ_w is the density of the fluid, g the gravitational constant. The space coordinate z is taken positive downwards. Frictional effects due to air flow in the unsaturated medium are neglected as well as evaporation.

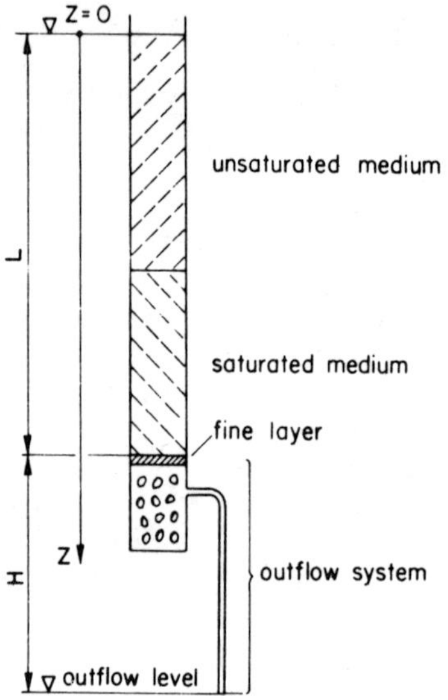

Figure 1 Sand column with outflow system

Under these conditions and taking into consideration the flow continuity, the unsaturated flow is governed by the Richard's equation

$$\phi = \frac{\partial S}{\partial t} = \frac{\partial}{\partial z} \left(k_w \cdot \frac{\partial}{\partial z} \left(- z + \frac{p_w}{\rho_w \cdot g} \right) \right) \tag{2}$$

where ϕ is the porosity, S is the saturation and t is the time. From the porosity and the saturation, the water content θ can be determined as

$$\theta = \phi \, . \, S \tag{3}$$

Since drainage processes of an initially saturated medium are considered, it is advantageous to replace S by the effective saturation S_e

$$S_e = \frac{S - S_r}{1 - S_r} \tag{4}$$

S_r is the residual saturation. For the description of the stationary relation between capillary pressure p_c and saturation and between saturation and conductivity, the approach of Brooks and Corey (1964) is used

$$S_e = \left(\frac{p_b}{p_c}\right)^{\lambda} \qquad p_c \geq p_b \tag{5}$$

and

$$k_w = k_s \cdot S_e^{(3 + \frac{2}{\lambda})} \tag{6}$$

p_b, λ and k_s are experimentally determined constants. Since air flow is neglected, the fluid pressure is

$$p_w = - p_b \cdot S_e^{(- \frac{1}{\lambda})} \tag{7}$$

The dynamic effects in relations (5) and (6) are taken into consideration by approximative corrections. The first correction concerns the difference in capillary pressure for transient and stationary flow conditions at the same water content

$$\Delta p_c = - \alpha \cdot \rho_w \cdot g \cdot \phi_e \cdot \frac{\partial S_e}{\partial t} \tag{8}$$

where α is an experimentally determined constant. ϕ_e is the effective porosity

$$\phi_e = \phi . (1 - S_r) \tag{9}$$

By considering correction (8), equation (7) is transformed into

$$p_w = - p_b . S_e^{(- \frac{1}{\lambda})} + \alpha . \rho_w . g . \phi_e . \frac{\partial S_e}{\partial t} \tag{10}$$

The dynamic effect in relation (6) is taken into account by changing the exponent, so that

$$k_w = k_s . S_e^{\varepsilon} \tag{11}$$

The exponent ε has to be determined experimentally. The value of ε does not appear to depend strongly on $\frac{\partial S_e}{\partial t}$ in the range of observation and is, therefore, taken as a constant.

Introducing these relations and the normalized length Z

$$Z = \frac{z . \rho_w . g}{p_b} \tag{12}$$

and the time T

$$T = \frac{t . k_s . \rho_w . g}{p_b . \phi_e} \tag{13}$$

one obtains Darcy's law in non-dimensional form

$$Q = S_e^{\varepsilon} . \left(- \frac{1}{\lambda} . \frac{\partial S_e}{\partial Z} . S_e^{(-1-\frac{1}{\lambda})} - \beta . \frac{\partial^2 S_e}{\partial Z . \partial T} + 1 \right) \tag{14}$$

and the Richard's equation

$$\left. \begin{aligned} \frac{\partial S_e}{\partial T} = \frac{\partial}{\partial Z} \left(\frac{1}{\lambda} . S_e^{(\varepsilon-1-\frac{1}{\lambda})} . \frac{\partial S_e}{\partial Z} + \beta . S_e^{\varepsilon} . \frac{\partial^2 S_e}{\partial Z . \partial T} \right) \\ - \varepsilon . S_e^{(\varepsilon-1)} \frac{\partial S_e}{\partial Z} \end{aligned} \right\} \tag{15}$$

with the constant β

$$\beta = \frac{\alpha . \rho_w^2 . g^2 . k_s}{p_b^2} \tag{16}$$

and the specific flux

$$Q = \frac{q}{k_s} \tag{17}$$

<u>Initial and boundary conditions</u>

The initial condition at the initial time T_A is given by the saturation profile S_e $(Z, T = T_A)$. The medium column may consist of a saturated and an unsaturated region (figure 1). The interface between saturated and unsaturated region is moving downwards, as shown as curve B-C in figure 2. Along curve B-C, the saturation is known (lower boundary condition)

$$S_e = 1$$

The initial location of the interface is point B. The subsequent motion of the interface is determined by the overall continuity condition

$$\int_0^L (S_e(T_A) - S_e(T)).dZ = \int_{T_A}^{T} Q(Z=L).dT \qquad (18)$$

Here, Q (Z = L) is the specific flux at the lower column end and is given by the flow characteristics of the outflow system

$$Q(Z=L) = \frac{k_H}{k_s} . \left\{ 1 + \frac{p_w(Z=L)}{\rho_w.g.h} \right\} \qquad (19)$$

L is the length of the sand column, h is the height of the outflow system. k_H is the conductivity of the outflow system related to the cross section of the column. Thus, Q (Z = L) represents not only the specific outflow, but also the specific flux in the saturated part of the column. When no saturated region exists, the continuity condition (18) is used to determine the saturation at the lower column end at time T (lower boundary condition). The continuity condition applied at the upper column end (along A-E) leads to the upper boundary condition.

$$S_e(Z=0,T_A) - S_e(Z=0,T) = \int_{T_A}^{T} Q(Z=0,T).dT \qquad (20)$$

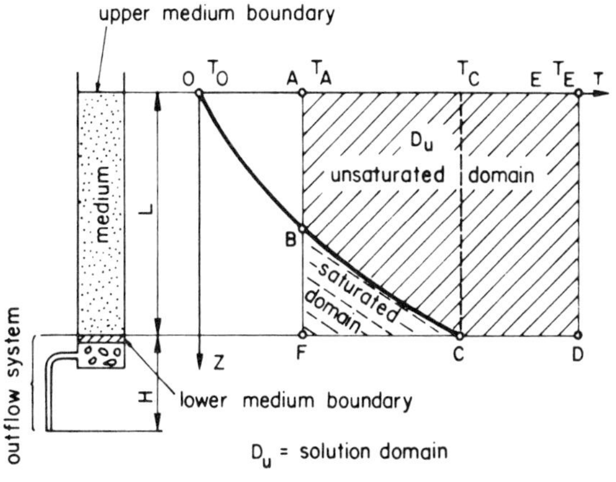

Figure 2 Representation of the solution domain of the function $S_e(Z,T)$

SIMULATION METHOD

Equation (15) has the form

$$\left.\begin{aligned} F(S_e(Z,T)) &= \frac{\partial}{\partial Z}\left(a.S_e^b.\frac{\partial S_e}{\partial Z} + c.S_e^d.\frac{\partial^2 S_e}{\partial Z.\partial T}\right) \\ &\quad + e.S_e^f.\frac{\partial S_e}{\partial Z} - \frac{\partial S_e}{\partial T} = 0 \end{aligned}\right\} \qquad (21)$$

a, b, c, d, e, f are constants. The equation is non-linear parabolic of second order in the main parts with a non-linear additional term of third order. This additional term is due to the consideration of the dynamic effects. For simulation, the finite element method and the Galerkin method are used. The method has been applied to problems of saturated groundwater flow (Pinder and Frind, 1972) and of unsaturated flow in porous media (Zyvoloski and Bruch, 1973).

Application of the finite element and the Galerkin method

The solution function S_e (Z,T) with the described initial and boundary conditions has to be found within the domain D_u in space and time (figure 2). The domain D_u is divided into n partial domains D_i. The approximate solution within such a domain D_i will be of the form

$$S_e(Z,T) = \sum_{j=1}^{r} N_j(Z,T).S_{ej}(Z_j,T_j) \qquad (22)$$

r is the number of nodes within the domain D_u. S_{ej} is the discrete representation of the unknown solution funcion at node j. N_j is a linearly independent function which is defined in the domain belonging to node j. N_j is equal to one at node j and vanishes at the other nodes in D_u. For each node k the Galerkin requirement

$$\iint_{D_u} N_k(Z,T).F(S_e(Z,T)).dZ.dT = 0 \qquad (23)$$

has to be formulated. Substituting equation (21) into (23) yields

$$\left.\begin{aligned}
&\iint_{D_u} \left[N_k \cdot \frac{\partial}{\partial Z}(a.S_e^b) \cdot \frac{\partial S_e}{\partial Z}\right] .dZ.dT \\
&+ \iint_{D_u} \left[N_k .a.S_e^b \cdot \frac{\partial^2 S_e}{\partial Z^2}\right] .dZ.dT \\
&+ \iint_{D_u} \left[N_k \cdot \frac{\partial}{\partial Z}(c.S_e^d) \cdot \frac{\partial^2 S_e}{\partial Z.\partial T}\right] .dZ.dT \\
&+ \iint_{D_u} \left[N_k .c.S_e^d \cdot \frac{\partial}{\partial Z}\left(\frac{\partial^2 S_e}{\partial Z \partial T}\right)\right] .dZ.dT \\
&+ \iint_{D_u} \left[N_k .e.S_e^f \cdot \frac{\partial S_e}{\partial Z}\right] .dZ.dT \\
&- \iint_{D_u} \left[N_k \cdot \frac{\partial S_e}{\partial T}\right] .dZ.dT = 0
\end{aligned}\right\} \quad (24)$$

By the application of partial integration on parts of the integral, it is possible to reduce the order of the derivatives.

$$\left.\begin{aligned}
&\iint_{D_u} \left[N_k .a.S_e^b \cdot \frac{\partial^2 S_e}{\partial Z^2}\right] .dZ.dT \\
&= - \iint_{D_u} \left[N_k \cdot \frac{\partial}{\partial Z} (a.S_e^b) \cdot \frac{\partial S_e}{\partial Z}\right] .dZ.dT \\
&- \iint_{D_u} \left[\frac{\partial N_k}{\partial Z} \cdot a.S_e^b \cdot \frac{\partial S_e}{\partial Z}\right] .dZ.dT \\
&+ \int_{T_1}^{T_2} \left[N_k .a.S_e^b \cdot \frac{S_e}{Z}\right]_{Z_1}^{Z_2} .dT
\end{aligned}\right\} \quad (25)$$

$$\iint_{D_u} \left[N_k . c.S_e^d . \frac{\partial}{\partial Z}\left(\frac{\partial^2 S_e}{\partial Z.\partial T}\right)\right] .dZ.dT$$

$$= - \iint_{D_u} \left[N_k . \frac{\partial}{\partial Z}(c.S_e^d) . \frac{\partial^2 S_e}{\partial Z.\partial T}\right] .dZ.dT$$

$$- \iint_{D_u} \left[\frac{\partial N_k}{\partial Z} .c.S_e^d . \frac{\partial^2 S_e}{\partial Z.\partial T}\right] .dZ.dT \qquad (26)$$

$$+ \int_{T_1}^{T_2} \left[N_k .c.S_e^d . \frac{\partial^2 S_e}{\partial Z.\partial T}\right]_{Z_1}^{Z_2} .dT$$

where T_1, T_2, Z_1, Z_2 are related to the boundary of the domain D_u. Substituting these terms, the Galerkin equation (23) becomes

$$F = - \iint_{D_u} \left[\frac{\partial N_k}{\partial Z} .a.S_e^b . \frac{\partial S_e}{\partial Z}\right] .dZ.dT + \int_{T_1}^{T_2} \left[N_k .a.S_e^b . \frac{\partial S_e}{\partial Z}\right]_{Z_1}^{Z_2} .dT$$

$$- \iint_{D_u} \left[\frac{\partial N_k}{\partial Z} .c.S_e^d . \frac{\partial^2 S_e}{\partial Z.\partial T}\right] .dZ.dT + \int_{T_1}^{T_2} \left[N_k .c.S_e^d . \frac{\partial^2 S_e}{\partial Z.\partial T}\right]_{Z_1}^{Z_2} .dT$$

$$+ \iint_{D_u} \left[N_k .c.S_e^f . \frac{\partial S_e}{\partial Z}\right] .dZ.dT - \iint_{D_u} \left[N_k . \frac{\partial S_e}{\partial T}\right] .dZ.dT = 0 \qquad (27)$$

The expressions S_e, $\frac{\partial S_e}{\partial Z}$, $\frac{\partial S_e}{\partial T}$, $\frac{\partial S_e}{\partial Z.\partial T}$ are obtained from equation (22).

Elements

Two sorts of elements are used. For the 8-point rectangular element (figure 3a), the shape function is

$$S_e(\zeta,\eta) = \alpha_1 + \alpha_2.\zeta + \alpha_3.\eta + \alpha_4.\zeta^2 + \alpha_5.\zeta.\eta + \alpha_6.\eta^2 + \alpha_7.\zeta^2\eta + \alpha_8.\zeta.\eta^2 \qquad (28)$$

where $\alpha_i = f_i(\zeta_1, \ldots \zeta_8; \eta_1, \ldots \eta_8; S_{e1}, \ldots S_{e8})$

For the 6-point triangular element (figure 3b), the shape function is

$$S_e(\zeta,\eta) = \beta_1+\beta_2.\zeta + \beta_3.\eta + \beta_4.\eta + \beta_5.\zeta.\eta+\beta_6.\eta^2 \qquad (29)$$

where $\beta_i = f_i(\zeta_1,\ldots\zeta_6;\ \eta_1,\ldots\eta_6;\ S_{e1},\ldots S_{e6})$

The coordinates ζ_i, η_i are related to the node i of the element with the function value S_{ei}. The functions N_i can be determined as follows:

$$\lfloor N \rfloor = \lfloor \Omega \rfloor . \lfloor A^{-1} \rfloor$$

where for the 8-point element for example

$$\lfloor N \rfloor = \lfloor N_1,\ldots N_8 \rfloor$$

$$\lfloor \Omega \rfloor = \lfloor 1,\zeta,\eta,\zeta^2,\zeta.\eta,\eta^2,\zeta^2.\eta,\zeta.\eta^2 \rfloor$$

$$\lfloor A \rfloor = \begin{bmatrix} \lfloor \Omega(\zeta_1,\eta_1) \rfloor \\ \lfloor \Omega(\zeta_8,\eta_8) \rfloor \end{bmatrix}$$

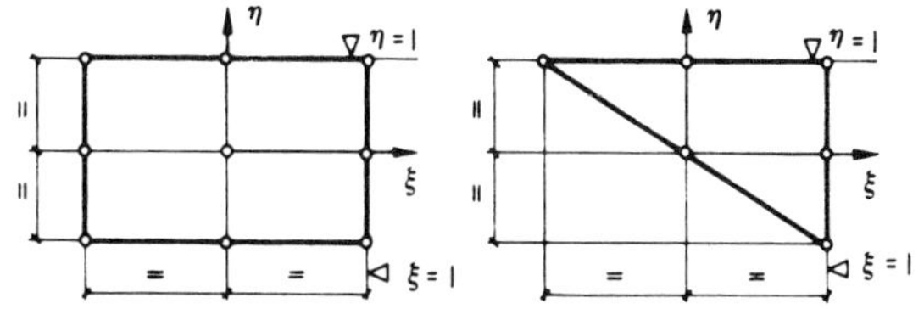

a) b)

Figure 3

8-node-element 6-node-element

Computation steps

The numerical model is realized in the programme VDRAIN. After reading of the input variables, the programme carries out a given number of computation cycles. A cycle corresponding to a time step ΔT consists of the determination of the function $S_e(Z,T)$ in a defined solution domain D_u. The method distinguishes between completely unsaturated columns (no saturated region) and between partial unsaturated columns (existence of a saturated and an unsaturated region).

The programme steps are (see figure 4)

1. The column length L is divided into N_E partial sections of equal length ΔZ.

2. At the beginning of a cycle, the interface between saturated and unsaturated region (if existant) has to be located at the end of a section (point B in figure 4).

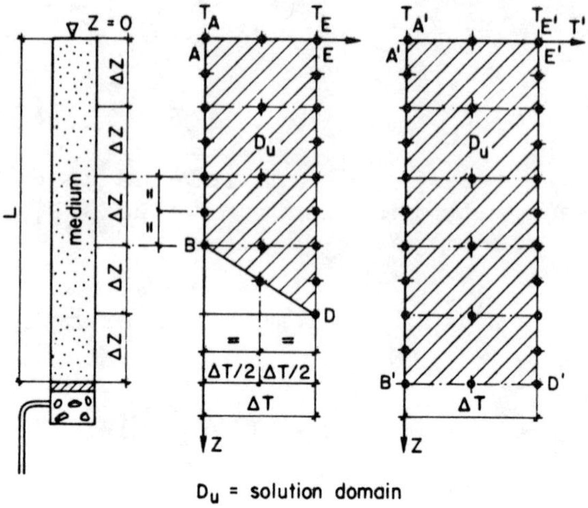

Figure 4 Solution domain for a computation cycle

3. The time step ΔT is given. The time between the interval

$$T_A \leq T \leq T_A + \Delta T$$

is considered. T_A is the initial time of the cycle. In the case of a partially unsaturated column, it is assumed that the interface between saturated and unsaturated region is moving downwards with a constant velocity

$$V = \frac{\Delta Z}{\Delta T}$$

during the considered time.

4. The solution domain S_e (Z, T) is divided into rectangles (figure 4b) and, in case of a trapezoidal domain, a triangle (figure 4a). The dimensions of these time - space - elements are ΔT and ΔZ. The elements correspond to those defined in figure 3.

5. The initial values S_{ei} at the nodes i along the boundary A-B or A'-B' respectively (figure 4) are given. For a partially unsaturated column, the values of the nodes along B-D are known as

$$S_{ei} = 1$$

6. The upper boundary condition is formulated for each node i along A-E or A'-E' respectively excluding point A or A' respectively.

$$F_i = S_e(Z=0,T=T_A) - S_e(Z=0,T_i) - \int_{T_A}^{T_i} Q(Z=0,T).dT \quad (30)$$

F_i is computed numerically with assumed values of the unknown function S_e at the nodes. For the integration of the integral in equation (3), the Gaussian quadrature formula using three terms is applied.

7. In the case of a completely unsaturated column the lower boundary condition is formulated at the nodes i along the boundary B'-D' (figure 4) excluding point B'

$$F_i = \int_0^L (S_e(Z,T_A) - S_e(Z,T_i)).dZ - \int_{T_A}^{T_i} Q(Z=L,T).dT \quad (31)$$

For the computation of Q equation (19) is used. The pressure p_w is expressed with the aid of equation (10). F_i is also computed numerically, again using assumed values of S_{ei}. The first integral is computed with the application of the Simpson rule

$$\int_0^L f(z).dz = \sum_{j=1}^{N_E} \left[\frac{\Delta z}{6} .(f(z_o(j) - \frac{\Delta z}{2}) + 4.f(z_o(j)) + f(z_o(j) + \frac{\Delta z}{2}))\right] \quad (32)$$

with $z_o = \Delta Z.(j - \frac{1}{2})$

$f(z) = S_e(Z,T_A) - S_e(Z,T_i)$

The application of the Simpson rule is useful, because the function f(Z) is of second degree within

$$Z_o - \frac{\Delta Z}{2} \leq Z \leq Z_o + \frac{\Delta Z}{2}$$

The second integral of equation (31) is computed in the same manner as outlined in step 6.

8. For the remaining nodes i the Galerkin equation (2) is formulated. The integrals are computed numerically with the aid of the Gaussian quadrature formula using three terms in each direction of an element. Assumed values of S_e are needed at the nodes.

9. For the N_U nodes i with unknown function values S_{ei} the corresponding non-linear equations are composed to a non-linear equation system

$$F_i = 0, \quad i = 1,N_U$$

For the solution of the equation system a standard routine using the Newton-Raphson method is applied. The function values S_{ei} at the nodes i are found interatively using the assumed values.

10. In the case of a partially saturated column, the overall mass conservation equation (18) is checked for time T_E. The integral

$$\int_{T_A}^{T_E} Q(Z=L,T).dT$$

is computed by the Simpson method with three terms. If the continuity condition

$$\Delta V = \int_{0}^{L} (S_e(T_A) - S_e(T)).dZ - \int_{T_A}^{T_A+\Delta T} Q(Z=L).dT \qquad (33)$$

is not satisfied within a given tolerance

$$|\Delta V| \leq \delta$$

a better approximation for the time step ΔT is computed with the aid of equation (33). In this case, steps 6 to 10 are repeated using the new time step ΔT.

11. The computation cycle is terminated and the results are printed. The computed function values S_{ei} at the nodes i along D-E resp. D'-E' are the initial values for a following cycle which involves steps 3 to 11.

Application remarks

The numerical model approximates the solution function S_e (Z, T). The result can only be accepted when the result is insensitive to the number of elements. A good approximation is of special importance, because second order derivatives of the solution function are needed. Generally, the accuracy of the model increases with the number of elements used.

The number of elements is limited by the capacity of the computer. The main computational capacity is required by the routine for solving the non-linear equation system.

The simulation of the first drainage phase of a fully saturated column (near Z = 0) has to be treated with a few elements only. Since this is the phase of the greatest relative changes of the solution function, difficulties may arise using the method presented here. In such a case, excessive gradients of the solution function occur which cannot be approximated by the numerical model. The occurance of these difficulties depends strongly on the input parameters. The difficulties manifest themselves by the fact that the iterative procedure for finding the time step ΔT does not converge.

SIMULATION

Two drainage experiments are simulated using the measured experimental parameters.

Experiment 1
The experimental parameters are

l = 16.9 cm = length of the column
ϕ_e = .30
k_s = .0142 cm/sec
$\frac{p_b}{\rho_w \cdot g}$ = 36.51 cm
λ = 3.975
h = 50.0 cm
k_H = .00514 cm/sec
α = 2400 cm· sec
ε = 3.5

Number of Elements: 5
Measurement location S1 : z = 8.4 cm
P1 : z = 3.4 cm
P2 : z = 13.4 cm

The drainage process of an initially fully saturated column is simulated beginning with the initial time T_0. Figures 5 and 6 show experimental data compared with the numerical simulation.

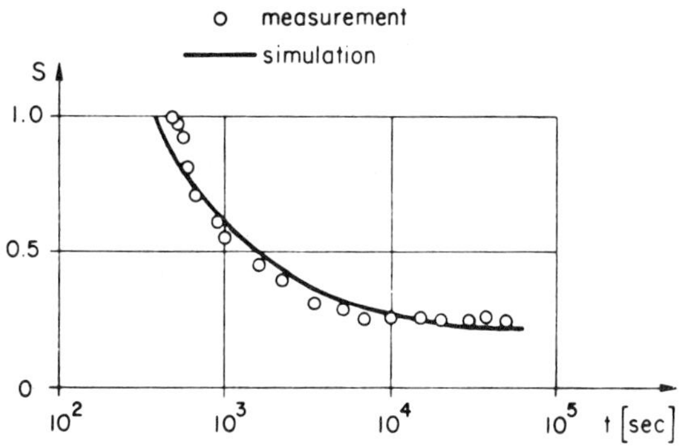

Figure 5 Function S(t) at location S1
Experiment 1 and simulation

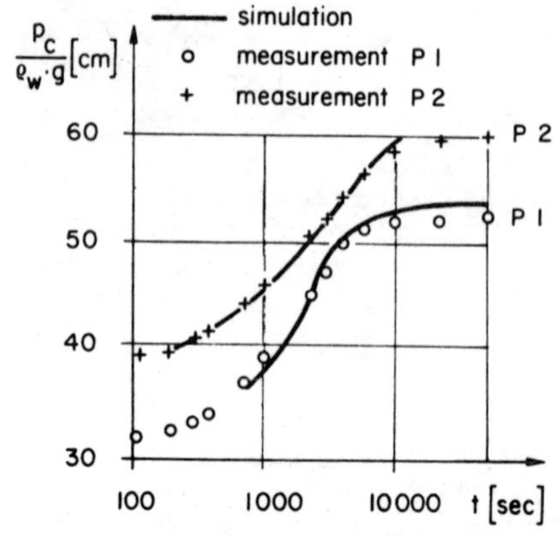

Figure 6 Function $p_c(t)$ at locations P1 and P2.
Experiment 1 and simulation

Experiment 2 (see figures 7 to 9)
The experimental parameters are

l = 63.8 cm = length of the column
ϕ_e = 0.291
k_s = 0.0211 cm/sec
$\frac{p_b}{\rho_w \cdot g}$ = 32.3 cm
λ = 6.0
h = 36.7 cm
k_H = 0.064 cm/sec
α = 820 cm· sec
ε = 2.4

Measurement locations	S1 : z	= 7.0 cm
	S2 : z	= 17.0
	S3 : z	= 27.0
	S4 : z	= 37.0
	S5 : z	= 47.0
	S6 : z	= 57.0

Number of elements for simulation: 10

A fully saturated medium column is drained. The process could not be simulated in the first drainage phase due to the difficulties outlined in the application remarks. The simulation begins with a partial desaturation of the column. The initial values of the saturation S (z) are taken from the experimental data.

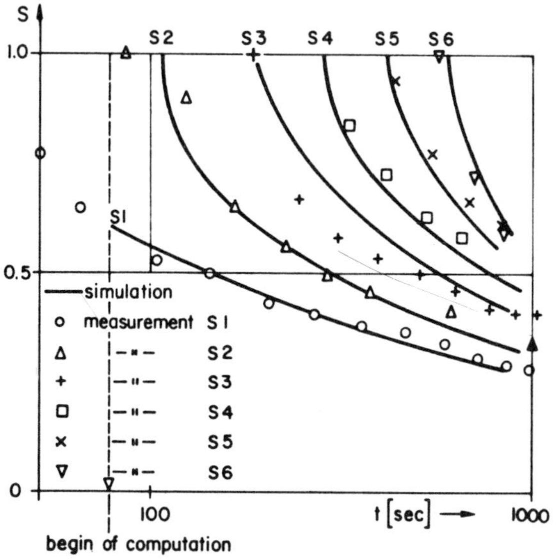

Figure 7 Function S(t) of experiment 2 and simulation

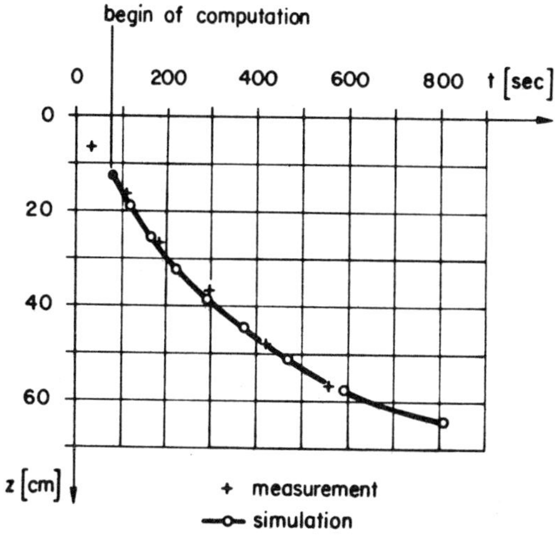

Figure 8 Function z(t) of the interface between saturated and unsaturated region.
Experiment 2 and simulation

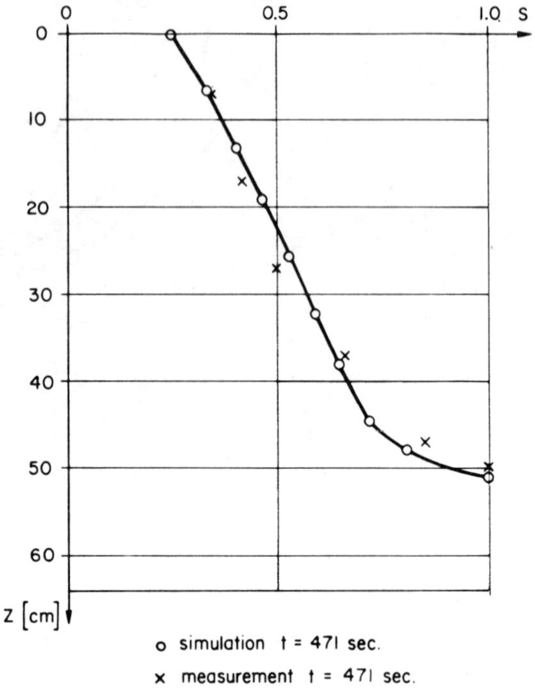

Figure 9 Function S(z) from experiment 2 and simulation

CONVERGENCE

The basic equation of the numerical model is of third order. By applying partial integration, the order of the highest occuring derivatives in the Galerkin equation is reduced to two. This derivative is of a mixed type:

$$\frac{\partial^2 S_e}{\partial Z . \partial T}$$

To be sure that the convergence criteria of Zienkiewicz are fulfilled, the first order derivatives should be continuous within the solution domain D_u. This requires continuity conditions at the boundary between two elements. Through these conditions, higher order elements would be required, so that the number of equations in the non-linear equation system is increasing by the factor of about 1.6. The presented numerical model does not include such additional continuity conditions. The elements used can thus be termed as non-conforming. The mathematical proof of the convergence of the method has not be given. Nevertheless, it is thought that the convergence criteria of Zienkiewicz can be applied because only interconnecting boundaries in the T-direction occur (figure 4). The consequence is that the derivative $\frac{\partial S_e}{\partial T}$ is continuous within the solution domain, so that only the derivatives $\frac{\partial^2 S_e}{\partial Z \cdot \partial T}$ and $\frac{\partial S_e}{\partial Z}$ can be discontinuous at the boundary between two elements. The agreement of the numerical model with the experimental data shows that the model can be applied with high accuracy to similar flow problems. By omitting the continuity conditions between two elements, the number of non-linear equations to solve can be kept at a minimum.

REFERENCES

Brooks, R.H. and Corey, A.T.(1964), Hydraulic Properties of Porous Media. Hydrology Papers, Colorado State University

Pinder, G.F. and Frind, E.O. (1972), Application of Galerkin's Procedure to Aquifer Analysis. Water Res. Res., 8, Nr. 1: 108-120

Stauffer, F. (1977), Einfluss der kapillaren Zone auf instationäre Drainagevorgänge. Bericht R 13-77, IHW, ETH Zürich

Topp, G.C., Klute, A. and Peters, D.B. (1967), Comparison of Water Content - Pressure Head Data obtained by Equilibrium, Steady-State and Unsteady-State Methods. Soil Sci. Soc. Amer. Proc. 31: 312-314

Zienkiewicz, O.C. and Cheung, Y.K. (1967), The Finite Element Method in Structural and Continuum Mechanics. McGraw-Hill, London

Zyvoloski, G. and Bruch, J.C. (1973), Finite Element Weighted Residual Solution of One-Dimensional, Unsteady and Unsaturated Flows in Porous Media. Dept. Mech. Eng. Univ. California, Santa Barbara,73-4

FLOW TO LARGE DIAMETER BOTTOM ENTRY WELLS

C.R. Dudgeon and R.J. Cox

University of New South Wales, Water Research Laboratory

INTRODUCTION

Prior to the development and general availability of drilling equipment suitable for the construction of water wells of small diameter (commonly called bores, boreholes or tube-wells), all wells were necessarily hand dug and of a size large enough for a man to work in. Square or circular openings of the order of 1 to $1\frac{1}{2}$ metres across were normal. The shift towards small diameter holes, normally between 150 and 300 mm in diameter has occurred partly as a result of the limitations of drilling equipment and partly as a result of such unqualified statements as "...doubling the diameter of a water-table well will increase its yield about 11 percent." (Ref.6, p 107). Statements such as this are based on the assumption that flow to the well obeys the Dupuit assumptions - i.e. radial one dimensional Darcy flow in a homogeneous isotropic aquifer. While these conditions may adequately cover most cases of well flow, there is a range of practical cases for which the statements may be in significant error. One such case is that of flow to large diameter wells in unconfined gravel aquifers where the effects of finite diameter, partial penetration and/or non-Darcy flow may cause the relationship between discharge and well diameter to differ greatly from that predicted using the Dupuit assumptions.

The availability of the finite element method has allowed the drawdown-discharge relationship to be investigated for cases such as this.

This paper describes an investigation of the characteristics of flow to bottom entry partially penetrating wells in unconfined aquifers. Wells of this type are still dug by hand in relatively undeveloped parts of the world and may be the most economical type of groundwater extraction facility in shallow aquifers even in highly developed countries. Machine augering of holes between one and two metres in diameter is now possible to depths of the order of 20 metres and, where suitable drilling equipment is available, the overall cost of extracting water from large diameter augered holes may be less than that of producing the same flow from a number of smaller diameter drilled holes. Economic comparisons should be based on local cost data and the drawdown-discharge characteristics of the various types of well considered practicable under the local conditions. Provided the hydraulic characteristics of the aquifer material are known it should be possible to use the results presented in this paper in such an economic analysis.

FLOW EQUATION

Comparison of the performance of wells may be based on steady flow conditions since relatively long term pumping may be assumed. The analysis of the flow requires a field equation which results from combining the steady continuity equation with an appropriate flow equation relating velocity and piezometric gradient. The Forchheimer equation

$$I = aV + bV^2 \tag{1}$$

which directly relates the flow velocity V and piezometric gradient I has been chosen for the flow equation. This equation has been subjected to theoretical (Stark and Volker, 1967) and experimental (Sunada, 1966) validation and found to fit permeameter data with sufficient accuracy using a single pair of coefficients a and b for a given aquifer material (Cox, 1976) over the range of velocities encountered in flow towards wells. It covers both "Darcy" and "non-Darcy" flow regimes. At low "Darcy flow" velocities, the bV^2 term becomes insignificant and the equation reduces to

$$I = aV = \frac{1}{K} V \tag{2}$$

where K is the hydraulic conductivity.

For three dimensional groundwater flow in a homogeneous isotropic aquifer, the Forchheimer equation may be written as

$$\frac{\partial h}{\partial x_i} = -(a + b\,|V|)\,v_i \tag{3}$$

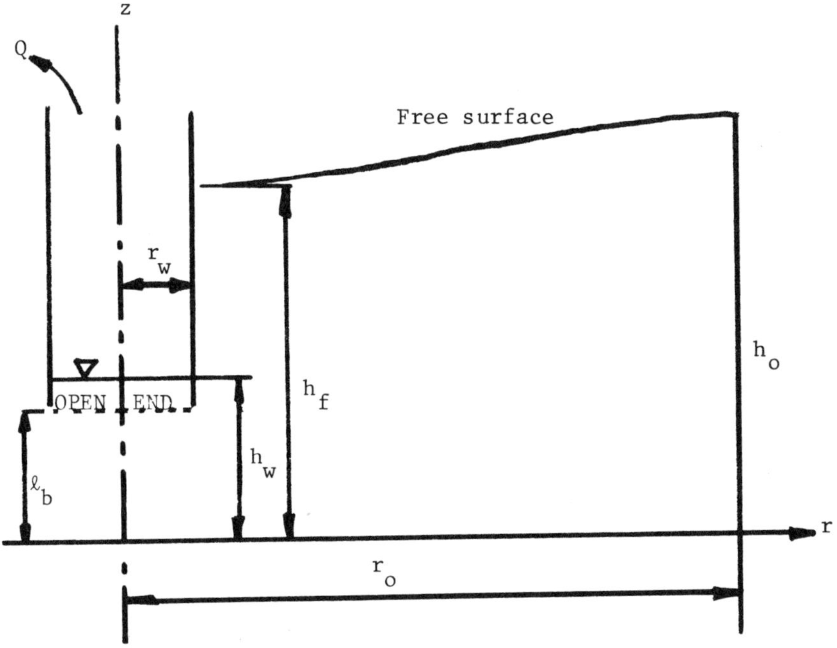

FIGURE 1 BOTTOM ENTRY LARGE DIAMETER WELL IN AN UNCONFINED AQUIFER

where h = piezometric head

v_i = the components of velocity in the x_i cartesian co-ordinate system

$|V|$ = magnitude of velocity vector

The steady flow continuity equation is

$$-\frac{\partial v_i}{\partial x_i} = 0 \qquad (4)$$

Combination of equations (3) and (4) yields the field equation

$$\frac{\partial}{\partial x_i}\left[\left(-\frac{a}{2b} + \sqrt{\left(\frac{a}{2b}\right)^2 + \left|\frac{\partial h}{\partial l}\right| / b}\right)\left(\frac{\partial h}{\partial x_i}\right) \Big/ \left|\frac{\partial h}{\partial l}\right|\right] = 0 \qquad (5)$$

where $\left|\frac{\partial h}{l}\right| = \left(\frac{\partial h}{\partial x_i}\frac{\partial h}{\partial x_i}\right)^{\frac{1}{2}}$

NUMERICAL METHOD

Figure 1 defines the geometry of the well-aquifer configuration investigated. Equation (5) was solved for this case with the appropriate boundary conditions by the method of finite elements. Full details of the variational formulation, solution procedures, finite element meshes and convergence studies have been presented in the PhD thesis by Cox (1976). Only some of the features of particular interest are outlined here.

A finite element network typical of those used to investigate the problem is shown in Figure 2. The critical region of the flow occurs near the open bottom end, particularly at the foot of the casing. In this zone non-Darcy flow may occur (i.e. the bV^2 term in the Forchheimer equation may become large) and high piezometric gradients can develop. As shown in the figure, triangular ring elements were used throughout the flowfield with their size graded from a minimum under the well to a maximum at $r = r_o$, the radius of influence. Automatic generation of the mesh reduced the amount of work required in iterating for the free surface.

A trial free surface was assumed and subsequently adjusted using an over-relaxation iteration scheme until the boundary condition $h = z$ was satisfied along the free surface. Trials indicated that the final solution of the complete flow field was only negligibly affected by the originally assumed

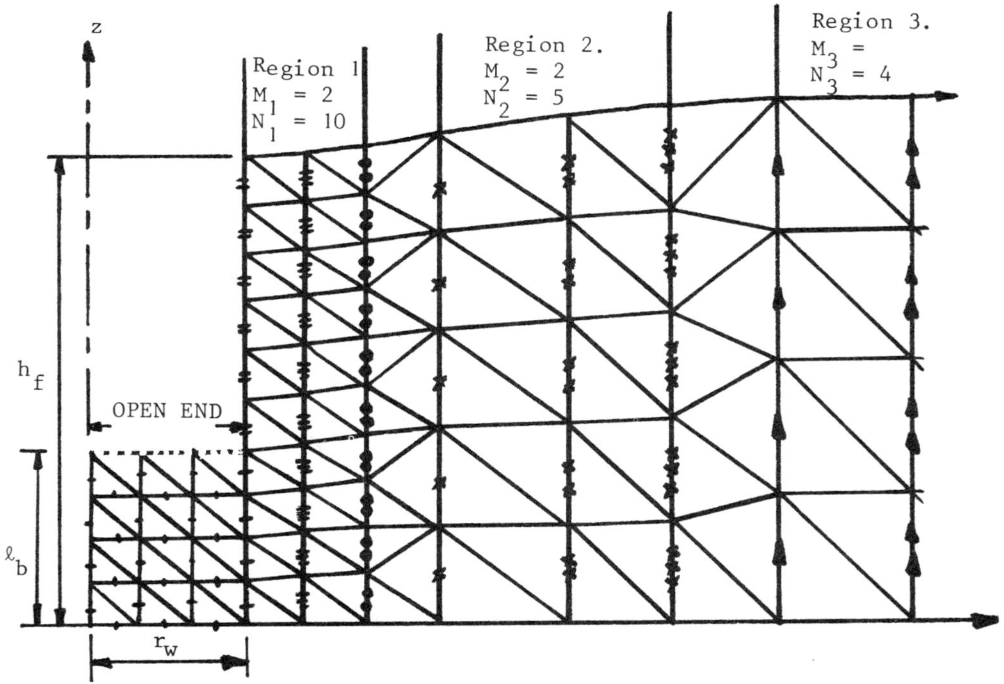

FIGURE 2. FINITE ELEMENT NETWORK

position of the free surface. Various alternatives for the initial trial location of h_f and the free surface were investigated but the most rapid convergence was obtained using the equations presented by Hall (1955)

$$\frac{h_f - h_w}{h_o - h_w} = \left[1 - \left(\frac{h_w}{h_o}\right)^{2.4}\right] \Big/ \left[1 + \frac{\ln(r_o/r_w)}{50}\right]\left[1 + \frac{5\,r_w}{h_o}\right] \qquad (6)$$

$$\frac{z - h_f}{h_o - h_f} = 2.5\left(\frac{r - r_w}{r_o - r_w}\right) - 1.5\left(\frac{r - r_w}{r_o - r_w}\right)^{1.5} \qquad (7)$$

The symbols are as shown on Figure 1.

The non-linearity introduced by the Forchheimer equation was also handled by an over-relaxation iteration procedure. An over-relaxation factor of between 1.5 and 1.9 gave satisfactory convergence within 5 iterations.

Solution accuracy was investigated to allow the minimum solution time to be achieved consistent with the required accuracy. Since the solution was to be repeated a large number of times to produce well design data it was necessary to avoid unnecessarily long iterations so that a greater number of combinations of well and aquifer variables could be covered within the computer time available. It was found that the accuracy of a particular solution depended mainly on the number of nodes along the bottom entry section of the well. For a given refinement of the network, designated by N_1, the number of radial "tubes" in the first network region, the solution accuracy was effectively independent of all variables other than the dimensionless ratio h_o/r_w. Table 1 gives the percent over-estimates in discharge for a number of selected networks.

Table 1: Finite Element Network Errors

Percentage Overestimate in Well Discharge

$r_o/h_o \geqslant 2$ $\quad h_w \geqslant l_b$

h_o/r_w	Network Refinement, N_1				
	16	24	32	48	64
5	9	4	2	0	
10		14	9	4	1
20			16	7	3

The data shown were used to adjust the solutions obtained for the wider range of design variables for which solutions were subsequently obtained using values of N_1 = 32, 48 and 64 for $\frac{h_o}{r_w}$ = 5, 10, 20 respectively. The overall solution accuracy

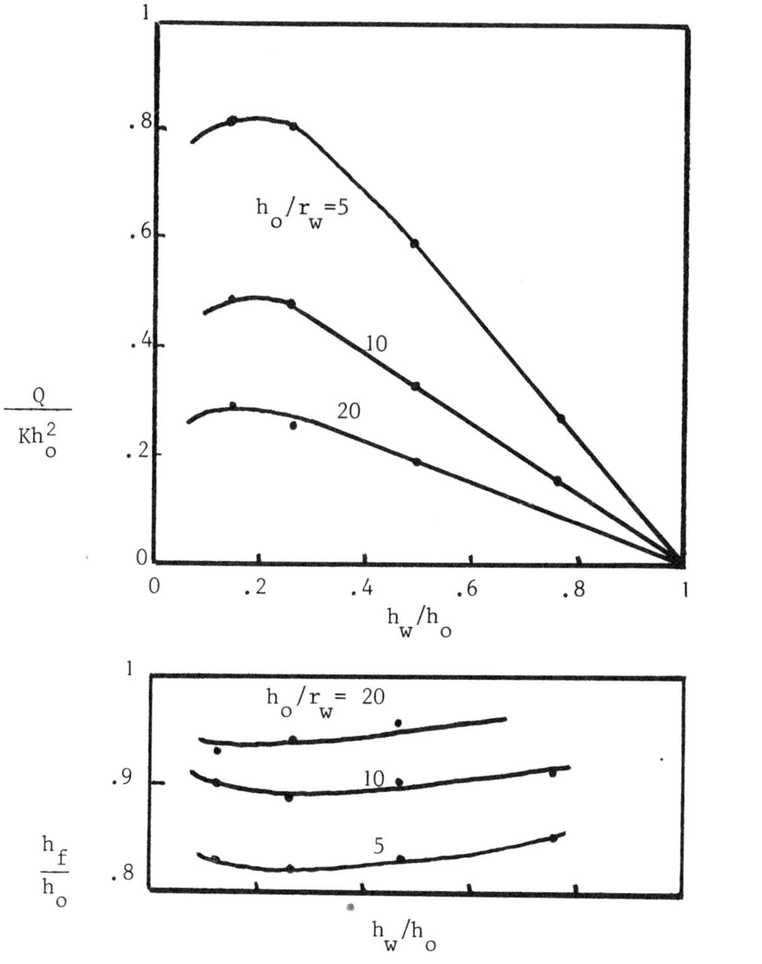

$r_o/h_o = 2$ $\ell_b = h_w$ Darcy Flow

FIGURE 3 DISCHARGE AND FREE SURFACE DRAWDOWN AT THE WELL - DARCY FLOW thus achieved was considered to be high in relation to the accuracy with which the well and aquifer variables would be known in particular practical applications.

VERIFICATION

Although no experimental verification of the large diameter bottom entry well flow analysis was carried out, it should be pointed out that the theoretical and numerical procedures employed were verified for side entry wells in confined and unconfined aquifers by

(i) comparing numerical solutions with analytical solutions for cases where both were available, and

(ii) by comparing numerical solutions with experimental results from a large sand tank and an electrolytic tank.

The experiments in the sand tank covered both Darcy and non-Darcy flows. In all cases the comparison between the results of the numerical analysis and experiments was found to be very good. Full details of the verification have been given by Cox (1976).

RESULTS

Table 2 shows the results of the investigation in dimensionless form. The table covers the range of variables found to occur in practice in Australia and considered to be generally relevant throughout the world. Figure 3 is a plot of the results for cases for which wholly Darcy flow occurs (i.e. the effect of the bV^2 term in the Forchheimer equation is negligible). Figure 4 is a plot of the results incorporating the effects of non-Darcy flow for particular values of r_o/h_o and h_o/r_w. Graphs for other values of these parameters may be prepared from Table 2 if required. Both Figures 3 and 4 apply to the limiting case of drawdown to the bottom of the well (i.e. $l_b = h_w$). This is considered to be the case most relevant in the comparison of the performance of various well designs.

In all Darcy flow cases examined in which the full available drawdown was not utilised, both the well discharge and free surface water level at the well varied linearly with h_w between $h_w = l_b$ and $h_w = h_o$. For any well water level $h_w > l_b$, the discharge and free surface position may be estimated by linear proportioning from the solution for the limiting condition of drawdown to the bottom of the well, $h_w = l_b$.

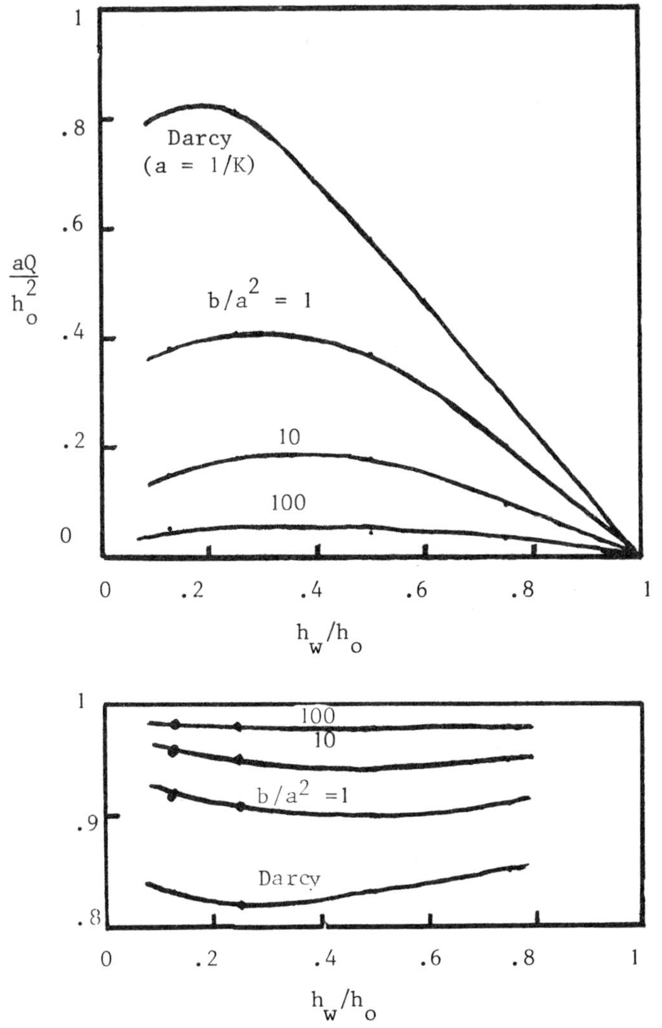

$r_o/h_o = 2$ $h_o/r_w = 5$ $\ell_b = h_w$

FIGURE 4 DISCHARGE AND FREE SURFACE DRAWDOWN AT THE WELL - NON DARCY FLOW

Table 2: Discharge and Free Surface Drawdown at the Well

$\frac{r_o}{h_o}$	$\frac{h_o}{r_w}$	$\frac{b}{a^2}$	$\frac{l_b}{h_o}$	h_w/h_o				
				1/8	1/4	1/2	3/4	1
(a) Well Performance - Tabulated Values of Q/Kh_o^2 (K=1/a)								
2	5	0	h_w/h_o	.81	.81	.58	.29	0
	10			.48	.48	.32	.16	0
	20			.29	.27	.19		0
	5	1	h_w/h_o	.38	.41	.37	.20	0
		10		.15	.18	.18	.093	0
		100		.052	.057	.048	.034	0
	10	1	h_w/h_o	.18	.18	.16	.088	0
		10		.070	.068	.060	.038	0
		100		.023	.023	.019	.014	0
	10	0	1/8	.48	.41	.27	.14	0
			1/4		.48	.31	.15	0
			1/2			.32	.17	0
	5	0	1/8	.81	.69	.46	.23	0
			1/4		.81	.54	.28	0
			1/2			.58	.34	0
4	5	0	h_w/h_o	.72	.71	.50	.25	0
8	5	0	h_w/h_o	.63	.62	.45	.22	0
(b) Free Surface Location at the Well - Tabulated Values of h_f/h_o								
2	5	0	h_w/h_o	.83	.82	.83	.85	1
	10			.90	.89	.90	.91	1
	20			.93	.94	.95		1
	5	1	h_w/h_o	.92	.91	.90	.91	1
		10		.96	.95	.94	.95	1
		100		.98	.98	.98	.98	1
	10	1	h_w/h_o	.96	.96	.95	.95	1
		10		.99	.99	.98	.98	1
		100		.99	.99	.99	.99	1
	10	0	1/8	.89	.91	.94	.97	1
			1/4		.88	.93	.97	1
			1/2			.90	.95	1
	5	0	1/8	.83	.86	.91	.96	1
			1/4		.82	.88	.94	1
			1/2			.83	.91	1
4	5	0	h_w/h_o	.74	.73	.75	.84	1
8	5	0	h_w/h_o	.66	.65	.71	.83	1

Note: $b/a^2 = 0$ indicates wholly Darcy flow solution

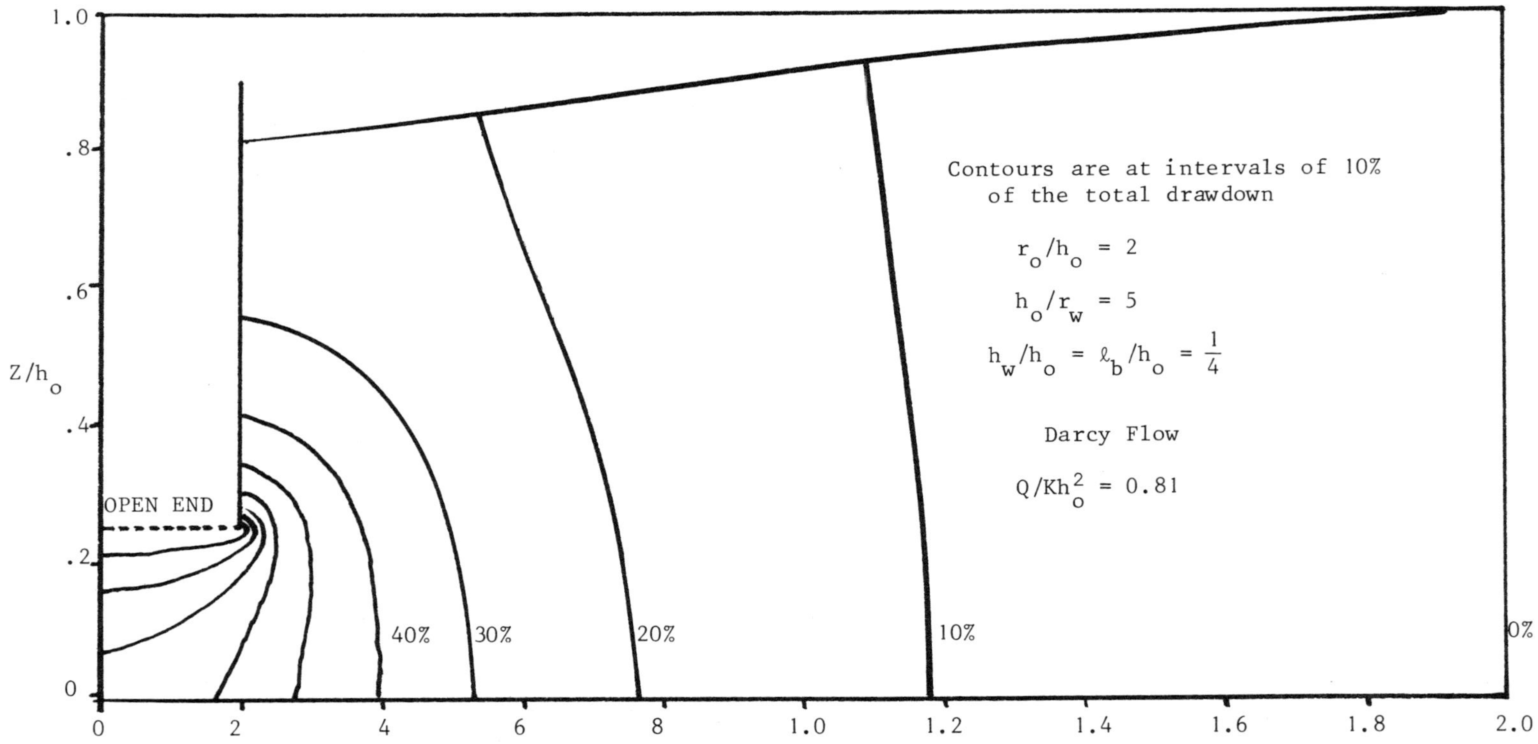

FIGURE 5 TYPICAL HEAD DISTRIBUTION

The computed drawdown distribution for a typical Darcy flow problem ($r_o/h_o = 2$, $h_o/r_w = 5$, $l_b/h_o = h_w/h_o = \frac{1}{4}$) is shown in Figure 5. This gives a guide to the region of greatest head loss.

For a specified degree of non-linearity (value of b/a^2), the discharge flux distribution along the open well bottom was found to be approximately independent of r_o/h_o, h_o/r_w and l_b/h_o within the range of values investigated (i.e. $r_o/h_o \geqslant 2$, $5 \leqslant h_o/r_w \leqslant 20$, $l_b/h_o > 1/8$). The discharge flux distributions are shown in Figure 6a for cases of Darcy flow and non-linear flow with $b/a^2 = 1$ and 100. Figure 6b shows the radial distribution of velocity along the bottom of the well for Darcy flow. It should be noted that high entrance velocities occur as r approaches r_w. This is very significant in relation to possible sand movement into the well. The cumulative well discharge distribution is shown in Figure 6c for Darcy flow.

APPLICATION OF RESULTS

Application of the data given in Table 2 and Figures 3 and 4 requires a knowledge of the hydraulic characteristics (Forchheimer coefficients a and b) of the aquifer material and the saturated thickness, h_o. The drawdown-discharge characteristics of a number of possible well designs can be determined from the data once the appropriate dimensionless ratios have been calculated. An economic comparison allowing for both capital and running costs can then be made for a given flow rate.

The data on the location of the free surface may also be used with advantage in engineering dewatering applications.

A field investigation to compare wells of various diameters in a shallow aquifer was carried out by the Queensland (Australia) Irrigation and Water Supply Commission. Analysis of part of the pumping test data by Huyakorn (Dudgeon et al 1973) yielded the Forchheimer coefficients a = 6 secs/m and b = 40 $(\text{sec/m})^2$ at one site. The corresponding value of $\frac{b}{a^2}$ is 1.1. The aquifer thickness was found to be very variable but, for the example, a saturated unconfined thickness of 5m is assumed. For $\frac{1}{2}$ and 1 metre diameter bottom entry wells Table 2 predicts maximum discharges of approximately $.18\frac{h_o^2}{a}$ and $.41\frac{h_o^2}{a}$ respectively i.e. 7.5 and 17 litres/second respectively. The expected increase in maximum

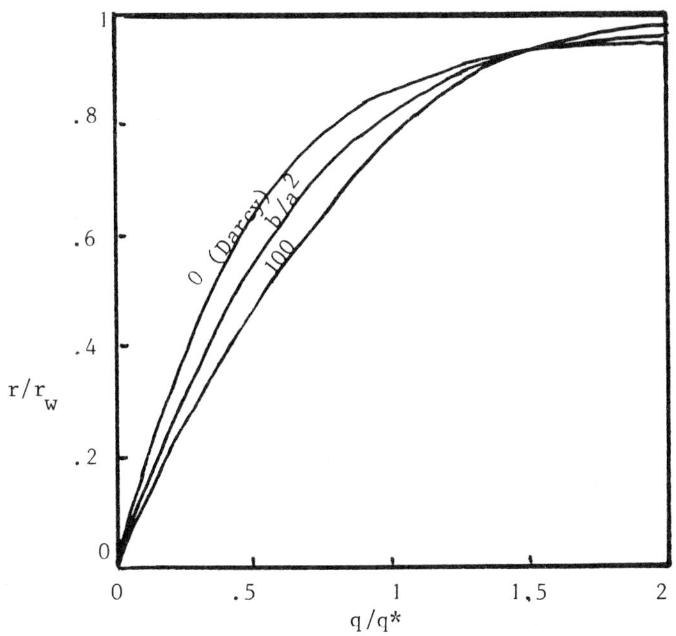

(a) $q^* = Q/r_w$

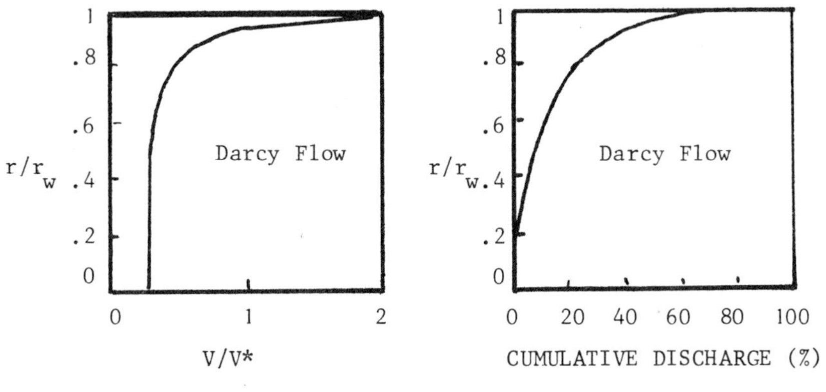

(b) $V^* = Q/\pi r_w^2$

(c)

$r_o/h_o = 2$, $5 \leq h_o/r_w \leq 20$, $\ell_b/h_o > 1/8$

FIGURE 6 DISTRIBUTION OF DISCHARGE FLUX, VELOCITY AND CUMULATIVE WELL DISCHARGE ALONG BOTTOM ENTRY SECTION discharge obtained by doubling the diameter of the well is thus of the order of 100%. A 100mm diameter screened well at this site should provide a maximum discharge of approximately $.24\frac{h_o^2}{a}$ i.e. 10.4 litres/second when half the aquifer is screened (Dudgeon and Cox, 1977).

A more comprehensive set of calculations with local cost factors introduced would allow the optimum unconfined well to be determined for this site. Such a set of calculations based on the assumption of confined flow to side entry wells has been made by Keshavarz (1974).

Actual costs of 200 to 1200 mm diameter experimental wells constructed on this site have been reported by the Queensland Irrigation and Water Supply Commission (Ref. 5). In the field investigation it was found difficult to give an exact interpretation of the results of the economic comparison because of the variability of the aquifer and doubt about the degree of confinement of the flow. However, the results showed that large diameter augered wells may be the most economical solution if all cost factors are considered.

CONCLUSION

The results of the investigation demonstrate how the finite element method can be used to prepare design data suitable for use in water well design studies. Computation costs involved in analysing non-Darcy free surface flow to a well are considerable because of the double iteration procedure required to cope with the free surface and the non-linear flow equation. It is considered that for the case investigated the preparation of design graphs and charts provides a more economical solution than the direct solution of particular problems on the computer.

REFERENCES

1. Cox, R.J. A study of near-well groundwater flow and the implications in well design. Uni. New South Wales, Ph.D. thesis, 1976.

2. Dudgeon, C.R. et al "Hydraulics of flow near wells in unconsolidated sediments. Vol. 2. Field studies". Uni. New South Wales, Water Research Laboratory, Report No. 126, 1973.

3. Dudgeon, C.R. and Cox, R.J. "Effect of partial screening and non-Darcy flow on the performance of wells in unconfined aquifers". Institution of Engineers, Australia, 6th Australasian Hydraulics and Fluid Mechanics Conference, Adelaide. Proc. 80-83, 1977.

4. Hall, H.P. "An investigation of steady flow toward a gravity well." La Houille Blanche, 10, 8-35, 1955.

5. Irrigation and Water Supply Commission, Queensland, Report to Australian Water Resources Council on Research Project 68/8, Extraction of water from unconsolidated sediments (and supplementary report), (undated).

6. Johnson Division, Universal Oil Products Co., "Ground water and wells", Johnson Divn., U.O.P. 1972.

7. Keshavarz, M.H. "Optimal design of wells in unconsolidated sediments". Uni. New South Wales, M.Eng.Sc. project, 1974.

8. Stark, K.P., Volker, R.E. Non-linear flow through porous materials. Some theoretical aspects. Univ. College of Townsville, Dept. of Engineering, Bull. No.1, 1967.

9. Sunada, D.K. Turbulent flow through porous media. Univ. of California, Berkeley, Water Resources Centre, Contribution No. 103, 1965.

FREE SURFACE SEEPAGE FLOW IN EARTH DAMS

I. Kazda

Technical University of Prague

INTRODUCTION

Percolating water affects the stability of earth dam slopes to a considerable extent. Attention to the solution of seepage through earth dams by the Finite Element Method has been paid since 1965, when Zienkiewicz and Cheung published the first paper on the ground water seepage analysis under hydraulic structures. This paper with another work written by Zienkiewicz, Mayer and Cheung in 1967 has fully proved the advantage of using the Finite Element Method for the numerical solution of complex seepage flow problems.

The seepage analysis carried out by the Finite Element Method is considered very simple for confined flow. In the case of unconfined flow the solution is considerably complicated by the fact that a part of the region boundary is created by the free surface the position of which is not known beforehand. The problem of determining the free surface was first dealt with by Taylor and Brown (1967) and Finn (1967) who had used a simple iterative procedure by which the problem could be substituted by the solution of problem sequence without free surface. Neuman and Witherspoon (1970) stated that in some cases the Taylor and Brown method diverged, and suggested a more complicated two step iterative technique. As it was stated further, this technique could not be regarded as reliable in all cases either. Taylor, France and Zienkiewicz (1973) solved a great number of typical flow problems of steady and transient seepage by the Finite Element Method.

The solution of a great number of seepage problems in earth cores of rockfill dams has shown that for practical use it is of great advantage to apply the Taylor and Brown method adapted by the way described further.

THEORETICAL CONSIDERATIONS

Steady potential flow in an incompressible, non-homogeneous anisotropic porous continuum is described in the two-dimensional case by the following self-adjoint partial differential equation

$$\frac{\partial}{\partial x}\left(k_x \frac{\partial h}{\partial x}\right) + \frac{\partial}{\partial y}\left(k_y \frac{\partial h}{\partial y}\right) = 0 \qquad (1)$$

where for the permeability coefficients is true that $k_x = k_x(x,y)$, $k_y = k_y(x,y)$ and total head h is defined as a sum of elevation above datum plane and pressure head

$$h = y + p_h \qquad (2)$$

In Equation 1 the permeability coefficients in horizontal and vertical directions are functions of coordinates x,y. It is assumed that the seepage continuum is generally non-homogeneous and anisotropic in such a manner that axes of anisotropy are parallel to coordinate axes. Unless the latter assumption is fulfilled, a transform of the compoments of the permeability tensor will be carried out by turning the axes of anisotropy in the direction of global axes. Further it is assumed that the water flows into or out of the considered domain R only accros its boundary S. In the case of earth dams those assumptions are fulfilled in all respects.

Let us search for the solution of Equation 1 when prescribed: Dirichlet's boundary condition on the part of the boundary S_D

$$h = h_D\,(x,y) \tag{3}$$

Neumann's boundary condition on the part of the boundary S_N

$$k_x \frac{\partial h}{\partial x} n_x + k_y \frac{\partial h}{\partial y} n_y + q = 0 \tag{4}$$

where n_x and n_y are direction cosines of the outward normal to S_N and $q = q(x,y)$ is the prescribed flux.

The boundary value problem defined in such a manner may be transposed in a variational problem

$$I(h) = \frac{1}{2} \int_R \left\{ k_x \left(\frac{\partial h}{\partial x}\right)^2 + k_y \left(\frac{\partial h}{\partial y}\right)^2 \right\} dR + \int_{S_N} qh \; dS \tag{5}$$

Equation 1 is Euler's equation for the functional I(h) so that the function h minimizing I(h) gives the solution of the boundary value problem. It is advisable to search for the minimum of the functional I(h) by means of a numerical method, namely by the Finite Element Method.

We have not yet taken into account the problem how the existence of free surface in the dam will affect the solution. Let us consider the following example: the domain R (Figure 1) limited by the boundary $S = S_1 + S_2 + S_3 + S_4$ representing an earth core. The part of the boundary S_1 represents the upstream face, whereas S_2 is the seepage face,

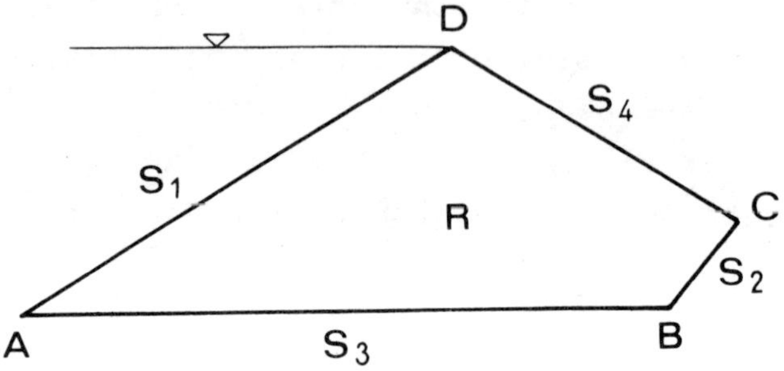

Figure 1 Region boundaries for potential flow with free surface

the exact lenght of which is not known beforehand, S_3 is the surface of the impervious layer and S_4 is an approximation of unknown position of the free surface. Point D of the free surface is known and point C must lie on S_2. In that case the steady potential flow in closed domain $\overline{R} = R + S$ represents the boundary problem

$$A(h) = \frac{\partial}{\partial x}\left(k_x \frac{\partial h}{\partial x}\right) + \frac{\partial}{\partial y}\left(k_y \frac{\partial h}{\partial y}\right) = 0 \qquad (6a)$$

$$h = H = \text{const} \qquad \text{on } S_1 \qquad (6b)$$

$$h = y \qquad \text{on } S_2 \qquad (6c)$$

$$k_x \frac{\partial h}{\partial x} n_x + k_y \frac{\partial h}{\partial y} n_y = 0 \qquad \text{on } S_3 \qquad (6d)$$

$$\left.\begin{array}{l} h = y = \xi(x) \qquad (6e) \\ k_x \dfrac{\partial h}{\partial x} n_x + k_y \dfrac{\partial h}{\partial y} n_y = 0 \qquad (6f) \end{array}\right\} \text{simultaneously on } S_4$$

As the position of the free surface S_4 is not known beforehand it is necessary to substitute the solution of the boundary problem by an iterative process enabling to find the free surface. A simple technique of creating the iterative process has been used by Taylor and Brown (1967). According to their method an initial approximation of the free surface $\xi^{(0)} = \xi^{(0)}(x)$ is chosen and then a new boundary value problem is solved

$$A_1(h) = \frac{\partial}{\partial x}\left(k_x \frac{\partial h}{\partial x}\right) + \frac{\partial}{\partial y}\left(k_y \frac{\partial h}{\partial y}\right) = 0 \qquad (7a)$$

$$h = H = \text{const} \qquad \text{on } S_1 \qquad (7b)$$

$$h = y \qquad \text{on } S_2 \qquad (7c)$$

$$k_x \frac{\partial h}{\partial x} n_x + k_y \frac{\partial h}{\partial y} = 0 \qquad \text{on } S_3 \text{ and } S_4^{(0)} \qquad (7d)$$

in the domain $\bar{R}^{(0)} = R^{(0)} + S_1 + S_2 + S_3 + S_4^{(0)}$ where $S_4^{(0)}$ is that part of the boundary formed by the first approximation of the free surface. This problem can be transposed to a variational problem and be solved by the FEM. The resulting solution will be the function $h^{(1)} = h^{(1)}(x,y)$, $x,y \in \bar{R}^{(0)}$. This solution will not generally satisfy the condition $h^{(1)} = y$ on $S_4^{(0)}$. Therefore that condition may be used for the correction of the free surface approximation:

$$\xi^{(1)} = \beta h^{(1)} \qquad (8)$$

where β is a coefficient close to 1, the value of which may be changed during iteration. $\xi^{(1)}$ defines the new position $S_4^{(1)}$ of the free surface approximation and for the new domain $\bar{R}^{(1)}$ changed in this manner the boundary problem (Equation 7) is again being solved. Thereby the iteration procedure is defined for the completion of which the criterion

$$\sum \left| h^{(i+1)} - \xi^{(i)} \right| \leq \varepsilon \tag{9}$$

may be used, where ε is the the required accuracy of the approximation of the free surface.

As shown by Neuman and Witherspoon (1970) the mentioned iterative procedure is of poor numerical stability and can diverge mainly in point C (Figure 1) in which the approximation of the free surface intersects the seepage face. Neuman and Witherspoon have made an attempt to eliminate this difficulty by using a more complicated two step iterative procedure. If the infiltration on the free surface is not taken into account, the procedure may be described as follows. In the first step the boundary value problem is solved:

$$A_2(h) = \frac{\partial}{\partial x}\left(k_x \frac{\partial h}{\partial x}\right) + \frac{\partial}{\partial y}\left(k_y \frac{\partial h}{\partial y}\right) = 0 \tag{10a}$$

$$h = H = \text{const} \qquad \text{on } S_1 \tag{10b}$$

$$h = y \qquad \text{on } S_2^{(0)} \text{ and on } S_4^{(0)} \tag{10c}$$

$$k_x \frac{\partial h}{\partial x} n_x + k_y \frac{\partial h}{\partial y} n_y = 0 \qquad \text{on } S_3 \tag{10d}$$

where $S_4^{(0)}$ is again the first approximation of the free surface. The resulting values of h may be used for the determination of the flux $q^{(1)}$ on the seepage face $S_2^{(0)}$. In the second step the region $\bar{R}^{(0)}$ remains without change and a new boundary problem is solved

$$A_3(h) = \frac{\partial}{\partial x}\left(k_x \frac{\partial h}{\partial x}\right) + \frac{\partial}{\partial y}\left(k_y \frac{\partial h}{\partial y}\right) = 0 \tag{11a}$$

$$h = H = \text{const} \qquad \text{on } S_1 \tag{11b}$$

$$k_x \frac{\partial h}{\partial x} n_x + k_y \frac{\partial h}{\partial y} n_y + q^{(1)} = 0 \qquad \text{on } S_2^{(0)} \tag{11c}$$

$$k_x \frac{\partial h}{\partial x} n_x + k_y \frac{\partial h}{\partial y} n_y = 0 \qquad \text{on } S_3 \text{ and on } S_4 \qquad (11d)$$

The values of h on $S_4^{(o)}$ may be used for verifying the approximation of the free surface again according to Equation 8 and then a new iteration step follows. The iterative procedure will continue until the Equation 9 is fulfilled.

The iterative procedure introduced by Taylor and Brown is very simple and may be well coded. It is sufficient for the programme used for the solution of potential confined seepage flow to be completed by a subroutine correcting the free surface approximation on the base of the results of the preceding iteration step. The Neuman and Witherspoon method is more complicated and necessitates considerable modifications of the basic programme.

Both mentioned techniques for determining the free surface were being tested by the author on a great number of practical problems. In Figure 2 as an example the comparison of the results of determining the free surface in a homogeneous earth dam with a toe drain (founded on a layer of impervious soil) for the solution by Kozeny (solid line) and the method by Taylor and Brown (crosses) has been presented. The results deduced by the method by Taylor and Brown have been gained after only 8 iterations. The results of both solutions are in good correspondence. In solving the same problem by means of the Neuman and Witherspoon method the computation diverged and it was not possible to determine the free surface. The Neuman and Witherspoon method converged very good, when the domain was short in the direction of the flow and the length of the free surface was small.

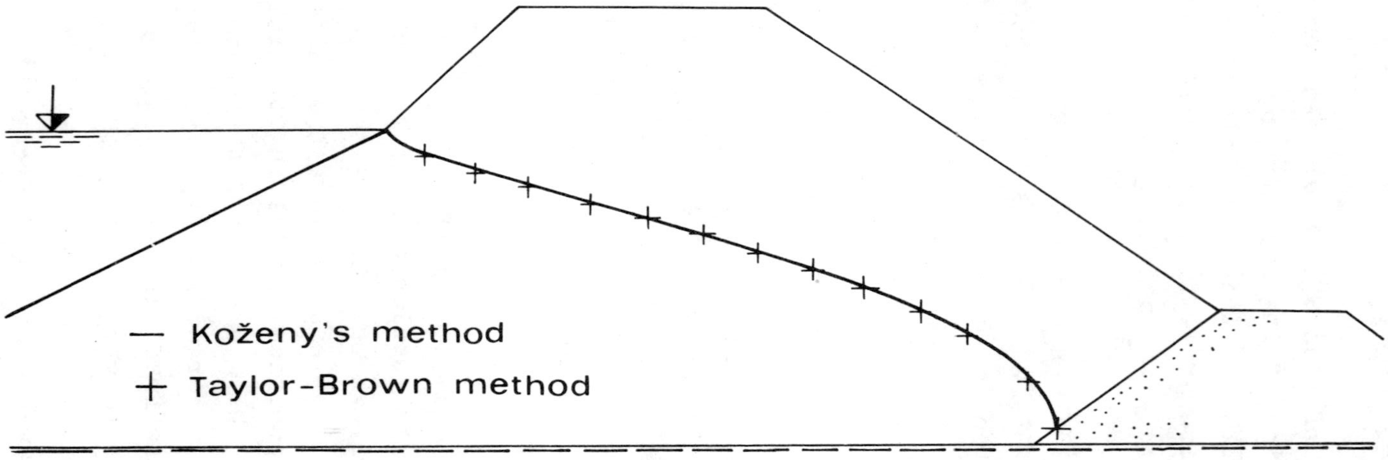

Figure 2 Results comparison for the Kozeny's method and the method by Taylor and Brown

CONCLUSIONS FROM RESULTS OF NUMERICAL EXPERIMENTS

A detailed analysis of the results of numerical experiments deduced from the method by Taylor and Brown and that of Neuman and Witherspoon has shown that in the latter the convergence of the iteration procedure mainly depends on the following factors:

- the division of the net into elements in the vicinity of the exit point C in which the free surface crosses the seepage face,
- the position of the initial approximation of the free surface,
- the length of the free surface with regard to the length of other parts of the boundary.

It may be easily shown how the form of the elements in the vicinity of the exit point C affects the value of the flux q_C in this nodal point. In Figure 3 is the CST element ijk with the area A, for which i ≡ C is valid. For the flux q_i in the nodal point i we may obtain (for simplicity reason let us assume first the isotropic permeability k_x=k_y=k):

$$\frac{4}{k} A q_i = m_{ii} h_i + m_{ij} h_j + m_{ik} h_k = q_i^* \qquad (12)$$

where m_{ii}, m_{ij} and m_{ik} are the members of the characteristic matrix of the element M. The values of the head h_i and h_j may be expressed by means of h_k (Figure 3), so that

$$q_i^* = (m_{ii} + m_{ij} + m_{ik})\, h_k - m_{ii} n_j + m_{ij} n_i \qquad (13)$$

It is easy to make sure that the sum in the brackets at h_k is equal zero and after a simple modification we may obtain

$$q_i = \frac{k}{2} (x_k - x_j) \qquad (14)$$

According to the form of the element ijk this expression may be both positive or negative (Figure 3), even if, in fact, in the node i ≡ C the water can only flow out, i.e. $q_i < 0$. So as not to intro-

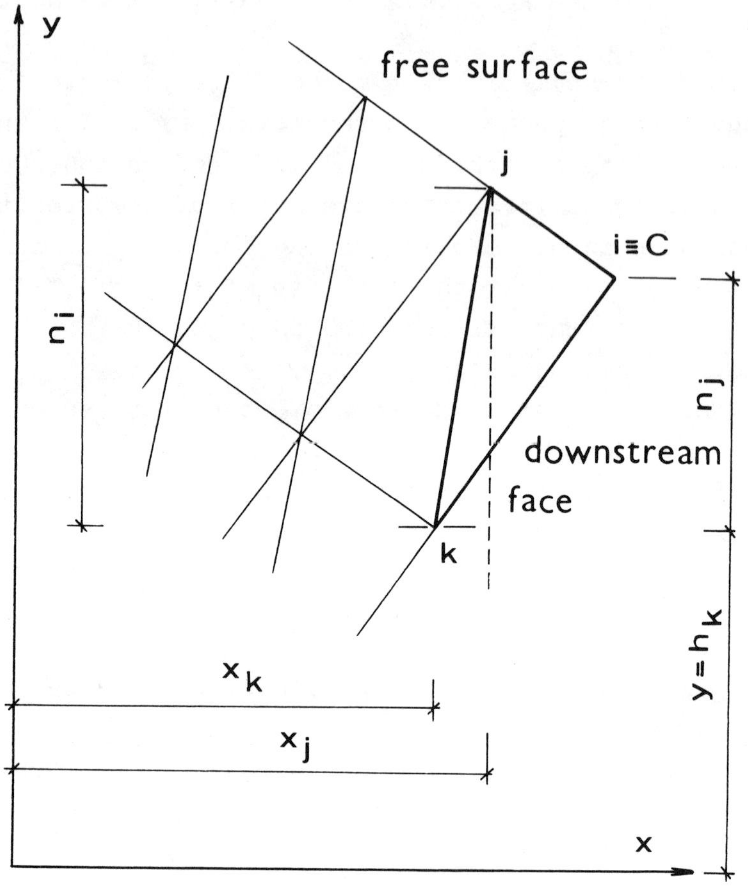

Figure 3 Influence of the form of the element with the exit nodal point C

duce in the computation an error beforehand that will cause that $q_i > 0$ and the convergence of the iterative procedure will be endangered, in dividing into elements it is necessary to fulfill the condition

$$x_j > x_k \tag{15}$$

It may be easily shown that the criterion is valid for the anisotropic permeability as well. By similar modifications as the above mentioned ones we may obtain

$$q_i = \frac{1}{2} k_y (x_k - x_j) \qquad (16)$$

In a number of problems solved by the Neuman-Witherspoon method there exists a dependence of the convergence of the free surface approximation to the correct shape on the selection of the initial approximation. It has been shown that the most suitable first approximation is of straight line character and it must be so positioned that it lies below the actual free surface. For such an initial approximation the convergence to the correct shape has been fast and continuous. In the selection of the initial approximation above the true position value of the criterion given by Equation 9 has also been continuously decreased though, but in the vicinity of the exit point C a significant deformation of the free surface shape has occurred.

Even the approximation formed by two line segments has been compared with the straight line approximation in the Neuman-Witherspoon method. Even if this estimate is closer to the shape of the true free surface, it has not proved to be advantageous for the iterative procedure. The breaking point has appeared as a further singularity that has inconveniently affected the convergence to the correct shape of the free surface. In the node of breaking the difference between h and y has not been succssfully minimized, not even then when in other nodes on the approximation it has been very close to zero.

The influence of the free surface length on the convergence of the Neuman-Witherspoon method can also be explained. In the first step of each iterative cycle Dirichlet´s boundary condition (Equation 10c) is being introduced on the free surface approximation. In that way on a part of the boundary S_4 representing the free surface the unknown flux q_S is implicitly introduced. The smaller this flux will be, the nearer will be the approximation to the correct shape. The longer S_4 will be in comparison with the upstream face S_1 and the seepage face S_2, the greater will be the influence of the flux q_S on the iteration convergence. If the first approximation is still different from the true position of the free surface, a divergence of the iterative procedure may occur.

The analysis of the computations carried out by the Taylor-Brown method has shown that the iteration convergence is mostly affected by setting the boundary condition in the node C which is a singular point. In this point the change of the boundary shape always occurs and at the same time the type of the boundary condition changes. The point C may be included in the part of the noundary S_4 on which, during iteration, Neumann´s boundary condition is given or in the part S_2 on which Dirichlet´s boundary condition is given.

Neumann´s boundary condition being given in the node C, the divergence may occur considerably fast during the iterative procedure.A typical example of such a divergence is shown in Figure 4. A similar instance is given by Neuman and Witherspoon (1970).This divergence may be removed, if the node C is taken as the point of the seepage face S_2 and during iterationDirichlet´s boundary condition is given in it. Generally, it is not possible to guarantee even in that case that the free surface approximation will always converge to the correct shape in the vicinity of the node C.

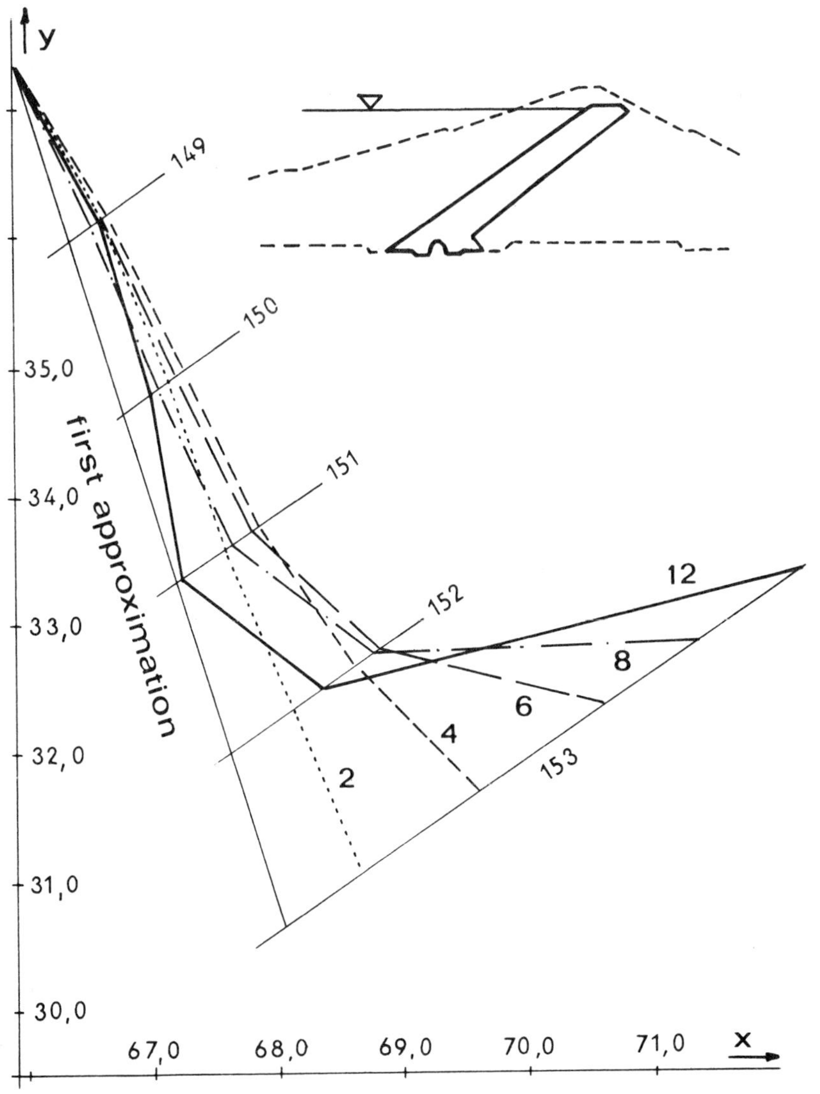

Figure 4 Divergence of the Taylor-Brown method for Neumann´s boundary condition in the node C

TAYLOR-BROWN MODIFIED METHOD

The computations carried out have shown that Dirichlet´s boundary condition being given in the node

C, all the other nodes with the exception of that node will converge on the free surface approximation in relation to its correct position. This fact has brought about the conclusions that the Equation 8 is generally not suitable for the correction of the node C position. From the examined correction methods a simple axtrapolation has proved to be the most successful since the new node C position on the seepage face may be obtained from the new position of the adjacent nodes on S_4 in any iterative step. During extrapolation full use has been made of the fact that the free surface shall be theoretically close to the part of the parabola with horizontal axis.

This way of modifying the iterative procedure has proved to be satisfactory (Kazda, 1977). The simplicity of the Taylor-Brouwn method has been preserved, the number of the needed iterations has been reduced and the convergence of the free surface approximation to the correct shape and position has been secured not only in the mean, but also in all the nodes.

The total procedure may be summarized as follows:

- by neans of the Equation 8 a new approximation of the free surface may be found at the end of each iterative step,
- from the new position of the nodes adjacent on the approximation to the node C its corrected position may be found by axtrapolation and a new value of Dirichlet´s condition given,
- a new iterative cycle is being solved.

The whole procedure is repeated until the Equation 9 is fulfilled. Figure 5 presents examples of the approximation of the free surface which have

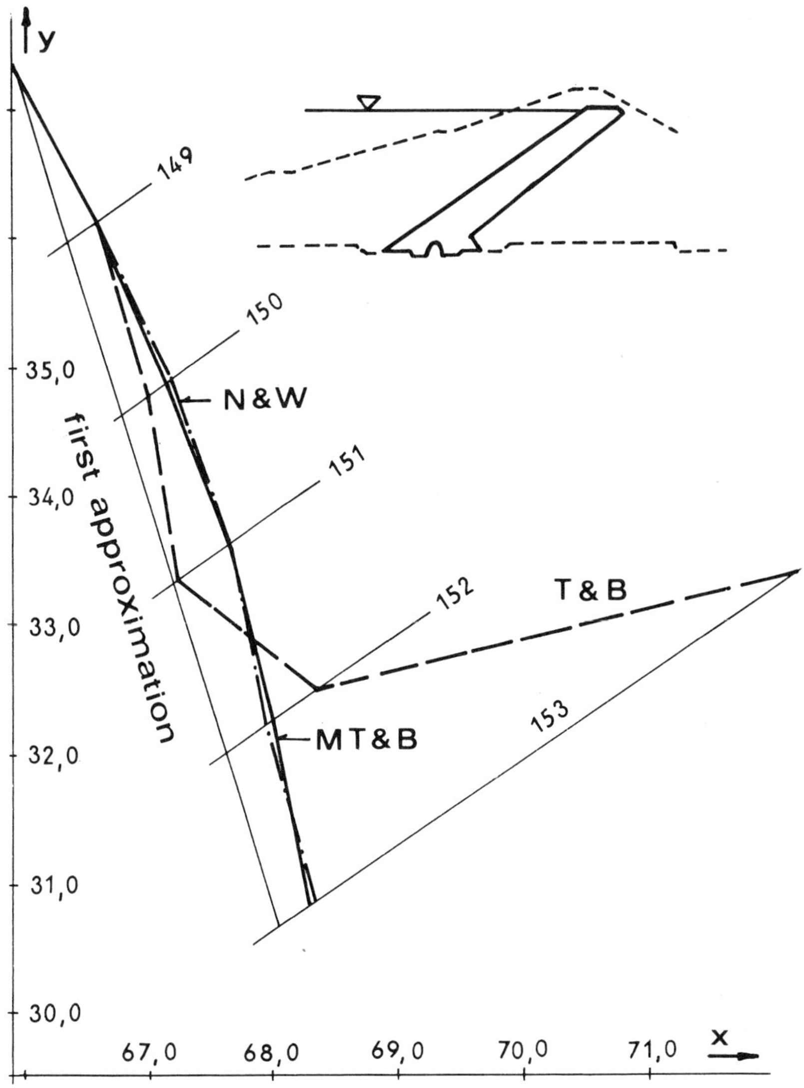

Figure 5 Results comparison for the Neuman-Witherspoon method, the Taylor-Brown method and the modified Taylor-Brown method

been obtained after 12 cycles of iteration by the Neuman-Witherspoon method (N&W), by the Taylor-Brown method (T&B) and finally by the modified Taylor-

Brown method (MT&B) with extrapolation.

In addition, it is necessary to mention that the great accuracy of the free surface position has practical importance only when the aim of the computation is to find the very position of the free surface. If the distribution of the head in the given region is mainly the matter, then the small shape deviations of the free surface approximation play an important role only in the immediate vicinity of the free surface.In the direction from the free surface into the region their influence disappears very quickly. E. g. the position of the free surface determined by the Neuman-Witherspoon method and represented in Figure 5 would be completely satisfactory for practical reasons.

CONCLUSIONS

For practical computations of the seepage flow with free surface in earth dams and earth cores of rockfill dams the modified Taylor-Brown method has proved to be successful. In this method the free surface is searched for by the iterative procedure of Taylor and Brown, but in the exit point C of the free surface approximation Dirichlet's boundary condition is always given and its new position on the seepage face is searched for after any iterative step by extrapolation from the position of the adjacent nodes on the free surface approximation.

The modified Taylor-Brown method is very easy to code and the iteration converges quickly not only in the mean, but also in any node of the approximation. With regard to the experience gained until now it is necessary to apply from 7 to 12 iteration cycles for the Equation 9 to be satisfied with precision of 0.05 m for a typical problem. The simplicity of the procedure enables to consume little computer time: in the case of the use of ICL 4/72 50-80 seconds of the CPU time were needed for typical computations.

REFERENCES

Finn, W. D. Liam (1967) Finite-element analysis of seepage through dams. Journal of the Soil Mechanics and Foundations Division, 93, SM6:41-48.

Kazda, I. (1977) Long-term pore pressure values in rockfill dam cores. Proceedings of the 5th Danube European Conference on Soil Mechanics and Foundation Engineering, vol. II, pp. 103-116.

Neuman, S. P. - Witherspoon, P. A. (1970) Finite element method of analyzing steady seepage with a free surface. Water Resources Research, 6, 3:889-897.

Taylor, R. L. - Brown, C. B. (1967) Darcy flow solution with a free surface. Journal of the Hydraulics Division, 93, HY:25-33.

Taylor, C. J. - France, P. W. - Zienkiewicz, O. C. (1973) Some free surface transient flow problems of seepage and irrotational flow. The Mathematics of Finite Elements and Applications (J. R. Whiteman ed.), Academic Press, London, pp. 313-325.

Zienkiewicz, O. C. - Cheung, Y. K. (1965) Finite elemnts in the solution of fields problems. The Engineer, 220, 5722:507-510.

Zienkiewicz, O. C. - Mayer, P. - Cheung, Y. K. (1966) Solution of anisotropic seepage by finite elements. Journal of the Engineering Mechanics Division, 92, EM1:111-120.

UNSTEADY GROUNDWATER REGIME DURING DRY-DOCK BUILDING

I. Seteanu, R. Popa

Polytechnical Institute of Bucharest, Rumania.

INTRODUCTION

The seepage discharge prediction from an aquifer through the dock walls is of great interest during the construction stage to selecting a sure, as well as economic capacity of drainage pumping station.

Some experimental results using electrohydrodynamic models (Davidenkoff and Franke, 1965) are now available for both homogeneous and isotropic soils under steady groundwater regime. However, for the most actual cases, the soil layers having quite anisotropic ("sandwich") structure, a more thorough analysis is to be done.

To determine the amount of groundwater table drawdown because of dock drainage, the groundwater, unsteady flow regime is analysed in the present paper, a particular attention being paid to separation zone between saturated and unsaturated domains, where the capillary pressure effect is prevailing.

A mathematical model is herein developed, consisting of a sequence of two-dimensional boundary value problems on a time-variable domain that are solved by finite element method (a Galerkin's scheme), and time-iterative integrations (a "backward difference" scheme). The free groundwater level is assumed stationary at far distances.

MATHEMATICAL MODEL

The continuity equation for the liquid phase may be written as:

$$- \frac{\partial}{\partial x} (\rho.u) - \frac{\partial}{\partial y} (\rho.v) = \frac{\partial}{\partial t} (\rho.m.S) \qquad (1)$$

where: ρ is the water density; u and v are the filtration (Darcy) velocity components along the x and y axes respectively; m is the soil porosity; and S is the relative volumetric water saturation (water volume/void volume).

The dynamic equations (Darcy's law) are:

$$u = - K_o . (K_{xx} . \frac{\partial H}{\partial x} + K_{xy} . \frac{\partial H}{\partial y})$$

$$v = - K_o . (K_{yy} . \frac{\partial H}{\partial x} + K_{yy} . \frac{\partial H}{\partial y}) \qquad (2)$$

where

$$H = y + \frac{p - p_a}{\gamma}$$

is the hydraulic head, while p and p_a are the pore and atmospheric pressures, respectively. In the above formulae, the symmetric permeability tensor $K_o(S).K_{ij}(x,y)$ is introduced by assuming the separation of the unsaturated effect, where K_o is the relative hydraulic conductivity ($0 \leq K_o \leq 1$), and K_{ij} representing the conductivity at saturation (Neuman, 1973).

By neglecting the soil deformability effect but taking into account the liquid phase compressibility following the relation:

$$\frac{\Delta \rho}{\rho} = \gamma . c_a . \Delta H \qquad (3)$$

where c_a is the bulk water compressibility coefficient, and combining the relations (1) and (2), one gets:

$$\frac{\partial}{\partial x} \left[K_o . (K_{xx} . \frac{\partial h}{\partial x} + K_{xy} . \frac{\partial h}{\partial y} + K_{xy}) \right] +$$

$$+ \frac{\partial}{\partial y} \left[K_o . (K_{yx} . \frac{\partial h}{\partial x} + K_{yy} . \frac{\partial h}{\partial y} + K_{yy}) \right] = (C + \frac{\theta}{m} . A) . \frac{\partial h}{\partial t} \qquad (4)$$

In this relation h is the pressure head ($h = (p-p_a)/\gamma$); C is the specific moisture capacity ($= \partial\theta/\partial h$); $\theta = S.m$ represents the volumetric moisture content, and $A = \gamma.m.c_a$ is the specific storage; ($\gamma = \rho.g$).

The formula (4) consists of a parabolic, quasilinear, partial differential equation of second order governing the pressure head field for unsteady regime through an anisotropic and nonhomogeneous porous medium with both saturated and unsaturated zones.

Regarding the initial conditions, one assumes to be given the pressure head field:

$$H(x,y,0) = h_o(x,y) \qquad (5)$$

throughout the initial flow domain.

The boundary conditions may be:

- imposed pressure head surface (on the porous medium-aquifer interfaces or on the seepage surfaces);

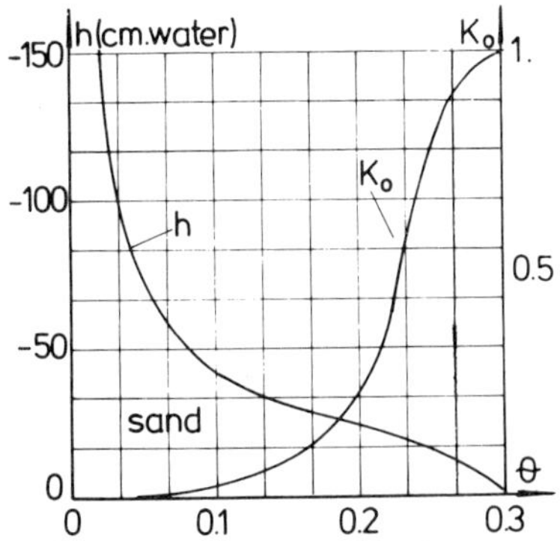

Figure 1 Material laws of the porous medium

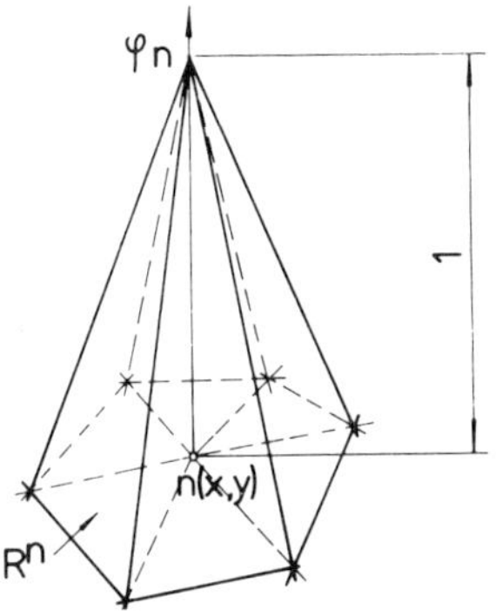

Figure 2 Unitary function of coordinates

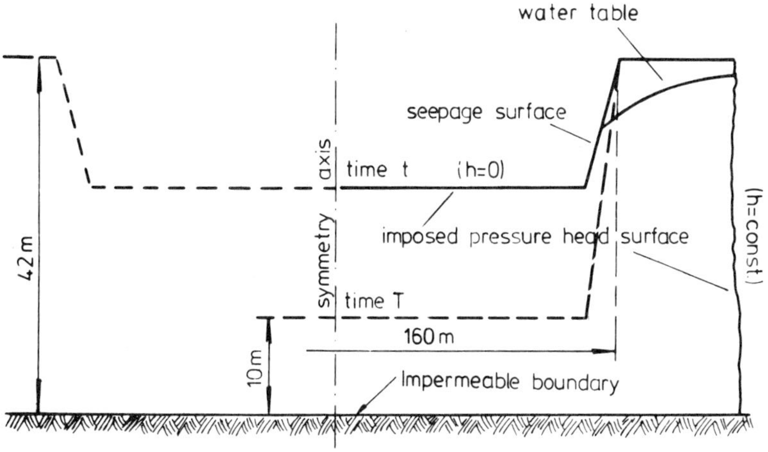

Figure 3 Sketch of excavation for computational model

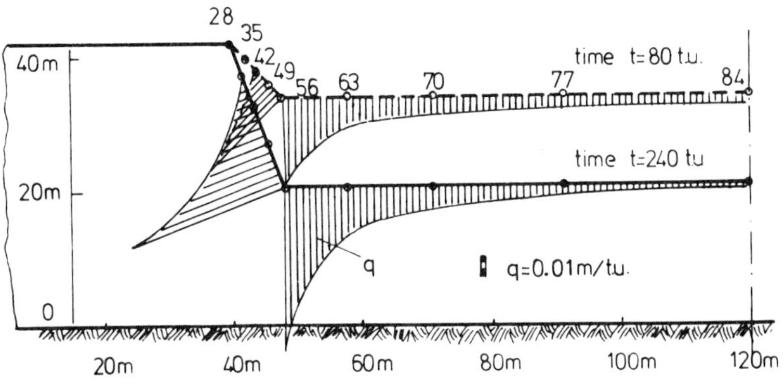

Figure 4 Specific flow rates on the excavation contour

$$- V_n(x,y,t) = K_o.(K_{xx}.\frac{\partial h}{\partial x} + K_{xy}.\frac{\partial h}{\partial y} + K_{xy}).n_x + $$
$$+ K_o.(K_{yx}.\frac{\partial h}{\partial x} + K_{yy}.\frac{\partial h}{\partial y} + K_{yy}).n_y \qquad (7)$$

where n_x and n_y are the outer normal unit vector components.

The material relationships $h = h(\theta)$ and $K_o = K_o(\theta)$ specifically denoted for any particular porous medium, are derived experimentally as indicated, for instance in the Figure 1.

Drawing the system of axes along the principal directions of the permeability tensor, the following relations may be inferred:

$$L(h) = \frac{\partial}{\partial x}(K_o.K_x.\frac{\partial h}{\partial x}) + \frac{\partial}{\partial y}\left\{K_o.(K_y.\frac{\partial h}{\partial y} + K_y)\right\} -$$
$$- (C + \frac{\theta}{m}.A).\frac{\partial h}{\partial t} = 0 \qquad (8)$$

$$K_o.K_x.\frac{\partial h}{\partial x}.n_x + K_o.(K_y.\frac{\partial h}{\partial y} + K_y).n_y = - V_n(x,y,t) \qquad (9)$$

instead of the formulae (4) and (7), where $K_x = K_{xx}$, $K_y = K_{yy}$, and $K_{xy} = K_{yx} = 0$.

FINITE ELEMENT METHOD (GALERKIN'S SCHEME)

By discretising the flow domain into finite elements, one defines the coordinate unit functions (as in the Figure 2):

$$\phi_n(x,y) \begin{cases} 1 & \text{for } x = x_n,\ y = y_n \\ \text{linear within } R^n & \\ 0 & \text{for the rest} \end{cases} \qquad (10)$$

where R^n denotes the assembly of the elements neighbouring the n-th nodal point $(n = 1,2, \ldots,N)$, N being the total number of grid points.

The solution is then found as having the following form:

$$h^N(x,y,t) = \sum_{n=1}^{N} h_n(t).\phi_n(x,y) \qquad (11)$$

where $h_n(t)$ are the unknowns, and by imposing the condition that the generalised scalar product vanishes for the chosen g grid points:

$$\int_R L(h^N).\phi_n.dx.dy = 0 \qquad (n = 1,2,\ldots,N) \tag{12}$$

Assuming that the weighted average time-derivatives are expressed as (Neuman, 1973):

$$\frac{\partial h_n}{\partial t} = \frac{\int_R (C + \frac{\theta}{m}.A).\frac{\partial h}{\partial t}.\phi_n.dx.dy}{\int_R (C + \frac{\theta}{m}.A).\phi_n.dx.dy}, \tag{13}$$

one gets the ordinary differential equations system:

$$\sum_{j=1}^{N} \left(A_{ij}.h_j(t) + F_{ij}.\frac{\partial h_j}{\partial t} \right) = Q_i - B_i \tag{14}$$

$$(i = 1,2,\ldots,N)$$

where A_{ij} is a sparse, symmetrical matrix, F_{ij} is a diagonal matrix, while Q_i and B_i are vectors (Neuman, 1972).

The nodal flow rates, Q_i, are equal to zero along the impermeable boundary, as well as for the internal points and the boundary points near to the unsaturated domain. For the boundary points where the flow is directed to the inside of the domain one considers $Q_i > 0$.

INTEGRATION SCHEME OVER TIME

For numerical integrating the ordinary differential equations system (14), one splits up the total time-interval T into small steps t. Then the system (14) may be written again in finite differences by using an implicit "backward scheme" as follows:

$$\sum_{j=1}^{N} \left(A_{ij}^{(k+\frac{1}{2})} + \frac{1}{\Delta t_k}.F_{ij}^{(k+\frac{1}{2})} \right).h_j^{(k+1)} = $$

$$= Q_i^{(k+\frac{1}{2})} - B_i^{(k+\frac{1}{2})} + \frac{1}{\Delta t_k}.\sum_{j=1}^{N} F_{ij}^{(k+\frac{1}{2})}.h_j^{(k)} \tag{15}$$

$$(i = 1,2,\ldots,N)$$

The solution is iteratively derived with $A_{ij}^{(k+\frac{1}{2})}$ and the other coefficients determined for the time-moment $t_k + \Delta t_k/2$, assuming that the piezometric head is

$$h_j^{(k+\frac{1}{2})} = \frac{1}{2} \cdot \left[h_j^{(k)} + h_{jo}^{(k+1)} \right]$$

where $h_{jo}^{(k+1)}$ represents the previous iteration solution (for the first iteration one takes $h_{jo}^{(k+1)} = h_j^{(k)}$).

The Gauss elimination algorithm is finally used for solving the linear equations system (15).

The chosen computation method assures the damping of the oscillation tendency specific to the numerical schemes of parabolic equations often mentioned by various authors (Mikhlin, 1964).

BOUNDARY CONDITIONS SPECIFICATION

One solves the system (15) by disregarding the i-th order equation corresponding to i-th boundary grid points where the piezometric head is assumed to be given. These equations are then used to derive the filtration flow rates in the same grid points.

For the i-th boundary grid points where the filtration flow rates are given, one obtains the nodal discharge values, Q_i (e.g. along the impermeable boundary: $Q_i = 0$).

For the unsaturated i-th boundary grid points, oneimpos imposes $Q_i = 0$ and one then checks the required restriction $h_i < 0$, when the computations of the given iteration are finished. If contrary, one infers that the respective boundary grid points have become saturated.

For the i-th boundary grid points placed on the seepage surface, the condition $h_i = 0$ (direct contact with the atmospheric pressure) is imposed and one tries to verify when the computations of the given iteration are finished, if the required condition $Q_i < 0$ is satisfied. If not, one concludes that the now mentioned boundary grid points are unsaturated (Neuman, 1973).

One proceeds the process, by successive iterations, for the given time-moment, until the absolute error begins to be less than a specified precision level, over the whole flow domain. This way, the actual flow domain is obtained (for which the saturation S = 1). Meantime, the unsaturated zone where the capillary force effects are dominant, is finally found.

NUMERICAL RESULTS

An almost parallelepipedic excavation, (Figure 3), with the length long enough to accept that the flow regime might be two-

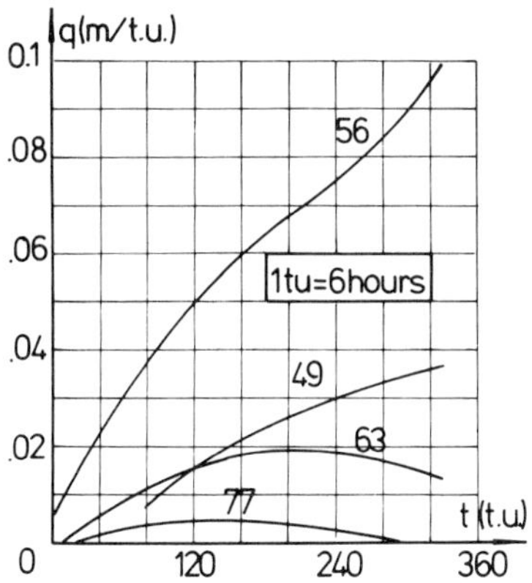

Figure 5 Specific flow rates on the excavation contour

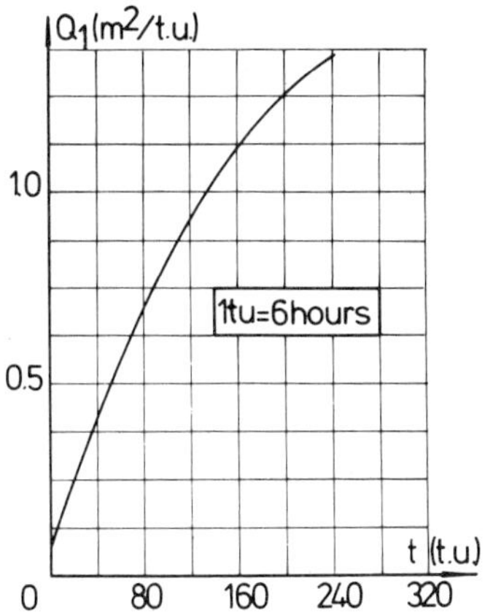

Figure 6 Seepage discharge on the unit length

dimensional, is analysed. Because of the assumed linear time-variation of the depth during the dry-dock building, a time varying network of nodal points with a specified geometric structure was chosen, to more easily introduce the boundary conditions.

The assumed time-step is $\Delta t = 6 \div 60$ hours, while the total time of analysis was T = 85 days. An anisotropic soil ($K_x = 1.2K_y$) with $K_y = 0.33 \times 10^{-4}$ cm.s^{-1} is considered.

The specific discharges for the excavation contour grid points were obtained (Figure 4), corresponding to various time moments. The time-variation of the specific discharges for some specified points are presented in the Figure 5.

The seepage discharge, Q_1, on the unit length, was obtained with the relation

$$Q_1 = \int q(s).ds$$

the integration being performed along the whole contour of the excavation cross-section.

The time-variation of the discharge Q_1 is represented in the Figure 6, and an almost parabolic increasing of it may be observed from the picture.

By tightly analysing the numerical results, a concentration tendency of the specific discharges may be remarked near to the excavation cross-section corners, while a decreasing of the discharge contributions is observed near the central inside of the excavation.

REFERENCES

Davidenkoff, R. and Franke, O. (1965) Untersuchung der Raumlichen Sicherstromung in line Limspundete Baugrube in offenen Gewassern, Bautechnik, Sept.

Mikhlin, S.G. (1964) Variational Methods in Mathematical Physics, Pergamon Press, Oxford.

Neuman, S.P. (1972) Theory of Flow in Unconfined Aquifers Considering Delayed Response of the Water Table, Water Resources Research, 8, 4.

Neuman, S.P. (1973) Saturated-Unsaturated Seepage by Finite Elements. Proc. ASCE, J. Hydraulics Division, 99, HY 12.

A QUANTITATIVE COMPARISON BETWEEN FINITE ELEMENT SOLUTION AND EXPERIMENTAL RESULTS OF TRANSIENT, UNCONFINED GROUNDWATER FLOW

G. Ōhashi

Dept. of Agricultural Eng., Ehime University, Matsuyama, 790, Japan

INTRODUCTION

In modeling hydrology systems,unconfined groundwater flows have indirect effects upon the hydraulic behavior of both ground-surface and subsurface. However, they play many important roles in hydraulic problem or technical design concerning to not only water resources but also the multiple assessments.

On the other hand, a wide range of groundwater flow problems have been handled by the approximate numerical techniques. Moreover after going through the necessary discussions, much of this activity has centered on the numerical analysis of groundwater flow with a phreatic surface in particular the more difficult problem of transient seepage.

Recently, Neuman(1972,1973) proposed a linear model that treated the unconfined aquifer as a compressible system and the phreatic surface as a moving boundary. The transient phenomenon can be simulated mathematically by using constant values of specific storage and specific yield without recourse to unsaturated theory. Under this treatment, the basic equation of the flow has a parabolic differential type which involves the time derivative.

At the same time, a comprehensive review of the application of the finite element method were given by Neuman et al.(1973).

The purpose of this paper, therefore, is to consider the extention of the Neuman's model so that an unconfined groundwater flow system in vertical two-dimension with the boundary, which has an effect of periodic fluctuation, can be analized. First, an attention is paid to the dependent movement of the phreatic surface, because a simplified procedure based on the direct shift technique of free surface position is utilized for determining the height. That is, the problem of stability and convergence in the sense of the maximum norm for parabolic equation of second order has been analized by using Fujii's definition(1974). Thus, the reliability of computational stability for a pair of given parameter has been ascertained by performing various numerical illustrations.

Second is to find out restrictions to the validity of the above technique. After the solutions obtained by finite element method are compared with the experimental results obtained by means of both Hele-Shaw madel and Sand layer model, several restrictions in light of these comparisons have been presented.

DIRECT FORMULATION OF TIME DEPENDENT PROBLEM

Assuming the conditions such that fluid is uncompressible and porous media is homogeneous in an isothermal state, one can derive a partial differential equation from both Darcy's law and the equation of continuity

$$\rho K(\nabla^2\phi) = \rho S_* \partial\phi/\partial t, \quad \text{------------}(1)$$

where

$$S_* = \rho g S_w \lambda c_f, \quad \text{------------}(2)$$

$$c_f = -\alpha_b/\lambda , \quad \text{------------}(3)$$

in which S_* = the elastic storativity of unconfined aquifer; K = the hydraulic conductivity; ρ = the density of water; S_w = the degree of volumetric water saturation($0<S_w\leq1$); λ = the porosity; g = the gravity acceleration; and α_b = the coefficient of bulk compressibility defined such as the fractional change in pressure under the constant external stress.

In this elastic treatment, the constant value of elastic storativity(S_*) in the right hand side of Equation (1) is considered to be the same as well as the specific storage in confined aquifer. The basic equation of the vertical flow concerned(Figure 1) can be represented in two-dimensional form by

$$K(\partial^2\phi/\partial x^2 + \partial^2\phi/\partial y^2) = S_*\partial\phi/\partial t, \quad \text{------------}(4)$$

where x and y are horizontal and vertical cartesian coordinate, ϕ is an unknown harmonic function.

The initial conditions for the flow can be specified as

$$\phi(x,y,t) = \phi_0(x,y) \qquad \text{on } R, \quad \text{------------}(5)$$

$$\psi(x,0) = \psi_0(x) \qquad \text{on } FS, \quad \text{------------}(6)$$

where *R* is the flow region, *FS* designates the free surface, $\psi(x,t)$ represents a equation of free surface, and suffix 0 denotes the initial stage.

The boundary condition for a prescribed head is given by

$$\phi(0,y,t) = H(t) \qquad \text{on } \Gamma_1, \quad \text{------------}(7)$$

and for the impermeable boundary is by

$$K\cdot\partial\phi/\partial n = 0 \qquad \text{on } \Gamma_2, \quad \text{------------------(8)}$$

where Γ_1 and Γ_2 represent the portions of boundary and n designate the normal direction along the impermeable boundary.

In unconfined groundwater flow, the other conditions must be satisfied on the free surface;

$$\psi(x,t) = \phi(x,y_{FS},t), \quad \text{-----------------------------(9)}$$

$$K\cdot(\partial\phi/\partial x)\cdot n_x + K\cdot(\partial\phi/\partial y)\cdot n_y = S_y\cdot(\partial\phi/\partial t), \quad \text{------(10)}$$

where S_y is the specific yield and n_x and n_y are the x and y directional cosines of the unit out-ward normal along the free surface. In this compressible treatment, the basic equation takes a form in which time derivative of unknown function occurs.. However the time dependent problem of Equation (4) is precisely reduced to those treated in the steady flow problem at a particular instant.

The derivation of finite element solution and discretization through the solution domain have already been well known. Let the unknown function ϕ of each element with this situation be approximated as

$$\phi = [N(x,y)]\{\phi\}^e , \quad \text{-----------------------------(11)}$$

in which N(x,y) is the appropriate interporation function defined picewise element by element and $\{\phi\}^e$ is the nodal value of ϕ in the discretised domain. A standard form of assembled equation for the overall matrices in the whole region can be written as

$$[H]\{\phi\} + [C]\{\dot{\phi}\} + \{F\} = 0, \quad \text{----------------------(12)}$$

where $\{\dot{\phi}\}$ is the derivative of $\{\phi\}$ with time t.

In order to obtain overall matrices and the total heads of all nodes, the terms which are the element along the free surface, in the inside regionand along the impermeable boundary, whould be evaluated in terms of the detailed formulation with respect to each element matrix.

On the other hand, over the time interval(t, t-Δt) the nodal point potential head will be approximated by a linear expression as

$$\{\phi\}_\xi = (\xi-t')(\{\phi\}_t-\{\phi\}_{t'})/\Delta t + \{\phi\}_{t'} , \quad \text{-----------(13)}$$

where $t>\xi>t'$, $t'=t-\Delta t$, $\xi=t'+\theta\Delta t$, and $\theta\neq 0$ in which θ is the coefficient of a difference scheme and Δt is an arbitrary time increment.

Using the simillar expression in terms of θ-scheme constant a finite difference approximation will be represented by

$$(\{\phi\}_t-\{\phi\}_{t'})/\Delta t = \theta\{\dot{\phi}\}_t + (1-\theta)\{\dot{\phi}\}_{t'} , \quad \text{---------}(14)$$

where $\theta \neq 0$. Upon insuring continuity of the potential field with respect to time variable, the above expression will give a distribution specified at a discrete number of nodal point along the boundary. In addition to this, Equation (12) is rewritten to reflect the solution at time t in terms of pseude-initial values at time t-Δt. Therefore, substituting Equation (13) and (14) into Equation (12) results in the matrix equation for the step by step solution procedure

$$(\theta[H]+[C]/\Delta t)\{\phi\}_t = -\left[(1-\theta)[H]-[C]/\Delta t\right]\{\phi\}_{t'}$$

$$-\theta\{F\}_t-(1-\theta)\{F\}_{t'} , \quad \text{----------}(15)$$

where $\theta \neq 0$. This equation is written in a more convenient form for the purpose of the solution technique as

$$[D]\{\phi\}_\xi + \{R\} = 0, \quad \text{------------------------------}(16)$$

where $[D] = ([H] + [C]/\theta\Delta t)$, $\quad$ -------------------------(17)

$$\{R\} = -([C]/\theta\Delta t)\{\phi\}_{t'}+\left[\theta\{F\}_t+(1-\theta)\{F\}_{t'}\right], \text{---------}(18)$$

$$\{\phi\}_\xi = \theta\{\phi\}_t + (1-\theta)\{\phi\}_{t'} , \quad \text{--------------------}(19)$$

and $\theta \neq 0$. In addition to this, one can obtain the time variation in potential head by starting at t=0, that is

$$\{\phi\}_{t'} = \{\phi\}_{t=0} , \quad \text{------------------------------}(20)$$

Upon evaluating the values $\{\phi\}_\xi$ by Equation (16), the new total potential heads at time t are calculated by

$$\{\phi\}_t = \{\phi\}_\xi/\theta - \left[(1-\theta)/\theta\right]\{\phi\}_{t'} , \quad \text{---------------}(21)$$

where $\theta \neq 0$. According to the well known procedure, at each time step the iteration method should be chosen for relocating the position of the free surface and then recalculating its element matrices.

NUMERICAL METHOD AND PROPOSED DEFINITIONS

Assuming the replacement of the original transient problem by a discrete number of steady ones, the movement of the phreatic surface can be approximated as a series of discrete surfaces. In order to satisfied the above assumption, the changes in shape of the flow domain should be each of slightly varying and each separated by a small interval of time. Under normal condition, the surface configuration and boundary values are known at the beginning of each time step. And, the iterative process is repeated for several chosen time interval until the full range of profiels has been investigated or until steady state condition prevail.

In this paper, however, the shifting of the free surface is restricted to vertical direction and the position of the free surface for each time step is estimated approximately by

$$\psi(x,t) = \phi(x,y_{FS},t-\Delta t), \qquad (22)$$

and the boundary condition prescribed in Equation (7) is given by

$$H(t) = A\cdot\sin \omega t, \qquad (23)$$

where A is amplitude and ω ($=2\pi/T$) is an angular frequency.

In other words, the above mentioned procedure is a direct relocation of free surface without using the iteration method, because the flow concerned has a steady regulated motion with a constant period(T). Later, the experimental results will give the evidence that the proposed method is satisfactly utilized for a repeating method in stead of the iteration one.

For the purpose of comparison with numerical results, a theoretical solution of the one-dimensional flow corresponding to the situation of Equation (4) is presented by

$$\psi(x,t) = A\exp(-m_A x)\cdot\sin(\omega t-m_L x), \qquad (24)$$

where $-m_A$ is the damped coefficient of amplitude, $-m_L$ is the coefficient of phase shift, and A is the amplitude given. Furthermore, the relation between these coefficients and the parameters concerning to the measurement and properties of aquifer is expressed by

$$m_A = m_L = \sqrt{\lambda\omega/2Kh}, \qquad (25)$$

where h is a mean depth of aquifer. As for the length of aquifer, it is necessary to bring into use a finite value in practical objectives.

Then, a new parameter in stead of these coefficients is proposed by

$$P = \sqrt{\pi\lambda\ell^2/TKh}, \qquad (26)$$

where ℓ is a finite length of aquifer. In addition to this a non-dimensional parameter of time is defined by

$$t' = (Kh/\lambda\ell^2)t. \qquad (27)$$

Using this parameter, one can transform P-value into a non-dimensional frequency in terms of ω' as

$$P = \sqrt{\omega'/2}, \qquad (28)$$

where $\omega'=2\pi/T'$ and T' is a non-dimensional period.

Finally, the physical meaning of P-value can be clarified by means of the following definitions;

$$P = \sqrt{1/2R_{e\pi}} \ , \qquad \text{------------------------------(29)}$$

where $R_{e\pi} = V_{\pi}\cdot L_{\pi}/\nu$, ------------------------------(30)

$$\left.\begin{array}{ll} V_{\pi} = gT/2\pi, & L_{\pi} = kh/\lambda\ell^2, \\ K = k\rho g/\mu, & \nu = \mu/\rho \ , \end{array}\right\} \text{-------------------(31)}$$

in which, μ = the coefficient of viscosity, ν = the coefficient of kinematic viscosity, k = the intrinsic permeability, and V_{π} = the phase velecity.

As for the above relationship, $R_{e\pi}$ is defined in the same manner as Reynold's Number of fluid flow. Then, the proposed expression of $R_{e\pi}$ is called as " pi (π)-Reynold's Number" with respect to seepage flow in porous media.

STABILITY AND CONVERGENCE

The success of the procedure concerning to free surface shifting depends upon both the accuracy and the stability of the calculus schemes which contain both time difference scheme and triangulation parameters.

Fujii(1974) showed that a parabolic scheme problems constructed in Finite-Element-Galerkin manner could be analized mathematically with a view to obtain the discrete maximum principle property as well as stability in the mean square sense. The main concerns in his argument were the notion of triangulation of acute type and a priori estimation of the discrete approximation for the parabolic equation. The conditional representation to guarantee the maximum principle stability(L^{∞}-stability) and the mean square sense stability(L^2-stability) were expressed in terms of several parameters for the lumped mass type and the consistent mass type. In his paper some numerical illustrations of the stabilities were only shown in one-dimensional exsamples.

Now, the extension of his evaluating criteria to the problems in two-dimensional flow is considered and critical examinations of these evaluations are carried out.
Figure 2 represents the relation with the acuteness of a simple triangular and the definition of η_i and δ_i. Then, the expression $\sigma = \max_i\cdot\cos(\delta_i)$ results $\sigma = -\min_i\cdot\cos(\eta_i)$, because $\delta_i+\eta_i = \pi$. Hence, the condition which $\sigma \leqq 0$ reduces equivalently to the one which a simple triangular is acute type.

In a similar way, considering a triangulation with regular simplices and putting $\zeta = \gamma_1 K\Delta t/\kappa_{min}^2$, in which γ_1 is constant(>0) and κ_{min} is the minimum perpendicular length of all the simplices triangulation, the computational stable conditions for two type schemes are reduced to the following expressions as:

(i) Stability for Lumped Mass Type Scheme ($\gamma_1=\sqrt{12}$)

L^{∞}-Stable : $\zeta \leqq 1/3(1-\theta)$ $\quad (0 \leqq \theta < 1)$ ----------(32)

L^2-Stable : $\zeta \leqq 1/3(1-2\theta)$ $\quad (0 \leqq \theta < 1/2)$ --------(33)

(ii) Stability for Consistent Mass Type Scheme ($\gamma_1=\sqrt{48}$)

L^{∞}-Stable : $1/6\theta \leqq \zeta \leqq 1/6(1-\theta)$ $\quad (0 < \theta < 1)$ ---------(34)

L^2-Stable : $\zeta \leqq 1/12(1-2\theta)$ $\quad (0 \leqq \theta < 1/2)$.--------(35)

Figure 3 gives the corresponding stability regions for both lumped mass type and consistent one. It is found out from both figures that the region of maximum principal property are more restricted than those of one-dimensional example. Especially, when $\theta = 1/2$ in consistent mass type, only that $\zeta = 1/3$ is allowable. The ranges of L^2-stable in both types are always satisfied under condition which $\theta \geqq 1/2$, and the smaller the value of K is, the more wide the region of L^2-stable is.

Since the mesh spacing in the above argument are assumed to be uniform, the more restricted condition should be required in the case of non-uniform spacing.

Let ϕ_i^{nT} represent the value of ϕ at time $i\Delta t$ in the n-th period, where i is the i-th increment of the time interval in each steady period nT. For the calculus scheme expressed by Equation (16), a proposed norm corresponding to the well known norm may be defined as follow:

$$\|\Delta\phi\| = |\ \phi_i^{(n+1)T} - \phi_i^{(n)T}\ | \qquad \text{----------------------(36)}$$

where $\|\cdot\|$ and $|\cdot|$ denote norm and absolute value respectively.

Similary, let Λ a constant parameter, which is equivalent to the Lipshitz constant, represent the convergence index, it may be defined by

$$\Lambda = |\phi_i^{(n+1)T} - \phi_i^{(n)T}| \ / \ |\phi_i^{(n)T} - \phi_i^{(n-1)T}| \ , \qquad \text{------(37)}$$

If for several trial choices of Δt, Λ is found to be very small compared 1.0, the sequential time step can be follow to arrive at eventually a condition satisfying that the Lipshitz constant is less than one.

EXPERIMENTAL METHOD

Laboratry experiments were set up to measure transient water movement for the prescribed boundary conditions. A middle soil tank, 300.7cm long, 63.2cm deep, and 44.8cm wide was used to represent a section of the aquifer profile extending from a lateral ditch at one end of the tank to the midpoint between ditches or the impervious boundary at the other end. The soil used was a fine sand which has the properties such as

d_m = 0.57mm, K = $7.26 \cdot 10^{-3}$cm/sec(in steady state), and was uniformly packed in the tank. An apparatus of Hele-Shaw model consists of parallel glass plates spaced 1.44mm in apart. The plates are 25.0 high and 42.0 long. The fluid used in the model was the mixture of both glycerine(ν=8.302cm^2/sec, ρ=1.252 gr/cm^3, and 95.0% purity) and water in several ratios.

The experimental setup of two models are shown in Figure 4. For both models the water table was initially horizontal at a distance h above the impermeable layer. Soil water pressure heads were measured by porous pipes connecting to the piezometers, located at suitable distances from the end of layer. The water table profiles of Hele-Shaw model were measured indirectly by making photographs for each sampling time. Among many experiments, the representative cases of both models, which were conducted for different initial water table, periodic time, and amplitude combinations, are selected and summarised in Table 1 and 2.

RESULTS AND DISCUSSION

The spatial configuration of the finite element nodes is shown in Figure 5. The region of flow concerned was divided into 110 triangulars. The nodes on free surface and with given head were numbered from 1 to 19 and from 70 to 74 respectively. Thus, the other nodes were unknown heads. The spacing near the end of aquifer was chosen so that the large gradient could be approximated with a fair degree of accuracy.

Typical plot for the hydraulic head distribution as measured experimentally by Hele-Shaw model of No.14' and as predicted by solution of finite element method is given in Figure 6. As shown in example , the two-dimensional finite element method has provided an accurate solution for the complicated water table fluctuations of groundwater flow. Especially , all results measured by means of Hele-Shaw model have shown that close agreement was obtained between the predicted and the measured.

On the basis of this experimental fact, it is infered that the proposed numerical method has been acceptable for the most purposes. Then, numerical examinations concerning to stability and convergence were made by using the above mentioned model.

First, the differences of the convergence depending on the θ-scheme are shown in Figure 7 and 8, which are given by the parameters $|\Delta\phi|$ and Λ respectively. Further example , as shown in Figure 9, clarify the relative tendencies of the convergences with several time increments. The feasibility of approximating the groundwater flow by using finite difference scheme for the time domain are evaluated relatively from these exsamples. In this sence, although three θ-schemes provide resemble results, it is apparent that the backward difference scheme is the most desirable and the Galerkin scheme is second.

As shown in Figure 3, these tendencies agree with the fact that the stability region of which θ-scheme value is near one has been more spacious than the others. For the time increment Δt, the most suitable value can be chosen from the economical point of view. Hence, a restrictive expression of the time increment is given relatively such that when $\theta = 1/2$, $\Delta t/T \leq 1/30$.

On the other hand, the experimental results obtained by means of Sand model has two groups of tendencies such that the former cases are in good agreement with the numerical solution of finite element method as well as Hele-Shaw model and the others have apparent lags with the predicted. These tendencies are evident in Figure 10 and 11, which show the phase lag in the sharp motion cases. Namely , the cases which their cycle time T are shorter than 120sec have not good agreement with the numerical results. And their discrepancy in phase will be increase with the intensification of fluctuations.

To sum up these results, another explanation is given in Figure 12, which displays the relationship among the one-dimensional results(m_A,m_L), the two-dimensional ones($\tilde{m}_A$,$\tilde{m}_L$), and the experimental ones($\hat{m}_A$,$\hat{m}_L$). While agreement among m_A, $\tilde{m}_A$, and $\hat{m}_A$ is acceptable for the most purposes, the fact that the experimental results($\hat{m}_L$) which encircled by broken line in Figure 12 have lag to both m_L and $\tilde{m}_L$ indicate that soil properties used may have some upper bound against the sharp fluctuation. In fact, the phase difference shown in Figure 10 could be calculated in terms of hydraulic conductivity(K). That is, it seems likely that the fitted value is about 30 times larger than the K-value obseved in laboratory under steady state.

In other words, the difference mentioned in the above examples is particularly important because it implies that the assumption of linearity in the theory to the flow concerned has not been satisfied effectively. Consequently, it is necessary to introduce such a restrictive index as the kinematic property to the concerned flow could be estimated.

For this purpose, a new parameter P-value defined in Equation (26) is utilized for taking into account effectively. At the same time, the kinematic meaning of this parameter is explained by the relation depending on the physical characterristics as well as Reynold's Number. Then, P-value concerning to all cases of both models are calculated and shown in Table 1 and 2. Results presented in two Tables indicate that the case having good agreement with the numerical solution keeps such values as P-values are smaller than 5.0.

As a result, the effect of P-value display is in accord with the consequence from the comparisons between two approximate solutions and the experimental ones.
When reliable experimental results and theoretical consideration concerning to P-value are provided, this parameter will continue to be useful for predicting water table movement in view of significant practical validity.

CONCLUSION

Based on the study reported in this paper, the following conclusion can be drawn:

1. A numerical method has been developed in which the shifting values of free surface can be estimated very efficiently using the explicit vales of potential head on the free surface.
2. The results presented show that the 'simplified shift' method gives satisfactory agreement with the experimental ones.
3. A check has been made on the convergence and stability of difference schemes for the time dependent part. When the scheme constant becomes near 1.0, they have good ends.
4. The restrictive index as to the application of linear theory of the concerned flow has been proposed by 'P-value' or 'Pi -Reynold's Number', which consists of a non-dimensional parameter as similar as Reynold's Number and is related with physically measurable values.
5. When p-value is smaller than 5.0, three results have good agreement with each other. This expression means that a kinematic characteristics for the flow is briefly described.

ACKNOWLEDGEMENT

This research is partly supported by the National Science Foundations through Grant No.156157(1976), No.102023(1976), and No.202024(1977). Digital computation works were carried out on the F-230-28 at Ehime Univ. and M-190 at Kyoto Univ.

REFFERENCES

Fang,C.S.,S.N.Wang, and W.Harrison (1972) Groundwater Flow in a Sandy Tidal Beach (2).,Water Resour.Res., 8 ,1:121-128.

Fujii,H (1974) A Note on Finite Element Approximation of Evolution Equation., Tech. Rep.202, Math. Sci. Res. Inst, Kyoto Univ., 96-117.

Neuman,S.P. and P.A.Witherspoon (1971) Analisis of Non-Steady Flow with Free Surface using the Finite Element Method.,Water Resour.Res., 7 ,3:611-623.

Neuman,S.P. (1972) Theory of Flow in Unconfined Aquifer considering Delayed Response of the Water Table.,Water Resour.Res. , 8 , 4:1031-1045.

Ohashi,G. (1977) Studies on Analysis of Unconfied Groundwater Flow in Alluvial Fan.(in Japanese), Ph.D.thesis,Kyoto Univ., Kyoto, 26-55.

Pinder,G.F. and E.O.Frind (1972) Application of Galerkin Procedure to Aquifer Analysis., Water Resour.Res., 8, 1:108-120.

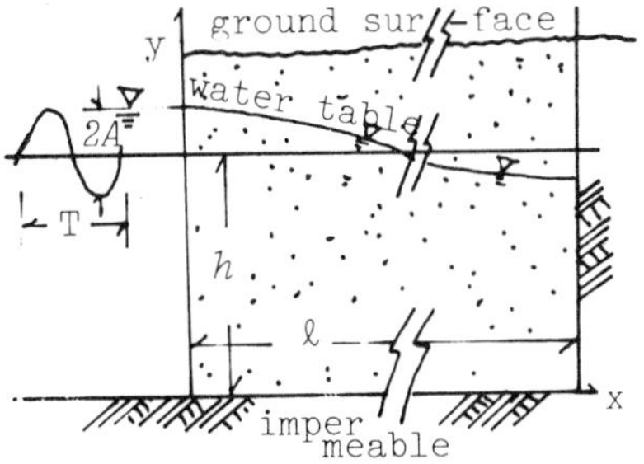

Fig.1 Schematicdiagrm of unconfined aquifer.

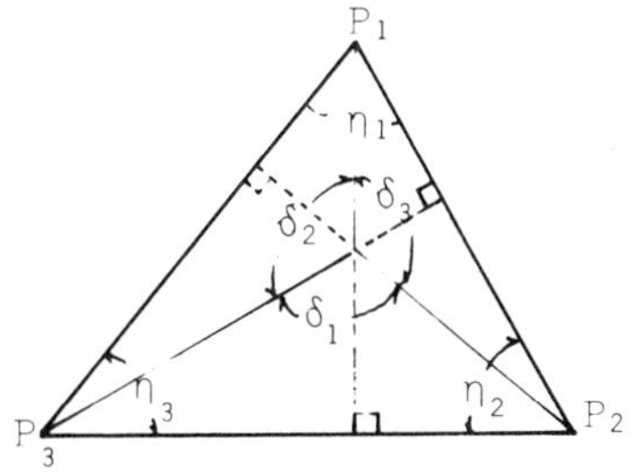

Fig.2 Definition of acute type triangular.

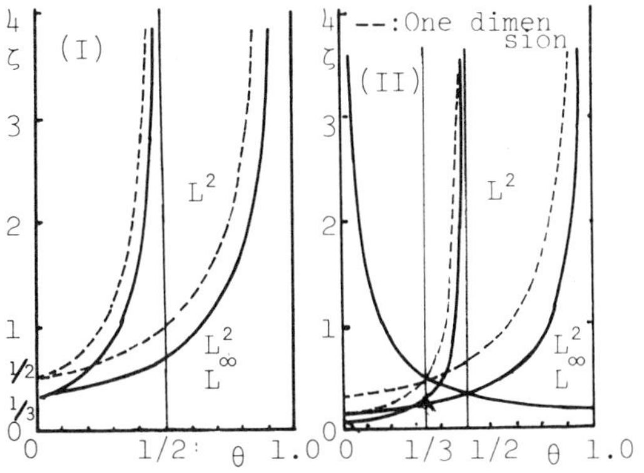

Fig.3 Stability regions for (I):Lumped mass type and (II) Consistent mass type.

Table 1 Hele-Shaw model

No.	T	2A	h	K	P	Δt	%
3′	30	2.92	11.15	6.253	1.63	2	70
8′	60	3.72	11.30	6.160	1.15	2	70
10′	30	2.30	11.50	1.377	3.42	2	85
14′	60	2.32	11.30	1.446	2.38	2	85
17′	120	2.84	11.38	1.446	1.67	4	85
20′	120	1.92	11.48	0.204	4.40	4	100
19′	240	2.76	11.13	0.204	3.19	8	100
18′	480	3.69	11.33	0.204	2.23	16	100

(cm, sec)

Table 2 Sand-model (P_1:K=0.034, P_2:K=0.34)

No	T	2A	h	P_1	P_2	Δt	
19	30	4.49	43.09	25.4	8.04	1	σ
12	60	1.73	42.71	18.1	5.71	2	
13	60	3.66	42.81	18.0	5.70	2	
14	60	5.62	43.09	18.0	5.68	2	○
15	60	7.70	43.15	18.0	5.68	2	
16	60	9.53	43.02	18.0	5.69	2	
2	120	5.74	43.25	12.7	4.01	4	Δ
8	240	5.62	43.09	8.99	2.84	8	□
11	480	5.51	42.74	6.38	2.02	16	▽

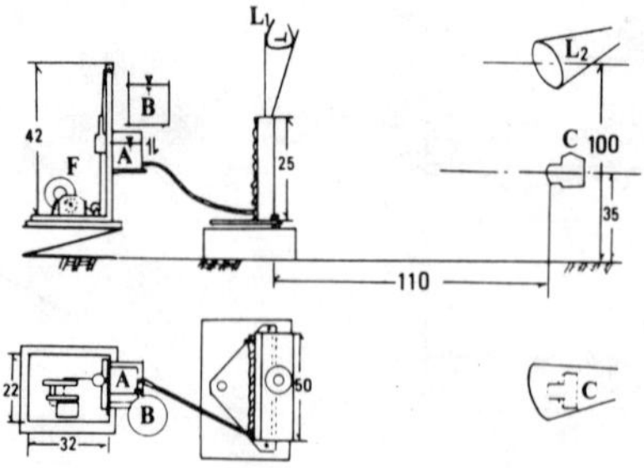

Hele-Shaw Model

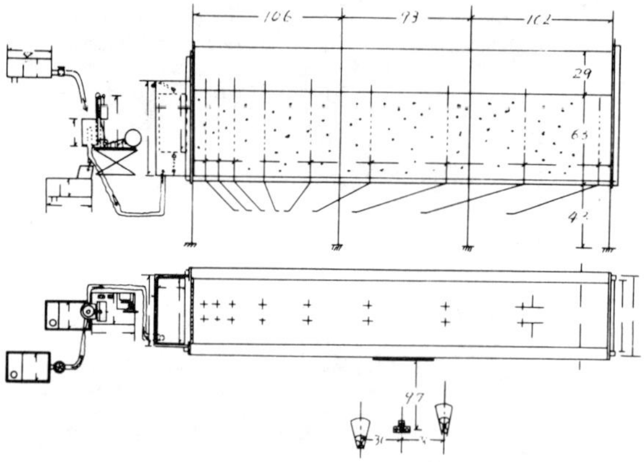

Sand Tank Model

Fig. 4. Apparatus used for periodic flow.

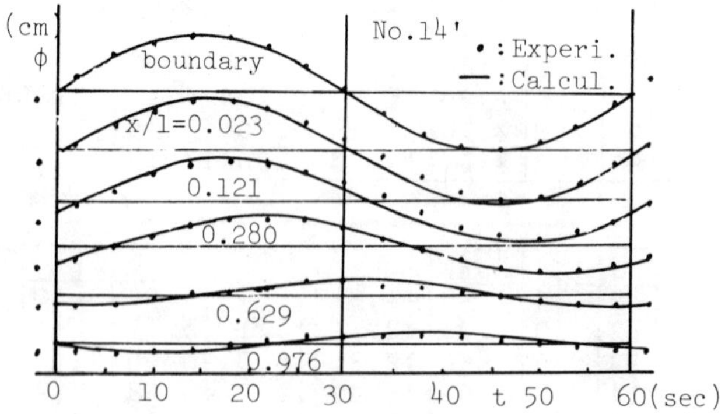

Fig.6 Comparison between Experimental result of Hele-shaw model and Finite Element solution.

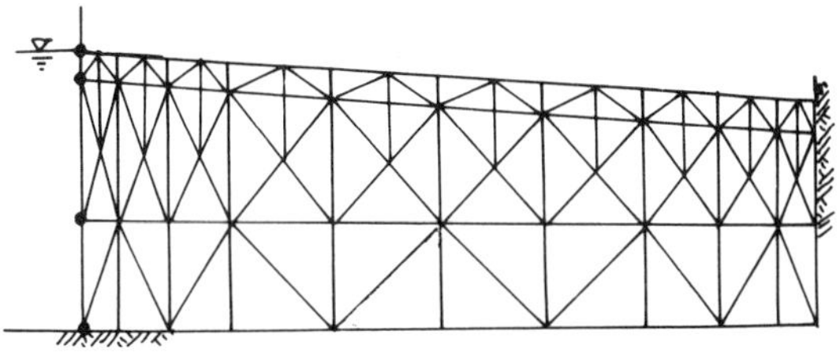

Fig. 5 The finite element display of the flow region in unconfined aquifer.

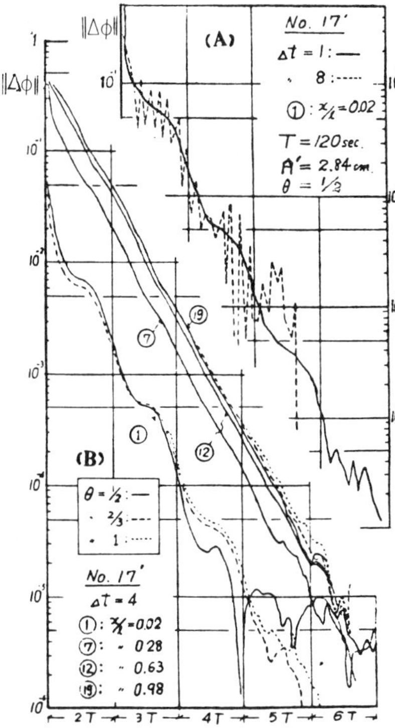

Fig.7 Examples of convergence at the free surface nodes with several values of both Δt and θ.

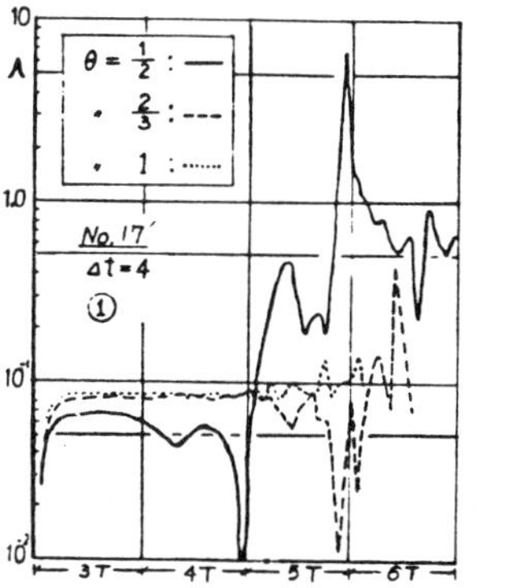

Fig.8 Examples of convergence for several values of θ.

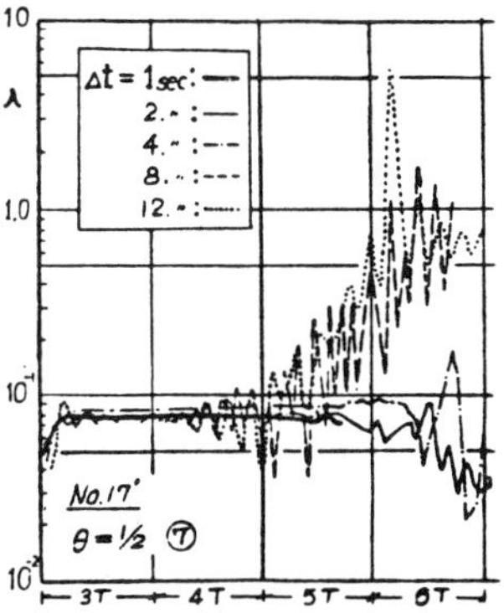

Fig.9 Example of convergence for several time increments.

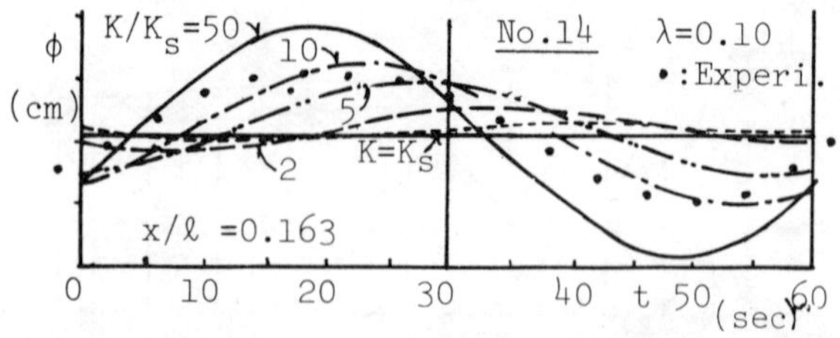

Fig. 10 Comparison between experimental result of Sand model and the calculated with several hydraulic conductivities, where K_s is value of laboratry test.

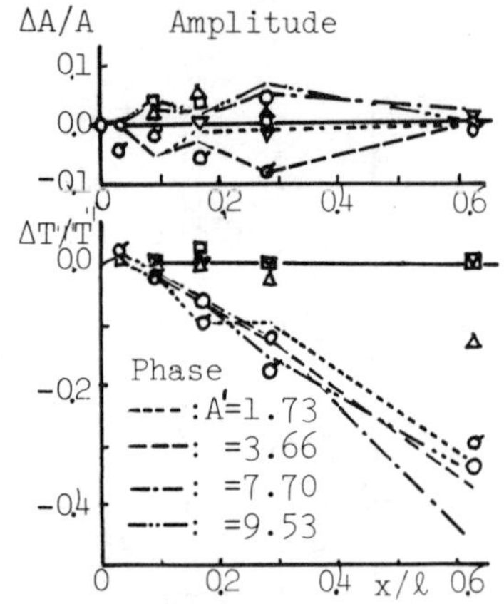

Fig. 11 Comparisons of differences between the experimental and the solution with amplitude and phase, where ΔA and ΔT are differences of two results respectively. (A'=2A).

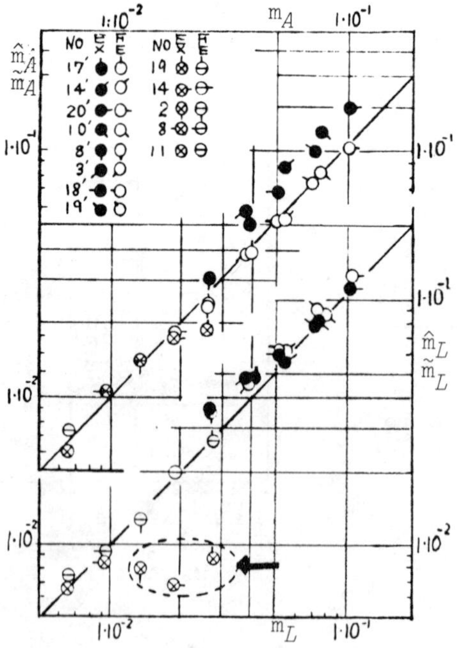

Fig. 12 Comparisons of three results, in which their properties are expressed by the dampped coefficient of amplitude and the coefficient of phase shift.

m_A, m_L: One-dimension

$\tilde{m}_A$, $\tilde{m}_L$: Two-dimension

$\hat{m}_A$, $\hat{m}_L$: Experimental

GROW1: A PROGRAM FOR ANISOTROPIC SOIL FLUID FLOW WITH FREE OR ARTESIAN SURFACE. APPLICATION TO EXCAVATIONS AT LOCK OF ZEEBRUGGE (BELGIUM)

J.P. RAMMANT, E. BACKX
Catholic University of Louvain, Belgium

ABSTRACT

A finite element program for plane or axisymmetric soil fluid flow in anisotropic media is developed. The element used to model the subsoil water flow is a two-dimensional isoparametric element with 4 to 8 variable-number-nodes. The input to the program consists of: the geometry, the subdivision of the model in finite elements, the selection of finite elements which can be adjusted to the free surface, permeability coefficients, boundary conditions, sources and sinks. The output yields: the nodal hydraulic head values, the streamfunction values and the velocities. Subroutines for graphical display of streamlines and of potential lines are included. The application of the program to a practical case is illustrated: the comparison of in situ piezometric measurements with computed values for excavations at the lock of Zeebrugge is presented.

NOMENCLATURE

F^e "force" vector for an element
g gravity acceleration
k^e "stiffness" matrix of an element
k_r permeability coefficient in r-direction
k_x permeability coefficient in x-direction
k_y permeability coefficient in y-direction
k_z permeability coefficient in z-direction
N matrix of shape functions
N_i shape function for node i
p pressure
q number of nodes in an element
r radius
S boundary of domain
S_2 boundary where flow is imposed
u_r velocity component in r-direction
u_x velocity component in x-direction
u_y velocity component in y-direction
u_z velocity component in z-direction
V volume of complete domain
V^e element volume
x,y,z coordinates
ϕ hydraulic head
ϕ_i nodal value of head
ϕ_e matrix of element head values
ξ,η normalized coordinates
ψ streamfunction
ρ density

INTRODUCTION

Seepage problems often occur in relation to the construction of dams, pile walls and excavations. Large settlements of constructions situated in the influence zone of groundwater lowering have to be feared. This phenomenon occured during the construction of the locks of Zeebrugge during 1972. The works slowed down till 1977. An extensive study was made to model efficiently the subsoil fluid flow and to anticipate the difficulties for future more important excavations.

At that stage the finite element program GROW1 was introduced. Pumping of wells in a small excavation was experimented and the observed piezometer heights were compared with the computer predictions. Due to the good agreement between prediction and measurements the program has been used to study the feasability of larger projects, including the study of injection possibilities, projected pile walls, horizontal and vertical impermeable screens and the use of recycled groundwater.

BASIC EQUATIONS

The basic differential equations governing the flow through porous media are deduced from the general Navier-Stokes equation, taking into account the following simplifying assumptions. The flow is assumed to be incompressible, laminar and the macroscopic velocity is related to the pressure by Darcy's law. The head, or energy per unit mass, is defined as (Lambe, 1969):

$$\phi = z + p/(\rho g)$$

The steady flow equations are then given in table 1. Using the Galerkin weighted residual method, the finite element technique is applied; for the axi-symmetric flow minimization of the error gives:

$$\iiint_V \left\{\frac{\partial}{\partial r}\left(k_r\frac{\partial\phi}{\partial r}\right) + \frac{\partial}{\partial z}\left(k_z\frac{\partial\phi}{\partial z}\right)\right\}\delta\phi r d\theta dr dz = \iint_{S_2}(v_n - \bar{v}_n)\delta\phi d\theta dS \qquad (3)$$

TABLE 1 Basic equations	Plane flow	Axi-symmetric flow
(1)	$\frac{\partial}{\partial x}(k_x\frac{\partial \phi}{\partial x})+\frac{\partial}{\partial y}(k_y\frac{\partial \phi}{\partial y})=0$	$\frac{\partial}{\partial r}(k_r\frac{\partial \phi}{\partial r})+\frac{\partial}{\partial z}(k_z\frac{\partial \phi}{\partial z})=0$
(2)	$u_x = -k_x \frac{\partial \phi}{\partial x}$ $u_y = -k_y \frac{\partial \phi}{\partial y}$	$u_r = -k_r \frac{\partial \phi}{\partial r}$ $u_z = -k_z \frac{\partial \phi}{\partial z}$

V stands for the whole domain which will be subdivided into subvolumes V^e namely the finite elements; $\bar{v}_n$ is the imposed flow velocity normal to the boundary S_2.
Integration of (3) by parts gives:

$$\sum_e \iiint_{V_e} (k_r\frac{\partial \phi}{\partial r}\frac{\partial \delta\phi}{\partial r} + k_z\frac{\partial \phi}{\partial z}\frac{\partial \delta\phi}{\partial z})dvol - \sum_e \iint_{S_2^e} \bar{v}_n \delta\phi d\theta dS=0 \qquad (4)$$

A 4 to 8 noded isoparametric element is constructed. The head is interpolated upon the head values at the nodes:

$$\phi = \sum_i^q N_i \phi_i \qquad q = 4,5,6,7 \text{ or } 8$$

The N_i are given in table 2; (Bathe, 1976)

TABLE 2
Shape functions for isoparametric element

	Add if node i is defined			
	i=5	i=6	i=7	i=8
$N_1 = \frac{1}{4}(1+\xi)(1+\eta)$	$-\frac{1}{2}N_5$			$-\frac{1}{2}N_8$
$N_2 = \frac{1}{4}(1-\xi)(1+\eta)$	$-\frac{1}{2}N_5$	$-\frac{1}{2}N_6$		
$N_3 = \frac{1}{4}(1-\xi)(1-\eta)$		$-\frac{1}{2}N_6$	$-\frac{1}{2}N_7$	
$N_4 = \frac{1}{4}(1+\xi)(1-\eta)$			$-\frac{1}{2}N_7$	$-\frac{1}{2}N_8$
$N_5 = \frac{1}{2}(1-\xi^2)(1+\eta)$				
$N_6 = \frac{1}{2}(1-\xi)(1-\eta^2)$				
$N_7 = \frac{1}{2}(1-\xi^2)(1-\eta)$				
$N_8 = \frac{1}{2}(1+\xi)(1-\eta^2)$				

Equation (4) simplifies then to the following stiffness relationship:

$$\sum_e [k^e][\phi^e] = \sum_e [F^e] \tag{5}$$

with

$$[k^e]_{qxq} = \iint_{\Omega^e} \left(rk_r \frac{\partial[N]^T}{\partial r} \frac{\partial[N]}{\partial r} + rk_z \frac{\partial[N]^T}{\partial z} \frac{\partial[N]}{\partial z}\right) 2\P drdz$$

$$[F^e]_{qx1} = 2\P \iint_{S_2^e} [N]^T \bar{v}_n \, dS$$

$$[N]_{1xq} = [N_1 \; N_2 \; .. \; .. \; .. \; .. \; N_q]$$

For a plane analysis, we put simply r=1 and replace r by x and z by y.

At the free surface ϕ should be equal to z. This is a boundary condition achieved by iterations: one assumes a form for the free surface at first, the equation (5) is solved and z at the free surface is adjusted till convergence; (Zienkiewicz, 1971; Connor, 1976).

The iterations, in which the finite elements are moved, converge readily as shown later in an example. The streamfunction ψ is defined by:

$$\psi = - 2\pi r \int_{z_1}^{z_2} u_r \, dz$$

A simple numerical integration scheme is adopted in the program to find ψ out of u_r.

APPLICATION

Figure 1 shows the general lay-out of the construction of the lock at Zeebrugge. The excavation at which the pumping was tested is indicated at (4) of Fig.1. (1) represents the lock; (2) and (3) are the lock doorrooms for which large excavations (and thus large ground settlements) are forecasted; (4) and (5) represent the machinery rooms of which the excavations are considered in this paper; (6) stands for the docks; (7) is the sea. At the location H of Fig.1 are the nearest houses of Zeebrugge which suffer from extensive cracking damage.

Although the sub-soil water flow following pumping at excavation (4)(Fig.1) is really a threedimensional problem, a good approximation can be found by considering an axi-symmetric flow, thus replacing the rectangular excavation by a circular one with radius r=18.9m. The considered geometrical configuration is shown in figure 2.

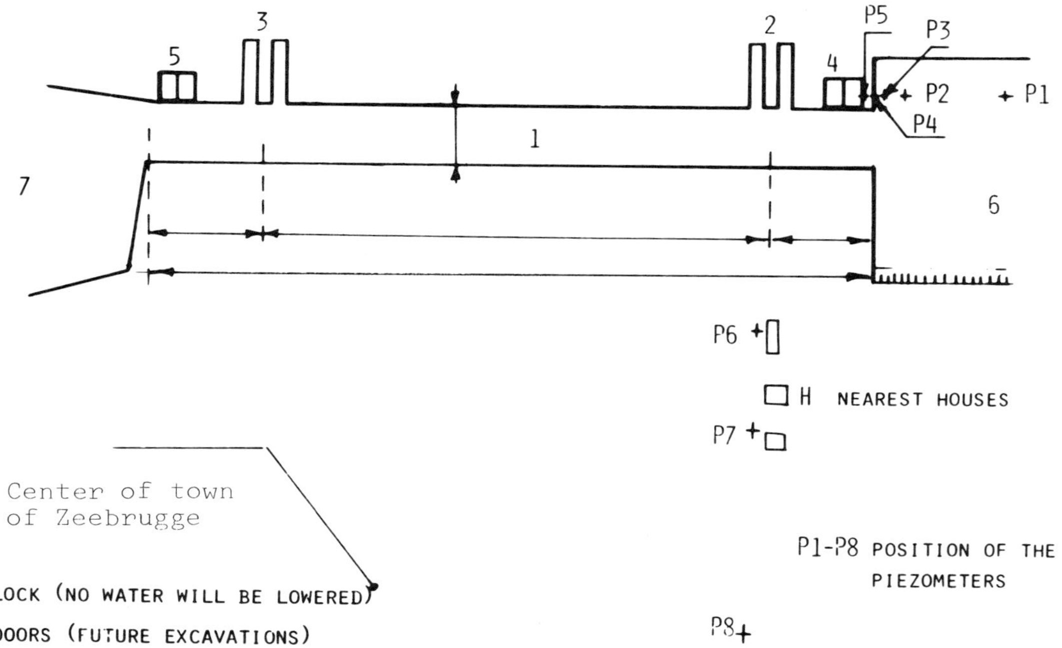

1 LOCK (NO WATER WILL BE LOWERED)

2,3 DOORS (FUTURE EXCAVATIONS)

4,5 MACHINERY ROOMS (PRESENT EXCAVATIONS CONSIDERED IN THIS PAPER)

6 DOCKS

7 SEA

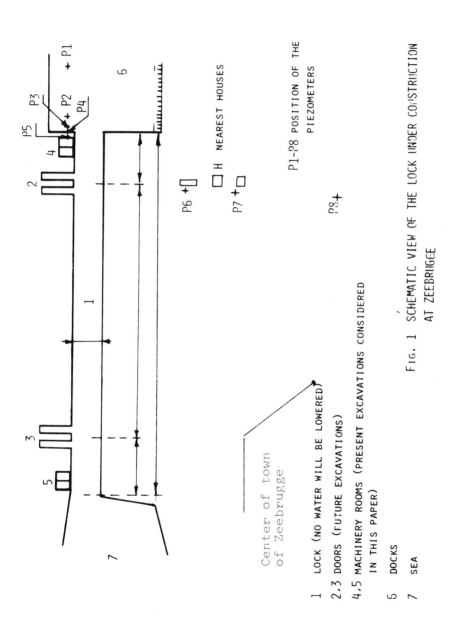

FIG. 1 SCHEMATIC VIEW OF THE LOCK UNDER CONSTRUCTION AT ZEEBRUGGE

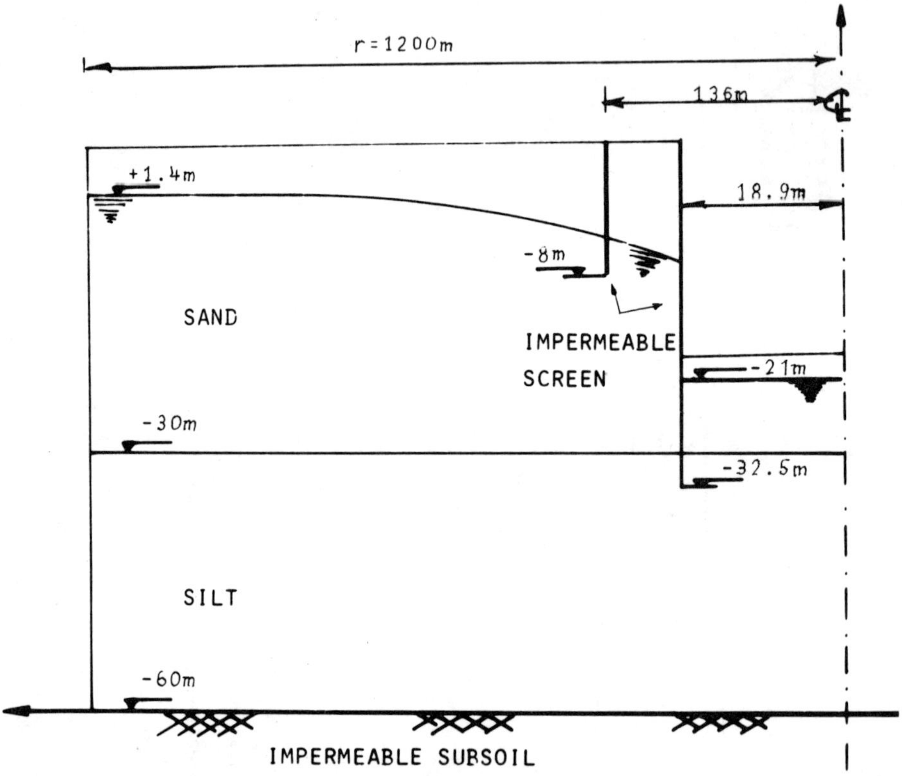

FIG. 2. GEOMETRY OF THE IDEALIZED SUBSOIL

An impermeable subsoil is located at -60m. Between -60m and -30m the subsoil is silt, while between -30m and the surface the subsoil is quaternary sand. At large distance (r=1200m) the subsoil water surface is at +1.4m. In the excavation the water is removed at -21m. An impermeable screen surrounds the excavation.

The soil permeabilities are given in table 3.

TABLE 3

Soil permeability coefficients

	k_r (m/sec)	k_z (m/sec)
Quaternary sand	$1.5\ 10^{-4}$	$0.79\ 10^{-4}$
Ledian silt	$0.3\ 10^{-4}$	$0.20\ 10^{-4}$

The subdivision of the soil in finite elements is plotted in figure 3. One remarks the small vertical screen at nodes 40 to 44 (the nodes have each two numbers) in the neighbourhood of which quadratic elements are choosen.

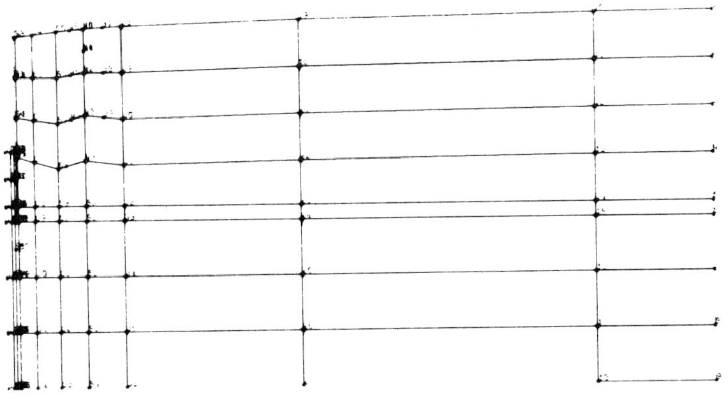

FIG. 3 PLOT OF THE FINITE ELEMENT MESH

The boundary conditions used for the considered problem are:

nodes 1 till 9: ϕ= 61.4m (right side of Fig.3)

nodes 89,92,100,107: ϕ=39m (horizontal in excavation of Fig.3)

no sinks or sources are present in this case.

Figure 4 compares the computed free surface position with the in situ observed piezometric heights. The rate of flow is also indicated. Two field measurements are compared with the computed values. The agreement is satisfying.

The rapid convergence of the iterations to find the free surface is illustrated in figure 5 where the z-position of boundary point 82 (upper left point of Fig.3) is drawn.
The iteration is achieved by moving the nodes of the elements above line z=31.5 into the z-direction. For an artesian flow no iterations have to be performed. The total computing time for the 3 iterations on a IBM 370/158-3 is 26 seconds.
The water velocity components are also given in the computer output. The potential lines and the streamlines are plotted in figure 6. In Fig.6.A the loss of potential head at the screen around the excavation is illustrated. In Fig.6.B the streamlines indicate the way in which the flow rate is formed. Of this overall picture detailed drawings can be obtained.

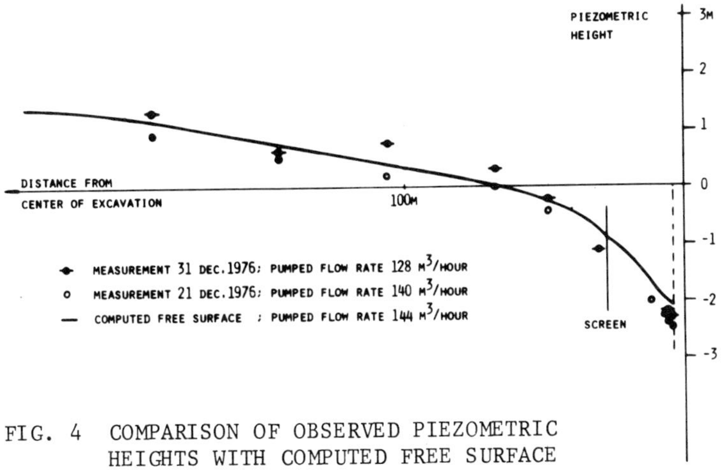

FIG. 4 COMPARISON OF OBSERVED PIEZOMETRIC HEIGHTS WITH COMPUTED FREE SURFACE

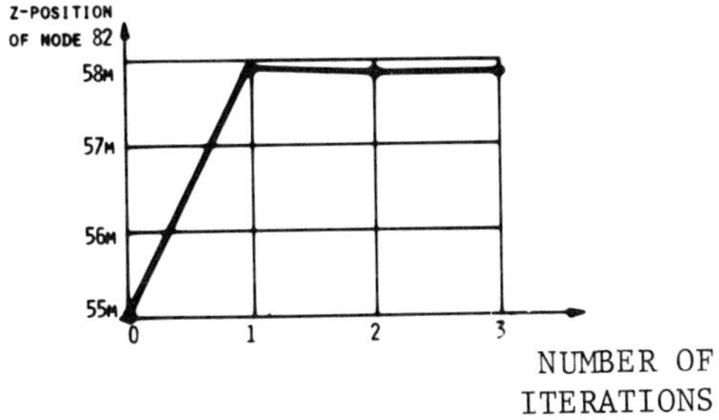

FIG. 5 ILLUSTRATION OF THE CONVERGENCE OF THE ITERATIONS FOR SEARCHING THE FREE SURFACE

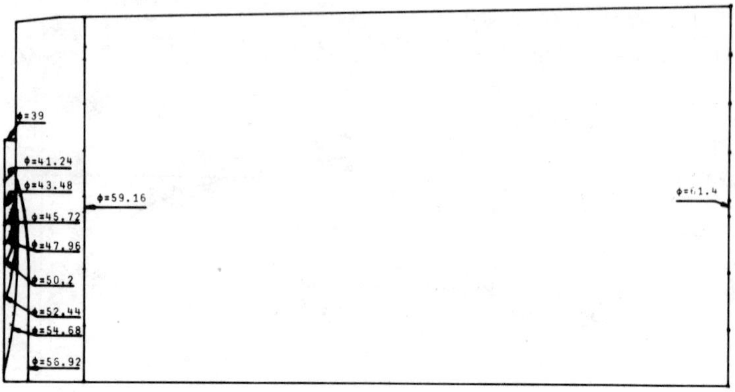

FIG. 6.A POTENTIAL LINES

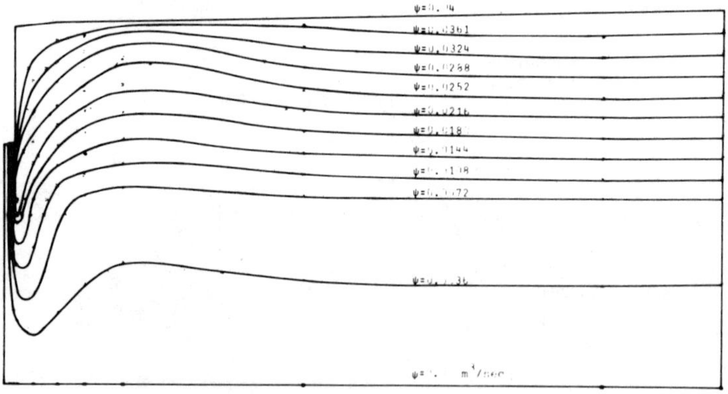

FIG. 6.B STREAMLINES

CONCLUDING REMARKS

A computer program for solving the Poisson equation for fluid flow has been presented. It can be applied to groundwater flow with free or artesian surface. The seepage may be plane or axisymmetric. The program proved to be a valuable tool in practical applications. For the excavations at Zeebrugge at (2) and (3) of figure 1 various proposed execution models were tested numerically: i.e. deep and thick vertical nearly impermeable screens, a horizontal injected nearly impermeable layer, use of recycled subsoil water and the combination of different proposals. The cheapest solution, i.e. use of recycled water at specified locations, proved to be also the most effective and is followed by the contractor.

Many other problems governed by the same equations can be solved using the same program: e.g. heat transfer, magnetic fields, etc. The subroutines for plotting the results are a valuable tool for checking coordinates and output.

ACKNOWLEDGEMENTS

The authors wish to express their heartly thanks to Professor M. Van Laethem for valuable discussions, to the firm Société Belge du Béton and its engineers A. Van Bruwaene, E. Berleur and S. Vanmarcke for their support.

REFERENCES

[1] K.J. Bathe, E.L. Wilson, Numerical Methods in Finite Element Analysis, Prentice Hall, Inc., New Jersey, 1976.

[2] J.J. Connor, C.A. Brebbia, Finite Element Techniques for fluid flow, Newnes-Butterworths, London-Boston, 1976.

[3] T.W. Lambe, R.V. Whitman, Soil Mechanics, John Wiley and Sons, Inc., N.-Y., 1969.

[4] O.C. Zienkiewicz, The Finite Element Method in Engineering Science, McGraw-Hill, London, 1971.

AUTOMATIC COMPUTING OF A TRANSMISSIVITY DISTRIBUTION USING ONLY PIEZOMETRIC HEADS.

A. Yziquel, J.C Bernard

Coyne et Bellier, Bureau d'Ingénieurs Conseils, Paris, France

POSITION OF PROBLEM

Introduction

As part of the planning studies for a large dam, it may be important to have a mathematical model of the ground water flow under the structure, in order to examine the leakage rates from the reservoir. Except where the geology is particularly unfavourable, the physical law describing ground water flow is the local diffusivity equation (generalized Darcy law) :

$$- \frac{\partial}{\partial x} \left(a(x) \frac{\partial y(x,t)}{\partial x} \right) = q(x,t) + S(x) \frac{\partial y(x,t)}{\partial t} \qquad (1)$$

with $a(x)$ transmissivity at x
$y(x,t)$ piezometric head at x at time t
$q(x,t)$ inflow field at x at time t
$S(x)$ porosity at x

With a discrete finite element model of this law, one can build a ground water flow model for computer analysis. This model, using the local transmissivities, will determine the piezometric heads at any point in the domain. This is known as the"direct problem" approach. With a piezometric heads boundary condition at boundary Γ , the problem is said to be "properly formulated" (existence and uniqueness) - fig.1.

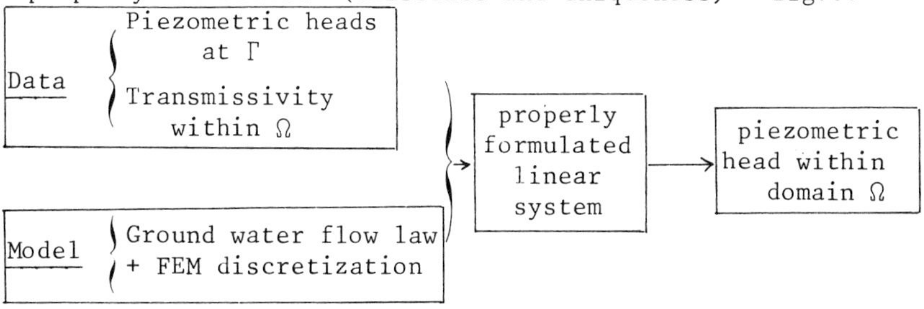

Fig.1 - Properly formulated piezometric head problem

Unfortunatly, transmissivity measurements in the field require costly pumping tests whereas piezometric heads are more accessible, e.g. in farm wells. If it is possible to solve the "inverse problem", i.e. determine local transmissivities from the piezometric heads, the leakage rate can be calculated by solving an inverse problem (transmissivity calibration) followed by a direct problem (leakage) - fig.2.

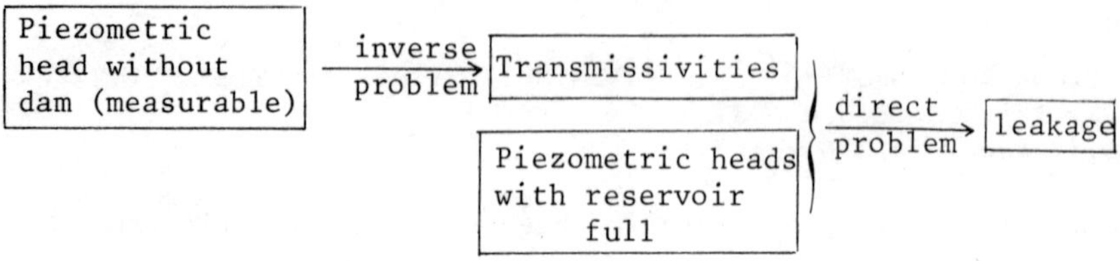

Fig.2 - Leakage calculation

Solution of inverse problem

The inverse problem is said to be "improperly formulated" (there is more than one solution) if only the piezometric heads at boundary are used as the sole boundary condition. There are three methods of solving this improperly formulated problem :

1) Change the improperly formulated problem to a properly formulated problem by introducing further data (transfer flow boundary conditions) and simplifying assumptions (structure of transmissivity function, isotropy, etc.)(Emsellem and de Marsily, 1971) (Clouet d'Orval, 1971) - Fig.3.

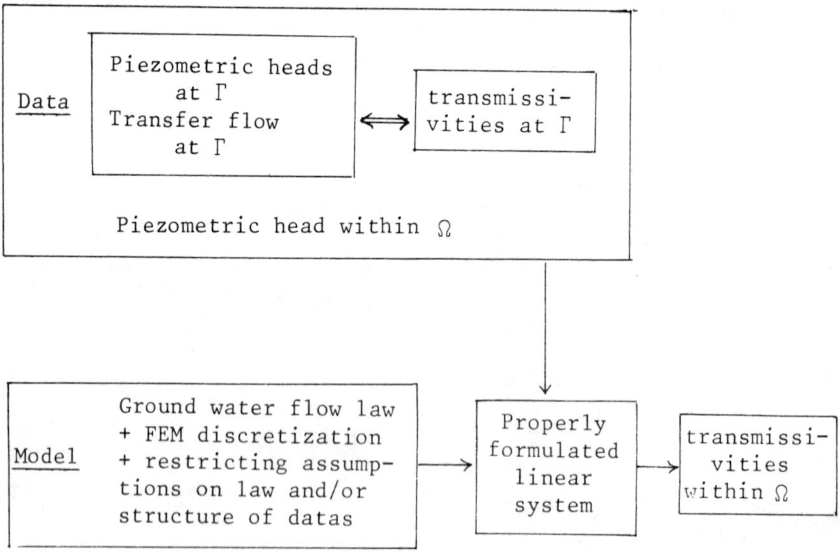

Fig.3 - Properly formulated transmissivity problem

2) Take a set of transmissivities, solve a direct properly formulated problem and then alter the transmissivity values "manually" to reduce the deviation between the calculated piezometric heads and observed piezometric heads in the field. This is known as the "manual calibration method".
3) Automate the above procedure by means of an algorithm which minimises a deviation functional between the calculated and observed piezometric heads without any manual trial, by iteration on the direct properly formulated piezometric head problem. The object being not to use unreliable data (transfer flow at boundary Γ), keep assumptions to a minimum (flow law structure and data structure) and free the calculation from the subjective influence of manual calibration.

Description of method

The third method described was chosen, using an optimal control formulation and a gradient method. The theory has been generated by G. Chavent (Chavent, 1971) (Chavent, 1973), who describes it as follow :
"The method uses the modern theory of function space control
"and consists of minimising an error criterion (non quadratic
"in its parameters) by an ordinary gradient method. The advan-
"tages of this procedure are that :
"- it is not necessary to assume any algebraic form of the
"parameter function to calculate the gradient with respect to
"this function,
"- calculating the gradient only requires solving two equations
"with partial derivatives, even when the unknown is a function,
"and
"- the method can be used even if only a limited number of
"measurement points in space are available."
The chart for the method is shown in fig.4.

MATHEMATICAL FORMULATION OF CONTINUUM PROBLEM

The notation is as follows :

Ω	geometric domain
Γ	boundary of Ω
$x = (x_1, x_2)$	geometric coordinates of a point within Ω
a(x)	transmissivity field over Ω
y(a,x)	piezometric field obtained by solving direct problem with Darcy's law and associated transmissivity field a(x)
z(x)	observed piezometric field over Ω
g(x)	piezometric field on boundary Γ
q(x)	inflow field (well or spring, infiltration) within domain Ω

y and z are such that their restriction on the boundary is equal to g(x), that is : $y(a,x)|_\Gamma = z(x)|_\Gamma = g(x)$ (2)

The problem to be solved is : Find a(x) such that

$$y(a,x) = z(x) \quad \forall \ x \in \Omega \tag{3}$$

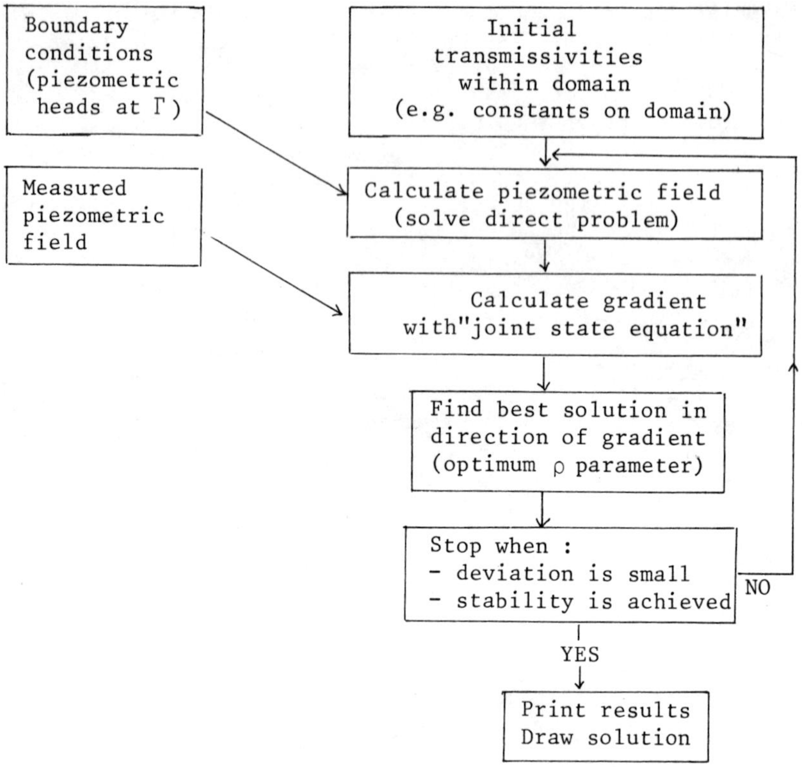

Fig.4 - Chart of method

Deviation functional

The object is to find a convergent iterative algorithm which , by modification of the initial transmissivity field, minimises the mean square deviation functional J between the calculated piezometric head y(a,x) and the observed piezometric head z(x). The deviation functional is defined as :

$$J(a) = \int_{\Omega} \{y(a,x) - z(x)\}^2 \, dx \qquad (4)$$

In the optimal control terminology, a(x) is the control, y(a,x) is the state and z(x) is the desired observation. a(x) and y(a,x) are related by the state equation, which is the integral variational form of the local Darcy law (see description of the state equation below). The object is to find the control a(x) which minimises the deviation functional J(a).

<u>Gradient algorithm</u>

Let $a_i(x)$ be a control at iteration i. We must find a

$a_{i+1} \stackrel{\perp}{=} a_i + \delta a$ such that $J(a_{i+1}) \leqslant J(a_i)$ (5)

If it is possible to differentiate J, we can write

$$\delta J(a_i) = \int_\Omega \frac{\partial J}{\partial a}(a_i,x) \quad \delta a(x)\ dx \tag{6}$$

This defines the gradient function $\frac{\partial J}{\partial a}(a_i,x)$ at a_i

If we make the choice $\delta a = -\rho \frac{\partial J}{\partial a}(a_i,x)$ with $\rho > 0$ (7)

then $\delta J(a_i) = -\rho \int_\Omega \left[\frac{\partial J}{\partial a}(a_i,x)\right]^2 dx < 0$ (8)

Therefore $J(a_{i+1}) \leqslant J(a_i)$ (9)

A proper choice for ρ makes the process convergent.

<u>Gradient calculation and "joint state equation"</u>

The following didactic formulation (which was introduced by Abadie in his mathematical programming course) is quite convenient to avoid the usual "guess" about the joint state equation. Let us differentiate equation (4)

$$\delta J(a_i) = 2\int_\Omega \{y(a_i x) - z(x)\}\delta y(x)\ dx \tag{10}$$

Let us put equation (10) in a form similar to equation (6)
For that purpose, let us differentiate equation (1)

$$\frac{\partial}{\partial x}\left\{\delta a(x)\frac{\partial y(a,x)}{\partial x} + a(x)\frac{\partial \delta y(x)}{\partial x}\right\} = 0 \ \forall x \in \Omega \tag{11}$$

Let $\psi(x)$ be a function over Ω. For any $\psi(x)$ we can write :

$$\int_\Omega \psi(x)\frac{\partial}{\partial x}\left[\delta a(x)\frac{\partial y(a,x)}{\partial x} + a(x)\frac{\partial \delta y(x)}{\partial x}\right] dx = 0 \tag{12}$$

Using the following Green's formula

$$\int_\Omega \psi(x)\frac{\partial\ \phi(x)}{\partial x}\ dx = \int_\Gamma \psi(x)\ \ \phi(x)\ d\gamma - \int_\Omega \frac{\partial\psi(x)}{\partial x}\phi(x)dx \tag{13}$$

with the choice

$$\begin{cases} \phi(x) = \delta a(x)\dfrac{\partial y(a,x)}{\partial x} + a(x)\dfrac{\partial \delta y(x)}{\partial x} & (14) \\ \psi(x)\big|_\Gamma = 0 & (15) \end{cases}$$

we obtain

$$\int_\Omega \frac{\partial\psi(x)}{\partial x}\left\{\delta a(x)\frac{\partial y(a,x)}{\partial x} + a(x)\frac{\partial\delta y(x)}{\partial x}\right\} dx = 0 \tag{16}$$

Then we add equation (16) to equation (10) to obtain

$$\begin{aligned} \delta J(a_i) = & \int_\Omega \frac{\partial\psi(x)}{\partial x}\frac{\partial y(a_i,x)}{\partial x}\delta a(x)\ dx \\ & + \int_\Omega a_i(x)\frac{\partial\psi(x)}{\partial x}\frac{\partial\delta y(x)}{\partial x} + 2\{y(a_i,x)-z(x)\}\delta(x)\ dx \end{aligned} \tag{17}$$

The first part of the second member of equation (17) is similar to the second member of equation (6), and the second part of the second member of equation (17) can be made equal to zero with a particular choice for $\psi(x)$. Let $\psi(x)$ be defined by the following properly formulated problem (i.e. solution is unique)

$$\psi(x)|_\Gamma = 0 \text{ and } \quad \xi(x), \quad \xi(x)|_\Gamma = 0$$
$$\int_\Omega a_i(x) \frac{\partial\psi(x)}{\partial x} \frac{\partial\xi(x)}{\partial x} dx = -2\int_\Omega \{y(a_i,x)-z(x)\}\,\xi(x)\,dx \tag{18}$$

With $\xi(x) = \delta y(x)$ equation (17) and (18) lead us to

$$\delta J(a_i) = \int_\Omega \frac{\partial\psi(x)}{\partial x} \frac{\partial y(a_i,x)}{\partial x} \delta a(x)\,dx \tag{19}$$

Equation (19) defines the gradient (see equation (6))

$$\frac{\partial J}{\partial a}(a_i,x) = \frac{\partial\psi(x)}{\partial x} \frac{\partial y(a_i,x)}{\partial x} \tag{20}$$

Equation (18) is known as the joint state equation and $\psi(x)$ as "joint state".

<u>State equation</u>

With a set of test functions $\xi(x)$ we can write the integral variational form of local equation (1) (without porosity)

$$\int_\Omega a_i(x) \frac{\partial y(a_i,x)}{\partial x} \frac{\partial\xi(x)}{\partial x} dx = \int_\Omega \phi(x)\,\xi(x)\,dx \tag{21}$$

We derive this equation by mean of the Green's formula (13) with the choice :

$$\phi(x) = -a_i(x)\frac{\partial y(a_i,x)}{\partial x}, \quad \psi(x) = \xi(x), \ \xi(x)|_\Gamma = 0 \tag{22}$$

By adding a boundary condition, we obtain the following second properly formulated problem

$$y(a_i,x)|_\Gamma = g(x) \text{ and } \quad \xi(x), \quad \xi(x)|_\Gamma = 0$$
$$\int_\Omega a_i(x) \frac{\partial y(a_i,x)}{\partial x} \frac{\partial\xi(x)}{\partial x} dx = \int_\Omega q(x)\,\xi(x)\,dx \tag{23}$$

Equation (23) is known as state equation (properly formulated piezometric head problem).

<u>Remarks</u>

We compute the gradient in three steps :
- solve equation (23) to obtain $y(a_i,x)$
- solve equation (18) to obtain $\psi(x)$
- obtain gradient using equation (20)

It is important to notice the similarity between equation (18) and (23) :

	state equation	joint state equation
Function	$y(a_i, x)$	$\psi(x)$
Boundary condition	$y(a_i, x)\|_\Gamma = g(x)$	$\psi(x)\|_\Gamma = 0$
Second member	$q(x)$	$-2\{y(a_i, x) - z(x)\}$

The operator $a_i(x) \frac{\partial \xi(x)}{\partial x} \frac{\partial}{\partial x}$ is the same in both equations. This feature of G. Chavent's method leads to time saving in computing and programming because we use the same assembly and solve procedure for both state and joint state equations.

NUMERICAL IMPLEMENTATION

Discretization

By reason of the continuous formulation, discretization of the state and joint state equations are identical. A finite element method was chosen of the following type :

1) The mesh over domain Ω is built with triangular or quadrilateral elements with parabolic edges
2) The elements were chosen to minimise the Laplacian interpolation error :
   - isoparametric Lagrange triangles with six nodes and seven integration points
   - isoparametric serendipity quadrilaterals with eight nodes and nine integration points.
3) It was decided to express the control $a(x)$ only at the element integration points without imposing any structure on the transmissivity field (continuity, algebraic structure)

This requires two discretizations :

(a) One discretization at the nodes of the observed piezometric field, the calculated piezometric field and the joint state which are written $\{z\}^N$, $\{y\}^N$ and $\{\psi\}^N$

(b) One discretization of the transmissivity field at the integration points of the elements, written $\{a\}^I$

Subscript i is used for the iteration number of the gradient algorithm.

The discretized state equation is then written :

$$[A(\{a\}^I)] \{y\}^N = \{f\}^N \quad ; \quad \{y\}|_\Gamma^N = \{g\}^{N'}, \quad N' < N$$

with $[A]$ global flow matrix

$\{g\}^{N'}$ observed piezometric heads at the boundary Γ (N' nodes)

$\{f\}^N$ flow rate vector after discretization of $\int_\Omega q(x)\, \xi(x)\, dx$

The discretized joint state equation is written :

$$[A(\{a\}^I)]\{\psi\}^N = \{h\}^N; \quad \{\psi\}^N|_\Gamma = \{0\}^{N'}$$

with $\{0\}^{N'}$ null vector at the boundary Γ (there is no deviation at the boundary)

$\{h\}^N$ deviation vector after discretization of $-2\int_\Omega [y(x)-z(x)]\ \xi(x)\,dx$

Gradient algorithm

We choose the transmissivity field $\{a_0\}^I$ for the iteration 0.
We write for the iteration i the following iteration scheme :

1 Compute state $\{y_i\}^N$ which solves

$$[A(\{a_i\}^I)]\ \{y_i\}^N = \{f\}^N \text{ with } \{y_i\}^N|_\Gamma = \{g\}^{N'}$$

2 Compute joint state $\{\psi_i\}^N$ which solves

$$[A(\{a_i\}^I)]\ \{\psi_i\}^N = \{h_i\}^N \text{ with } \{\psi_i\}^N|_\Gamma = \{0\}^{N'}$$

$\{h_i\}^N$ being computed from $\{y_i\}^N$ and $\{z\}^N$

3 Compute the derivatives $\frac{\partial y_i}{\partial x_j}$ and $\frac{\partial \psi_i}{\partial x_j}$ at the integration points and build vectors $\{\frac{\partial y_i}{\partial x_j}\}^I$ and $\{\frac{\partial \psi_i}{\partial x_j}\}^I$, $j = 1,2$

4 Compute gradient $\frac{\partial J}{\partial a}(a_i,x)$ at integration point :

$$\{\frac{\partial J}{\partial a}\ i\}^I = \{\frac{\partial y_i}{\partial x_1}\ \frac{\partial \psi_i}{\partial x_1} + \frac{\partial y_i}{\partial x_2}\ \frac{\partial \psi_i}{\partial x_2}\}^I$$

5 Compute $\{a_{i+1}\}^I_{\rho_0}$ by means of the gradient formula (5) with a small positive ρ_0 :

$$\{a_{i+1}\}^I_{\rho_0} = \{a_i\}^I - \rho_0\ \{\frac{\partial J}{\partial a}\ i\}^I$$

6 Linearize and project on a transmissivity convex (minimum and maximum values allowed for transmissivities) to obtain ρ_L which is the initialvalue for the minimum search algorithm of the deviation functional in the gradient direction.

7 Minimum search algorithm to obtain $\{a_{i+1}\}^I$ such that

$$J(\{a_{i+1}\}^I) = \underset{\rho>0}{\text{Min}} \; J(\{a_{i+1}\}^I_\rho)$$

The search algorithm multiplies or divides ρ by two and is a modification of an algorithm used by Cea (Cea, 1971). We use quadratic intérpolation to obtain the final value.

8 If the deviation functional value is sufficently small and stable, stop; else go back to step 1 with value i equal to i+1.
Note that iterations on a real project showed the non-convexity of the deviation functional J.

EXTENSION OF THE BASIC MODEL

We introduced three extensions in the basic model for a better approximation of a real aquifer.

Pseudo-transient term
We added a pseudo-transient term in the second member of state equation. This term, already mentioned in equation (1), is related to the observed piezometric head rate of change on the day after the reference date used for the identification.

Infiltration and porosity identification
The infiltration and porosity datas supplied were constant over the whole domain. It was thought not to be realistic and we decided to modify those datas according to the same process used for transmissivity identification (gradient computation and functional minimum search). This improvement lowers the deviation and allows for better identification of transmissivities in the next iterations.

"Zone isolation" process
The use of basic method shows two limitations : the possibility for the model to represent the real aquifer in certain sub-domains and the local effect of gradient algorithm (i.e. the deviation in some elements induces transmissivity modifications in the same or next elements only).
It is therefore important to be able to :
- either isolate some elements to perform a local identification in a sub-domain,
- or isolate some other elements to neglect their contribution in gradient computation.

The last case occurs when gradient components are and stay large in absolute value without improvements during the iterative process. This prevents effective reduction of the deviation in other elements.
In both cases, we have to keep the isolated zone for flow calculation, since flow rates through isolated elements may change with iterations. The new gradient must be computed

properly with respect to the joint state equation homogeneity (it is not sufficient to set some gradient components to zero). Both isolation processes were implemented in program, with good results in overcoming local minima.
The manual choice for elements to be isolated means a loss in objectivity, but we are now working on an algorithm including memory over the past iterations. This memory should allow for automatic detection of elements to be isolated by both processes.

APPLICATION OF THE METHOD

Tests

Before applying the method to an actual project, it was condidered prudent to run a few tests combining various different domains and transmissivity fields. The procedure was as follows : the transmissivity field on the domain and the boundary condition were imposed, and the direct problem was first solved to obtain the piezometric heads on that domain. Using this data, it was checked that the iterative process described produced the initial transmissivity function.

Application to project

The details of the aquifer involved were as follows : the domain covered 300 Km^2 in a chalk formation. The known river levels were used as boundary conditions and there were 76 piezometer measurements at various points in the domain. Two calculations were made, one being fairly summary with a quitecoarse mesh and with no allowance for infiltration or porosity. The second used a more refined mesh with infiltration and porosity datas.

First model

The domain was discretized with 130 elements and 390 nodes. The measured piezometric records were extended to the whole domain by a simple "nearest neighbour" procedure.
This calculation yielded an average deviation per node reducing from 2.37 m to 1.10 m in 17 iterations, requiring 60 seconds of CDC 7600 time. It revealded an impervious barrier north-east of the reservoir, which could not be adequatly represented by the mesh used. An assumption of a perched water table, incompatible with the planar aquifer model, was considered. Results are plotted on fig.5.

Second model

In order to obtain more precise results in the dam zone and better results in the zones containing high piezometric gradients, the problem was repeated with a 335 element, 978 node model, which included for infiltration and porosity (pseudo-transient term).
The piezometric records were extended statistically (Matheron, 1970) by the Hydrogeology Section of the Ecole des Mines, Paris.

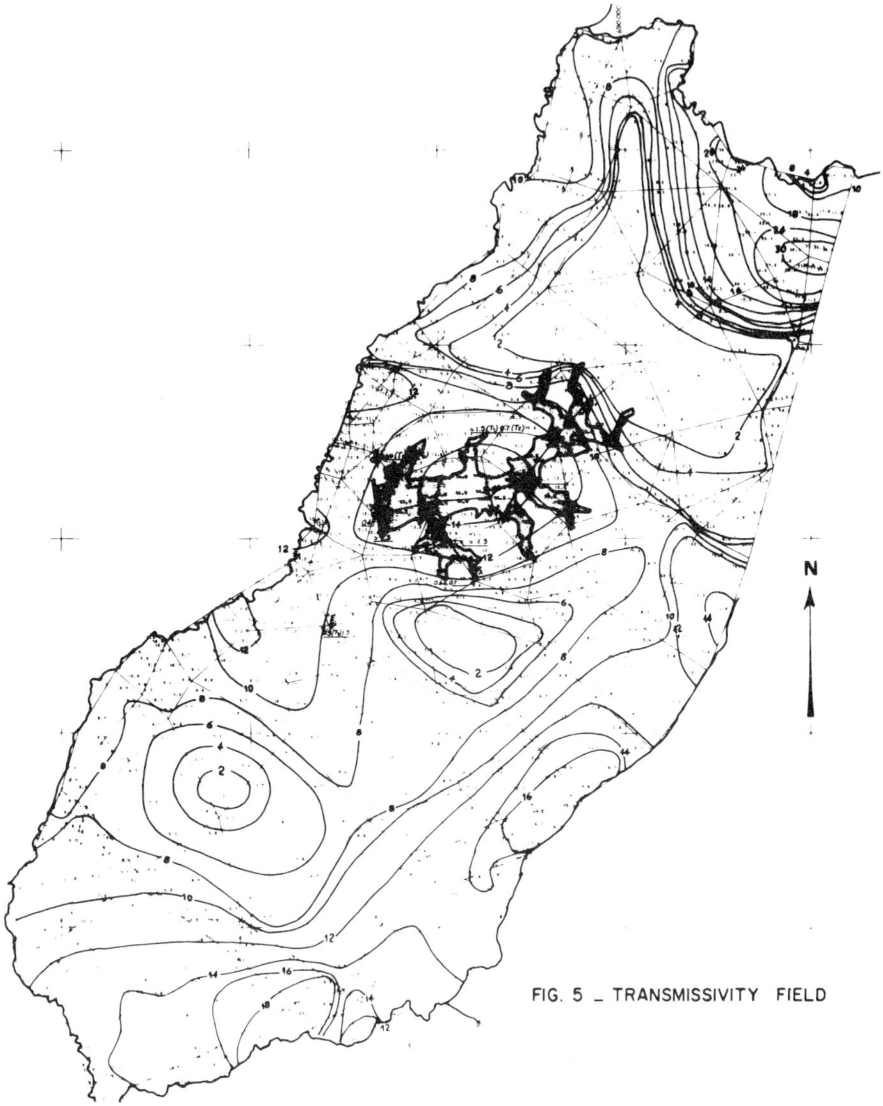

FIG. 5 _ TRANSMISSIVITY FIELD

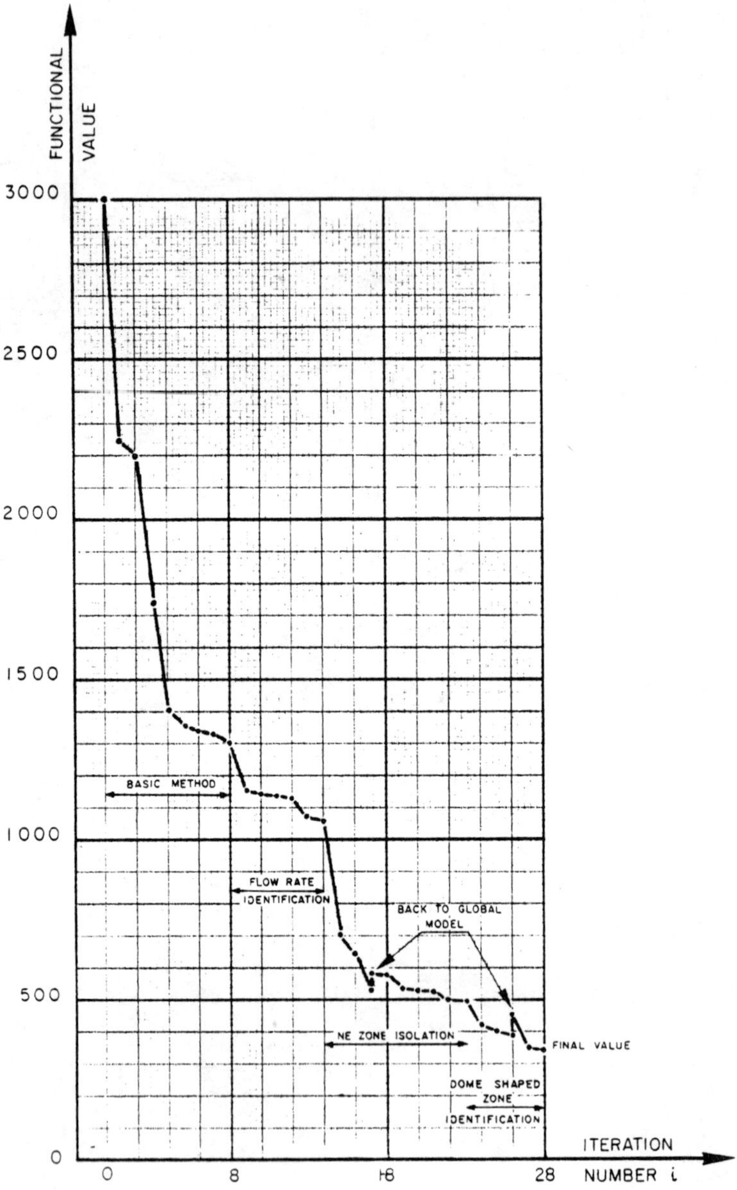

FIG. 6 – DEVIATION FUNCTIONAL $\int_{\Omega} \left[y\,(a_i\,,x) - z\,(x) \right]^2 dx$

This method constructs a rather stiff piezometric field. The calculation was made in four stages :

1) The basic method was first applied, reducing the deviation from 3.16m to 2.08m in 8 iterations.

2) A gradient was then used to search better values for infiltration rates and porosity. With an average rainfall of 6 litres/km2/sec, values ranged from 3 to 15 litres/km2/sec. The porosity coefficient ranged from 1 % to 20 %, with a 5 % average. This extra identification reduced the average deviation per node to 1.87m in 6 iterations.

3) The zone isolation process was then used for the north-east zone of the reservoir, which introduces steep gradients which could not be reduced by local transmissivity changes. The improvement over the rest of the aquifer brought the average deviation per node down to 1.30m in 9 iterations.

4) The last improvement was obtained by isolating the southern zone of the reservoir where the piezometric lines are dome-shaped. This dome was first solved on a small domain, and then transferred to the full model.

The ultimate value of the average deviation was 1.07m as shown on fig.6.

REFERENCES

Cea, J. (1971) Optimisation. Théorie et algorithmes. Dunod, Paris.

Chavent, G. (1971) Analyse fonctionnelle et identification de coefficients répartis dans les équations aux dérivées partielles - Thèse d'Etat, Mathématiques Paris VI, 158 pages.

Chavent, G. (1973) Estimation des paramètres distribués dans les équations aux dérivées partielles - in "Computing Methods in Applied Sciences and Engineering" Part 2 Springer Verlag, Berlin.

Clouet d'Orval, M. (1971) Détermination automatique des transmissivités dans un horizon aquifère isotrope d'après les données piezométriques. Thèse de Docteur Ingénieur, Nancy I.

Emsellem, Y.and de Marsilly, G. (oct. 1971) An automatic solution for the inverse problem. Water ressources research vol.7, N°5.

Matheron, G. and Huijbrechts, C. (June 1970) Universal Kriging An optimal method for estimating and contouring in trend surface analysis, paper presented at International Symposium on Techniques for Decision Making in the Mineral Industry, Inst. of Mining and Met., Montreal Can., June 1970.

GROUNDWATER FLOW SIMULATION USING COLLOCATION FINITE ELEMENTS

G.F. Pinder, E.O. Frind, and M.A. Celia

Princeton University, Princeton, N.J. 08540

ABSTRACT

A new numerical method is presented for the solution of two-dimensional potential flow problems. The method combines the most attractive features of orthogonal collocation and finite elements. It is particularly suited to obtaining C^1 continuous solutions in irregular domains because it is much more efficient for problems of this type than a comparable Galerkin scheme. This makes it an interesting method for simulating groundwater flow and mass transport problems. A comparison of the accuracy of the resulting solution with that of the Galerkin finite element method using linear triangles is given, and the convergence properties of the new method are investigated.

INTRODUCTION

The purpose of this paper is to present what appears to be a new numerical method for the solution of partial differential equations of the form commonly encountered in water resources. This scheme incorporates the computational efficiency and conceptual simplicity of orthogonal collocation while maintaining the flexibility of the finite element method (FEM). We denote this the collocation finite element method (CFM).

The classical collocation method dates back at least to the work of Lanczos (1938). The method has been used widely in chemical engineering where the governing equations often exhibit a strong nonlinearity (see, for example, Finlayson, 1972; Villadsen and Michelsen, 1978). The mathematical basis of the collocation method is described in detail by Prenter (1975).

CFM is particularly attractive when continuous velocities, and therefore continuous potential gradients, are required. The most commonly used method, that of solving first for the potentials and then differentiating, yields velocities with an accuracy one order lower than that of the potentials. Alternatively, one can use Hermite finite elements (Frind, 1977; Van Genuchten et al, 1977), which give potentials and gradients at the same time, both having the same order of accuracy. Unfortunately, these Hermitian elements have sixteen degrees of freedom each, which is relatively large, while the classical Galerkin finite element formulation works most efficiently with elements having a small number of degrees of freedom. In CFM on the other hand, elements with a large number of degrees of freedom are not penalized and the formulation can be very efficient. As a result, CFM with bicubic Hermite elements requires only a fraction of the computational effort necessary with FEM using the same elements (Frind and Pinder, 1978). It also yields equal or better accuracy.

In this paper, we will compare CFM using bicubic Hermite elements with FEM using linear triangles. We will also investigate the convergence of CFM in the case of distorted grids.

THEORETICAL BACKGROUND

We will include here only those elements of the theory that are necessary to present the basic concept of CFM. Let us consider an equation of the form

$$Lu = f \tag{1}$$

with L being an operator in a bounded domain α. We can find an approximate solution to Equation (1) in the form of a finite series

$$u \simeq \hat{u} = \sum_{j=1}^{N} a_j \phi_j(\underset{\sim}{x}) \tag{2}$$

where the a_j are undetermined coefficients and the $\phi_j(\underset{\sim}{x})$ are linearly independent basis functions. To determine a_j, and thus the approximate solution $\hat{u}$, we invoke the general form of the method of weighted residuals which can be written for L as

$$\int_{\alpha} (L\hat{u}-f) w_i(\underset{\sim}{x}) d\alpha = 0 \quad i = 1,2,\ldots N \tag{3}$$

The FEM is obtained when the weighting functions $w_i(\underset{\sim}{x})$ are chosen to be the basis functions $\phi_j(\underset{\sim}{x})$. The collocation method results when the Dirac delta function $\delta_i(\underset{\sim}{x})$ is used as the weighting function (Finlayson and Scriven, 1965),

$$\int_{\alpha} (L\hat{u}-f) \delta_i(\underset{\sim}{x}) d\alpha = 0 \quad i = 1,2,\ldots N \tag{4}$$

Equation (4) can be rewritten as

$$(L\hat{u}-f)\Big|_{\underset{\sim}{x}_i} = \sum_{j=1}^{N} a_j L\phi_j(\underset{\sim}{x}_i) - f(\underset{\sim}{x}_i) = 0 \quad i = 1,2,\ldots N \tag{5}$$

where i designates a number of points in a known as collocation points. Provided the ϕ_j are specified, these equations may be solved for the N coefficients a_j.

The accuracy of the collocation method is known to be quite sensitive to the location of collocation points. It can be shown that the optimal locations of these points are at the roots of the Legendre polynomials (Villadsen and Stewart; 1967, DeBoor and Swartz; 1973; Prenter, 1975). When the collocation points are selected in this way, the method is known as orthogonal collocation. In the case of regular subspaces, such as rectangles, the location of the optimal collocation points is easily determined (see Figure 1). By using an isoparametric transformation, an accurate collocation scheme can also be achieved using irregular subspaces such as illustrated in Figure 2. The procedure involves the following steps:

- specification of a set of bases defined in local $\underset{\sim}{\xi}$ co-ordinates
- selection of collocation points defined in $\underset{\sim}{\xi}$
- transformation of the governing partial differential equation from cartesian $\underset{\sim}{x}$ to curvilinear $\underset{\sim}{\sigma}$ coordinates
- formulation and subsequent solution of a set of approximate algebraic equations.

The approximating function $\hat{u}$ can be formulated using Hermite polynomials defined in local $\underset{\sim}{\xi}$ coordinates. The relationship is given for two dimensions as

$$\hat{u} = \sum_{j=1}^{N} \phi_{00j}(\underset{\sim}{\xi})U_j + \phi_{10j}(\underset{\sim}{\xi})\frac{\partial U_j}{\partial \xi_1} + \phi_{01}(\underset{\sim}{\xi})\frac{\partial U_j}{\partial \xi_2} + \phi_{11j}(\underset{\sim}{\xi})\frac{\partial^2 U_j}{\partial \xi_1 \partial \xi_2} \tag{6}$$

where the Hermite polynomials are defined as:

$$\phi_{00j} = \frac{1}{16}(\xi_1+\xi_{1j})^2(\xi_1\xi_{1j}-2)(\xi_2+\xi_{2j})^2(\xi_2\xi_{2j}-2) \tag{7a}$$

$$\phi_{10j} = -\frac{1}{16}\xi_{1j}(\xi_1+\xi_{1j})^2(\xi_1\xi_{1j}-1)(\xi_2+\xi_{2j})^2(\xi_2\xi_{2j}-2) \tag{7b}$$

$$\phi_{01j} = -\frac{1}{16}(\xi_1+\xi_{1j})^2(\xi_1\xi_{1j}-2)\xi_{2j}(\xi_2+\xi_{2j})^2(\xi_2\xi_{2j}-1) \tag{7c}$$

$$\phi_{11j} = \frac{1}{16}\xi_{1j}(\xi_1+\xi_{1j})^2(\xi_1\xi_{1j}-1)\xi_{2j}(\xi_2+\xi_{2j})^2(\xi_2\xi_{2j}-1) \tag{7d}$$

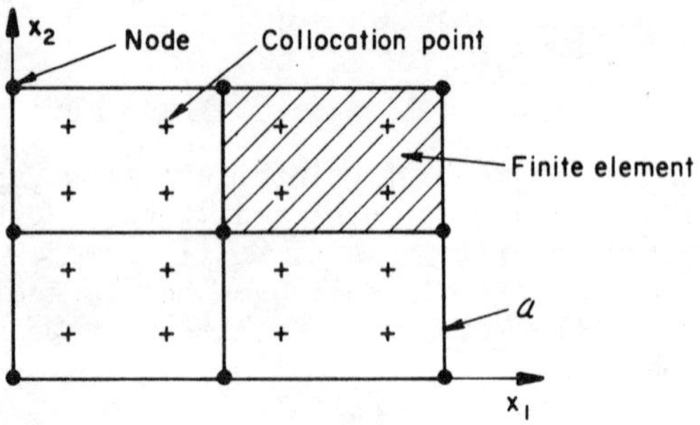

Figure 1 Orthogonal collocation points specified for rectangular subspaces.

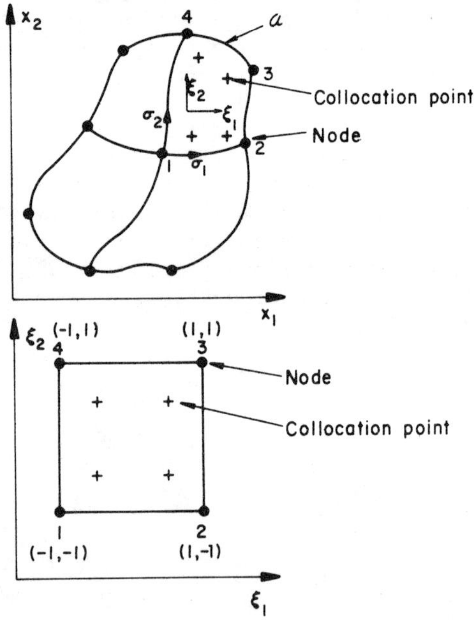

Figure 2 Collocation points for a deformed subspace (isoparametric element) in global ($\tilde{x}$) and local ($\tilde{\xi}$) coordinates.

The unknown parameters in Equation (6) are defined in the $\underset{\sim}{\xi}$ coordinate system, but we require the derivatives at the nodes $\left(\frac{\partial U_j}{\partial \xi_1}, \frac{\partial U_j}{\partial \xi_2}, \frac{\partial^2 U_j}{\partial \xi_1 \partial \xi_2} \right)$ in terms of the curvilinear coordinates $\underset{\sim}{\sigma}$. Thus, we transform Equation (6) using the relationship (Frind, 1977):

$$\begin{Bmatrix} \dfrac{\partial u}{\partial \xi_1} \\ \dfrac{\partial u}{\partial \xi_2} \\ \dfrac{\partial^2 u}{\partial \xi_1 \partial \xi_2} \end{Bmatrix}_j = \begin{bmatrix} \dfrac{\partial \sigma_1}{\partial \xi_1} & 0 & 0 \\ 0 & \dfrac{\partial \sigma_2}{\partial \xi_2} & 0 \\ \dfrac{\partial^2 \sigma_1}{\partial \xi_1 \partial \xi_2} & \dfrac{\partial^2 \sigma_2}{\partial \xi_1 \partial \xi_2} & \dfrac{\partial \sigma_1 \partial \sigma_2}{\partial \xi_1 \partial \xi_2} \end{bmatrix}_j \begin{Bmatrix} \dfrac{\partial u}{\partial \sigma_1} \\ \dfrac{\partial u}{\partial \sigma_2} \\ \dfrac{\partial^2 u}{\partial \sigma_1 \partial \sigma_2} \end{Bmatrix}_j \qquad (8)$$

Substitution of Equation (8) into Equation (6) yields

$$\hat{u}(\underset{\sim}{\sigma}) = \sum_{j=1}^{N} \phi_{0j} U_j + \phi_{\sigma_1 j} \frac{\partial U_j}{\partial \sigma_1} + \phi_{\sigma_2 j} \frac{\partial U_j}{\partial \sigma_2} + \phi_{\sigma_1 \sigma_2 j} \frac{\partial^2 U_j}{\partial \sigma_1 \partial \sigma_2} \qquad (9)$$

where the transformed basis functions are defined as

$$\phi_{0j} = \phi_{00j} \qquad (10a)$$

$$\phi_{\sigma_1 j} = \phi_{10j} \left(\frac{\partial \sigma_1}{\partial \xi_1} \right)_j + \phi_{11j} \left(\frac{\partial^2 \sigma_1}{\partial \xi_1 \partial \xi_2} \right)_j \qquad (10b)$$

$$\phi_{\sigma_2 j} = \phi_{01j} \left(\frac{\partial \sigma_2}{\partial \xi_2} \right)_j + \phi_{11j} \left(\frac{\partial^2 \sigma_2}{\partial \xi_1 \partial \xi_2} \right)_j \qquad (10c)$$

$$\phi_{\sigma_1 \sigma_2} = \phi_{11j} \left(\frac{\partial \sigma_1}{\partial \xi_1} \right)_j \left(\frac{\partial \sigma_2}{\partial \xi_2} \right)_j \qquad (10d)$$

It is now possible to combine Equation (4) and Equation (9) to give

$$\sum_{e=1}^{E} \sum_{j=1}^{4} U_j \int_{\tilde{\xi}} L\phi_{0j}\delta_i(\tilde{\xi})\det J d\tilde{\xi} + \frac{\partial U_j}{\partial \sigma_1} \int_{\tilde{\xi}} L\phi_{\sigma_1 j}\delta_i(\tilde{\xi})\det J d\tilde{\xi}$$

$$+ \frac{\partial U_j}{\partial \sigma_2} \int_{\tilde{\xi}} L\phi_{\sigma_2 j}\delta_i(\tilde{\xi})\det J d\tilde{\xi} + \frac{\partial^2 U_j}{\partial \sigma_1 \partial \sigma_2} \int_{\tilde{\xi}} L\phi_{\sigma_1\sigma_2 j}\delta_i(\tilde{\xi})\det J d\tilde{\xi}$$

$$= 0 \qquad i = 1,2,\ldots M \tag{11}$$

where $M = 4E$, J is the Jacobian matrix, E is the number of elements, and for simplicity in presentation, we have set f=0. One can alternatively write

$$\sum_{j=1}^{N} \left[U_j L\phi_{0j} + \frac{\partial U_j}{\partial \sigma_1} L\phi_{\sigma_1 j} + \frac{\partial U_j}{\partial \sigma_2} L\phi_{\sigma_2 j} + \frac{\partial^2 U_j}{\partial \sigma_1 \partial \sigma_2} L\phi_{\sigma_1 \sigma_2 j} \right]_i$$

$$= 0 \qquad i = 1,2,\ldots M \tag{12}$$

The coefficient matrix of Equation (12) is easily obtained provided the derivatives in L(•) can be transformed into ξ coordinates. This is achieved using chain rule expansions and a classical isoparametric transformation

$$x_\alpha(\tilde{\xi}) = \sum_{j=1}^{4} \left(\phi_{00j} X_{\alpha j} + \phi_{10j} \frac{\partial X_{\alpha j}}{\partial \xi_1} + \phi_{01j} \frac{\partial X_{\alpha j}}{\partial \xi_2} \right) \tag{13}$$

$$\alpha = 1,2$$

where we have omitted the cross-derivative from the transformation function. Note that in the regular ξ system, the optimal collocation points are easily found (they are, in fact, the Gauss points commonly used in numerical integration).

Equation (13) contains, in addition to the global node coordinates $X_{\alpha j}$, the nodal derivatives $\frac{\partial X_{\alpha j}}{\partial \xi_1}$ and $\frac{\partial X_{\alpha j}}{\partial \xi_2}$.

These can be obtained from the nodal derivatives $\frac{\partial\sigma_\alpha}{\partial\xi_\alpha}$ (see Equation 8) by rotation. The latter express the coordinate transformation at the corners of the element and they must be specified independently of the isoparametric transformation. The simplest approach is to use the ratio of the corresponding side lengths L_{σ_α} of the distorted element and the 2x2 basic square (Figure 2), that is

$$\frac{\partial\sigma_\alpha}{\partial\xi_\alpha} = \frac{L_{\sigma_\alpha}}{2} \qquad \alpha = 1,2 \tag{14}$$

These quantities are, however, discontinuous at the nodes. The discontinuity of the transformation has the effect of reducing the continuity of the solution from C^1 to C^0 except for the nodal values themselves. To achieve a continuous transformation, the nodal derivatives can be expressed as some weighted average of the lengths of the adjoining sides. We experimented with various averaging schemes and obtained best results with the harmonic mean, which we use here. The nodal derivatives then become

$$\frac{\partial\sigma_\alpha}{\partial\xi_\alpha} = \frac{L^{e-}_{\sigma_\alpha} L^{e+}_{\sigma_\alpha}}{L^{e-}_{\sigma_\alpha}+L^{e+}_{\sigma_\alpha}} \qquad \alpha = 1,2 \tag{15}$$

where e- and e+ refer to the two elements adjoining the node in the α-direction. Extrapolation must be used at the boundary nodes in order to obtain a consistent transformation. We will show in the examples that the continuous transformation does not necessarily always give better results.

Because the collocation equations are written at the collocation points, of which there are four per element, while the degrees of freedom are associated with the nodes, there will in general be more unknowns than equations in the most general form of the collocation scheme. It can be shown, however, that at least for elliptic problems such as the steady state groundwater flow problem, the number of unknowns can be reduced to exactly the number of equations by utilizing the boundary conditions. In this way, no additional collocation points on the boundaries are needed.

The above exposition illustrates the basic principles of formulating the CFM. For a more detailed discussion of the theory, the reader is referred to Frind and Pinder, 1978.

NUMERICAL EXPERIMENTS

The purpose of the numerical experiments is fourfold:

1. To investigate the behavior of the solution as the grid is distorted.
2. To compare the discontinuous and the continuous transformation of the CFM Hermite element.
3. To compare CFM using Hermite elements with FEM using linear triangles.
4. To investigate the convergence of the CFM solution as the grid is refined.

We will consider the well-known problem of steady radial flow to a well located in an isotropic, homogeneous aquifer. The governing equation written in cartesian coordinates is

$$L(u) = \frac{\partial^2 u}{\partial x^2} + \frac{\partial^2 u}{\partial y^2} = 0 \tag{16}$$

and the boundary conditions are of either the Dirichlet or the Neumann type. A wedge-shaped region (Figure 3) is convenient for this problem and we use it as a basis for our experiments.

Experiment A: Distorting the Grid

In this experiment we will investigate the effect of the grid distortion on the error produced by the numerical solution. We will also compare the CFM using isoparametric bicubic Hermite elements (Figure 3a) with FEM using linear triangular elements (Figure 3b). The FEM grid was designed to give a linear discretization equivalent to that of the Hermite cubic; this was done by placing two additional nodes along each of the Hermite element sides in the radial direction. No additional nodes were placed in the tangential direction where the solution is invariant. After accounting for the boundary conditions, the Hermite grid has a total of 16 degrees of freedom while the triangle grid has 15. Although the degrees of freedom are nearly the same, it must be kept in mind that in the linear grid, they are utilized more effectively because a larger proportion is located along the radial sides where the solution is variable.

The Hermite grid is distorted by increasing the radial length of the larger elements from 1 to 10, while keeping the radial length of the smaller elements constant at 1. The side length ratio is thus defined (see Figure 3(a)) as $(r_3-r_2)/(r_2-r_1)=(r_3-2)$. The triangle grid is distorted in a similar way.

Two cases will be considered, according to the type of boundary conditions specified at the left side.

Case A1: Dirichlet boundary

$$u = c_1 \text{ at } r = 1 \tag{17a}$$

$$u = 0 \text{ at } r = r_3 \tag{17b}$$

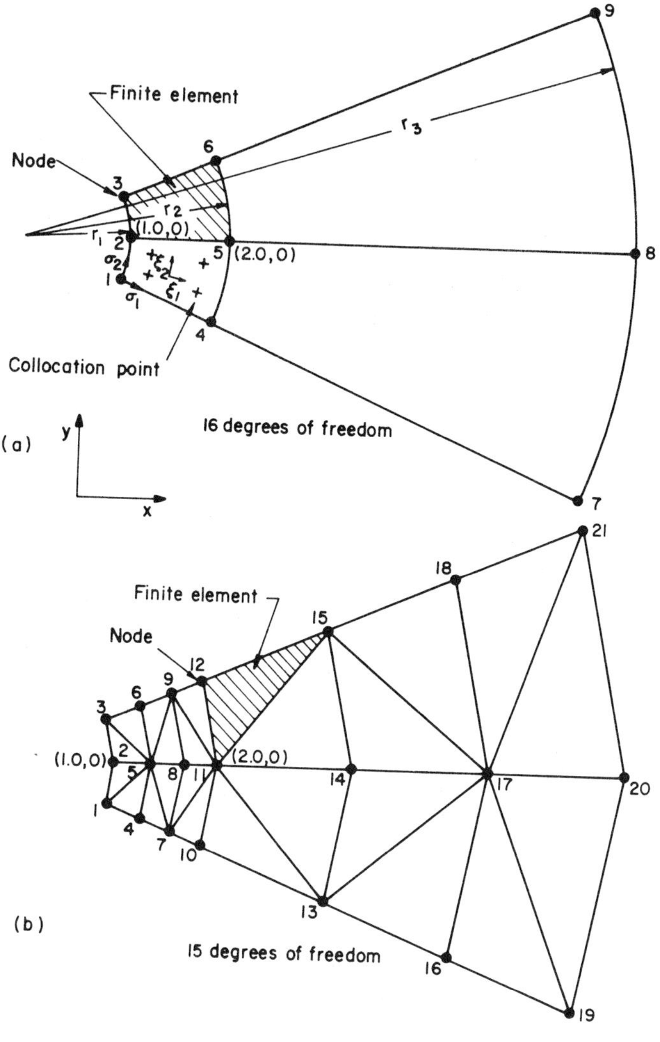

Figure 3 Finite element configuration for CFM using Hermite polynomials (a) and FEM using triangles (b).

Case A2: Neumann boundary

$$\frac{du}{dr} = c_2 \quad \text{at} \quad r = r_1 \tag{18a}$$

$$u = 0 \quad \text{at} \quad r = r_3 \tag{18b}$$

For the Hermite CFM grid, we will, in addition, use the two options available for the isoparametric transformation: the discontinuous transformation as per Equation (14), and the continuous transformation using the harmonic mean as per Equation (15).

For each case, the error of the numerical solution is determined by comparison with the exact solution. Figures 4 and 5 show the resulting error plotted versus the logarithm of the side length ratio. Three sets of curves are shown in each figure: (a) the remaining degree of freedom at the boundary r=1, (b) the potential at r=2, and (c) the radial gradient at r=2. Plot (a) is representative of the behavior of the solution at the boundary, while plots (b) and (c) are representative of the behavior in the interior of the grid.

An examination of Figures 4 and 5 reveals the following:

1. The error, as expected, increases as the grid distortion increases. (The apparent decrease in the error in the gradient for the linear triangle in Figure 4(a) is merely a consequence of the way the grid was set up.)

2. For moderate grid distortion (side length ratio of up to about 4), CFM with a harmonic mean transformation gives somewhat better accuracy than CFM with a discontinuous transformation.

3. For large grid distortion (side length ratio greater than 4), the error in CFM with harmonic mean increases faster than that in the discontinuous version.

4. FEM with linear triangles generally gives better potentials, while CFM with Hermite elements gives better gradients.

5. Gradients produced by CFM are of about the same accuracy at the boundary and in the interior, while those produced by linear FEM are of lower accuracy at the boundary.

6. Potentials produced by linear FEM are less affected by large grid distortions than those produced by CFM.

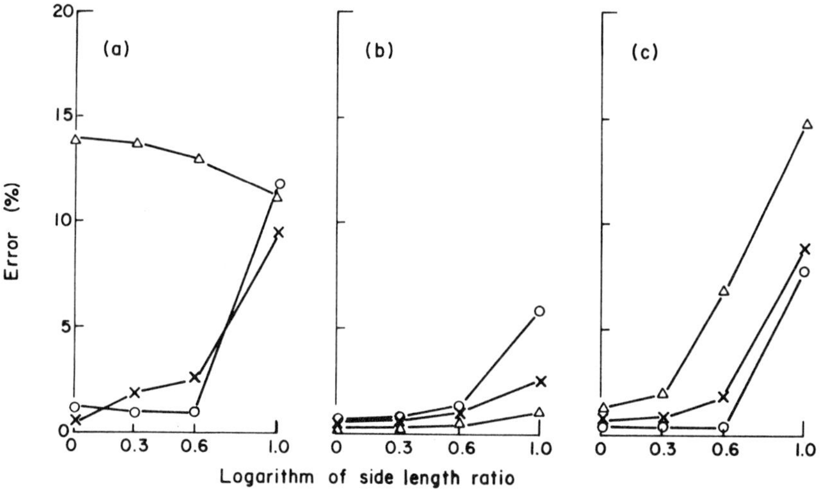

Figure 4 Effect of grid distortion: Case A1, Dirichlet boundary. (a) Radial gradient at r=1, (b) Potential at r=2, (c) Radial gradient at r=2. FEM with linear triangles (Δ), CFM with discontinuous transformation (x), CFM with harmonic mean transformation (o).

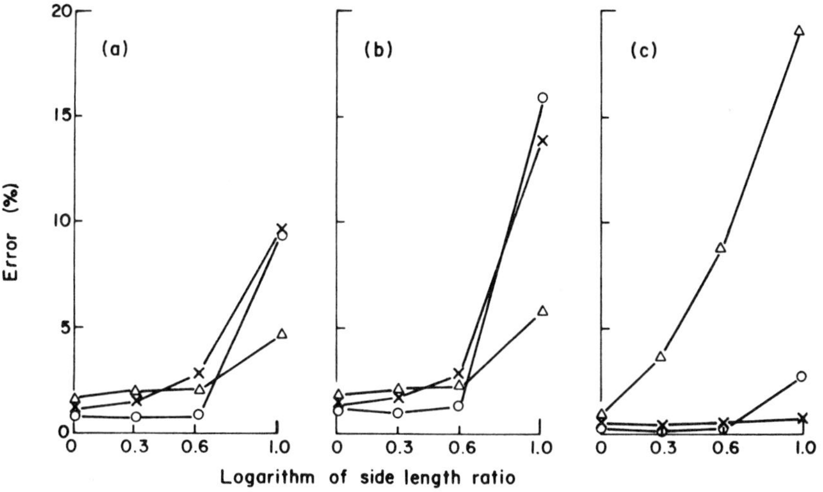

Figure 5 Effect of grid distortion: Case A2, Neumann boundary. (a) Potential at r=1, (b) Potential at r=2, (c) Radial gradient at r=2. FEM with linear triangules (Δ), CFM with discontinuous transformation (x), CFM with harmonic mean transformation (o).

Experiment B - Convergence Study

Here we will investigate the convergence of CFM using both the discontinuous and the continuous transformation. We use the basic grid configuration of Figure 3(a), but with the nodes placed logarithmically in the radial direction. The boundary conditions are those given by Equation (17). In order to show the pattern of convergence under increasing grid distortion, we will consider two cases, according to the geometry of the initial grid:

Case B1: Sidelength ratio 10^1: 1

$$r_1 = 10^0 = 1$$

$$r_2 = 10^1 = 10$$

$$r_3 = 10^2 = 100$$

Case B2: Sidelength ratio $10^{1.5}$: 1

$$r_1 = 10^0 = 1$$

$$r_2 = 10^{1.5} = 31.6$$

$$r_3 = 10^3 = 1000$$

Each of the two grids is successively refined by subdividing the elements logarithmically in the radial direction only. The minimizing sequence consists of four grids having 4,8,16, and 32 elements in each case. Figures 6 and 7 show the results of the convergence study plotted logarithmically as percent error (absolute) versus the number of elements. To distinguish between boundary and interior nodes, three sets of curves are again shown. These curves reveal some interesting properties of the CFM:

1. CFM with discontinuous transformation converges essentially monotonically.

2. CFM with harmonic mean transformation converges more rapidly at first than the discontinuous version; eventually however both converge to the same point.

3. For very large grid distortions (greater than 10: 1), the continuous version converges in an oscillatory pattern, while the discontinuous version still converges nearly monotonically. However, the continuous version almost always produces a smaller error.

In terms of computational effort, a comparison between CFM and FEM is not easy because of the different ways in which

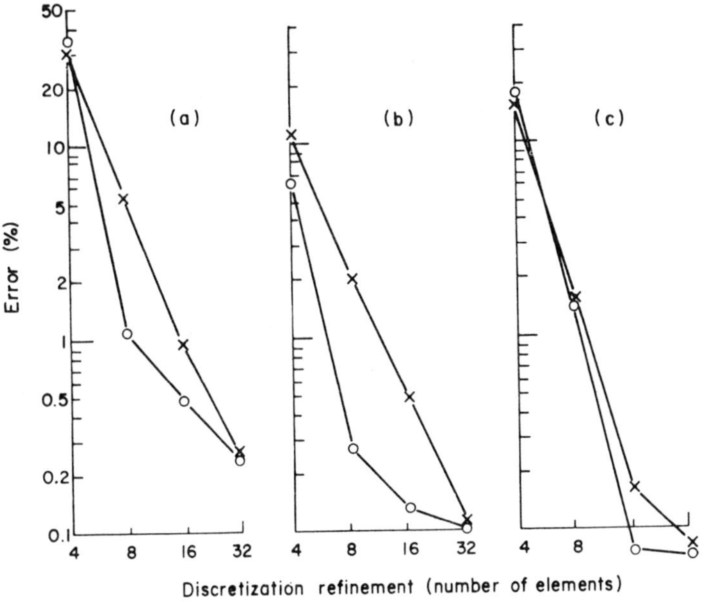

Figure 6 Convergence of CFM: Case B1, Initial sidelength ratio 10^1:1. (a) Radial gradient at r=1, (b) Potential at r=10, (c) Radial gradient at r=10. CFM with discontinuous transformation (x), CFM with continuous transformation (o).

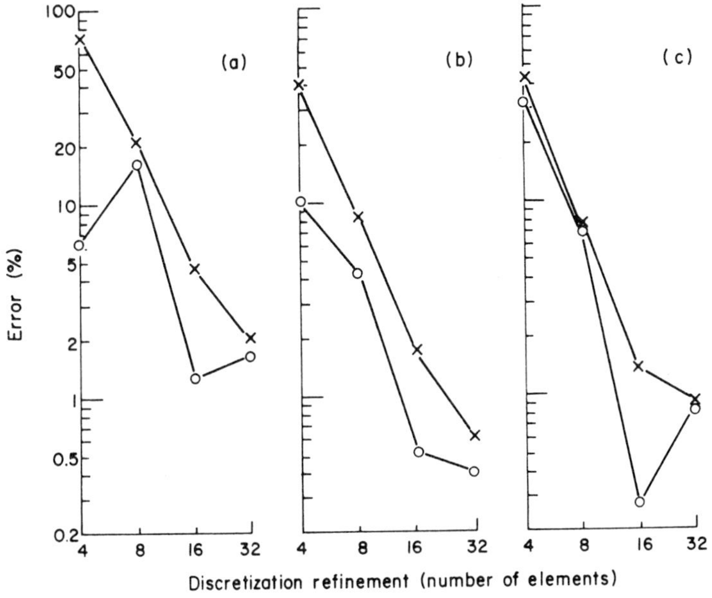

Figure 7 Convergence of CFM: Case B2, Initial sidelength ratio $10^{1.5}$:1. (a) Radial gradient at r=1, (b) Potential at r=31.6, (c) Radial gradient at r= 31.6. CFM with discontinuous transformation (x), CFM with continuous transformation (o).

the degrees of freedom are utilized. For example, it can be shown by means of an operations count that the FEM grid of Figure 3(b) is somewhat less costly than the CFM grid of Figure 3(a). As mentioned before, the linear elements here are at an advantage because the degrees of freedom can be located in a more effective way. If however the FEM grid were designed such that the linear discretization is equivalent to the bicubic Hermite discretization in both coordinate directions, the CFM grid would be the less costly one in terms of computational effort. In this case, the degrees of freedom of the Hermite will be more efficiently utilized computationally because all of them (that is, all the interior ones) are shared by four elements. Also the work of preparing the grid will be less because there are fewer nodes. On average, the two methods appear to be computationally equivalent.

CONCLUSION

It appears that CFM may be used to good advantage in groundwater flow problems where accurate velocities are required. The method uses isoparametric Hermite elements and is therefore suited to regions with irregular boundaries. It provides a C^1 continuous solution and the gradients obtained are of the same order of accuracy as the potentials. Comparing the method with C^0 continuous FEM, it is found that FEM will produce better potentials, while CFM will produce better gradients, particularly at the boundary. CFM using Hermite elements is competitive in terms of cost with FEM using linear triangles. It can also be shown that CFM is much more efficient than C^1 continuous FEM using Hermite elements.

ACKNOWLEDGEMENTS

We appreciate the helpful comments provided by D.H. Tang and M.T. Van Genuchten of Princeton University. The study was supported by the U.S. National Science Foundation, NSF-RANN, grant #NSF-AER74-01765 and the University of California through sub-contract P.O.3143202 under Energy Research & Development Administration Contract No. W-7405-ENG.48.

REFERENCES

De Boor, C., and B. Swartz (1973), Collocation at Gaussian Points, SIAM J. Numerical Analysis, 10(4), pp. 582-606.

Finlayson, B.A. (1972), The Method of Weighted Residuals and Variational Principles, Academic Press, New York.

Finlayson, B.A. and L.E. Scriven (1965), The Method of Weighted Residuals and its Relation to Certain Variational Principles for the Analysis of Transport Processes, Chem. Eng. Science, 20, pp. 395-404.

Frind, E.O. (1977), An Isoparametric Hermitian Finite Element for the Solution of Field Problems, Int. J. for Numerical Methods in Engineering, 11, pp. 945-962.

Frind, E.O., and G.F. Pinder (1978), A Collocation Finite Element for Potential Problems in Irregular Domains, Submitted to Int. J. for Numerical Methods in Engineering.

Lanczos, C. (1938), Trigonometric Interpolation of Empirical and Analytical Functions, J. Math. Physics, 17, pp. 123-199.

Prenter, P.M. (1975), Splines and Variational Methods, Pure and Applied Mathematics Interscience Series, John Wiley and Sons, New York.

Van Genuchten, M.T., G.F. Pinder and E.O. Frind (1977), Simulation of Two-dimensional Contaminant Transport with Isoparametric Hermitian Finite Elements, Water Resources Research, 13(2), pp. 451-458.

Villadsen, J.V. and F. Michelsen (1978), Solution of Differential Equation Models by Polynomial Approximation, Prentice-Hall, Englewood Cliffs, N.Y.

Villadsen, J.V. and W.E. Stewart (1967), Solution of Boundary Value Problems by Orthogonal Collocation, Chem. Eng. Science, 22, pp. 1485-1501.

UNSTEADY FLOW IN POROUS MEDIA SOLVED BY COMBINED FINITE ELEMENT-METHOD OF CHARACTERISTICS MODEL

A.A. Hannoura, J.A. McCorquodale

Civil Engineering Department, University of Windsor, Windsor, Ontario, Canada

INTRODUCTION

There are several flow problems in which the inertia terms in the governing equations are significant. The method of characteristics has been a popular solution technique for this type of problem. The one dimensional formulation of this method is very simple; however, in a number of problems, including the example of this paper, the real phenomenon has two of three dimensional variations which cannot be neglected.

The concept of the hybrid model presented here is to treat the time integration by the Method of Characteristics (MOC) and the space integration by the Finite Element Method (FEM). Thus the major inertial effects are included in the MOC and the effects of non-homogeneity, velocity and pressure variation in space are computed by the FEM and used to update the MOC.

Unsteady flow in coarse granular media results from the interaction of water waves with coastal structures and rock-fill dams. This flow may be Darcy or non-Darcy and one or two phase. Furthermore, the porous matrix is usually nonhomogeneous and the structural boundaries are irregular. The flow phenomenon is described by a set of hyperbolic partial differential equations involving coefficients of pressure distribution, inertial and hydraulic conductivity. These coefficients may vary in time and space. A hybrid method of characteristics-finite element model is used to solve the governing equations.

FORMULATION OF THE PROBLEM

The problem considered in this study is the analysis of the interflow due to wave interaction on a porous structure with an impervious core as shown schematically in Figure 1. The problem is formulated in two stages: (a) the development of the hyperbolic differential equations which will be solved by the characteristics method in order to compute the instantan-

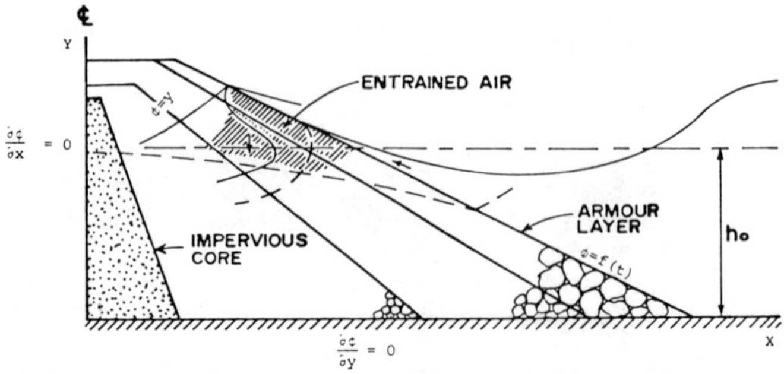

Figure 1 Wave Interaction with a Porous Structure

eous water surface; (b) the introduction of the finite element method to estimate representative values of pressure distribution correction factor (1), and the hydraulic conductivity of porous matrix.

The characteristic formulation

Applying the one dimensional momentum equation to the internal flow in the embankment shown in Figure 1, the following system of equations for unsteady non-Darcy flow in a porous structure is obtained:

$$\begin{bmatrix} C_m & u/m & 0 & mgC_p \\ 0 & (h_o+\eta)/m & 1 & u/m \\ dt & dx & 0 & 0 \\ 0 & 0 & dt & dx \end{bmatrix} \begin{Bmatrix} \frac{\partial u}{\partial t} \\ \frac{\partial u}{\partial x} \\ \frac{\partial \eta}{\partial t} \\ \frac{\partial \eta}{\partial x} \end{Bmatrix} = \begin{Bmatrix} -gFum \\ 0 \\ du \\ dt \end{Bmatrix} \qquad (1)$$

in which C_m = inertia coefficient, u = horizontal Darcy velocity, m = porosity, C_p = pressure distribution correction factor, t = time, g = acceleration due to gravity, F = non-Darcy friction term, h_o = mean water level, η = internal wave amplitude, and x is the horizontal axis as shown in Figure 2. From Equation 1, the characteristic directions can be written as

$$\alpha = \left.\frac{dx}{dt}\right|_+ = (1+\frac{1}{C_m})\frac{u}{2m} + \sqrt{(\frac{u}{m})^2(\frac{C_m+3}{4C_m}-1)+g(h_o+\eta)\frac{C_p}{C_m}} \qquad (2)$$

$$\text{and } \beta = \left.\frac{dx}{dt}\right|_- = (1+\frac{1}{C_m})\frac{u}{2m} - \sqrt{(\frac{u}{m})^2(\frac{C_m+3}{4C_m}-1)+g(h_o+\eta)\frac{C_p}{C_m}} \qquad (3)$$

The characteristic solution proceeds by solving the governing equations along the characteristic directions α and β. The

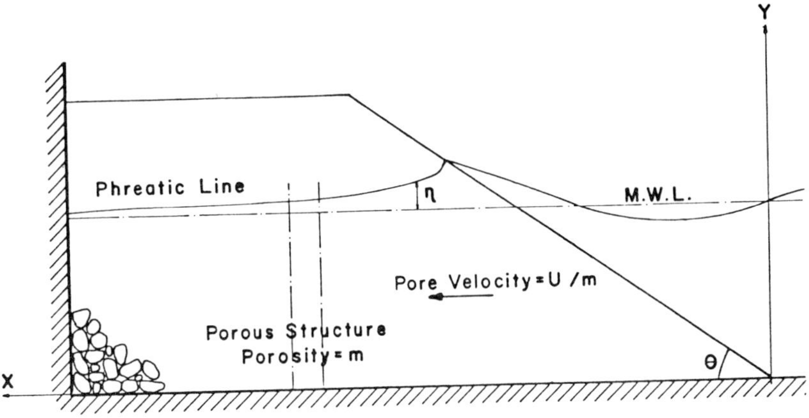

Figure 2 Defining Sketch of Example Problem

equations of motion and continuity can be obtained from Equation 1 and rewritten in term u and c, wave celerity = $\sqrt{gy}$ as follows:

$$c_m \frac{\partial u}{\partial t} + \frac{u}{m}\frac{\partial u}{\partial x} + 2cc_p m \frac{\partial c}{\partial x} = -gFmu \tag{4}$$

and

$$\frac{\partial c}{\partial t} + \frac{u}{m}\frac{\partial c}{\partial x} + \frac{c}{2m}\frac{\partial u}{\partial x} = 0 \tag{5}$$

An explicit finite difference scheme is used to discretize Equations 4 and 5 in order to obtain u and c at the internal grid points (see Figure 3), i.e.

$$\begin{aligned} u(i,j+1) = u(i,j) - \frac{\Delta t}{2\Delta x c_m}\{&\frac{u(i,j)}{m}[u(i,j+1)-u(i-1,j)] + \\ &mc(i,j)c(i,j)[c_p(i+1,j)-c_p(i-1,j)] + \\ &2mc_p(i,j)c(i,j)[c(i+1,j)-c(i-1,j)] + \\ &2gF(i,j)mu(i,j)\Delta x\} \end{aligned} \tag{6}$$

and

$$\begin{aligned} c(i,j+1) = c(i,j) - \frac{\Delta t}{2m\Delta x}\{&u(i,j)[c(i+1,j)-c(i-1,j)] + \\ &\frac{c(i,j)}{2}[u(i+1,j)-u(i-1,j)]\} \end{aligned} \tag{7}$$

Equations 2 and 3 are used to control the discretization of the x-t plane as shown in Figure 3.

<u>The finite element formulation</u>

The momentary two dimensional internal flow can be described by

$$\frac{\partial}{\partial x}\left(K(|\nabla\phi|)\frac{\partial \phi}{\partial x}\right) + \frac{\partial}{\partial y}\left(K(|\nabla\phi|)\frac{\partial}{\partial y}\right) = 0 \tag{8}$$

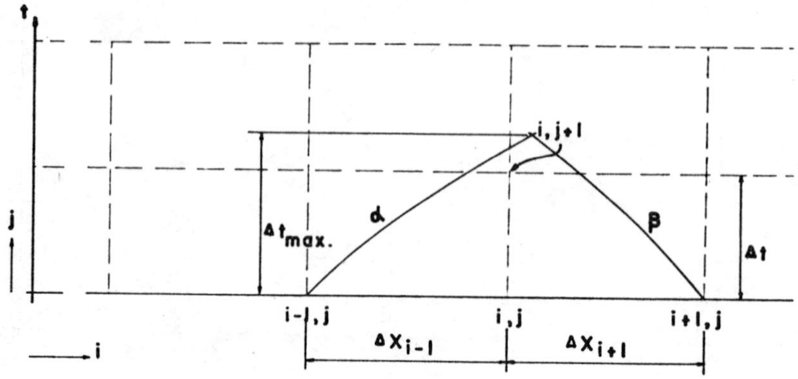

Figure 3 Discretization of the x-t Plane

in which $K(|\nabla\phi|)$ = non-linear hydraulic conductivity and ϕ = piezometric head. McCorquodale (2) gives the corresponding functional form of Equation 4 as

$$\chi = \int_A\int [K(|\nabla\phi|)(\phi_x^2+\phi_y^2) + G(|\nabla\phi|)]dydx \tag{9}$$

where $G(|\nabla\phi|)$ is a continuity function. The solution of Equation 8 is now reduced to finding the distribution of the piezometric function, ϕ, which minimizes χ, subjected to certain boundary conditions. The finite element grid for the example problem is shown in Figure 4. The integral of Equation 9 is given by

$$\sum_e \sum_j K(|\nabla\phi|)^e S_{ij}^e \phi_j = 0 \tag{10}$$

in which $S_{ij}^e = (b_i b_j + c_i c_j)/A^e$ = the element stiffness matrix

$$b_i = y_i - y_k$$

$$c_i = x_k - x_j$$

and i,j and k are the element node numbers (see Figure 4). Equation 10 gives a set of non-linear simultaneous equations in the unknown values of ϕ.

NUMERICAL SOLUTION PROCEDURES

Initial conditions

At t = 0,

$u = 0$, $c = \sqrt{gh_o}$ and $c_p = 1$

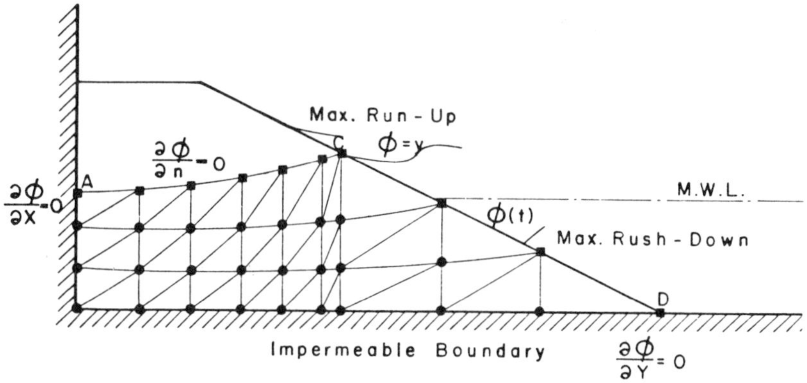

Figure 4 Discretization of the X-Y Plane

Boundary conditions

At the core boundary and base of the embankment, u = 0, i.e. $\frac{\partial \phi}{\partial x}$ = 0 (see Figure 4). An experimental study was undertaken to obtain representative boundary conditions for the outcrop point movement and the pressure distribution along the sloping face bc (see Figures 2 and 4). Figures 5 and 6 show the experimental boundary conditions for the outcrop point movement and the pressure distribution along bc for large amplitude waves.

Since the outcrop point location and the macroscopic velocity at the core are prescribed, Equations 4 and 5 can be manipulated to obtain expressions to compute the macroscopic velocity at the interface and the celerity at the core.

The numerical solution is carried out in two stages: (a) the finite difference solution (FD) guided by Equations 2 and 3 is used to advance the solution from t = 0 to $t + \varepsilon$; (b) at $t + \varepsilon$, the finite element method (FEM) is called to solve Equation 8 for the interior space domain (see Figure 4). The finite element solution is used to compute the hydraulic conductivity $K(|\nabla\phi|)$ and the pressure distribution correction factor. A weighted average value of the hydraulic conductivity was calculated and then used in the next run of the finite difference solution, i.e., from time $t + \varepsilon$ to $t + 2\varepsilon$. The pressure distribution correction factor is obtained by comparing the computed pressure distribution to the assumed hydrostatic distribution. Since the two grids of the FD and the FEM do not always coincide, an interpolation scheme is used to transfer the values of the hydraulic conductivity and the pressure distribution correction factor from the FEM to the FD grids.

CONCLUDING REMARKS

The combined finite element-method of characteristics model

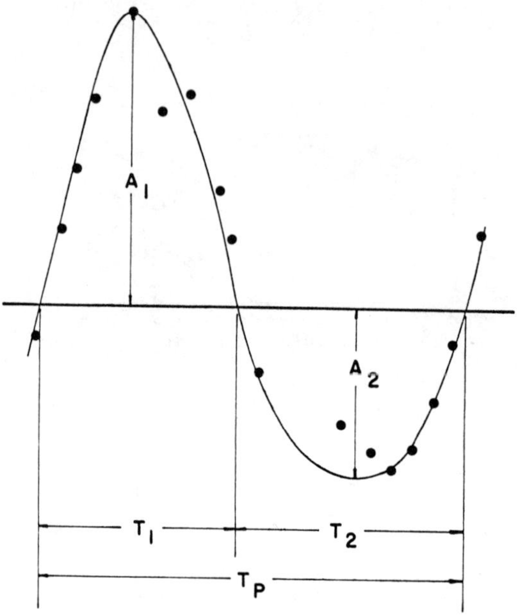

Figure 5 Interface Boundary Condition (Outcrop Point Movement)

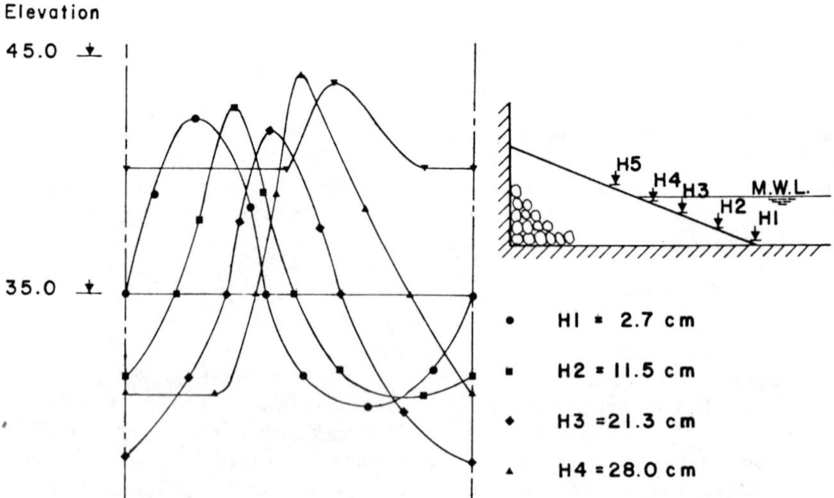

Figure 6 Interface Boundary Condition (Piezometric Head Variation along the u/s Slope)

was applied to the wave transmission in coarse granular porous media (see Figure 1). Figure 7 compares typical results obtained from the hybrid model with experimental results.

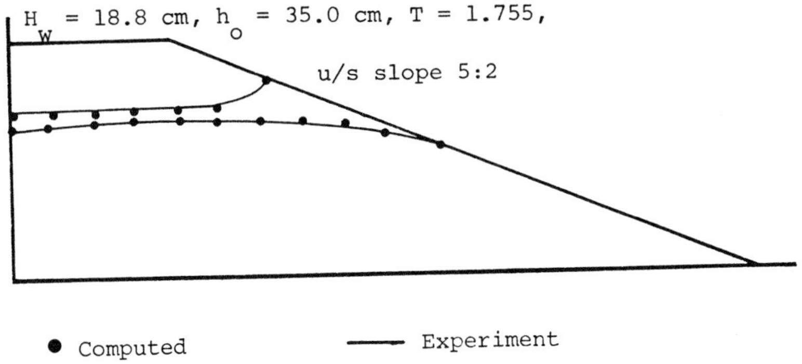

Figure 7 Computed and Experimental Wave Transmission for 4.36 cm Crushed Rock

The hybrid programme requires about twice the CPU time of a one dimensional method of characteristics model (3) but only 40% of the time required for a complete finite element model. Furthermore, the hybrid model can treat non-homogeneous embankments with irregular boundaries which is not convenient with the normal method of characteristics model. The model provides a simple treatment of wave transmission in porous structures which should permit design engineers to improve the stability analysis of these structures.

ACKNOWLEDGEMENT

This research was supported by the National Research Council of Canada. This support is gratefully acknowledged.

REFERENCES

1. Chow, V.T. (1959) Open Channel Hydraulics. McGraw-Hill, New York.

2. McCorquodale, J.A. (1970) "Variational Approach to Non-Darcy Flow", Journal of Hydraulics Division, ASCE, Vol. 96, No. HY11, Proc. Paper 7694, pp. 2265-2278.

3. McCorquodale, J.A. and Nasser, M.S. (1974) "Numerical Methods for Unsteady Non-Darcy Flow", The International Symposium on Finite Element Methods in Flow Problems, Swansea, United Kingdom, pp. 545-557.

A FINITE ELEMENT "DISCRETE KERNEL GENERATOR" FOR EFFICIENT GROUNDWATER MANAGEMENT

Tissa Illangasekare, H. J. Morel-Seytoux

Colorado State University, Fort Collins, Colorado 80523

INTRODUCTION

The constantly increasing demand for water in regions with already limited supply makes it imperative to manage the surface and groundwater supplies efficiently. In water scarce regions such as the Western United States, the water regulating agencies are facing the problem of regulating the water usage on a day-by-day basis during periods of high water demand. A typical case of a day-by-day controlled diversion of stream water to an irrigation district in the South Platte basin is shown on Figure 1. In the conventional approaches, for each set of decision variables such as aquifer pumping a simulation run has to be made and one must check whether the defined management objectives are met. Such an approach will be very inefficient in problems where large stream-aquifer systems are involved and a large number of decision variables has to be regulated on a day-by-day basis.

Most of the existing mathematical models of stream-aquifer systems are designed to predict the hydrologic behavior of the system in response to a particular set of numerical values of the excitations. An approach which makes use of the functional relation between the responses of the system to the excitation was presented by Morel-Seytoux (1973). The concept of the response function and the applicability of the approach to hydrologic modeling and simplified management problems were illustrated in a paper by Morel-Seytoux (1975). The use of these basic aquifer response functions in combination with a linearized stream routing model to predict the water table and river stage evolution was described briefly by Morel-Seytoux (1975).

Figure 2 schematically represents the operation of a management model using the influence coefficient approach. Figure 3 compares the conventional simulation approach and the suggested influence coefficient approach.

The models discussed by Morel-Seytoux (1975) and

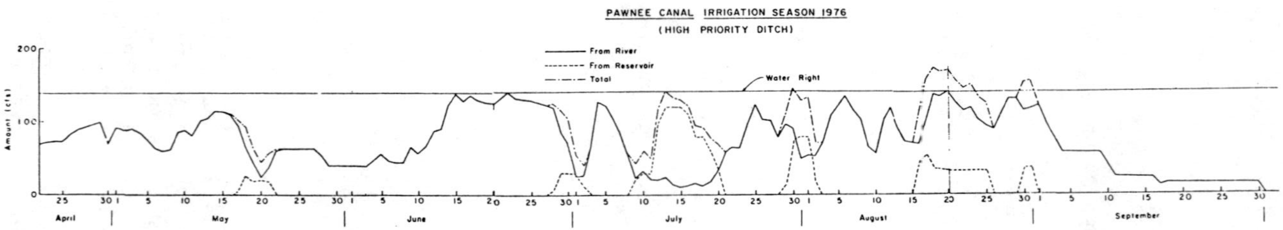

Figure 1 Daily diversion from stream, releases from reservoirs and total water delivery.

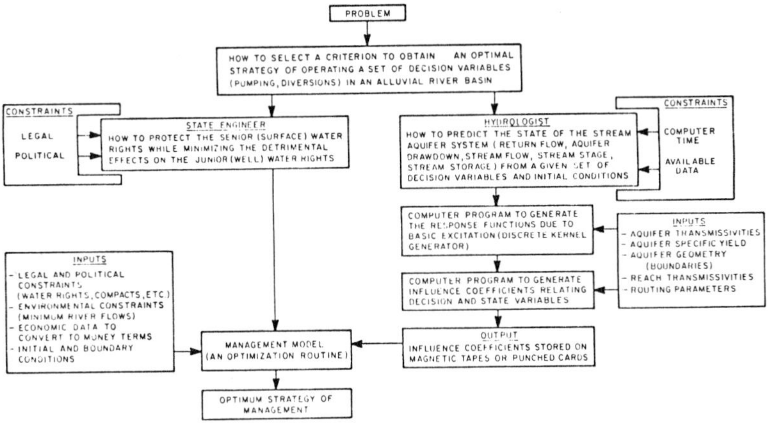

Figure 2 Schematic representation of the operation of management model using the influence coefficient approach.

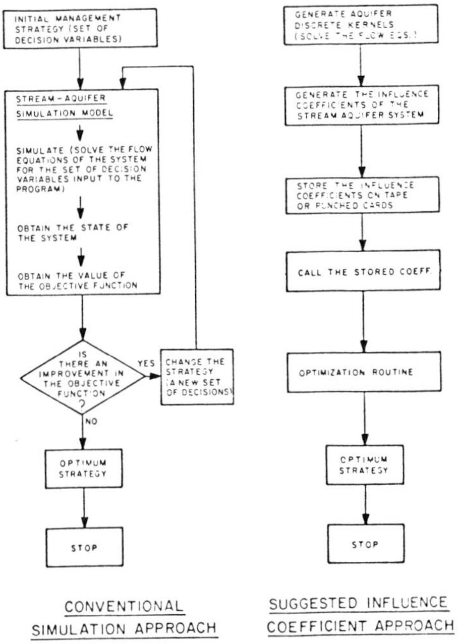

Figure 3 Comparison of conventional and new approach in deciding optimal management strategy.

Rodriguez-Amaya (1976) use two versions of the finite difference method to solve the basic saturated flow equation. A user oriented, (minimum input decisions taken by the user) storage efficient finite element model which generates the basic response functions without being limited by the size of the aquifer is presented in this paper.

DISCRETE KERNEL OF AQUIFER DRAWDOWN DUE TO PUMPING EXCITATION

The basic saturated flow equation is the Boussinesq equation:

$$\phi \frac{\partial s}{\partial t} - \frac{\partial}{\partial x}\left(T \frac{\partial s}{\partial x}\right) - \frac{\partial}{\partial y}\left(T \frac{\partial s}{\partial y}\right) = q_e \qquad (1)$$

where ϕ is the drainable (or effective) porosity, s is the drawdown measured positive downward from a (high) horizontal datum, t is time, x and y are the horizontal cartesian coordinates, T is the transmissivity, and q_e is the instantaneous pumping rate per unit area at excitation point e in the aquifer (chosen algebraically positive for a withdrawal excitation). It has been shown by Morel-Seytoux and Daly (1975) that the solution to Equation (1) can be expressed generally in the form:

$$s_{we}(n) = \sum_{\nu=1}^{n} \delta_{we}(n-\nu+1) Q_e(\nu) \qquad (2)$$

It is clear that once the discrete kernel coefficients $\delta_{we}(\nu)$, $\nu=1,2...n$ have been obtained, the drawdown response $s_{we}(n)$ can be obtained for any type of pumping schedule $Q_e(\nu)$, $\nu=1,2...n$ from Equation (2), whereas in the traditional simulation approach the total right-hand side of Equation (2) has to be computed for each given pumping schedule.

To generate the "discrete kernel" coefficients for an aquifer with given boundary conditions, Equation (1) has to be solved for a unit pulse excitation q_e on the right-hand side.

A USER ORIENTED COMPUTER-EFFICIENT MODEL

One of the shortcomings of existing numerical models of aquifer simulation is related to the decisions which has to be taken by the user with respect to the data inputs. The basic inputs needed for the numerical solution of Boussinesq's equation are the aquifer geometry, distribution of transmissivity, distribution of specific yield and the locations and values of net pumping excitations. In the "discrete kernel generator" the net pumping excitation is fixed as a unit pulse applied at a known node point. Thus all the inputs needed could be extracted from the basic data sources of maps defining aquifer boundaries, transmissivities and specific yields. A "user oriented" program is one such that the user does not have to make the decisions related to the type of mesh (or grid) system to be used in the numerical procedure, the spacing of nodes, the estimation of nodal transmissivity and specific yield values from data maps and time increment parameters, etc.

There are two aspects of efficiency which have to be considered in computer modeling. The efficiency with respect to the computer memory storage needed and the central processing time used. In solving problems associated with large stream-aquifer systems, the memory storage becomes a limiting factor. Even though the computing time which decides the computing cost becomes high for the generation of the discrete kernel coefficients, once they have been calculated and saved, simulation of aquifer behavior to any pumping excitation pattern can be obtained without ever making use any longer of the costly numerical model.

The "discrete kernel generator" developed in this study has the following features which makes it user oriented and storage efficient:

(1) The program uses the geometry of aquifer boundaries, contours of equal transmissivity and specific yields as inputs.

(2) The program generates a finite element mesh system to fit the given aquifer geometry.

(3) It defines a sub-mesh system to scan the total aquifer.

(4) The built-in time-parameters in the program guarantee numerical stability.

Moving sub-aquifer

The idea of the moving sub-aquifer is based on the fact that the aquifer drawdown (response) due to a pumping excitation at a node is significant only locally in the aquifer. Figure 4 shows the maximum value of the discrete kernels ($\delta_{we}(n)$) generated at different distances away from the excitation point e, for a homogeneous aquifer. For this case the response is significant only up to about the 8th node space. Hence for this particular case the width of the sub-aquifer within which the excitation is assumed to be felt is taken as 16 node spaces.

By assuming that the response is only significant locally in a region close to the excitation we are assuming that the initial zero gradient of the water table is not changed outside this region. Hence the boundary of this region which is also the boundary of the sub-aquifer (Figure 5) is assumed to have no flow boundary condition. The finite element equations are formulated for the sub-aquifer to solve Equation (1) for an excitation point on the excitation grid line EE (Figure 6). Once all the nodes on EE have been excited and the discrete kernels generated, the system moves by one grid space, making the next vertical grid line the excitation grid line. The sub-system scans the total aquifer by moving one grid space at

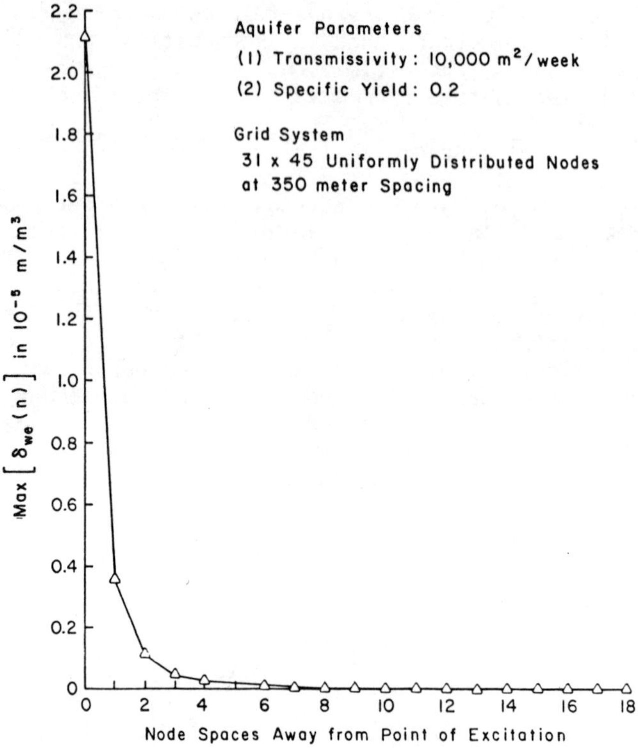

Figure 4 Maximum excitation response in a homogeneous aquifer.

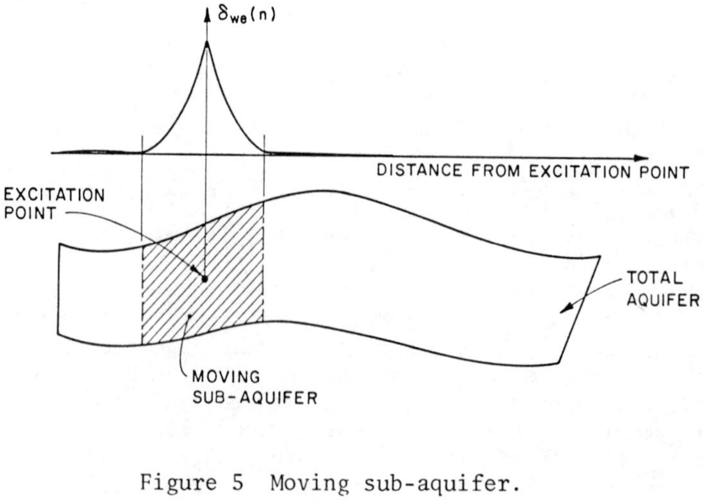

Figure 5 Moving sub-aquifer.

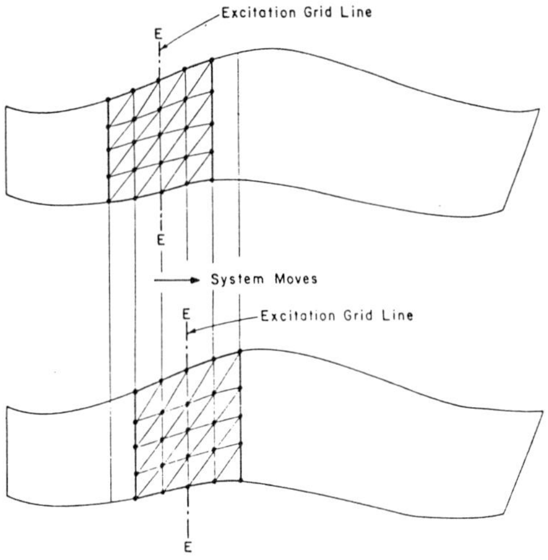

Figure 6 Moving sub-mesh.

a time till all the nodes have been excited.

FINITE ELEMENT FORMULATION

Using the Galerkin method on the operator defined by Equation (1) and following procedures described in various textbooks (e.g., Gray and Pinder (1974)) a matrix equation for the unknown drawdowns at the nodes is obtained. This equation is of the form:

$$[A]\ \{s\} = \{R\} \tag{3}$$

An element of the matrix [A] is given by,

$$A_{(i,k)(j,\ell)} = P_{ij}\ U_{k\ell} + Q_{ij}\ V_{k\ell} \tag{4}$$

where i,j are nodes defined on the x-y plane and k,ℓ are time nodes (Figure 7). The general expressions for the elements of matrices [P] and [Q] defined on the x-y plane and the elements of the matrices [U] and [V] defined on the time axis are given by:

$$P_{ij} = \sum_{m=1}^{K_i} \left\{ T_m \iint_{D^m_{xy}} \left(\frac{\partial N_i}{\partial x}\ \frac{\partial N_j}{\partial x} + \frac{\partial N_i}{\partial y}\ \frac{\partial N_j}{\partial y} \right) dxdy \right\} \tag{5}$$

$$Q_{ij} = \sum_{m=1}^{K_i} \left\{ \phi_m \iint_{D^m_{xy}} N_i\ N_j\ dxdy \right\} \tag{6}$$

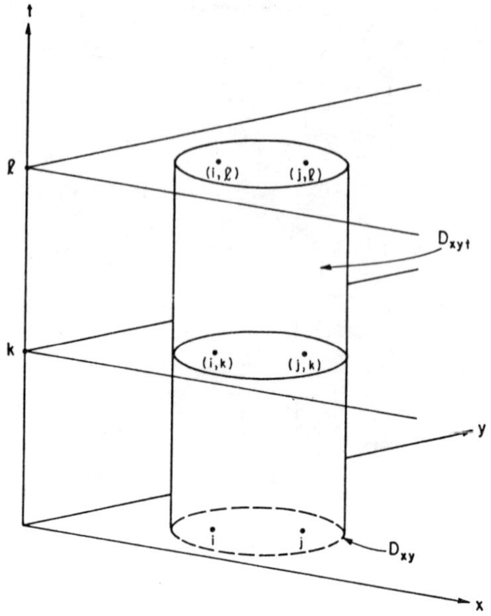

Figure 7 Definition of space and time nodes

$$U_{k\ell} = \int_{D_t} \gamma_k \gamma_\ell dt \tag{7}$$

$$V_{k\ell} = \int_{D_t} \gamma_k \frac{\partial \gamma_\ell}{\partial t} dt \tag{8}$$

where N_i and γ_k are the basis functions on space and time respectively, K_i is the number of elements sharing the space node i, T_m and ϕ_m are the constant transmissivity and specific yield in the mth element, D_{xy}^m is the space domain of mth element and D_t is the time domain.

An element of the right-hand side vector is given by,

$$R_{(i,\ell)} = \left\{ q_e \sum_{m=1}^{K_i} \iint_{D_{xy}^m} N_i \, dxdy \right\} \left\{ \int_{D_t} \gamma_\ell dt \right\} \tag{9}$$

The set of Equations (3) can be solved for $s_{(i,\ell)}$, where (i,ℓ) is the node whose projection is the ith node on the x-y plane and ℓ is the time level (Figure 7).

NUMERICAL SCHEME AND THE COMPUTER PROGRAM

The computer program developed for the generation of "discrete kernels" has three main components:

(1) a program which estimates the parameters and the size of the moving sub-mesh,
(2) a program which generates the mesh system and estimates the nodal values of transmissivity and specific yield from input data, and
(3) a program which, (a) computes and updates the "space" matrices [P] and [Q] as the system moves, (b) computes and updates the "time" matrices [U] and [V] as the time nodes move along the time axis, and (c) solves the system of Equations (3).

Algebraic structure of the finite element equations

The structure of the matrices defined by Equations (5) to (9) is determined by the type of elements and the basis functions used. In this study, simple triangular elements with linear basis functions $N_i(x,y)$ were used to generate the matrices [P] and [Q]. Along the time axis, two nodes per element with linear basis functions $\gamma_i(t)$ were used to generate the matrices [U] and [V].

In Equations (5), (6) and (9) the domains of integration D^m_{xy} become the triangular elements of the mesh system. The double integration was performed for each finite element to obtain the following expressions for an elemental setup shown in Figure 8.

$$[p] = \begin{bmatrix} \frac{-T_m(y_2-y_3)^2}{8A} & \frac{-T_m y_3(y_2-y_3)}{8A} & \frac{T_m y_2(y_2-y_3)}{8A} \\ \frac{-T_m y_3(y_2-y_3)}{8A} & \frac{-T_m(y_3^2 + \Delta L^2)}{8A} & \frac{T_m(y_2 y_3 + \Delta L^2)}{8A} \\ \frac{T_m y_2(y_2-y_3)}{8A} & \frac{T_m(y_2 y_3 + \Delta L^2)}{8A} & \frac{-T_m(y_2^2 + \Delta L^2)}{8A} \end{bmatrix} \quad (10)$$

$$[q] = \begin{bmatrix} \phi_m \frac{A}{6} & \phi_m \frac{A}{12} & \phi_m \frac{A}{12} \\ \phi_m \frac{A}{12} & \phi_m \frac{A}{6} & \phi_m \frac{A}{12} \\ \phi_m \frac{A}{12} & \phi_m \frac{A}{12} & \phi_m \frac{A}{6} \end{bmatrix} \quad (11)$$

$$\{r\} = \begin{bmatrix} \frac{A}{3} \\ \frac{A}{3} \\ \frac{A}{3} \end{bmatrix} \quad (12)$$

where [p], [q] and {r} are the components of elements of [P], [Q] and {R} contributed by each finite element and A is the area of the triangle.

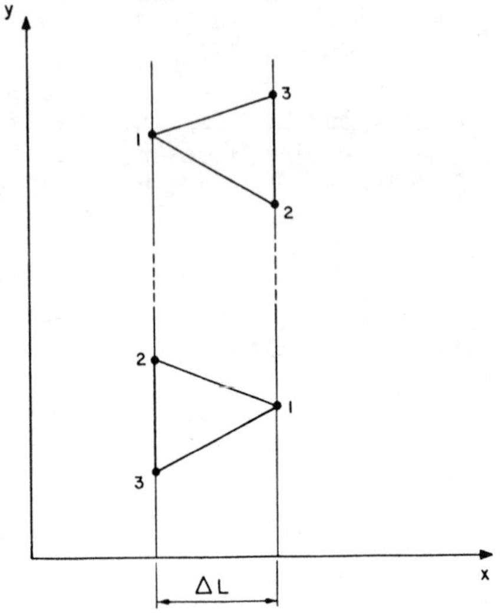

Figure 8 General element setup.

Using two nodes per element spaced at Δt, the following expressions were obtained for the "time" matrices.

$$[U] = \begin{bmatrix} \Delta t/3 & \Delta t/6 \\ \Delta t/6 & \Delta t/3 \end{bmatrix} \quad (13)$$

$$[V] = \begin{bmatrix} -1/2 & 1/2 \\ -1/2 & 1/2 \end{bmatrix} \quad (14)$$

Parameters of the moving sub-system

At all stages of the computation the "space" matrices [P] and [Q] hold only the information of the sub-mesh system. Therefore the mesh system is generated only for the sub-aquifer. Two parameters define totally the configuration of the sub-aquifer mesh (Figure 9), namely the vertical grid line spacing (ΔL) which is kept constant all along the length of the aquifer and the number of vertical grid lines (NSUBV) in a sub-aquifer. The number of nodes (NW) on a vertical grid line is estimated using a user supplied average total aquifer width and the parameter ΔL. The NW value used will make the distance between nodes on a vertical, approximately equal to ΔL.

An empirical relationship which guarantees stability of solution was obtained using computer runs made for a homogeneous square aquifer whose nodes were spaced equally at a

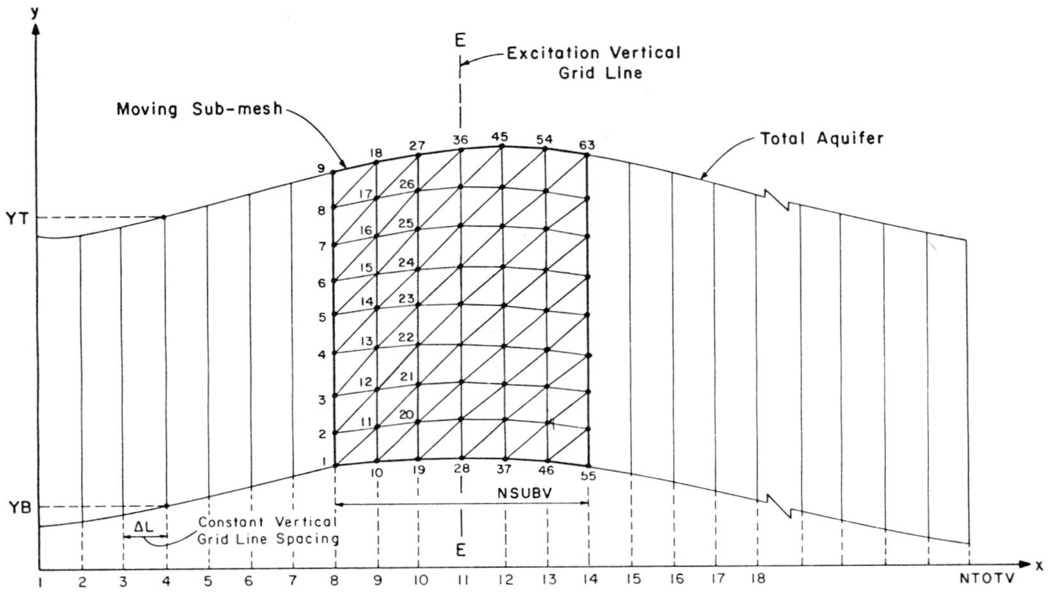

Figure 9 Finite element mesh system.

distance ΔL. The routine which solves the finite element equations was used for different combinations of T/ϕ, ΔL and time increments Δt and it was found that the following condition should be satisfied for stability of the numerical solution.

$$\frac{\phi}{T} \frac{(\Delta L)^2}{\Delta t} \leq 2.0 \tag{15}$$

Using the extreme case of $\Delta t = 0.5$ used during the first few periods,

$$\Delta L \leq \sqrt{\frac{T}{\phi}} \tag{16}$$

From user supplied values of average transmissivity and specific yield for the total aquifer, ΔL can be estimated from Equation (16). Using the same values of T and ϕ and a user supplied average aquifer width the analytical solution for a homogeneous aquifer is used to determine the number of grid spaces away from the excitation point at which the maximum value of the "discrete kernel" coefficients become insignificant. The above information is used to define the length of the sub-aquifer (NSUBV).

Finite element mesh generator

With a knowledge of the vertical grid line spacing ΔL the user can fit a vertical grid line system onto the total aquifer as shown in Figure 9. By superimposing the vertical grid line system onto the appropriate map the user prepares for each grid line vertical grid line information cards with the following information:

(a) the y-coordinates of the points at the intersection of the bottom and top boundaries of the aquifer. (YB and YT as shown on Figure 9),

(b) the contour values and the y-coordinates of the points of intersection of the contours of equal transmissivity,

(c) same as (b) for contours of equal specific yield.

Using the fixed number of nodes on a vertical (NW) the program distributes nodes at equal spaces on the vertical grid line between aquifer boundaries and the nodal values of transmissivity and specific yield are interpolated using the information given in (b) and (c). Each time the sub-aquifer moves a new vertical grid line information card is read and information on the leftmost grid line is dropped to minimize the computer storage requirements.

Finite element solution routine

Using the node numbering scheme shown on Figure 9, the symmetric "space" matrices [P] and [Q] becomes banded with a band width of NW + 2, where NW is the number of nodes on a vertical grid line. In addition to being banded the matrices are systematically sparse. That is, irrespective of the NW used, out of the NW+2 columns of the band all the columns have zero entries except columns 1, 2, NW+1 and NW+2 (Figure 10). This property makes it possible to save storage

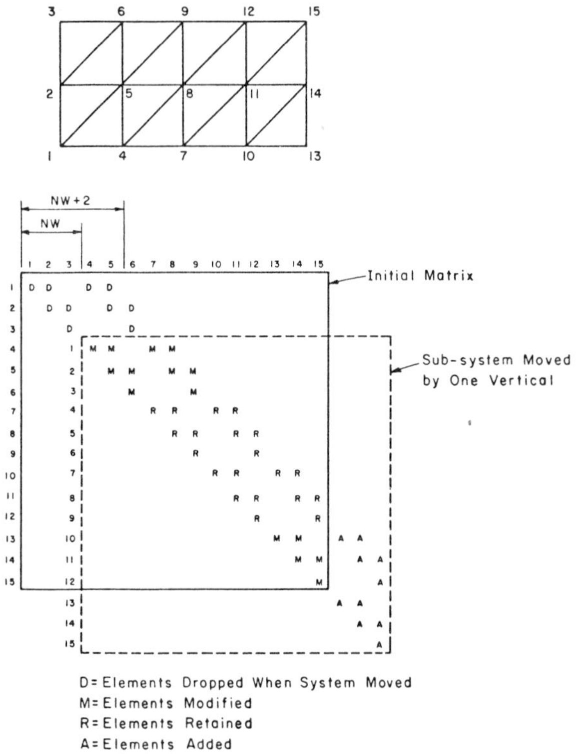

Figure 10 Structure of "space" matrices [P] and [Q]

as only these four columns have to be stored. In solving the system of equations given by Equation (3), the elements of matrix [A] are generated using Equation (4). Each time a node is excited the vector {R} is generated using Equation (9). The system of equations is solved using the Gauss-Seidel iterative scheme.

Once all the nodes on the excitation vertical grid line have been excited the sub-system is moved by one vertical. For the new sub-system the "space" matrices [P] and [Q] are modified by dropping the matrix elements corresponding to the set of nodes dropped and adding elements of the nodes of the added vertical. The elements to be dropped, modified, retained from the original matrix and added are shown in Figure 10 for an example problem.

The "discrete kernel" values generated are printed out (or stored on magnetic tape) at the end of each excitation.

RESULTS

The program developed was used to generate the discrete kernel coefficients for a homogeneous aquifer for which an analytical solution exists. A square aquifer of size 350 meters with no-flow boundary conditions and uniformly spaced nodes 350 meters apart was excited at the central node e (Figure 11). The analytical solutions and the program

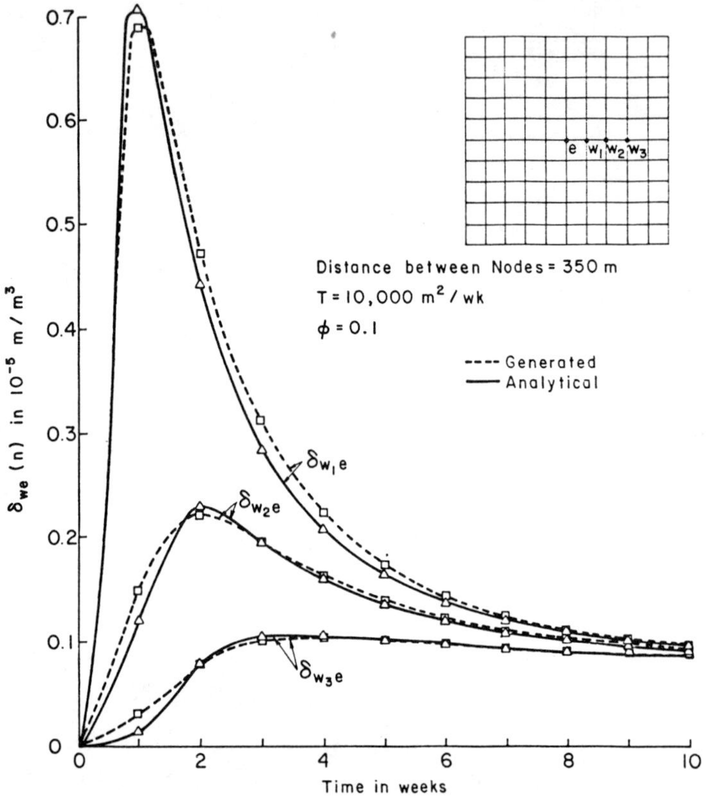

Figure 11 Comparison of program generated coefficients and analytical solutions - Case 1.

generated response functions $\delta_{we}(n)$ observed at the nodes w_1, w_2 and w_3 are compared on Figure 10. The comparison was done for a range of values of specific yield and the results are shown on Figures 12 and 13. The comparison shows that the analytical solutions and the program generated values are in very good agreement for a range of values of T/ϕ.

To demonstrate the steps involved in applying the model to real stream-aquifer systems, a sample reach was selected from the South Platte River (Figure 14).

Step 1: Using maps of aquifer geometry, transmissivity

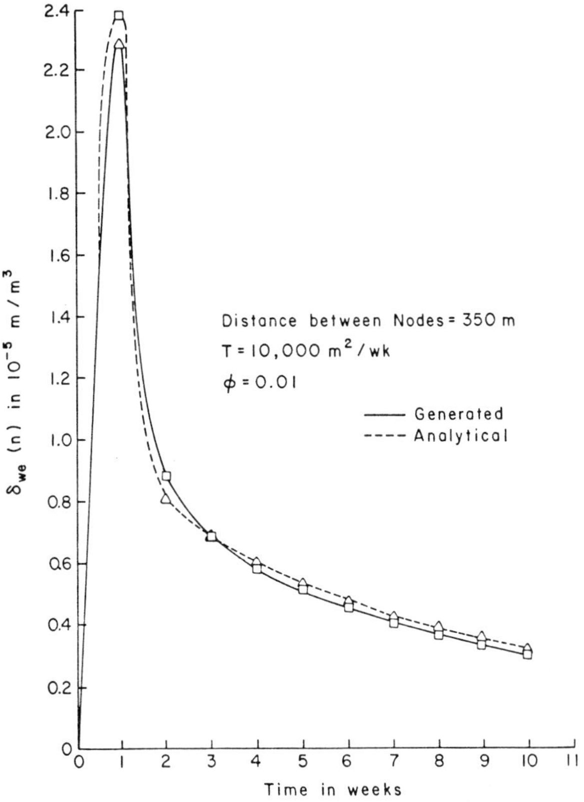

Figure 12 Comparison of program generated coefficients and analytical solutions - Case 2.

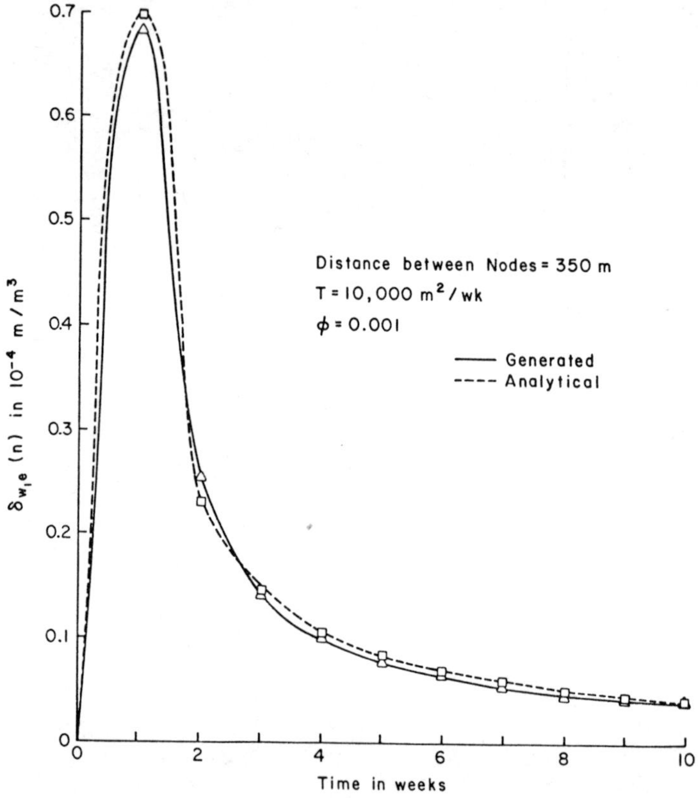

Figure 13 Comparison of program generated coefficients and analytical solutions - Case 3.

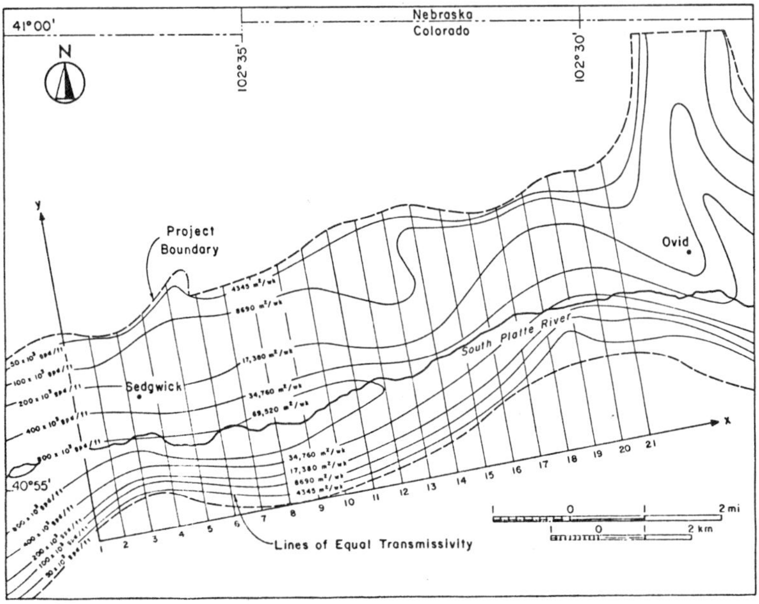

Figure 14 Sample reach from South Platte River.

and specific yield, estimate average values of aquifer width, transmissivity and specific yield.

average width = 2600 meters

$T = 24{,}600\ m^2/wk$

$\phi = 0.2$

From the relationship (16), using the average values of T and ϕ, the vertical grid line spacing ΔL is estimated.

$\Delta L = 350$ meters

Step 2: Using the average values of aquifer width, T, ϕ and the estimated ΔL the program computes the moving-aquifer parameters.

length of the sub-system = 16 grid line spaces

number of nodes on a vertical = 9.

Step 3: The user defines the x,y axes (Figure 14) and fits a vertical grid line system spaced at 350 meters. The vertical grid line information cards are prepared for the 21 grid lines to be used as the inputs to the "discrete kernel generator" program.

CONCLUSION

A cost efficient method which has the potential to be used effectively in taking day-by-day short term management decisions in stream-aquifer management problems was presented. The user oriented "kernel generator" program discussed simplifies the application of the model to real field problems. The agreement of the model generated values with analytical solutions shows that the model can be used to generate the "discrete kernels" to a reasonably good accuracy for different ranges of values of transmissivity and specific yield.

REFERENCES

Morel-Seytoux, H. J., R. A. Young and G. E. Radosevich (1973) Systematic Design of Legal Regulations for Optimal Surface-Groundwater Usage. OWRR Completion Rep. Ser. 53, Environ. Resour. Cntr., Colo. State Univ., Fort Collins, Colo., 81 p.

Morel-Seytoux, H. J. (1975) Optimal Legal Conjunctive Operation of Surface and Ground Waters. Paper presented at 2nd World Congress on Water Resources, New Delhi, India.

Morel-Seytoux, H. J. (1975) A Combined Model of Water Table and River Stage Evolution. Water Resour. Res., Vol. 11, No. 6, 968-972.

Morel-Seytoux, H. J. and C. J. Daly (1975) A Discrete Kernel Generator for Stream-Aquifer Studies. Water Resour. Res., Vol. 11, No. 2, 253-260.

Gray, W. G. and G. F. Pinder (1974) Galerkin Approximation of the Time Derivative in the Finite Element Analysis of Groundwater Flow. Water Resour. Res., Vol. 10, No. 4, 821-828.

Rodriguez-Amaya, C. (1976) A Decomposed Aquifer Model Suitable for Management. Ph.D. dissertation, Colorado State University, Fort Collins, Colorado.

ACKNOWLEDGMENTS

The work upon which this paper is based was supported in part by funds provided by the U.S. Department of Interior, Office of Water Research and Technology, as authorized under the Water Resources Research Act of 1964, and pursuant to Grant Agreement No. 14-34-0001-6006. The OWRT support is gratefully acknowledged.

The authors also wish to express their appreciation for many fruitful discussions with Mr. Charles Daly, Research Associate, Dept. of Civil Engineering, Colorado State University, and Greg Peters, Exxon Production Research Co., Houston, Texas.

FINITE ELEMENT DESCRIPTION OF FLOWFIELD IN GROUNDWATER MANAGEMENT MODELS

K. Elango, H. Suresh Rao

Indian Institute of Technology, Madras, India

INTRODUCTION

One way of classifying the mathematical models developed for water resources systems analysis, as suggested by Meta Systems Inc. (1976), is to identify whether the model is 'descriptive' or 'prescriptive' in nature. A perusal of the literature indicates that the use of the finite element method in groundwater problems has been limited to descriptive models. This paper presents the results of a preliminary study on the performance of a prescriptive model, incorporating the finite element description of groundwater flow within a mathematical programming framework. The feasibility of such a model was first suggested by Aguado and Remson (1974).

The efficient management of the piezometric surface of a confined aquifer under steady-state condition is the problem chosen for study. The situation is simple enough to possess an analytical solution for the flow field, so that the results of the prescriptive finite element model can be compared with those of the prescriptive analytical model. Secondly the solution of the state of the piezometric surface obtained through the simplex algorithm is compared with that obtained through the use of a descriptive finite element model and one based on analytical solution.

RELATED STUDIES

The review herein is limited to linear models. Aguado and Remson (1974) presented the method of including groundwater variables as decision variables in linear programming (LP) management models, with the finite difference approximations of the governing differential equation as constraints in the LP formulation. They concluded that any other method of approximation of the differential equation that yields a set of algebraic linear equations could be used in place of the finite difference approximations. They cited the finite element method as an example. Aguado et al (1974) reported the utility of a LP model with the finite difference scheme describing groundwater hydraulics in the study of a major dewatering programme to facilitate construction of a dock. Suresh Rao et al (1975) indicated that a LP model based on closed form (i.e. analytical) solution for the groundwater flow field was useful in the study of aquifer-depressurization through a system of wells in a large lignite mine area. The details of this model were presented by Elango et al (1976). Schwarz (1976) discussed the possibility of representing aquifer hydraulics in the form of either an influence matrix or a transformation model and imbedding the same in a LP formulation. The influence matrix in effect implies the use of analytical solutions and the transformation model is a multicell aquifer model using finite difference scheme. A hypothetical case of management of quality of groundwater was formulated by Willis (1976) as a mixed integer programming problem, with the flow field described by finite difference method. Futagami et al (1976) presented a LP model for pollution control in a water body and the governing 'field' equations were handled by the finite element method. Elango and Suresh Rao (1977) presented a unified view of these various attempts at the incorporation of 'field' equations as constraints in mathematical programming models.

FORMULATION OF THE PROBLEM

The problem-situation chosen for this study is indicated in Figure 1. A confined aquifer, circular in shape with radius, R_o, has a constant value of transmissibility, T, throughout. The boundary is maintained at a constant potential of H_o. The potential surface inside the region is to be properly controlled to facilitate some engineering activity, by pumping through wells in an efficient manner. Let it be necessary that at the two points P and R certain specified levels of depressurization be achieved under steady-state condition. These points are denoted by the index values j = 1 and 5 respectively. Pumping is possible at three locations A, B and C, denoted by i = 3, 7 and 14 respectively. The problem is to determine the minimum total pumpage and its allocation between the wells, that makes the potentials at P and R less than or equal to $\bar{h}_1$ and $\bar{h}_5$ respectively. There exists restriction on pump capacity at each well; that is, pumpages from A, B and C should be less than or equal to $\bar{Q}_3$, $\bar{Q}_7$ and $\bar{Q}_{14}$ respectively.

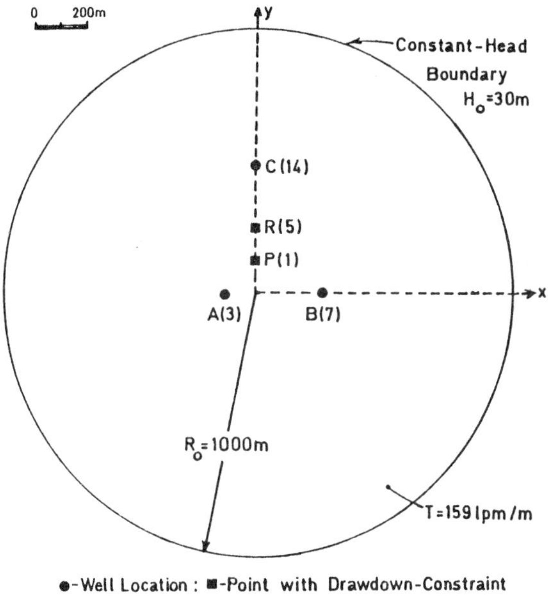

FIGURE 1 PLAN VIEW OF AQUIFER SYSTEM

Prescriptive Model based on Analytical Solution

For the situation described above, the hydraulic condition can be described by an analytical solution for the potential surface, as given by Polubarinova-Kochina (1962). The drawdown, s_{ij}, at a point with coordinates (x_j, y_j) due to a pumpage, Q_i, at the well at point (x_i, y_i) is given by

$$s_{ij} = \frac{Q_i}{2\pi T} \ln \frac{\{(R^2 - x_i x_j - y_i y_j)^2 + (x_i y_j - y_i x_j)^2\}^{0,5}}{R\{(x_i - x_j)^2 + (y_i - y_j)^2\}^{0,5}} \qquad (1)$$

The specific drawdown β_{ij} due to unit pumpage is given by

$$\beta_{ij} = \frac{s_{ij}}{Q_i} \tag{2}$$

The 'principle of superposition' can be adopted to obtain the resultant drawdown due to more wells at other points. Thus the resultant potential at point j, h_j is

$$h_j = H_o - \sum_i Q_i \beta_{ij} \tag{3}$$

The management problem can now be formulated as a LP problem, in terms of the specific response function, as follows:

(It may be mentioned that the following formulation is purposely made specific to the problem considered and the index notation is so chosen to be common for both the analytical and finite element models.)

$$\text{Minimize } Q_T = \sum_i Q_i \quad , \quad i = 3, 7 \text{ and } 14 \tag{4}$$

subject to
(a) the 'field' equations expressed as

$$h_j + \sum_i \beta_{ij} Q_i = H_o \quad , \quad i = 3, 7 \text{ and } 14 \text{ and } j = 1 \text{ and } 5 \tag{5}$$

(b) the management aspects

(1) the required depressurization should be achieved;

$$h_j \leq \bar{h}_j \quad , \quad j = 1 \text{ and } 5 \tag{6}$$

(11) the pumpage limits should not be exceeded;

$$Q_i \leq \bar{Q}_i \quad , \quad i = 3, 7 \text{ and } 14 \tag{7}$$

(111) the non-negativity requirements for LP formulation;

$$Q_i \geq 0 \quad , \quad i = 3, 7 \text{ and } 14 \tag{8a}$$

and

$$h_j \geq 0 \quad , \quad j = 1 \text{ and } 5 \tag{8b}$$

The computer programm written to solve this version is refered as LPCF in subsequent discussions.

Prescriptive Model based on Finite Element Method
The same problem is now modelled using the finite element description of the flow field, as shown in Figure 2. The governing differential equation is

$$T \nabla^2 h + Q \, \delta(x_i, y_i) = 0 \qquad (9a)$$

with the boundary condition as

$$h = H_o \quad , \quad \text{along the outer boundary.} \qquad (9b)$$

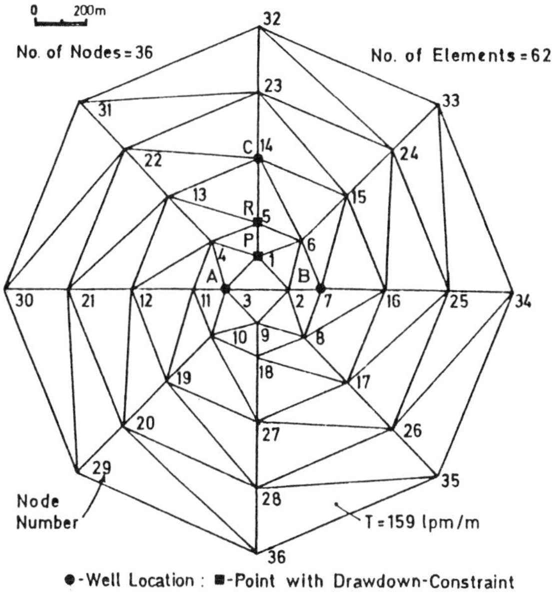

FIGURE 2 FINITE ELEMENT DISCRETIZATION OF FLOW FIELD

The field is discretized into simple triangles, with nodes only at the vertices. The node configuration is chosen as usual, to include points of specific interest, namely wells and points of prescribed minimum depressurization. Linear interpolation functions are adopted. The handling of concentrated pumpage at the wells is done in the usual manner, as explained for instance by Pinder and Gray (1977). The node numbering is so done that nodes 1 to 28 are internal nodes and 29 to 36 are boundary nodes.

The essential difference in this version is that the field equations now take a form different from that of Equation-set 5. The corresponding finite element equations are

$$\sum_j a_{ij} h_j + \Delta_i Q_i = b_i \quad , \quad j = 1,\ldots,28 \qquad (10)$$
$$i = 1,\ldots,28$$

In the present case $\Delta_i = 1 \quad , \quad i = 3,\ 7$ and 14 (11)

$= 0$ other $1 \leq i \leq 28$

The coefficients a_{ij} and b_j are obtained through the following steps:

(1) Evaluation of the matrices of the elements

(11) Formulation of the global matrix

(111) Introduction of the boundary condition and determination of the reduced form of the matrix and revised values of the right hand side vector.

It should noted at this stage, that the quantity Q_i in the second term in the Equation-set 10 is as such an unknown in the prescriptive model and has to be obtained as part of the solution of the LP problem, that simultaneously satisfies these field equations. The finite element program segment to evaluate a_{ij} and b_j requires the following as input data: number of nodes, number of elements, node-connectivity matrix, the x- and y-coordinates of each node, number of nodes with specified boundary values, their node numbers and the corresponding specified value of potential and the program is called FESEG.

With the evaluation of a_{ij} and b_j, the following LP problem can be enunciated:

Minimize $\quad Q_T = \sum_i Q_i \qquad i=3,\ 7$ and 14 (4)

subject to $\quad \sum_j a_{ij} h_j + \Delta_i Q_i = b_i, j=1,\ldots,28$ (10)

$i=1,\ldots,28$

$h_j \leq \bar{h}_j, j=1$ and 5 (6)

$Q_i \leq \bar{Q}_i, i=3,\ 7$ and 14 (7)

$$Q_i \geq 0, \ i = 3, 7 \text{ and } 14 \qquad (8a)$$

$$\text{and} \quad h_j \geq 0, \ j = 1,...,28 \qquad (12)$$

The program to solve this version is referred as LPFE.

Check using Descriptive Models

In order to check the values of the potential at internal nodes as calculated by the simplex method, used for solving the LP models, two other computer programs of the descriptive type are used. The first one DCF calculates the cumulative drawdown at internal nodes (other than well-nodes) for specified values of pumpages, using Equation 1 for given values of T, R_o and nodal coordinates. The second program DFE is an extended version of FESEG. The additional segment reads-in the pumpages at specified nodes and solves the resulting set of simultaneous equations for h_j, through the gaussian elimination procedure.

COMPUTATIONAL ASPECTS

The linear programming problems are solved using the package-program MPOS (version 3.2), on a CYBER 175 digital computer. The procedure of solution adopted is called REGULAR. Cohen and Stein (1976) state that the algorithm used is the original two-phase simplex method. First negative slack variables are introduced. Positive slack variables are introduced next and finally, the artificial variables. In Phase I, the infeasibility form, namely the sum of the artificial variables, is driven to zero to produce a feasible solution. In Phase II, an optimal solution is sought.

The finite element program DFE is an adapted version of that given by Connor and Brebbia (1976) for the Laplace equation.

Values of parameters, used in the computations are: R_o = 1000 m; T = 159 litres/minute/metre width; H_o = 30 m; $\bar{h}_1$ = 24 m; $\bar{h}_5$ = 24·5m; $\bar{Q}_3$ = 3000 litres/min (i.e. lpm); $\bar{Q}_7$ = 4000 lpm and $\bar{Q}_{14}$ = 5000 lpm. The coordinates in metres of selected nodes like 1,3,5,7,14,16,25 and 34 are (0,125), (-125,o), (0,250), (250,0), (0,500), (500,0), (750,0) and (1000,0) respectively.

RESULTS AND DISCUSSION

The results consist of the minimum pumpage required, its optimal allocation between the wells and the actual potentials developed due to these pumpages at points with constraints on the potential values. The solution from LPFE in addition contains the potential values at other internal nodes. A comparison of potential surface according to the LP solution with those obtained through direct evaluation by DCF and DFE is then presented. Certain general observations are also mentioned.

Optimal Pumpage

Table 1 indicates the optimal pumpages at points A, B and C according to LPCF and LPFE and the corresponding value of minimum total pumpage.

Table 1: Details of Optimal Pumpage

Location of Well	Pumpage in lpm according to	
	LPCF	LPFE
A	3000	2588
B	0	0
C	1337	2062
Total Pumpage	4337	4650

The choice of the wells to be pumped is seen to be similar in both cases. Largest rate is at A (node 3) and the well at C (node 14) is pumped at a lower rate. Both versions indicate that depressurization capacity of well at B (node 7) is the least, so it need not be pumped at all as the pumping of the wells at A and C needed to effect sufficient depressurization is within the individual limits. However the optimal values are seen to differ for each well as also the gross pumpage. The minimum pumpage according to the analytical solution based model is less than that indicated by the finite element based model by 313 lpm, a difference of about 6·7 %. This is considered to be due to the inherent approximations in the finite element method. These include the fineness of discretization, simplified assumptions of the variation of the potential over the element etc.

Thus a different node configuration may not lead to the same set of values as obtained herein, when the discretization is not fine enough. This convergence characteristic observed in descriptive finite element models thus appears in the prescriptive models also.

Values of Potential

Table 2 presents the values of potential at the internal nodes computed by different procedures. LPCF yields the values at nodes 1 and 5. DCFCF refers to results, corresponding to pumpages of LPCF based on DCF. Similarly DCFFE results are based on DCF using pumpages from LPFE. Here the drawdowns at the pumped-well locations (nodes 3 and 14) are not evaluated, as the form of the analytical solution is such that the results are very sensitive to the assumed value for the radius of the well. Based on the optimal pumpages of LPCF and LPFE, the program DFE yields, two sets of values at all the internal nodes. The former is denoted by DFECF and the latter by DFEFE.

The following are observed when comparing the results of the finite element model with the analytical model. In the case of the LPCF solution the constraint on potential only at the node 5 is a binding one. However in the case of LPFE, both constraints at the nodes 1 and 5 are binding constraints. Comparison of DCFCF with DCFFE and DFECF with DFEFE indicates that in general that the potentials based on the analytical model are slightly lower than those based on the finite element model. Evaluations by DFE indicate that the maximum deviations occur at the pumped-well nodes, because of the differences in the pumpages given by the two models. For instance, DFECF yields potentials of 19·18 m and 25·13 m at the nodes 3 and 14, as against 20·10 m and 23·76 m by DFEFE. Similar differences are reflected in the values, at nodes, surrounding the pumped well nodes. The number of nodes for which the difference is less than $\pm$ 0·2 m is 20 for DFE and 19 for DCF.

Comparison of the results of LPCF with DFECF and those of LPFE with DFECF indicates slight differences between the nodal values evaluated by the simplex algorithm and those by the gaussian elimination procedure. LPCF indicates 23·58 m against 23·63 by DFECF at the node 1 for the same values of pumpage. It is 24·50 m against 24·76 m for the node 5. The same trend of lower values from the descriptive model is noticed while comparing results from LPFE and DFEFE. At the nodes 1 and 5, the values from

Table 2: Values of Potential at Nodes

Node Number	Potential in metres according to					
	LPCF	LPFE	DCFCF	DCFFE	DFECF	DFEFE
1	23.58	24.00	23.58	23.63	23.78	23.85
2	-	25.55	24.91	25.01	25.32	25.40
3	-	20.24	-	-	19.18	20.10
4	-	24.51	23.79	23.82	24.35	24.38
5	24.50	24.50	24.50	24.19	24.76	24.37
6	-	25.84	25.57	25.36	25.88	25.70
7	-	26.53	25.94	26.00	26.42	26.41
8	-	26.63	26.19	26.32	26.40	26.50
9	-	24.74	24.09	24.42	24.26	24.59
10	-	25.30	24.41	24.77	24.87	25.18
11	-	24.22	23.08	23.50	23.77	24.12
12	-	27.05	26.79	26.91	26.92	27.01
13	-	26.58	26.52	26.30	26.71	26.48
14	-	23.85	-	-	25.13	23.76
15	-	27.18	27.29	26.97	27.40	27.07
16	-	28.03	27.95	27.91	28.01	27.95
17	-	28.19	27.99	28.05	28.08	28.12
18	-	26.18	25.63	25.87	25.86	26.07
19	-	27.58	27.22	27.38	27.39	27.53
20	-	28.99	28.88	28.96	28.92	28.97
21	-	28.79	28.75	28.82	28.72	28.75
22	-	28.52	28.57	28.47	28.60	28.48
23	-	27.65	27.95	27.40	28.09	27.59
24	-	28.76	28.87	28.73	28.87	28.71
25	-	29.17	29.19	29.21	29.15	29.11
26	-	29.24	28.18	29.21	29.19	29.20
27	-	28.00	27.71	27.82	27.83	27.93
28	-	29.18	29.05	29.09	29.09	29.13

LPFE are 24·00 m and 24·50 m against 23·85 m and 24·37 m from DFEFE. This is so for almost all the nodes. However the difference is well within 0·2 m for most of the nodes.

Some Observations

In the case of descriptive finite element models, the existence of symmetry and bandedness of the global matrix makes possible substantial reduction in computations. However such an advantage is not apparent for prescriptive models, with the usual optimization algorithms.

The finite element based prescriptive models are more versatile in handling heterogeneity, complicated boundary geometry and conditions. But the possible variability in the solutions due to the freedom in the choices of nodal configuration, shape of elements and interpolation functions have to be kept in mind. The knowledge gained on these aspects from the performance of descriptive models may be usuful in studying their influence in prescriptive models.

In the application of this type of management model to large size problems, an interface program to link the DFESG with the LP package would be useful in considerably reducing the data preparation and handling.

The effect of relaxing the strict equality requirement of the finite element equations into inequalities of the form

$$b_i - \varepsilon \leq \sum_j a_{ij} \cdot h_j + \Delta_i Q_i \leq b_i + \varepsilon \qquad (13)$$

has been studied as a function of the value of ε. This is with a view to make possible computationally favourable reductions in the density of the constraint matrix by forcing to zero, all coefficients with magnitude less than prescribed value. The possible utility and implications of this modification are being further studied.

SUMMARY

Comparison of the prescriptive finite element model with the analytical model for the simple groundwater management problem in this preliminary study indicates that its performance is satisfactory. Some aspects that need further systematic studies to improve the accuracy, efficiency and capacity in dealing with large size problems by such a model have been identified.

ACKNOWLEDGEMENT

The first author thanks o.Prof. Dr.Ing. G. Rouvé and the staff at the Institut für Wasserbau und Wasserwirtschaft, RWTH, Aachen, for their kind help in the preparation of this paper. He thanks also the Alexander von Humboldt-Foundation for making possible his stay in West-Germany, where most of the work reported herein has been carried out.

REFERENCES

Aguado, E., and Remson, I. (1974) Ground-Water Hydraulics in Aquifer Management. Proc. ASCE, 100, No. HY1

Aguado, E., Remson, I., Fikul, M.F. and Thomas, W.A. (1974) Optimal Pumping for Aquifer Dewatering. Proc. ASCE, 100, No.HY7

Cohen, C., and Stein, J. (1976) Multi Purpose Optimization System-User's Manual No.320, Computing Centre, Northwestern University, Illinois

Connor, J.J., and Brebbia, C.A. (1976) Finite Element Techniques for Fluid Flow. Newnes-Butterworths, London

Elango, K., Jothishankar, N.J., Suresh Rao, H., and Sethuraman, V. (1976) Optimal Pumping for Control of Groundwater Pressure in an Open-cast Mine. Proc. Symp. on Water Resources and Fossil Fuel Production, Düsseldorf

Elango, K., and Suresh Rao, H. (1977) Discussion on FEM Coupled with LP for Water Pollution Control. Prod. ASCE 103, No.HY6

Futagami, T., Tamai, N., and Yatsuzuka, M. (1976) FEM Coupled with LP for Water Pollution Control. Proc. ASCE 102, No. HY7

Meta Systems Inc. (1975) Systems Analysis in Water Resources Planning. Water Information Centre Inc., New York

Pinder, G.F., and Gray, W.G. (1977) Finite Element Simulation in Surface and Subsurface Hydrology, Academic Press

Polubarinova-Kochina (1962) Theory of Groundwater Movement. Princetown University Press

Schwarz, J. (1976) Linear Models for Groundwater Management. Jl.of Hydrology, 28, 2/4

Suresh Rao, H., Jothishankar, N.J., Elango, K. and Sethuraman, V. (1975) Discussion on Optimal Pumping for Aquifer Dewatering. Proc. ASCE 101, No.HY7

Willis, R. (1976) Optimal Groundwater Quality Management: Well Injection of Waste Waters. Water Resources Research, 12, 1

TWO TECHNIQUES ASSOCIATED WITH THE GALERKIN METHOD FOR SOLVING GROUNDWATER FLOW PROBLEMS

S.P. Kjaran, S.T. Sigurdsson

National Energy Authority, Reykjavík, Iceland

INTRODUCTION

The linear parabolic differential equation describing unsteady groundwater flow in an anisotropic, nonhomogenous aquifer has been solved by various authors using the finite element Galerkin method for the space variables. This paper deals with two techniques connected with the implementation of this method.

The timederivative can successfully be dealt with using a mode superposition technique instead of a finite difference or Galerkin method. The technique involves computing the major eigenvalues and eigenvectors of the matrices arising from the Galerkin method. This can be done effectively with the subspace iteration algorithm which has been developed for stress-strain problems. An advantage of this technique in the present context is that the eigenvalues can be associated with the time constants of linear reservoirs. From a hydrological standpoint it is in turn useful to interpret the aquifer system as a system of parallel linear reservoirs. The inflow to each reservoir is determined from the eigenvectors. Whereas the aquifer system can be shown to be equivalent to infinitely many parallel reservoirs with definite time constants it is in practice sufficient to know the constants of the 1-10 most delayed reservoirs and treat the remaining inflow without time-delay. A further computational advantage is that the groundwater level can be recalculated by a simple convolution if the input changes.

It may sometimes be of equal importance to obtain good approximations to flow values across external or internal boundaries as to actual groundwater levels. If these flow values are obtained directly from the calculated groundwater gradients continuity with inflow will in general not be retained. It is, however, possible to make use of more accurate difference approximations to normal gradients, that are inherent to the Galerkin method, to overcome this problem and obtain more reliable flow values. A similar technique is known to be successful in electro-magnetic field problems.

BASIC THEORY

In this section we present the basic equation for groundwater flow which we are concerned with in this paper and develop, in particular, the analogy between the aquifer system and system of parallel linear reservoirs.

The differential equation governing the groundwaterlevel, h, in an anisotropic, nonhomogeneous aquifer is given by:

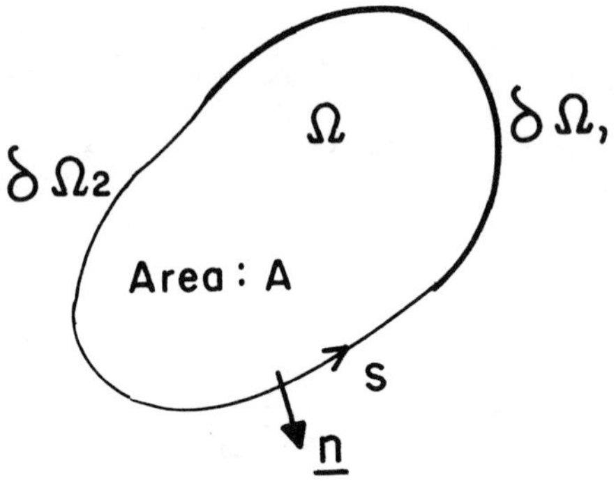

Figure 1

$$S\frac{\partial h}{\partial t} = Lh + R(x,y,t) \text{ in} \Omega \quad (1)$$

$$h = g \quad \text{on} \quad \partial\Omega_1 \quad (2)$$

$$-Nh = f \quad \text{on} \quad \partial\Omega_2$$

$$h(x,y,t_o) = \hat{h}\,(x,y) \quad \text{in} \quad \Omega \quad (3)$$

where the operators L and N are defined as:

$$L = \frac{\partial}{\partial x}\left(T_{xx}\frac{\partial}{\partial x} + T_{xy}\frac{\partial}{\partial y}\right) + \frac{\partial}{\partial y}\left(T_{yy}\frac{\partial}{\partial y} + T_{yx}\frac{\partial}{\partial x}\right) \quad (4)$$

$$N=\underline{n}\cdot(T_{xx}\frac{\partial}{\partial x}+T_{xy}\frac{\partial}{\partial y},T_{yy}\frac{\partial}{\partial y}+T_{yx}\frac{\partial}{\partial x}) \qquad (5)$$

In the isotropic case equation 4 and equation 5 reduce to

$$L = \nabla \cdot (T\ \nabla) \qquad (6)$$

$$N = T\frac{\partial}{\partial n} \qquad (7)$$

R is some source function (infiltration, pumping, recharge) and g and f are given timeindependent functions. S is the storage coefficient and T the transmissivity. It is convenient to divide the groundwaterlevel into a stationary and a transient part:

$$h(x,y,t) = h_1(x,y,t) + h_o(x,y) \qquad (8)$$

where the stationary level satisfies the following equation:

$$L\ h_o = -\ \overline{R}(x,y) \quad \text{in } \Omega \qquad (9)$$

$$h_o = g \quad \text{on } \partial\Omega_1 \qquad (10)$$

$$-\ Nh_o = f \quad \text{on } \partial\Omega_2$$

$\overline{R}$ is the mean source level. The transient part must then satisfy:

$$S\frac{\partial h_1}{\partial t} = L\ h_1 + (R-\overline{R}) \quad \text{in } \Omega \qquad (11)$$

$$h_1 = o \quad \text{on } \partial\Omega_1$$

$$-\ Nh_1 = o \quad \text{on } \partial\Omega_2 \qquad (12)$$

$$h_1(x,y,t_o) = \hat{h}(x,y) - h_o(x,y) \quad \text{in } \Omega \qquad (13)$$

We now define the eigenproblem:

$$L\ \phi_n + (\frac{\lambda_n T_o}{AS_o})\ S\ \phi_n = 0 \quad \text{in } \Omega$$

$$\phi_n = 0 \quad \text{on } \partial\Omega_1 \qquad (14)$$

$$-\ N\phi_n = 0 \quad \text{on } \partial\Omega_2$$

We have normalized the eigenvalues with respect to the area A, and some reference transmissivity, T_o, and storage coefficient, S_o. Now using the eigenfunction expansion we get:

$$h_1(x,y,t) = \sum_{n=1}^{\infty} a_n(t)\,\phi_n(x,y) \tag{15}$$

$$R(x,y,t) - \overline{R}(x,y) = \sum_{n=1}^{\infty} b_n(t)\,\phi_n(x,y)\,S(x,y) \tag{16}$$

where $b_n(t)$ is given by:

$$b_n(t) = \int_{\Omega} (R(x,y,t)-\overline{R}(x,y))\,\phi_n(x,y)\,dx\,dy \tag{17}$$

Similar eigenfunction expansion is used for h_1 in equation 13.

Inserting equation 15, equation 16 and equation 17 in the differentialequation for h_1 and solving for the unknowns $a_n(t)$ we get:

$$h_1(x,y,t)=\sum_{n=1}^{\infty} \phi_n(x,y) \int (f(\zeta,\eta)-h_o(\zeta,\eta))\,d\zeta\,d\eta$$

$$e^{-(t-t_o)/K_n} + \sum_{n=1}^{\infty} \phi_n(x,y) \int \phi_n(\zeta,\eta)$$

$$\int_{\tau_o}^{t} (R(\zeta,\eta,\tau)-\overline{R}(\zeta,\eta))\,e^{-(t-\tau)/K_n}\,d\zeta\,d\eta \tag{18}$$

Now taking $t_o = -\infty$ we get:

$$h_1(x,y,t) = \sum_{n=1}^{\infty} \phi_n(x,y) \int_{\Omega} \phi_n(\zeta,\eta)$$
$$\int_0^{\infty} e^{-\tau/K_n} (R(\zeta,\eta,t-\tau)-\overline{R}(\zeta,\eta))\,d\tau)\,d\zeta\,d\eta \tag{19}$$

where K_n is defined as: $K_n = \dfrac{AS_o}{\lambda_n T_o}$ (20)

We now proceed to calculate the boundary outflow, which is given by:

$$q(t) = - \int_{\partial\Omega} Nh \, ds \tag{21}$$

By using Gauss' theorem we get:

$$q(t) = \sum_{n=1}^{\infty} \frac{1}{K_n} \int_{\Omega} \phi_n dxdy \int_{\Omega} \left\{ \phi_n(\zeta,\eta) \int_0^{\infty} e^{-\tau/K_n} R(\zeta,\eta,t-\tau)\, d\tau \right\} d\zeta d\eta \tag{22}$$

By differentiating the n'th term in equation 22 we find:

$$q_n + K_n \frac{dq_n}{dt} = \int_{\Omega} \phi_n dxdy \int_{\Omega} \phi_n(\zeta,\eta)\, R\,(\zeta,\eta,t)\, d\zeta d\eta$$

$$q(t) = \sum_{n=1}^{\infty} q_n(t) \tag{23}$$

This is a linear, ordinary differential equation, and is the differential equation for a linear reservoir with timeconstants K_n. The boundary outflow is therefore the outflow from a system of infinitely many parallel, simple linear reservoirs as illustrated in figure 2.

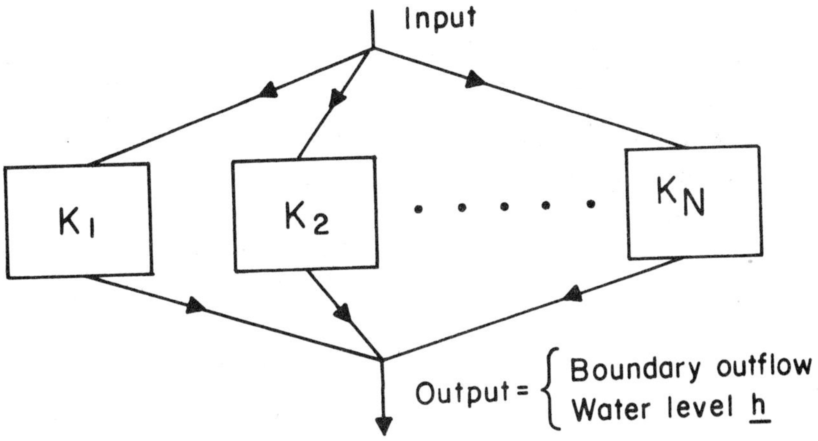

Figure 2

The eigenvalues have the following properties:

$$0 \leq \lambda_1 \leq \lambda_2 \leq \cdots\cdots\cdots \leq \lambda_n \cdots$$

and $\lambda_n \to \infty$, when $n \to \infty$. We, therefore, have that $K_n \to o$, when $n \to \infty$. The higher order inflow terms are therefor delayed very little in time. If the source term is not very rich in the higher harmonics, that is to say it does not fluctuate very much over the area, just a few of the inflow terms give any significant value to the total inflow. In that case just the few first terms in the eigenfunction expansion need to be evaluated.

THE GALERKIN FINITE ELEMENT METHOD

The Galerkin finite element method has been used by many authors, in one form or another, for solving groundwater problems, see f.ex. Zienkiewicz (1971), Neuman and Witherspoon (1970), Pinder and Frind (1972) and Pinder and Gray (1974).

In this section we, therefore, restrict ourselves to a brief development of the method in the form that is relevant for the special techniques to be presented in the following sections.

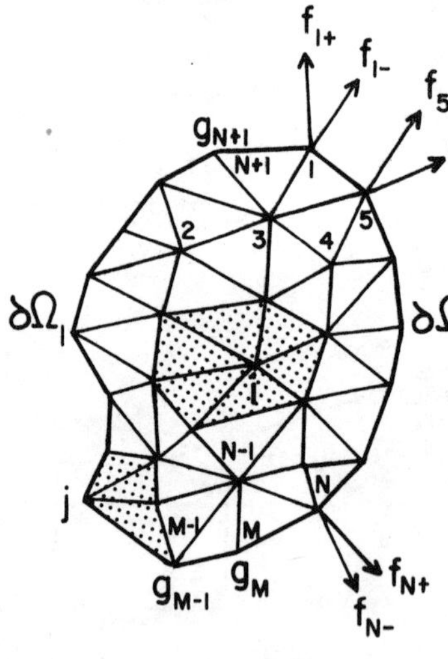

Figure 3

The region Ω is subdivided into triangular elements (see figure 3) and the nodal points are numbered, starting with points in Ω and on $\partial\Omega_2$ (assume the total number of these to be N), and finishing with points on $\partial\Omega_1$ (assume the total number of these to be M-N). We associate with nodal point i the piecewise linear pyramid function $\psi_i(x,y)$, which takes the value 1 at the point and the value 0 outside the adjacent triangles (the shaded region in figure 3). We then seek to obtain a continuous and piecewise linear approximate solution to the problem given by equation 1 to 5 in the form:

$$\bar{h}(x,y,t) = \sum_{i=1}^{N} h_i(t)\psi_i(x,y) + \sum_{i=N+1}^{M} g_i\psi_i(x,y) \qquad (24)$$

The coefficients $h_i(t)$ or g_i are the values of $\bar{h}$ at point i, the constant coefficients g_i are known from the prescribed boundary condition on $\delta\Omega_1$, whereas the timedependent coefficients $h_i(t)$ are to be determined. We further make the assumption that the second term on the RHS of equation 24 satisfies the prescribed flow conditions on $\partial\Omega_2$ and the first term the corresponding homogenous condition of zero flow.

Considering first the case of isotropic flow, the Galerkin conditions for determining $h_i(t)$ are that:

$$\int_{\Omega} \{S\frac{\partial \bar{h}}{\partial t} - \nabla(T\nabla\bar{h}) - R\}\psi_i dxdy = 0 \quad i = 1,\ldots,N \qquad (25)$$

However, since $\bar{h}$ is only piecewise once differentiable in x and y, these conditions have no proper meaning unless we integrate by parts using Green's theorem. By doing so, and including the assumption of the satisfaction of the flow conditions equation 25 can be rewritten as:

$$C\frac{d\underline{h}}{dt} = - B\underline{h} + \underline{b} \qquad (26)$$

where $\underline{h}(t) = \{h_i(t)\}_N$

$$C = \{C_{ij}\}_{NxN} = \{ \int_{\Omega} S\psi_j\psi_i dxdy\}_{NXN} \qquad (27)$$

$$B = \{b_{ij}\}_{NxN} = \{ \int_{\Omega} T\nabla\psi_j\nabla\psi_i \; dxdy\}_{NxN} \qquad (28)$$

$$\underline{b} = \{b_i\}_N = \{ \int_{\Omega} R\psi_i dxdy$$

$$- \int_{\Omega} T \sum_{j=N+1}^{M} g_j\nabla\psi_j)\nabla\psi_i \; dxdy - \oint_{\partial\Omega_2} f\psi_i ds\}_N \qquad (29)$$

Equation 26 represents a system of N linear ordinary differential equations that can be solved for given initial conditions $\underline{h}(o) = \underline{\hat{h}}$ (cf.equation 3).

For the steady state problem with $\frac{d\underline{h}}{dt} = 0$ and $\underline{b}(t) = \overline{\underline{b}}$ (cf. equation 9) it reduces to a system of N linear equations. In the evaluation of the integrals in equation 27-29 we use the approximations that S and T are constant within each triangular element and that R is continous in Ω and linear within each triangular element, taking prescribed values at the nodal points, referred to as R_i. We can also include in the infiltration term, R, Dirac delta functions, $-Q\delta(x,y;\zeta,\eta)$, corresponding to a pumping at the point (ζ,η). Finally, we use the approximation that f is linear between nodal points on $\partial\Omega_2$, taking on prescribed flow values at the nodal points, referred to as f_{i-} or f_{i+} depending on whether we are considering the flow value just before or just after the nodal point as we pass along $\partial\Omega_2$ anticlockwise (see figure 3).

The system in equation 26 is most readily obtained by assembling it from the contributions of each triangular element. These contributions can, in turn, be obtained by applying the Galerkin method to such an element and imposing arbitrary flow conditions round the whole boundary (see figure 4).

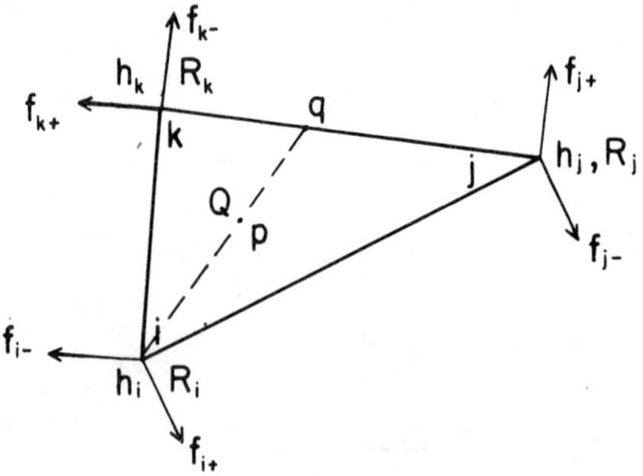

Figure 4

The resulting system corresponding to equation 26 is, if we assume that we have a pumping Q at point p as well as a continuous infiltration:

$$\frac{SA}{12}\begin{bmatrix}2&1&1\\1&2&1\\1&1&2\end{bmatrix}\frac{d}{dt}\begin{bmatrix}h_i\\h_j\\h_k\end{bmatrix}=$$

$$-\frac{T}{8A}\begin{bmatrix}2l_{jk}^2 & l_{ij}^2-l_{jk}^2-l_{ki}^2 & l_{ki}^2-l_{ij}^2-l_{jk}^2\\ l_{ij}^2-l_{jk}^2-l_{ki}^2 & 2l_{ki}^2 & l_{jk}^2-l_{ki}^2-l_{ij}^2\\ l_{ki}^2-l_{ij}^2-l_{jk}^2 & l_{jk}^2-l_{ki}^2-l_{ij}^2 & 2l_{ij}^2\end{bmatrix}\begin{bmatrix}h_i\\h_j\\h_k\end{bmatrix} \tag{30}$$

$$+\frac{A}{12}\begin{bmatrix}2&1&1\\1&2&1\\1&1&2\end{bmatrix}\begin{bmatrix}R_i\\R_j\\R_k\end{bmatrix}$$

$$-Q\begin{bmatrix}\alpha_i\\\alpha_j\\\alpha_k\end{bmatrix}-\frac{1}{6}\begin{bmatrix}l_{ki}(f_{k+}+2f_{i-})+l_{ij}\ (2f_{i+}+f_{j-})\\ l_{ij}(f_{i+}+2f_{j-})+l_{jk}\ (2f_{j+}+f_{k-})\\ l_{jk}(f_{j+}+2f_{k-})+l_{ki}\ (2f_{k+}+f_{i-})\end{bmatrix}$$

Here A denotes the area of the element and l_{ij} the length of the edge between points i and j.

$\alpha_i = \frac{lpq}{l_{iq}}$, similarly for α_j and α_k.

When assembling the element contributions we use the assumption that flow terms across common boundaries of adjacent elements must cancel out. In the final assembled system we are thus only left with flow terms on the outer boundaries $\delta\Omega_1$ and $\delta\Omega_2$. On $\delta\Omega_2$ the flow terms are prescribed, whereas on $\delta\Omega_1$, h_i, are known and we can drop the corresponding equations from the system. The resulting system will then be identical to that in equation 26. It should further be noted that if we sum the equations 30 we get that:

$$SA\frac{d}{dt}\left(\frac{h_i+h_j+h_k}{3}\right) = A\frac{R_i+R_j+R_k}{3} - Q$$

$$-l_{ij}\left(\frac{f_{i+}+f_{j-}}{2}\right) - l_{jk}\left(\frac{f_{j+}+f_{k-}}{2}\right) - l_{ki}\left(\frac{f_{k+}+f_{i-}}{2}\right)$$

Thus, continuity of mass is preserved within the element and the same will hold true for any assembled subregion and in the end for the total region Ω.

Considering finally the case of anisotropic flow, the change in equations 26-29 is that in equation 28 we now have that:

$$b_{ij} = \iint_{\Omega} \left\{ T_{xx} \frac{\partial \psi_j}{\partial x} \frac{\partial \psi_i}{\partial x} + T_{yy} \frac{\partial \psi_j}{\partial y} \frac{\partial \psi_i}{\partial y} + T_{xy} \frac{\partial \psi_j}{\partial y} \frac{\partial \psi_i}{\partial x} + T_{yx} \frac{\partial \psi_j}{\partial x} \frac{\partial \psi_i}{\partial y} \right\} dxdy$$

(cf. equation 4). However, if we approximate T_{xx}, T_{yy}, T_{xy} and T_{yx} with constants within each triangular element, this can be brought within the framework of equation 30 if we determine the eigenvalues, λ_1 and λ_2, and corresponding unit eigenvectors (principal directions of the anisotropy) (c_{11}, c_{12}) and (c_{21}, c_{22}) for the matrix

$$\begin{bmatrix} T_{xx} & T_{xy} \\ T_{yx} & T_{yy} \end{bmatrix}$$

Assuming that $T_{xy} = T_{yx}$ and $T_{xx} T_{yy} > T_{xy}^2$, so that the eigenvalues are real and positive and the eigenvectors orthonormal, we introduce the following change of coordinates:

$$\zeta = \frac{1}{\lambda_1}\{c_{11}x + c_{12}y\}, \gamma = \frac{1}{\lambda_2} \{c_{21}x + c_{22}y\} \tag{31}$$

If the lengths l_{ij}, l_{jk} and l_{ki} in the first matrix on the RHS of equation 30 are recalculated in this new coordinate system and T is replaced by $\sqrt{\lambda_1 \lambda_2}$ the triangular element contribution to the final system still remains valid.

THE MODESUPERPOSITION TECHNIQUE FOR THE TIMEDEPENDENT PROBLEM

In this section we develop the modesuperposition technique for solving the timedependent problem in equation 26. In analogy with the treatment of the differentialequation we split equation 26 into two equations a stationary part and a transient part. We have for the stationary part:

$$\underline{B}\underline{h}_o = \underline{\bar{b}} \tag{32}$$

where $\underline{\bar{b}}$ means long time average values of the $\underline{b}$-vector. For the transient part we get:

$$C \frac{d\underline{h}_1}{dt} = - B \underline{h}_1 + (\underline{b}(t) - \overline{\underline{b}}) \tag{33}$$

$$\underline{h} = \underline{h}_o + \underline{h}_1 \tag{34}$$

We now define the eigenvalue problem:

$$B\underline{\phi} = \lambda C\underline{\phi} \tag{35}$$

where we have normalized B with respect to some reference transmissivity, T_o, and normalized C with respect to some reference storage coefficient, S_o, and the area, A. By writing $(\underline{b} - \overline{\underline{b}})$ as an eigenvector series expansion we get the solution to equation 33:

$$\underline{h}_1(t) = \frac{1}{AS_o} \sum_{n=1}^{N} \underline{\phi}_n^T \int_0^\infty \underline{\phi}_n^T (\underline{b}(t-\tau) - \overline{\underline{b}}) e^{-\tau/K_n} d\tau \tag{36}$$

where $\underline{\phi}_n^T$ means $\underline{\phi}_n$ transposed and K_n is defined from:

$$K_n = \frac{AS_o}{\lambda_n T_o} \tag{37}$$

N is the number of unknown triangular mesh points. By comparing equation 36 with equation 19 we see that we again have our linear reservoirs, but in this case we have a finite number of reservoirs, that is to say N. A computational advantage of equation 36 is the easy evaluation of the convolution integral, especially if we use the approximation that $\underline{b}(t)$ is piecewise constant. If we wish to calculate $\underline{h}(t)$ for a new input function (infiltration, pumping), the eigenfunctions and eigenvalues remain the same and we just have to perform the simple convolution integral to get a new solution. As mentioned in the section on basic theory, only the first, say M, eigenfunctions and eigenvalues have to be computed, because the timeconstants for the rest become rapidly very small. Equation 36 can then be approximated by:

$$\begin{aligned} \underline{h}_1(t) &= \frac{1}{AS_o} \sum_{n=1}^{M} \underline{\phi}_n \int_0^\infty \underline{\phi}_n^T (\underline{b}(t-\tau) - \overline{\underline{b}}) \bar{e}^{\tau/K_n} d\tau \\ &+ \frac{1}{AS_o} \sum_{n=M+1}^{N} \underline{\phi}_n^T (\underline{b}(t) - \overline{\underline{b}}) \underline{\phi}_n K_n + \varepsilon \\ &= \frac{1}{AS_o} \sum_{n=1}^{M} \underline{\phi}_n \int_0^\infty \underline{\phi}_n^T (\underline{b}(t-\tau) - \overline{\underline{b}}) \bar{e}^{\tau/K_n} d\tau \end{aligned} \tag{38}$$

$$+ B^{-1} (\underline{b}(t) - \overline{\underline{b}})$$

$$- \frac{1}{AS_o} \sum_{n=1}^{M} \underline{\phi}_n^T (\underline{b}(t) - \overline{\underline{b}}) \underline{\phi}_n K_n + \varepsilon$$

where ε is the measure of the truncation error. The middle term on the RHS is most effectively dealt with by obtaining the Cholesky factor of B at the start of the calculation. Since the other terms now only involve the eigenvectors corresponding to the M smallest eigenvalues of the problem in equation 35, it is advantageous to use the socalled subspace iteration algorithm. The algorithm, which can be viewed as an extension of the well-known power method, obtains the eigenvalues and eigenvectors simultaneously and involves only at each iterationstep a solution of a reduced eigenproblem of size k as well as k linear systems of the form $B\underline{x} = \underline{y}$ where $k \simeq M + 8$. The convergence is in general very rapid. It has been used successfully in solving stress-strain problems. See Bathe et al. (1974) and Strang and Fix (1973).

Let us consider a small example. We take our area Ω as a rectangle and the source term to be given in the form of constant pumping. We let the aquifer be isotropic and homogeneous. See figure 5 and 6. From equation 38 we have the approximate solution for the drawdown S:

$$\underline{S}(t) = \frac{Q}{AS_o} \sum_{n=1}^{M} K_n \underline{\phi}_n \underline{\phi}_n^T \underline{q} (1 - e^{-t/K_n})$$

$$+ Q B^{-1} \underline{q} \qquad (39)$$

$$- \frac{Q}{AS_o} \sum_{n=1}^{M} K_n \underline{\phi}_n \underline{\phi}_n^T \underline{q} + \varepsilon$$

where $\underline{q}$ is defined as:

$$\underline{q} = \begin{Bmatrix} 0 \\ \cdot \\ \cdot \\ 1 \\ \cdot \\ \cdot \\ 0 \end{Bmatrix} \qquad (40)$$

with zero everywhere, except in the point where the pumping takes place. We insert the following numerical values. Pumping well has the coordinates $(\zeta,\eta) = (\frac{1}{2}, \frac{1}{2})$, the observation well has the coordinates $(x,y) = (\frac{1}{4}, \frac{1}{2})$,

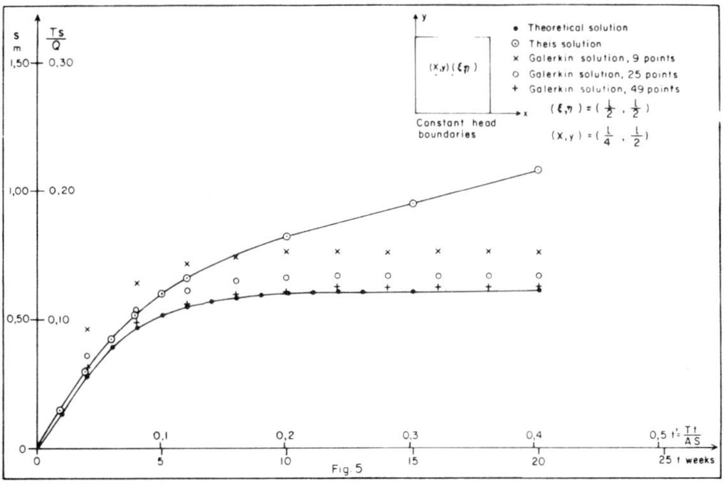

Figure 5

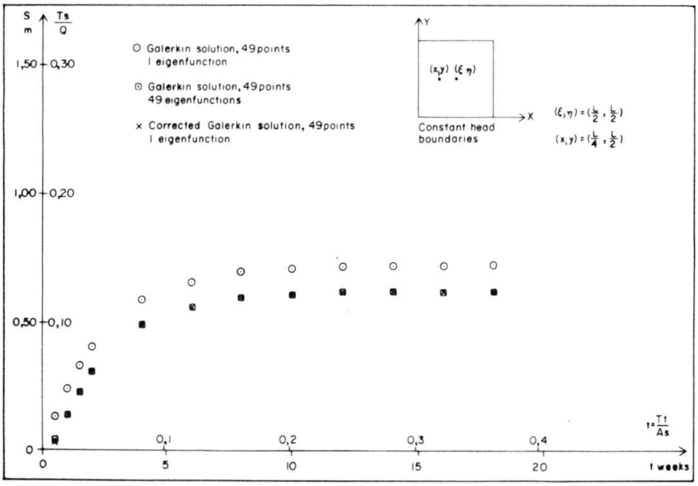

Fig. 6

Figure 6

$A = 10^8 m^2$, $l = 10^4 m$, $T = 0{,}2\ m^2/s$, $S = 0{,}06$,

$Q = 1{,}0\ m^3/s$. Firstly we take M = N in equation 39, we then get:

$$\underline{s}(t) = \frac{Q}{AS_o} \sum_{n=1}^{N} K_n\, \underline{\phi}_n\, \underline{\phi}_n^T\, \underline{q}\, (1-e^{-t/K_n}) \qquad (41)$$

Equation 41 is plotted in figure 5 for 9, 25 and 49 internal points together with the theoretical solution and Theis solution for an infinite aquifer. There are two scales on the axis. On the vertical axis we have the actual drawdown and the dimensionless drawdown $s' = \frac{Ts}{Q}$. On the horizontal axis we have the actual time in weeks and the dimensionless time $t^1 = \frac{Tt}{AS}$. Then the diagram can be used for different values of Q,T,S and A by using the dimensionless scale on the axis. Now we take just one eigenfunction but 49 internal points, that is to say, M = 1, N = 49. Equation 39 then reduces to:

$$\underline{s}(t) = \frac{Q}{AS_o} K_1 \underline{\phi}_1\, \underline{\phi}_1^T\, \underline{q}\, (1-e^{-t/K_1}) \qquad (42)$$
$$+ Q\, B^{-1} \underline{q} + \frac{Q}{AS_o} K_1 \underline{\phi}_1 \underline{\phi}_1^T \underline{q} + \varepsilon$$

which can be written as:

$$\underline{s}\ (t) = QB^{-1}\, \underline{q} - \frac{Q}{AS_o} K_1\, \underline{\phi}_1\, \underline{\phi}_1^T\, \underline{q}\; e^{-t/K_1} + \varepsilon \qquad (43)$$

The first term in equation 42 is the drawdown corresponding to the first eigenfunction and it is plotted in figure 6 together with s for 49 eigenfunctions and equation 43, which is called the corrected solution in figure 6. The accuracy becomes less for smaller times and thus there is a lower time limit for the accuracy.

Finally, it could be mentioned that if we have a leaky aquifer with the leakage proportional to the drawdown, then the eigenproblem does not have to be solved again. Then the eigenfunctions remain the same and the new eigenvalues are simply given by:

$$\lambda_{new} = \lambda + \frac{A \cdot \gamma}{T_o} \tag{44}$$

where γ is the leakage factor. Now just the convolution integral has to be performed again.

The modesuperposition technique we have presented seems to be suitable, when many utilization alternatives have to be checked. When calibrating the model, that is to say determine the T and S values the eigenproblem has to be solved many times, in that case the finite difference scheme for the time derivative might be less time consuming. For small pumping times care must be taken that the relative error is not too big, when using modesuperposition.

CALCULATION OF FLOW ACROSS BOUNDARIES

In this section we restrict our attention to the steady state problem unless otherwise specified. We assume that we have determined the level values, h_i, at all nodal points and that we are now interested in determining the flow distribution across internal as well as external boundaries f.ex. boundaries A1-A9 and B1-B7 in figure 7.

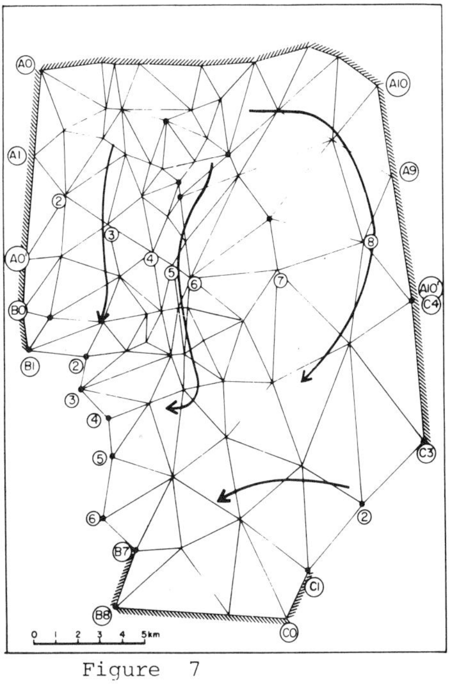

Figure 7

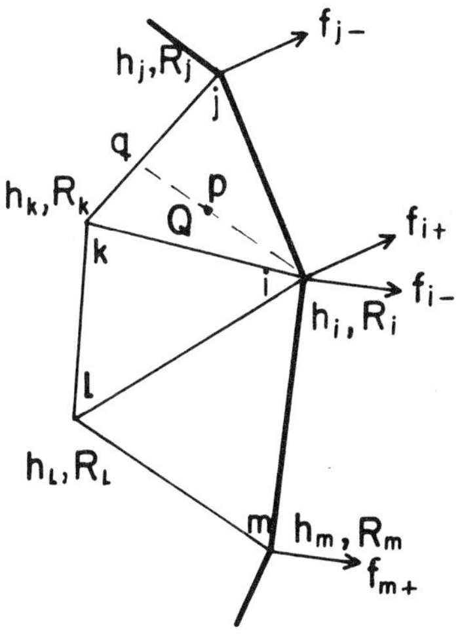

Figure 8

The most straightforward approach would be to obtain the flow values from the normal gradients of the approximate solution, $\bar{h}$, within the triangular elements next to the boundaries..In the isotropic case this amounts to using the following formula for obtaining the total flow across the edge ij in figure 8:

$$\frac{1}{2}l_{ij}(f_{i+} + f_{j-}) = \frac{T}{4A}\{2l_{ij}^2 h_k + (l_{jk}^2 - l_{ki}^2 - l_{ij}^2)h_j + (l_{ki}^2 - l_{ij}^2 - l_{jk}^2)h_i\} \qquad (45)$$

where A denotes the area of the element ijk, l_{ij} the length of the edge between points i and j and T the constant transmissivity within the element. The same formula remains valid in the anisotropic case if we transform the (x,y)-coordinates according to equation 31 and recalculate the lengths of the edges in the new coordinate system as well as replacing T by $\sqrt{\lambda_1\lambda_2}$ (cf. comments before and after equation 31).

The above formula, however, has the following disadvantages:

i) When flow values are calculated across a closed boundary they will in general not match the infiltration within the boundary, indeed discrepancies of up to 50% are not uncommon in practical problems.

ii) Flow values based on it will not agree with specified flow values on the boundary $\partial\Omega_2$.

iii) At an internal boundary we obtain different flow values depending on, on which side the triangular element is that we base our calculations on.

We, therefore, propose that the following formula should be used, if we wish to calculate the flow across the boundary mij in figure 8:

$$\frac{1}{6}\{l_{mi}(f_{m+} + 2f_{i-}) + l_{ij}(2f_{i+} + f_{j-})\}$$

$$= -\sum_{\Delta} \frac{T}{8A}\{2l_{jk}^2 h_i + (l_{ij}^2 - l_{jk}^2 - l_{ki}^2)h_j + (l_{ki}^2 - l_{ij}^2 - l_{jk}^2)h_k\} + \frac{A}{12}\{2R_i + R_j + R_k\} - \alpha_i Q\} \qquad (46)$$

The LHS of this equation amounts to half the total flow across the boundary mij and may thus be interpreted as the total flow from the halfway point between points m and i to the halfway point between points i and j. On the RHS we have assumed that we have a pumping Q at the point p and define α_i as in equation 30. $\sum_{\Delta}$ denotes that we are including on the RHS contributions from the triangles ikl and ilm analogous to the one presented for triangle ijk. We further note that anisotropy is dealt with in exactly the same way as in equation 45 and in the case of a timedependent problem the only change that we have to make in equation 46 is to replace R_i by $R_i - S\frac{dh_i}{dt}$. Finally, it may be observed that the first term on the RHS of equation 46 is, after we have summed over the appropriate triangles, exactly half the gradient inflow through the boundary jklm (cf. equation 45).

Equation 46 is derived as part of the assembly process described in the section on the Galerkin method, i.e. it is derived by combining the equations associated with nodal point i in the element contributions for the triangular elements ijk, ikl and ilm (see equation 30), and by making use of the assumption that flowterms across common boundaries of adjacent elements cancel out. The consequence of this derivation is that the flow values obtained from equation 46 are such that if we were to solve the steady state problem with these flow values as boundary conditions we would obtain exactly the given h_i values. It further means that all the three drawbacks associated with the formula in equation 45 completely disappear.

It is of interest to note the form that the RHS of equation 46 takes in the regular case shown on the left in figure 9, where $l_{ij} = l_{ik} = l_{im} = l$, say,

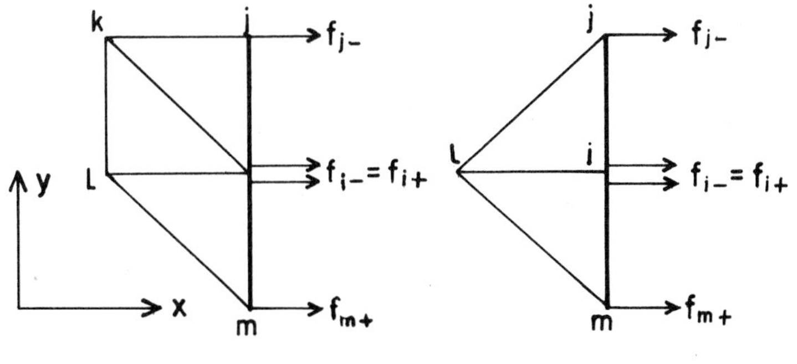

Figure 9

and we further assume that we have the same transmissivity, T, within all the triangles and omit the pumping term. The RHS becomes:

$$-T\ \{2h_i - \frac{1}{2}(h_j+h_m)-h_l\} + \frac{l^2}{24}\{6R_i+R_j+R_m+2(R_k+R_l)\} \tag{47}$$

By applying Taylor expansions round the point i it can be shown that this is in fact a second order difference approximation to $-lT\frac{\partial h}{\partial x}\big|_i$ for a function h that satisfies the differential equation

$$-\ \nabla\ (T\nabla h) = R$$

and indeed a third order difference approximation to

$$-lT\frac{1}{6}\ \{\frac{\partial h}{\partial x}\Big|_m + 4\frac{\partial h}{\partial x}\Big|_i + \frac{\partial h}{\partial x}\Big|_j\}$$

if $\frac{\partial R}{\partial y} = 0$. The difference approximation $-l(\frac{h_i-h_l}{l})$, corresponding to the calculated groundwatergradient is of course only a first order approximation. However, it should be mentioned that in the regular case shown on the right in figure 9 the RHS of equation 46 becomes identical to equation 47 except the term $2R_k$ is missing which in turn means that the improved order of the difference approximation is lost unless $R = 0$. The fact that the Galerkin finite element method may lead to high order difference approximations on the boundary has been observed by f.ex. Strang and Fix (1973), p. 33, in the case of ordinary differential equations.

As a demonstration of the difference between the flow formulas in equations 45 and 46 we present results from an isotropic steady-state problem associated with the region in figure 7. In this problem zero flow is specified across the shaded part of the boundary and a given groundwater level on the remaining part of it. The calculated level values within the region range from 280 m to 560 m, transmitting values from $1.26 \cdot 10^{-4}$ to $1.26 \cdot 10^{-1}$ m^2/s and infiltration from 500 to 1050 mm/year. The arrows indicate approximate flowlines. In figure 10 we show the flow per unit length across the internal boundary A1-A9 and the external boundaries B1-B7 and C1-C3. Downstream and upstream gradient values stand for flow values based on equation 45 depending on, on which side of the boundary the triangular elements are, and Galerkin gradient values stands for flow

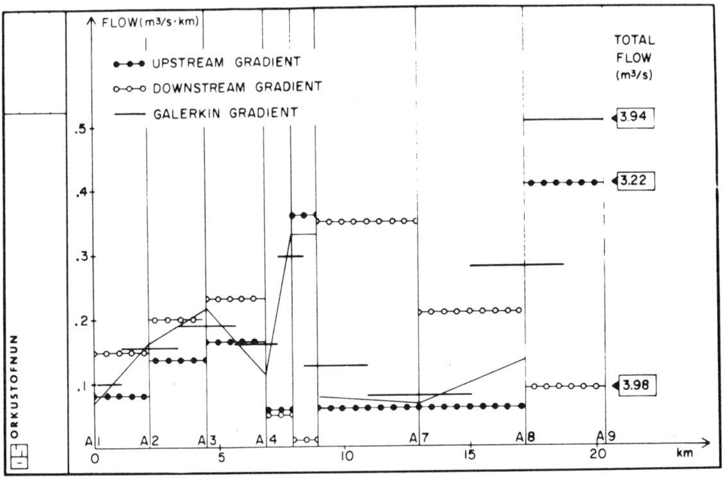
FLOW(m³/s·km)
UPSTREAM GRADIENT
DOWNSTREAM GRADIENT
GALERKIN GRADIENT
TOTAL FLOW (m³/s)
3.94
3.22
3.98
A1 A2 A3 A4 A7 A8 A9
0 5 10 15 20 km
ORKUSTOFNUN

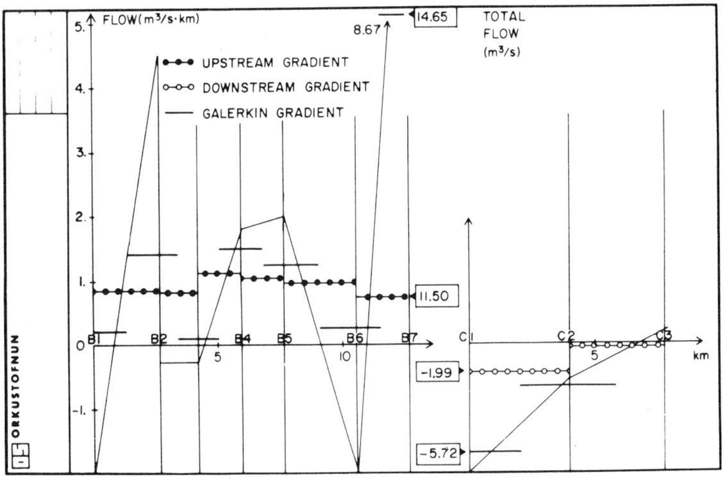
FLOW(m³/s·km)
UPSTREAM GRADIENT
DOWNSTREAM GRADIENT
GALERKIN GRADIENT
TOTAL FLOW (m³/s)
8.67
14.65
11.50
-1.99
-5.72
B1 B2 B4 B5 B6 B7
C1 C2 C3
5 10 km
ORKUSTOFNUN

Figure 10

values based on equation 46. In both cases we divide the total flow across the appropriate edge with the length of the edge in order to obtain the given values. As already mentioned, the flow values in the case of the Galerkin gradient do not depend on, on which side the elements are. In order to obtain the flow values at the end of the boundaries from equation 46 we make use of the fact that we have zero flow across the boundaries A0-A1, A9-A10, B0-B1, B7-B8, C0-C1 and C3-C4.

The total flow across A1-A9 according to the Galerkin gradient method is 3.94 m^2/s and this matches exactly the infiltration above that boundary. It is interesting to note that downstream gradient values which lead to almost correct total flow match the Galerkin gradient values in general considerably worse than the upstream gradient values. The difference between the total outflow through the boundary B1-B9 and the inflow through the boundary C1-C3 is 8.93 m^3/s according to the Galerkin gradient method which again matches exactly the infiltration within the whole region. Here it is interesting to note that, although the corresponding difference is only 6% greater for the upstream and downstream gradient values, the Galerkin gradient method gives ca. 30% greater outflow through B1-B9 and almost three times as great inflow through C1-C3, a result which is, in fact, in better agreement with hydrological evidence.

Rather than just interpreting the LHS of equation 46 as the total flow across the edge on both sides of point i, we can, of course, go on and determine the actual f values, provided we make assumptions to the effect that $f_{i-} = f_{i+}$ (which holds true on a straight boundary) or that f is constant along a given edge. As an example, we can assume along boundary A1-A9 that $f_{i-} = f_{i+}$, $i = 2,3,4,5,7$ and $f_{i+} = f_{(i+1)-}$ $i = 5,8$. Then the appropriate values, $f_{1+}, f_2, f_3, f_4, f_5, f_{6+}, f_7, f_{8-}$ and f_{8+}, are obtained by solving tridiagonal system.

The almost contionous flowline in figure 10 is obtained by solving this system and will of course result in the same total flow as the discontinous Galerkin gradient flowline. Similarly for boundaries B1-B7 and C1-C3.

Rodi (1976) has developed a method for calculating accurate gradients in electromagnetic field problems and demonstrated its usefulness. Although his motivation and derivation differs from ours, and rectangular elements are used in the solution of the problem, the basic formula of this method can be shown to be analogous to equation 46.

REFERENCES

Bathe, K.J.; Wilson, E.L.; Peterson, F.E. (1973) SAP IV. A Structural Analysis Program For Static and Dynamic Response of Linear Systems. College of Engineering University of California. Report No EEPC 73-11.

Neuman, S.P.; Whitherspoon,P.A. (1970) Finite Element Method of Analysing Steady Seepage with a Free Surface. Water Resources Research, 6, 3: 889-897.

Pinder, G.F.; Frind, E.O. (1972) Application of Galerkin's Procedure to Aquifer Analysis. Water Resources Research, 8, 1: 108-120.

Pinder, G.F.; Gray, W.G. (1974) Galerkin Approximation of the Time Derivative in the Finite Element Analysis of Groundwater Flow. Water Resources Research, 10, 4: 821-828.

Rodi, W.L. (1976) A Technique for Improving the Accuracy of Finite Element Solutions for Magnetotelluric Data. Geophysical Journal of the R.A.S. 44, 2.

Strang, G; Fix, G.J. (1973) An Analysis of the Finite Element Method. Prentice-Hall, Inc.

Zienkiewicz, O.C. (1971) The Finite Element Method in Engineering Science. (The second, expanded and revised, edition of The Finite Element Method in Structural and Continuum Mechanics). McGraw-Hill, London.

FINITE ELEMENT ANALYSIS OF GROUND WATER FLOW AND SETTLEMENTS IN AQUIFERS CONFINED BY CLAY

K. Runesson, H. Tägnfors and N-E. Wiberg

Chalmers University of Technology, Gothenburg, Sweden

SUMMARY

The finite element analysis of the ground water flow in aquifers confined by soft clay and resulting settlements of the ground surface is discussed. As a result of a disturbance of the hydraulic equilibrium in the aquifer transient flow and deformation (consolidation) of the confining clay occur. The analysis is based on the concept of clay as a two phase saturated porous medium with the displacements and the excess pore pressure as primary unknowns. A finite element method in space and a numerical integration method in time are utilized. Due to the rapidly decaying character of the time dependent process it is not efficient to maintain an unchanged time step throughout the process. An automatic prediction of the time step based on an accepted local truncation error is to prefer and is also a part of the method. The clay is assumed to be normally consolidated or slightly overconsolidated, thus calling for non-linear constitutive properties of the soil skeleton. An initially anisotropic yield condition of the Critical State type is proposed because in situ stresses are generally anisotropic.

A finite element computer program GEOFEM is described which is used for solution of the seepage problem in the aquifer as well as the coupled consolidation problem.

INTRODUCTION

Groundwater hydraulics is of great technical and economic importance both in water resource hydrology and in soil mechanics. Disturbances of the hydraulic equilibrium in an aquifer due to for example discharge of water, leakage into rock tunnels, or changes of the perkolation characteristics give rise to ground water flow.

In Sweden many aquifers are fairly thin sand layers, which are confined from above by a thick deposit of soft clay and from below by solid rock. An important consequence of a disturbance of the hydraulic equilibrium is the occurrence of pronounced settlements of the ground surface. It is obvious that a complete analysis must include both the transient seepage in the aquifer itself as well as the more delayed settlements and pore water flow in the clay.

Clearly, the solution of equations for groundwater flow in the aquifer is of primary interest to the hydrologist. However, it also provides information for an accurate analysis of deformations not only in the aquifer itself but also in the confining clay layer. Basic principles of groundwater hydraulics are well established and can be found in a number of text books, e.g. Li (1972) and Domenico (1972). The numerical solution has been discussed by, among others, Desai and Abel (1972) and Connor and Brebbia (1976).

In this paper the main emphasis will be put on the confining layer of saturated clay, which has to be treated as a two-phase material. The motion of the two phases (the clay skeleton and the pore water) is coupled due to interactive forces. Though established already by Terzaghi (1943) and Biot (1941), the mechanical relations for the coupled problems seem to have been firmly set first by Sandhu (1968).

Uncoupled theories, pseudo-theories, have been applied both to consolidation of clay and to seepage problems, e.g. aquifer analysis, by a number of authors The pore pressure is first determined and an effective stress analysis using the gradient of the known pore pressure values as a load is quite straight-forward.

In judging the magnitude and progress of the settlements of a clay layer, the choice of a reliable constitutive model for the soil skeleton is of outstanding importance. Of equal importance is the accessability to in situ stress values estimated with a high degree of accuracy. For soft clay it seems reliable to include plasticity in the constitutive equations, because clay (at least most Swedish clays) exhibits abrupt yield at the preconsolidation stresses, even at rapid loading. Appropriate yield criterions should be initially anisotropic when K_0^{NC}, the ratio between horizontal and vertical effective stresses at normally consolidated conditions, is not equal to unity, Larsson (1977).

Creep effects are important in many Swedish clays especially when the organic contents are significant. There seems to be experimental evidence for the use of viscoplasticity as observed from undrained triaxial tests, Larsson (1977). However, though the creep rate has been tested under fairly restrictive stress conditions, use will be made in this paper of an associated creep law according to the theory of Perzyna (1966).

MECHANICS FOR SATURATED SOIL

Effective stress hypothesis

Saturated soil may be idealized by a mechanical model consisting of basically two phases, a solid phase (the soil skeleton) and a fluid phase (the pore water), mixed together. From the effective stress principle by Terzaghi (1943) the total stress $\bar{\sigma}_{ij}$ is equal to the difference between the effective stress σ_{ij} in the soil skeleton and the pore water pressure p (tension stresses positive).

$$\bar{\sigma}_{ij} = \sigma_{ij} - p\delta_{ij} \qquad (1)$$

It is to be noticed that stresses are measured per bulk area. By definition all deformations in the soil are coupled to the effective stresses.

Field equations

Adopting Cartesian coordinates we may summarize the basic relations for saturated soil. Before any further approximations are introduced, the equations are valid both for the aquifer analysis and for the clay layer analysis.

Equilibrium equations:

$$-(\sigma_{ij,j} - \sigma^0_{ij,j}) + p_{,i} = 0 \qquad (2)$$

where

$$-\sigma^0_{ij,j} = U^0_i$$

The weight of the skeleton represented by the initial volume forces U^0_i is equilibrated by the initial stresses σ^0_{ij}. Further, p is the excess pore pressure (diffusive pressure), since the hydrostatic pressure equilibrates the pore water bulk weight. Additional loads are applied only on the boundary, thus makin (2) homogeneous.

Continuity conditions:

$$\dot{\varepsilon}_{kk} + q_{k,k} = 0 \qquad (3)$$

where $\dot{\varepsilon}_{kk}$ is the volumetric strain rate and q_i are the pore water velocities relative the soil skeleton. A do denotes differentiation with respect to time. Equation (3) arises from the assumption of the pore water being completely incompressible.

Kinematic equations (small strains):

$$\varepsilon_{ij} = \frac{1}{2}(u_{i,j} + u_{j,i}), \quad g_i = -p_{,i} \tag{4}$$

where displacements are denoted by u_i, strains by ε_{ij}, and diffusive forces by g_i.

Constitutive equations for the soil skeleton (including elastic-viscoplastic-plastic behaviour) and for the pore water flow (Darcy's law):

$$\begin{cases} \dot{\sigma}_{ij} = S^{ep}_{ijkl}(\dot{\varepsilon}_{kl} - h_{kl}), \\ S^{ep}_{ijkl} = S^{ep}_{ijkl}(\sigma_{mn}, \varepsilon^{p}_{kk}), \; h_{kl} = h_{kl}(\sigma_{mn}, t) \end{cases} \tag{5}$$

$$q_i = K_{ij} g_j \tag{6}$$

where S^{ep}_{ijkl} are components of the elastic-plastic tangential stiffness tensor, K_{ij} are components of the permeability tensor, and h_{kl} are functions expressing the viscoplastic strain rate. Explicit time hardening has to be contained in h_{kl} for an adequate description of the behaviour of many Swedish clays.

Boundary conditions

The boundary of a finite body of saturated soil is divided into two parts in two different ways. On disjunct parts relevant boundary conditons have to be imposed on displacements u_i and tractions T_i as well as on the excess pore pressure p and the drainaging flow Q

$$u_i = \bar{u}_i \tag{7a}$$

$$T_i = \bar{T}_i, \quad T_i = n_k(\sigma_{ki} - p\delta_{ki}) \tag{7b}$$

$$p = \bar{p} \tag{7c}$$

$$Q = \bar{Q} + \gamma(p - p_s), \quad Q = n_k q_k \tag{7d}$$

where n_i is the outward unit normal to the boundary, and where a bar denotes a prescribed value. The flow boundary condition contains a so called convective part (often occurring in diffusion type problems), proportional to the rise of the excess pore pressure from an initial level p_s. Free drainage means that the total pressure is zero (neglecting the atmospheric pressure), i.e. the excess part is equal to minus the hydrostatic pressure. In practice there are difficulties tied to the adequate choice of, not only the boundary conditions, but the region for analysis, at least when conventional finite element methods are used.

Special problems

Taking advantage of the certain physical behaviour of a soil one may obtain special formulations approximating the ones given above. Undrained and completely drained behaviour are associated with constant volume, $\varepsilon_{kk} \equiv C$, where the constant C may be given a priori or the determination of which is a part of the problem. Undrained behaviour is associated with C = 0.

The assumption of undrained conditions is justified in mainly two situations: at rapid loading (the water dissipation is negligable), and when the permeability is very small (the effect of water dissipation is negligable in the time interval of interest). In the latter case a state of hydraulic equilibrium is obtained, $q_{k,k} = 0$, when the permeability coefficients are zero. In practice, undrained behaviour is assured mostly during rapid loading, during which process creep strains are zero.

Completely drained behaviour is associated with a vanishing rate of volumetric strain, yielding hydraulic equilibrium (from the continuity equation) and with prescribed excess pore pressures along some part of the boundary. If (4) and (6) are used, a unique solution to $q_{k,k} = 0$ exists, giving the excess pore pressure. Displacements are calculated from the stresses obtained from

$$-(\sigma_{ij,j} - \sigma^{0}_{ij,j}) = -p_{,i}, \quad p = p(x,t) \tag{8}$$

where the right hand side of (8) now is known.

If prescribed pressures are zero everywhere, (8) is homogeneous, because the unique solution to $q_{k,k} = 0$ then is $p \equiv 0$. This is the case when the primary consolidation of a clay layer, subjected to a surface load, is finished (asymptotically). The greater the c_v-value, expressing the product of the permeability and the soil skeleton stiffness, the sooner a drained state is achieved.

It is possible to approximate the partly drained behaviour by assuming that the total mean stress is stationary, $\dot{\sigma}_{ii} = 0$, throughout the process. In the case of isotropic linear elasticity and isotropic permeability, this leads to the pseudo-theory defined by Terzaghi's classical parabolic pore pressure equation

$$-p_{,ii} + \frac{1}{c_v}\dot{p} = 0 \tag{9}$$

Known the pore pressure from (9) the displacements can be calculated from (8).

Aquifers confined by clay - integrated analysis

Both the permeability and the soil stiffness (expressed by the c_v-value) are very large for the aquifer material compared to clay. This means that, due to a disturbance of the hydraulic equilibrium, the consolidation process in the aquifer will be very rapid compared to that in the clay layer. A complete study of the behaviour of the integrated system aquifer-confining clay can, of course, always be performed by considering the true interaction. Due to the physical character, however, such an approach seems not to be computationally efficient. Much economy is gained if the aquifer and the clay layer are analyzed separately with the assumption that the two processes are separated in time. The following procedure is deviced, Figure 1, where a single well action is shown as an example.

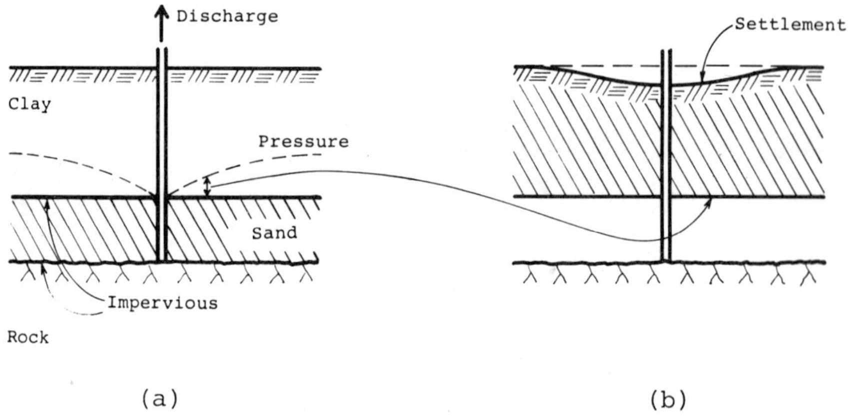

Figure 1 (a) aquifer analysis, (b) consolidation analysis

Using the pseudo-theory in (9) the pore pressure change in the aquifer is analyzed. For simplicity it is assumed that the pore pressure decrease is limited so that confined conditions are maintained, Figure 1a. The sand-clay interface is assumed to be impervious, i.e. the leakage from the low-permeable clay is neglected. Relevant boundary conditions are (7c) and (7d). As the c_v-value is large a simple stationary version of (9) is solved. This gives in particular the pore pressure at the sand-clay interface, Figure 1a.

It is noticed that it is possible to use the well-known approximate equations for confined aquifers, e.g. Connor and Brebbia (1976). Such an approach is, however, not used in this paper.

The consolidation analysis of the clay layer, Figure 1b, is performed by use of the equations (2), (3), (4), (5) and (6) with the boundary conditions (7). The pore pressures at the sand-clay interface are conveniently prescribed to the values obtained from the aquifer analysis. Another possibility is to use the convective type boundary condition in (7d), where p_s is chosen as the values calculated in the aquifer. A jump in pore pressure will be obtained at the interface. (This kind of boundary condition is not used in the calculated examples below.)

NON-LINEAR CONSTITUTION OF THE CLAY SKELETON

General

Because of the great variety in behaviour of clay depending on sedimentation environments, porosity, water contents, organic contents, cementing interactive forces, etc., there is little possibility to unify the behaviour of clay in one single constitutive model. Swedish clays are often soft with high porosity and are contractant under deviatoric loading beyond the preconsolidation stresses. Even under rapid, undrained loading a dramatic yield is obtained at the state of preconsolidation, Sällfors (1975). In some cases the behaviour is almost perfectly plastic. In addition, creep effects are significant in the overconsolidated region, even for moderate additional stresses. These properties indicate that the choice of an elastic-viscoplastic-plastic model is well motivated. Pure creep and inviscid plasticity are obtained as special cases of the more general model.

Elastic-viscoplastic-plastic model

The rheological model for the material model used is given in Figure 2a. Associated with the two sliders are two families of closed surfaces in stress space, Figure 2b, the inviscid yield surface ($\alpha = 1$) and the quasistatic yield surface ($\alpha = 2$), defined by

$$f_\alpha = 0, \quad f_\alpha = f_\alpha(\sigma_{ij}, \kappa_\alpha), \quad \alpha = 1,2 \qquad (10)$$

where κ_α are hardening (softening) parameters. In the present application the inviscid yield surface $f_1 = 0$ is assumed to be hardening (softening). The plastic part of the volumetric strain is chosen as the hardening measure κ_1, i.e. restriction is made to pure density hardening. The quasistatic yield surface $f_2 = 0$ is assumed to be perfectly plastic so that κ_2 is irrelevant. The functions f_α are similarly defined such that the surfaces $f_\alpha = 0$, i = 1,2, completely coincide for a certain amount of softening in $f_1 = 0$. While

$f_1 \leq 0$ must always hold, there is no such restriction on f_2. Thus, $f_2 = c$, $c > 0$, define subsequent loading surfaces.

Depending on the position of the stress point in stress space different characteristic stress states are obtained, Figure 2b.

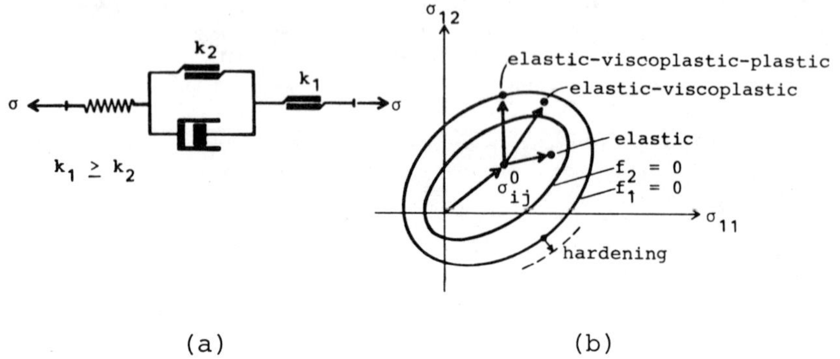

Figure 2 (a) 1D-rheological model, (b) loading surfaces in stress space

The kinematic decomposition of the strain rate is (small strains)

$$\dot{\varepsilon}_{ij} = \dot{\varepsilon}^{e}_{ij} + \dot{\varepsilon}^{c}_{ij} + \dot{\varepsilon}^{p}_{ij} \tag{11}$$

Using associated flow rules (which may be questionable for certain choices of f_α) both for the plastic part, $\dot{\varepsilon}^{p}_{ij}$, and for the creep part, $\dot{\varepsilon}^{c}_{ij}$, in accordance with Perzyna (1966), we obtain the constitutive law in stiffness form according to (5) with the explicit expressions

$$S^{ep}_{ijkl} = S^{e}_{ijkl} - \frac{1}{D} S^{e}_{ijmn} \frac{\partial f_1}{\partial \sigma_{mn}} S^{e}_{klpq} \frac{\partial f_1}{\partial \sigma_{pq}} \tag{12}$$

where

$$D = \frac{\partial f_1}{\partial \sigma_{ij}} S^{e}_{ijkl} \frac{\partial f_1}{\partial \sigma_{kl}} - \frac{\partial f_1}{\partial (\varepsilon^{p}_{kk})} \frac{\partial f_1}{\partial \sigma_{kk}}$$

and

$$h_{ij} = \gamma(t) < \Phi(F) > \frac{\partial f_2}{\partial \sigma_{ij}} \Big/ \left| \frac{\partial f_2}{\partial \sigma_{ij}} \right| \tag{13}$$

where γ and Φ are creep functions, F is a non-dimensional variable obtained by scaling f_2 with a reference parameter. The length of a second order tensor

T_{ij} is denoted $|T_{ij}| = (T_{kl}T_{kl})^{1/2}$. Further, for a scalar x, $\langle x\rangle = x$ for $x \geq 0$, $\langle x\rangle = 0$ for $x < 0$.

Actual loading surfaces

Two kinds of surfaces have been implemented, representing isotropic yield criterions. Denoting by q and p the second deviator invariant and the first invariant respectively, $[q = (3s_{ij}s_{ij}/2)^{1/2},\ p = -\sigma_{kk}/3]$, we may define a class of criterions by

$$f_\alpha = q - M^{(\alpha)}(p - p_b), \quad \alpha = 1,2 \qquad (14)$$

containing extended von Mises (compression cone, extension cone), Drucker-Prager, etc., Figure 3a. Surfaces associated with the Critical State (CS) theory are defined by, Figure 3b,

$$f_\alpha = q^2 + M^2(p - p_b)(p - p_c^{(\alpha)}), \quad \alpha = 1,2 \qquad (15)$$

where, in particular, $p_c^{(1)}$ is the isotropic preconsolidation pressure, p_b is a cohesion parameter, and M is the slope of the Critical State Line (CSL), defined as $h = q - M(p - p_b) = 0$.

An anisotropic extension of (15) has been developed recently, Runesson and Axelsson (1977), Runesson (1978). The degree of anisotropy, which is of stress-induced kind, is assumed to be governed by K_0^{NC}, which is the ratio of horizontal and vertical effective stresses at the state of normal consolidation. Compared to (15), the parameters $p_c^{(\alpha)}$ are substituted for $\bar{p}_c^{(\alpha)}$ associated with (anisotropic) loading along the K_0^{NC}-line, Figure 3b. For $K_0^{NC} = 1$ the extended condition coincides with (15), i.e. isotropy is maintained. Experimental evidence for the anisotropic extension has been provided by Berre (1975) and Larsson (1977).

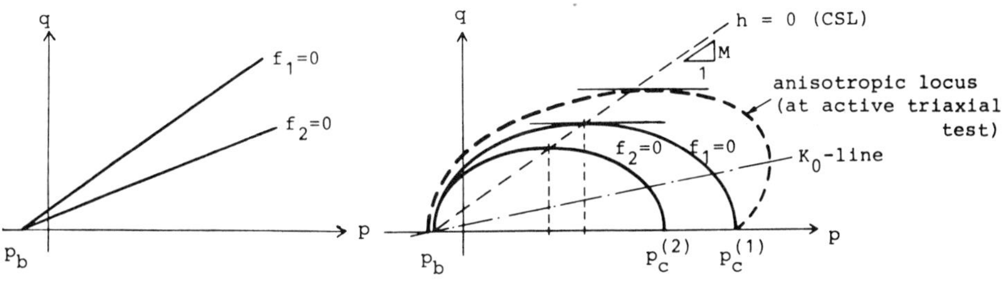

(a) (b)

Figure 3 Isotropic yield criterions. (a) Extended von Mises etc., (b) Critical State

A hardening rule suggested by Janbu (1970) is used in the post-yield region in combination with the inviscid yield surface in (15)

$$H = \frac{\partial p_c^{(1)}}{\partial \varepsilon_p^p} = k_m p_j \left(\frac{p_c^{(1)}}{p_j}\right)^{1-\beta}, \quad \varepsilon_p^p = -\varepsilon_{kk}^p \tag{16}$$

where β and k_m are non-dimensional constants and p_j is a reference stress. For simplicity $\beta = 0$ is chosen, which is reasonable for many Swedish clays.

Creep

Recent tests (undrained) on Swedish clays, Larsson (1977), indicate a linear creep law in the overconsolidated region while an exponential dependence similar to the one by Singh and Mitchell (1968) seems reliable for stresses in the post-yield region. In addition, explicit time hardening has to be taken into account in the viscosity parameter $\gamma(t)$. The details can be found in Runesson (1978).

APPROXIMATE SOLUTION

FE-discretization in space

A discretization in space of an appropriate variational formulation leads to the semi-discrete Galerkin method for problems with time dependence. A system of ordinary differential equations in time is obtained.

The space discretization used in the program GEOFEM, Runesson et.al. (1978), for calculation of the 2D-problems shown below, is based on a parametric finite element with a biquadratic displacement approximation and a bilinear excess pore pressure approximation in local coordinates. The element geometry is mapped from quadratic regions in the local coordinates by biquadratic functions. For coupled consolidation such an element may be termed quasi-isoparametric, while for the pseudo-theory (aquifer analysis) the element will be superparametric.

Numerical integration in time

The linear initial value problem arizing from the FE-discretization of the pore pressure equation in the pseudo-theory is formulated in matrix form

$$\begin{cases} G\dfrac{d\tilde{p}}{dt} = -(K + G_S)\tilde{p} + \tilde{P}(t) \\ \tilde{p}(0) = \tilde{p}_0 \end{cases} \tag{17}$$

where $\tilde{p}$ contains the nodal values of the excess pore pressures and $\tilde{P}$ contains contributions from known drainage on the boundary and from convective flow across the boundary. Further, K is the permeability matrix, G is a storage matrix, and G_S is associated with the convective flow boundary condition.

The non-linear initial value problem arizing from the FE-discretization of the consolidation problem can be formulated in matrix form analogeously with the formulation (17)

$$\begin{cases} A(\tilde{v})\dfrac{d\tilde{v}}{dt} = -B\tilde{v} + \tilde{h}(\tilde{v},t) + \tilde{V}(t) \\ \tilde{v}(0) = \tilde{v}_0 \end{cases} \tag{18}$$

The nodal values of the displacement $\tilde{u}$ and the excess pore pressures $\tilde{p}$ are collected in the matrix $\tilde{v}$, while the applied load $\tilde{U}$ differentiated with respect to time and the known drainage $\tilde{P}$ are collected in $\tilde{V}$. A pseudo-load due to creep, $\tilde{U}^c(\tilde{v},t)$, is contained in $\tilde{h}$.

$$\tilde{v}(t) = \begin{bmatrix} \tilde{u}(t) \\ \tilde{p}(t) \end{bmatrix}, \quad \tilde{V}(t) = \begin{bmatrix} \dfrac{d\tilde{U}}{dt}(t) \\ \tilde{P}(t) \end{bmatrix}, \quad \tilde{h}(\tilde{v},t) = \begin{bmatrix} \tilde{U}^c(\tilde{v},t) \\ 0 \end{bmatrix}$$

The square matrix A, which is state dependent, and B, which is constant, are defined by

$$A(\tilde{v}) = \begin{bmatrix} S(\tilde{v}) & C \\ C^t & 0 \end{bmatrix}, \quad B = \begin{bmatrix} 0 & 0 \\ 0 & -(K+G_S) \end{bmatrix}$$

where S is the tangential stiffness matrix for the skeleton and C is a rectangular coupling matrix.

In the program GEOFEM-C a single-step method of linear semi-explicit Runge-Kutta type is used to trace the solution in time. The method used is basically a so called W-method, a class of methods which has recently been proposed by Wolfbrandt (1977), Steihaug and Wolfbrandt (1977). W-methods are convenient for stiff systems like (17) and (18) because they are $L(\beta)$-stable, $0 \leq \beta \leq \pi/2$, when applied to linear systems, while they can be ensured to be $A(\beta)$-stable, $0 \leq \beta \leq \pi/2$, for non-linear systems. Another advantage is that no iterations are required in a non-linear problem. For linear problems the W-methods are closely connected to the Noersett methods, Noersett (1974) as they provide a rational approximation to the exponential with real and equal poles.

An automatic prediction of the next time step is based on the calculation of the principal error term, which is determined by a comparison of the results obtained from two methods of different order (imbedding technique). Hence, for small time steps a reliable estimate for the local truncation error is possible, provided the non-linear terms have the required regularity to match the theoretical order of accuracy. As the coefficient matrix will be constant during a time-interval, an extra stage in the method corresponding to the estimation of the truncation error only requires an extra-substitution (provided an elimination solution method is used, which is the case in the computer program GEOFEM.

The particular method adopted is called (2,4)-W. It is a 2-stage method of order 2. Applied to the problem

$$\begin{cases} \dfrac{d\tilde{y}}{dt} = \tilde{f}(\tilde{y},t), & t > t_0 \\ \tilde{y}(t_0) = \tilde{y}_0 \end{cases} \tag{19}$$

where $\tilde{f}$ is a column matrix of the same dimension as $\tilde{y}$, the method is defined by

$$\begin{cases} \tilde{y}_1 = \tilde{y}_0 + \dfrac{\Delta t}{4}(\tilde{k}_1 + 3\tilde{k}_2) \\ W(\Delta t,\alpha,P)\tilde{k}_1 = \tilde{f}(\tilde{y}_0,t_0) \\ W(\Delta t,\alpha,P)\tilde{k}_2 = \tilde{f}(\tilde{y}_0 + \frac{2}{3}\Delta t\tilde{k}_1,\ t_0 + \frac{2}{3}\Delta t) - \\ \qquad\qquad - \frac{4}{3}\Delta t\alpha P\tilde{k}_1 \\ W(\Delta t,\alpha,P) = I - \Delta t\alpha P \end{cases} \tag{20}$$

where P is an arbitrary real quadratic matrix and α is the inverse of the largest root to the second Laguerre-polynomial, i.e. $\alpha = 1 - 1/\sqrt{2}$. This choice ensures $L(\pi/2)$-stability for a linear problem.

A third order estimate $\tilde{T}$ of the local truncation error is given by

$$\tilde{T} = \frac{\Delta t}{8}(\tilde{k}_1 - 5\tilde{k}_2 + 5\tilde{k}_3 - \tilde{k}_4) \tag{21}$$

where the extra vectors $\tilde{k}_3$ and $\tilde{k}_4$ satisfy

$$\begin{cases} W(\Delta t,\alpha,P)\tilde{k}_3 = \tilde{f}(\tilde{y}_1,t_1) \\ W(\Delta t,\alpha,P)\tilde{k}_4 = \tilde{f}(\tilde{y}_1 + \frac{2}{3}\Delta t\tilde{k}_3,\ t_1 + \frac{2}{3}\Delta t) + \\ \qquad\qquad + \Delta t\alpha P(\frac{2}{3}\tilde{k}_1 + 6\tilde{k}_2) \end{cases} \tag{22}$$

Due to the certain kind of non-linearity introduced in (18) at the adoption of an elastic-viscoplastic-plastic model, some modification is made at the practical application to the problem of consolidation, Runesson (1978). This is motivated from a solution economy point of view. The main modification is that the tangent stiffness matrix S is evaluated for $\tilde{v}_0$ only and is kept unchanged through all stages during a single time step. Physically, such a modification corresponds to a linear material response throughout the time interval. Though the theoretical order of the method is no longer achieved the engineering applicability is not influenced and in the authors' experience the method has worked well.

Analogeously to solid analysis, the fullfillment of the yield condition in the end of the time interval results in a lack of equilibrium, which is compensated for by applying a fictitious residual load. This load gives rise to additional displacements and pressures. An iterative procedure leads to the simultaneous satisfaction of the yield condition and the variational equilibrium equation.

Computer program GEOFEM

A computer program, GEOFEM, based on the assumptions discussed in this paper is under development. Some special versions containing the same basic subroutines are:

GEOFEM S	Solid (drained, plasticity)
GEOFEM G	Groundwater flow (confined, linear)
GEOFEM C	Consolidation (partly drained or undrained, plasticity, creep)

The number of element variables are shown in Figure 4.

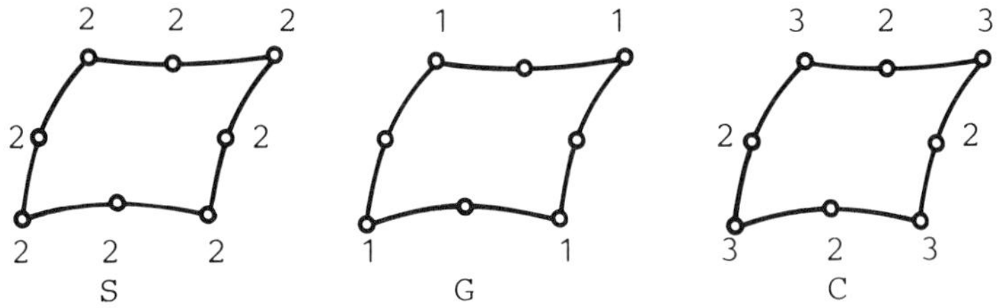

Variables 1:(p), 2:(u_1,u_2), 3:(u_1,u_2,p)

Figure 4 Element variables for different versions of GEOFEM

The solution of the equation system with a positive definite or indefinite coefficient matrix is performed by a direct Crout method, Wiberg and Tägnfors (1976). Input data are given in free format. The program has plotting facilities.

NUMERICAL EXAMPLES

Basic data

The ground profile in Figure 5a is considered under two different loading alternatives. Compared to the layer of soft clay, sand and rock are considered as very stiff. The initital pore pressure distribution is assumed to be hydrostatic.

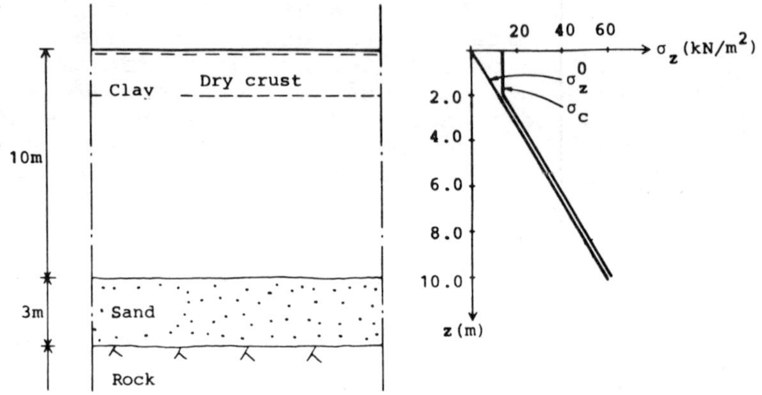

(a) (b)

Figure 5 (a) ground profile, (b) initial stress and preconsolidation stress distribution with depth

The vertical effective pressure (compression stress) in situ is assumed as

$$\sigma_z^0 = \rho' g z, \quad \rho' = \rho_m - \rho_w$$

A normal value for Swedish clay is $\rho_m g = 16$ kN/m^3 giving $\rho' g = 6$ kN/m^3.

It is assumed that the preconsolidation pressure $\bar{p}_c^{(1)}$, which is the equivalent to $p_c^{(1)}$ for the anisotropic extension of (15), can be estimated as

$$\bar{p}_c^{(1)} = \sigma_c(1 + 2K_0)/3, \quad K_0(\text{in situ}) \approx K_0^{NC}$$

where, in the present example, $K_0 = 0.8$ is chosen. The distribution of σ_c, which is the vertical preconsolidation stress obtained at an oedometer test, is assumed as, Figure 5b,

$$\sigma_c = \begin{cases} 2.0\alpha \text{ kN/m}^2, & 0 \leq z \leq 2\text{m} \\ z\alpha \text{ kN/m}^2, & 2 \leq z \leq 10\text{m} \end{cases}$$

where α is a constant. The constant value of σ_c in the upper part of the clay layer corresponds to a dry crust.

The quasi-static preconsolidation pressure $\bar{p}_c^{(2)}$ is assumed to be equal to p^0, the mean effective pressure in situ. It can also be calculated as

$$\bar{p}_c^{(2)} = \bar{p}_c^{(1)}/OCR$$

where the in situ value of the overconsolidation ratio OCR is

$$OCR = \frac{\sigma_c}{\sigma_z^0} = \begin{cases} 2.0\alpha/\rho' gz, & 0 \leq z \leq 2m \\ \alpha/\rho' g, & 2 \leq z \leq 10m \end{cases}$$

Beneath the dry crust a slight degree of overconsolidation is assumed by the choice of $\alpha = 6.6$ (OCR = 1.1).

The elastic constants used are $E = 3000$ kN/m^2 and $\nu = 0.30$. The internal friction angle is $\Phi = 30^o$ (giving M) and a small cohesion is assumed, $c = 5$ kN/m^2 (giving p_b). To define the hardening along the normal consolidation line at K_0-consolidation, the value $k_m = 4$ has been chosen. Isotropic permeability is assumed with $k = 2 \cdot 10^{-10} m^4/kNs$.

Point well discharge and rock tunnel leakage

Due to discharge in a point well, Figure 6a, the pressure is assumed to decrease to zero across the aquifer. Complete axisymmetry is assumed. The extremes of no drainage and free drainage along the well are compared. Due to leakage into a rock tunnel beneath a cracked zone, Figure 6b, there will be a decrease of pore pressure. Plain strain and flow are assumed. The effect of concrete injection of the cracked zone is investigated.

When the stationary pore pressure distribution at the sand-clay interface is calculated, hydrostatic distribution is assumed at a certain distance from the well or the cracked zone respectively.

The consolidation settlement and the total pore pressure development at the symmetry axis caused by point well discharge are shown in Figures 7 and 8. The larger settlements when drainage along the well is assumed are significant. It is noticed that the excess pore pressure gradient in general does not vanish at hydraulic equilibrium but there is a state of stationary flow. The finite element mesh is chosen so that a refinement is obtained in the lower part of the clay layer. This is because the true pore pressure gradient in this part is very steep for early times, which is demonstrated in Figure 8.

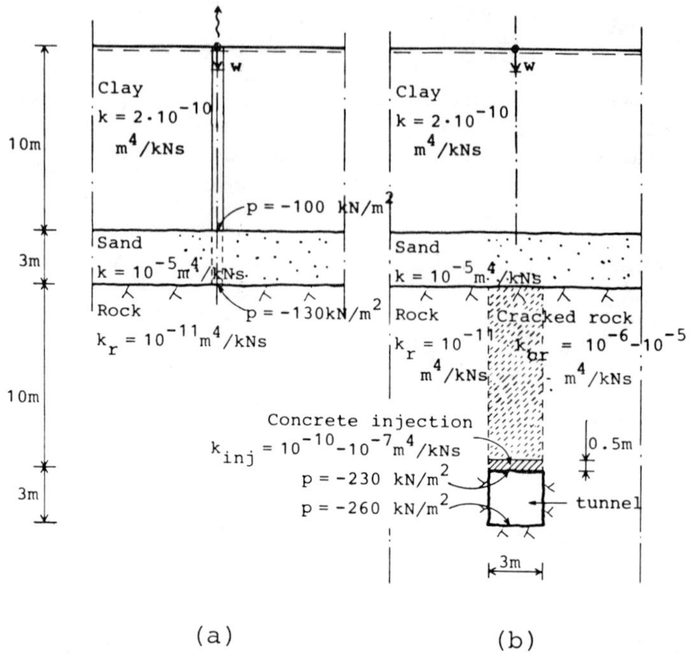

Figure 6 Hydraulic loading, (a) point well discharge, (b) rock tunnel leakage

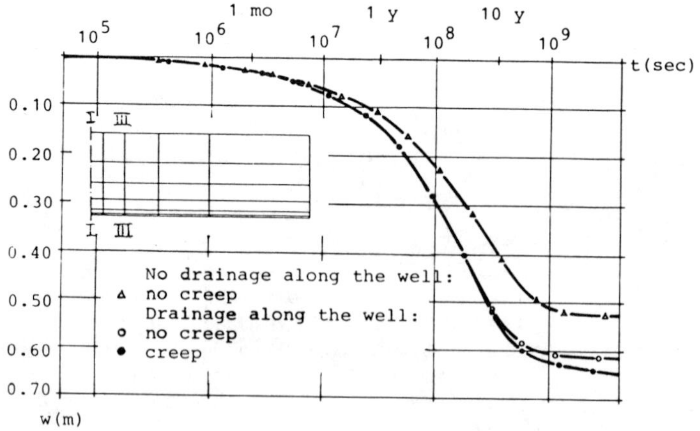

Figure 7 Settlement at symmetry axis due to point well discharge

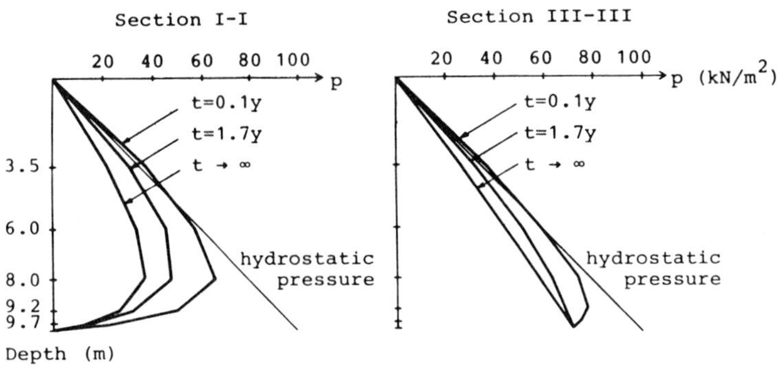

Figure 8 Total pore pressure profiles in the case of no drainage along the well

Similar calculations for the tunneling problem are performed. Due to the great uncertainty of the coefficient of permeability k_{cr} in the zone of cracked rock, a few values are tried: $k_{cr} = 10^{-5}$ and $k_{cr} = 10^{-6}$ m^4/kNs. The effect of concrete injecting of the tunnel is investigated for the case $k_{cr} = 10^{-5}$ m^4/kNs. It turns out that even a poor concrete has a very advantageous effect on the pore pressure decrease. Horizontal pore pressure decrease profiles just beneath the clay layer are shown in Figure 9 for various conditions. In the case of no injecting the consolidation process (without creep) is shown in Figure 10. As the larger settlement corresponds to considerable strains, a small deformation theory may be questionable.

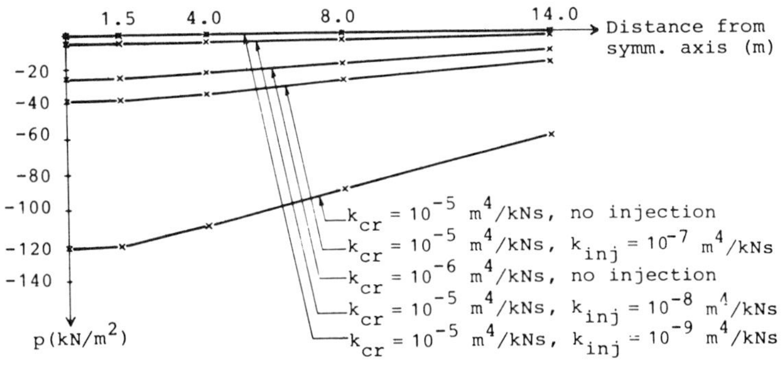

Figure 9 Pore pressure decrease beneath the clay layer for different permeabilities of cracked rock and injecting concrete

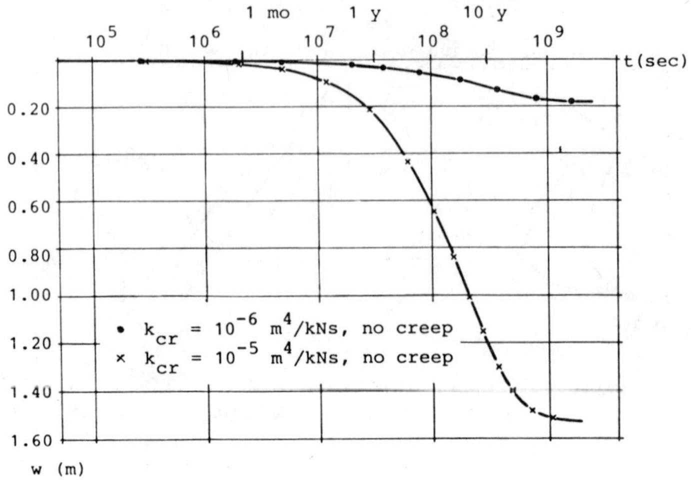

Figure 10 Consolidation due to tunnel leakage (no injecting)

CONCLUSION

The integrated analysis of sand-clay layer systems is efficiently treated if the pore pressure development in the two layers are assumed to be separated in time. To obtain reliable results from the time-dependent settlement analysis the constitutive model for clay must inclue plasticity and creep. The adopted time-integration method gives unconditionally stable solutions and the time steps are chosen automatically based on the estimated truncation error. Calculations performed on leakage into a rock-tunnel through cracked rock indicate dramatic differences in the result due to the choice of permeabilities.

ACKNOWLEDGEMENTS

Partial financial support has been obtained from the Swedish Board of Building Research, which is gratefully acknowledged.

REFERENCES

Berre, T. (1975) The Use of Triaxial and Direct Shear Tests for the Solution of Geotechnical Problems (in Norwegian). Nordisk Geoteknikermoede i Koebenhavn, Polyteknisk Forlag, Copenhagen.

Biot, M. (1941) General Theory of Three-dimensional Consolidation. J. Appl. Physics, 12, 155-164.

Connor, J., Brebbia, C.A. (1976) Finite Element Techniques for Fluid Flow. Newnes-Butterworth, London.

Desai, C.S., Abel, J.F. (1972) Introduction to the Finite Element Method. van Nostrand Reynhold Company, New York.

Domenico, P.A. (1972) Concepts and Models in Groundwater Hydrology. McGraw-Hill, New York.

Janbu, N. (1970) Foundations in Soil Mechanics (in Norwegian). Tapir, Trondheim.

Larsson, R. (1977) Basic Behaviour of Scandinavian Soft Clays. Swedish Geotech. Inst., Rep. No. 4, Linköping.

Li, W-H. (1972) Differential Equations of Hydraulic Transients, Dispersion and Groundwater Flow. Prentice-Hall, Englewood Cliffs.

Noersett, S.P. (1974) One-step Methods of Hermite Type for Numerical Integration of Stiff Systems. BIT, 14, 63-77.

Perzyna, P. (1966) Fundamental Problems in Viscoplasticity. Advances in Appl. Mech., 9, 243-377.

Runesson, K. (1978) On Non-linear Consolidation of Soft Clay. Chalmers Univ. of Techn., Dep. of Struct. Mech., Göteborg.

Runesson, K., Axelsson, K. (1977) An Anisotropic Yield Criterion for Clay. Proceedings of the Int. Conf. on Finite Elements in Nonlinear Solid and Struct. Mech. Geilo, 29.8-1.9 1977. Tapir, Trondheim.

Runesson, K., Tägnfors, H., and Wiberg, N-E. (1978) GEOFEM, A Computer Program for Finite Element Analysis of Geotechnical Problems. Chalmers Univ. of Techn., Dep. of Struct. Mech., Göteborg (in preparation).

Sandhu, R.S. (1968) Fluid Flow in Saturated Porous Elastic Media. Univ. of California, Berkeley.

Singh, A., Mitchell, J. (1968) General Stress-strain-time Functions for Soils. J. Soil Mech. and Found. Div., ASCE, 94, 21-46.

Steihaug, T., Wolfbrandt, A. (1977) An Attempt to Avoid both Iteration and Exact Jacobian in the Numerical Solution of Stiff Differential Equations. Chalmers Univ. of Techn., Dep. of Comp. Sci., Göteborg.

Sällfors, G. (1975) Preconsolidation Pressure of Soft, High-plastic clays. Chalmers Univ. of Techn., Geotechn. Dep., Göteborg.

Terzaghi, K. (1943) Theoretical Soil Mechanics. Wiley and Sons, New York.

Wiberg, N-E., Tägnfors, H. (1976) General Solution Routines for Symmetric Equation Systems. The Mathematics of Finite Elements and Applications II, MAFELAP 1975 (Ed. Whiteman), Academic Press, London, 499-509.

Wolfbrandt, A. (1977) A Study of Rosenbrock Processes with Respect to Order Conditions and Stiff Stability. Chalmers Univ. of Techn., Dep. of Comp. Sci., Göteborg.

S E S S I O N   2

SURFACE WATER FLOW

A TWO-DIMENSIONAL HYDRODYNAMIC MODEL OF A TIDAL ESTUARY

Roy A. Walters, Ralph T. Cheng

U.S. Geological Survey, Menlo Park, Calif. 94025

ABSTRACT

This paper describes a finite element model which is used in the computation of tidal currents in an estuary. This numerical model is patterned after an existing algorithm and has been carefully tested in rectangular and curve-sided channels with constant and variable depth. One of the common uncertainties in this class of two-dimensional hydrodynamic models is the treatment of the lateral boundary conditions. Special attention is paid specifically to addressing this problem. To maintain continuity within the domain of interest, "smooth" curve - sided elements must be used at all shoreline boundaries. The present model uses triangular, isoparametric elements with quadratic basis functions for the two velocity components and a linear basis function for water surface elevation. An implicit time integration is used and the model is unconditionally stable. The resultant governing equations are nonlinear owing to the advective and the bottom friction terms, and are solved at each time step iteratively by the Newton-Raphson method.

Further model test runs have been made in the southern portion of San Francisco Bay, California, USA (South Bay) where a two-dimensional model is justifiable. Due to the complex bathymetry, the hydrodynamic characteristics of South Bay are dictated by the generally shallow basin which contains deep, relict river channels. Great care must be exercised to ensure that the conservation equations remain locally as well as globally accurate. Simulations have been made over several representative tidal cycles using this finite element model and the results compared to existing data.

INTRODUCTION

Estuaries are generally characterized by complex interactions between physical, chemical, and biological processes. Because water circulation is one of the most important factors controlling these processes, considerable effort has been devoted to the study of tidal and long-term circulation patterns in estuaries.

Early work in tidal dynamics was largely confined to analytical studies using the linearized equations of motion (Lamb, 1932). With the growth of computer technology, there has been a shift to the use of numerical methods to solve the complete equations of motion. The finite difference method was applied first and is exemplified by the work of Leendertse (1967). Subsequently, this method has been expanded to include the transport equations and applied to various estuaries (Reid and Bodine, 1968; Leendertse, 1970; Hess, 1976).

More recently, numerous researchers have begun an examination of the finite element method, partly because this method handles the generally complex geometry of estuaries with less difficulty. Advances have been made in finite element simulations of tidal hydrodynamics using vertically uniform or vertically integrated approximations to the complete equations of motion; as yet, three-dimensional models are not practical. These simulations have used a variety of formulations and solution techniques, such as the use of various basis functions, different matrix storage, linear and nonlinear solution methods, and different time-stepping methods. For instance, Wang and Connor (1975) used linear basis functions along with the split-time differencing method. Partridge and Brebbia (1976), King et al. (1974), and Kawahara et al. (1966) used quadratic elements and a variety of explicit and implicit time-integration methods. In a somewhat different approach, Pearson and Winter (1977) used harmonic decomposition and linear basis functions to solve the tidal problem in an elliptical embayment.

This paper discusses a finite element model which computes the tidal and residual currents in an estuary. This numerical model is patterned after an algorithm described by King et al. (1974), and has been carefully tested in rectangular and curvilinear channels with constant and variable depth. One of the common uncertainties in this class of two-dimensional hydrodynamic models is the treatment of lateral boundary conditions. This problem is addressed by using smooth, curve-sided elements at all shoreline boundaries. Also of special interest is the use of mixed interpolation--quadratic basis functions for the two velocity components, and a linear basis function for water surface elevation.

Development of the mathematical formulation and the finite element model are summarized in the following two sections. Next, a summary of the numerical results for hypothetical basins is presented, including a discussion of the effects of various types of boundary conditions. Finally, simulations for the southern portion of San Francisco Bay are discussed.

MATHEMATICAL MODEL

The problem considered here is the calculation of the water surface elevation and the horizontal velocity field within a tidal basin that is connected to the open ocean. The governing equations are the conservation equations for momentum and mass. When the momentum equations are time-averaged over turbulent time scales, they are known as the Reynolds equations (Neumann and Pierson, 1966). Only changes in the dependent variables over a time period of the order of a tidal period are considered significant and treated.

The full three-dimensional governing equations are averaged over the depth of the water column to arrive at a system of two-dimensional, depth-integrated momentum and continuity equations. This system of equations is known as the shallow water equations. Details of this development can be found in Leendertse (1967), Connor and Wang (1974), and Nihoul (1975). The vertically integrated system of equations is written as follows:

the x-momentum equation,

$$\frac{\partial u}{\partial t} + u\frac{\partial u}{\partial x} + v\frac{\partial u}{\partial y} - fv$$

$$= -g\frac{\partial h}{\partial x} + \frac{1}{H}\left[\frac{\partial}{\partial x}(H\tau_{xx}) + \frac{\partial}{\partial y}(H\tau_{xy}) + \tau_x^s - \tau_x^b\right] \qquad (1)$$

the y-momentum equation,

$$\frac{\partial v}{\partial t} + u\frac{\partial v}{\partial x} + v\frac{\partial v}{\partial y} + fu$$

$$= -g\frac{\partial h}{\partial y} + \frac{1}{H}\left[\frac{\partial}{\partial x}(H\tau_{yx}) + \frac{\partial}{\partial y}(H\tau_{yy}) + \tau_y^s - \tau_y^b\right] \qquad (2)$$

and the continuity equation,

$$\frac{\partial h}{\partial t} + \frac{\partial}{\partial x}(Hu) + \frac{\partial}{\partial y}(Hv) = 0 \qquad (3)$$

where

density is assumed constant,

x,y are the Cartesian coordinates in the horizontal plane (m),

u,v are the depth-averaged velocity components in the x,y direction (m/sec),

t is time (sec),

h is the water surface elevation measured from mean lower low sea level (m),

H is total depth of the water column (m),

$f = 2\Omega \sin\theta$ is the Coriolis parameter (sec^{-1}),

Ω is the rotation rate of the earth (sec^{-1}),

θ is the latitude (deg),
g is the gravitational acceleration (m/sec^2),
τ_{xx}, τ_{xy}, τ_{yx}, and τ_{yy} are the combination of the molecular and Reynolds stresses, and the dispersion terms (m^2/sec^2),
and τ_x^s and τ_y^s are the surface wind stresses (m^2/sec^2) and
τ_x^b and τ_y^b are the bottom stresses (m^2/sec^2) in the x and y direction, respectively.

In order to solve equations 1 through 3, all the stress terms must be expressed as a function of the mean variables or be specified as inputs. These terms are dependent upon correlations between perturbation quantities of the dependent variables and are usually unknown. Attempts to compute these terms lead to the problem of closure which is discussed by Tennekes and Lumley (1972), Batchelor (1967), and Nihoul (1975). As an ad hoc approach, these terms are approximated as dispersion terms with the coefficients adjusted empirically, i.e.

$$\tau_{xx} = \varepsilon_{xx}\frac{\partial u}{\partial x}, \quad \tau_{xy} = \varepsilon_{xy}\frac{\partial u}{\partial y}, \quad \tau_{yx} = \varepsilon_{yx}\frac{\partial v}{\partial x}, \quad \tau_{yy} = \varepsilon_{yy}\frac{\partial v}{\partial y}. \qquad (4)$$

The surface stresses are specified by a quadratic form of the wind velocity (Pond, 1975),

$$\tau_x^s = C_D\left(\frac{\rho_a}{\rho}\right)W_x(W_x^2 + W_y^2)^{\frac{1}{2}}, \quad \tau_y^s = C_D\left(\frac{\rho_a}{\rho}\right)W_y(W_x^2 + W_y^2)^{\frac{1}{2}} \qquad (5)$$

where
ρ_a = density of air (kg/m^3),
ρ = density of water (kg/m^3),
C_D = drag coefficient (nondimensional),
W_x, W_y = wind speed (m/sec) in the x,y direction respectively.

The Manning-Chezy formulation for bottom stress in open channel flows is extended to two dimensions to give

$$\tau_x^b = \frac{g}{C^2}\,u(u^2 + v^2)^{\frac{1}{2}}, \quad \tau_y^b = \frac{g}{C^2}\,v(u^2 + v^2)^{\frac{1}{2}} \qquad (6)$$

where C is the Chezy Coefficient (m$^{\frac{1}{2}}$sec^{-1}) (Chow, 1959).

All that remains to complete the problem statement is a specification of the initial and boundary conditions. Initial conditions are somewhat arbitrary while the boundary conditions are complicated because they depend upon a wide range of external forcing conditions.

The treatment of the surface and bottom stresses has already been discussed. It can be shown by scaling that the surface and bottom stress terms dominate over the lateral stresses in most parts of an estuary, particularly in shoal areas and nearshore regions. Inclusion of lateral stresses in the mathematical formulation is largely to improve numerical stability. As a result, only parallel flow conditions are applied at shoreline boundaries. At open boundaries, the water surface elevation and (or) the current velocity can be specified so long as the system is not overconstrained. The application of these boundary conditions and their effects on accuracy of the solution are considered in the results section.

FINITE ELEMENT METHOD

The depth-integrated continuity and momentum equations (equations 1 through 3) along with the associated boundary conditions form a well posed initial-boundary value problem whose numerical solution is obtained by the finite element method. The Galerkin finite element formulation used here follows from the work of King et al. (1974), while general background information for the finite element method may be found in Pinder and Gray (1977), Huebner (1975), and Cheng (1978).

Finite element spatial discretization is used to approximate the spatial domain with six-node triangular elements. Values of the dependent variables in each element are interpolated from the nodal values to arrive at continuous approximations covering the entire domain of interest. The coefficients in the interpolation equations are the nodal values of the dependent variables.

Mixed interpolation is used in this study--quadratic functions are used to interpolate the velocity components u and v and linear basis functions are used to interpolate the water surface elevation h. This scheme is purported to have advantages over the use of the same interpolation functions for both velocity and elevation. The study of the Navier-Stokes equations by Hood and Taylor (1974) compared mixed interpolation using quadratic/linear and cubic/quadratic basis functions, with interpolation using quadratic/quadratic basis functions. They conclude that: (1) there is a great improvement in the accuracy of the pressure field (represented here by water surface elevation), (2) there is a slight degradation in the accuracy of the velocity field, and (3) there is a large improvement in computational efficiency due to the decreased number of nodal equations.

With the element network and basis functions specified, the Galerkin method is applied to the governing equations in order to reduce these equations to a discrete system which can be readily solved by computer. The residuals of the governing equations are formed when the approximations for the dependent variables are substituted into the momentum and the continuity equations. The weighted residuals are defined as the weighted mean of these residuals over the domain of interest. In the Galerkin finite element formulation, the weighting functions used are the basis functions for the respective equation being integrated. By requiring that the weighted residuals equal zero, the governing equations take an integral form. The dispersive stress terms are in a divergence form; the order of these terms is reduced by the use of integration by parts. As a result, the governing equations can be written as,

$$\sum_e \int_e \Bigg[\{\phi\} \Bigg\{ [\phi]\frac{d\{u\}}{dt} + ([\phi]\{u\})\frac{\partial[\phi]}{\partial x}\{u\} + ([\phi]\{v\})\frac{\partial[\phi]}{\partial y}\{u\} - f[\phi]\{v\} + g\frac{\partial[\psi]}{\partial x}\{h\} - \frac{1}{[\psi]\{H\}}\left[\tau_x^s - \tau_x^b\right]\Bigg\} + \frac{\partial\{\phi\}}{\partial x}\tau_{xx} + \frac{\partial\{\phi\}}{\partial y}\tau_{xy}\Bigg] dxdy - \oint\{\phi\}\left[\tau_{xx}dy - \tau_{xy}dx\right] = 0 \tag{7}$$

$$\sum_e \int_e \Bigg[\{\phi\} \Bigg\{ [\phi]\frac{d\{v\}}{dt} + ([\phi]\{u\})\frac{\partial[\phi]}{\partial x}\{v\} + ([\phi]\{v\})\frac{\partial[\phi]}{\partial y}\{v\} + f[\phi]\{u\} + g\frac{\partial[\psi]}{\partial y}\{h\} - \frac{1}{[\psi]\{H\}}\left[\tau_y^s - \tau_y^b\right]\Bigg\} + \frac{\partial\{\phi\}}{\partial x}\tau_{yx} + \frac{\partial\{\phi\}}{\partial y}\tau_{yy}\Bigg] dxdy - \oint\{\phi\}\left[\tau_{yx}dy - \tau_{yy}dx\right] = 0 \tag{8}$$

$$\sum_e \int_e \{\psi\} \Bigg\{ [\psi]\frac{d\{h\}}{dt} + \frac{\partial}{\partial x}\Big[([\psi]\{H\})([\phi]\{u\})\Big] + \frac{\partial}{\partial y}\Big[([\psi]\{H\})([\phi]\{v\})\Big]\Bigg\} dxdy = 0 \tag{9}$$

where

- $[\phi]$ are the quadratic basis functions,
- $[\psi]$ are the linear basis functions,
- [ ] denotes a row vector,
- { } denotes a column vector.

Note that the lateral stress terms represented by the line integrals along the boundary can be set equal to zero for reasons discussed previously.

After evaluating the coefficient integrals, equations 7 to 9 become a system of nonlinear first order differential equations in time for {u}, {v}, and {h}. In general, the integrations over the elements in the x-y plane are too complex to be carried out analytically. For this reason, the elements are mapped into a standard right triangle using an isoparametric transformation, and the integrations are performed numerically using a seven-point Gaussian quadrature (see Huebner, 1974).

Because isoparametric elements are used, any point (x,y) can be expressed in terms of the quadratic basis functions defined in the transformed space (ξ,η) and the nodal coordinates of the element. Along one side of the element, 3 of the 6 basis functions are zero so that the isoparametric transformation can be written in parametric form,

$$x = [\phi(\zeta)]\{x\}$$

$$y = [\phi(\zeta)]\{y\}$$

where ζ is the distance along the side of the element and the vectors contain three components. From these expressions, the slope dy/dx may be calculated. By specifying the slope and locations of the corner nodes, the location of the center node can be calculated from the expressions for the slopes evaluated at the two corner nodes. Following the above procedure, a smooth, curved boundary is created by specifying the locations and slopes at the corner nodes of each element and calculating the position and slope of the center node. More importantly, the smoothness of the side boundary between elements is assured because the slope is uniquely defined at the common corner node.

The system of ordinary differential equations is integrated with respect to time by using finite difference techniques. A fully implicit method is adopted because it is unconditionally stable and not subject to the generally stringent stability criterion associated with the explicit method. Large time steps may be used; however, the global coefficient matrix is time dependent and must be calculated at each iteration. With all things considered, the implicit method seems to be competitive with the explicit method for two-dimensional problems of small to moderate size due to the time step limitation for explicit methods.

After approximating the time derivatives, the governing equations are reduced to a nonlinear algebraic system which is solved by the Newton-Raphson iteration method (Newbery, 1974). At each iteration the original system is approximated by a system of linear algebraic equations which is solved by Gaussian elimination. Convergence of the solution depends upon the starting conditions for steady problems and the rate of change of the dependent variables for time-dependent problems. A steady problem usually requires 5 to 10 iterations for convergence; a time-dependent problem requries 2 to 3 iterations. Convergence occurs when the maximum change in the dependent variables between successive iterations is less than 1 percent.

RESULTS AND DISCUSSION

In this section, the results of two studies employing the model are discussed in detail. The first study examines the effects of boundary conditions on the accuracy of the model. These results are largely derived from a set of numerical experiments in regular-shaped basins and consider the effects of conditions specified at both lateral and open boundaries.

In the second study the model is applied to south San Francisco Bay, USA; boundary conditions and the general behavior of the model are discussed.

Boundary Conditions

A common difficulty in two-dimensional finite element hydrodynamic models is the specification of the lateral boundary conditions at the shoreline. As pointed out earlier, the lateral stresses are important only within the boundary layer along the shoreline (solid boundary). Because the size of an element is usually much greater than the thickness of the shore boundary layer, setting the velocity equal to zero at the boundary in many cases distorts the velocity field to an unrealistic extent. Alternatively, as is the case for this study, the velocity can be constrained to be tangent to the boundary so that no mass flux exists across the shoreline boundary.

When straight-sided elements are utilized, the direction of the boundary is undefined at those nodes where there is curvature to the boundary. Thus, no matter how the direction is specified, there will be a normal velocity near these nodes. A typical manifestation of this effect would include a flow through an island or peninsula. One of the primary reasons for using quadratic basis functions is to allow the use of curve-sided elements. When the domain of interest is represented by finite elements, the entire shoreline boundary is approximated by a smooth, curved boundary as described above. Thus the ambiguity of the boundary direction at any of the nodes is eliminated.

In a separate study, a series of numerical experiments was conducted to compare piecewise-linear and curve-sided elements, and to assess the effects of depth variations on the accuracy. As a simple index of error, simulations were made with a steady flow through various basins and computed discharges at several cross sections were compared. A steady flow was used since these comparisons are meaningless for time dependent flows (see the discussion in Norton et al., 1973). Two of the basin shapes which were used in the simulations are shown in figure 1. These basins had a constant depth and a decrease in width of up to 80 percent. In one case, the decrease was approximated linearly over an element; in the other, the slope was set at the center node and curved adjacent elements were used.

The results indicated that continuity errors were smaller when using curve-sided elements, especially for the larger reductions in width. The continuity accuracy was better in both a local and global sense. But more importantly, the flow field was much more realistic than that computed using straight-sided elements due to the fact that the parallel flow conditions cannot be satisfied exactly where there is an abrupt change in boundary direction. As a result, there were significant flows into and (or) out of

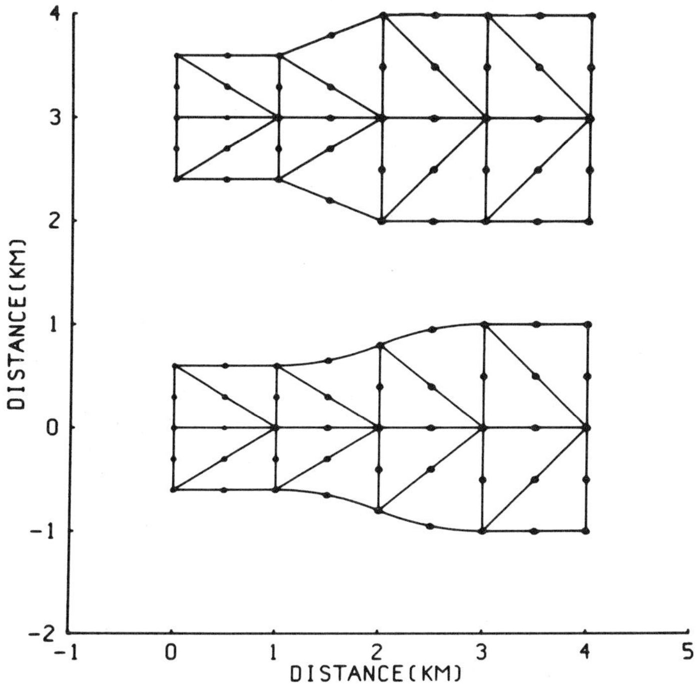

FIGURE 1. Straight-sided and curve-sided basins used in the numerical experiments. The outflow is on the left.

the lateral boundaries.

The seaward boundary conditions have proved to be difficult to specify both in a physical and a numerical sense. In a physical sense, the characteristics of the tidal wave at this boundary are determined by the incident oceanic wave and the reflected waves within the embayment. Only for simple basin shapes can the problem be solved (Taylor, 1920; Lamb, 1932; and a recent discussion by Pearson and Winter, 1977). Furthermore, the velocity field is sensitive to the lateral gradient in water surface elevation, which itself is time-dependent. In practice, data are usually available to specify the water surface elevation at only one station on an open boundary, while additional knowledge of the velocity field or the lateral gradient in surface elevation is required. Otherwise, there is insufficient information to specify the boundary condition at an open boundary for a two-dimensional flow field.

Other problems arise in the numerical implementation of these boundary conditions. As a result, numerical experiments were made where the model was tested with respect to a variety of boundary condition types. The results are briefly summarized where they apply to tidal basins.

Of all the boundary condition types tested, the least accurate is the specification of a spatially constant water surface elevation across an open boundary. If the surface elevation is specified, the corresponding equations are removed from the system of equations. Unfortunately, these are continuity equations; consequently, any velocity gradient in the set of elements along the open boundary is poorly approximated.

In the numerical experiments, a sinusoidal surface elevation was specified at one end of a rectangular basin. The results for the surface elevation agreed with the analytical solution (Lamb, 1932) everywhere; the velocity agreed everywhere except at the nodes on the seaward boundary. The velocity gradient in this region was under-approximated, resulting in 5 to 10 percent errors in the velocity. It should be noted that the use of linear (Wang and Connor, 1975) or quadratic (Partridge and Brebbia, 1976) basis functions for both elevation and velocity did not seem to suffer this problem. The increased number of nodes without continuity constraints when using mixed interpolation is probably responsible for these errors. As might be expected, the application of a velocity boundary condition resulted in small local errors everywhere; however, this condition is usually unknown and is one of the desired results.

While local errors at the seaward boundary may be reduced by network refinement, a different approach is adopted here. At this boundary, the surface elevation is specified at only one point while the direction of the velocity is specified at the remainder. In this way, the continuity constraint is retained everywhere except at one node on the boundary, and the need to specify the lateral head gradient is obviated. The lateral gradient automatically adjusts itself so that smooth flow conditions prevail at the open boundary. As a result, local errors in continuity are reduced to the same magnitude as exist throughout the remainder of the domain.

Simulations

San Francisco Bay (fig. 2) is a complex estuary which can be represented by basically two estuarine types. Because nearly all freshwater enters from the Sacramento-San Joaquin River delta, the northern reach is a partially mixed estuary during winter (the wet season) and vertically well mixed during summer, while South Bay is usually isohaline (McCulloch et al., 1970; Imberger et al., 1977). The salinity in South Bay varies seasonally and is primarily controlled by exchanges with the northern reach and Pacific Ocean. Some salinity stratification may be present in winter due to transport of delta-derived water into South Bay during wet years, while the water is otherwise nearly

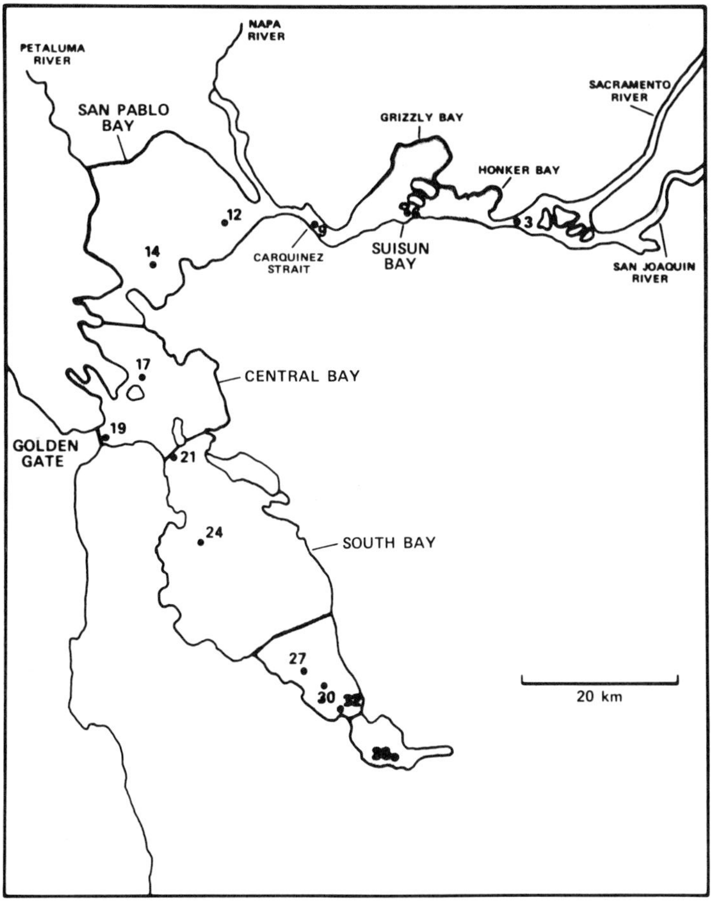

FIGURE 2. The San Francisco Bay system. The shaded areas denote depths of less than 2 meters below mean lower low sea level and the numbers indicate the U.S. Geological Survey routine monitoring stations.

isohaline due to low freshwater flows and wind-induced mixing. As a result, the use of a vertically integrated model in South Bay can be justified.

The bottom topography of South Bay is characterized by shallow depths with several deep, relict river channels. The mean depth is of the order of 5 meters while the channels are as deep as 15 meters. Near the center of the basin, the channel shoals, probably due to sedimentation from tidal scouring farther north. In the southern portion, there are extensive shoal areas with depths of the order of 1 meter. At the extreme southern end, the bay narrows into a number of sloughs and small creeks.

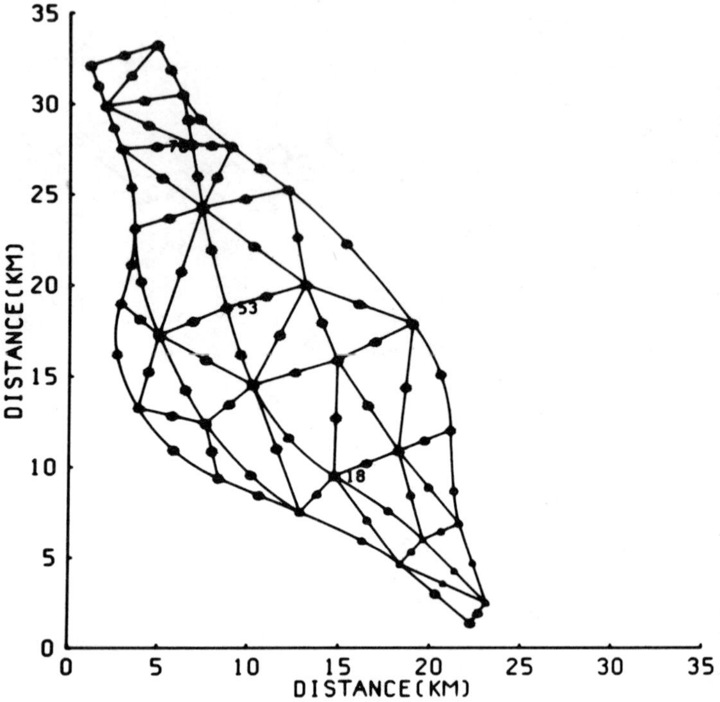

FIGURE 3. The finite element network used to represent South San Francisco Bay. u,v velocity components are specified at all nodes, water surface elevation is only specified at corner nodes.

Modeling difficulties in South Bay arise due to the extreme variations in depth. As a result, the modeling effort proceeded in two stages: (1) the deep channel was modeled with isoparametric elements, (2) elements were added to describe the shoal areas in the basin. Circulation in South Bay seems to follow the channel to a certain extent, probably due to the decreased effect of bottom friction. Thus smaller continuity errors were expected with smooth channel contours. The finite element network for South Bay is shown in figure 3; the channel can be recognized by the curved elements which lie along the western side of the bay. The southernmost portion of South Bay is modeled in only a crude sense as a large amount of network refinement would be required to model the flow here. This refinement was felt to be unwarrented due to the small size and shoal nature of this region.

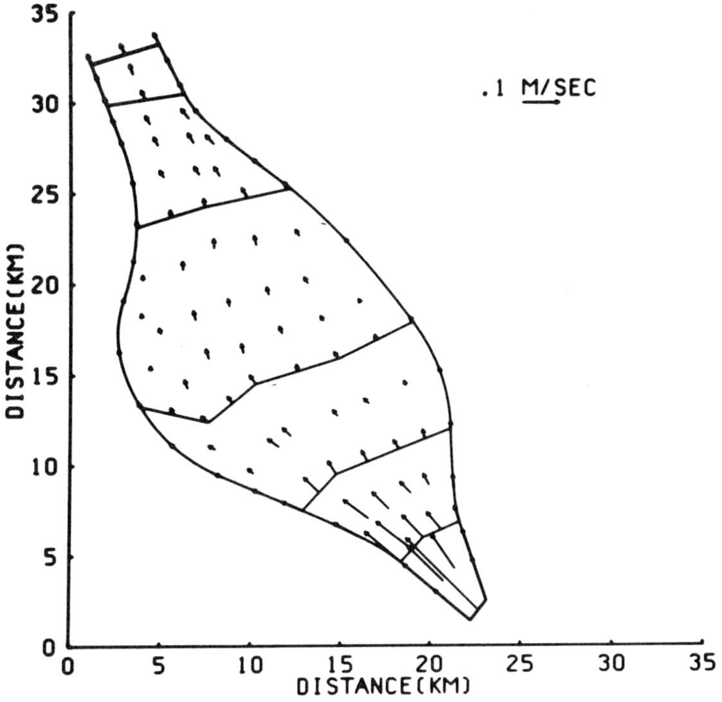

FIGURE 4. Velocity field for a simulated steady flow in South San Francisco Bay. Velocity boundary conditions are used at the inflow and a one point head condition at the outflow. Discharge is computed along the noted sections.

The continuity criterion was applied to the South Bay network by introducing a steady flow at the southern end of the basin and comparing the calculated discharges across several sections. These sections and the resultant velocity field are shown in figure 4. Refining the network can improve accuracy; however, continuity errors were kept under 10 percent by using the network in figure 3. The model was sensitive to the conditions specified at the seaward boundary such that if the lateral gradient in surface elevation was not taken into account, the local errors in velocity were usually large. Due to the pronounced asymmetry of the basin, this gradient could not be approximated in a simple way--such as with a geostrophic balance for instance. Specifying the surface elevation at one node and the velocity direction at all the nodes on the open boundary resolved this problem.

Until now, measurements of water movement in South Bay have been derived from drifter measurements and a few current-meter observations. Although presently being planned, there has been as yet no comprehensive program to define the spatial and temporal variations in the velocity field. Lacking field data, the model was examined using a semidiurnal tide and the detailed calibration was deferred until the necessary measurements have been made. For the purposes of this simulation, Chezy coefficients of 30 $m^{\frac{1}{2}}sec^{-1}$ were used in the shoal elements and 50 $m^{\frac{1}{2}}sec^{-1}$ in the channel elements. The dispersion coefficients were set equal to 24 m^2/sec (see Glenne, 1966). The model was started from rest at high water and reached equilibrium in about 1½ tidal cycles as is shown by the tidal ellipses in figure 5. Farther north, the ellipses become almost linear due to channeling effects at the seaward boundary (fig. 5b).

A representative flow field at three hours after high water is shown in figure 6 and three hours after low water in figure 7. In the southern portion of the basin, there is a tendency for the water in the shoal areas to flow towards the channel rather than directly toward the seaward boundary due to the effects of bottom friction. In these simulations, twenty time steps per tidal cycle were used. The results varied slightly for one-hour time steps.

This model can be compared to others through the use of the Courant number. Define the Courant number as

$$C_r = \sqrt{gH}\,\Delta t/\Delta x$$

where

Δt is the time step size,

and Δx is the length of a typical element side.

With half-hour time steps, C_r varies between approximately 1.3 and 11 over the South Bay network. If the explicit method is used, the usual upper limit on C_r is approximately 0.5.

Glenne (1966) has shown that the length of South Bay is approximately a quarter of the wavelength of the semidiurnal tide. Therefore, the basin is near resonance, resulting in a marked amplification of the tidal range and a correspondingly small phase lag between the northern and southern ends. The phase lag as computed by the model (45 to 55 minutes) agrees with the measured value of approximately 45 minutes (Glenne, 1966). On the other hand, the model shows a slightly smaller amplification of tidal height at the southern end: 25 percent compared to a measured value of about 35 percent. The tidal volumes as computed along the noted sections (fig. 4) agree reasonably well with data from Glenne (1966).

Summary

This paper describes a finite element model which has two features of special interest: (1) use of mixed interpolation

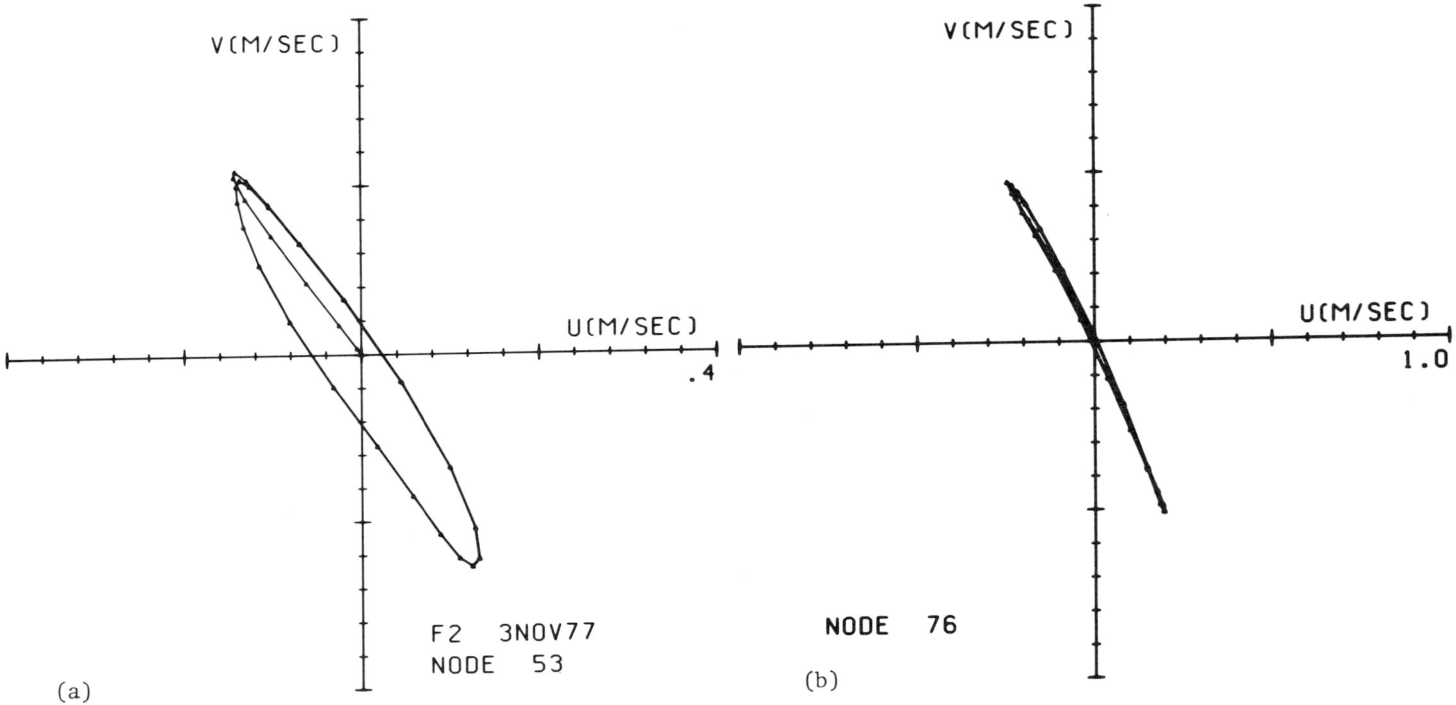

FIGURE 5. Tidal ellipse near the center (a) and in the northern portion (b) of South Bay. Node 53 and node 76 are indicated in figure 3.

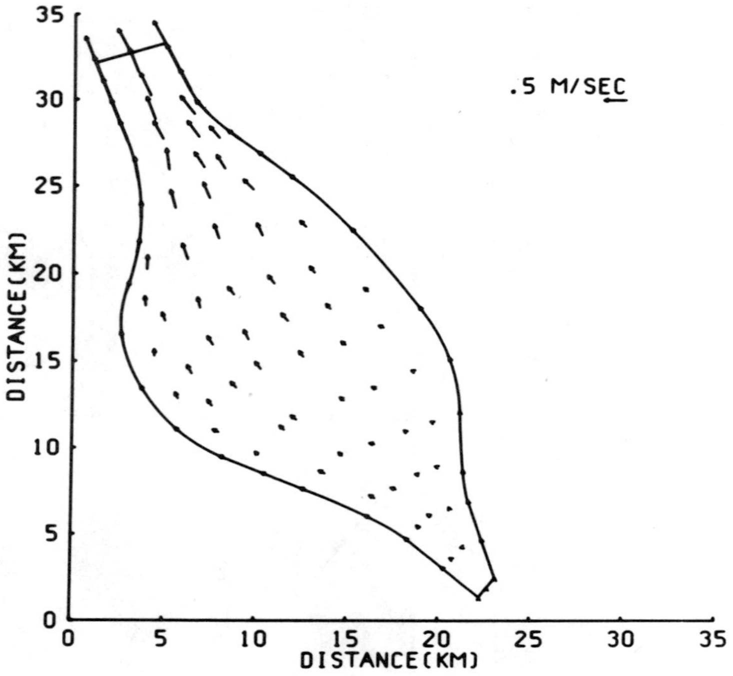

FIGURE 6. Velocity field 3.1 hours after high tide at the north entrance of South San Francisco Bay.

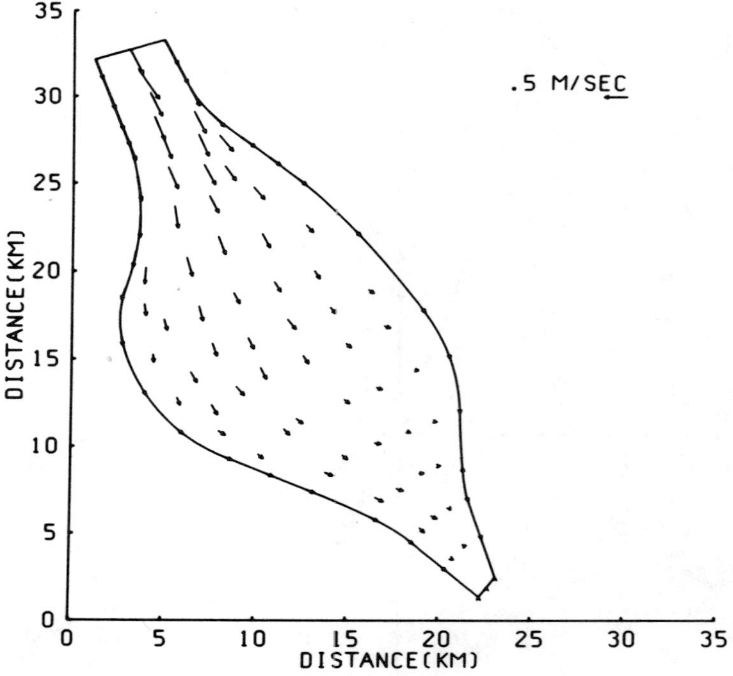

FIGURE 7. Velocity field 3.1 hours after low tide at the north entrance of South San Francisco Bay.

with quadratic basis functions for velocity and linear functions for surface elevation, and (2) use of smooth, curve-sided elements. The use of mixed interpolation improves computational efficiency but causes a reduction in accuracy of the velocity field when surface elevation conditions are specified at open boundaries. However, this problem can be remedied by a careful treatment of the boundary conditions. The use of curve-sided elements eases the specification of parallel flow conditions at lateral boundaries and results in a more accurate representation of the flow field.

The implicit method is used as the time integration scheme because it is unconditionally stable; the size of the time step is limited only by the rate of change of the forcing conditions. As a result, the model can be readily extended to a tidally-averaged seasonal model in the sense of residual circulation as described by Nihoul (1975).

REFERENCES CITED

Batchelor, G. K. (1967) An Introduction to Fluid Dynamics. Cambridge University Press, New York, 738 p.

Cheng, R. T. (1978) Modeling of Hydraulic Systems by Finite Element Methods *in* Chow, V. T., ed., Advances in Hydroscience, 11:207-284.

Chow, V. T. (1959) Open-Channel Hydraulics. McGraw-Hill Book Co., New York, 680 p.

Connor, J. J. and Wang, J. D. (1974) Finite Element Modelling of Hydrodynamic Circulation *in* Brebbia, C. A., and Connor, J. J., eds., Numerical Methods in Fluid Dynamics. Pentech Press, London, p. 355-411.

Glenne, Bard (1966) Diffusive Processes in Estuaries: Sanitary Eng. Research Lab. Rept. 66-6, California Univ., Berkeley, 78 p.

Hess, K. W. (1976) A Three-Dimensional Numerical Model of the Estuary Circulation and Salinity in Narragansett Bay. Estuarine and Coastal Marine Sci., 4:325-338.

Hood, P. and Taylor, C. (1974) Navier-Stokes Equations using Mixed Interpolation *in* Oden, J. T. et al, eds., Finite Element Methods in Flow Problems. UAH Press, Huntsville, Ala., p. 121-132.

Huebner, K. H. (1975) The Finite Element Method for Engineers. John Wiley and Sons, New York, 500 p.

Imberger, Jorg, Kirkland, W. B., Jr., and Fischer, H. B. (1977) The Effect of Delta Outflow on the Density Stratification in San Francisco Bay. H. B. Fischer, Inc., Berkeley, Calif., and Waterfront Design Assoc., Alameda, Calif., Prepared for the Assoc. of Bay Area Govts., Rept. HBF-77/02, 109 p.

Kawahara, M., Yoshimura, N., Nakagawa, K., and Ohsaka, H. (1976) Steady and Unsteady Finite Element Analysis of Incompressible Viscous Fluid. Internatl. Jour. Numerical Methods Eng. 10:437-456.

King, I. P., Norton, W. R., and Iceman, K. R. (1974) A Finite Element Model for Two-Dimensional Flow *in* Oden, J. T. et al, eds., Finite Element Methods In Flow Problems. UAH Press, Huntsville, Ala., p. 133-137.

Lamb, H. (1932) Hydrodynamics, 6th Edition. Dover Press, New York, 738 p.

Leendertse, J. J. (1967) Aspects of A Computational Model for Long-Period Water-Wave Propogation. Rand Corp., Santa Monica, Calif., RM-5294-PR, 165 p.

Leendertse, J. J. (1970) A Water-Quality Simulation Model for Well-Mixed Estuaries and Coastal Seas. Volume 1. Principles of Computation. Rand Corp., Santa Monica, Calif., RM-6230-RC, 71 p.

McCulloch, D. T., Peterson, D. H., Carlson, P. R., and Conomos, T. J. (1970) A preliminary Study of the Effects of Water Circulation in the San Francisco Bay Estuary. U.S. Geol. Survey Circ. 637A and 637B, 34 p.

Neumann, G. and Pierson, W. J., Jr. (1966) Principles of Physical Oceanography. Prentice-Hall, Inc., Englewood Cliffs, N.J., 545 p.

Newbery, A.C.R. (1974) Numerical Analysis *in* Pearson, C. E., ed., Handbook of Applied Mathematics. Van Nostrand Reinhold Co., New York, p. 1002-1057.

Nihoul, J. C. (1975) Hydrodynamic Models *in* Nihoul, J. C., ed., Modeling of Marine Systems. Elsevier, Amsterdam, p. 41-68.

Norton, W. R., King, I. P., and Orlob, G. T. (1973) A Finite Element Model for Lower Granite Reservoir. Water Resources Engineering Inc., Walnut Creek, Calif., 138 p.

Partridge, P. W., and Brebbia, C. A. (1976) Quadratic Finite Elements in Shallow Water Problems. Am. Soc. Civil Eng. Jour. Hydraulics Div. HY9:1299-1313.

Pearson, C. E. and Winter, D. F. (1977) On the Calculation of Tidal Currents in Homogeneous Estuaries. Jour. Phys. Oceanography 7:520-531.

Pinder, G. F. and Gray, W. G. (1977) Finite Element Simulation in Surface and Subsurface Hydrology. Academic Press, Inc., New York, 295 p.

Pond, S. (1975) The Exchanges of Momentum, Heat, and Moisture at the Ocean-Atmosphere Interface *in* Proceedings, Symposium on Numerical Models of Ocean Circulation. Washington, D.C., Natl. Acad. Sci., p. 26.

Reid, R. O. and Bodine, B. R. (1968) Numerical Model for Storm Surges in Galveston Bay. Am. Soc. Civil Eng. Jour. Waterways and Harbor Div., p. 33-57.

Taylor, G. I. (1920) Tidal Oscillations in Gulfs and Rectangular Basins *in* Proceedings, London Math. Soc. 20:148-181.

Tennekes, H. and Lumley, J. L. (1972) A First Course in Turbulence. MIT Press, Cambridge, Mass., 300 p.

Wang, J. D. and Connor, J. J. (1975) Mathematical Modeling of Near Coastal Circulation. Mass. Inst. Technology Rept. no. 200, 272 p.

FINITE ELEMENT SIMULATION OF SHALLOW WATER PROBLEMS WITH MOVING BOUNDARIES

Daniel R. Lynch and William G. Gray

Water Resources Program
Department of Civil Engineering
Princeton University

SUMMARY

The general problem of numerical solution of the non-linear shallow water equations is discussed. The usual formulation involves boundary conditions wherein either the depth or the mass flux is specified through time along fixed boundaries. However, many practical problems involving transient flooding of dry areas are properly formulated by specifying zero depth along a moving boundary whose position depends upon the time history of the flow rate. The boundary condition in this case is nonlinear, and numerical treatment requires a grid which deforms during the course of simulation.

The effects of various approximations to this boundary condition are discussed. An extension of the FE method is presented which allows direct integration of the equations and the boundary conditions, while explicitly accounting for the continuous deformation of the grid. General principles of node movement are discussed, application to two example problems is demonstrated, and directions for further research are indicated.

INTRODUCTION

The shallow water equations have found widespread application in the water resources field. To date, areas of investigation have included storm surge; coastal circulation and transport of coastal pollution; overland runoff response to rainfall; flood routing; and the spreading of oil slicks. In general, the behavior of these systems is dominated by the effects of irregular boundaries, nonlinear interactions, variable physical properties, and arbitrary forcing, the result being that solution of realistic problems requires a numerical approach. Interest in the application of the finite element method to these problems has grown due to its promise of superiority in three areas: 1) the representation of complex geometry; 2) the representation of variable coefficients, especially bathymetric information; 3) the inherent flexibility of arbitrary grid point locations, which allows the investigator to build in detail only where required by the physical problem. There has, however, been little practical motivation for a generalized, variable grid in the time dimension. This, coupled with the ease of interpretation of results offered by a grid which is regular in at least one dimension, has led most investigators to employ a generalized FE mesh in the spatial domain in conjunction with equal intervals in the time domain. The result, whether intentional or not, is effectively a finite difference scheme in time. Given a set of initial and boundary conditions, a continuous FE solution in space is typically obtained at discrete points in time by time stepping.

Boundary conditions for this problem are generally formulated on fixed boundaries, the necessary information being either the total depth of fluid (effectively the pressure force normal to the boundary) or the mass flux across the boundary of the domain, specified throughout time. There is, however, a third type of BC for shallow water problems, namely, the requirement of zero depth on a boundary which deforms continuously, following the motion of the fluid:

$$H = 0 \quad \text{on} \quad \underset{\sim}{X} = \underset{\sim}{X}_0 + \int_0^t \underset{\sim}{V}_f \, dt \qquad (1)$$

where $H(x,y,t)$ is the fluid depth,

$\underset{\sim}{X}(t)$ is the location of the boundary at time t,

$\underset{\sim}{X}_0$ is the initial location of the boundary, and

$\underset{\sim}{V}_f(x,y,t)$ is the fluid velocity at the boundary.

While the first two types of BC's are adequate to solve a variety of practical problems, the details of fluid motion in the vicinity of a moving boundary are often the most important aspects of the solution, and may in fact be the motivation for the modeling effort in the first place. Examples include coastal flooding due to storm surge; periodic tidal flooding of salt flats and wetland; coastal water quality effects on wetlands, beaches, etc.; rate of spreading of oil slicks; and general beach processes such as sedimentation, erosion, etc. If the details near the boundary are required for any of these problems, a method is required which accounts for the continuous deformation of the spatial domain, based on the fluid mechanics of the situation. In numerical terms, the position of at least the boundary nodes must be considered to be functions of time: $\underset{\sim}{X}_i = \underset{\sim}{X}_i(t)$, where the subscript i is a node designation. Since the rate of change of $\underset{\sim}{X}_i(t)$ -- i.e., the nodal velocity -- depends upon the solution itself, the moving boundary condition is nonlinear. Further, since the spatial FE grid is different at every point in time, any matrix manipulation will in general be nonstationary.

Owing to these difficulties, there are two common approaches which effectively circumvent the moving boundary problem. The simplest method is to construct an imaginary vertical barrier near boundaries which would otherwise require node motion, as in Fig. 1. The solution is then found, assuming no flux across the boundaries at the fixed artificial locations. If required the actual location of the fluid boundary at any time is approximated by a horizontal inland extrapolation. This is in fact the approximation which is always made at non-vertical, natural shoreline boundaries, and can be expected to be a good one provided that the actual shoreline motion is small compared to the domain under study. If, however, the boundary motion is significant compared to the spatial discretization, this approximation should fail in several ways:

1) The most obvious problem is the mass deficit involved in the extrapolation of the boundary inland. In this sense, continuity is violated.

2) A corresponding deficit in momentum is not accounted for. The specification of zero velocity at the barrier results in zero frictional loss on this boundary, the influence of which is felt throughout the boundary elements. In fact, since the depth is small near the boundary, frictional losses there could be quite large.

3) In connection with the friction considerations, kinetic energy of the fluid is converted to potential energy at the vertical boundary. The frictional loss which would in reality be incurred by the fluid while moving inland is not accounted for in this transformation. As a result, waves incident to the fixed boundary are reflected perfectly, while in reality the reflection would be im-

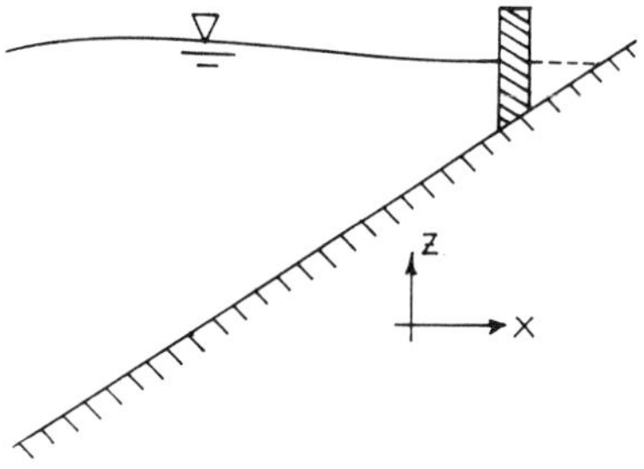

Figure 1. Artificial vertical barrier on flooding shoreline boundary.

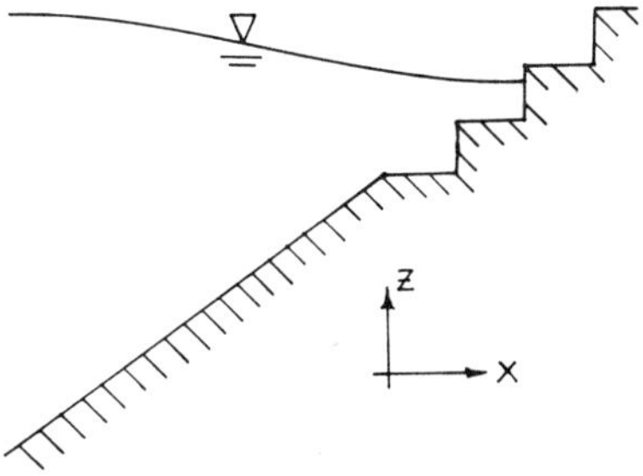

Figure 2. Staircase approximation of bathymetry at shoreline boundary.

perfect. Further, the excursion of the moving boundary is assumed to follow the depth response at the imaginary boundary, while in reality the friction losses result in both the attenuation and delay of the boundary motion. In transient situations this could cause substantial error.

A more refined approach than that discussed above would replace the continuous bathymetry with a discontinuous, "staircase" representation as shown in Fig. 2. Entire elements of the spatial domain are switched on or off, depending upon the computed water levels, assuming instantaneous wetting of a complete element in a manner which conserves mass. Thus, the non-stationary domain is approximated by a series of stationary ones, each of which may be in effect for several time steps with a corresponding economy of computing effort. While this is an improvement over the approach represented in Fig. 1, momentum is still not conserved, and the impulsive boundary motion accompanied by instantaneous wetting of an element can generate spurious disturbances which propagate from the boundary. The assumption of instantaneous wetting is the source of difficulty, for in reality the boundary must move with the velocity of the fluid, which will generally be small in shallow areas.

To illustrate this point, consider the distance which the true boundary moves in one time step, which is approximately

$$V\Delta t = \left(\frac{V}{\sqrt{gH}}\right)\left(\frac{\Delta t}{\Delta x}\sqrt{gH}\right)\Delta x \tag{2}$$

Most models operate with the ratio $\frac{\Delta t}{\Delta x}\sqrt{gH}$ in the range .5 to 1.0; further, velocities are generally much smaller than the wave speed, so typically $\frac{V}{\sqrt{gH}}$ is in the range .01 to .05. Together these imply that $V\Delta t = (.005 \text{ to } .05)\,\Delta x$, that is, it takes from 20 to 200 time steps for the boundary to move a distance of Δx. Thus the impulsive, instantaneous motion implied by the "staircase" method can lead to erroneous results in the vicinity of the boundary when dynamic or transient effects are important.

It is interesting to note that in both approaches outlined above, the exact BC (1), which specifies the depth, is replaced by one in which the normal mass flux is specified. The numerical implication of this approximation is that a component of the momentum equation is discarded at the boundary in favor of the BC on normal velocity, leaving the depth and the tangential velocity as unknowns which must be obtained. In contrast the exact BC (1) requires that the continuity equation be discarded at the boundary in favor of the condition on depth, leaving both components of fluid velocity as unknowns. It is the purpose of this paper to present a technique which uses the correct boundary condition and which computes the location of the domain boundary at each time step.

The simulation of a moving boundary problem can be decomposed into two parts, node movement and rezoning. Simulation begins with the specification of initial conditions on an initial finite element grid and marches forward in time, allowing nodes to move while maintaining the initial grid connectivity. If at any time the initial mesh becomes unacceptably distorted due to accumulated node movement, a rezoning of the domain is required. Generally, rezoning will be necessary when either of two problems arises:

1) Elements become distorted, or the node distribution is too sparse or dense to give the required level of solution detail in various parts of the domain.

2) The grid becomes incapable of accurately representing fixed spatial functions such as bathymetry, forcing functions or friction coefficients.

Once rezoning is accomplished, the simulation proceeds again as an initial value problem. Thus, grid deformation is achieved by the juxtaposition of two processes: incremental growth with fixed connectivity, and instantaneous rezoning with fixed boundaries. Both parts of the problem must be accomplished accurately and efficiently to obtain a good solution. Incremental growth occurs during each time step and requires an effective equation solution technique; this process is the major concern of this paper and will be discussed below. The rezoning process implies automatic recognition of mesh distortion as well as the rezoning itself, both of which are non-trivial programming problems. However these problems appear to be solvable by an efficient bookkeeping system similar to that used in automatic mesh generation and will not be discussed further here.

PRINCIPLES OF NODE MOVEMENT

Generally speaking, it is desirable to minimize unnecessary node motion, since it is computationally efficient to keep the grid as stationary as possible. In particular, for studies of bounded water bodies such as estuaries, lakes, or bays, it is reasonable to keep most interior nodes fixed. This has the advantage of allowing the bathymetry function, which is often a major determinant of initial node placement, to remain fixed in most of the domain. Furthermore, the numerical method reduces to its proven stationary equivalent throughout most of the region. (The case of an oil slick spreading in an unbounded domain might be an exception to these general principles.) Once interior node positions are fixed, tangential motion of boundary nodes could cause unnecessary computational difficulties due to the "shearing" of boundary elements. Thus it is generally desirable to eliminate this motion where possible.

Although the numerical method to be developed will account for arbitrary motion of any node, the physics of the problem as determined from the governing equations place a constraint on boundary node motion. The rationale for boundary node motion may be developed by examining the continuity equation:

$$\frac{\partial H}{\partial t} + \underset{\sim}{V} \cdot \underset{\sim}{\nabla} H + H\underset{\sim}{\nabla} \cdot \underset{\sim}{V} = 0 \tag{3}$$

At a moving boundary H = 0 and this equation reduces to

$$\frac{\partial H}{\partial t} + \underset{\sim}{V} \cdot \underset{\sim}{\nabla} H = \frac{DH}{Dt} = 0 \tag{4}$$

Further expansion of (4) yields

$$\frac{\partial H}{\partial t} + V_n \frac{\partial H}{\partial n} + V_\lambda \frac{\partial H}{\partial \lambda} = 0 \tag{5}$$

where the directions normal and tangential to the boundary are denoted by n and λ, respectively. Since H = 0 along the boundary, $\frac{\partial H}{\partial \lambda} = 0$ and (5) reduces to

$$\frac{\partial H}{\partial t} + V_n \frac{\partial H}{\partial n} = 0 \tag{6}$$

Thus it is sufficient that the boundary move with velocity equal to the normal component of fluid velocity, or more generally,

$$(\underset{\sim}{V}_b - \underset{\sim}{V}_f) \cdot \hat{n} = 0 \tag{7}$$

where the subscripts b and f refer to boundary and fluid, respectively and $\hat{n}$ is a unit normal to the boundary. Boundary condition (1) can now be restated in a less restrictive form:

$$H = 0 \quad \text{on} \quad \underset{\sim}{X} = \underset{\sim}{X}_0 + \int_0^t \underset{\sim}{V}_b \, dt \quad ; \quad (\underset{\sim}{V}_b - \underset{\sim}{V}_f) \cdot \hat{n} = 0 \tag{8}$$

where $\underset{\sim}{V}_b$, $\underset{\sim}{V}_f$, and $\hat{n}$ are all functions of space and time. One final detail must be taken care of: in general, the discretized FE domain will have a discontinuous normal at the junction of two boundary elements. It is therefore not possible to satisfy the normal velocity relation in (8) exactly at every point on the boundary. It is reasonable instead to require that the relation hold in an average sense such that

$$\int_S (\underset{\sim}{V}_b - \underset{\sim}{V}_f) \cdot \hat{n} \, dS = 0 \qquad (9)$$

where S is the boundary of the domain. Expression of the velocities in terms of the FE basis functions, yields

$$\underset{\sim}{V}_b = \sum_i \underset{\sim}{V}_{bi} \, \phi_i \qquad (10a)$$

$$\underset{\sim}{V}_f = \sum_i \underset{\sim}{V}_{fi} \, \phi_i \qquad (10b)$$

Substitution of these expressions into (9) gives the relation

$$\sum_i \int_S (\underset{\sim}{V}_{bi} - \underset{\sim}{V}_{fi}) \cdot \hat{n} \, \phi_i \, dS = 0 \qquad (11)$$

The most "local" way to satisfy (11) is to require each term in the summation to equal zero, or

$$(\underset{\sim}{V}_{bi} - \underset{\sim}{V}_{fi}) \cdot \int_S \hat{n} \, \phi_i \, dS = 0 \qquad (12)$$

which serves as a definition of the "nodal normal direction" $\hat{n}_i$:

$$\hat{n}_i \equiv \frac{\int_S \hat{n} \, \phi_i \, dS}{\left| \int_S \hat{n} \, \phi_i \, dS \right|} \qquad (13)$$

Gray (1977) and Pinder and Gray (1977) have defined the nodal normal direction in this manner for fixed boundaries where the normal flux is specified, so the evaluation of the normal direction for moving boundaries presents no new difficulty. Gray further points out that application of the divergence theorem to the surface integral in (13) results in

$$\int_S \hat{n}\,\phi_i\,dS = \iint_A \nabla\phi_i\,dA \qquad (14)$$

where A is the entire domain. Thus, the nodal normal direction is readily evaluated at any point in time using existing methods, regardless of the type of elements being used. It is useful to point out that for the special case of linear triangles, the nodal normal is perpendicular to the line AC in Fig. 3. Wang and Connor (1975) have also indicated that this choice of nodal normal for linear triangles conserves mass for a fixed no-flux boundary.

The final form of boundary condition (1) is thus

$$H_i = 0 \quad \text{on} \quad \underset{\sim}{X}_i = \underset{\sim}{X}_{i0} + \int_0^t \underset{\sim}{V}_{bi}\,dt \quad ; \quad (\underset{\sim}{V}_{bi} - \underset{\sim}{V}_{fi}) \cdot \hat{n}_i = 0 \qquad (15a)$$

where the subscript i ranges over all moving boundary nodes. It should be emphasized that in order to reduce element shearing, the tangential velocity of a node is arbitrarily required to be zero such that

$$\underset{\sim}{V}_{bi} \cdot \hat{\lambda}_i = 0 \qquad (15b)$$

DISCRETIZATION IN SPACE AND TIME

The governing shallow water equations can be expressed as

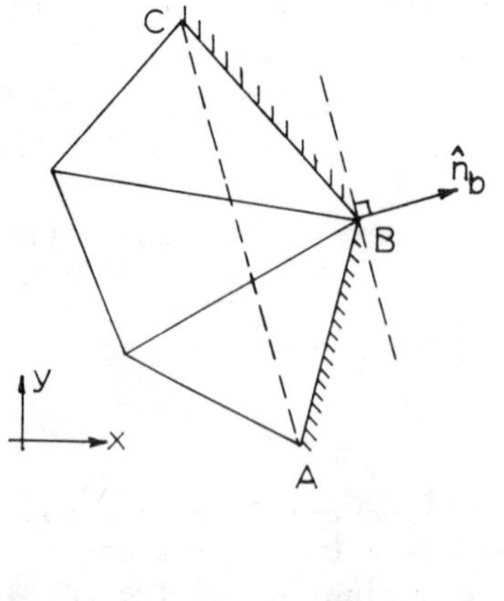

Figure 3. Approximate normal direction at node B.

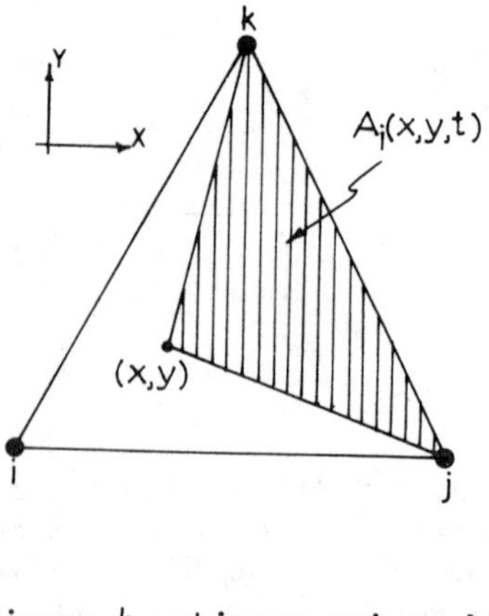

Figure 4. Linear triangular finite element.

$$\frac{\partial^2 H}{\partial t^2} + \tau \frac{\partial H}{\partial t} = \underset{\sim}{\nabla} \cdot (gH\underset{\sim}{\nabla}\zeta) + H\underset{\sim}{V} \cdot \underset{\sim}{\nabla}\tau + \underset{\sim}{\nabla} \cdot [\underset{\sim}{\nabla} \cdot (H\underset{\sim}{V}\underset{\sim}{V})$$

$$+ \underset{\sim}{f} \times H\underset{\sim}{V} - \underset{\sim}{W}] \tag{16a}$$

$$\frac{\partial \underset{\sim}{V}}{\partial t} = - \underset{\sim}{V} \cdot \underset{\sim}{\nabla}\underset{\sim}{V} - \underset{\sim}{f}\times\underset{\sim}{V} - g\underset{\sim}{\nabla}\zeta - \tau\underset{\sim}{V} + \frac{\underset{\sim}{W}}{H} \tag{16b}$$

where x,y,t are the independent variables in space and time,

$H(x,y,t)$	is the total fluid depth,
$\underset{\sim}{V}(x,y,t)$	is the vertically averaged fluid velocity,
$\zeta(x,y,t)$	is the elevation of the free surface above a reference datum,
$H_0(x,y)$	is the bathymetry,
$\underset{\sim}{f}$	is the coriolis parameter,
g	is gravity,
$\underset{\sim}{W}(x,y,t)$	is the wind stress,
$\tau\underset{\sim}{V}(x,y,t) \equiv \dfrac{g\lvert\underset{\sim}{V}\rvert\underset{\sim}{V}}{C^2 H}$	is the bottom stress, and
$C(x,y,t)$	is the Chezy coefficient.

Here the wave equation has been substituted for the usual continuity equation, since this formulation has led to superior results in studies by Lynch (1978) involving fixed boundaries. Following the usual weighted residual procedure one may express the dependent variables as combinations of predetermined basis functions:

$$H(\underset{\sim}{x},t) \simeq \sum_{j=1}^{N} H_j(t)\ \phi_j(\underset{\sim}{x},\underset{\sim}{X}_\ell(t)) \tag{17a}$$

$$\underset{\sim}{V}(\underset{\sim}{x},t) \simeq \sum_{j=1}^{N} \underset{\sim}{V}_j(t)\ \phi_j(\underset{\sim}{x},\underset{\sim}{X}_\ell(t)) \tag{17b}$$

$$\tau(\underset{\sim}{x},t) \simeq \sum_{j=1}^{N} \tau_j(t)\ \phi_j(\underset{\sim}{x},\underset{\sim}{X}_\ell(t)) \tag{17c}$$

where the $\underset{\sim}{X}_\ell(t)$ are the moving node coordinates, and N is the number of nodes. The functions ϕ_j are thus implicit functions of time, in addition to the usual spatial variation. (Note that this formulation ensures that H_j and $\underset{\sim}{V}_j$ are equal at all times to the approximate values of H and $\underset{\sim}{V}$ at node j regardless of nodal motion.) While the spatial gradients of H and $\underset{\sim}{V}$ take on their usual form in finite element analysis, the time derivatives will be slightly more complex because of the implicit dependence of the basis functions on time:

$$\frac{\partial H}{\partial t} \simeq \sum_{j=1}^{N} \frac{dH_j}{dt} \phi_j + \sum_{j=1}^{N} H_j \frac{\partial \phi_j}{\partial t} \tag{18a}$$

$$\frac{\partial^2 H}{\partial t^2} \simeq \sum_{j=1}^{N} \frac{d^2 H_j}{dt^2} \phi_j + 2 \sum_{j=1}^{N} \frac{dH_j}{dt} \frac{\partial \phi_j}{\partial t} + \sum_{j=1}^{N} H_j \frac{\partial^2 \phi_j}{\partial t^2} \tag{18b}$$

$$\frac{\partial \underset{\sim}{V}}{\partial t} \simeq \sum_{j=1}^{N} \frac{d\underset{\sim}{V}_j}{dt} \phi_j + \sum_{j=1}^{N} \underset{\sim}{V}_j \frac{\partial \phi_j}{\partial t} \tag{18c}$$

The terms in (18) which do not involve time derivatives of ϕ_j are the ones which usually arise in the approximation of a time derivative. The remaining terms are nonzero only in elements which are undergoing deformation.

Because (17) and (18) are approximations, their substitution into (16) will not exactly satisfy the equations. However, a set of N weighting functions ψ_i may be chosen such that the residuals in (16) are orthogonal to each of the functions. When this is done, the coefficients H_j and $\underset{\sim}{V}_j$ may be solved for and will approximate the exact solution. Because the spatial domain varies with time, the set of functions ψ_i must also vary with time in order to "fill" the space. Choice of $\psi_i = \phi_i$ will suffice and will result in the usual Galerkin formulation in the absence of node movement. Thus a set of nonlinear ordinary differential equations in the time domain is obtained:

$$\sum_{j=1}^{N} \Big[\frac{d^2 H_j}{dt^2} \langle\phi_j \phi_i\rangle + 2 \frac{dH_j}{dt} \langle\frac{\partial \phi_j}{\partial t}, \phi_i\rangle + H_j \langle\frac{\partial^2 \phi_j}{\partial t^2}, \phi_i\rangle + \frac{dH_j}{dt} \langle\tau\phi_j, \phi_i\rangle + H_j \langle\tau \frac{\partial \phi_j}{\partial t}, \phi_i\rangle\Big] = \langle R_W, \phi_i\rangle \tag{19a}$$

$$\sum_{j=1}^{N} \Big[\frac{d\underset{\sim}{V}_j}{dt} \langle\phi_j \phi_i\rangle + \underset{\sim}{V}_j \langle\frac{\partial \phi_j}{\partial t}, \phi_i\rangle\Big] = \langle \underset{\sim}{R}_M, \phi_i\rangle \qquad i = 1, 2, \ldots N \tag{19b}$$

where < > indicates integration over the entire spatial domain, and $R_W(x,y,t)$ and $R_{\sim M}(x,y,t)$ are the right-hand sides of equations (16a) and (16b) respectively, with approximations (17) substituted for the exact solution H and $\underset{\sim}{V}$. There are only two differences between equations (19) and their stationary counterparts:

1) The functions ϕ_j are implicit functions of time; thus in general all of the terms $<\phi_j,\phi_i>$, $<\frac{\partial\phi_j}{\partial x},\phi_i>$, etc. are time-dependent.

2) Extra terms appear in the left-hand side matrix in the forms $<\frac{\partial\phi_j}{\partial t},\phi_i>$, $<\frac{\partial^2\phi_j}{\partial t^2},\phi_i>$, etc.

Application of the finite difference approach in time requires equations (19) to be satisfied approximately at discrete points in time t_K. Use of a three-level scheme is required by the second derivative in (19a), and further makes possible an explicit formulation which is centered in time. Replacement of the time derivatives of the dependent variables with second-order correct FD approximations completes the discretization:

$$\left.\frac{d^2H_j}{dt^2}\right|_K \cong \left[\frac{H_{K+1} - 2H_K + H_{k-1}}{\Delta t^2}\right]_j \tag{20a}$$

$$\left.\frac{dH_j}{dt}\right|_K \cong \left[\frac{H_{K+1} - H_{K-1}}{2\Delta t}\right]_j \tag{20b}$$

$$\left.\frac{d\underset{\sim}{V}_j}{dt}\right|_K \cong \left[\frac{\underset{\sim}{V}_{K+1} - \underset{\sim}{V}_{K-1}}{2\Delta t}\right]_j \tag{20c}$$

where the K indices are used to denote time levels.

In the absence of node motion, this scheme reduces to the explicit wave equation scheme, which in linearized form is governed by the stability constraint [Gray and Lynch, 1977]

$$\left(\frac{\Delta t}{\Delta x}\right)^2 gH \leq \frac{1}{3} \tag{21}$$

To complete the discretization process, relations between node motion and the terms $\frac{\partial\phi_j}{\partial t}$, $\frac{\partial^2\phi_j}{\partial t^2}$ must be developed.

BASIS FUNCTIONS

Consider first the linear traingular elements, shown in Fig. 4.

For this case, the basis function associated with node i is given simply by the ratio

$$\phi_i(x,y,t) = \frac{A_i(x,y,t)}{A(t)} \tag{22}$$

where A(t) is the area of the triangle ijk. Equation (22) can be differentiated to yield

$$\frac{\partial \phi_i}{\partial t} = \frac{1}{A}\left[\frac{\partial A_i}{\partial t} - \phi_i \frac{dA}{dt}\right] \tag{23a}$$

$$\frac{\partial^2 \phi_i}{\partial t^2} = \frac{1}{A}\left[\frac{\partial^2 A_i}{\partial t^2} - \phi_i \frac{d^2 A}{dt^2}\right] - \frac{2}{A}\frac{dA}{dt}\frac{\partial \phi_i}{\partial t} \tag{23b}$$

Next consider an arbitrary triangle BCD in the (x,y) plane, as shown in Fig. 5, with all nodes in motion. The area of BCD is simply

$$A = \frac{1}{2}\,(\underset{\sim}{DB} \times \underset{\sim}{DC}) \cdot \hat{z} \tag{24}$$

where, for example $\underset{\sim}{DB}$ is the vector extending from point D to point B, and $\hat{z}$ is the unit normal vector pointing upward from the x-y plane. (In equation (24), vector cross product notation has been used.) Differentiation of (24) with respect to time gives

$$\frac{dA}{dt} = \frac{1}{2}\left\{\frac{d}{dt}(\underset{\sim}{DB}) \times \underset{\sim}{DC} + \underset{\sim}{DB} \times \frac{d}{dt}(\underset{\sim}{DC})\right\} \cdot \hat{z} \tag{25a}$$

Because the time derivative of a vector joining two points is the difference in the velocities of the two points, this equation becomes

$$\frac{dA}{dt} = \frac{1}{2}\left\{(\underset{\sim}{V}_B - \underset{\sim}{V}_D) \times \underset{\sim}{DC} + \underset{\sim}{DB} \times (\underset{\sim}{V}_C - \underset{\sim}{V}_D)\right\} \cdot \hat{z} \tag{25b}$$

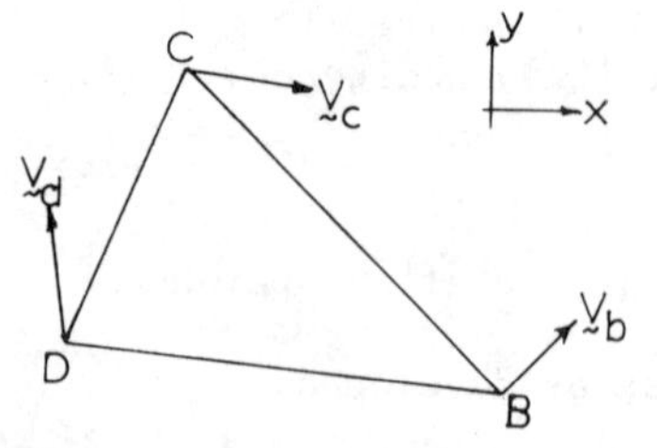

Figure 5. Triangle, non-zero nodal velocities.

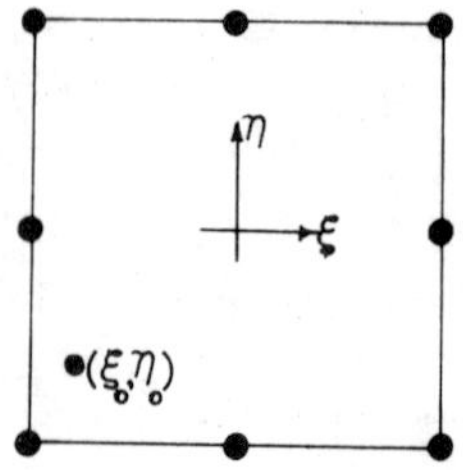

Figure 6. Quadratic isoparametric element in local coordinates.

or, after rearrangement,

$$\frac{dA}{dt} = \frac{1}{2}\{\underset{\sim}{V}_B \times \underset{\sim}{DC} + \underset{\sim}{V}_C \times \underset{\sim}{BD} + \underset{\sim}{V}_D \times \underset{\sim}{CB}\} \cdot \hat{z} \qquad (25c)$$

More generally, if the verticesi,j,k are numbered counter-clockwise, and

$$\Delta\underset{\sim}{X}_i \equiv \underset{\sim}{X}_j - \underset{\sim}{X}_k \qquad (26)$$

then

$$\frac{dA}{dt} = \{\frac{1}{2}\sum_{i=1}^{3} \underset{\sim}{V}_i \times \Delta\underset{\sim}{X}_i\} \cdot \hat{z} \qquad (27)$$

Further differentiation yields

$$\frac{d^2A}{dt^2} = \{\frac{1}{2}\sum_{i=1}^{3} \frac{d\underset{\sim}{V}_i}{dt} \times \Delta\underset{\sim}{X}_i + \sum_{i=1}^{3} \underset{\sim}{V}_i \times \underset{\sim}{V}_{i+1}\} \cdot \hat{z} \qquad (28)$$

Substitution of (27) into (23a) yields, after manipulation,

$$\frac{\partial\phi_i}{\partial t} = \{\frac{\Delta\underset{\sim}{X}_i}{2A} \times \underset{\sim}{V}_E\} \cdot \hat{z} \qquad (29)$$

where $\underset{\sim}{V}_E$, the "elemental velocity", is defined within each element in terms of the nodal velocities $\underset{\sim}{V}_{ni}$:

$$\underset{\sim}{V}_E(x,y,t) \equiv \sum_{i=1}^{3} \underset{\sim}{V}_{ni}\,\phi_i \qquad (30)$$

Observing the definition of ϕ_i and its spatial derivatives, (29) can be rearranged into the form

$$\frac{\partial\phi_i}{\partial t} = -\underset{\sim}{V}_E \cdot \underset{\sim}{\nabla}\phi_i \qquad (31)$$

which implies that moving with the elemental velocity, the total differential of ϕ_i is zero. Equation (31) provides a convenient means of determining $\frac{\partial\phi_i}{\partial t}$ in terms of its spatial gradients, which are used throughout any FE scheme and thus present no new complications.

Substitution of (27) and (28) into (23b), and considerable manipulation, yields the result

$$\frac{\partial^2 \phi_i}{\partial t^2} = \underset{\sim}{\nabla}\phi_i \cdot [\underset{\sim}{V}_E \sum_{\ell=1}^{3} 2\underset{\sim}{V}_{n,\ell} \cdot \underset{\sim}{\nabla}\phi_\ell - \Delta\underset{\sim}{V}_E]$$

$$+ \frac{1}{A} [\underset{\sim}{V}_{n,j} \times \underset{\sim}{V}_{n,k} - \phi_i \sum_{\ell=1}^{3} \underset{\sim}{V}_{n,\ell} \times \underset{\sim}{V}_{n,\ell+1}] \cdot \hat{z} \qquad (32)$$

where $\underset{\sim}{V}_{n,\ell}$ and $\underset{\sim}{V}_E$ are the nodal and elemental velocities as above, nodes i,j,k are numbered counterclockwise, and

$$\Delta\underset{\sim}{V}_E \equiv \sum_{\ell=1}^{3} [\frac{d}{dt} \underset{\sim}{V}_{n,\ell}]\phi_\ell \qquad (33)$$

Again, given the nodal velocities and accelerations, $\frac{\partial^2 \phi_i}{\partial t^2}$ can be computed in terms of the basis functions and their spatial gradients at any point in time.

The result (31) is perhaps not unexpected. Neuman and Witherspoon (1971) obtained the same result by a different procedure for linear quadrilaterials, and it can be derived for any isoparametric element as follows. Consider the transformed domain of the isoparametric basis functions, illustrated in Fig. 6 for the special case of 8-node rectangles. Since the basis functions ϕ_i depend upon ξ and η only (the ξ, η space does not deform in time), points of constant ξ and η such as the point (ξ_0, η_0) are points of constant $\phi_i = \phi_{i0}$. Since the element is deforming in the (x,y) space, the point (ξ_0, η_0) will be displaced during time δt from (x_0, y_0) to (x_1, y_1). From the definition of the isoparametric transformation,

$$\underset{\sim}{x}_1 = \sum_{i=1}^{M} \underset{\sim}{X}_i(t + \delta t)\ \phi_i\ (\xi_0, \eta_0) \qquad (34a)$$

$$\underset{\sim}{x}_0 = \sum_{i=1}^{M} \underset{\sim}{X}_i(t)\ \phi_i\ (\xi_0, \eta_0) \qquad (34b)$$

where M is the number of nodes in the element.

Since ξ_0 and η_0 are fixed, subtraction of (34b) from (34a) yields

$$\delta\underset{\sim}{x} = \sum_{i=1}^{M} \delta\underset{\sim}{X}_i\ \phi_i\ (\xi_0, \eta_0) \qquad (35)$$

and a point displaced by $\delta \underset{\sim}{x}$ experiences no change in the functions ϕ_i (i.e., $\delta\phi_i = 0$). Dividing by δt and taking the limit as δt approaches zero, one obtains

$$\frac{d}{dt}(\phi_i) = 0 \tag{36a}$$

when

$$\frac{d}{dt}(\underset{\sim}{x}) = \sum_{i=1}^{M} \left[\frac{d}{dt}(\underset{\sim}{X}_i)\right] \phi_i = \underset{\sim}{V}_E \tag{36b}$$

Now because

$$\frac{d\phi_i}{dt} = \frac{\partial\phi_i}{\partial t} + \frac{d\underset{\sim}{x}}{dt} \cdot \underset{\sim}{\nabla}\phi_i \tag{37a}$$

one obtains

$$\frac{\partial\phi_i}{\partial t} + \underset{\sim}{V}_E \cdot \underset{\sim}{\nabla}\phi_i = 0 \tag{37b}$$

which is the desired result.

Equation (37b) thus provides a simple determination of $\frac{\partial\phi_i}{\partial t}$ for any isoparametric element in terms of the nodal velocities and the spatial gradients of ϕ_i. The term $\frac{\partial^2\phi_i}{\partial t^2}$ can similarly be obtained by differentiation of (37b).

All of the relations required by the model are now available, and a typical time step would proceed as follows. At time t_K, the nodal normal directions are computed based on the existing grid, using equation (13); and the boundary node velocities are determined in accordance with (15). The node locations $\underset{\sim}{X}_{i,K+1}$ can be extrapolated by a finite-difference approximation to (15):

$$\underset{\sim}{X}_{i,K+1} \cong \underset{\sim}{X}_{i,K-1} + 2\Delta t\, \underset{\sim}{V}_{ni,K} \tag{38}$$

and the term $\frac{d\underset{\sim}{V}_i}{dt}$, required for computation of $\frac{\partial^2\phi_j}{\partial t^2}$, can likewise be approximated:

$$\frac{d\underset{\sim}{V}_{i,K}}{dt} = \frac{d^2\underset{\sim}{X}_{i,K}}{dt^2} \cong \frac{\underset{\sim}{X}_{i,K+1} - 2\underset{\sim}{X}_{i,K} + \underset{\sim}{X}_{i,K-1}}{\Delta t^2} \tag{39}$$

With this information, all of the required terms in (19, 20) can be evaluated. Thus the values of the dependent variables H_i and $\underset{\sim}{V}_i$ may be obtained at time t_{K+1}. At this point, the new nodal normals can be computed based on the deformed grid, and the entire procedure repeated.

APPLICATIONS

One numerical problem not mentioned in the above analysis is the characterization of bottom friction at the moving boundary. The most common formulation in shallow water problems, and the one used in this paper (equation 16) is the Manning-Chezy formula for the friction slope $\underset{\sim}{S}_f$:

$$g\underset{\sim}{S}_f = \frac{g|\underset{\sim}{V}|\underset{\sim}{V}}{C^2 H} \equiv \tau(x,y,t)\underset{\sim}{V} \qquad (40)$$

The term τ, although finite throughout the interior of the domain, approaches infinity at the moving boundary since $H = 0$ there. In the numerical scheme developed above, the function $\tau(x,y,t)$ is expanded in terms of its nodal values τ_j and the basis functions (equation 17c). The value of τ_j at a boundary node must be representative of the frictional effect throughout the boundary element. A rule is required whereby the determination of τ_j at boundary nodes is based on some small, positive depth. The importance of this determination is illustrated by a simple example.

Consider a one-dimensional canal with constant bathymetry, terminated in a beach as in Fig. 7 and subject to a surge at $x = 0$:

$$\zeta(x = 0\ ,\ t > 0) = 1. - \exp(-rt) \qquad (41)$$

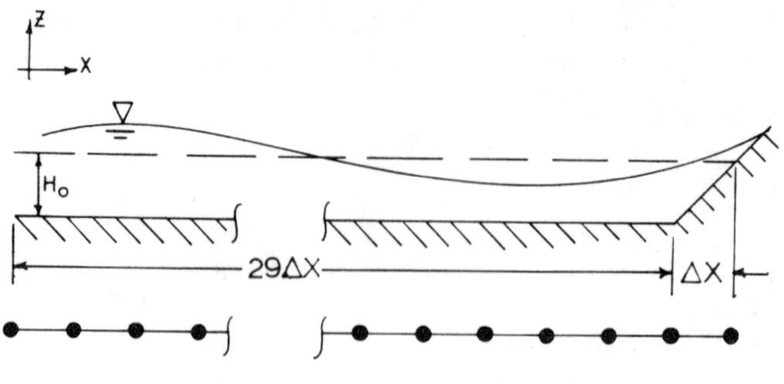

Fig. 7. Bathymetry and finite element grid for one-dimensional computation.

This problem has been solved with one-dimensional linear elements (Fig. 7), with $H_0 = 30.$ ft, $\Delta x = 50{,}000.$ ft, $\Delta t = 0.06$ hours, and $r = 0.333\ hr^{-1}$. (The exponential in (41) decays after about 200 time steps.) The moving boundary condition was applied at the landward boundary, and the solution compared to that obtained by applying a no-flux condition at $x = 29\Delta x$. While the detailed solution at the beach boundary should depend heavily on the choice of boundary condition, the solutions in the interior should be substantially the same. The surface elevation solution at $x = 14\Delta x$ is examined in Figs. 8a and 8b.

In Fig. 8a, the moving boundary solution was computed using the value H = 1. ft for the friction determination at the beach boundary. It is apparent that a significantly different reflection occurs when the surge first reaches the beach. In Fig. 8b, the boundary friction was based on H = 15. ft, roughly the average depth in the boundary element. The two solutions agree quite well. It is apparent that the solution even in the interior of the domain is very sensitive to the characterization of frictional resistance at the moving boundary.

A second example again involves a canal, this time with linear bathymetry (Fig. 9a) subject to periodic forcing at the seaward end: $\zeta(x = 0\ ,\ t > 0) = \sin\omega t$ (42)

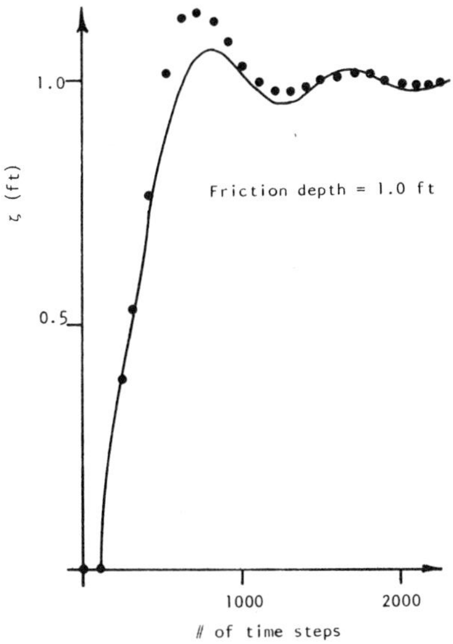

Fig. 8a. Surface elevation for one-dimensional computation at $x = 14\Delta x$.
- No flux boundary
• Moving boundary

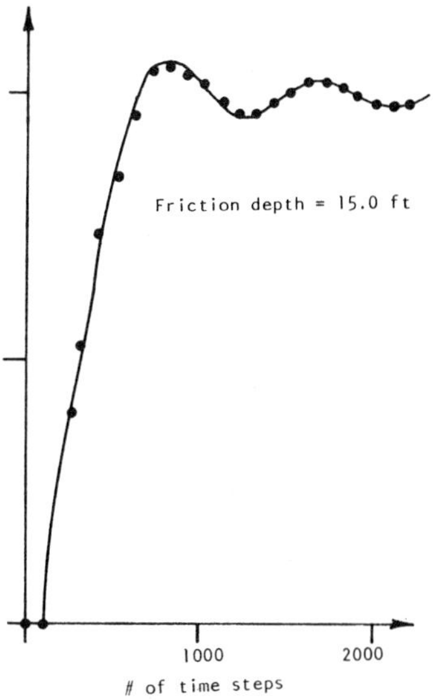

Fig. 8b. Surface elevation for one-dimensional computation for $x = 14\Delta x$. Moving and no-flux boundary.

This problem was solved using two-dimensional linear triangles as in Fig. 9b, with $\Delta x = 50{,}000.$ ft, $H_0 = 30.$ ft, $\Delta t = 0.062$ hours, and $\frac{2\pi}{\omega} = 12.4$ hours. The Coriolis effect was not considered. Again, the moving boundary results were compared with results obtained by requiring no flux at $x = 8\Delta x$. The frictional depth at the moving boundary was chosen to be 1. ft. The dynamic steady state responses at and near the boundary are shown in Figs. 10a and 10b.

Figure 10a shows the boundary surface elevation responses for the no-flux condition (node 45) and for the moving boundary (nodes 45 and 50). The two responses at node 45 are similar, being almost equal in phase, but differing by about a factor of two in amplitude. If the no-flux boundary solution (node 45) is extrapolated inland and compared with the moving boundary elevation (node 50), the discrepancy in both amplitude and phase between the two boundary conditions becomes severe. These results are in agreement with the qualitative discussion above which suggested that the use of the approximate no-flux condition would not provide sufficient damping or delay of the boundary motion.

Figure 10b shows the surface elevation responses further in from the boundary at nodes 40 and 35. The agreement in phase continues to be good, and the discrepancy in amplitude decreases with distance from the boundary. The responses at node 20, well removed from the boundary, are practically indistinguishable.

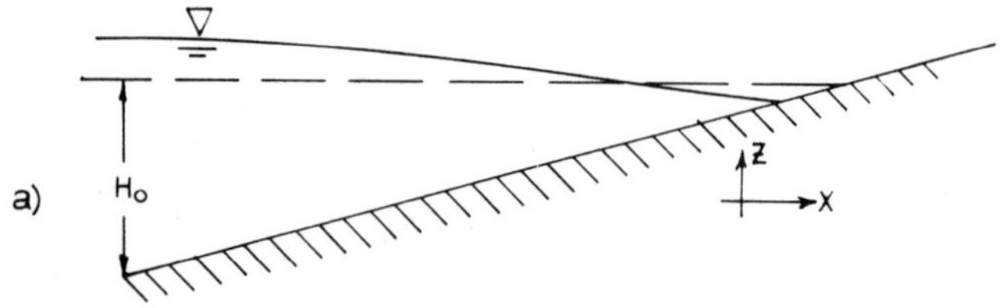

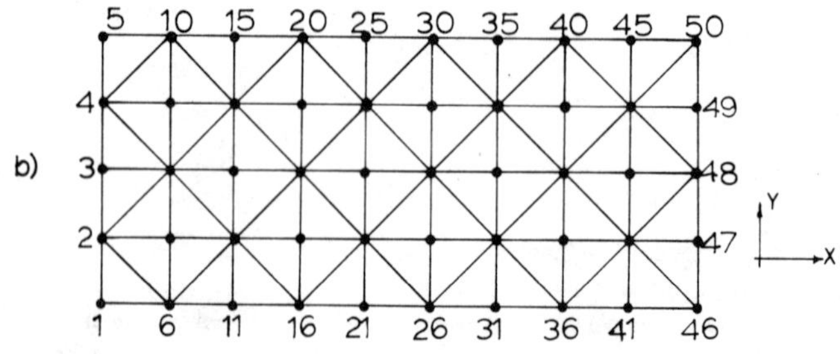

Fig. 9a. Linear bathymetry.

Fig. 9b. Two-dimensional computational grid.

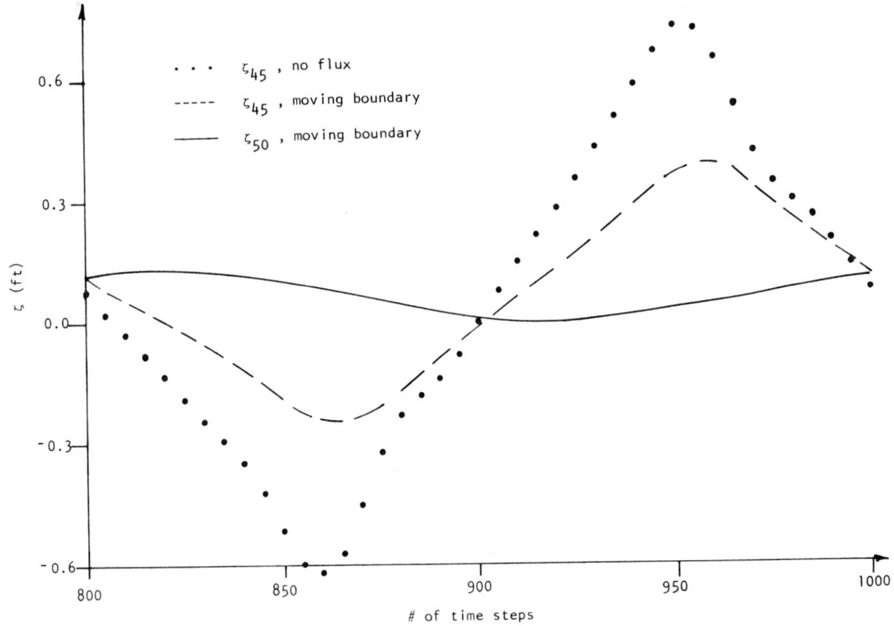

Fig. 10a. Surface elevation for two-dimensional computation at the boundary.

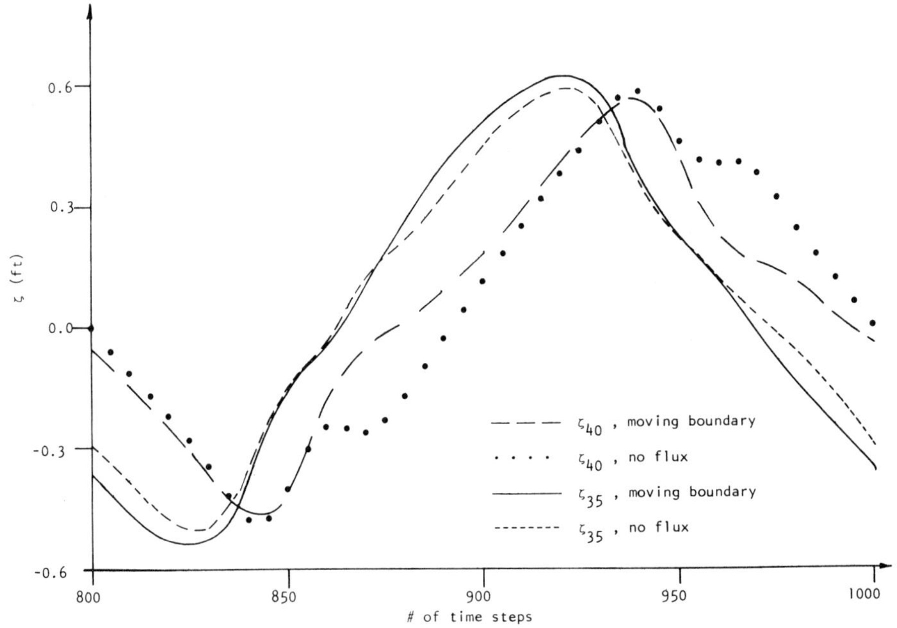

Fig. 10b. Surface elevation for two-dimensional computation near the boundary.

CONCLUSIONS

The method presented here incorporates the moving boundary condition into an established finite element shallow water model. Experiments to date show that the method introduces no new instabilities and is capable of computing reasonable solutions to some simple problems. The solutions near the boundary are markedly different from those generated by assuming a no-flux condition on a nearby fixed boundary, and the discrepancies decrease in importance in the interior of the domain. The characterization of frictional loss at the boundary has a dramatic effect on the solution even in the interior of the domain. Further basic research is required for the accurate description of boundary friction, and experimental data is needed in order to compare computed solutions with actual measurements.

ACKNOWLEDGMENT

This work has been supported in part by the United States Geological Survey.

REFERENCES

Gray, W.G., An Efficient Finite Element Scheme for Two-Dimensional Surface Water Computation, Proc. First Intl. Conf. on Finite Elements in Water Resources, Pentech Press, London, (1977).

Gray, W.G., and D.R. Lynch, Time Stepping Schemes for Finite Element Tidal Model Computations, Advances in Water Resources, 1, 2, pp. 83-95, (1977).

Lynch, D.R., Numerical Solution of the Shallow Water Equations--An Evaluation of the Galerkin-Finite Element Method, Ph.D. thesis, Department of Civil Engineering, Princeton University, (1978).

Neuman, S.P., and P.A. Witherspoon, Analysis of Nonsteady Flow with a Free Surface Using the Finite Element Method, Water Resources Research, 7, 3, pp. 611-623, (1971).

Pinder, G.F., and W.G. Gray, Finite Element Simulation in Surface and Subsurface Hydrology, Academic Press, New York, (1977).

Wang, J.D., and J.J. Connor, Mathematical Modeling of Near Coastal Circulation, M.I.T. Parsons Laboratory Report #200, (1975).

BEHAVIOUR OF A HYDRODYNAMIC FINITE ELEMENT MODEL

R.A. Harrington, N. Kouwen, and G.J. Farquhar

Dept. of Civil Eng., Univ. of Waterloo, Waterloo, Ont., Canada. N2L 3G1

SUMMARY

A finite element model which solves the vertically-integrated momentum and continuity equations is described. Linear triangular elements are used to describe the geometry and parameter variations. The Galerkin method of weighted residuals is employed to cast the equations in a form amenable to numerical solution. The model is based on a fully-implicit formulation using finite differences for the temporal derivatives.

Means of evaluating the non-linear terms of the governing equations are described, and model results are presented for a frictionless tidal channel. The example is chosen such that the non-linearities have a large influence on the solution, and as a result the linearization scheme significantly affects the model's behaviour.

Suppression of the non-linear instabilities generated by the convective terms in the momentum equations is examined for the case of flow around a 180° bend. Both the imposition of artificially high roughness coefficients and the use of an effective eddy viscosity are examined in terms of their ability to damp the oscillations which arise for this example.

Finally, model results are presented for a case study involving determination of remedial measures to improve flow conditions at a river outfall in Southern Ontario.

MODEL DEVELOPMENT

The vertically-integrated momentum and continuity equations may be written in Cartesian tensor notation as

$$\frac{\partial u_i}{\partial t} + u_j \frac{\partial u_i}{\partial x_j} + g \frac{\partial \eta}{\partial x_i} - e_{ij} f\, u_j - \frac{\partial}{\partial x_j} \varepsilon \frac{\partial u_i}{\partial x_j} - F_i = 0 \qquad (1a)$$

$$\frac{\partial \eta}{\partial t} + \frac{\partial\, h u_i}{\partial x_i} = 0 \qquad (1b)$$

where the summation is over i,j = 1,2. In the above equations, u_i is the depth-averaged velocity in the x_i direction, t is time, g is the acceleration of gravity, η is the water surface elevation above some datum, e_{ij} is the alternating tensor, f is the Coriolis parameter, ε is an effective eddy viscosity, h is the total depth of flow, and F_i is a force due to bed and wind stress. The force term is expressed as

$$F_i = - \frac{g n |u| u_i}{h^{4/3}} + \frac{\gamma^2 \rho_a |w| w_i}{\rho\, h} \qquad (2)$$

where n is Manning's roughness coefficient, $|u|$ is the magnitude of the average velocity, γ^2 is a wind stress coefficient (Pollack, 1960), ρ_a and ρ are the densities of air and water respectively, and w_i is the x_i component of the average wind velocity $|w|$.

Equations 1a and 1b are the well-known shallow water equations, in which the usual assumptions have been made (e.g. the omission of vertical accelerations other than that of gravity, the assumption of hydrostatic pressure). The initial conditions required for these equations are specified values of u_i and η at t = 0 over the whole domain. In many applications, for lack of better information these values are typically zero velocity and a flat water surface. The boundary conditions are of three types, namely

i) specified water levels at sea or open boundaries,

ii) zero normal velcoties at land boundaries, and
iii) specified inflows at flux boundaries such as rivers.

The finite element expression of Equations 1a and 1b is formulated by means of the Galerkin method of weighted residuals. Linear triangular elements are employed, primarily because of the resulting simplicity of integration afforded by these elements. In the subsequent development, the following notation is employed: $<\cdot>$ represents a row vector; $\{\cdot\}$ a column vector; $[\cdot]$ a square matrix; and a superscript T indicates the transpose.

Applying the Galerkin method to Equations 1a and 1b, and integrating by parts the second order terms in the momentum equations as well as the spatial derivatives in the continuity equation yields

$$\sum_{n_e} \iint \Big(<N>^T<N> \frac{\partial\{u_i\}}{\partial t} + <N>^T<N>\{\bar{u}_j\} \frac{\partial <N>}{\partial x_j} \{u_i\}$$
$$+ g <N>^T \frac{\partial <N>}{\partial x_i} \{\eta\} - e_{ij} f <N>^T<N>\{u_j\}$$
$$+ \frac{\partial <N>^T}{\partial x_j} <N>\{\varepsilon\} \frac{\partial <N>}{\partial x_j} \{u_i\} - <N>^T<N>\{F_i\} \Big) d\Delta$$
$$= \sum_{n_e} \int <N>^T \varepsilon \frac{\partial u_i}{\partial x_j} \ell_j \, ds \tag{3a}$$

$$\sum_{n_e} \iint \Big(<N>^T<N> \frac{\partial\{\eta\}}{\partial t} - \frac{\partial <N>^T}{\partial x_i} <N>\{\bar{h}\}<N>\{u_i\} \Big) d\Delta$$
$$= \sum_{n_s} \int <N>^T q_n \, ds \tag{3b}$$

where n_e is the number of elements, n_s is the number of flux boundaries, $<N>$ is the linear interpolation function, $d\Delta$ indicates integration over an element, ds signifies integration over an element boundary, ℓ_j is the direction cosine of the outward normal, and q_n is the inflow per unit width due to rivers etc. The overbars in the preceeding equations indicate quantities which must be linearized in some way. Assuming for the present that linearization has been carried out, evaluation of the various integrals, summation of the contribution of each element, and application of suitable initial and boundary conditions results in the expression

$$[P]\{\dot{\Psi}\} + [K]\{\Psi\} = \{F\} \tag{4}$$

where $\{\Psi\}$ is the vector of nodal values of u_i and η, $\{\dot{\Psi}\}$ is the time derivative of $\{\Psi\}$, $\{F\}$ contains the wind stress and flux terms, $[P]$ is the mass matrix, and $[K]$ is the system matrix.

Solution of Equation 4 is achieved by means of a fully-implicit formulation using finite differences in time for the approximation of $\{\dot{\Psi}\}$. The result is

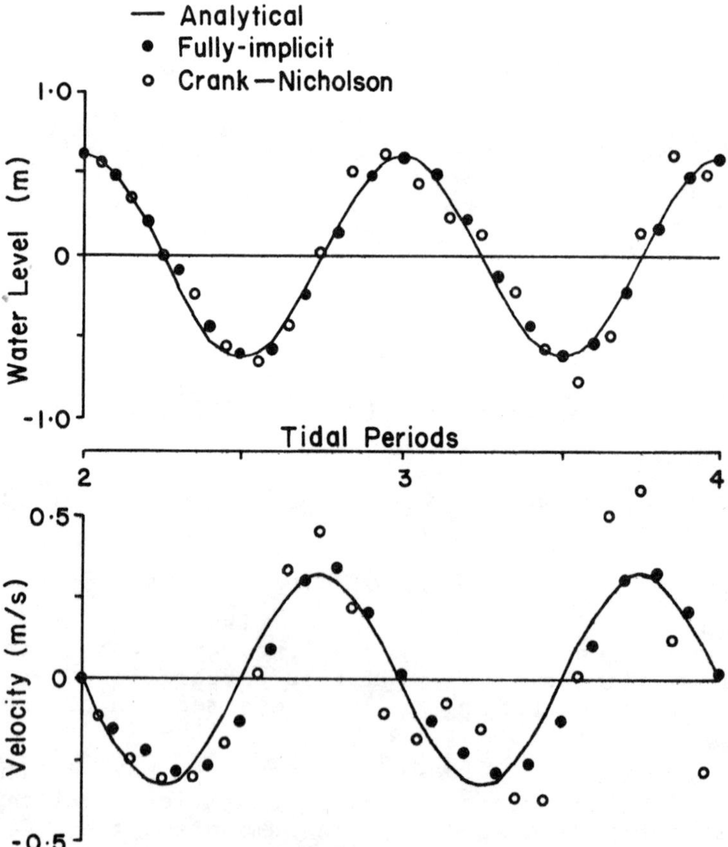

FIGURE 1
Model Results for Tidal Channel

$$[A]\{\Psi\}_{t+\Delta t} = \{F^*\} \tag{5a}$$

with $[A] = [P] + \Delta t\ [K]$, and (5b)

$$\{F^*\} = \Delta t\{F\} + [P]\{\Psi\}_t \tag{5c}$$

where Δt is the time step. The inherent stability of the fully-implicit scheme is illustrated in Figure 1, which shows the model result for a frictionless channel closed at one end and subject to a sinusoidal tidal variation at the other. The fully-implicit formulation is compared to the Crank-Nicholson approximation, and Figure 1 shows that the implicit scheme gives results which compare very well to the analytical solution, while the Crank-Nicholson scheme deteriorates very rapidly. For this case the amplitude to depth ratio was small, so the effects of the convective terms in the momentum equations may be ignored. The time step used in the model was the same for both the implicit and Crank-Nicholson formulations, and although it may be possible to obtain stability for the Crank-Nicholson scheme by employing a smaller time step, the increased computational effort is probably not warranted.

LINEARIZATION SCHEMES

When the linearization indicated in Equations 3a and 3b is carried out at the unknown time level (referred to herein as implicit linearization), an iterative technique is required to advance the solution. It is usually advantageous to solve for the correction to be applied to each estimate of $\{\Psi\}$ rather than for $\{\Psi\}$ itself, so the equations actually solved are

$$[A]\{\Delta\Psi\} = \{\Delta F^*\} \tag{6}$$

where Δ indicates the difference between the current and previous iterate. The convergence criterion used in the model is given by

$$e = \left| \frac{\Sigma\ \Delta\Psi^2}{\Sigma(\Psi_{t+\Delta t})^2_{\text{current}}} \right|^{1/2} \times 100\% \tag{7}$$

where a typical tolerance for e is 0.1%. The coefficient matrix [A] is updated for each iteration, so the iterative technique is effectively a full Newton-Raphson method. Of course, if the linearization is carried out at the known time level (referred to here as explicit linearization), no iteration is required.

However, it may be argued that the linearization scheme should be compatible with the overall method of temporal integration. For example, implicit time integration coupled with explicit linearization results in the continuity equation being satisfied by means of a velocity at one time level and a depth at another. In this case some phase shift would be expected if the depth or velocity changed significantly over the time step. The same type of argument may be applied to the contribution of the convective terms in the momentum equation. Figure 2 illustrates this effect

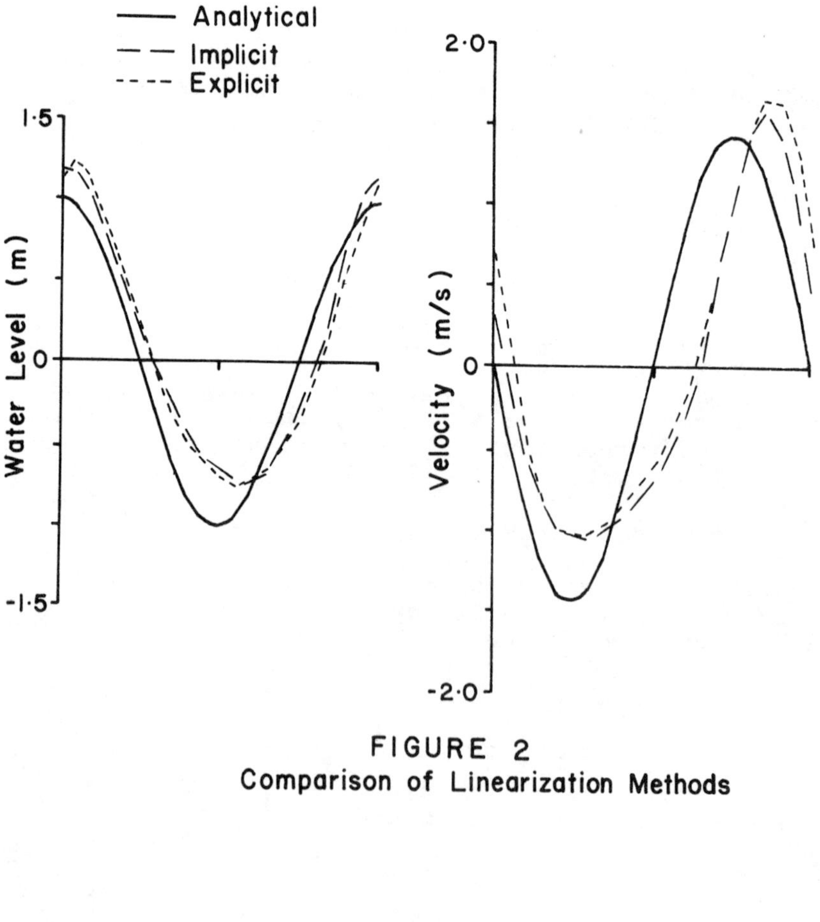

FIGURE 2
Comparison of Linearization Methods

for the case of a frictionless tidal channel. The amplitude to depth ratio was increased from that of the previouse case in order that the contribution of the non-linear terms would be magnified. Figure 2 shows the model results compared to the first-order analytical solution. As expected, low water is delayed and is somewhat deeper than the analytical solution, and the velocities are no longer exactly 90° out of phase with the water surface elevations. The time for high water should not change appreciably, and Figure 2 indicates that implicit linearization results in no change in the time of high water, although the depth is somewhat larger, as would be expected. However, explicit linearization results in high water being delayed and being raised the same amount that low water is raised. Both of these phenomena are contrary to what would be expected physically. The differences for this example are not large, but for more complex geometries the linearization method may strongly influence the numerical solution. As the time step is reduced, the differences between the two schemes becomes smaller since the variation over the time step is decreased. With very small time steps, the advantages of implicit time integration are outweighed by increased running costs. Also, if the problem being considered is not strongly time dependent, the nature of the linearization scheme becomes less important.

One of the most troublesome aspects of numerical solution of the equations of motion is suitable treatment of the convective terms in the momentum equations. While in some cases these terms have been neglected altogether, it has been shown (van de Ree, 1975) that their influence cannot be neglected in all applications. The previous discussion indicated that in general the linearization scheme should be compatible with the overall integration procedure of the model, consequently two treatments of the convective terms will be examined. The first is the formal linearization indicated in Equation 3a, and the second employs only element-averaged velocities; in both cases the velocities used are those at the advanced time level.

The effect of the convective terms on the numerical solution is to generate motion too small to be resolved on the computational grid. If this high-frequency energy is not dissipated in some way, the solution becomes unstable and deteriorates. These non-linear instabilities may be suppressed in effectively two ways, either the imposition of a large roughness coefficient (Partridge and Brebbia, 1976), or application of some type of eddy viscosity (Wang, 1976). Both of these methods are examined for the case of flow around a 180° bend as described in Chow (1959). This example was chosen because the high velocity gradients developed ensure that the convective terms play a dominant role in the solution. The system was discretized using 65 nodes and 96 elements as shown in Figure 4. Direct solution for the steady state flow pattern is not possible since the coefficient matrix becomes non-positive definite when the time derivative is removed from the continuity equation. Consequently the solution is obtained by marching in time until a quasi-steady state is achieved.

The first series of tests involved formal linearization of the convective terms and use of various roughness coefficients. Figure 3 shows the normalized relative error of the solution field between successive time steps as the solution proceeds. It is seen that for relatively small rugosity (which would be the case for this example) the solution very quickly become unstable and failed to converge, the computations eventually being terminated due to unrealistic flow situations such as negative depths. For larger roughness coefficients, the solution converged with no instabilities being generated, but the steady state flow pattern achieved was physcally unrealistic, depths being too large and velocities being too small. A second series of tests used element-averaged velocities for the convective terms, and the results were very much like those shown in Figure 3, the only difference being that somewhat more time was required. The instabilities were still generated for smaller roughnesses, and convergence to an erroneous result still occurred for the larger roughnesses. The lag in this case was due to the fact that the element-averaging caused a reduction in the instantaneous contribution of the convective terms, but the overall effect was virtually unchanged.

The next series of tests comprised imposition of a large roughness coefficient during the initial stages of the computations and a gradual reduction in the coefficient as the solution progressed. In this case no instabilities were generated initially, but as the roughness decreased, and as the model began to approach the measured flow situation, instabilities set in and the simulations were eventually terminated. The same result was obtained for both element-averaging and for formal linearization. In all of the previous tests, the size of the time step was also varied, but it had no effect on the results other than to delay (or accelerate) the behaviour described.

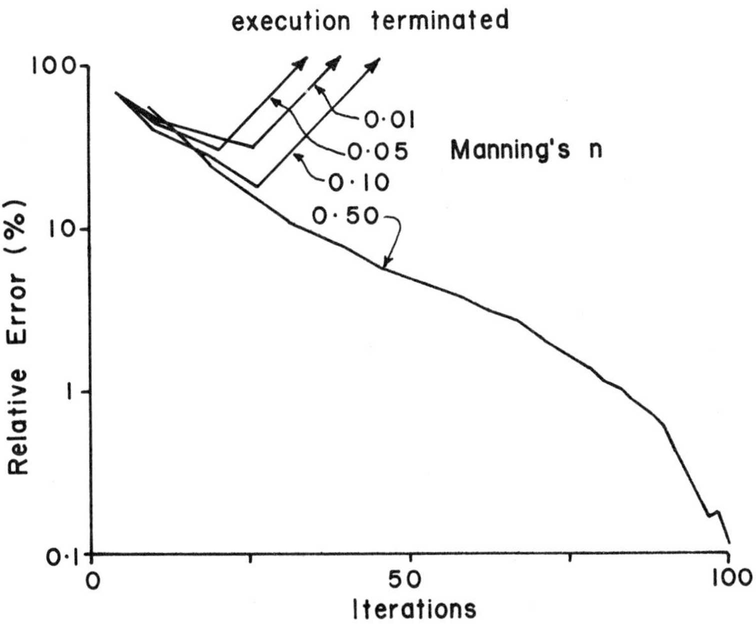

FIGURE 3
Effect of Roughness Coefficient

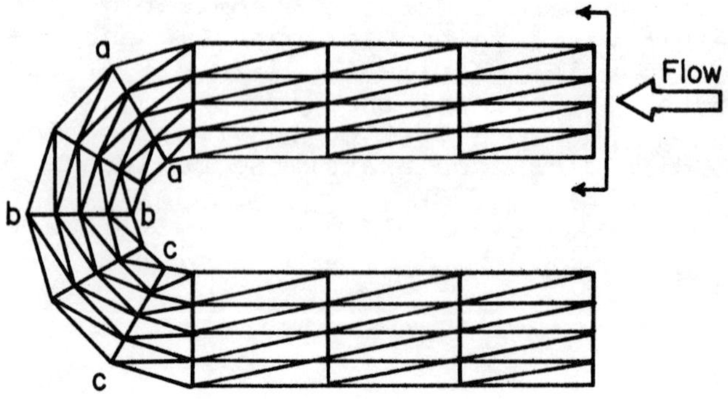

Finite Element Grid

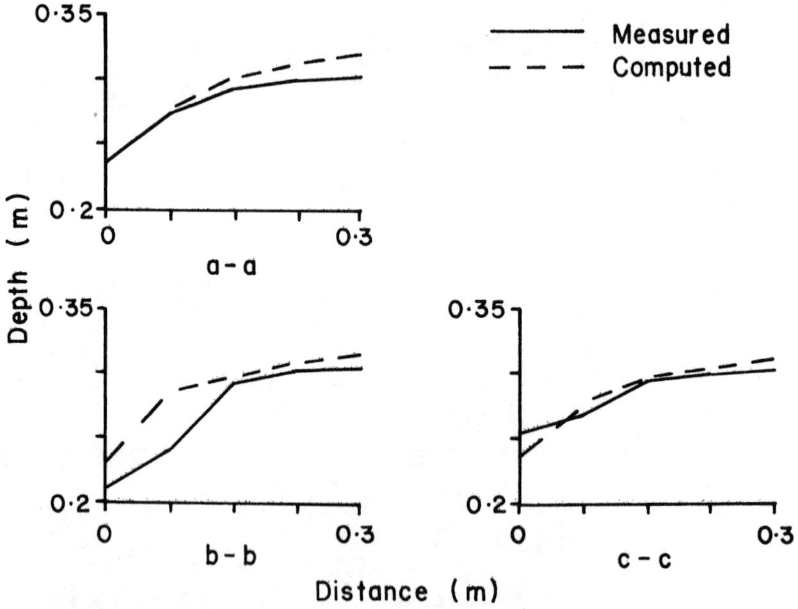

FIGURE 4
Results for 180° Bend

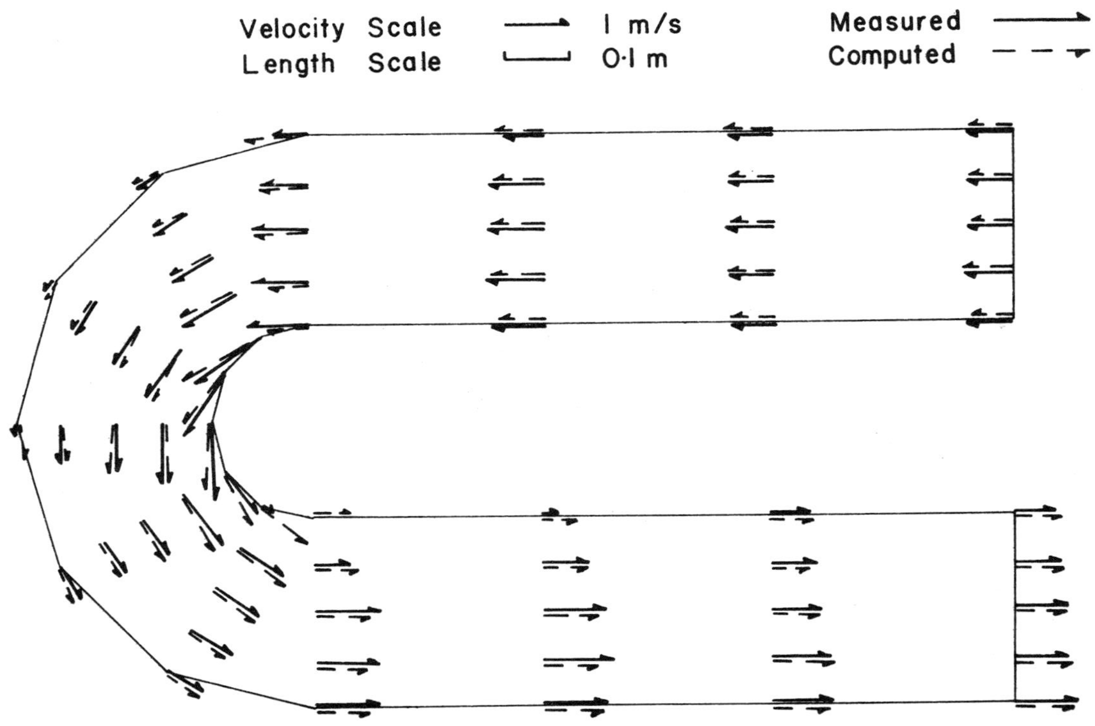

FIGURE 5
Velocity Distribution

The final series of tests were carried out using an eddy viscosity to damp the high-frequency oscillations. It was found that for this case sufficient damping was provided to overcome the non-linear instabilities when the dimensionless eddy viscosity ε/uh was of the order of 0.01. Figures 4 and 5 show the water levels and velocities produced by the model for this case. The agreement between the model results and the measured values is fairly good, although the grid is too coarse to permit resolution of the stagnation point downstream of the bend, and this is reflected in the water levels.

MODEL APPLICATION

The Humber River discharges into Lake Ontario at Toronto as shown in Figure 6. A previous paper (Garret et al., 1976) described the techniques used to assess alternatives proposed to alleviate water quality problems at the Humber outfall. One such problem was the diversion of river water behind a series of breakwaters. The resulting siltation rendered the protected area behind the breakwaters unsafe for recreational users such as small craft operators. The proposed remedy was the construction of a jetty which would deflect the river water past the breakwater. A hydraulic model study was carried out to assess this alternative, and the optimum length of jetty which provided both access of small craft and deflection of river water was determined.

The area of interest was discretized using 68 nodes and 91 elements. Figure 7 shows the model results when existing conditions without the jetty were simulation, and the deflection of river water behind the breakwater is clearly seen. Figure 8 shows the model results for a short jetty. Significant quantities of river water are still being entrained behind the breakwater, and an eddy is produced behind the jetty. Finally, Figure 9 shows the results for a longer jetty. The river water is almost completely deflected past the breakwater, and an eddy is again formed behind the jetty. These results are all consistent with the results of the hydraulic model study.

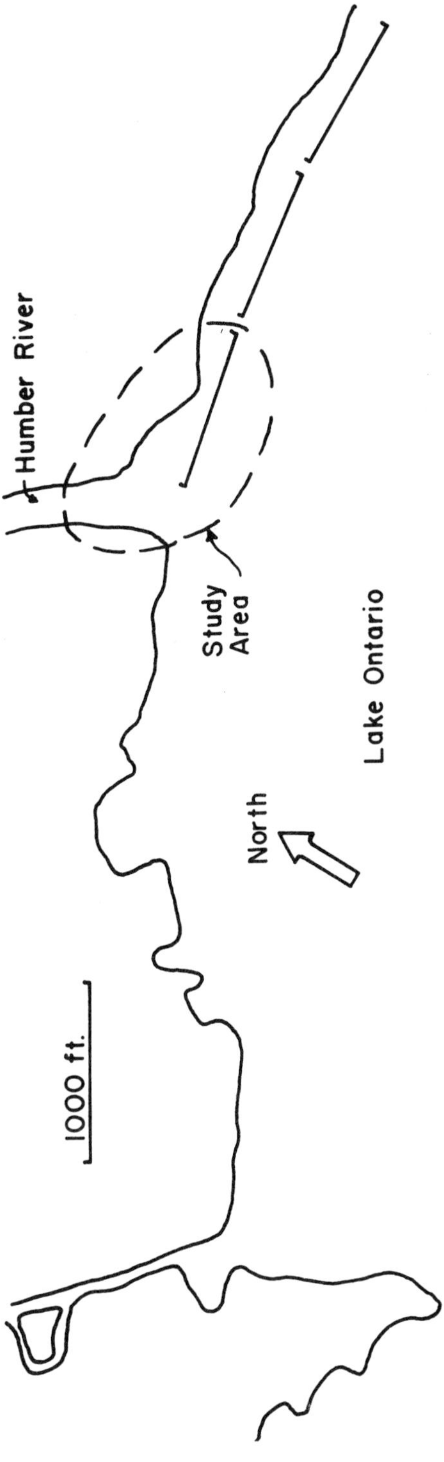

FIGURE 6
Site Plan

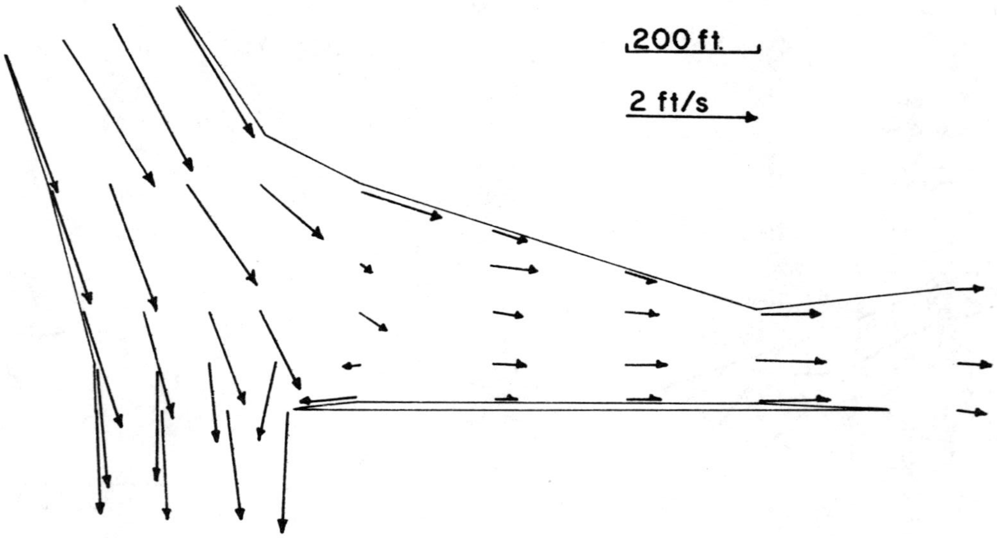

FIGURE 7
Velocity Distribution - Existing Conditions

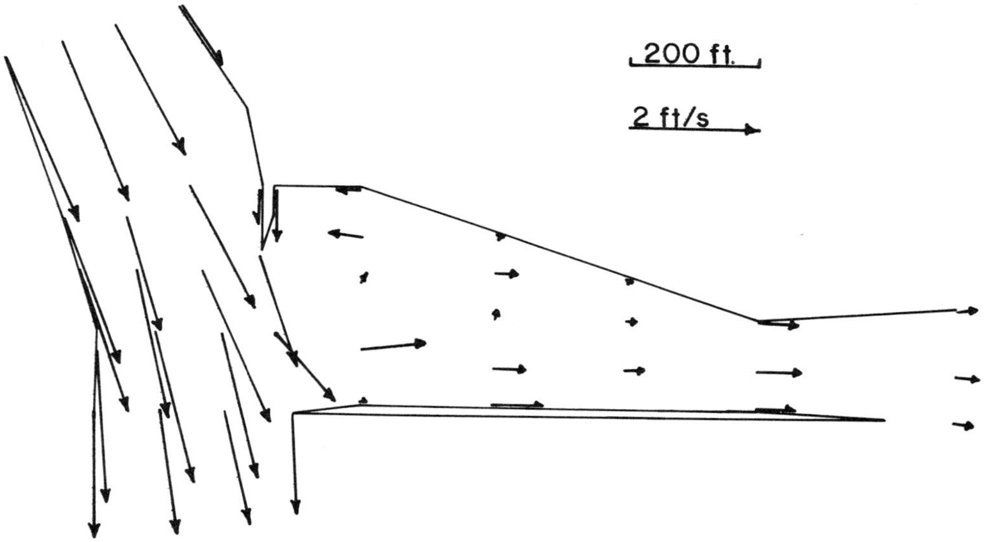

FIGURE 8
Velocity Distribution - Short Jetty

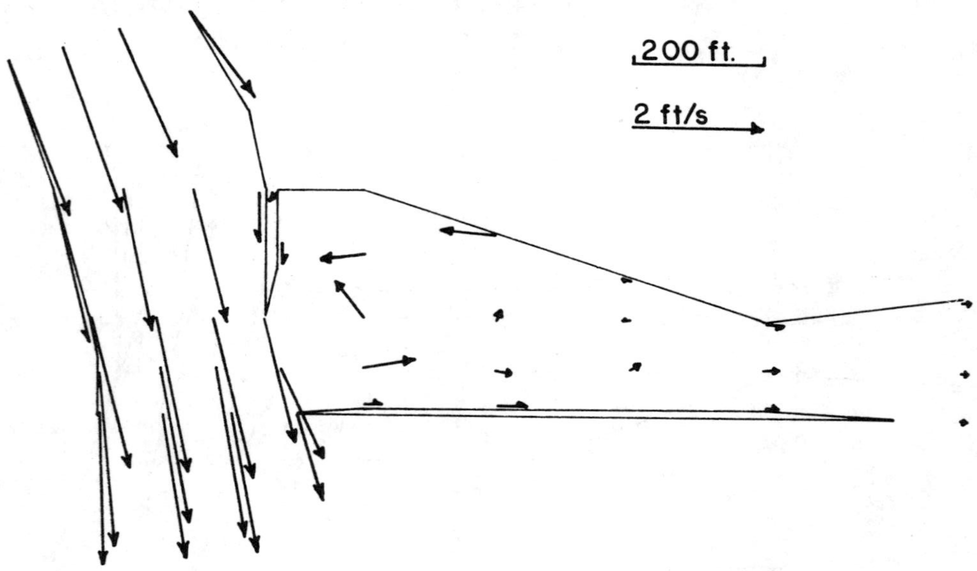

FIGURE 9
Velocity Distribution - Long Jetty

CONCLUSIONS

This paper has described a finite element model developed to solve the vertically-integrated momentum and continuity equations. The model employs a fully-implicit solution scheme which was shown to be more stable than the Crank-Nicholson approximation. The example of a closed frictionless channel with a tidal boundary was used to demonstrate that the linearization scheme should be compatible with the overall integration method of the model for those cases in which there are significant variations in parameters over the time step. The case of flow around a 180° bend was used to illustrate the fact that imposition of a large roughness coefficient had little effect on damping the non-linear instabilities generated for this example. Use of an eddy viscosity was quite effective in suppressing these components. Finally, model results were presented for a case study involving the assessment of measures to ameliorate flow conditions at a river outfall in Southern Ontario.

ACKNOWLEDGEMENTS

The first author was supported by a scholarship from the Central Mortgage and Housing Corporation, Ottawa, Canada. The research was funded by the University of Waterloo and the National Research Council of Canada.

REFERENCES

Chow, V.T., Open Channel Hydraulics, McGraw-Hill Book Co., 1959

Garret, M.R., Harrington, R.A., Kouwen, N., Roake, A.F., and Wisner, P.E., "Methodology for Sedimentation and Water Quality Modelling at the Humber River Mouth", Rivers '76, ASCE, pp. 650-666

Partridge, P.W., and Brebbia, C.A., "Quadratic Finite Elements in Shallow Water Problems", Journal of the Hydraulics Division, ASCE, Vol. 102, No HY9, Sept., 1976, pp. 1299-1313

Pollack, J.J., "Wind Set-up and Shear Stress Coefficient in Chesapeake Bay", Journal of Geophysical Research, Vol. 65, No 10, Oct., 1960, pp. 3383-3389

van de Ree, W.J., and Schaap, H.Y., "Measured Contribution of the Terms of the Vertically Integrated Hydrodynamic Equations", Modelling '75, ASCE, pp. 1237-1248

Wang, H.P., "Multi-Levelled Finite Element Hydrodynamic Model of Block Island Sound", Finite Elements in Water Resources, Pentech Press, July, 1976, pp. 4.69-4.93

DEVELOPMENT OF GENERALIZED FREE SURFACE FLOW MODELS USING FINITE ELEMENT TECHNIQUES

D. Michael Gee, Robert C. MacArthur

The Hydrologic Engineering Center, U.S. Army Corps of Engineers, Davis, California

INTRODUCTION

The Corps of Engineers' Hydrologic Engineering Center is involved in the development, evaluation, and application of mathematical models. Two finite element hydrodynamic models, one for two-dimensional free surface flow in the horizontal plane and one for the vertical plane are being evaluated. Although the models are formulated to solve dynamic flow problems, all work to date has been with steady state solutions. Recent research has focused on mass continuity performance of the models, proper boundary condition specification, and comparison with finite difference techniques. The objective of this research is to develop generalized mathematical models for routine use by the engineering community. This paper presents recent results of evaluation and application of the models.

THE MODEL FOR TWO-DIMENSIONAL FREE SURFACE FLOW IN THE HORIZONTAL PLANE

The model for two-dimensional free surface flow in the horizontal plane solves the governing equations in the following form:

Continuity

$$\frac{\partial h}{\partial t} + \frac{\partial}{\partial x}(uh) + \frac{\partial}{\partial y}(vh) = 0 \qquad (1)$$

Momentum

$$\frac{\partial u}{\partial t} + u\frac{\partial u}{\partial x} + v\frac{\partial u}{\partial y} + g\frac{\partial h}{\partial x} + g\frac{\partial a_o}{\partial x} - \frac{\varepsilon_{xx}}{\rho}\frac{\partial^2 u}{\partial x} - \frac{\varepsilon_{xy}}{\partial}\frac{\partial^2 u}{\partial y^2} - 2\omega v \sin\phi$$

$$+ \frac{gu}{C^2 h}\left(u^2 + v^2\right)^{1/2} - \frac{\zeta}{h} V_a^2 \cos\psi = 0 \qquad (2)$$

$$\frac{\partial v}{\partial t} + u\frac{\partial v}{\partial x} + v\frac{\partial v}{\partial y} + g\frac{\partial h}{\partial y} + g\frac{\partial a_o}{\partial y} - \frac{\varepsilon_{yx}}{\rho}\frac{\partial^2 v}{\partial x^2} - \frac{\varepsilon_{yy}}{\rho}\frac{\partial^2 v}{\partial y^2} + 2\omega u \sin\phi$$

$$+ \frac{gv}{C^2 h}\left(u^2 + v^2\right)^{1/2} - \frac{\zeta}{h} V_a^2 \sin\psi = 0 \qquad (3)$$

where

u, v = x and y velocity components respectively
t = time
h = depth
a_o = bed elevation
ε = turbulent exchange coefficients
g = gravitational acceleration
ω = rate of earth's angular rotation
ϕ = latitude
C = Chezy roughness coefficient
ζ = empirical wind stress coefficient
V_a = wind speed
ψ = angle between wind direction and x - axis
ρ = fluid density

Before solution, the equations are recast with flow (velocity times depth) and depth as the dependent variables. A linear shape function is used for depth and a quadratic function for flow. The Galerkin method of weighted residuals is used and the resulting non-linear system of equations solved with the Newton-Rapheson scheme. Details of the solution have been published previously by Norton, et al (1973) and King, et al (1975). General discussions of finite element techniques have been published by Zienkiewicz (1971), Hubner (1975), and Strang & Fix (1973).

Evaluation of Continuity Errors

The finite element method yields a solution which approximates the true solution to the governing partial differential equations. The approximate nature of this solution becomes evident when mass continuity is checked at various locations in the solution domain for a steady state simulation. Although overall continuity is maintained (inflow equals outflow over the boundary), calculated flows across internal sections deviate somewhat from the inflow/outflow values. A study was made to evaluate errors in continuity as a function of network density. Poor continuity approximation is important of itself if water quality simulation is the goal. In the present applications, however, water surface elevations and velocities are the variables of interest. Therefore, the impact of continuity errors on these parameters was also investigated.

Flows on the Rio Grande de Loiza flood plain were simulated using several networks. This flood plain was selected because of its complex flow field and a prior study by the U.S. Army Corps of Engineers (1976) had made the data readily available. Model performance had previously been evaluated for simple hypothetical and laboratory flows by Norton et al (1973) and King et al (1975). The Loiza flood plain is about 10 by 10 km (6 by 6 miles) in extent and is characterized by variable bottom topography, one inlet and two outlets, and several islands. Three of the networks used in the study are shown in Figs. 1 to 3 illustrating progressive increase in network detail.

The solution was considered acceptable if flow at all continuity check lines deviated from inflow by less than $\pm$ 5%. Continuity is checked by integrating the normal component of velocity times depth along lines specified by the modeler. The continuity check lines used in this study are indicated by dark lines on Figs. 1 to 3. Note that, because the flow divides around the islands, in some cases the sum of flows across two check lines (such as 5 and 6) should be compared with inflow. Various parameters of the problem are summarized in Table 1. No attempt was made to calibrate the coefficients used.

The continuity approximation improved with increasing network detail, as expected. Flow at the worst check line in the coarsest network (7 + 8) improved from 79.3% to 98.2% of inflow as network detail was increased. Network characteristics, computer execution times, and results of the simulations with these three networks are summarized in Table 2. Average depths and velocities along the continuity check lines are given in Table 3. The check line numbers in Tables 2 and 3 refer to the lines indicated on Figs. 1-3.

Table 1 Data for Loiza Flood Plain Simulation

1. Boundary conditions:
   a. Inflow (line 1) = 8200 cms (290,000 cfs)
   b. Outlets (lines 11 & 12), water surface elevation = 2.5 m (8 ft) MSL
   c. All other boundaries; either tangential flow or stagnation points
2. Bed roughness: Chezy C spatially varied from 5.5 to 22 $m^{1/2}$/sec (10 to 40 $ft^{1/2}$/sec)
3. Turbulent exchange coefficients: varied with element size from 24 to 48 m^2/sec (260 to 500 ft^2/sec)

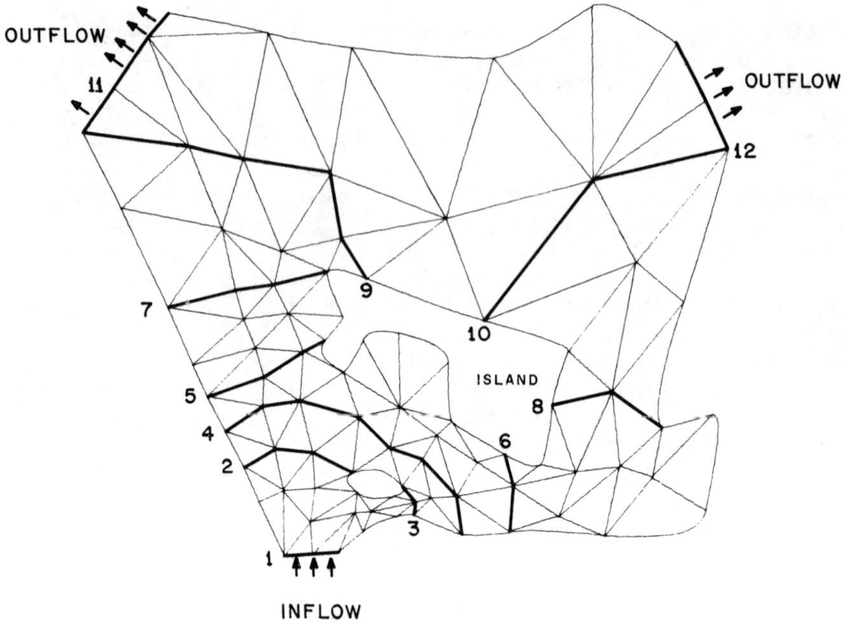

Figure 1 Continuity Check Network 3.1
(Dark Lines Indicate Continuity Check Lines)

Table 2 Continuity Performance of the Networks

Network	3.1	3.3	3.5
No. of Nodes	310	375	432
No. of Elements	131	162	189
CDC 7600 Execution Time (sec)	22	31	45
Check Line	Percent of Inflow		
1 (inflow)	100.0	100.0	100.0
2 + 3	89.2	90.8	96.2
4	114.9	106.8	104.9
5 + 6	87.5	92.0	96.4
7 + 8	79.3	90.1	98.2
9 + 10	99.8	99.4	98.7
11 + 12 (outflow)	100.0	100.0	100.0

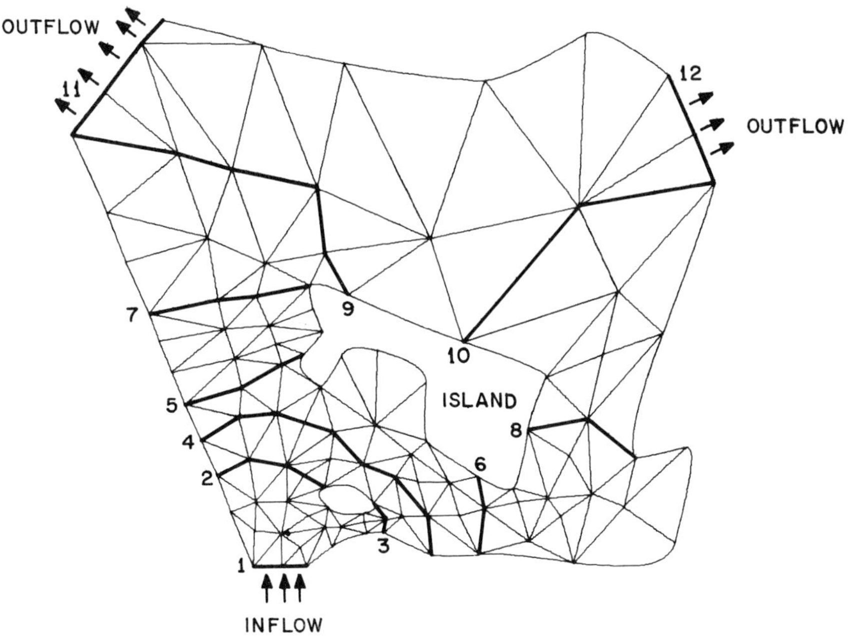

Figure 2 Continuity Check Network 3.3

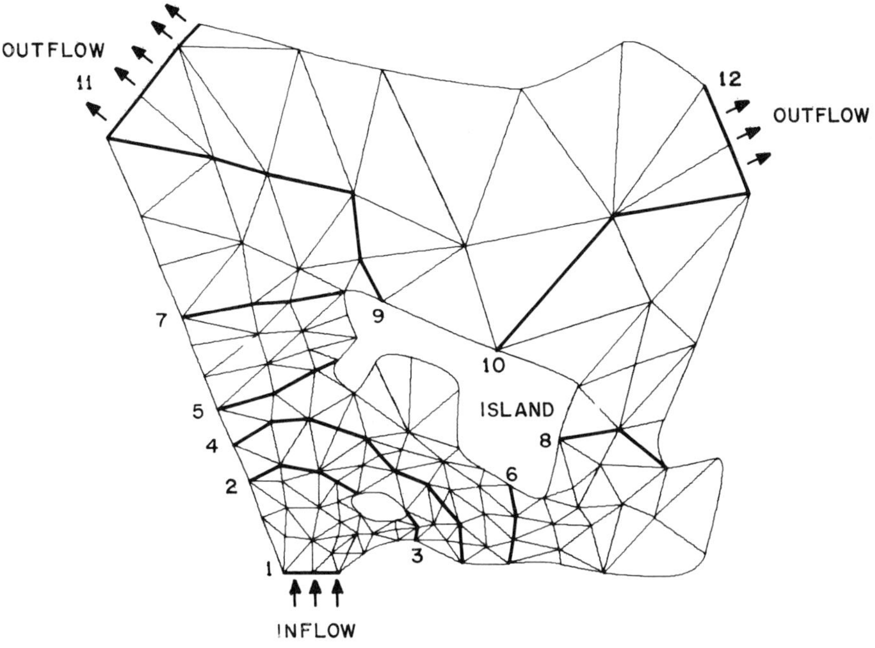

Figure 3 Continuity Check Network 3.5

Table 3 Flows (as percent of inflow), depth, and velocities for the networks

Line	NETWORK	3.1	3.3	3.5
1	%	100	100	100
	Y(m)	5.09	5.34	5.36
	V(mps)	1.92	1.83	1.83
2	%	50.4	50.1	53.7
	Y(m)	3.02	2.84	2.87
	V(mps)	.70	.74	.79
3	%	38.8	40.7	42.5
	Y(m)	2.48	2.51	2.63
	V(mps)	2.52	2.62	2.61
4	%	114.9	106.8	104.9
	Y(m)	2.91	2.78	2.78
	V(mps)	.66	.65	.63
5	%	36.0	37.9	39.7
	Y(m)	1.91	1.98	2.02
	V(mps)	.75	.76	.78
6	%	51.5	54.1	56.7
	Y(m)	3.82	3.90	3.76
	V(mps)	.90	.93	1.01
7	%	36.8	37.9	40.9
	Y(m)	1.86	1.88	1.91
	V(mps)	.62	.63	.67
8	%	42.5	52.2	57.3
	Y(m)	2.26	2.37	2.37
	V(mps)	.80	.94	1.03
9	%	42.1	41.0	42.6
	Y(m)	2.48	2.48	2.50
	V(mps)	.24	.23	.24
10	%	57.7	58.4	56.1
	Y(m)	3.09	3.07	3.05
	V(mps)	.31	.31	.30
11	%	45.8	46.0	47.0
	Y(m)	2.36	2.37	2.37
	V(mps)	.67	.68	.69
12	%	54.2	54.0	53.0
	Y(m)	2.31	2.31	2.30
	V(mps)	1.06	1.06	1.05

For most of the check lines, the improvement in continuity obtained with increasing network detail was associated with changes in both velocity and depth. In two cases, lines 6 & 8, the velocity changes were substantial. This region of the flow field is characterized by a rapid change of direction. The results reinforce the caveat that increased network detail is important in such regions. Furthermore, it appears that depth is somewhat less sensitive to errors in continuity than is velocity. Therefore, if one is interested in water surface elevations only, a less stringent continuity performance criterion could be accepted than if velocities are of interest.

<u>Application to McNary Dam Second Powerhouse Study</u>

An example of a "production" type application of the horizontal flow model is the second powerhouse site selection study for McNary lock and dam on the Columbia River. Flow fields downstream of the dam were simulated for several possible locations of the second powerhouse. Of interest were velocities, both magnitudes and directions, in the vicinity of the approach channel to the navigation lock. The study area and several of the possible second powerhouse locations are shown on Fig. 4. Finite element networks for the existing condition and for the south shore powerhouse with excavated discharge channel are shown in Figs. 5 and 6. Data are summarized in Table 4. The roughness coefficient was calibrated to reproduce an observed condition.

This study was greatly facilitated by an automatic reordering algorithm (Collins (1973)) which has been incorporated into the model. This algorithm makes modification of a network (compare Figs. 5 and 6) straightforward in that the entire network need not be re-numbered. The existing numbering scheme is utilized for input/output and the system of equations internally re-ordered to reduce storage.

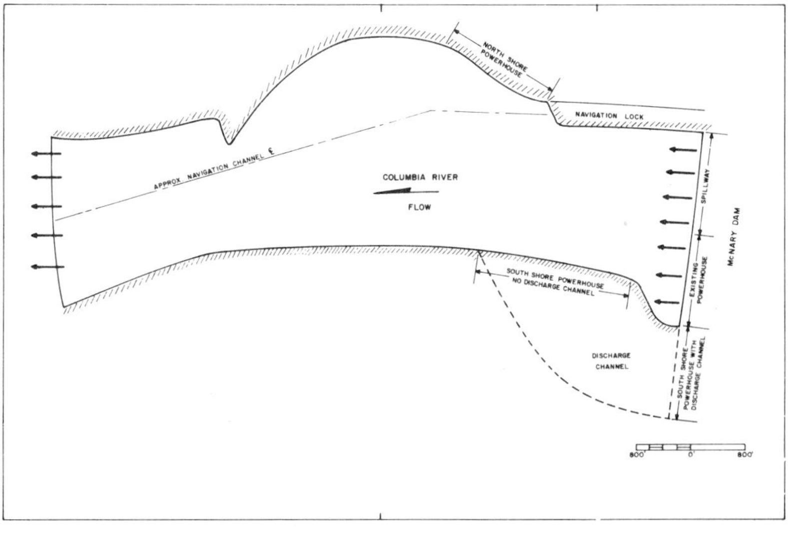

FIG. 4 STUDY AREA SHOWING POSSIBLE SECOND POWERHOUSE LOCATIONS

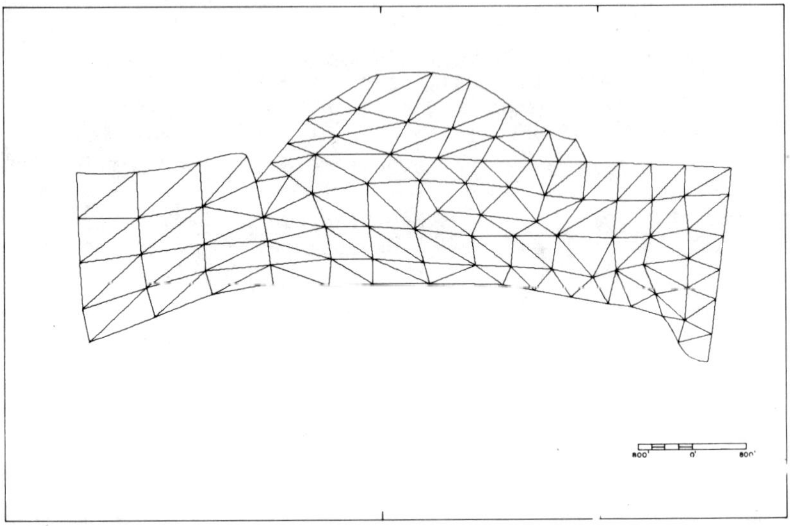

FIG. 5 FINITE ELEMENT NETWORK FOR RUNS 1-4

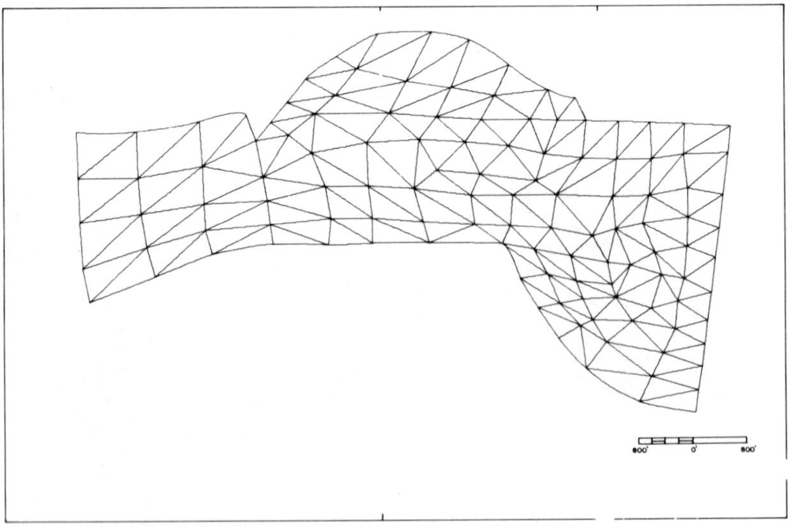

FIG. 6 FINITE ELEMENT NETWORK FOR RUN 5

Table 4 Data for McNary Second Powerhouse Study

1. Upstream boundary condition:
   a. Spillway: Q = 7000 cms (250,000 cfs) for calibration runs
      Q = 0 for production runs
   b. Existing powerhouse: Q = 6500 cms (230,000 cfs)
   c. Second powerhouse: Q = 7000 cms (250,000 cfs)
2. Downstream boundary condition: Water surface elevation = 82.4 m (270.3 ft) MSL*
3. All other boundaries: Either tangential flow or stagnation points
4. Roughness: Chezy C = 55 $m^{1/2}$/sec (100 $ft^{1/2}$/sec)
5. Turbulent exchange coefficients: Varied with element size from 4.8 to 14.4 m^2/sec (50 to 150 ft^2/sec)

*For production runs in which total river discharge was 13600 cms (480,000 cfs). This elevation was varied according to a known stage-discharge relationship for other discharges.

A vector plotting routine was used to display simulated flow fields. Two such plots are shown on Figs. 7 and 8. Plots of this type are considered essential for interpreting and analyzing complex flow fields.

Continuity errors were generally less than ±5% with the exception of the constriction near the downstream boundary where errors were on the order of -15%. If future detailed studies are made, and velocities in that area become important, more network detail will be provided.

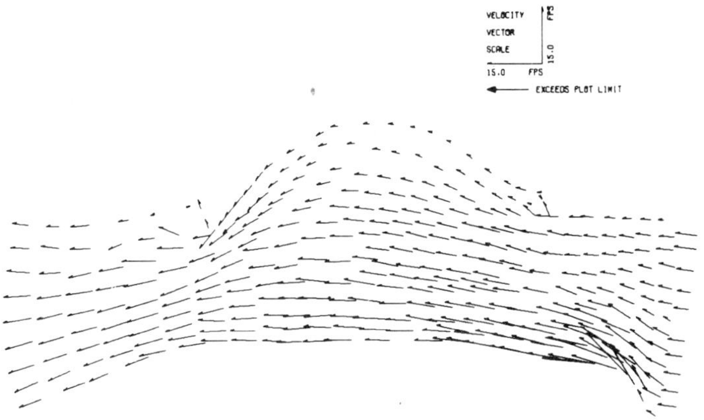

Figure 7 Velocities for Spillway Q = 7000 cms (250,000 cfs), Existing Powerhouse Q = 6500 cms (230,000 cfs), Slip Boundary Conditions

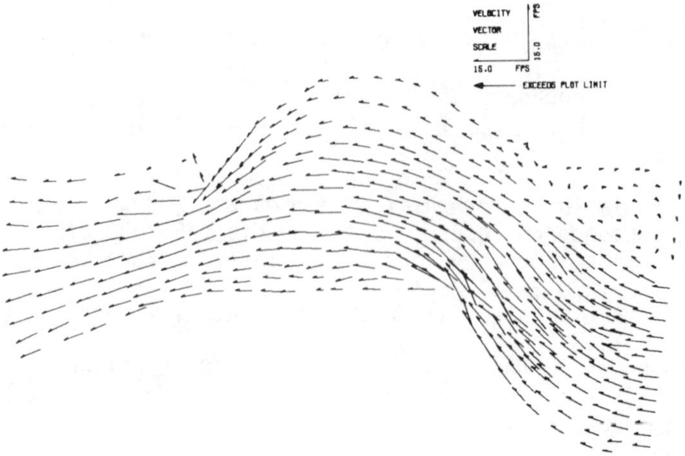

Figure 8 Velocities for Spillway Q = 0, Existing Powerhouse Q = 6500 cms (230,000 cfs), Second Powerhouse Q = 7000 cms (250,000 cfs), Slip Boundary Conditions

The model allows two valid types of boundary conditions at boundaries where no flow enters or leaves the system. One is the stagnation point where both components of velocity are zero; the other is the slip boundary condition where the velocity on the boundary is tangential to the boundary. The slip condition requires use of curved-sided elements on the boundaries. Use of curved boundaries with tangential flow is favored. Use of stagnation points along the boundaries results in a substantially different solution as shown in Fig. 9. Not only is the velocity distribution altered, but calculated head loss in the reach is about 0.21 m (0.7 ft.) greater than with the slip boundary condition. Continuity performance for the two simulations was similar, though in other problems analyzed by Resource Management Associates (1977), the slip condition was superior. Use of different boundary conditions should be investigated in an attempt to identify under what conditions the modeler should choose slip or stagnation point boundaries.

It is encouraging to note that the McNary study required no code changes to the model.

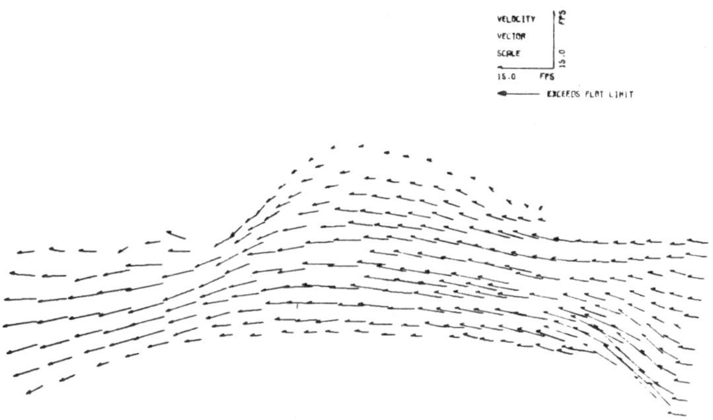

Figure 9 Velocities for Spillway Q = 7000 cms (250,000 cfs), Existing Powerhouse Q = 6500 cms (230,000 cfs), Stagnation Point Boundary Conditions

TWO-DIMENSIONAL MODELS IN THE VERTICAL PLANE

Two-dimensional (longitudinal and vertical) hydrodynamic models have been developed to aid the Corps in the description and analysis of reservoir water quality. The importance of the longitudinal as well as the vertical exchange in long, relatively narrow and deep impoundments has been studied by Pritchard (1971), Anthony and Drummond (1973) and the Tennessee Valley Authority (1969). Investigations such as these have shown that the hydrodynamics of a stratified reservoir influences the water quality and, therefore, the biological productivity of deep impoundments. Additional objectives for the development of multidimensional models are to be able to predict the effects that outlet type and location, degree of stratification, and reservoir operation have on the water quality in downstream rivers and streams.

As well as the general interest in simulating flows in the vertical plane, this research has provided the opportunity to compare the performance of an implicit finite difference method (FDM) model with that of a finite element method (FEM) model. The FDM model was developed by Edinger and Buchak (1977) and is named LARM (Laterally Averaged Reservoir Model). The FEM vertical model was developed by Norton et al (1973) and King et al (1975). Although initial development of the vertical FEM model was accomplished at the same time as that of the horizontal model previously discussed, further refinement and use of the vertical model has lagged considerably.

The primary objectives of the comparison of these two hydrodynamic models were: (1) to compare the relative ease with which the required data and boundary conditions could be prepared and coded; (2) to compare the overall performance of the two different approaches with respect to stability, convergence, accuracy and practicality; and finally, (3) to compare relative run times and simulation costs between the two methods for similar problems. The following paragraphs present the fundamental equations used by the two models.

Governing Equations

Both models incorporate similar forms of the so-called phenomenological equations for momentum, along with the continuity equation and a form of the convective-diffusion equation for thermal or material transport in the vertical. Note, however, that the FEM model retains the vertical momentum equation, which is replaced by the hydrostatic pressure distribution in the FDM model. Both models utilize a Cartesian coordinate system with the longitudinal x dimension positive downstream. The vertical z dimension is referenced positive upward from the x-axis in the FEM model, while it is positive downward from the x-axis in LARM. Both models allow for a variable width in the lateral y direction.

FEM Hydrodynamic Model

Momentum Equation:

$$\rho b \left(\frac{\partial u}{\partial t} + u\frac{\partial u}{\partial x} + w\frac{\partial u}{\partial z}\right) + \frac{\partial (pb)}{\partial x} - \varepsilon_{xx}\frac{\partial}{\partial x}\left(b\frac{\partial u}{\partial x}\right) - \varepsilon_{xz}\frac{\partial}{\partial z}\left(b\frac{\partial u}{\partial z}\right)$$

$$+ \rho g C^{-2} u|u| A_b - \zeta V_a^2 \cos\psi \, A_s = 0 \quad (4)$$

$$b\rho \left(\frac{\partial w}{\partial t} + u\frac{\partial w}{\partial x} + w\frac{\partial w}{\partial z}\right) + \frac{\partial (pb)}{\partial z} + b\rho g - \varepsilon_{xz}\frac{\partial}{\partial x}\left(b\frac{\partial w}{\partial x}\right)$$

$$- \varepsilon_{zz}\frac{\partial}{\partial z}\left(b\frac{\partial w}{\partial z}\right) = 0 \quad (5)$$

Continuity Equation:

$$\frac{\partial}{\partial x}(bu) + \frac{\partial}{\partial z}(bw) = 0 \quad (6)$$

Convective-Diffusion Equation for Density:

$$\frac{\partial b\rho}{\partial t} + b\left(u\frac{\partial \rho}{\partial x} + w\frac{\partial \rho}{\partial z}\right) - D_x\frac{\partial}{\partial x}\left(b\frac{\partial \rho}{\partial x}\right) - D_z\frac{\partial}{\partial z}\left(b\frac{\partial \rho}{\partial z}\right) = 0 \quad (7)$$

where

u, w = fluid velocity in the x and z directions respectively
b = breadth
p = pressure
D_x, D_z = eddy diffusion coefficients in the x and z directions respectively
A_b = area over which bottom stress is effective
A_s = the area over which the wind stress is effective

Other variables have previously been defined.

FDM Hydrodynamic Model LARM

Momentum Equation:

$$\frac{\partial}{\partial t}(ub) + \frac{\partial}{\partial x}(u^2b) + \frac{\partial}{\partial z}(uwb) + \frac{1}{\rho}\frac{\partial}{\partial x}(pb) - \frac{\partial}{\partial x}\left(b\varepsilon_x \frac{\partial u}{\partial x}\right) - \frac{\partial}{\partial z}(\tau_z b) = 0 \qquad (8)$$

Boundary stresses are found using the following expressions:
at the surface:

$$\tau_z = \zeta\frac{\rho_a}{\rho} V_a^2 \text{ Cos } \psi \qquad (9)$$

at the bottom:

$$\tau_z = \frac{\rho g}{C^2} u|u| \qquad (10)$$

Hydrostatic Pressure distribution:

$$\frac{\partial p}{\partial z} - \partial g = 0 \qquad (11)$$

Continuity Equation:

$$\frac{\partial}{\partial x}(ub) + \frac{\partial}{\partial z}(wb) = qb \qquad (12)$$

Thermal Convective-Diffusion Equation:

$$\frac{\partial(Tb)}{\partial t} + \frac{\partial(uTb)}{\partial x} + \frac{\partial(wTb)}{\partial z} - \frac{\partial}{\partial x}\left(D_x b \frac{\partial T}{\partial x}\right) - \frac{\partial}{\partial z}\left(D_z b \frac{\partial T}{\partial z}\right) = \frac{fb}{\rho c_p} \qquad (13)$$

Equation of state:

$$\rho = \rho(T) \qquad (14)$$

where

ρ = fluid density
ρ_a = density of air
τ_z = boundary shear stress
q = lateral inflow per unit volume
T = temperature
D_x, D_z = heat transport dispersion coefficients
f = heat inflow per unit volume
c_p = specific heat

Other variables have been previously defined.

Description of the Test Problem

Data collected by the Tennessee Valley Authority (TVA)(1969) were used to test and compare the two vertical models in a reservoir simulation. These data were for the Fontana Reservoir in North Carolina. The models were applied to the first 23 km (14.5 miles) of the reservoir upstream from the dam. To simplify geometric requirements, a uniform reservoir breadth of 638 m (2095 ft) was used. This breadth was selected to conserve reservoir volume. The bottom profile and elevations were determined from sediment investigation cross sections. Conditions that existed in the reservoir during the last week of March, 1966 were used to provide the boundary conditions for the simulation. Water temperature profiles, water surface elevations, and flows into and out of the test reach were obtained from the TVA (1969) data. The reservoir was stratified and was approximately 108 m (353 ft.) deep at the dam. Surface heat exchange, wind velocity, and tributary inflows were all assumed to be zero for the purposes of this investigation; a steady inflow and outflow of 140 cms (5000 cfs) was used.

Discussion

The time and effort necessary to describe the reservoir geometry for both the FDM and FEM models were comparable. To achieve calculated results at comparable locations in space, optional quadrilateral elements were used so that the finite element network (Fig. 10) was almost identical to FDM grid (not shown). It is recognized that this network does not exploit the capability of the FEM model to allow increased geometric resolution where desired, such as near the reservoir outflow point, but this simplification was useful for comparison of results.

The convergence of the FEM solution was noted to be somewhat more sensitive to the magnitude of the turbulent exchange coefficients than the FDM model. The ranges of values of the coefficients over which convergent solutions can be obtained for the two models have not yet been firmly established. Additional sensitivity investigations shall be undertaken at a later time. Ariathurai, et al (1977) examined similar equations and found that stability and convergence of the solution could be related not only to spatial and temporal step sizes but also to the Peclet number which is the ratio of convective transport to diffusive transport.

The flow fields calculated with the FDM and FEM models are shown in Figs. 11 and 12 respectively. The vertical scale of Figs. 10-12 is exaggerated by a factor of 100. Coefficients used (refer to equations 4-7) were: ε_{xx}=24, ε_{xz}=4.8x10^{-3}, ε_{zz}=240, D_x=23, D_z=9.3x10^{-7} m^2/sec (260,0.05, 2600, 250, 10^{-5} ft^2/sec). Although the models have numerous detailed differences, particularly in the description of boundary conditions, the calculated flow fields are similar and reasonable. For the test application, the reservoir was thermally stratified, with the incoming fluid cooler and more dense than the fluid in the surface layers. The stable density gradient in the region of the thermocline tends to inhibit vertical momentum and material transport, yet circulation appears in the upper layers. The circulation in the surface layers is driven by internal horizontal shearing between the cool water flowing toward the outlet and the warmer water above. A similar flow pattern is also observed in the bottom region below the main flow in the

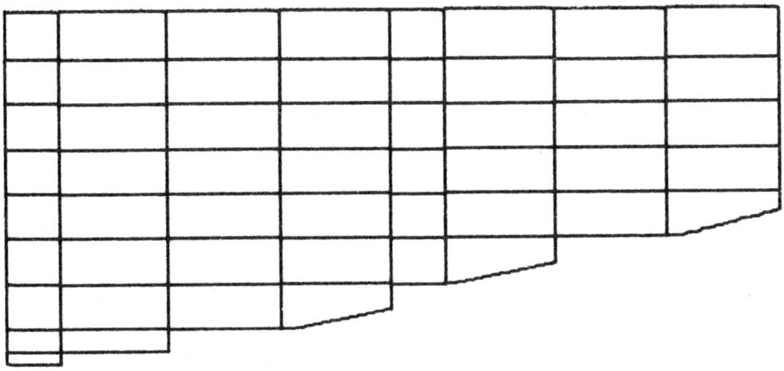

Figure 10 Finite Element Network for Reservoir Simulation

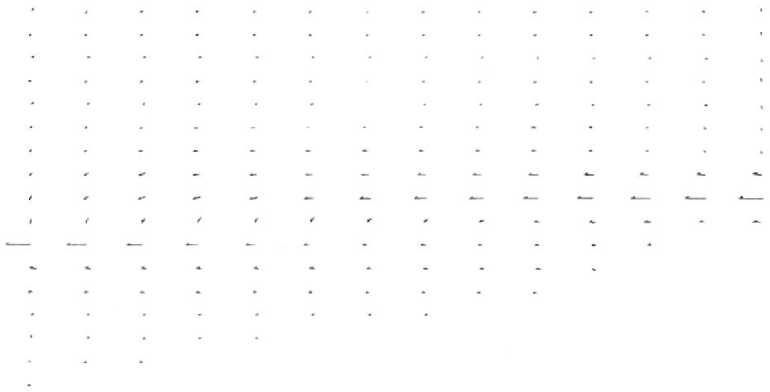

Figure 11 Reservoir Velocities Calculated with the Finite Difference Model

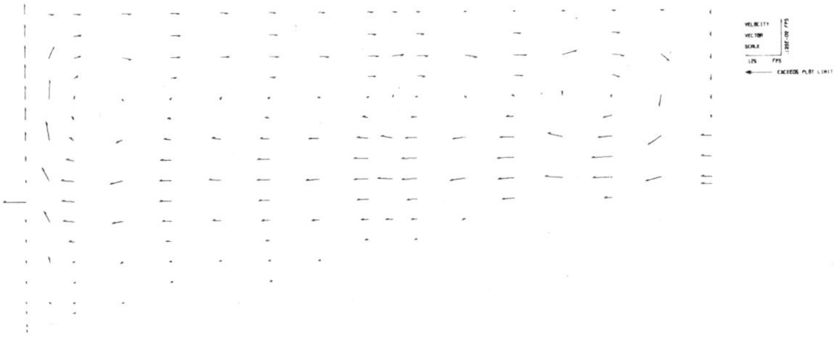

Figure 12 Reservoir Velocities Calculated with the Finite Element Model

FDM model (Fig. 11). Generally, the FEM solution predicts larger vertical velocity components, perhaps due to the retention of the vertical momentum equation. Comparison of the solutions with available field data will be undertaken once general performance characteristics of the two models are further defined.

For these steady state simulations, the FDM took about 6 times more CDC 7600 computer time than the FEM. The primary reason is that, to achieve a steady state solution, the FDM model must be run through pseudo-time with constant boundary values until transients from initial conditions die out (about 75-100 days in this case). The FEM model, however, has the capability of solving the system once with zero time derivatives to arrive at a steady state solution. Comparative costs for dynamic simulations will depend primarily upon length of time step and number of elements used to define the study region.

SUMMARY

The work to date with the horizontal flow model indicates the following:

(1) Internal continuity errors can be reduced to acceptable levels by increasing network detail, particularly in areas of large curvature of the velocity field.

(2) Errors in continuity tend to be reflected more strongly in the velocity than the depth.

(3) General application of the model to steady state simulations is feasible at present.

The preliminary work with the vertical flow models indicates the following:

(1) The finite element method model is less costly than the finite difference model for steady state solutions.

(2) Simulation of flows in which density gradients are important requires careful selection of turbulent exchange and eddy diffusion coefficients.

(3) The finite element model predicts larger vertical velocities than the finite difference model, perhaps due to the retention of the vertical momentum equation.

(4) More experience with, and development of, the vertical models will be required before "production" applications can be easily made.

Indicated areas of further work are:

(1) Verification of models' performance when an adequate data set becomes available.

(2) Development of guidance on selection of turbulent exchange coefficients, relationship to flow properties etc.

(3) Investigate models' behavior for dynamic simulations.

(4) Evaluate use of stagnation vs. slip boundary conditions in the finite element models.

(5) Extend simulations with the vertical models to variable breadth problems.

ACKNOWLEDGEMENTS

The evaluation of the mass continuity performance was funded by the U.S. Federal Highway Administration. The Walla Walla District, U.S. Army Corps of Engineers funded the McNary second powerhouse study. The comparison of finite element and finite difference techniques is being funded by the U.S. Army Corps of Engineers Waterways Experiment Station. Continued advice of the finite element model developers: Ian King and Bill Norton of Resource Management Associates, and the finite difference model developers: John Edinger and Ed Buchak of J. E. Edinger and Associates, is gratefully acknowledged.

REFERENCES

Anthony, M. and Drummond, G. (1973) "Reservoir Water Quality Control," in Man-Made Lakes: Their Problems and Environmental Effects, Geophysical Monograph 17, American Geophysical Union, pp. 549-551.

Ariathuri, R., MacArthur, R.C., and Krone, R.B. (1977) "Mathematical Model of Estuarial Sediment Transport," Tech Report D-77-12, Office, Chief of Engineers, U.S. Army Corps of Engineers.

Collins, R.J. (1973) "Bandwidth Reduction by Automatic Renumbering," International Journal for Numerical Methods in Engineering, Vol. 6, 345-356.

Edinger, J.E. and Buchak, E.M. (1977) "A Hydrodynamic Two-Dimensional Reservoir Model: Development and Test Application to Sutton Reservoir, Elk River, West Virginia," Report to U.S. Army Corps of Engineers, Ohio River Division.

Gray, W.G., Pinder, G.F., and Brebbia, C.A. (1977) Finite Elements in Water Resources, Proceedings of the First International Conference on Finite Elements in Water Resources, Pentech Press.

Huebner, K. (1975) The Finite Element Method for Engineers, John Wiley & Sons.

King, I.P., Norton, W.R., and Iceman, K.R. (1975) "A Finite Element Solution for Two-Dimensional Stratified Flow Problems," in Finite Elements in Fluids, Vol. 1, John Wiley & Sons.

Norton, W.R., King, I.P., and Orlob, G.T. (1973) "A Finite Element Model for Lower Granite Reservoir," Report to U.S. Army Corps of Engineers, Walla Walla District.

Pritchard, D.W. (1971) "Hydrodynamic Models: Two-Dimensional Model," in Estuarine Modeling: An Assessment, Environmental Protection Agency, Stock No. 5501-0129, U.S. G.P.O., Washington D.C.

Resource Management Associates (1977) "Case Studies of Continuity Satisfaction with an Improved 2-D Hydrodynamic Model," Report to the U.S. Army Corps of Engineers, Hydrologic Engineering Center.

Strang, G. and Fix, G. (1973) An Analysis of the Finite Element Method, Prentice-Hall.

Tennessee Valley Authority (1969) "Evaluation of Fontana Reservoir Field Measurements," Laboratory Report No. 17-90.

U.S. Army Corps of Engineers, Jacksonville District (1976) "Flood Hazard Information (Technical Appendix), Rio Grande de Loiza, Puerto Rico."

Zienkiewicz, O.C. (1971) _The Finite Element Method in Engineering Science_, McGraw Hill.

RECENT APPLICATION OF RMA'S FINITE ELEMENT MODELS FOR TWO DIMENSIONAL HYDRODYNAMICS AND WATER QUALITY

I. P. King, W. R. Norton

Resource Management Associates, Lafayette, CA

INTRODUCTION

In 1971-1972, the authors developed for the U.S. Army Corps of Engineers (Norton et al, 1973) two dimensional finite element models for fluid flow; these original models were developed to be as general as possible for the class of problem described. They were, however, new in concept and over the last 5 years a series of refinements have been incorporated both to improve model performance and computer speed. The formulation of the two dimensional vertical model has been previously reported (King et al, 1975). In this paper we plan to review recent simulations using the model. Space does not permit a description of all applications and so several have been selected to demonstrate the model not as a test vehicle but as a practical tool being applied to make engineering decisions.

FORMULATION AND MODEL STRUCTURE

Hydrodynamics Model

The model is constructed to solve the Navier Stokes equation reduced to two dimensional form. The Boussinesq approximation is used to represent turbulence effects through an equivalent eddy viscosity coefficient.

The governing equations may thus be written as:

Motion Equations

$$\frac{\partial u}{\partial t} + u\frac{\partial u}{\partial x} + v\frac{\partial u}{\partial y} + g\frac{\partial h}{\partial x} + g\frac{\partial z_o}{\partial x} - \frac{\varepsilon_{xx}}{\rho}\frac{\partial^2 u}{\partial x^2} - \frac{\varepsilon_{xy}}{\rho}\frac{\partial^2 u}{\partial y^2}$$

$$- 2\omega v \sin\phi + \frac{gu}{C^2 h}(u^2 + v^2)^{1/2} - \frac{\zeta}{h} V_a^2 \cos\psi = 0 \qquad (1)$$

$$\frac{\partial v}{\partial t} + u\frac{\partial v}{\partial x} + v\frac{\partial v}{\partial y} + g\frac{\partial h}{\partial y} + g\frac{\partial z_o}{\partial y} - \frac{\varepsilon_{yx}}{\rho}\frac{\partial^2 v}{\partial x^2} - \frac{\varepsilon_{yy}}{\rho}\frac{\partial^2 v}{\partial y^2}$$

$$+ 2\omega u \sin\phi + \frac{gv}{C^2 h}(u^2 + v^2)^{1/2} - \frac{\zeta}{h} V_a^2 \sin\psi = 0 \qquad (2)$$

Continuity Equation

$$\frac{\partial h}{\partial t} + \frac{\partial}{\partial x}(uh) + \frac{\partial}{\partial y}(vh) = 0 \tag{3}$$

where ρ = fluid density; u,v = velocity components in the x and y directions; h = water depth; C = Chezy coefficient; g = gravitational constant; z_o = bottom elevation; ε_{xx}, ε_{xy}, ε_{yx} and ε_{yy} = turbulent eddy coefficients; x,y = Cartesian coordinates; t = time; ω = rate of earths angular rotation; ϕ = local latitude; ζ = empirical coefficient; V_a,ψ = local wind velocity and direction. The dependent variables of equations (1), (2) and (3) are u, v, and h. In the finite element model formulation, it is convenient to transform the velocity components to flow components, r and s, such that r = uh, s = vh.

This transformation has a number of virtues

(a) More accurate representation of uniform flow conditions.

(b) Ease of representation of flow boundary conditions.

and (c) Linearization of equation (3).

The Galerkin method of weighted residuals is used to the finite element representation of the system. Quadratic shape functions are used to represent flow components r and s, linear functions to represent h. This mixed form of representation has been found to be the best representation of the various order of drivatives existing in the basic equations.

In the original version of the model, elements were limited to be straightsided, however, this presented considerable difficulty when flow parallel to a boundary was desired. In what has been the most important change to the original model, the methodology was changed in the current version to iso-parametric elements (quadratic sides) and thus by adjustment of location of the "mid-side node" smooth parallel flow boundaries between elements could be maintained with no loss of flow. The impact of this change will be demonstrated clearly in our example problems of the right angle bend and Talahalla Creek.

A full implicit scheme has been utilized for reduction of the time dependent term to a finite difference representation. In this scheme reported previously (King, 1977) for one dimensional problems the variations are a time step of a typical dependent variable to have the form

$$\phi = a + bt + ct^{\alpha} \tag{4}$$

where a, b, c and α are constants from which it may be deduced that at the end of a time interval Δt

$$\dot{\phi} = (\phi - \phi_o)\alpha/\Delta t - (\alpha - 1)\dot{\phi}_o \tag{5}$$

where dot represents derivative with respect to time and

subscript zero refers to the beginning of the time step.

In practical applications for both one and two dimensional models we have found that α = 1.5 develops a stable solution without unnecessary damping. It should be noted that although this procedure is similar to conventional intermediate evaluation using the so called "theta" the implications are different for the response of non-linear problems.

Equations (1) and (2) remain non-linear after transformation and so the finite element formulation is completed by using a Newton Raphson scheme to reduce the non-linear equations into linear form and iteration is used to generate a solution.

The model as structured allows both steady state and time dependent solutions. The ability to run steady state simulations has proved a valuable tool when networks are first established. It allows rapid evaluation of the adequacy of the mesh and appropriateness of the elements.

Water Quality Model

The water quality model has been developed to solve the two dimensional convection-diffusion equation which may be written as:

$$h\frac{\partial c}{\partial r} + u\frac{\partial}{\partial x}(ch) + u\frac{\partial}{\partial y}(ch) - Dx\frac{\partial}{\partial x}(h\frac{\partial c}{\partial x}) - Dy\frac{\partial}{\partial y}(h\frac{\partial c}{\partial y})$$

$$- h\sigma + khc = 0 \qquad (6)$$

where c = the constituent concentration; σ = any local source of constituent mass; and k = the rate of mass decay.

The finite element model has been developed for triangular and quadrilateral iso-parametric elements using quadratic shape functions. The Galerkin Method of Weighted Residuals is again used to form the set of linear equations for this problem; time dependence is treated in the same way as for the hydrodynamic model.

The model was developed to be compatible with the hydrodynamic model, and velocities and depths developed in that model may be used directly as input files to the water quality model. As with the hydrodynamic model, quality may be simulated either in a fully dynamic mode or in a steady state form.

Model Structure

Both models are written in FORTRAN and have been executed on many computer configurations. The programs are designed in versions that are either entirely within fast core storage or are such that for smaller machines they utilize file storage during solution of the equations. Pre- and post-processors have been developed to simplify network preparation and allow automatic plotting of networks, velocity vectors, contours of water depth and constituent concentration, and isometric three dimensional views of quality constituents.

MODEL TESTING AND VERIFICATION

The basic formulation of the hydrodynamic model does not ensure overall satisfaction of mass continuity at all points. The continuity equation is only one of three equations that must be approximated by the finite element solution. Overall mass continuity along outer boundaries is however maintained. This apparent shortcoming can however be turned to advantage because the satisfaction of continuity across internal element boundaries is readily calculated by integrating the total flow crossing these boundaries. Therefore it is possible to use continuity as a measure of accuracy of the solution at various areas of the system.

In this context it was found that the limitation to straight sided elements presented a number of conceptional difficulties in practical application because it was impossible to specify parallel flow conditions at all locations on the boundary of an irregular system without "leakage" into or out of the system. Under such conditions the total flow in and out of the system would not balance and overall conservation of flow could not be maintained. The initial solution to this problem was to specify zero flow at all such locations and to insert a narrow boundary layer element with low eddy viscosity near the boundary; the result of this construction was to allow high velocity gradients between interior points and the boundary. While this method maintained overall satisfaction of flow continuity, it created a notable lack of satisfaction of continuity in the vicinity of the zero flow points. It appears that this lack of continuity was caused by forcing the flow regime to zero and thus creating unrealistically high velocity gradients that the model has difficulty representing.

A second potential solution to the problem of irregular boundaries is to construct the system network such that the angle of flow is forced to be exactly orthogonal to perpendicular bisector of the boundaries corner angle (Wang and Connor, 1975). This type of construction relieves the necessity of having a zero flow condition and exactly as much flow "leaks" out of the system as leaks in (or vice versa). While this solution works reasonably well for flow modeling, it causes numerous problems in water quality simulation and special techniques must be used to keep track of the apparent mass transport in and out of the system. The drawbacks associated with this type of network construction seem to outweigh its merits, and it has been dropped from RMA's models.

RMA's current approach to both the problem of parallel flow and continuity satisfaction has been to develop iso-parametric elements which have continuous slopes at interelemental boundary connections.

The curved elements have been designed so that for most problems flow parallel to the boundary can be maintained without loss of continuity or the need to specify zero flow at the boundary. This change was achieved by reforming the element coefficient matrix in iso-parametric form. The transformations necessary to achieve this formulation raise the order of the approximating functions and it is necessary to use a higher order numerical integration scheme than was previously used for straight-sided elements. For example, sixteen point schemes are now used for curved triangles and quadrilaterals.

As a side effect of the iso-parametric elements, the surface integrals that result from the partial integration of 2nd derivatives cannot be treated as zero on curved boundaries and thus must be explicitly evaluated. Also, when curved sides are used on boundaries, the location of "mid-side nodes" must be adjusted so that the slopes of the boundary at both of the corner nodes exactly achieve the desired values. This problem is solvable and a computer program has been developed to perform this computation and to plot the resulting curved-sided network. We have found that extreme care must be used in specifying these slopes or an overall flow continuity will be lost due to slight leakage.

Model Testing

Two examples are presented to demonstrate the improved results which have been obtained from the most recent version of the model. The first example is the hypothetical problem of flow around a right angle bend in a rectangular channel. The second problem is an application of curved sided elements to a river crossing which was originally simulated with all straight sided elements (Tseng, 1975).

Example 1 The first problem consists of horizontal flow around a right angle bend of nominal dimensions. The networks constructed for this problem represent a prototype that is six feet along each of its outside edges (see Figure 1) with a nominal water depth of ten feet. In all cases flow is introduced along the top right face of the system (nodes 28, 31 and 33) and removed at the lower left boundary (nodes 1, 2 and 4). A head loss in the order of a few hundredths of a foot was calculated with the values of eddy viscosity of 1.0 $lb/sec/ft^2$, Chezy C of 100 and flow equal to 20 cfs which were used.

Four separate levels of network detail were prepared for this problem as is shown in Figure 1. The only difference in Levels 1-3, and their resulting computer simulations, is the amount of detail in the area of the bend. Each level of network has approximately twice the detail at the bend as it's predecessor. The network indicated as Level 4 has the same number of nodes and elements as Level 2, but uses a short radius curve and parallel flow at nodes 16 and 18, and thus has no corners or zero flow boundary conditions.

The results obtained with each of the four networks is summarized in Tables 1 and 2. Please note that the definition of the lines across which mass continuity was evaluated is taken from the Level 1 network and is at the same physical location in each example.

The results obtained with each of the four networks is summarized in Tables 1 and 2. The results for the first three

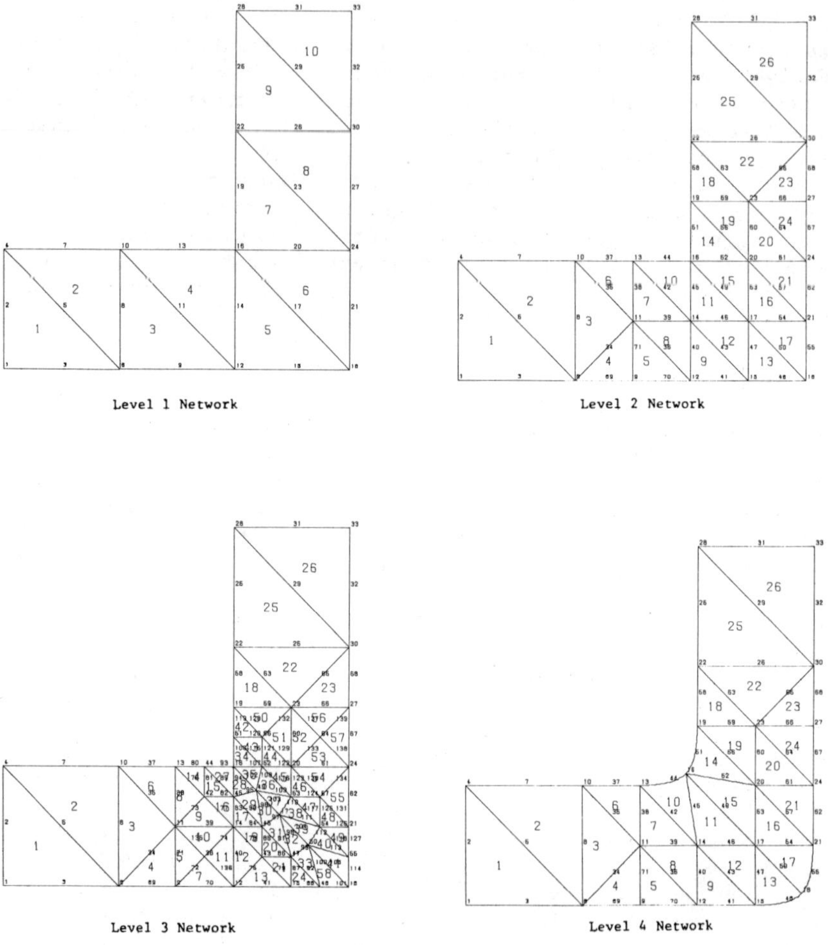

FOUR LEVELS OF DETAIL USED IN EVALUATING
THE INFLUENCE OF NETWORK CONSTRUCTION ON
CONSERVATION OF FLOW AROUND A RIGHT ANGLE BEND

Figure 1

networks are about as to be expected. As network detail increased, the apparent error in continuity is decreased, with the most detailed network giving the most consistent answers. One can see, however, that even with the relatively extreme detail of the Level 3 network an apparent continuity violation of just over six percent is present at the network's "corner".

Table 1 COMPARISON OF CONSERVATION OF MASS FOR VARIOUS LEVELS OF NETWORK DETAIL

Line Number	Defined By Nodes Of Level 1	Percentage of Flow Relative to Line G			
		Level 1	Level 2	Level 3	Level 4
A (outflow)	1, 2, 4	100.0	100.0	100.0	100.0
B	4, 5, 6	93.8	100.0	--	99.8
C	10,11,12	114.2	96.2	98.4	98.6
D	16,17,18	71.4	84.9	93.6	97.9
E	22,23,24	106.7	98.7	100.4	99.7
F	28,29,30	99.3	100.0	99.9	100.0
G (inflow)	28,31,33	100.0	100.0	100.0	100.0

Table 2 COMPARISON OF CENTER LINE VELOCITIES FOR VARIOUS LEVELS OF NETWORK DETAIL

Node Number	Network Level	X-Velocity Component(fps)	Y-Velocity Component(fps)	Vector Velocity(fps)
14	1	-0.773	-0.313	0.834
14	2	-0.990	-0.367	1.056
14	3	-1.031	-0.263	1.064
14	4	-0.859	-0.315	0.915
17	1	-0.517	-0.550	0.755
17	2	-0.541	-0.624	0.826
17	3	-0.554	-0.622	0.833
17	4	-0.442	-0.457	0.636
20	1	-0.224	-0.856	0.885
20	2	-0.187	-1.000	1.017
20	3	-0.126	-1.020	1.028
20	4	-0.244	-0.877	0.910

It seem clear that this violation is due mainly to the fact that zero velocity has been specified at nodes 16 and 18 to keep flow from crossing the network's boundaries. The imposition of such specifications creates large velocity gradients which the model has difficulty approximating. The result is a relatively poor representation of the local flow field with the apparent "continuity violation" across line D of the network.

A comparison of the center line velocities in Table 2 shows some interesting results. The zero velocity specification at the corners (nodes 16 and 18) forces more flow to the center of the channel and shows higher velocity gradients across the section. It is interesting to note, however, that there is less difference in the predicted velocities among the various levels of network detail than there is in the total flow comparisons. This is especially true in values of the velocity vectors and suggests that individual velocities may have more or less error than the flow integral.

Example 2 The second example problem is taken from a flow simulation of Tallahalla Creek in Mississippi. The physical setting is a convergent-divergent flow regimen in a natural flood plain constricted by a bridge crossing. This problem was first analyzed by Franques and Yannitell (1974) with a non-linear potential flow type finite element model and later by Tseng (1975) with the original finite element model. The network developed by Tseng is used as the reference point as shown in Figure 2.

In the process of evaluating the Tallahalla Creek problem, a total of seven levels of network detail were developed, three of which are reported herein. Each of the networks was prepared with increasing detail to evaluate the impact of smooth boundaries and network refinements on the simulation's apparent violation of flow continuity. Table 3, below, describes each of the various networks, and Figure 2 shows the geometric form of each construction.

Table 3 NETWORKS USED FOR SIMULATION OF TALLAHALLA CREEK

Level	Nodes	Elements	Description
1	199	86	Basic network with zero flow sharp corners
2	221	96	One row of elements added immediately upstream of bridge, curved boundaries
3	283	124	Additional detail in the areas of the convergent flow, curved boundaries

Each of the simulations was performed with a total flow of 22,000 cfs and a downstream head condition along the right hand edge of the network specified at elevation 309.9 ft; seven different element eddy viscosities were used from the original verification which ranged from 75-750 lb-sec/ft^2.

The results from Tallahalla Creek with the three different networks are presented in Tables 4, 5 and 6. The first table shows the apparent errors in continuity satisfaction (as a percentage to total inflow) at seven cross sections along the system; the exact location of each cross section is indicated in Figure 2. Please note that not all cross sections existed for

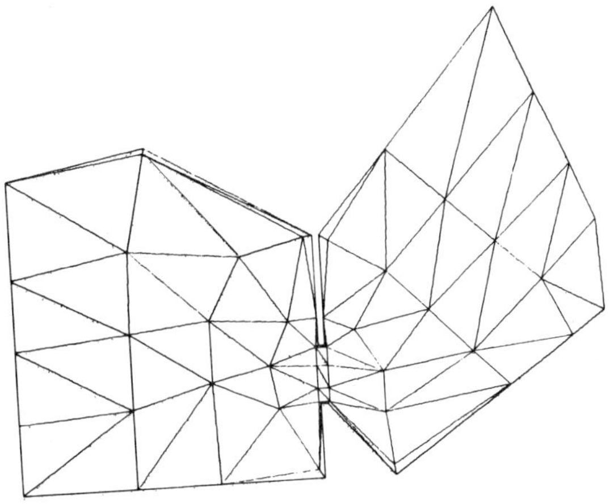

Original Tallahalla Creek Network - Level 1

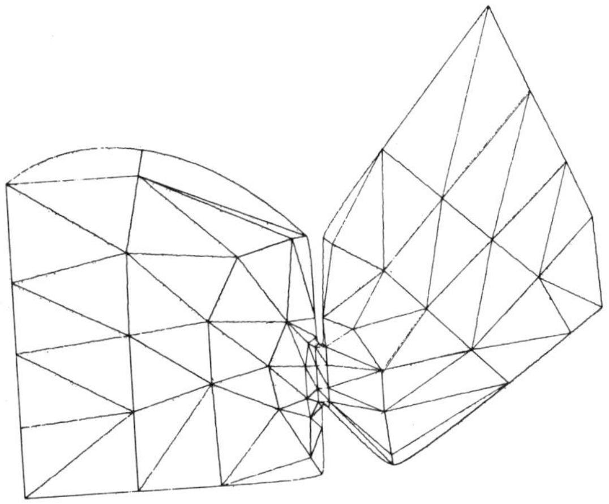

Intermediate Tallahalla Creek Network - Level 2

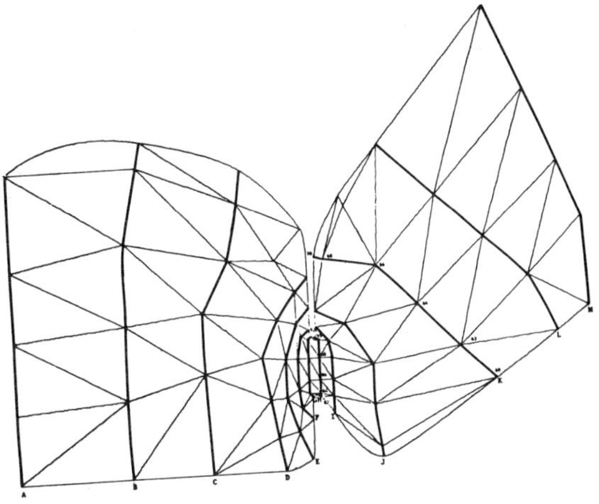

Final Tallahalla Creek Network - Level 3

NETWORK LAYOUTS FOR TALLAHALLA CREEK
Figure 2

each simulation. This table shows the improvement achievement by various levels of refinement. The error indicated in Line L is associated with the network configuration and specification of uniform water surface elevation downstream. No attempt was made in these networks to correct this problem, thus the error remains relatively consistent. Particularly important is the improvement achieved by the first all-parallel flow case (Level 2) where flow singularities were eliminated. In this case the maximum error close to the bridge was reduced by about an order of magnitude. The next table, Table 5, shows the predicted elevation of the center line water surface at each of ten cross sections for each network. It is interesting to note that position of the water surface is nearly the same for networks 1 and 3, and somewhat lower for network 2. This is thought to be a result of the high head losses in the first case due to the stagnation points and in the third case to relatively large velocities of large friction losses; the second case has no stagnation points and its velocities are relatively lower than the third case as shown in Table 5.

Table 4 APPARENT VIOLATION IN FLOW CONTINUITY FOR THREE DIFFERENT TALLAHALLA CREEK NETWORKS INDICATED VALUES ARE PERCENTAGES OF UPSTREAM FLOW

Cross Section	Level 1	Level 2	Level 3
A	0	0	0
B	- 4.2	- 4.4	- 4.6
C	5.5	6.1	- 2.5
D	- 6.5	- 7.2	6.5
E	--	--	- 7.4
F		-10.5	0.0
G	-54.3	- 4.9	- 7.4
H	-28.8	-17.5	- 0.7
I	--	--	- 4.4
J	- 1.2	3.2	0.7
K	8.1	6.6	5.6
L	-19.3	-21.5	-21.0
M	0	0	0

As a final demonstration of the influence of mesh refinement, the velocity distribution at the bridge cross section is shown in Table 6. Note that during the network changes to parallel flow, most of the velocity changes are close to the bridge pier and, in fact, velocity at mid-span is reduced. However, as network refinement continues, velocities at midspan steadily rise. As the bottom line of Table 6 shows, there is a considerable increase in net flow with refinement. When examining Table 6, it should be remembered that the bottom profile through the bridge opening is not regular and that the nodes at the piers are not at the center of the cross section.

For those reasons one should not expect a uniform velocity distribution at the opening.

Table 5 WATER SURFACE ELEVATIONS (FEET) ALONG CENTER LINE*

Cross Section	Level 1	Level 2	Level 3
M	309.84	309.82	309.82
L	310.29	310.28	310.29
K	312.03	312.01	312.06
J	313.30	313.20	313.29
H	314.62	314.37	314.70
G	314.71	314.41	314.76
D	315.64	315.21	315.58
C	316.14	315.83	316.16
B	316.38	316.11	316.42
A	316.82	316.61	316.87

*See Figure 2 for location of values

Table 6 VELOCITY DISTRIBUTION AT BRIDGE SECTION
VALUES ARE FEET/SEC

Node	Level 1	Level 2	Level 3
82	0	3.95	3.66
83	1.28	3.70	3.62
84	2.19	2.89	3.58
85	2.80	2.85	3.28
86	2.79	2.57	3.01
87	2.68	2.38	2.75
88	2.76	2.36	2.70
89	3.19	2.74	3.57
90	2.52	2.51	3.40
91	1.60	3.14	3.28
92	0	3.99	3.25
% Cont.	71.2	82.6	99.3

Model Verification

In the final analysis the value of any model can only be determined from its applications and how well it fulfills its intended function. Serious questions must be asked as to whether or not the model produces a reasonably accurate representation of the prototype at an acceptable cost and without resorting to a large number of special "fixes" for every problem attempted.

In the paragraphs which follow, a comparison is made between field observations of water surface elevation and velocities . and the same parameters as predicted by simulation with RMA's hydrodynamic model. The objective of this work is to demonstrate how well the model reproduces prototype behavior and to what extent it can be verified.

The physical location for the comparison of model/prototype behavior was the Suisun Bay Area of the San Francisco Bay Delta System (see Figure 3). The fraction of the total Bay-Delta System included in the model verification extended from Benicia on the west to Rio Vista on the Sacramento River and the intersection of Three Mile Slough and the San Joaquin River on the east. The period chosen for calibration was a twenty-five hour tidal cycle observed September 14-15, 1967; the net delta outflow was estimated at 23,800 cfs and the total delta demands at 6,600 cfs for this period. All data used for calibration were collected and reported by Corps of Engineers (1974). These data are part of the information used by the Corps in connection with the calibration of its Sausalito-based physical model of the same area.

Model calibration was accomplished by a trial and error approach, and the adequacy of model validation was determined judmentally. The procedure used was basically as follows. First, a finite element grid of the system was constructed over a set of standard NOAA Nautical Charts. Bottom elevations were taken from the same source, with spatial averages used where high bottom relief was indicated.

Next, a set of boundary conditions were prepared from the observed data. Tidal elevations were used at the Benicia location and on the San Joaquin River, while estimated flows were used at Rio Vista. The Rio Vista flows were calculated as the product of the average observed velocity and the stream cross section.

The final activity in model verification was determining the parametric values for the empirical expressions representing system energy losses. The model includes two such relationships, eddy viscosity and bottom friction. The eddy viscosity values were set at nominal values based on results reported in the literature and from previous model applications, and the calibration work focused on a determination of bottom friction coefficients. Nominal values were selected for bottom friction and a series of simulations conducted to determine the agreement between model predictions and observed values, and the response of the system to changes in specified values. This procedure was repeated until such time as it was judgmentally determined that basic processes observed in the prototype were also observed in the model and that point by point agreement between observed and simulated values of tidal elevation, stream velocity and salinity were sufficiently coincident that conclusions drawn from simulated system behavior would be the same as those drawn from observed prototype experience.

For the purposes of simulation, a finite element grid was prepared directly as an overlay on the NOAA Charts. The network, also shown in Figure 3, was configured to represent the highly variable geometry of the system to a very close degree, with virtually all lengths and widths realistically accounted for. In its final state, the network contained a combination of 116 triangular and quadrilateral elements and 488 node

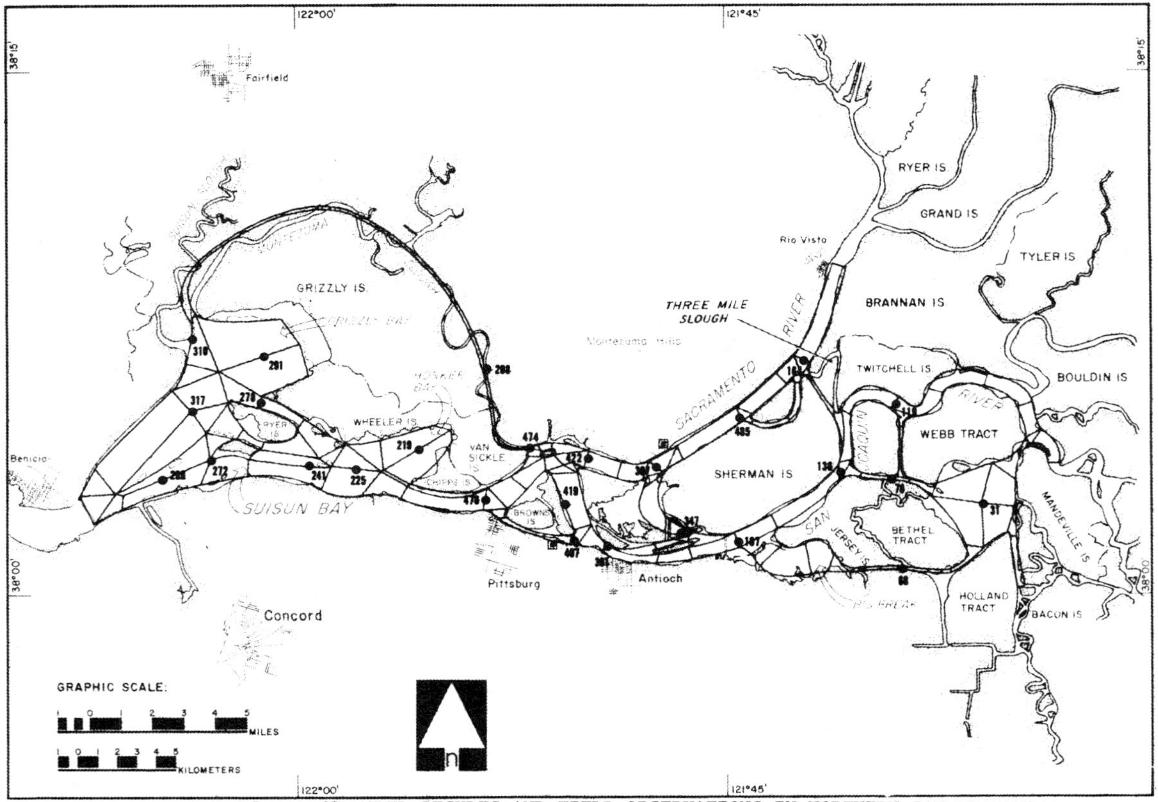

LOCATION MAP--COMPARISON OF MODEL RESULTS AND FIELD OBSERVATIONS IN NORTHERN SAN FRANCISCO BAY
Figure 3

points. The elements ranged in size from 0.04 mi^2 to 1.71 mi^2, and encompassed a tidal prism of just over 68 mi^2.

Bottom elevations, also estimated from the Nautical Charts, ranged from a minimum of -40 feet to a maximum of -7 feet. In those areas where high bottom relief was indicated, bottom elevations were estiamted at values which gave hydraulically equivalent sections in the area of a particular node point. In all cases the bottom elevations were set at values below lower low water and there were no instances of alternately "wet" or "dry" nodes.

The values of eddy viscosity were found to lie in the range of 30 to 1500 lb-sec/ft^2 and were scaled approximately in relation to the size of the network's element sizes. Bottom friction, using Chezy's "C", ranged from 30 to 105, with the lower values (high friction) assigned to constricted, rough areas and the higher number used in the main channel regions of more uniform flow.

Comparisons between field observations, numerically simulated values and observations of the Corps' physical model for stream velocity, and surface elevation for the 25-hour tide of September 14-15, 1967, are presented in Figure 4; the location of each station can be found on Figure 3.

The model/prototype comparisons are presented as a series of plots, with the "continuous" records constructed from field measurements shown as a solid line and the results from the mathematical model indicated as solid dots. Also shown, as a dashed line, are the observations made in the Corps' physical model. In general, there seems to be adequate agreement between model and prototype at all locations for all observations, with the dominant features of the tidal cycle apparent in all predictions. It is also interesting to note that there appears to be little to choose from between the results produced by the mathematical or the physical model for this particular problem.

MODEL APPLICATION

To demonstrate the combined operation of RMA's quantity and quality models, representative results from a recent outfall study are summarized below. The objective of the study was to evaluate the water quality impacts of proposed alternative waste water outfall locations in the San Pablo Bay region of northern San Francisco Bay. The study included fully dynamic simulation of both flow and water quality, with output produced in both tabular and graphic form.

In plan, the prototype system extended from Bluff Point in the south to Sonoma Creek in the north and the Petaluma River in the west to the Carquinez Straight in the east with a total surface area of roughly 150 mi^2; the network used for this study is reproduced as Figure 5. The bottom topography is such that water depths ranged from a maximum of 60 feet in narrowest part of the system to an average of 25 feet along the south central part of the network to alternately wet and

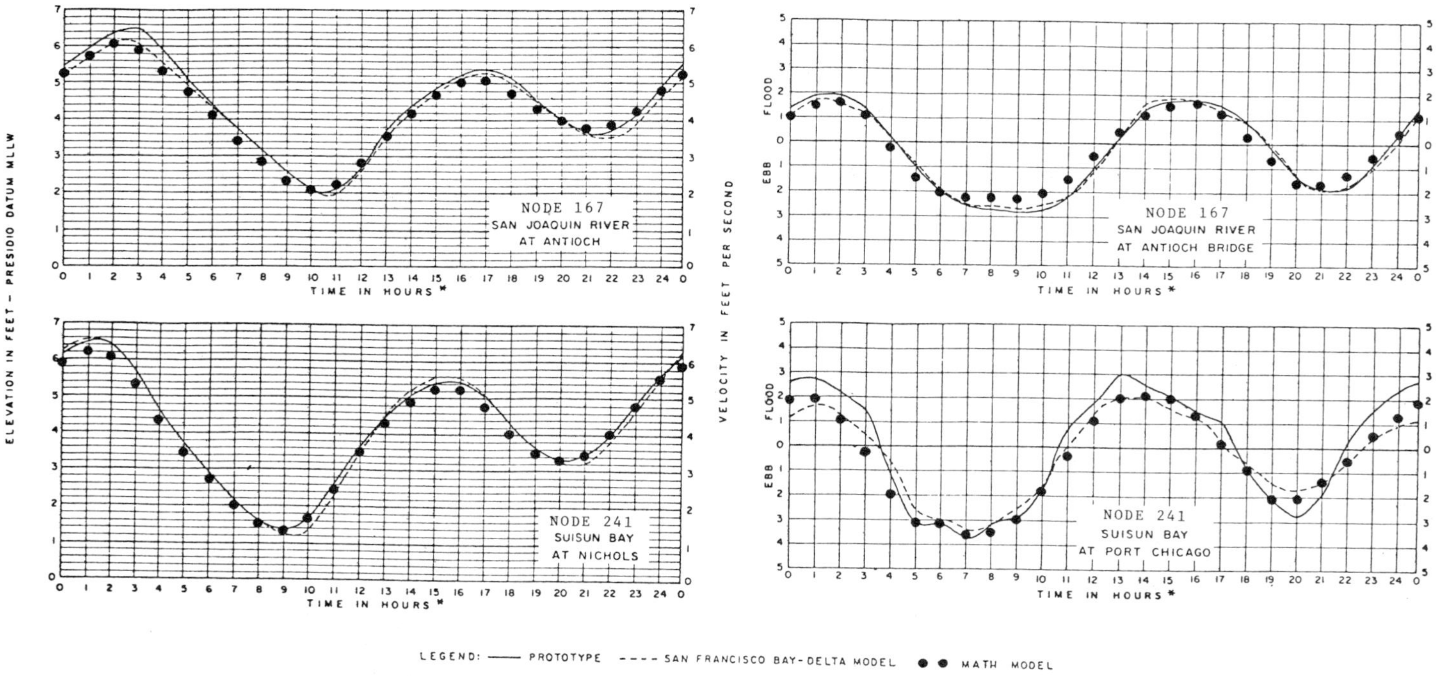

COMPARISON OF MODEL RESULTS AND FIELD OBSERVATIONS IN NORTHERN SAN FRANCISCO BAY
Figure 4

dry mud flats along the west and northwestern parts of the upper bay. During development of these networks it was found that the sharp variation of bottom topography, especially in shallow areas can create difficulties due to rapid gradients of flow vectors and as a result network detail in these areas must be increased to accomodate the rapid changes.

The system was run with a long term average tide as the boundary condition at the seaward (south) end of the network and a flow boundary condition representative of a net fresh-water flow of 4400 cfs at its upstream (eastern) limit. A number of different outfall locations were investigated as suggested by the variable levels of detail in Figure 5.

The simulation procedure was carried out as follows. The hydraulic system was started from rest and run to dynamic equilibrium with the repeating, average tide; this proceudre took about one and a quarter tidal cycles and the results were stored on magnetic tape. The wet and dry mud flat areas in the western and northern portions of the network were automatically included or deleted from the network depending on local tidal height and caused no apparent problems.

Subsequent to the hydraulic analysis, the water quality model was run for the various outfall locations. Mass emission rates were specified at the outfall sites, and up to six independent analyses of both conservative and non-conservative substances simulated in a single computer run. All quality simulations used the same description of the hydraulic field, and the assumption was made that the small amount of flow introduced by the individual outfall (< 10 cfs) was insignificant when compared to the tidal flows. Time steps of one hour were used both for the hydrodynamic and water quality models.

Representative results from the study are shown in Figure 6 as plots of iso-concentration lines after approximately three days of quality simulation for an outfall in the south-western corner of the upper part of the bay. The plot at 66 hours shows the concentrations at slack water after high water, and indicates how the wastewater is confined largely to the western edge of the system with relatively little mixing out into the main channel to the east. The plot at hour 73 shows the influence of the same outfall at slack water low tide and shows a more pronounced seaward movement of the wastewater, although it is still confined to the western edge of the network in the north bay. Please note that in this plot the concentration contours end somewhat to the east of the other plot, and reflect the boundary of the mud flat area at low tide.

A substantial number of plots of the type shown here were prepared for this study, and proved to be quite useful in evaluating the various alternative locations. Field studies are now underway to refine the topography in the area of the final alternatives and dye studies are planned to verify the model prior to final selection of construction alternatives.

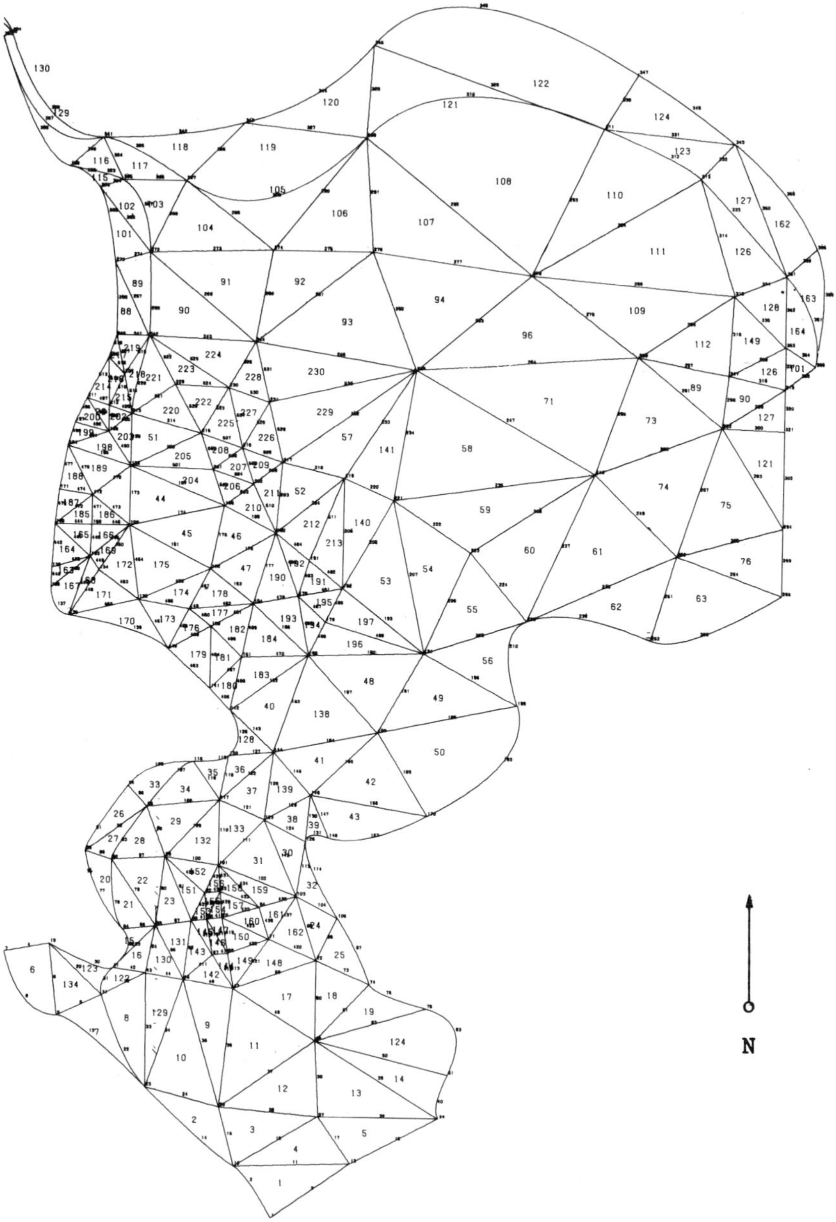

HYDRODYNAMIC AND WATER QUALITY NETWORK FOR SAN PABLO BAY
Figure 5

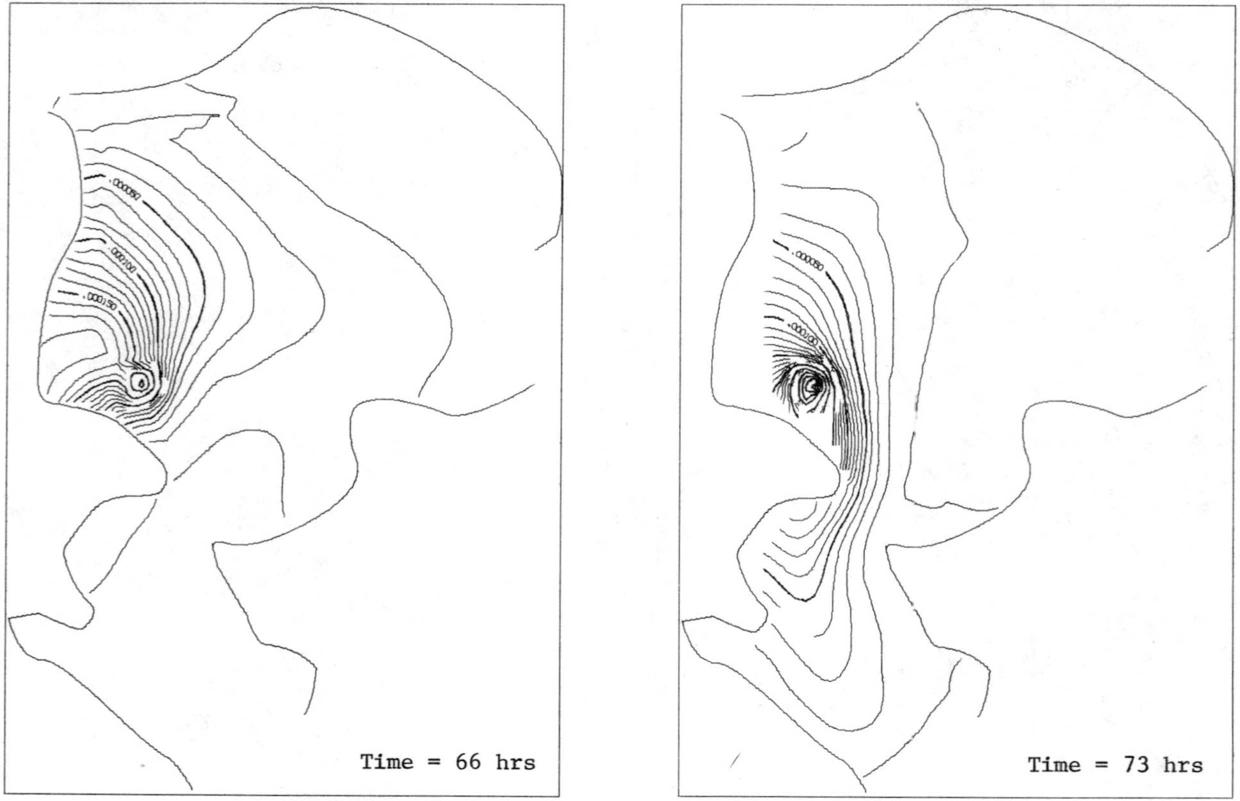

ISOCONCENTRATION LINES FOR WASTE LOAD IN SAN PABLO BAY
Figure 6

CONCLUDING REMARKS

This paper has concentrated on showing the applicability of two dimensional finite element models. The evaluation of continuity lines has demonstrated not only that the model results improve sharply with the use of curved elements, but that with network refinement overall performance can be improved.

The verification and application sections have demonstrated that the model can be economically used for study of outfall locations both from the hydrodynamic and water quality perspective.

Finally it should be pointed out that useful application of these models still requires considerable judgment in laying out networks and are most suitable only for those experienced in their application.

REFERENCES

Corps of Engineers, San Francisco District (1974), "Results of Sensitvity Tests," San Francisco Bay and Sacramento-San Joaquin Delta Water Quality and Waste Disposal Investigation. San Francisco Bay-Delta Model," Technical Memo. No. 1, U.S. Army.

Franques, J. T. and D. U. Yannitell (1974), "Two-Dimensional Analysis of Backwater at Bridges," Journal of Hydraulic Division ASCE, Vol. 100, Hy3, pp. 379-392.

King, I. P., Norton, W. R., and K. R. Iceman (1975), "A Finite Element Solution for Two-Dimension Stratified Flow Problems," Finite Elements in Fluids, Vol. 1, pp. 133, John Wiley and Sons, London.

King, I. P. (1977), "Finite Element Models for Unsteady Flow Routing Through Irregular Channels," Finite Elements in Water Resources, 4.165, Pentech Press, London.

Norton, W. R., King I. P. and G. T. Orlob (1973), "A Finite Element Model for Lower Grande Reservoir," Report to U.S. Corps of Engineers, Walla Walla District.

Tseng, M. T. (1975), "Finite Element Model for Bridge Backwater Computation," Evolution of Flood Risk Factors in the Design of Highway Stream Crossings, Vol. 3, Report to Federal Highway Administration.

Wang, J. E. and J. J. Connor (1975), "Mathematical Modeling of Near Coastal Circulation," Report to Sea Grant Office, National Oceanic and Atmospheric Administration, Department of Commerce.

A WATER POLLUTION PREDICTION SYSTEM BY THE FINITE ELEMENT METHOD

Y. Matsuda

IBM Japan Ltd., 1-10-10, Utsubo Honmachi, Nishi-ku, Osaka, Japan 550

INTRODUCTION

Traditionally, the Finite Difference Method (FDM) has been used for tidal computations by Dronkers (1964) and Leendertse (1967), but it suffers from lack of mesh flexibility and difficulty in the satisfaction of the boundary conditions and difficulty in the composition of the general purpose program.

To overcome these faults in FDM, the finite element method (FEM), which has been successfully used for structural analysis, has been applied by Grotkop (1973), Connor (1976) and Taylor (1975). But on the other hand, there is a paper by Weare (1976) suggesting that the FEM needs more computation time than the FDM. So, it was difficult to apply the FEM in larger seas, such as the Seto Inland Sea, in Japan.

For example, in the sample models, the number of nodes were always less than 70 and the number of elements were always less than 100 (Table 1).

Considering these circumstances, we developed a new simulation system (IBM Japan Ltd., 1978) using the newly created "Explicit Finite Element Method" by Matsuda (1977) and the steady dispersion method (IBM Japan, Ltd., 1975).

Table 1. Size of Simulation Models

	Grotkop (1973)	Taylor & Davis (1975)	Connor & Brebbia (1976)	This Paper
No. of Nodes	69	64	53/58	183/596
No. of Elements	97	9*	74/86	306/906

(*—isoparametric element)

ANALYSIS OF TIDAL CURRENT

The equations of tidal current are formulated as follows:

$$\frac{\partial U}{\partial t} = -g\cdot(h+\zeta)\cdot\frac{\partial \zeta}{\partial x} - u\cdot\frac{\partial U}{\partial x} - v\cdot\frac{\partial U}{\partial y} + f\cdot V - \gamma^2\cdot u\cdot\sqrt{u^2+v^2} \quad (1)$$

$$\frac{\partial V}{\partial t} = -g\cdot(h+\zeta)\cdot\frac{\partial \zeta}{\partial y} - u\cdot\frac{\partial V}{\partial x} - v\cdot\frac{\partial V}{\partial y} - f\cdot U - \gamma^2\cdot v\cdot\sqrt{u^2+v^2} \quad (2)$$

$$\frac{\partial \zeta}{\partial t} = -\frac{\partial U}{\partial x} - \frac{\partial V}{\partial y} \quad (3)$$

where

U: = $u\cdot(h+\zeta)$ (m^2/sec)

V: = $v\cdot(h+\zeta)$ (m^2/sec)

u,v: x-y component of vertically averaged velocity (m/sec)

h:	the height of a mean water level with respect to the bottom	(m)
ζ:	the height of a mean water level with respect to the bottom	(m)
t:	time	(sec)
γ^2:	dimensionless friction coefficient	(-)
g:	the acceleration due to gravity	(m/sec^2)
f:	the Coriolis parameter	(1/sec)
	$= 2\omega \sin\phi$	
	$\omega = 7.292 \times 10^{-6}$	(rad/sec)
	ψ: the latitude	(degree)

The boundary conditions are

(1) $u=u_B$, $v=v_B$, $\zeta=\zeta_B$

(Specified velocity and the height of the water level)

(2) $u.\ell_x + v.\ell_y = 0$ at land and island

where ℓ_x, ℓ_y : components of exterior unit vector normal to the boundary.

The initial condition defined through the domain under considera tion is,

$$\zeta(x,\ y,\ t=0)= \zeta_o(x,y)$$
$$u(x,\ y,\ t=0)= u_o(x,y)$$
$$v(x,\ y,\ t=0)= v_o(x,y).$$

Formulation by Implicit Method

In the element domain V^e, approximating certain variables by trial function takes the form:

$$U= \sum_1^{ne} Ni\ (x,y)\cdot Ui(t), \qquad V= \sum_1^{ne} Ni(x,y)\cdot Vi(t)$$

$$\zeta = \sum_1^{ne} Ni(x,y)\cdot \zeta i(t)$$

where Ni: shape functions

ne: number of nodes per element

Ui(t), Vi(t), ζi(t): nodal values of U, V, ζ.

After taking Ni as a weighting function in the process of the Galerkin method and adopting Green's theorem with boundary conditions, n algebraic differential equations are formulated as follows:

$$[P]\cdot\{\dot{U}\} + \{F\}_1 = 0 \tag{4}$$

$$[P]\cdot\{\dot{V}\} + \{F\}_2 = 0 \tag{5}$$

$$[P]\cdot\{\dot{\zeta}\} + \{F\}_3 = 0 \tag{6}$$

where

$$[Pij] = \sum_1^m \{\int_{ve} Ni\cdot Nj\cdot dx\cdot dy\}$$

$$\{Fi\}_k = -\sum_1^m \{\int_{ve} Ni\cdot Q_k\cdot dx\cdot dy\} \qquad (k = 1,\ 2,\ 3)$$

$$Q_1 = -g\cdot(h+\zeta)\cdot\frac{\partial\zeta}{\partial x} - u_e\cdot\frac{\partial U}{\partial x} - v_e\cdot\frac{\partial U}{\partial y} + f\cdot V - \gamma^2\cdot u\cdot\sqrt{u^2+v^2} \tag{7}$$

$$Q_2 = -g\,(h+\zeta)\cdot\frac{\partial \zeta}{\partial y} - u_e\cdot\frac{\partial V}{\partial x} - v_e\cdot\frac{\partial V}{\partial y} - f\cdot U - \gamma^2\cdot v\cdot\sqrt{u^2+v^2} \tag{8}$$

$$Q_3 = -\frac{\partial U}{\partial x} - \frac{\partial V}{\partial y} \tag{9}$$

ue, ve: velocity at the gravity point of an element
m : number of elements
n : number of nodes
V^e: domain of an element
Equations (4), (5), (6) are expressed as follows:

$$[P]\cdot\{\dot{C}\} + [D]\cdot\{C\} + \{F\} = 0 \tag{10}$$

Considering the time scheme:

$$\frac{\{C\}_t - \{C\}_{t-\triangle t}}{\triangle t} = \theta\cdot\{\dot{C}\}_t + (1-\theta)\cdot\{\dot{C}\}_{t-\triangle t}$$

where $0 \leq \theta \leq 1$.

Inserting this time scheme into equation (10) yields the following equation:

$$\frac{[P]}{\triangle t}\cdot\{C\}_t = \frac{[P]}{\triangle t}\cdot\{C\}_{t-\triangle t} - (1-\theta)\cdot\{F\}_{t-\triangle t} - \theta\cdot\{F\}_t \tag{11}$$

Approximating $\{F\}_t$ in the right hand side of equation (11) as $\{F\}_{t-\triangle t}$ results in the following expression:

$$\frac{[P]}{\triangle t}\cdot\{C\}_t = \frac{[P]}{\triangle t}\cdot\{C\}_{t-\triangle t} - \{F\}_{t-\triangle t} \tag{12}$$

<u>Formulation by Explicit Method</u>
(In the case of Linear Shape Functions)

<u>One-Step Explicit Method</u> Converting the matrix $[P]$ at the left hand side of equation (12) into consistent matrix $[\bar{P}]$ results in the following expression:

$$\frac{[\bar{P}]}{\triangle t}\cdot\{C\}_t = \frac{[P]}{\triangle t}\cdot\{C\}_{t-\triangle t} - \{F\}_{t-\triangle t} \tag{13}$$

<u>Two-Step Explicit Method</u> Coupling eq. (13) and eq. (11) results in the following expression:

$$\frac{[\bar{P}]}{\triangle t}\cdot\{C\}^*_t = a_1\cdot\frac{[P]}{\triangle t}\cdot\{C\}_{t-\triangle t} - a_2\cdot\{F\}_{t-\triangle t} \tag{14}$$

$$\frac{[\bar{P}]}{\triangle t}\cdot\{C\}_t = a_3\cdot\frac{[P]}{\triangle t}\cdot\{C\}_{t-\triangle t} + a_4\cdot\frac{[P]}{\triangle t}\cdot\{C\}^*_t - a_5\cdot\{F\}_{t-\triangle t} - a_6\cdot\{F\}^*_t \tag{15}$$

- Estimation of Coefficient ($a_1 \sim a_6$):

Considering the full matrix $[P]$ at the left hand side of eqs. (14) and (15) results in the following expression:

$$\{C\}_t = (a_3+a_1\cdot a_4)\cdot\{C\}_{t-\triangle t} + \triangle t\cdot(a_2\cdot a_4+a_5+a_1\cdot a_6)\cdot\{\dot{C}\}_{t-\triangle t} + \triangle t^2\cdot a_2\cdot a_6\cdot\{\ddot{C}\}_{t-\triangle t} \tag{16}$$

So, considering the Taylor expansion of $\{C\}_t$, the following relations are preferable as the suitable value of $(a_1 \sim a_6)$ because of lesser numerical errors:

$$\left.\begin{array}{l} a_3 + a_1 \cdot a_4 = 1 \\ a_2 \cdot a_4 + a_5 + a_1 \cdot a_6 = 1 \\ a_2 \cdot a_6 = 0.5 \end{array}\right\} \qquad (17)$$

Solving the equation (17) in the case of the three given coefficients among $(a_1 \sim a_6)$ retults in the definitions of value of all coefficients. As a special case, coupling expressions (13) and (11) results in the following expression:

$$a_1 = 1 \quad a_2 = 1, \quad a_3 = 1, \quad a_4 = 0, \quad a_5 = 1-\theta, \quad a_6 = \theta .$$

In the case of $\theta = 0.5$ (these relations satisfy the expression (17).), we call this the Two-Step Explicit Method (θ =0.5).

INVESTIGATION OF CALCULATION SCHEME BY ERROR ANALYSIS

After changing the equation (1) - (3) into one-dimensional form, we simplify these equations as follows:

$$\begin{cases} \dfrac{\partial u}{\partial t} = -g \cdot \dfrac{\partial \zeta}{x} & (18) \\[2ex] \dfrac{\partial \zeta}{\partial t} = -h \cdot \dfrac{\partial u}{\partial x} & (19) \end{cases}$$

In the following formulation, we consider the linear shape function.

Implicit Method

From equation (15), we can get the following equation:

$$u_{n-1,\,m} + 4 \cdot u_{n,\,m} + u_{n+1,\,m} = u_{n-1,\,m-1} + 4 \cdot u_{n,\,m-1} + u_{n+1,\,m-1} - \frac{3g \cdot \triangle t}{\triangle x} \cdot (1-\theta) \cdot (\zeta_{n+1,\,m-1} - \zeta_{n-1,\,m-1}) - \frac{3 \cdot g \cdot \triangle t}{\triangle x} \cdot \theta \cdot (\zeta_{n+1,\,m} - \zeta_{n-1,\,m})$$

where

n: suffix of the position

m: suffix of the number of the time step.

The method of the Fourier series is used for the check of stability. Inserting $u^{n,m} = A \cdot \xi^m \cdot e^{i\beta \cdot n \cdot \triangle x}$ and $\zeta^{n,m} = B \cdot \xi^m \cdot e^{i\beta \cdot n \cdot \triangle x}$ into the above expression results in the following expression:

$$A \cdot (\xi - 1) \cdot (2 + \cos(\beta \cdot \triangle x)) = -\frac{3 \cdot g \cdot \triangle t}{\triangle x} \cdot B \cdot i \cdot \sin(\beta \cdot \triangle x) \cdot (1 - \theta + \theta \cdot \xi) \qquad (20)$$

where A, B: auxiliary constant

β : wave number.

Similarly, we can get the following expression:

$$B \cdot (\xi - 1) \cdot (2 + \cos(\beta \cdot \triangle x)) = -\frac{3 \cdot h \cdot \triangle t}{\triangle x} \cdot A \cdot i \cdot \sin(\beta \cdot \triangle x) \cdot (1 - \theta + \theta \cdot \xi) \qquad (21)$$

Eliminating A and B in equation (17), (18) results in the following expression:

$$\xi = \frac{2+\cos(\beta\cdot\triangle x) \pm i\cdot 3\cdot(1-\theta)\cdot\sqrt{h\cdot g}\cdot\frac{\triangle t}{\triangle x}\cdot\sin(\beta\cdot\triangle x)}{2+\cos(\beta\cdot\triangle x) \mp i\cdot 3\cdot\theta\cdot\sqrt{h\cdot g}\cdot\frac{\triangle t}{\triangle x}\cdot\sin(\beta\cdot\triangle x)} \tag{22}$$

$$|\xi| = \sqrt{1+\frac{(1-2\theta)\cdot 9b^2\cdot\sin^2(\beta\cdot\triangle x)}{(2+\cos(\beta\triangle x))^2+9\cdot\theta^2\cdot b^2\cdot\sin^2(\beta\cdot\triangle x)}} \tag{23}$$

where $b=\sqrt{g\cdot h}\cdot\triangle t/\triangle x\cdot$ (24)

So, from equation (23), we can get following relations:

$$\begin{cases}\theta \geq \frac{1}{2}: & |\xi| \leq 1 \text{ ; stable,}\\ \theta < \frac{1}{2}: & |\xi| > 1 \text{ ; divergency.}\end{cases}$$

In case of formulation of equation (12),

$$|\xi| = \sqrt{1+\frac{9b^2\cdot\sin^2(\beta\cdot\triangle x)}{(2+\cos(\beta\cdot\triangle x))^2}} \geq 1 \text{ ; divergency}$$

Explicit Method

One-step Explicit Method - -

$$\zeta = \frac{2}{3}+\frac{1}{3}\cdot\cos(\beta\cdot\triangle x) \pm i\cdot b\cdot\sin(\beta\cdot\triangle x) \tag{25}$$

$$|\xi| = \sqrt{(\frac{2}{3}+\frac{1}{3}\cdot\cos(\beta\cdot\triangle x))^2 + b^2\cdot\sin^2\cdot(\beta\cdot\triangle x)} \tag{26}$$

Two-Step Explicit Method - -

$$\xi = \frac{Q_3}{6}+\frac{Q_1}{3}\cdot\cos(2\cdot\beta\cdot\triangle x)+\frac{Q_2}{3}\cdot\cos(\beta\cdot\triangle x)$$
$$\pm i\cdot\frac{1}{3}(Q_4\cdot\sin(2\cdot\beta\cdot\triangle x)+Q_5\cdot\sin(\beta\cdot\triangle x)) \tag{27}$$

where

$$\begin{aligned}
Q_1 &= \frac{3}{2}b^2\cdot a_2\cdot a_6 + \frac{a_1\cdot a_4}{6}\\
Q_2 &= a_3+\frac{4}{3}\cdot a_1\cdot a_4\\
Q_3 &= 4\cdot a_3-3b^2\cdot a_2\cdot a_6+3a_1\cdot a_4\\
Q_4 &= \frac{b}{2}\cdot(a_1\cdot a_6+a_2\cdot a_4)\\
Q_5 &= b\cdot(2\cdot a_1\cdot a_6+2a_2\cdot a_4+3\cdot a_5)
\end{aligned} \tag{28}$$

If the pairs of a_3 and a_5 are the same, the numerical error of the scheme is the same considering the equation (17) and the expression (28). So, if we solve the equation (17) giving the different kinds of combinations of (a_3, a_5, a_2), we can find ($a_1 \sim a_6$) in case of minimum numerical error.

Error Analysis of Explicit Method

In the case of the Implicit Method (equation (11)), the computing time is long. So, here, we investigate only the Explicit Method.

In case of g = h = c at equation (18) and (19), we can get an analytical solution

$$u=\zeta=A\cdot e^{i\beta\cdot(x-C\cdot t)} \tag{29}$$

where A: auxiliary constant
B: wave number
$b=C\cdot\triangle t/\triangle x$ (from expression (24))

So, the analytical phase change after the time Δt is as follows:

$$\psi A = -\beta \cdot C \cdot \Delta t = -b \cdot \beta \cdot \Delta x = -\frac{2\pi \cdot b}{k}$$

where $k = \frac{2\pi}{\beta \cdot \Delta x} = \frac{\lambda}{\Delta x}$ (integral number ≥ 2)

λ: wave length.

Considering that the numerical phase change is $\tan^{-1}$ (Imaginary part of ξ/Real part of ξ), the relative phase change error is as follows:

$$\frac{\psi}{\psi A} = -\frac{k}{2\pi \cdot b} \cdot \tan^{-1}\left(\frac{Im(\xi)}{Re(\xi)}\right).$$

From $|\xi|$, we can check the stability of the scheme.

In case of analytical solution (29), we can get the next relations:

$$|\xi| = 1, \quad \psi/\psi A = 1.$$

So, in order to estimate the error of the scheme, we consider the following standard deviations.

$$E_P = \sqrt{\frac{1}{M} \cdot \sum_1^M \left(\frac{\psi}{\psi_A} - 1\right)^2} \qquad E_A = \sqrt{\frac{1}{M} \cdot \sum_1^M (|\xi| - 1)^2}$$

Total Error is expressed as:

$$E_t = \sqrt{E_P^2 + E_A^2}$$

where M: number of k,
k = 2, 3, 4, 6, 8, 10.

Here, there are two values of Et from expression (25) and (27). So, arranging the numerical error by the average $\overline{Et_1}$, $\overline{Et_2}$ and Et $(=(\overline{Et_1}+\overline{Et_2})/2)$ within the stability condition ($|\xi| \leq 1$), Table 2 and Figure 1 show the results of the error analysis.

In the case of the Two-Step Explicit Method, changing the coefficients ($a_3=0\sim1$, $a_5=0\sim1$, $a_2=0.05\sim1$, $b=0.1\sim2$) results in the following values in case of minimum Et:

$a_1=0.78$, $a_2=0.5$, $a_3=1$, $a_4=0$, $a_5=0.22$, $a_6=1$

Table 2 shows that the Two-Step Explicit Method is the best method.

The stability condition of this method is as follows:

$$\Delta t \leq \frac{1 \cdot 3 \cdot \Delta x}{\sqrt{g \cdot h}} \tag{30}$$

Table 2 Summary of Numerical Error

Numerical Method	bmax. at $|\xi| \leq 1$	$\overline{E_{t1}}$	$\overline{E_{t2}}$	$\overline{E_t}$
Two-Step E.M.	1.3	0.58	13.42	7.00
Two-Step E.M. (θ=0.5)	1.1	0.56	15.22	7.89
One-Step E.M.	0.6	0.55	23.40	11.98

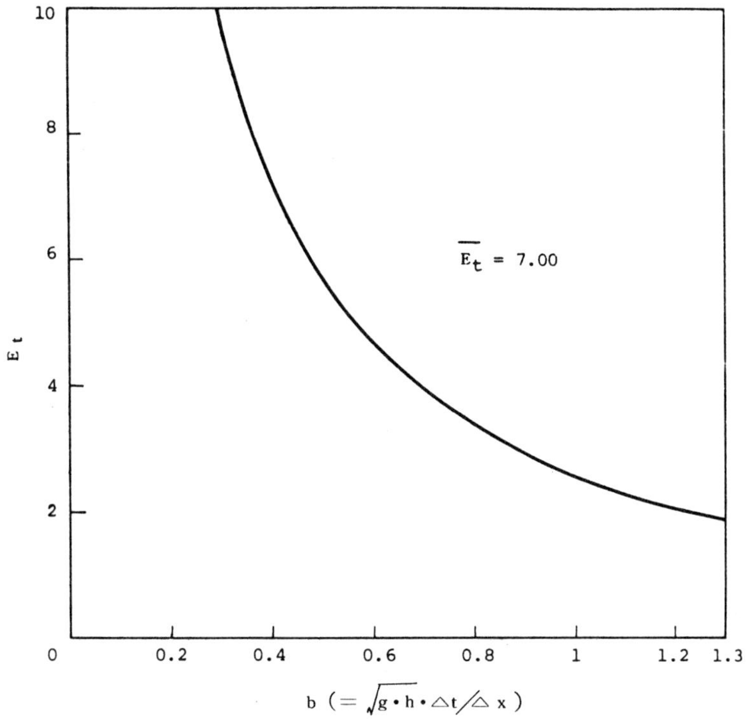

Fig. 1 Numerical Error of Two-Step Explicit Method

ANALYSIS OF DISPERSION PHENOMENA

The steady Dispersion Equation is as follows:

$$\overline{U}\cdot\frac{\partial C}{\partial x}+\overline{V}\cdot\frac{\partial C}{\partial y}=\frac{\partial}{\partial x}\left\{(Kx+Kxv)\cdot\overline{H}\cdot\frac{\partial C}{\partial x}\right\}+\frac{\partial}{\partial y}\left\{(Ky+Kyv)\cdot\overline{H}\cdot\frac{\partial C}{\partial y}\right\}+\overline{H}\cdot Q \qquad (31)$$

where

C: concentration	(p.p.m.)
$\overline{U}$,$\overline{V}$: average U, V during one tidal cycle	(m^2/sec)
$\overline{H}$: average ($h+\zeta$) during one tidal cycle	(m)
Kx,Ky: x-y component of eddy diffusivity	(m^2/sec)
Kxv,Kyv: x-y component of diffusivity dependent on velocity	(m^2/sec)

$$Kxo=\alpha\cdot\triangle L\cdot\sqrt{\frac{1}{M}\sum_{1}^{M}(u-\overline{u})^2} \qquad (m^2/sec)$$

$$Kyo=\alpha\cdot\triangle L\cdot\sqrt{\frac{1}{M}\sum_{1}^{M}(v-\overline{v})^2} \qquad (m^2/sec)$$

α : coefficient of representative length of diffusivity (-)

$\triangle L$: representative length of each element (m)

M: number of flow patterns

The boundary conditions are as follows:

$$\left.\begin{array}{l} C = C_o \text{ (prescribed value of C)} \\ \text{or} \\ \partial c/\partial n = 0 \\ n: \text{ component of exterior unit vector normal to the boundary.} \end{array}\right\} \quad (32)$$

Formulation by FEM is omitted here.

In order to improve the accuracy of this steady dispersion **analysis,** the following coefficient Gi of diffusivity by Spalding (1972) is adopted.

$$G_{xi} = \frac{2}{R_{xi}} \cdot \frac{e^{R_{xi}} - 1}{e^{R_{xi}} + 1}, \quad R_{xi} = \left(\frac{\bar{U}}{K_{xo}}\right)_i \cdot \triangle L$$

$$G_{yi} = \frac{2}{R_{yi}} \cdot \frac{e^{R_{yi}} - 1}{e^{R_{yi}} + 1}, \quad R_{yi} = \left(\frac{\bar{V}}{K_{yo}}\right)_i \cdot \triangle L$$

NUMERICAL EXAMPLES

Arrangement of the Calculations are as follows:

1. Ut from equation (4),
2. Vt from equation (5),
3. Boundary treatment,
4. ζ_t from equation (6),
5. return to 1.

Table 3 shows the size of numerical examples and Table 4 shows the time and core size needed to secure the numerical results.

Table 3 Size of Numerical Examples

	Range (km x km)	Depth (m)	No. of nodes	No. of elements
Simple Model	0.015x0.09	10	63	96
Osaka Bay	50 x 50	7 - 94	183	306
Seto Inland Sea	130 x 300	4 - 104	596	906

Table 4 Results of Numerical Examples

		Time Step (sec.)	Simulation Time (hour)	$Kx=Ky/\alpha$ (m^2/sec) (-)	Computing Time (min.)	Core Size (Kbytes)
Osaka Bay	Tidal Flow	72	24	*	8.5	222
	Dispersion	*	*	10/20	0.7 (5 cases)	276
Seto Inland Sea	Tidal Flow	200	36	*	14.2	364

(IBM S/370-M168)

Simple Model

In order to check the results of error analysis, we consider the simple model.
Fig. 2 shows the configuration of this model.

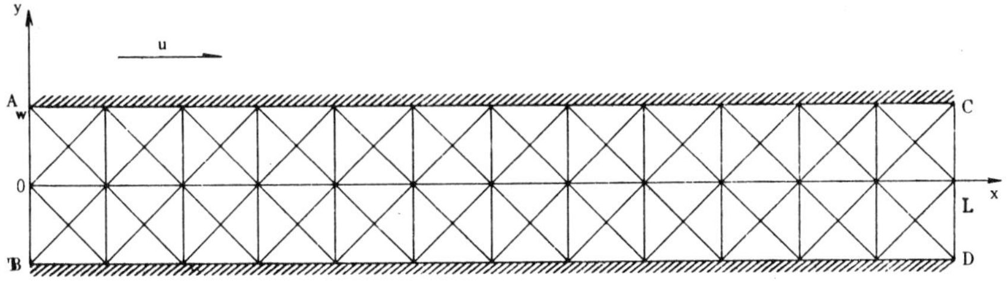

Fig. 2 Configuration of the Simple Model

Basic equations are the same as equations (18) and (19).

Boundary conditions are:

$$u, \zeta = \sin(t) \text{ at } x = 0 \quad (t > 0), \quad v = 0 \text{ at } y = \pm W$$

Initial conditions are:

$$u, \zeta = 0 \text{ at } t = 0 \quad (0 \leq x \leq L)$$

Considering g = 10 (m/sec^2) and h = 10(m), we get an analytical solution:

$$u, \zeta = \sin\left(t - \frac{x}{10}\right).$$

Here, W=7.5(m), L=90(m), $\triangle L = \left(\frac{L}{12} + \frac{L}{24}\right) / 2 = 5.625$ (m)

Results Fig. 3 shows the results of the One-step Explicit Method ($\triangle t$ =0.3182) and the Two-step Explicit Method (θ=0.5, $\triangle t$=0.6364) and Fig. 4 shows the result of the Two-step Explicit Method ($\triangle t = 0.7, \triangle t = 0.35$).

1. These results agree substantially with the error analyses of the numerical scheme in Table 2.
2. Fig. 1 shows that Et becomes large in case the value of b becomes small. This result agrees in large degree with the numerical one in Fig. 4.
3. Consequently, the maximum time step within the stability condition is preferable.
4. Table 5 shows the results of the numerical schemes. These results show the appropriateness of the stability condition in Table 2.
5. Judging by these results, the Two-step Explicit Method is the best method from the standpoint of accuracy and the computing time (the time step is long).

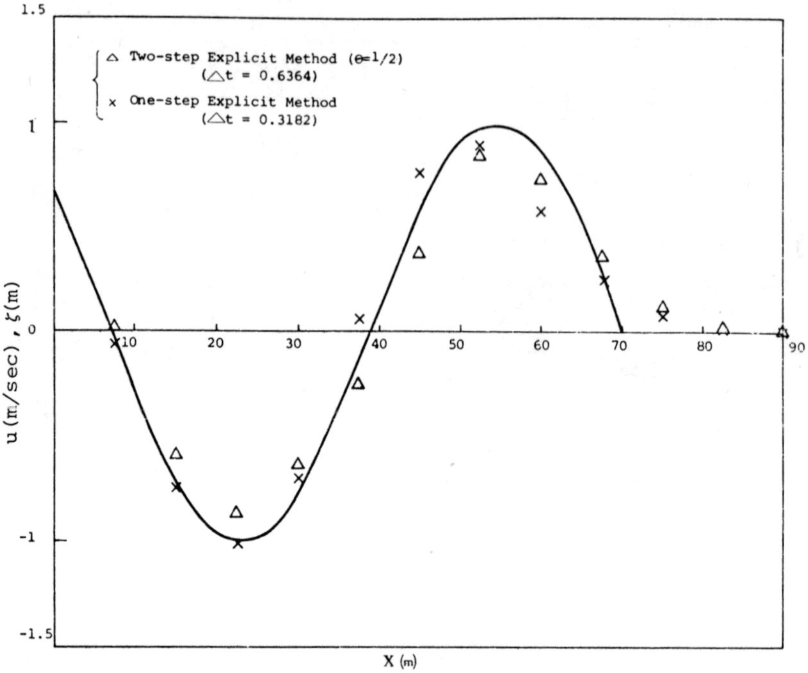

Fig. 3 Numerical Results of Simple Model (1)

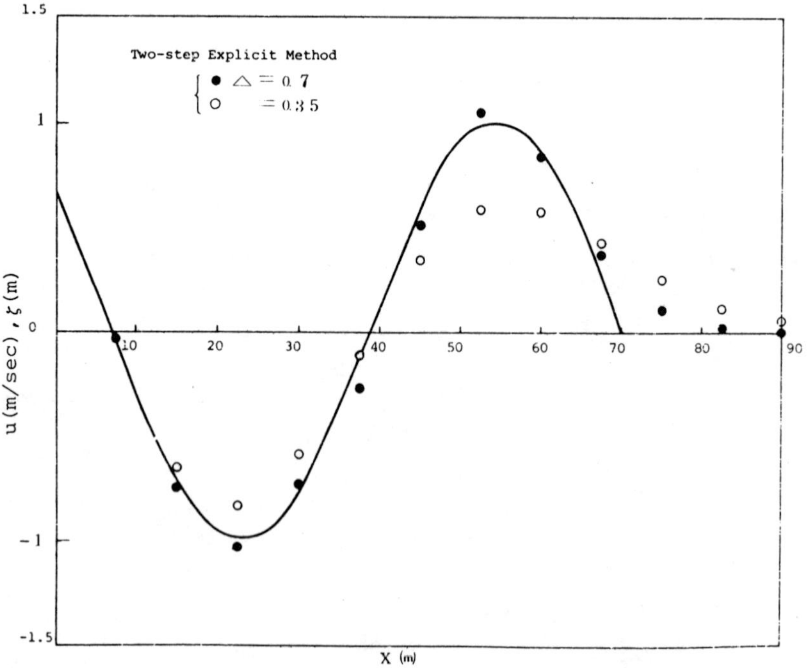

Fig. 4 Numerical Results of Simple Model (2)

Table 5 Comparison of Time Step Size

	$\triangle L$ (m)	hmax. (m)	Theoretical results $\triangle t$ (sec)	Numerical results $\triangle t$ (sec)	(Diverge)
Simple Model (One-step)	5.625	10	0.338	0.3182	(0.35)
Simple Model (Two-step, θ=0.5)	5.625	10	0.619	0.6364	(0.7)
Simple Model (Two-step)	5.625	10	0.731	0.7	(0.7778)
Osaka Bay (Two-step)	1800	94	77	72	(80)
Seto Inland Sea (Two-step)	5000	104	204	200	(225)

Osaka Bay

Fig. 5 shows the configuration of Osaka Bay.

Computational Conditions Observed values of velocity and water level are given at the boundaries A and B in Fig. 5.

Computational Results Fig. 6 shows the computed currents and Fig. 7 shows the time history of currents. These results show good agreement with the observed ones. Fig. 8 shows the concentration distribution.

The Seto Inland Sea

Fig. 9 shows the configuration of the Seto Inland Sea.

Computational Conditions Observed values of velocity and water level are given at the boundaries A-B-C and D-E in Fig. 9 and observed

values of velocity are given at 17 points in the Sea.

Computational Results Fig. 10 shows the computed currents.

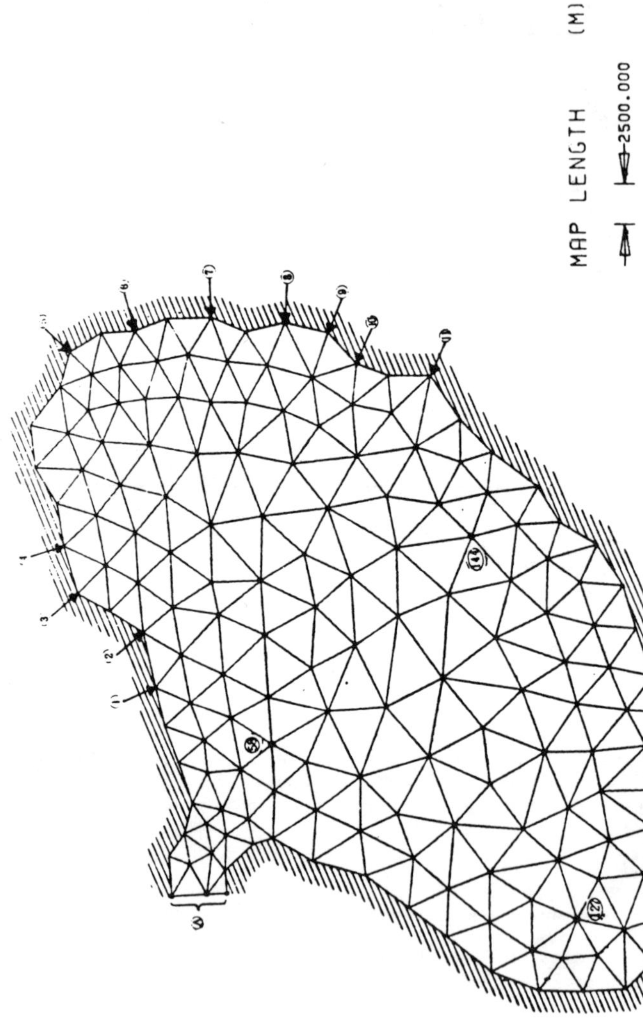

Fig. 5 Configuration of Osaka Bay

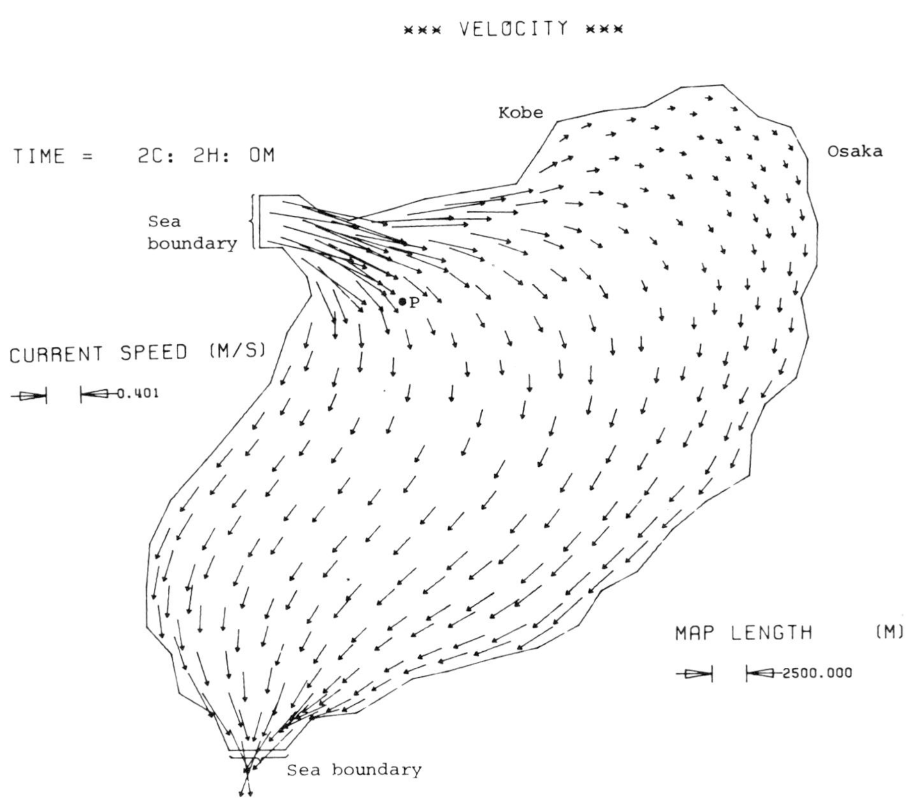

Fig. 6 Computed currents after 93,600 sec (2.167 tidal cycles)

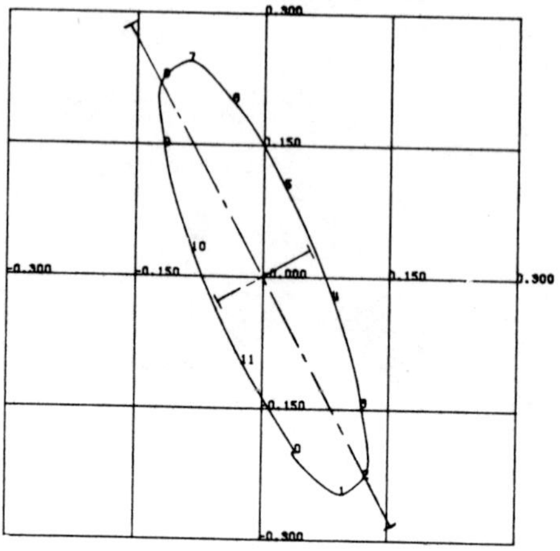

Fig. 7 Time History of Tidal Current at 'P'

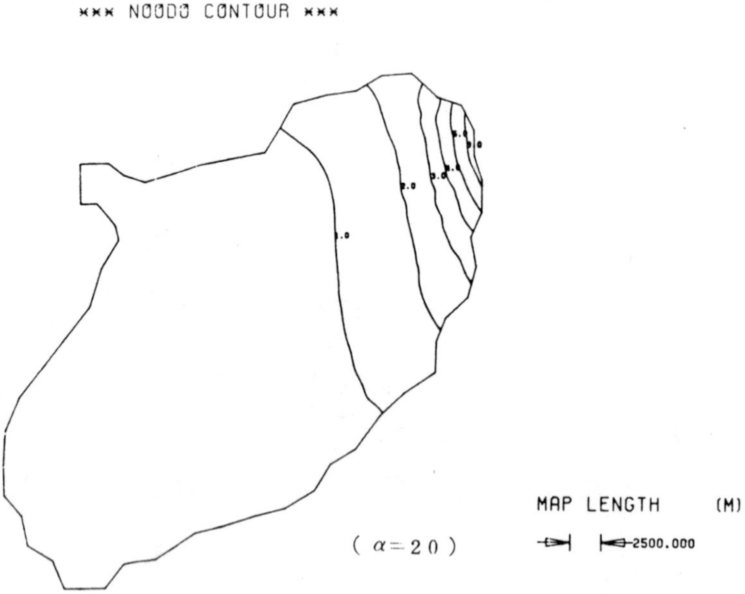

Fig. 8 Concentration distribution of COD (in P.P.M.)

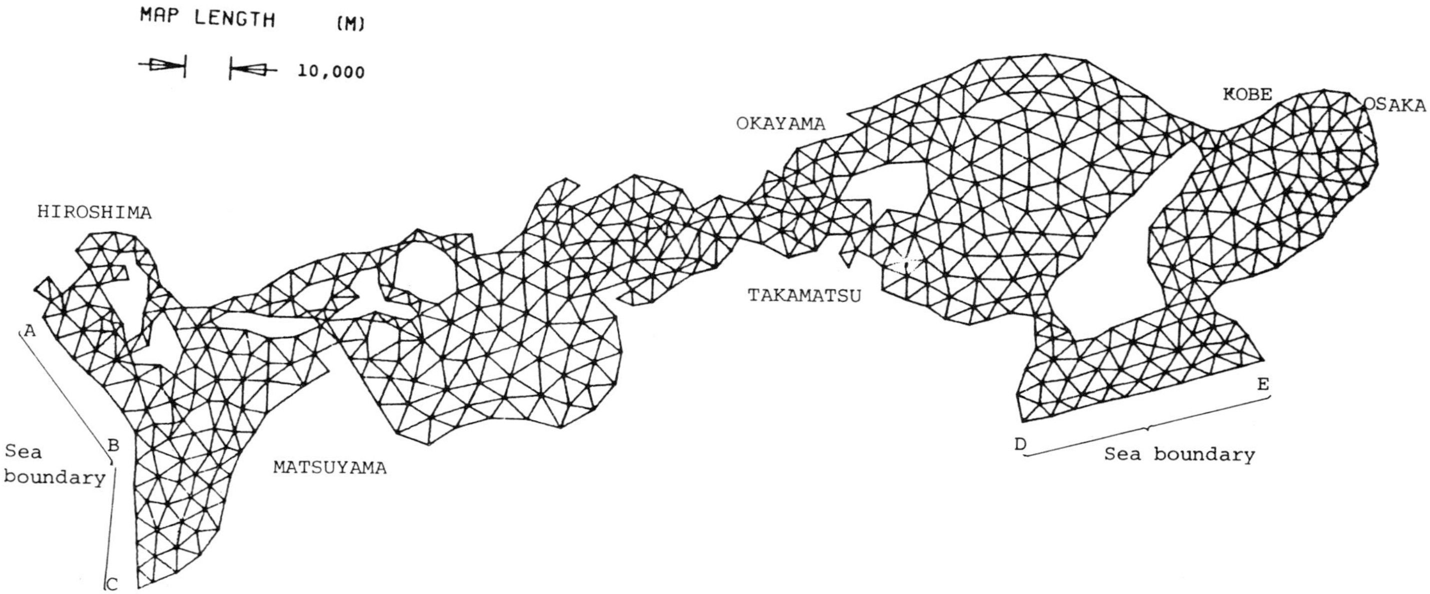

Fig. 9 Configuration of the Seto Inland Sea

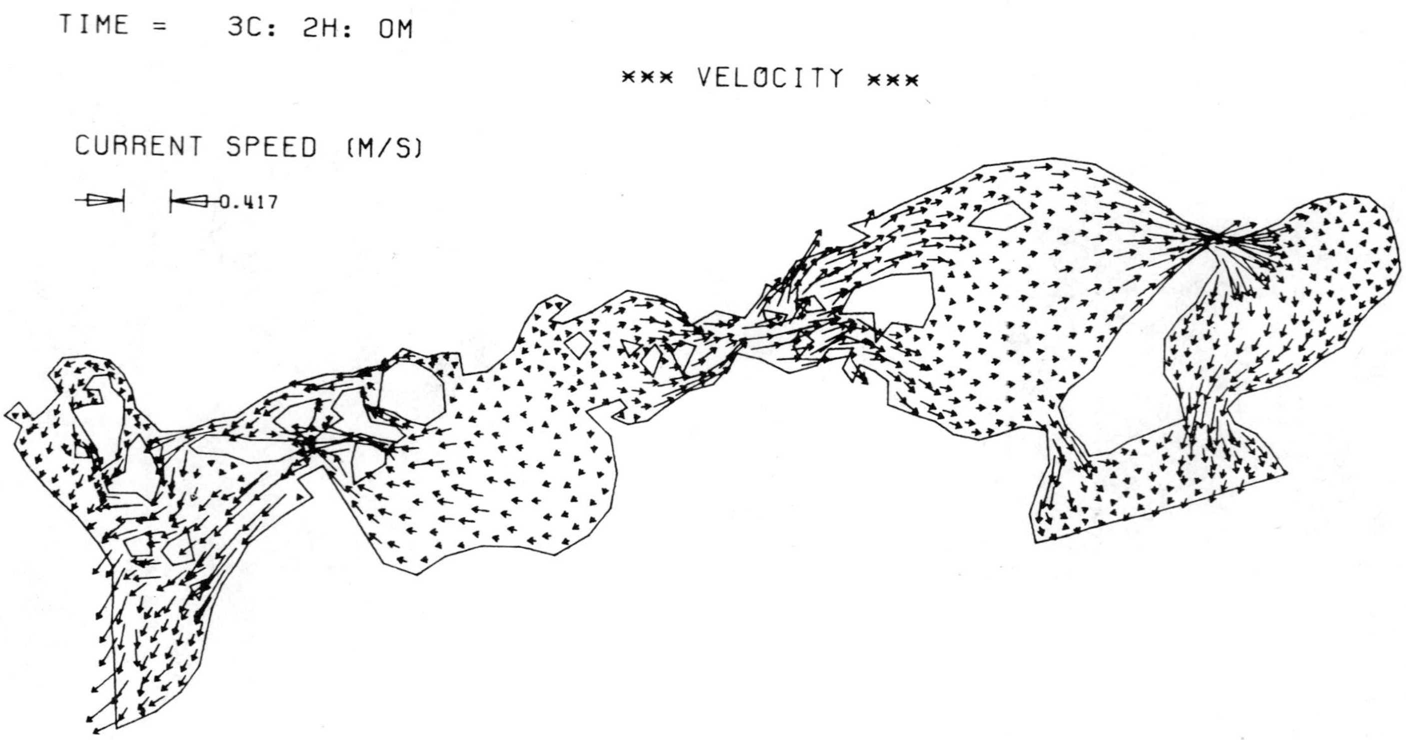

Fig. 10 Computed currents after 93,600 sec (2.167 tidal cycles)

CONCLUSION

Developing the simulation system using FEM, we found the following results:

1. this system can treat complex geometry;

2. we can use nonuniform meshes for economy;

3. the convergence of tidal flow is rapid — this is because of the consistency of the direction of velocity near land boundaries;

4. we can give the water level or the velocity or the water level and velocity simultaneously as forced conditions instead of only the water level or the velocity in the case of FDM. So, it is easy to simulate the flow pattern;

5. there is no significant difference between the computing time of FEM and that of FDM, because the type of computation by FEM is explicit.

Therefore, the numerical method of this paper is effective as a practical method for the simulation of water pollution problems, compared to the usual method by FDM or the other method using FEM.

REFERENCES

Connor, J. J. and Brebbia, C. A. (1976), Finite Element Techniques for Fluid Flow, Newnes-Butterworths: 233.

Dronkers, J. J. (1964), Tidal Computations, North Holland Publishing Company, Amsterdam.

Grotkop, G. (1973), Finite Element Analysis of Long-period Water Waves, Comp. Meth. Appl. Mech. Engng., 2:147.

IBM Japan (1978), FEM-SEAPOL, N: GE18-5134-0 (in Japanese).

IBM Japan (1975), WPS-FEM, N: GE18-5062-1 (in Japanese).

Leendertse, J. J. (1967), Aspects of a Computational Model for Long-period Water Wave Propagation, RAND Memorandum, RM-5294-PR, Santa Monica, California.

Matsuda, Y., Noda, S., Mizoguchi, M. and Iwasaki, M. (1977), A Development of a Water Pollution Prediction System by the Finite Element Method, 5th Symposium on Environmental Problem, JSCE (in Japanese).

Spalding, D. B. (1972), A novel finite difference formulation for differential expressions involving both first and second derivatives, Int. J. Num. Meth. Engng., 4:551.

Taylor, C. and Davis, J. M. (1975), Tidal and Long Wave Propagation — A Finite Element Approach, Comp. Fluids, 3:125.

Weare, T. J. (1976), Finite Element or Finite Difference Methods for the Two-dimensional Shallow Water Equations?, Comp. Meth. Appl. Mech. Engng., 7:351.

RAYLEIGH-RITZ FORMULATION OF PROBLEMS IN SEDIMENTOLOGY

Anand Prakash
Bechtel Inc., San Francisco, CA 94119

INTRODUCTION

A group of classical problems in potamology pertains to the investigation of the distribution of suspended sediment in erodible channels. Specific examples are: distribution of suspended sediment in estuaries, fluvial streams, earthen irrigation channels, open channels excavated in alluvial flood plains connecting perennial or ephemeral streams to the wet well of a pumping station, axisymmetric flow towards offshore intakes located in a lake and settling basins and sedimentation ponds of hydro and fossil-fuel power plants. A common task associated with all these problems is to estimate the amount of isoluble material that would settle out of suspension or be entrained with the flow and its spatial distribution at a particular instant of time.

In the general curvilinear orthogonal system of coordinates, the mass conservation of suspended sediment is given by the equation,

$$\frac{\partial C}{\partial t} = \frac{1}{h_1 h_2 h_3} \left\{ \frac{\partial}{\partial x_1} \left(\frac{h_2 h_3}{h_1} \varepsilon_1 \frac{\partial C}{\partial x_1} \right) + \frac{\partial}{\partial x_2} \left(\frac{h_1 h_3}{h_2} \varepsilon_2 \frac{\partial C}{\partial x_2} \right) + \frac{\partial}{\partial x_3} \left(\frac{h_1 h_2}{h_3} \varepsilon_3 \frac{\partial C}{\partial x_3} \right) - \frac{\partial}{\partial x_1} (u_1 C h_2 h_3) - \frac{\partial}{\partial x_2} (u_2 C h_1 h_3) - \frac{\partial}{\partial x_3} (u_3 C h_1 h_2) \right\} + S \qquad (1)$$

in which C = mass concentration of suspended sediment; u_1, u_2, u_3 = components of flow velocity; ε_1, ε_2, ε_3 = local turbulent mass transfer coefficients; t = time; x_1, x_2, x_3 = space coordinates; h_1, h_2, h_3 = linearizing scale factors and S =

contribution of a source or sink.

PROBLEM STATEMENT

For the formulation to be described herein, a typical earthen channel is simulated which takes off from an alluvial stream with predetermined sediment transport characteristics. In the lateral direction, the velocity, turbulent mass transfer coefficients and sediment concentrations are assumed to be invariant with position. The resulting two-dimensional equation for mass conservation is (Ariathurai and Krone, 1976; Sayre, 1969),

$$\frac{\partial C}{\partial t} = \frac{\partial}{\partial x}\left(\varepsilon_x \frac{\partial C}{\partial x}\right) + \frac{\partial}{\partial z}\left(\varepsilon_z \frac{\partial C}{\partial z}\right) - u\frac{\partial C}{\partial z} - w\frac{\partial C}{\partial z} + (\lambda_2 - \lambda_1 C) \quad (2)$$

in which, x, z = longitudinal and vertical coordinates; ε_x, ε_z = longitudinal and vertical turbulent mass transfer coefficients; u, w = longitudinal and vertical (fall) velocities; and $\lambda_1 C$, λ_2 = rates of deposition and erosion respectively. Assuming fluid velocity and turbulence to be low enough for the particles to settle, Karman-Prandtl velocity distribution, isotropic turbulence and Reynold's analogy for the transfer of mass and momentum by turbulence, it can be shown (Schlichting, 1955; Sayre, 1969; Ariathurai and Krone, 1976) that,

$$u = \bar{u} + \frac{u_*}{\kappa}\left[1 + \ell n \frac{z}{z_0}\right] \quad (3)$$

$$w = \frac{2d}{C_D}\,\frac{K_2}{K_1}\left[\frac{\rho_s}{\rho} - 1\right] \quad (4)$$

$$\varepsilon_x = \varepsilon_z = \beta' \kappa\, u_* z \left[1 - \frac{z}{z_0}\right] \quad (5)$$

$$\lambda_1 = \frac{w}{z_0}\left[1 - \frac{\tau_b}{\tau_{cd}}\right] \quad (6)$$

$$\lambda_2 = \frac{M}{z_0}\left[\frac{\tau_b}{\tau_{ce}} - 1\right] \quad (7)$$

in which d = mean particle diameter; C_D = coefficient of drag; ρ_s = particle density; ρ = fluid density; $\bar{u}$ = depth averaged flow velocity; κ = Von Karman's velocity coefficient; z_0 = flow depth; u_* = shear velocity; β' = coefficient relating the mass transfer coefficient with the momentum exchange coefficient; M = erodibility constant; τ_b = bed shear stress; and τ_{cd}, τ_{ce} = critical shear stresses for deposition and erosion respectively. K_1 and K_2 are shape factors such that the projected area, A, and volume, V, of the particle are $K_1 d^2$ and $K_2 d^3$ respectively.

RAYLEIGH-RITZ FORMULATION

Equation 2, as such, is in a non-self-adjoint form and therefore, is not directly derivable from a variational functional (Prakash, 1977 a; 1977 b). However, it can be transformed to a self-adjoint form using a reducing factor, exp (β), in which,

$$\beta = - \left[\frac{ux}{\varepsilon_x} + \frac{\omega z}{\varepsilon_z}\right] \tag{8}$$

The corresponding variational functional is,

$$J = \iint_R \exp(\beta) \left[\frac{\varepsilon_x}{2}\left(\frac{\partial C}{\partial x}\right)^2 + \frac{\varepsilon_z}{2}\left(\frac{\partial C}{\partial z}\right)^2 + \left(\frac{\partial C}{\partial t} + \lambda_2 - \lambda_1 \frac{C}{2}\right)C\right]$$

$$dx\, dz - \int_S q_s\, C_s\, C \exp(\beta)\, dS \tag{9}$$

in which $q_s C_s$ = outward flux of the dispersant across the boundary, S of the region, R. Following the usual steps of the finite-element method, a shape function [A] is now introduced such that C = [A] {C}, in which, [A] = [A_i A_j A_k ...]; i, j, k, ... = nodal elements of the shape function; {C} = a column vector specifying the values of the dispersant concentration on the nodes, i, j, k, ... of an element defined by the selected shape function.

Minimizing the variational functional of Equation 9 with respect to the nodal concentrations of an element, m, yields the following set of equations,

$$[\frac{\partial J}{\partial C_p}]^m = \iint_{R^m} \{ \varepsilon_x (\frac{\partial A_i}{\partial x} C_i + \frac{\partial A_j}{\partial x} C_j + \ldots) \frac{\partial A_p}{\partial x} +$$

$$\varepsilon_z (\frac{\partial A_i}{\partial z} C_i + \frac{\partial A_j}{\partial z} C_j + \ldots) \frac{\partial A_p}{\partial z} + [A_i \frac{\partial C_i}{\partial t} + A_j \frac{\partial C_j}{\partial t} +$$

$$\ldots] A_p + \lambda_2 A_p - [A_i C_i + A_j C_j + \ldots] \lambda_1 A_p \} e^\beta dxdz$$

$$- \int_S q_s C_s A_p e^\beta dS = 0 \qquad p = i, j, k, \ldots \tag{10}$$

Written in the form of elemental matrices, these equations are,

$$[T]\{\frac{\partial C}{\partial t}\}^m + [K] \{C\}^m = \{F\}^m \tag{11}$$

in which the elements of [T], [K] and {F} are,

$$T_{ij} = \iint_{R^m} A_i A_j e^\beta dx\, dz \tag{12}$$

$$K_{ij} = \iint_{R^m} (\varepsilon_x \frac{\partial A_i}{\partial x} \frac{\partial A_j}{\partial x} + \varepsilon_z \frac{\partial A_i}{\partial z} \frac{\partial A_j}{\partial z} + \lambda_1 A_i A_j) e^\beta dxdz \tag{13}$$

$$F_i = - \iint_{R^m} \lambda_2 A_i e^\beta dx\, dz + \int_S q_s C_s A_i e^\beta dS \tag{14}$$

Evaluating the integrals and assembling all the elemental matrices over the entire domain yields,

$$[TL] \{\frac{\partial C}{\partial t}\} + [KL] \{C\} = [FL] \qquad (15)$$

Note that the matrices [T], [K], [TL] and [KL] are symmetric and therefore, permit a large saving in computer storage and operation time as compared to the alternative formulation using the Galerkin approach where the corresponding matrices are nonsymmetric (Prakash, 1977 a and 1977 b).

INITIAL AND BOUNDARY CONDITIONS

For the simulation considered herein, the prescribed ambient concentration for each nodal point in the simulated region constitutes the initial condition i.e. $C(x, z, o) = C_{oi}$; $o < x < L$, $o \leq z \leq z_o$ in which L = length of the simulated region. The free surface is a Neumann boundary with no mass flux across it, i.e. $\varepsilon_z \frac{\partial C}{\partial z} - wC = o$; $z = z_o$. For cases where there is inflow of suspended sediment across a portion, S, of this boundary, the flux, $q_s C_s$, is specified and the resulting modified flow velocities u and w are prescribed as input to the program. The channel bottom is treated as an absorbing boundary if $\varepsilon_z \frac{\partial C}{\partial z} = o$; $w \neq o$, $z = o$ or a Neumann (reflecting) boundary $\varepsilon_z = \frac{\partial C}{\partial z} - wC = o$; $z = o$ (Sayre, 1969). The boundaries at the mouth and tail of the channel are treated as constant flux boundaries if the fluxes across the channel depth at x = o and x = L are known apriori. Alternatively, any one or both of these can be treated as Dirichlet (geometric) boundaries with prescribed nodal point concentrations. For any specific study, these concentrations should be determined by field measurements. In the absence of measured values, the vertical distribution of concentrations can be estimated by the equation (U. S. Dept. of Trans., 1975),

$$C = C_a \left[\frac{z_o - z}{z} \cdot \frac{a}{z_o - a} \right]^{\frac{w}{\beta' \kappa u_*}} \qquad (16)$$

in which C_a = predetermined concentration at a depth, a, above the bed.

SOLUTION TECHNIQUE

The expressions in Equations 3, 4, 5, 6 and 7 for u, w, ε_x, ε_z, λ_1 and λ_2 are fairly complex. Therefore, to avoid further complexity associated with the use of quadratic or isoparametric elements, a simpler formulation with triangular elements is attempted which can be extended to higher order elements without much difficulty. A local system of coordinates, with the centroid of the element as origin, is introduced such that,

$\varepsilon = x - \bar{x}^m$; $\eta = z - \bar{z}^m$ and $\bar{x}^m$, $\bar{z}^m$ = global coordinates of the centroid of mth element. Details of the corresponding elemental matrices are given in Appendix I.

The complex integrals over the elemental areas, $\iint \exp(\beta)\, \varepsilon_x \, d\xi \, d\eta$, $\iint \exp(\beta) \varepsilon_z \, d\xi \, d\eta$, $\iint \exp(\beta)\, A_i \, A_j \, d\xi \, d\eta$ and $\iint \exp(\beta)\, A_i \, d\xi \, d\eta$ are evaluated using the special weighting constants devised by Radau (Zienkiewicz, 1971). For simpler cases with elements as right angled triangles, direct integration can be used.

For programming efficiency, it is desirable that the symmetric matrices of Equation 15 have the smallest possible bandwidth. For this purpose, a special algorithm for mesh generation with an optimized node numbering scheme is used (Prakash, 1977 a). Assemblage of the elemental matrices with the optimized node numbering scheme results in matrix equations with the smallest possible bandwidth in which all nonzero elements are concentrated within a narrow band about the diagonal. All matrix operations are performed on the contracted forms of these matrices comprising only of the upper diagonal elements within the bandwidth.

The Neumann or natural boundary conditions are satisfied by the variational principle itself. Contribution of the constant flux boundaries, wherever applicable, is included in the column vector [FL]. For the Dirichlet or geometric boundary conditions, the equations corresponding to the nodes with prescribed nodal point concentrations are eliminated from the set of Equations 15 by deleting the rows and columns corresponding to each of these nodes from the component matrices. Concurrently, the right hand sides of the remaining equations are modified as follows (Prakash, 1976):

$$f_n^* = f_n - \sum_{\ell=1}^{L'} k_{n,\ell}\, c_\ell \qquad n \neq \ell'$$

$$n = 1, 2, 3, \ldots N \; ; \quad \ell = 1, 2, 3, \ldots L' \qquad (17)$$

in which f_n^* = modified values of the matrix elements on the right hand side; L' = total number of nodes with prescribed nodal point concentrations; N = total number of nodes in the domain; and f, k and c = individual elements of the matrices [FL], [KL] and the column vector {C} respectively. Thus, the total number of simultaneous equations to be solved is reduced by L'. The remaining $(N-L')$ equations are solved for $(N-L')$ values of $\frac{\partial C}{\partial t}$ by the Gauss elimination procedure for banded and symmetric matrices resulting in a set of (N-L´) first order differential equations. These equations are numerically integrated using the fourth order Runge-Kutta method as the starter and the Adams-Moulton multistep predictor-corrector method with Adams-Bashford corrector as the follower (Conte, 1965).

To reduce computer time, maximum and minimum error criteria for the difference in the values between the applications of a predictor and a corrector are specified. If the error is greater than the prescribed maximum, the time step size is halved automatically. If, on the other hand, the error is smaller than the minimum, the time-step size is doubled. After the solution for one time step attains convergence within the prescribed accuracy, the nodal point concentrations are produced as output and the solution is continued for successive time increments until the required time interval is completed (Guymon, 1970; Nalluswami, 1972).

Once the nodal point concentrations at every time-step throughout the domain of interest are determined, the amount of suspended sediment that has settled out of suspension, R(L), during a time period, t_o and over a settling length, L for a constant flux boundary at the tail can be obtained as follows:

$$R(L) = \int_0^{t_o} \int_0^{z_o} u(z)\, C(o,z,t)\, dz\, dt + \int_0^{L} \int_0^{z_o} C(x,z,o)\, dz\, dx - \int_0^{L} \int_0^{z_o} C(x,z,t_o)\, dz\, dx - \int_0^{t_o} \int_0^{z_o} (q_s\, C_s)\, dz\, dt \qquad (18)$$

The removal ratio, r, is given by,

$$r = R(L) / \left[\int_0^{t_o} \int_0^{z_o} u(z)\, C(o,z,t)\, dz\, dt + \int_0^{L} \int_0^{z_o} C(x,z,o)\, dz\, dx \right] \qquad (19)$$

The nodal point quantities, C (x,z,t) and u(z) are discrete values and not continuous functions. Therefore, the integrations of Equations 18 and 19 are performed as summations of the corresponding products of discretized values.

PROGRAM VALIDATION

Extensive testing of the formulation and solution technique is required to establish convergence and stability criteria for the numerical scheme. At present only limited testing has been conducted for simplified cases of one-dimensional transport with and without deposition. Analytical solution for the one-dimensional form of Equation 2 (hereinafter referred to as case (i)) with $\varepsilon_z = w = \lambda_1 = \lambda_2 = o$; ε_x = constant; $C(o,t) = C_o$; $C(\infty,t) = o$; $C(x, o) = o$, $o < x < \infty$ is given by (Bear, 1972; Carslaw and Jaeger, 1959),

$$\frac{C}{C_o} = \frac{1}{2}\left[\operatorname{erfc}\frac{x-ut}{\sqrt{4\varepsilon_x t}} + \exp\frac{ux}{\varepsilon_x}\operatorname{erfc}\frac{x+ut}{\sqrt{4\varepsilon_x t}}\right] \tag{20}$$

Analytical solution for the same case but with non zero values of λ_1 (hereinafter referred to as case (ii)) is given by (Bear, 1972),

$$\frac{C}{C_o} = \frac{1}{2}\exp\frac{ux}{2\varepsilon_x}\left[\exp(-x\alpha)\operatorname{erfc}\frac{x - \sqrt{u^2 + 4\varepsilon_x\lambda}\,t}{\sqrt{4\varepsilon_x t}} + \right.$$

$$\left.\exp(x\alpha)\operatorname{erfc}\frac{x + \sqrt{u^2 + 4\varepsilon_x\lambda}\,t}{\sqrt{4\varepsilon_x t}}\right] \tag{21}$$

in which $\alpha^2 = \dfrac{u^2 + 4\lambda\varepsilon_x}{4\varepsilon_x^2}$

The cases represented by Equations 20 and 21 imply semi-infinite domains of transport which are impossible to simulate in a numerical model. As an approximation, a fairly long (21,120 ft.) finite region was modeled with preassigned constant concentrations at the boundaries. The region of transport was divided into 128 triangular elements with 81 nodal points. Using the optimized node numbering scheme, a reduced bandwidth of 6 was obtained. Input parameters for the test problems were as follows:

$$\frac{C}{C_o} = 1,\ x = 0;\ \frac{C}{C_o} = 0.1482,\ x = 21{,}120 \text{ ft};$$

$$\varepsilon_x = 43{,}200 \text{ ft}^2/\text{day};\ u = 1.45 \text{ ft/day};\ w = \varepsilon_z = \lambda_2 = 0;$$

$\lambda_1 = 0$ for case (i) and $\lambda_1 = 0.35$ for case (ii). The results of the analytical and numerical solutions for the two cases at t = 1825 days are shown in Figure 1. As can be seen from the plots in Figure 1, the results of the numerical scheme are in reasonable agreement with those obtained from the corresponding analytical solutions. The discrepancy towards the tail end of the channel in both cases can be traced to the dissimilar boundary conditions at this end implied in the respective analytical and numerical solutions. The agreement can be improved further by mesh refinement and by enlargement of the transport region simulated in the numerical scheme so as to provide a closer approximation to the semi-infinite region represented by Equations 20 and 21.

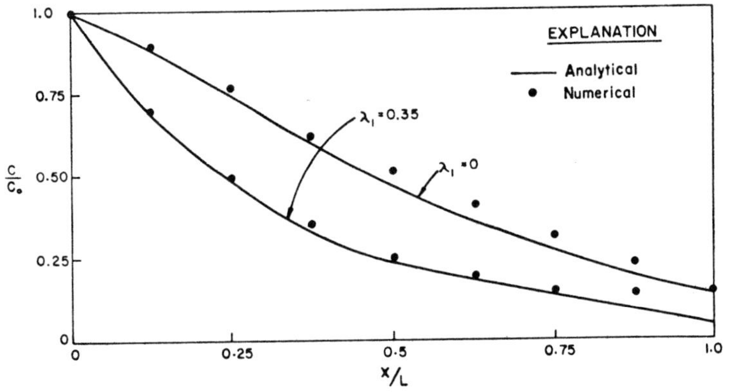

FIGURE 1 COMPARISON OF ANALYTICAL AND NUMERICAL RESULTS

CONCLUSION

A numerical simulator is developed which is based on the Rayleigh-Ritz formulation of the two-dimensional equation for the mass conservation of suspended sediment in an open channel. Contributions of the rates of local deposition and erosion of sediments are included in this equation. The formulation encompasses Neumann, Dirichlet, constant flux and absorbing boundary conditions. The validity of the solution technique is tested by comparing the results with known analytical solutions for simplified cases. Further testing with field and laboratory results is required before the simulator can be used for the solution of practical problems.

ACKNOWLEDGMENT

Special thanks are due to Dr. John J. Cassidy, Bechtel Inc., San Francisco, for the encouragement and helpful suggestions provided by him during the preparation of the manuscript of this paper.

REFERENCES

Ariathurai, R. and Krone, R. B. (1976) Finite Element Model for Cohesive Sediment Transport, Journal of the Hydraulics Division, ASCE, Vol. 102, No. HY3: 323-338.

Bear, J. (1972) Dynamics of Fluids in Porous Media, American Elsevier, Environmental Science Series: 764.

Carslaw, H. S. and Jaeger, J. S. (1959) Conduction of Heat in Solids, Oxford University Press, Oxford, England: 510.

Conte, S. D. (1965) Elementary Numerical Analysis, McGraw-Hill Book Company, New York: 278.

Guymon, L. R. G. (1970) Mathematical Modeling of Movement of Dissolved Constituents in Groundwater Aquifers by the Finite-Element Method, thesis presented to the University of California, at Davis, California, in partial fulfillment of the requirements of the degree of Doctor of Philosophy.

Nalluswami, M. (1972) Numerical Simulation of General Hydrodynamic Dispersion in Porous Media, thesis presented to Colorado State University at Fort Collins, Colorado, in partial fulfillment of the requirements for the degree of Doctor of Philosophy.

Prakash, Anand (1976) Discussion of Sparsity Oriented Analysis of Large Pipe Networks by M. Chandrashekar and K. H. Stewart, Journal of the Hydraulics Division, ASCE, Vol. 102, No.HY1: 109-112.

Prakash, Anand (1977 a) Finite Element Solutions of the Non-Self-Adjoint Convective-Dispersion Equation, International Journal for Numerical Methods in Engineering, Vol. 11: 269-287.

Prakash, Anand (1977 b) Discussion of Finite Element Model for Cohesive Sediment Transport, Journal of the Hydraulics Division, ASCE, Vol. 103, No. HY2: 200-203.

Sayre, W. W. (1969) Dispersion of Silt Particles in Open Channel Flow, Journal of the Hydraulics Division, ASCE, Vol. 95, No. HY3: 1009-1038.

Schlichting, H. (1955) Boundary Layer Theory, Pergamon Press Ltd., London: 647.

Sumer, B. M. (1977) Settlement of Solid Particles in Open-Channel Flow, Journal of the Hydraulics Division, ASCE, Vol. 103, No. HY11: 1323-1337.

U. S. Department of Transportation (1975), Federal Highway Administration, Highways in the River Environment, Hydraulic and Environmental Design Considerations, Training and Design Manual.

Zienkiewicz, O. C. (1971) The Finite Element Method in Engineering Science, McGraw-Hill Book Co., Inc., London, England: 521.

APPENDIX I

Constituents of the elemental matrices for a triangular element are as follows:

$$T_{mn} = \iint \exp(\beta)\, A_m\, A_n\, d\xi\, d\eta \qquad m=i,\ j,\ k;\ n=i,\ j,\ k$$

$$K_{mn} = \frac{a_{2m}\, a_{2n}}{4A^2} \int\int \exp(\beta)\, \varepsilon_x\, d\xi\, d\eta + \frac{a_{3m}\, a_{3n}}{4A^2} \iint \exp(\beta)$$

$$\varepsilon_z\, d\xi\, d\eta + \lambda_1 \iint \exp(\beta)\, A_m\, A_n\, d\xi\, d\eta$$

$$F_m = -\lambda_2 \iint \exp(\beta)\, A_m\, d\xi\, d\eta + \int_s \exp(\beta)\, q_s\, C_s\, A_m\, dS$$

$$A_m = [a_{1m} + a_{2m}\, \xi + a_{3m}\, \eta]/2A$$

$$a_{1i} = \xi_j\, \eta_k - \xi_k\, \eta_j;\ a_{2i} = \eta_j - \eta_k;\ a_{3i} = \xi_k - \xi_j$$

i, j, k follow in cyclic sequence.

A = Area of the element

$$u = \bar{u} + \frac{u_*}{\kappa} \left[1 + \ln \frac{\eta + \bar{z}^m}{z_0} \right]$$

$$\varepsilon_x = \varepsilon_z = \beta' \kappa u_* (\eta + \bar{z}^m) \left[1 - \frac{\eta + \bar{z}^m}{z_0} \right]$$

TSUNAMI WAVE PROPAGATION ANALYSIS BY THE FINITE ELEMENT METHOD

Mutsuto Kawahara, Shohei Nakazawa, Shunsuke Ohmori

Department of Civil Engineering, Chuo University, Kasuga, Bunkyo-ku, Tokyo, JAPAN.

and

Ken'ichi Hasegawa

Technical Research Division, Unic Corporation, Shinjuku, Shinjuku-ku, Tokyo, JAPAN.

INTRODUCTION

There has been a tremendous number of damages of human lives and properties caused by tsunami along the Pacific Coast especially in Japan. The sudden rise in sea level above the epicenter due to tectonic displacements of the sea bed by an earthquake generate a tsunami. The generated tsunami wave propagates toward coastal areas. As the waves approach nearshore of the coast, the amplitudes increase to form huge masses of water. If these waves pass near the nearly closed bays, the natural frequency of these basins is excited so that large waves are produced. Predictions of tsunami generations and propagation is, therefore, a very important means for coastal civilizations.

Once the ground displacement is specified, tsunami wave propagation is governed by the shallow water wave theory. Tsunami wave propagation has already been analyzed numerically by Aida [1969-a,b], Hwang [1973] and Hwang and Divoky [1970] using the finite difference method. In recent years, several finite element methods of the shallow water wave equation have been published. Grotkop [1973] has presented the method of discretizing both space and time functions using the finite element method. Cullen [1973] has employed leap-frog scheme to discretize time function. The isoparametric finite element method has been used by Taylor and Davis [1975-a,b], in which several numerical integration schemes have been compared. In the proceedings of the first symposium on Finite Elements in

Water Resources, Princeton University (Pinder, Gray, and Brebbia [1977]), a lot of finite element schemes have been presented. Wang [1977] has discussed multi-level hydrodynamic model and analyzed the tidal currents in the Block Island Sound. Sündermann [1977] has presented the finite element method based on the support method and solved tides in a channel with a barrier. Ferrante [1977] has dealt with general purpose program for hydrodynamic model. Lagrangian quadratic isoparametric elements have been used by Gray [1977]. Baker [1977] has established the finite elements of steady and/or transient three dimensional hydrodynamical environmental flow fields. The work by Holtz and Hennlich [1977] has been concerned with finite elements on explicit scheme. Cullen [1977] and Pinder and Gray [1977] have discussed the complex propagation factor. Kawahara, Takeuchi and Yoshida [1977] and Kawahara [1977] have presented the two step explicit finite element method and analyzed the Tokachi-Oki Tsunami wave propagation. Recent finite element techniques have been summarized in the books by Brebbia and Connor [1976] and Pinder and Gray [1977].

In this paper, a numerical analysis of tsunami wave propagation is presented by using the finite element method based on the shallow water wave equation. Following the conventional Galerkin procedures, the finite element method is applied to discretize space function of velocity and water elevation. As the interpolation function, linear polynomial function is used based on the three node triangle finite element. To discretize time function, a two step explicit schemes are employed. The schemes employed in this paper are the extensions of the Runge-Kutta scheme. Two step 1/2 formula and 2/3 formula are utilized. The scheme based on the 1/2 formula corresponds to the basic idea of the two step Lax-Wendroff scheme.

This paper also discusses the improved schemes of the above formula based on the smoothing technique and referred to the advanced formulae. For simple one dimensional long wave equation, stability conditions are examined by using von Neumann stability condition. Several numerical test examples are carried out to obtain the basic data for the practical application. The present finite element method has been applied to the prediction of tsunami wave propagation in North Japan.

FINITE ELEMENT EQUATIONS

According to the shallow water wave theory, the basic equations of tsunami propagation are from the Navier-Stokes equation using the assumption of hydrostatic pressure distribution i.e., equation of motion:

$$\frac{\partial u_i}{\partial t} + u_j u_{i,j} + g\zeta_{,i} - A_1(u_{i,j}+u_{j,i}) - f\varepsilon_{ij}u_j + \tau_i = 0 \qquad (2.1)$$

and equation of continuity:

$$\frac{\partial}{\partial t}(\zeta-b) + \{(H+\zeta-b)u_i\}_{,i} = 0 \qquad (2.2)$$

where U_i and ζ represent velocity and water elevation and H,g, A_1, f and τ_i are sea depth, gravity acceleration, eddy viscosity, Coriolis parameter, and bottom friction. As is shown in figure (1), the tectonic displacement of the sea bed in the epicenter region is denoted by b. Here and henceforth, the indecial notation is employed and the usual summation convention with repeated indices is used.

As the boundary conditions, the following three conditions are considered, i.e., the velocity is assumed to be known on the boundary S_1,

$$u_i = \hat{u}_i \qquad \text{on } S_1 \qquad (2.3)$$

the gradient of velocity is given on the boundary S_2,

$$r_i = (u_{i,j}+u_{j,i})n_j = \hat{r}_i \qquad \text{on } S_2 \qquad (2.4)$$

and the water elevation is prescribed on the boundary S_3,

$$\zeta = \hat{\zeta} \qquad \text{on } S_3 \qquad (2.5)$$

where superposed ^ denotes the prescribed value on the boundary and n_j is the components of the unit normals to the boundary. As the initial conditions, the appropriate initial state is assumed:

$$u_i = \hat{u}_i^0 \qquad \text{at } t = 0 \qquad (2.6)$$

$$\zeta = \hat{\zeta}^0 \qquad \text{at } t = 0 \qquad (2.7)$$

Using u_i and ζ^* as weighting functions, the weighted residual equations are derived according to the conventional procedures of the finite element Galerkin method.

$$\int_V (u_i^* \frac{\partial u_i}{\partial t}) dV + \int_V (u_i^* u_j u_{i,j}) dV + g\int_V (u_i^* \zeta_{,i}) dV + A_1 \int_V (u_{i,j}^* u_{i,j}) dV$$
$$+ A_1 \int_V (u_{i,j}^* u_{j,i}) dV - f\int_V (u_i^* \varepsilon_{ij} u_j) dV$$
$$+ \int_V (u_i^* \tau_i) dV = A_1 \int_S (u_i^* \hat{r}_i) dS \qquad (2.8)$$

$$\int_V [\zeta^* \{\frac{\partial}{\partial t} (\zeta-b)\}] dV + \int_V [\zeta^* \{(H+\zeta-b)u_i\}_{,i}] dV = 0 \qquad (2.9)$$

It is assumed that the wave propagation field to be analyzed is divided into small regions called finite elements. Let the interpolating equations for u_i and ζ in each finite element be expressed as:

$$u_i = \Phi_\alpha u_{\alpha i} \tag{2.10}$$

$$\zeta = \Phi_\alpha \zeta_\alpha \tag{2.11}$$

where Φ_α represents the interpolation function, $u_{\alpha i}$ is the nodal value of u_i at αth node of each finite element and ζ_α is the nodal value of ζ at αth node. For weighting functions u_i^* and ζ^*, the relations similar to equations (2.10) and (2.11) are introduced as follows.

$$u_i^* = \Phi_\alpha u_{\alpha i}^* \tag{2.12}$$

$$\zeta^* = \Phi_\alpha \zeta_\alpha^* \tag{2.13}$$

In the numerical computation in this paper, linear interpolation function based on three node triangle finite element is used for Φ_α. Introducing equations (2.10) ~ (2.13) into equations (2.8) and (2.9) and considering the arbitrariness of $u_{\alpha i}^*$ and ζ_α^* lead to the following finite element governing equation.

$$M_{\alpha i\beta j}\dot{u}_{\beta j} + K_{\alpha\beta\gamma j}u_{\beta j}u_{\gamma i} + H_{\alpha i\beta}\zeta_\beta + S_{\alpha i\beta j}u_{\beta j} + F_{\alpha i} = 0 \tag{2.14}$$

$$M_{\alpha\beta}(\dot{\zeta}_\beta - \dot{b}_\beta) + G_{\alpha\beta\gamma i}(H_\beta + \zeta_\beta - b_\beta)u_{\gamma i} = 0 \tag{2.15}$$

In equations (2.14) and (2.15), the depth of the sea and the bottom displacements are also interpolated as:

$$H = \Phi_\alpha H_\alpha \tag{2.16}$$

$$b = \Phi_\alpha b_\alpha \tag{2.17}$$

where H_α and b_α denote nodal values at each nodal point of finite element. Using the conventional superposition procedure, the finite element governing equation for the whole wave propagation field can be written in the following form.

$$M_{\alpha\beta}\dot{v}_\beta + K_{\alpha\beta\gamma}v_\beta v_\gamma + H_{\alpha\beta}h_\beta + S_{\alpha\beta}v_\beta + F_\alpha = 0 \tag{2.18}$$

$$M_{\alpha\beta}\dot{h}_\beta + G_{\alpha\beta\gamma}(H_\beta - b_\beta)v_\gamma + G_{\alpha\beta\gamma}h_\beta v_\gamma = 0 \tag{2.19}$$

where v_β and h_β are unknown field variables at the whole nodal points of the wave propagation field.

NUMERICAL INTEGRATION IN TIME

Finite element equations (2.18) and (2.19) in the whole propagation field can be rewritten in a more concise form as follows.

$$M_{\alpha\beta}\dot{u}_\beta + F_\alpha(u_\beta) = 0 \tag{3.1}$$

where u_β represents unknown field variables v_β and h_β. To

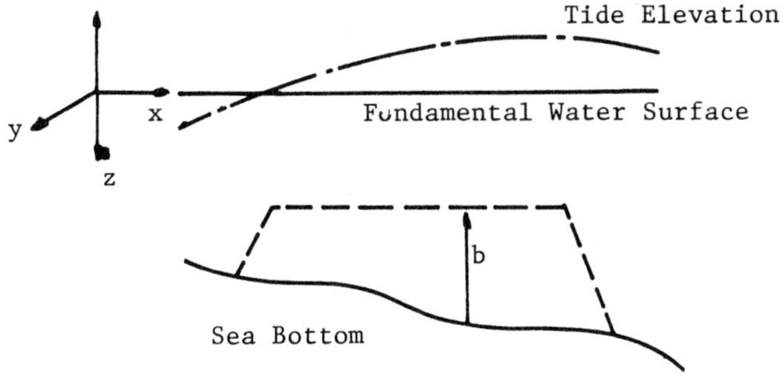

Fig.1 Coordinate System of Shallow Water Wave Equation

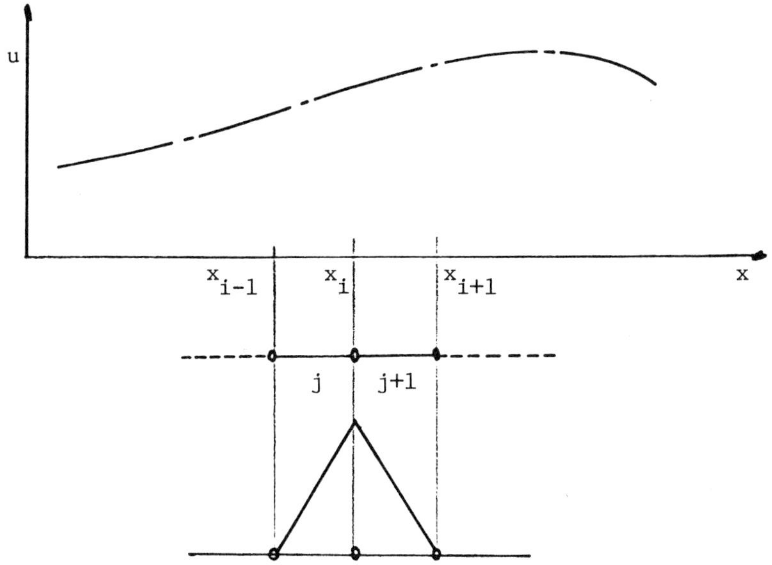

Fig.2 One Dimensional Finite Element Discretization for Derivation of the Stability Criteria

solve equation (3.1), numerical integration scheme in time needs be introduced. This paper employs the two step explicit time integration scheme.

Let T be total time interval to be analyzed. The total time is divided into short time increments, one of which is represented by Δt. Denoting the value of u at nth time point as u^n, general explicit two step scheme can be expressed in the following form.

$$u^{n+\alpha} = u^n + \Delta t(\alpha \dot{u}^n) \tag{3.2}$$

$$u^{n+1} = \beta_1 u^n + \beta_2 u^{n+\alpha} + \Delta t(\beta_3 \dot{u}^n + \beta_4 \dot{u}^{n+\alpha}) \tag{3.3}$$

where $0 \leqq \alpha \leqq 1$, $\beta_1 \sim \beta_4$ are appropriate constants. Substitution of equation (3.3) into equation (3.2) leads to:

$$u^{n+1} = (\beta_1+\beta_2)u^n + \Delta t(\beta_2\alpha+\beta_3+\beta_4)\dot{u}^n + \Delta t^2\beta_4\alpha\ddot{u}^n \tag{3.4}$$

This equation must be identical to the Taylor series expansion as follows.

$$u^{n+1} = u^n + \Delta t\dot{u}^n + \frac{\Delta t^2}{2}\ddot{u}^n + \ldots\ldots \tag{3.5}$$

Comparison of equation (3.4) with equation (3.5) yields:

$$\left.\begin{aligned} &\beta_1 + \beta_2 = 1 \\ &\beta_2\alpha + \beta_3 + \beta_4 = 1 \\ &\beta_4\alpha = \frac{1}{2} \end{aligned}\right\} \tag{3.6}$$

choosing such constants as to satisfy equation (3.6), several numerical integration schemes are obtained.
Taking that

$$\alpha = \frac{1}{2}\,,\ \beta_1 = \beta_4 = 1\,,\ \beta_2 = \beta_3 = 0 \tag{3.7}$$

two step Lax-Wendroff scheme (or Runge-Kutta 1/2 scheme) can be derived in the following form.

$$\begin{aligned} u^{n+1/2} &= u^n + \Delta t\dot{u}^n \\ u^{n+1} &= u^n + \Delta t\dot{u}^{n+1/2} \end{aligned} \tag{3.8}$$

Similarly, taking that

$$\alpha = \frac{2}{3}\,,\ \beta_1 = 1\,,\ \beta_2 = 0\,,\ \beta_3 = \frac{1}{4}\,,\ \beta_4 = \frac{3}{4} \tag{3.9}$$

two step Runge-Kutta scheme (2/3 formula) can be obtained.

$$u^{n+2/3} = u^n + \frac{2\Delta t}{3}\dot{u}^n$$
$$u^{n+1} = u^n + \frac{\Delta t}{4}(\dot{u}^n + 3\dot{u}^{n+2/3}) \tag{3.10}$$

Equation (3.8) indicates the basic idea of the two step Lax-Wendroff finite difference method. Similarly, equation (3.10) establishes the basic idea of the two step Runge-Kutta scheme.

In order to stabilize the formerly presented schemes, this paper employs a smoothing technique to improve. These schemes are referred as the advanced formulae and are expressed as follows.

The advanced two step Lax-Wendroff scheme:

$$u^{n+1/2} = \tilde{u}^n + \frac{\Delta t}{2}\dot{u}^n$$
$$u^{n+1} = \tilde{u}^n + \Delta t \cdot \dot{u}^{n+1/2} \tag{3.11}$$

The advanced two step Runge-Kutta scheme:

$$u^{n+2/3} = \tilde{u}^n + \frac{2\Delta t}{3}\dot{u}^n$$
$$u^{n+1} = \tilde{u}^n + \frac{\Delta t}{4}(\dot{u}^n + 3\dot{u}^{n+2/3}) \tag{3.12}$$

where $\tilde{u}^n$ has its components as

$$\tilde{u}_i^n = \mu u_i^n + (1-\mu)\frac{1}{N}\sum_{p=1}^{N} u_p^n \tag{3.13}$$

in which μ is an appropriate constant and N is the total number of nodal points connected to node i by the finite elements surrounding the node i. The nodal function $\tilde{u}^n$ is referred as the smoothed function in this paper.

Applications of equations (3.11) and (3.12) to equations (2.18) and (2.19) lead to the following two step explicit algorism.

The two step Lax-Wendroff scheme:

First step:

$$\bar{M}_{\alpha\beta} \cdot v_\beta^{n+1/2} = M_{\alpha\beta}\tilde{v}_\beta^n - \frac{\Delta t}{2}\{K_{\alpha\beta\gamma}v_\beta^n v_\gamma^n + H_{\alpha\beta}h_\beta^n + S_{\alpha\beta}v_\beta^n + F_\alpha^n\} \tag{3.14}$$
$$\bar{M}_{\alpha\beta}h_\beta^{n+1/2} = M_{\alpha\beta}\tilde{h}_\beta^n - \frac{\Delta t}{2}\{G_{\alpha\beta\gamma}(H_\beta + h_\beta^n)v_\alpha^n\} + G_{\alpha\beta\gamma}(b_\beta^{n+1/2} - b_\beta^n)v_\gamma^n \tag{3.15}$$

Second step:

$$\bar{M}_{\alpha\beta}v_\beta^{n+1} = M_{\alpha\beta}\tilde{v}_\beta^n - \Delta t\{K_{\alpha\beta\gamma}v_\beta^{n+1/2}v_\gamma^{n+1/2} + H_{\alpha\beta}h_\beta^{n+1/2} + S_{\alpha\beta}v_\beta^{n+1/2} + F_\alpha^{n+1/2}\} \tag{3.16}$$

$$\bar{M}_{\alpha\beta}h_\beta^{n+1} = M_{\alpha\beta}\tilde{h}_\beta^n - \Delta t\{G_{\alpha\beta\gamma}(H_\beta+h_\beta^{n+1/2})v_\gamma^{n+1/2}\} + G_{\alpha\beta\gamma}(b_\beta^{n+1}-b_\beta^n)v_\alpha^n \tag{3.17}$$

Two step Runge-Kutta scheme:

First step:

$$\bar{M}_{\alpha\beta}v_\beta^{n+2/3} = M_{\alpha\beta}\tilde{v}_\beta^n - \frac{2\Delta t}{3}\{K_{\alpha\beta\gamma}v_\beta^n v_\gamma^n + H_{\alpha\beta}h_\beta^n + S_{\alpha\beta}v_\beta^n + F_\alpha^n\} \tag{3.18}$$

$$\bar{M}_{\alpha\beta}h_\beta^{n+2/3} = M_{\alpha\beta}\tilde{h}_\beta^n - \frac{2\Delta t}{3}\{G_{\alpha\beta\gamma}(H_\beta+h_\beta^n)v_\alpha^n \;+ G_{\alpha\beta\gamma}(b_\beta^{n+2/3}-b_\beta^n)v_\alpha^n \tag{3.19}$$

Second step:

$$\begin{aligned}\bar{M}_{\alpha\beta}v_\beta^{n+1} = {} & M_{\alpha\beta}v_\beta^n - \frac{\Delta t}{4}\{(K_{\alpha\beta\gamma}v_\beta^n v_\alpha^n + H_{\alpha\beta}h_\beta^n + S_{\alpha\beta}v_\beta^n + F_\alpha^n) \\ & + 3(K_{\alpha\beta\gamma}v_\beta^{n+2/3}v_\gamma^{n+2/3} + H_{\alpha\beta}h_\beta^{n+2/3} + S_{\alpha\beta}v_\beta^{n+2/3} + F_\alpha^{n+2/3})\end{aligned} \tag{3.20}$$

$$\begin{aligned}\bar{M}_{\alpha\beta}\cdot h_\beta^{n+1} = {} & M_{\alpha\beta}h_\beta^n - \frac{\Delta t}{4}[\{G_{\alpha\beta\gamma}(H_\beta+h_\beta^n)v_\gamma^n + \frac{3}{2\Delta t}G_{\alpha\beta\gamma}(b_\beta^{n+2/3}-b_\beta^n)v_\gamma^n\} \\ & + 3\{G_{\alpha\beta\gamma}(H_\beta+h_\beta^{n+2/3})v_\gamma^{n+2/3} + \frac{3}{\Delta t}G_{\alpha\beta\gamma}(b_\beta^{n+1}-b_\beta^{n+2/3})v_\gamma^n\}]\end{aligned} \tag{3.21}$$

where $\bar{M}_{\alpha\beta}$ denotes lumped coefficient matrix and $\tilde{v}_\beta^n$ and $\tilde{h}_\beta^n$ are smoothed functions of the v_β and h_β respectively. The use of lumped coefficient matrix leads to the purely explicit scheme and moreover numerically stable computational algorisms.

STABILITY CONDITIONS

Von Neumann stability condition is investigated for one dimensional linear equation system. Consider linear shallow water wave equation:

$$\frac{\partial u}{\partial t} + g\frac{\partial h}{\partial x} = 0 \tag{4.1}$$

$$(0 \leqq t \leqq T, \quad 0 \leqq x \leqq L)$$

$$\frac{\partial h}{\partial t} + H\frac{\partial u}{\partial x} = 0 \tag{4.2}$$

where u and h denote velocity and water elevation and g and H are gravity acceleration and water depth respectively. Referring to figure (2), finite element equation is derived as follows.

$$\begin{pmatrix} \frac{\ell}{3} & \frac{\ell}{6} \\ \frac{\ell}{6} & \frac{\ell}{3} \end{pmatrix} \begin{pmatrix} \dot{u}_a \\ \dot{u}_b \end{pmatrix} + \begin{pmatrix} -\frac{g}{2} & \frac{g}{2} \\ -\frac{g}{2} & \frac{g}{2} \end{pmatrix} \begin{pmatrix} h_a \\ h_b \end{pmatrix} = \begin{pmatrix} 0 \\ 0 \end{pmatrix} \tag{4.3}$$

$$\begin{pmatrix} \frac{\ell}{3} & \frac{\ell}{6} \\ \frac{\ell}{6} & \frac{\ell}{3} \end{pmatrix} \begin{pmatrix} \dot{h}_a \\ \dot{h}_b \end{pmatrix} + \begin{pmatrix} -\frac{H}{2} & \frac{H}{2} \\ -\frac{H}{2} & \frac{H}{2} \end{pmatrix} \begin{pmatrix} u_a \\ u_b \end{pmatrix} = \begin{pmatrix} 0 \\ 0 \end{pmatrix} \tag{4.4}$$

The conventional superposition procedure leads to the finite element equation system, of which ith components are written as:

$$(\frac{\ell}{6})\dot{u}_{i-1} + (\frac{2\ell}{3})\dot{u}_i + (\frac{\ell}{6})\dot{u}_{i+1} + (-\frac{g}{2})h_{i-1} + (\frac{g}{2})h_{i+1} = 0 \tag{4.5}$$

$$(\frac{\ell}{6})\dot{h}_{i-1} + (\frac{2\ell}{3})\dot{h}_i + (\frac{\ell}{6})\dot{h}_{i+1} + (-\frac{H}{2})u_{i-1} + (\frac{H}{2})u_{i+1} = 0 \tag{4.6}$$

where ℓ represents the length of the nodal point interval. Application of two step Lax-Wendroff scheme to equations (4.5) and (4.6) leads to the following equations:

First step:

$$u_i^{n+1/2} = \frac{1}{6}u_i^n + \frac{2}{3}u_i^n + \frac{1}{6}u_{i+1}^n - \frac{\mu}{2}(-\frac{g}{2}h_{i-1}^n + \frac{g}{2}h_{i+1}^n) \tag{4.7}$$

$$h_i^{u+1/2} = \frac{1}{6}h_{i-1}^n + \frac{2}{3}h_i^n + \frac{1}{6}h_{i+1}^n - \frac{\mu}{2}(-\frac{H}{2}u_{i-1}^n + \frac{H}{2}u_{i+1}^n) \tag{4.8}$$

Second step:

$$u_i^{n+1} = \frac{1}{6}u_{i-1}^n + \frac{2}{3}u_i^n + \frac{1}{6}u_{i+1}^n - \mu(-\frac{g}{2}h_{i-1}^{n+1/2} + \frac{g}{2}h_{i+1}^{n+1/2}) \tag{4.9}$$

$$h_i^{n+1} = \frac{1}{6}h_{i-1}^n + \frac{2}{3}h_i^n + \frac{1}{6}h_{i+1}^n - \mu(-\frac{H}{2}u_{i-1}^{n+1/2} + \frac{H}{2}u_{i+1}^{n+1/2}) \tag{4.10}$$

where

$$\mu = \frac{\Delta t}{\ell} \tag{4.11}$$

and u_i^n and h_i^n denote the values of u and h at the ith node and at nth time step. Assume that the solution of equations (4.7) ~ (4.10) can be expressed as:

$$u_i^n = R^n e^{jwi} \tag{4.12}$$

$$h_i^n = S^n e^{jwi} \tag{4.13}$$

where $j = \sqrt{-1}$. Introducing equations (4.12) and (4.13) into

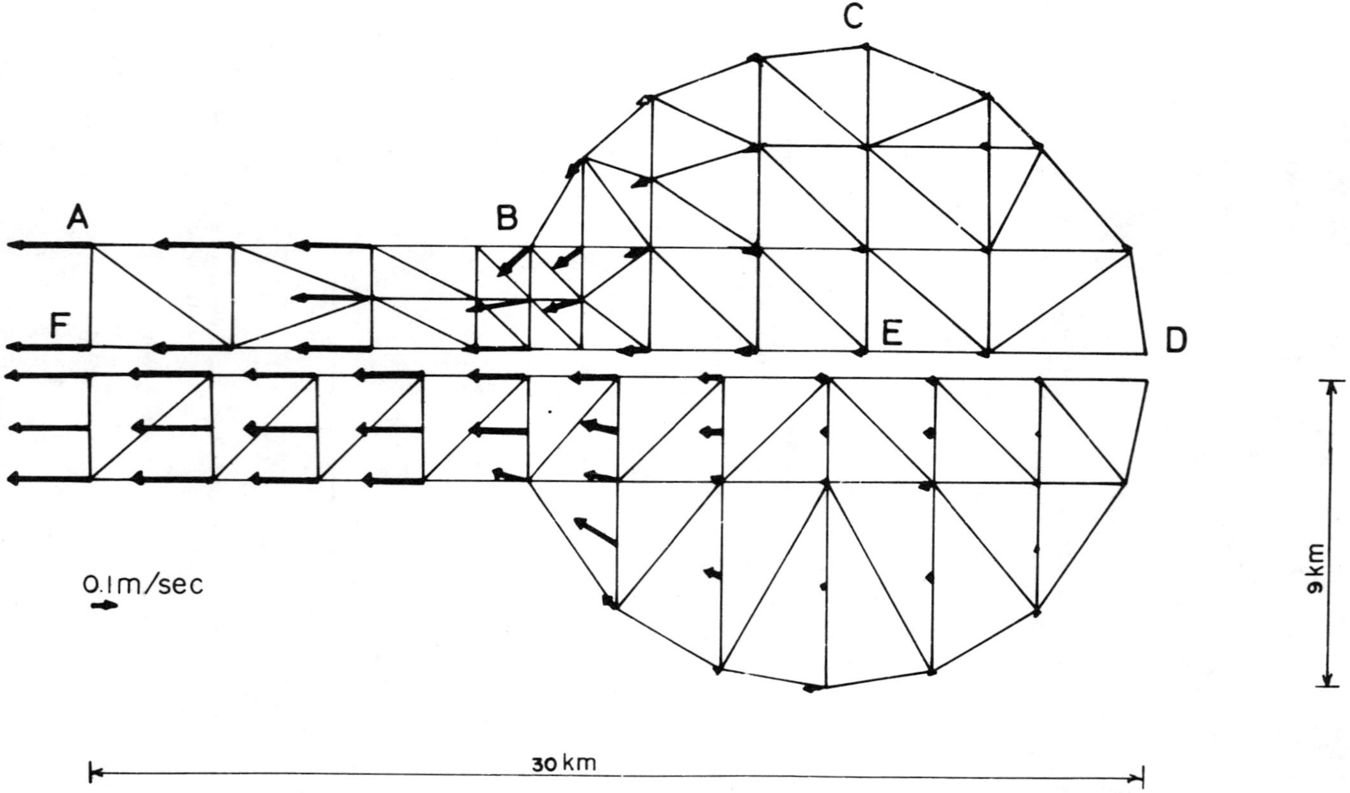

Fig.3 Computed Velocity Distribution Two Step Explicit Method and Periodic Method

equations (4.7) ~ (4.10) and rearranging the terms, the amplification relation can be derived as follows.

$$\begin{pmatrix} R^{n+1} \\ S^{n+1} \\ \hline R^{n+1/2} \\ S^{n+1/2} \end{pmatrix} = \left(\begin{array}{cc|cc} & gb & a & \\ Hb & & & a \\ \hline & & a & \frac{g}{2}b \\ & & \frac{H}{2}b & a \end{array}\right) \begin{pmatrix} R^{n+1/2} \\ S^{n+1/2} \\ \hline R^{n} \\ S^{n} \end{pmatrix} \tag{4.14}$$

where

$$a = \frac{2}{3} + \frac{1}{3}\cos\omega \qquad\qquad b\ -j\,\mu\,\sin\omega$$

Considering that the characteristic value of equation (4.14) should be less than or equal to 1, stability condition of two step Lax=Wendroff scheme can be derived as:

$$\mu \leqq \frac{\sqrt{2}}{3}\frac{1}{\sqrt{gH}} \tag{4.15}$$

In almost the same way, application of the two step Runge-Kutta scheme to equations (4.5) and (4.6) leads to the following amplification equations,

$$\begin{pmatrix} R^{n+1} \\ S^{n+1} \\ \hline R^{n+2/3} \\ S^{n+2/3} \end{pmatrix} = \left(\begin{array}{cc|cc} & \frac{g}{4} & a & \frac{3g}{4}b \\ \frac{H}{4} & & \frac{3H}{4}b & a \\ \hline & & a & \frac{2g}{3}b \\ & & \frac{2H}{3}b & a \end{array}\right) \begin{pmatrix} R^{n+2/3} \\ S^{n+2/3} \\ \hline R^{n} \\ S^{n} \end{pmatrix} \tag{4.16}$$

Using equation (4.16), stability condition of two step Runge-Kutta scheme can be obtained as:

$$\mu \leqq \frac{3}{2}\cdot\frac{1}{\sqrt{gH}} \tag{4.17}$$

To illustrate the adaptability of the present finite element schemes, several numerical investigation has been carried out. The comparison of the numerical results obtained by the Lax-Wendroff scheme and the periodic Galerkin method (Kawahara and Hasegawa [1977-a,b]) is studied. Figure (3) shows the finite element idealization and the computed velocities by both methods. At the entrance A - F, the water elevation is assumed to be given by:

$$h_\lambda = 0.5^m \sin\left(\frac{2\,\Pi}{12\text{ hour}}\,t\right) \qquad (4.18)$$

As the shape function, the quadratic polynomials for velocity and the linear polynomials for water elevation have been employed in the periodic Galerkin method, whereas, the linear polynomials are used for both velocity and water elevation in the computation by the two step explicit methods. In the figure, the upper half illustrates the results obtained by the two step Lax-Wendroff schemes and the lower half by the periodic Galerkin method. The computed numerical results are reasonably well in agreement.

In figures (4) and (5), numerical results by using various time increment Δt are compared. The computed velocity and water elevation by using two step Runge-Kutta scheme are shown in figure (4). The left half illustrates the results by the ordinary scheme (equation 3.10). The right half shows the results by the advanced scheme (equation 3.12). For comparison, the computed velocity and water elevation by periodic Galerkin method are plotted in figure (4). As is shown in the figure, one can use longer time increment Δt in case of the calculation by the advanced scheme than in case of the one by the ordinary scheme. However, it is noted that the computed velocity and water elevation by using Δt = 30 sec are resulted in too smaller values than that of the periodic Galerkin method. This seems to come from the artificial damping effects of the advanced scheme.

In figure (5), the velocity and water elevation are illustrated computed by the two step Lax-Wendroff scheme. The left half illustrates the results by the ordinary scheme and the right half represents the results by the advanced scheme. The same computed results by the ordinary scheme with Δt = 60 are obtained by the advanced scheme with Δt = 120. This shows that one can use twice longer time increment in case of the advanced scheme.

COMPUTATIONAL ASPECTS OF THE METHOD

The numerical treatment of the shallow water wave equation including the bottom upheaval was well established in the preceding chapters. In this chapter, the problems and difficulties of the practical applications of the method is discussed.

One is the computational aspect of the method. In this paper, the finite element method is constructed based on the triangulation of the given domain and the linear interpolation functions. For the practical applications, the finite element mesh subdivision of the domain consists the large amount of the nodal point coordinate and element connectivity data. For the simultaneous nonlinear differential equation in time, it should be stored on some storage units that the element co-

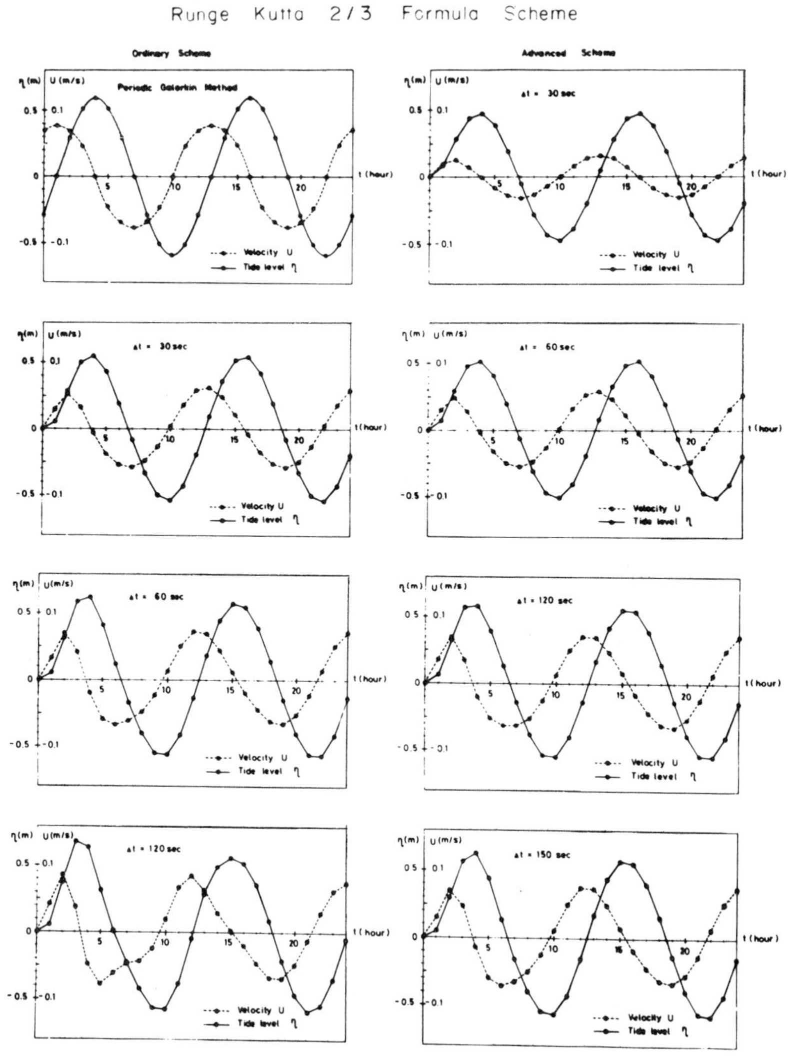

Fig.4 Computed Velocity and Tide Elevation
(Two Step Runge-Kutta Method)

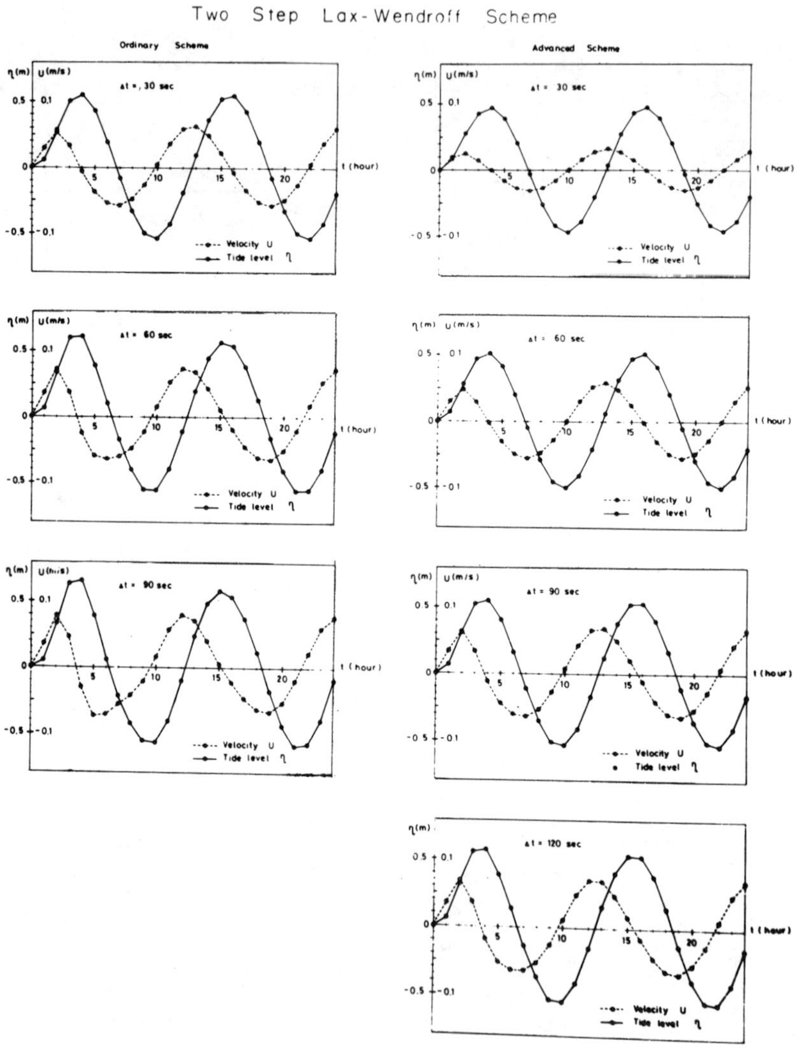

Fig.5 Computed Velocity and Tide Elevation
(Two Step Lax-Wendroff Method)

efficient matrices and element connectivity data. To keep the computation time in the executable range, the whole coefficient matrices should be stored on the core storage rather than the auxiliary storage devices. On the other hand, the restriction of the size of the core storage should be considered. The virtual storage facility allows the sufficient area for the whole matrices, but the effect of paging defects the numerical performance of the method in the computational sense. For the reduction of the requirement of the size of data area and computation time, a classification algorithm of the linear triangular elements is used for the identical elements. The data structure of the method is shown in Fig. (6).

The finite element model, as an example shown in Fig. (7), contains 6327 elements and 3396 nodes but has only 77 types of the identical element, which is used for the numerical simulation of the tsunami caused by the Great Kanto Earthquake in 1923.

For the explicit time integration, the spatial discretization is constrained by the stability condition of the scheme. In the case of linear elastic analysis, the mesh subdivision of the finite element affects the accuracy of the solution. The effect of spatial discretization is more significant in the flow analysis. Similar criteria is established for the nonlinear shallow water equation, as the linear equation derived in preceding chapter, and that implies the spatial discretization formula.

From the numerical experiments, the stability criteria for the element is

$$\Delta t \leq c \frac{\Delta x}{\sqrt{gH}} \tag{5.1}$$

and the value of c is less than 0.5. To keep the stability of the computation, the possible choice of Δt is to hold the value λ constant throughout the domain. For the two dimensional computation, the right triangle is mainly used and Δx is defined as the maximum length of it. Then, stability criteria is expanded as

$$\Delta t \leq c \min_{\Delta x \varepsilon \Omega} \left(\frac{\Delta x}{\sqrt{gH}}\right) \tag{5.2}$$

and c = 0.5 is mainly used.

For the finite element mesh, shown in Fig. (7), the minimum value of λ is obtained at the south edge of the field, and its value is about 30 sec. In the case of Δt = 20 sec the numerical solution is slightly unstable and in the case of Δt = 15 sec stable solution become obtainable. The computing time is

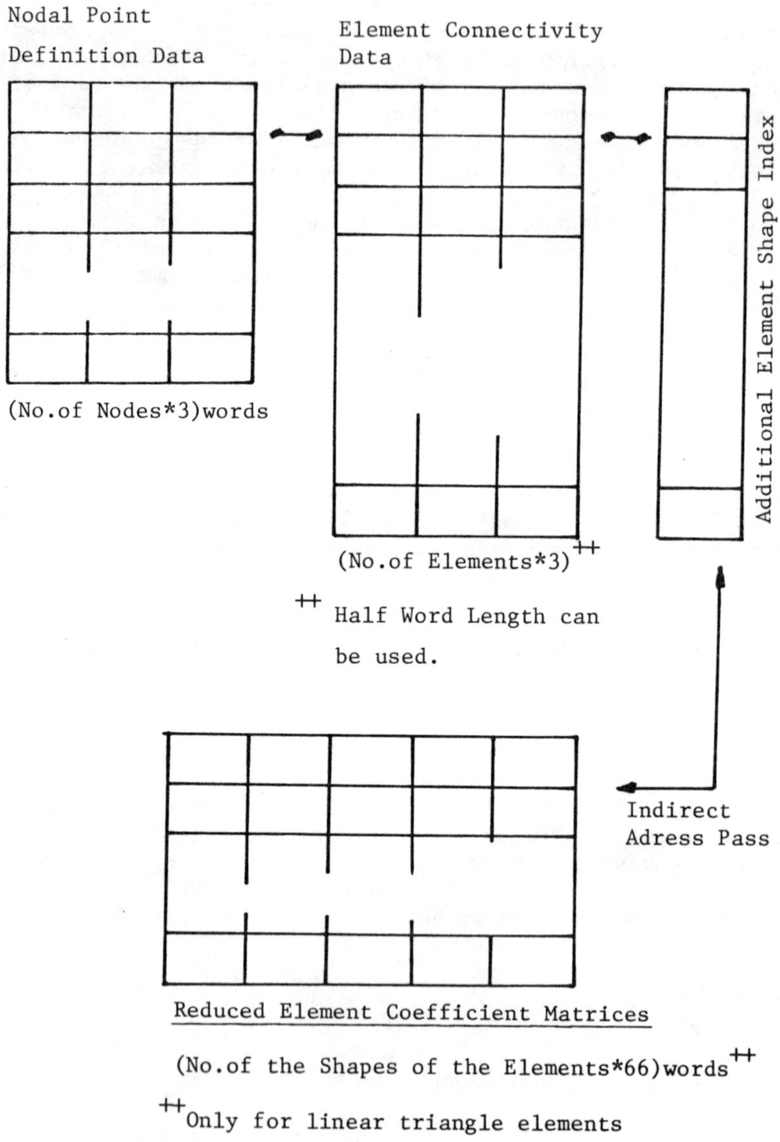

Fig.6 Proposed Data Structure for Explicit Finite Element Computation

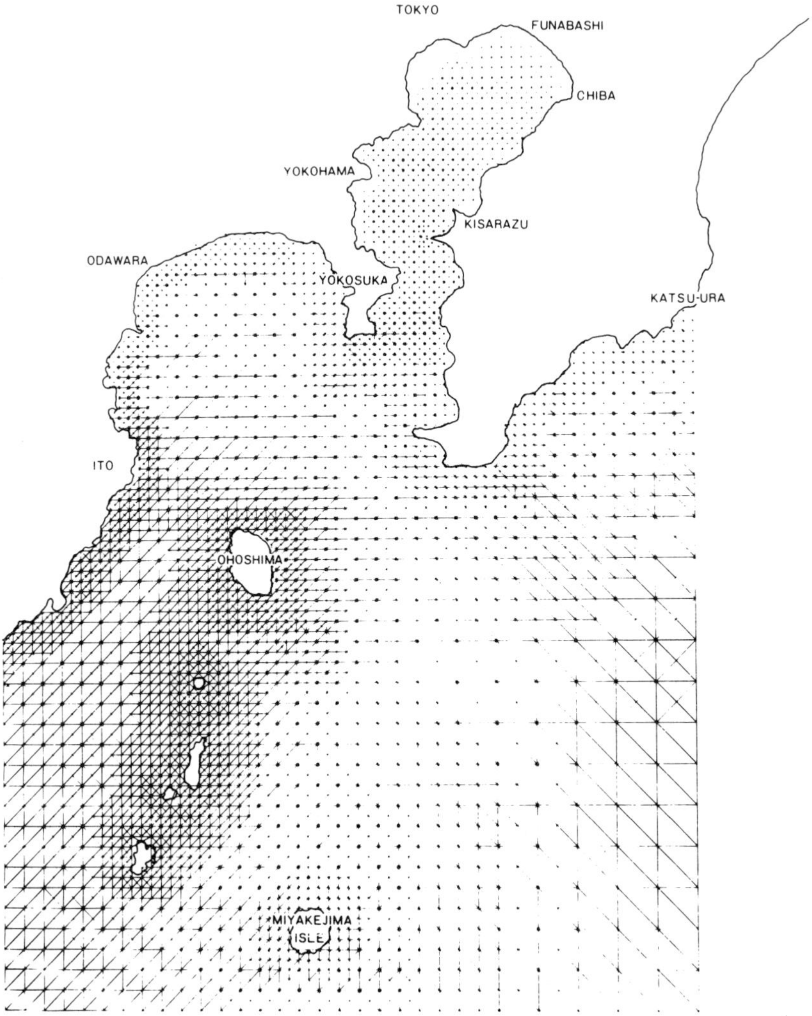

Fig.7 Finite Element Idealization for the Tsunami Wave Propagation Analysis of Great Kanto Earthquake

also examined by the Great Kanto Earthquake simulation. Two step Lax-Wendroff finite element algorithm is employed for it as the discretization of the nonlinear shallow water wave equation, and 100 step computation has taken 987 sec in CPU time by using IBM System 370-168. It is evident that the computing time of the explicit time integration depends almost linearly on the number of the element and it can be predicted by the criteria.

$$T = n \cdot m \ \Delta T \tag{5.3}$$

where T is total computing time, n, m denote the number of integration steps and the number of elements respectively, and ΔT is the computation time per element per time step. For the case of Fig. (7) by IBM 370-168, ΔT is almost 0.87 m sec. The finite element models used in this investigation is spanned from 2000 elements to 6000 element and the criteria shows good agreement with predicted computing time.

CONCLUDING REMARKS

The explicit time integration of semi-discrete Galerkin approximation of the shallow water wave equation is considered in this paper. The numerical examples, up to 6000 element finite element model, shows the numerical performance of the present method.

The stability criteria is derived for the one dimensional linearized equations, and extended to the nonlinear two dimensional problems successfully.

The present methods have been applied for the tidal computation and show good agreement with the field data. The evaluation of the computed wave height and the observed one is now in progress, and it may become possible to predict the damage of tsunami on the human lives and properties by the present method.

ACKNOWLEDGEMENT

The authors express their gratitude to Professor O.C. Zienkiewicz for his valuable discussions. The author is also thankful to Mr. Y. Shiratori, research engineer of UNIC Corporation for the painful data preparation. This work is partially supported by the Computer Center of JUSE.

LITERATURE CITED

Aida, I. [1969-a]: "Numerical Experiments for the Tsunami Propagation - 1964 Niigata tsunami and the 1968 Tokachi-oki tsunami," Bull, Earth. Res. Inst. Univ. Tokyo, Vol. 47, pp. 673-700

Aida, I. [1969-b]: "Numerical Experiments for Tsunamis Caused by Moving Deformations of Sea Bottom," Bull. Earth. Res. Inst. Univ. Tokyo, Vol. 47, pp. 849-862

Baker, A.J. [1975]: "Predictions in Environmental Hydrodynamics Using the Finite Element Method," I. Theoretical Development," AIAA J. Vol. 13, No. 1, pp. 36-42

Brebbia, C.A. and J.J. Connor [1976]: "Finite Element Techniques for Fluid Flow," Newnes-Butterworth, London

Cullen, M.J.P. [1973]: "A Simple Finite Element Method for Meteorological Problems," J. Inst. Math. Appls. Vol. 11, pp. 15-31

Cullen, M.J.P. [1977]: "The Application of Finite Element Methods to Primitive Equation of Fluid Motion," Finite Elements in Water Resources (eds. Pinder, et al.) Pentech Press, pp. 4.231-4.245

Ferrante, A.G. [1977]: "A General Purpose System for Computational Hydraulics," Finite Elements in Water Resources (eds. Pinder, et al.) Pentech Press, pp. 4.267-4.286

Gray, G.W. [1977]: "An Efficient Finite Element Schemes for Two Dimensional Surface Water Computation," Finite Elements in Water Rhsources (eds. Pinder, et al.) Pentech Press, pp. 4.33-4.49

Grotkop, G. [1973]: "Finite Element Analysis of Long-Period Water Waves," Comp. Meth. Appl. Mech. Engng. Vol. 2, pp. 147-157

Holtz, K.P. and H. Hennlich [1977]: "Numerical Experiences from the Computation of Tidal Waves by the Finite Element Method," Finite Elements in Water Resources (eds. Pinder, et al.), Pentech Press, pp. 4.19-4.31

Hwang, L.S. and D. Divoky [1970]: "Tsunami Generation," J. Geoph. Res. Vol. 75, No. 33, pp. 6802-6817

Hwang, L.S. [1976]: "Numerical Modeling of Tsunamis - Forecasting Heights in a Warning System," Topics in Ocean Engineering Vol. 3, (ed. C.H. Bretscheneider) Gulf Publishing Company

Kawahara, M., N. Takeuchi and Y. Yoshida [1977]: "Two Step Explicit Finite Element Method for Tsunami Wave Propagation Analysis," Int. J. Num. Meth. Engng. (to be published)

Kawahara, M. [1977]: "Steady and Unsteady Finite Element Analysis of Incompressible Viscous Fluid" Finite Elements in Fluids (to be published)

Kawahara, M. and K. Hasegawa [1977]: "Periodic Galerkin Finite Element Method for Tidal Flow Analysis," Int. J. Num. Meth. Engng. (to be published)

Kawahara, M., K. Hasegawa and Y. Kawanago [1977-b]: "Periodic Tidal Flow Analysis by Finite Element Perturbation Method," Comp. Fluid. (to be published)

Pinder, G.F. and W.G. Gray [1977]: "Finite Element Simulation in Surface and Subsurface Hydrology," Academic Press, New York

Pinder, G.F., W.G. Gray and C.A. Brebbia [1977]: "Finite Elements in Water Resources," Pentech Press, London

Sündermann, J. [1977]: "Computation of Barotoropic Tides by

the Finite Element Method," Finite Elements in Water Resources (eds. Pinder, et al.) Pentech Press, pp. 4.51-4.67

Taylor, C. and J.M. Davis [1975]: "Tidal and Long Wage Propagation - A Finite Element Approach," Comp. Fluid, Vol. 3, pp. 125-148

Taylor, C. and J.M. Davis [1975]: "Tidal Propagation and Dispersion in Estuaries," Finite Elements in Fluids, John Wiley & Sons, pp. 95-118

Wang, H.P. [1977]: "Multi-level Finite Element Hydrodynamic Model of Block Island Sound," Finite Elements in Water Resources (eds. Pinder, et al.) Pentech Press, pp. 4.69-4.93

WAVES GENERATED BY LANDSLIDE IN LAKES OR BAYS

C.G. Koutitas, T.S. Xanthopoulos

National Technical University of Athens

INTRODUCTION

The prediction of the generation and propagation of the waves due to a major landslide in an initially still water mass of a lake or bay is a combined soil mechanics and hydraulics problem. For reasons of simplicity the landslide is examined initially as a geological phenomenon and the deriving soil motion is subsequently applied on the flow field as a forcing factor generating the free surface disturbances. The case of a long narrow channel where the lateral landslide causes a sudden change of its cross section was previously examined [4] as a 1-D phenomenon. The present study deals with the creation and propagation of waves in finite three dimensional flow domains due to inflow of solid masses through a portion of their perimeter. The solution of the mathematical model is achieved by the Galerkin finite elements technique applied directly on the flow equations.

THE MATHEMATICAL MODEL

The mathematical model consists of the equilibrium and mass continuity equations which under the assumptions of the theory of long waves have the form,

$$F_1(u,v,\eta) = u_t + uu_x + vu_y + g\eta_x + \frac{\tau_x}{\rho H} = 0 \qquad (1)$$

$$F_2(u,v,\eta) = v_t + uv_x + vv_y + g\eta_y + \frac{\tau_y}{\rho H} = 0 \qquad (2)$$

$$F_3(u,v,\eta) = \eta_t + (uH)_x + (vH)_y = 0 \qquad (3)$$

where according to the notations of Fig. 1 u, v are the depth mean velocity components, η the free surface elevation, H the total depth and τ_x, τ_y the bottom shear stress components given by

$$\tau_x = C\cdot u\cdot\sqrt{u^2 + v^2} \qquad (4a)$$

$$\tau_y = C\cdot v\cdot\sqrt{u^2 + v^2} \qquad (4b)$$

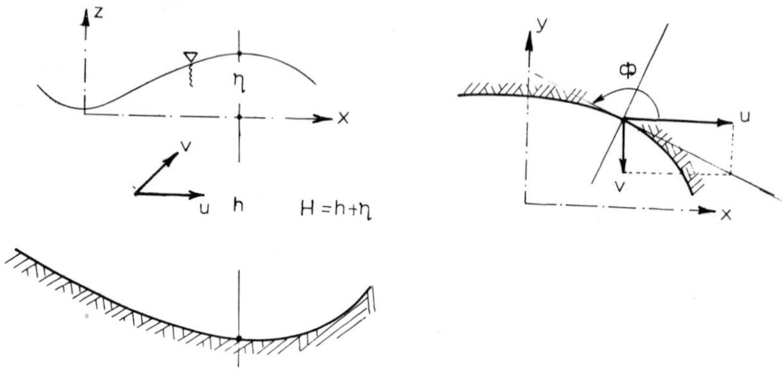

Figure 1 Notations for flow domain and boundaries.

The boundary conditions of the problem can be classified as follows:
a. Coastal boundaries. The supression of the velocity component normal to the boundary is imposed on them; thus and according to Fig. 1 a relation between the velocity components derive,

$$v = u\cdot tg\varphi \qquad (5)$$

b. Land slide boundary. On this boundary the landslide is described in an Eulerian frame by means of a solid dicharge (in magnitude and direction) diagram deriving from the edaphologic study of the unstable bank. The water depth is calculated by means of the continuity equation and the velocity by the relation

$$\upsilon_s = q_s/H \qquad (6)$$

where q_s the solid dicharge per unit length of the boundary.
c. Dam boundary. In case of a reservoir upstream of a dam as soon as the water depth becomes greater than the dam height H_D, the dam crest functions as a weir. The difference $H\text{-}H_D = h_w$ is used as the head to calculate the over flowing dicharge per unit length of crest

$$q = K\cdot h_w^n \qquad (7)$$

where K, n are coefficients depending on the dam shape.
d. Open sea boundary. In case of a landslide in a bay there is a ficticious line beyond which the waves must be left free to propagate towards the open sea without reflection. On the open

sea boundary the mass continuity principle provides a simple sufficient relation between H and υ_n the normal velocity

$$H\cdot\upsilon_n = \sqrt{gH}\cdot\eta \tag{8}$$

APPLICATION OF GALERKIN - F.E. METHOD

The flow field is discretised in triangular linear elements. The unknowns are the u, v, η nodal values. For an element (e) of nodes i, j, k the linear forms of the u, v, η functions are

$$\hat{u} = [N]^{(e)}\ \{u\}^{(e)} \tag{9}$$

$$\hat{v} = [N]^{(e)}\ \{v\}^{(e)} \tag{10}$$

$$\hat{\eta} = [N]^{(e)}\ \{\eta\}^{(e)} \tag{11}$$

where [N] the row matrix of the shape functions $[N] = [N_i, N_j, N_k]$ and {u}, {v}, {η} the column matrices of nodal values. The substitution of (9) (10) (11) in (1) (2) (3) gives raise to residuals:

$$F_m(\hat{u},\ \hat{v},\ \hat{\eta}) = R_m \qquad m = 1,\ 2,\ 3 \tag{12}$$

The Galerkin approach to the minimisation of R_1, R_2, R_3, [5] leads to the relations

$$\int_{\Omega^{(e)}} R_m\ N_\ell\ d\Omega = 0 \tag{13}$$

where m = 1, 2, 3 and ℓ = i, j, k and $\Omega^{(e)}$ is the surface of element (e). The time derivatives are replaced by forward differemces,

$$\frac{\partial A}{\partial t} = \frac{A - A_o}{\Delta t} \tag{14}$$

where the index o denotes value at the previous time step. The analytical form of equations (13) becomes:

$$\int_{\Omega^{(e)}} [\frac{1}{\Delta t}\ [N]\ N_\ell\ \{u\} - \frac{1}{\Delta t}\ [N]\ N_\ell\ \{u_o\} + [N]\ N_\ell\ \{u_o\}\ \frac{\partial}{\partial x}\ [N]\cdot\{u\}$$

$$+ [N]\ N_\ell\ \{v_o\}\ \frac{\partial}{\partial y}\ [N]\ \{u\} + g\ N_\ell\ \frac{\partial}{\partial x}\ [N]\{\eta_o\} + [N]\ N_\ell\ \{\frac{\tau_{xo}}{\rho H_o}\}]$$

$$d\Omega = 0 \tag{15}$$

$$\int_{\Omega^{(e)}} [\frac{1}{\Delta t}\ [N]\ N_\ell\ \{v\} - \frac{1}{\Delta t}\ [N]\ N_\ell\ \{v_o\} + [N]\ N_\ell\ \{u_o\}\ \frac{\partial}{\partial x}\ [N]\ \{v\}$$

$$+ [N]\ N_\ell\ \{v_o\}\ \frac{\partial}{\partial y}\ [N]\ \{v\} + g\ N_\ell\ \frac{\partial}{\partial y}\ [N]\{\eta_o\} + [N]\ N_\ell\ \{\frac{\tau_{yo}}{\rho H_o}\}]$$

$$d\Omega = 0 \qquad (16)$$

$$\int_{\Omega^{(e)}} [\frac{1}{\Delta t} [N] N_\ell \{\eta\} - \frac{1}{\Delta t} [N] N_\ell \{\eta_o\} + [N] N_\ell \{u_o\} \frac{\partial}{\partial x} [N]\{\eta\} +$$

$$[N] N_\ell \{\eta\} \frac{\partial}{\partial x} [N] \{u_o\} + [N] N_\ell \{v_o\} \frac{\partial}{\partial y} [N] \{\eta\} + [N] N_\ell \{\eta\}\frac{\partial}{\partial y}$$

$$[N]\{v_o\}+ [N]N_\ell\{u_o\} \frac{\partial}{\partial x} [N]\{h\}+ [N]N_\ell\{v_o\} \frac{\partial}{\partial y}[N]\{h\}+ [N]N_\ell\{h\}$$

$$(\frac{\partial}{\partial x} [N]\{u_o\} + \frac{\partial}{\partial y} [N] \{v_o\})] \, d\Omega = 0 \qquad (17)$$

where $\ell = i, j, k$. The relations (15) (16) (17) are algebraic systems of the form

$$[A]^{(e)} \{u\}^{(e)} = \{f\}^{(e)} \qquad (18)$$

$$[A]^{(e)} \{v\}^{(e)} = \{g\}^{(e)} \qquad (19)$$

$$[C]^{(e)} \{\eta\}^{(e)} = \{p\}^{(e)} \qquad (20)$$

holding for the element (e)
where,

$$A_{ij} = \int_{\Omega^{(e)}} [\frac{N_i N_j}{\Delta t} + N_m N_i u_{om} \frac{\partial}{\partial x} N_j + N_m v_{om} N_i \frac{\partial}{\partial y} N_j] \, d\Omega \qquad (21)$$

$$C_{ij} = \int_{\Omega^{(e)}} [\frac{N_i N_j}{\Delta t} + N_m u_{om} N_i \frac{\partial}{\partial x} N_j + N_m v_{om} N_i \frac{\partial N_j}{\partial y}$$

$$+ N_i N_j \frac{\partial}{\partial x} N_m u_{om} + N_i N_j \frac{\partial}{\partial y} N_m v_{om}] \, d\Omega \qquad (22)$$

$$f_i = \int_{\Omega^{(e)}} [\frac{N_m N_i u_{om}}{\Delta t} - gN_i \frac{\partial}{\partial x} N_m \eta_{om} - N_m N_i \frac{\tau_{xom}}{\rho H_{om}}] \, d\Omega \qquad (23)$$

$$g_i = \int_{\Omega^{(e)}} [\frac{N_m N_i v_{om}}{\Delta t} - gN_i \frac{\partial N_m}{\partial y} \eta_{om} - N_m N_i \frac{\tau_{yom}}{\rho H_{om}}] \, d\Omega \qquad (24)$$

$$p_i = \int_{\Omega^{(e)}} [\frac{N_i N_m \eta_{om}}{\Delta t} - N_m N_i u_{om} \frac{\partial}{\partial x} N_m h_m -$$

$$N_m N_i v_{om} \frac{\partial}{\partial y} N_m h_m + N_m h_m N_i (\frac{\partial}{\partial x} N_m u_{om} +$$

$$\frac{\partial}{\partial y} N_m v_{om})] \, d\Omega \qquad (25)$$

where m is summation index. The global matrices derive from the synthesis of relations (18) (19) (20) for all the elements of the flow field [3]. The global {u}{v}{η} vectors contain all the

unknown nodal values of u, v, η. The deriving algebraic systems are solved by Gauss Seidel iteration according to the following procedure:
If n is the time index ($t_n = n \cdot \Delta t$), the integration of the systems giving u, v values for time n+1 (as functions of u, v values for time n and η values for time n+1/2) are solved first, succeeded by the solution of the system giving η values for time n+3/2 (as functions of u, v values for time n+1 and η values for time n+1/2). As the u_o v_o η_o are used as a first approximation for the computation of u, v, η, the number of iterations is small (10 at most for a relative error 1 °/oo).

APPLICATION ON A REAL FLOW FIELD

For a long narrow reservoir upstream of Polifiton earth dam in Greece existed a previous study for one dimensional wave propagation due to a possible landslide at a distance of some hundred meters upstream of the dam. Due to the vicinity to the dam of the landslide area there is no space available for the formation and propagation of a one dimensional wave and the two dimensional wave study was consideral indispensable. The previous solid dicharge - time diagram was adopted, as the assumptions on which it was based were considered as adequate. Its form is given in Fig. 2.

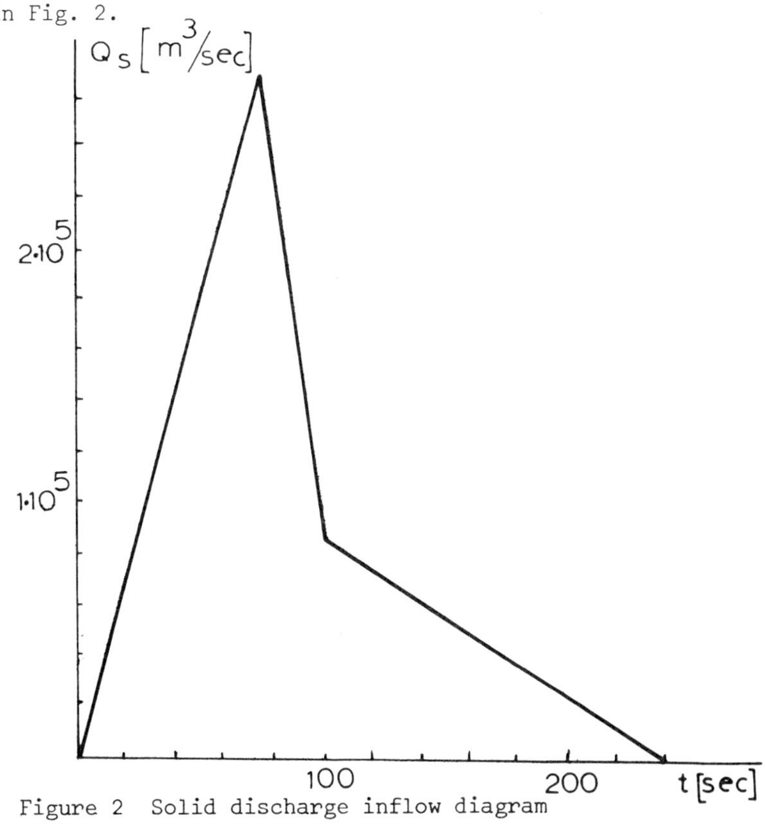

Figure 2 Solid discharge inflow diagram

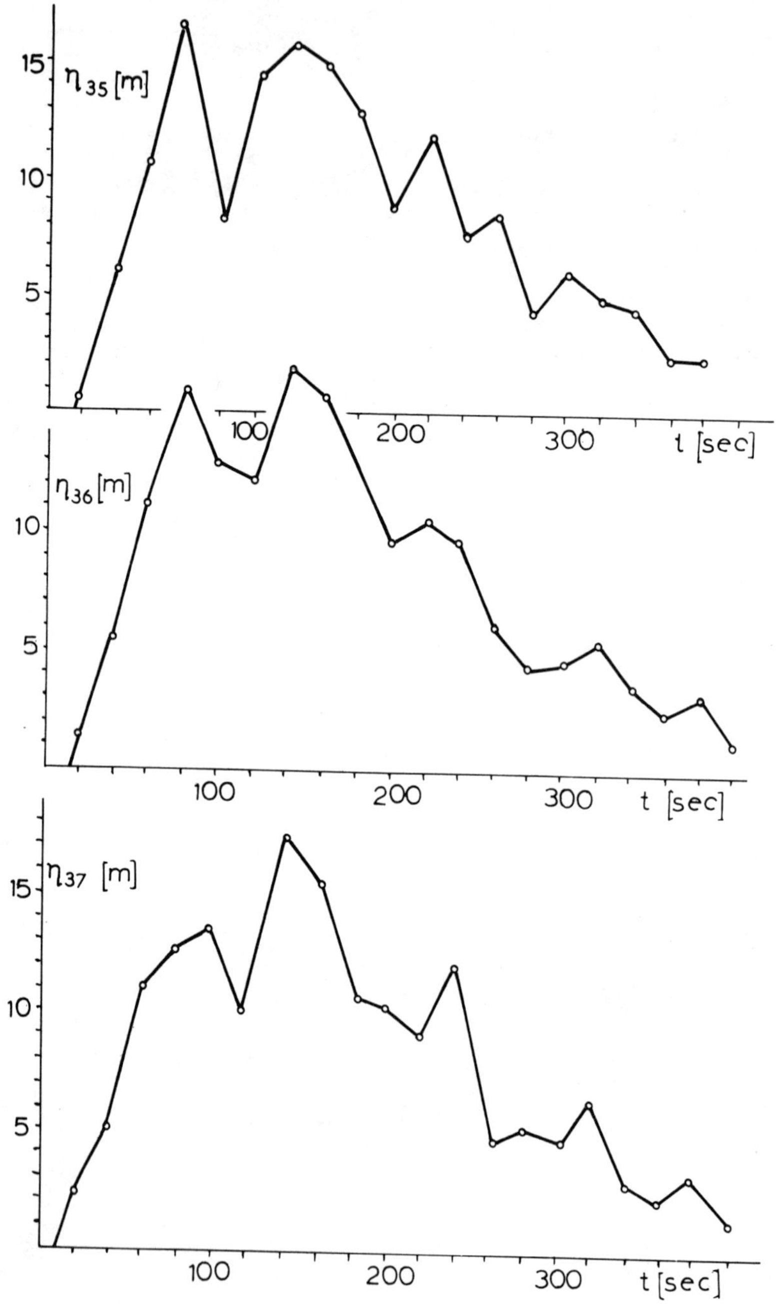

Figure 4 Free surface variations at points 35, 36, 37

The flow domain was discretised in triangular elements as is shown in Fig. 3.

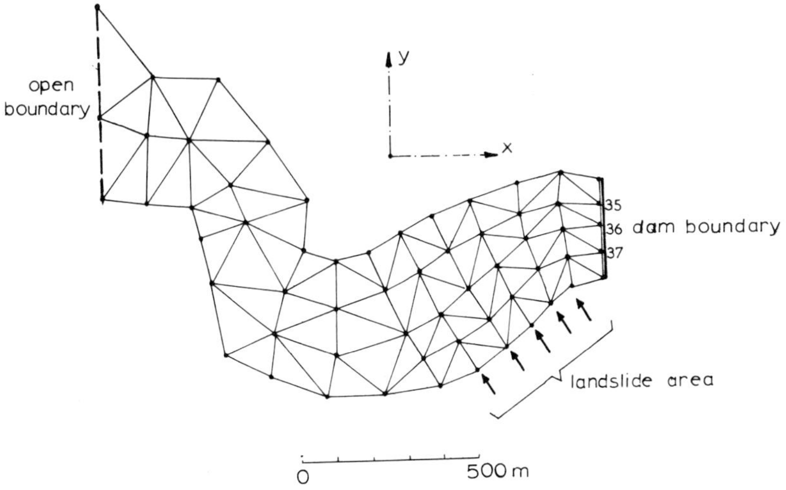

Figure 3 Flow field discretisation in triangular.

The used time step was 2 sec and C = 0,1. The free surface variations at points 1, 2, 3 along the dam are depicted in Fig.4. The computer time required for 60 nodes and 200 time steps on the UNIVAC 1106 of Thessaloniki University was 3 minutes.

CONCLUSIVE REMARKS

The developed implicit algorithm for the investigation of the two dimensional waves generated by landslides combines the simplicity of formulation that characterises the Galerkin method with the well known capability of description of flow domains of irregular geometry that characterises the finite elements method. Although it is a simple and effective mathematical tool for engineering practice it is subject to the accuracy limitations preimposed by the assumptions regarding the landslide procedure. The control of the results by comparison to the laboratory reproduction of the combined geologic - hydraulic phenomenon is consider necessary.

REFERENCES

1. Davidson D. Mc Cartney B "Water waves generated by landslides in reservoirs" ASCE J. Hydr. Div. 101, Dec. 1975.
2. Gray W., Pinder G. "On the relationship between the finite element and finite differences method" Int. J. for num. methods in Eng. Vol. 12, No 9, 1976.

3. Huebner K. "The finite element method for engineers" Wiley Interscience 1975.
4. Koutitas C. "Finite element approach to waves due to landslides" ASCE J. Hydr. Div. HY 9, Sept. 1977.
5. Pinder G., Frind E. "Application of Galerkin's procedure to awuifer analysis" Water Resources Res. 8, 108-120 (1972).
6. Zienkiewicz O. "The finite element method in engineering science" Mc Graw Hill, 1971.

LONG WAVE SIMULATIONS USING A FINITE ELEMENT MODEL

J. R. HOUSTON, D. G. OUTLAW

Waterways Experiment Station, Vicksburg, MS 39180

INTRODUCTION

Tsunamis and harbor resonance are two long-wave phenomena for which it is especially advantageous to use finite element numerical methods. Tsunamis have exceptionally long wavelengths in the deep ocean (several hundred kilometers) and when they enter shallow water their wavelengths decrease by orders of magnitude. Since finite element techniques allow elements in a numerical grid to change size and shape rapidly, a constant resolution of tsunami waves can be maintained during propagation from deep to shallow water by telescoping elements of the numerical grid. The arbitrary size and shape of elements of finite element grids also allow accurate representation of complicated land-water boundaries and, thus, the finite element method can be used to simulate the resonant response to long wave excitation of complicated harbors.

Tsunamis interacting with islands that have short continental shelves and waves exciting resonant responses of harbors are both governed by linear long-wave equations. A short continental shelf limits the time available for nonlinear and dissipative effects to develop in the case of tsunami propagation from the deep ocean to shore. Long waves that excite resonant responses of harbors and subsequent ship surging motions are known to have amplitudes of only a few centimeters. Nonlinear effects thus are negligible, although dissipative effects may be important.

This paper illustrates the use of a finite element solution of the linear long wave equations to investigate both tsunami interactions with the Hawaiian Islands and the resonant response to long wave excitation of the harbors complex of Los Angeles and Long Beach, California. Simulations of two historical tsunamis using the finite element numerical model are compared with four tide gage recordings of these tsunamis in the Hawaiian Islands. Numerical model simulations of the response of the Los Angeles and Long Beach (LA-LB) Harbors complex to long-wave excitation are compared with hydraulic model tests for this harbor.

NUMERICAL MODEL

The finite element model described in this paper is a modification of a model developed by Chen and Mei (1974) for harbor oscillation studies. The model solves the following generalized Helmholtz equation:

$$\nabla \cdot [h(x,y)\nabla\phi(x,y)] + \frac{\omega^2}{g}\phi(x,y) = 0 \qquad (1)$$

where $\phi(x,y)$ is the velocity potential defined by $U(x,y) = -\nabla\phi(x,y)$, with $U(x,y)$ being a two-dimensional velocity vector, ω an angular frequency, $h(x,y)$ the water depth, and g the acceleration due to gravity. Equation (1) is derived from the linear long-wave equation under the assumptions of irrotational flow and time-periodic motions and governs small amplitude undamped long waves in a region of variable water depth. Beyond the variable depth region covered by the finite element grid, it is assumed that there is a constant-depth ocean of infinite extent for the case of tsunami interaction with islands and of semi-infinite extent for the case of long-wave interaction with a harbor located along a straight and infinite coast. In the constant depth region the following Helmholtz equation governs propagation:

$$\nabla^2\phi(x,y) + \frac{\omega^2}{gh}\phi(x,y) = 0 \qquad (2)$$

Chen and Mei used a calculus of variations approach and obtained a Euler-Lagrange formulation of the boundary value problem. The following functional with the property that it is stationary with respect to arbitrary first variations of $\phi(x,y)$ was constructed by Chen and Mei:

$$F(\phi) = \iint 1/2\,[h(\nabla\phi)^2 - \frac{\omega^2}{g}\phi^2]\,dP + 1/2\,\S[h(\phi_R-\phi_I)\frac{\partial(\phi_R-\phi_I)}{\partial n_a}]\,da - \S[h\phi_a\frac{\partial(\phi_R-\phi_I)}{\partial n_a}]\,da - \S[h\phi_a\frac{\partial\phi_I}{\partial n_a}]\,da + \S[h\phi_I\frac{\partial(\phi_R-\phi_I)}{\partial n_a}]\,da \qquad (3)$$

where

P = the region containing the islands
$\S$ = line integral
ϕ_R = far-field velocity potential
ϕ_I = velocity potential of the incident wave
n_a = unit normal vector outward from region P

a = boundary of region P

ϕ_a = total velocity potential evaluated on boundary a

Proof was given by Chen and Mei that the stationarity of this functional is equivalent to the original boundary value problem.

The integral equation obtained from extremizing the functional is solved by using the finite element method. A linear interpolation function is used in the present study within the variable depth region because the element sizes are much smaller than local wavelengths. In the constant depth infinite region outside the variable depth region containing the islands, the scattered waves produced by the waves incident upon the islands satisfy Equation (2) and, therefore, have a velocity potential given by

$$\phi = \sum_{n=0}^{\infty} H_n(kr)(\alpha_n \cos n\theta + \beta_n \sin n\theta) \tag{4}$$

where α_n and β_n are unknown coefficients, $H_n(kr)$ are Hankel funtions of the first kind of order n, k is a wave number, and r and θ are spherical coordinates. In the constant depth semi-infinite region outside the variable depth harbor, the scattered waves satisfy Equation (2) and have a velocity potential given by

$$\phi = \sum_{n=0}^{\infty} H_n(kr)\alpha_n \cos n\theta \tag{5}$$

The velocity potential ϕ satisfies the Sommerfeld radiation condition that the scattered wave must behave as an outgoing wave at infinity. This condition may be expressed mathematically as

$$\lim_{r\to\infty} \sqrt{r}\left(\frac{\partial}{\partial r} - ik\right)\phi = 0 \tag{6}$$

where i represents an imaginary number.

The constant depth infinite or semi-infinite region is considered to be a single element with an "interpolation function" given by either Equation (4) or (5). The infinite series is terminated at some finite value such that the addition of further terms does not significantly influence the calculated values of $\phi(x,y)$. The resulting equation is combined with the system of equations that arise from the finite element formulation for unknown parameters at node points within the region containing the islands, and this complete system is solved using Gaussian elimination matrix methods.

The free surface elevation, $\eta(x,y)$, is related to $\phi(x,y)$ through the linearized dynamic free surface boundary condition expressed as

$$\eta(x,y) = \frac{1}{g}\left(\frac{\partial\phi(x,y)}{\partial t}\right) \tag{7}$$

The finite element method described in this paper is a time-harmonic solution of the boundary value problem. The interation of an arbitrary tsunami with a group of islands can easily be determined within the framework of a linearized theory. For example, an arbitrary tsunami in the deep ocean can be Fourier decomposed as follows:

$$b_o(t) = \int_{\infty}^{\infty} b(\omega)\, e^{-i[\omega t+\rho(\omega)]}\, d\omega \tag{8}$$

where

b_o = incident wave amplitude
$b(\omega)$ = amplitude of frequency component ω
$\rho(\omega)$ = phase angle

If $\eta(x,y,\omega)$ is the response amplitude an any point (x,y) along the island coasts due to an incident plane wave of unit amplitude and frequency ω, then the response to the arbitrary tsunami time history $b_o(t)$ is given by

$$\xi(x,y,t) = R_e\left[\int_{\infty}^{\infty} b(\omega)\eta(x,y,\omega)\, e^{-i\{\omega t+\rho(\omega)\}}\, d\omega\right] \tag{9}$$

where the operation $R_e[\ ]$ takes the real part of the braced quantity.

Therefore, as soon as $\eta(x,y,\omega)$ is known for all ω, the response to an arbitrary tsunami can be calculated. Of course, it is not feasible to calculate the integrals of Equations (8) and (9) over all frequencies. Instead, the frequency range is discretized and the integrals replaced by sums over a frequency range containing most of the energy of the tsunami.

The numerical model of Chen and Mei (1974) was modified before the investigations reported in this paper were conducted. Slight modifications were necessary to allow the model to handle problems with variable depths, since Chen and Mei had applied their numerical model only to constant depth problems. The subroutine that solves the large system of algebraic equations arising from the finite element formulation was modified to be more efficient by taking advantage of the sparseness (many zero terms) of the system of equations. This modification resulted in a decrease in the computational time of the numerical model by an order of magnitude for the grid covering the Hawaiian Islands and by a factor of approximately 2 or 3 for the grids covering the LA-LB Harbors complex. The solution of the algebraic equations also had to be modified so that calculations involved only small blocks of equations at a time since there were more than 800,000 terms in the system of equations for the case of tsunami interaction with the Hawaiian Islands.

NUMERICAL GRID FOR TSUNAMIS

The finite element grid used to study the interaction of tsunamis with the Hawaiian Islands is shown in Figure 1. Elements of the grid telescope from large sizes in the deep ocean to triangles with areas as small as 0.5 square kilometers in shallow coastal waters. The grid covers a region that has an area of over 400,000 square kilometers. In general, elements may be any arbitrary shape (e.g. quadrilateral) provided that the ratio of the lengths of the smallest and largest sides is not extreme. Triangular shapes were used for convenience in the present study.

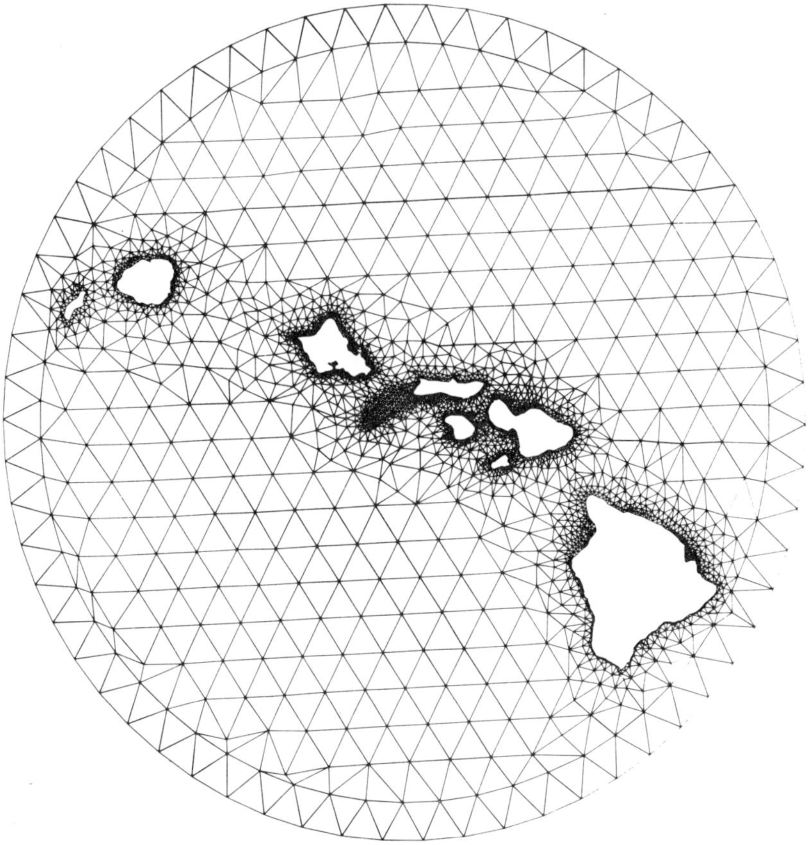

Figure 1. Finite element grid for Hawaiian Islands.

The geometric shapes of the eight islands comprising the Hawaiian Island chain obviously are modeled very precisely by the finite element grid of Figure 1. Extremely rapid bathymetric changes also are modeled very accurately by the grid. Furthermore, the number of node points along the shorelines of the islands is very dense. Since wave heights are calculated at node points by the finite element model, the model can adequately resolve the rapid wave-height variations along coastlines that are known to occur during tsunami activity in the Hawaiian Islands.

The computational time requirements of the finite element model for the grid of Figure 1 are very modest, making it economically feasible to determine the interaction of arbitrary tsunamis with the Hawaiian Islands, as discussed in the next section. The reason that the computational time required by the finite element model is small is that the grid uses small elements only in areas where they are necessary. The grid of Figure 1 has approximately 2,500 node points, whereas the finite difference grid used by Bernard and Vastano (1977) to study tsunami interaction with the Hawaiian Islands had 26,000 grid points. Even so, some of the elements of Figure 1 are as much as 60 times smaller than the finite difference grid cells used by Bernard, thus, allowing the first quantitative numerical simulations of nearshore propagation of actual tsunamis.

Figure 2 shows blow-ups of the finite element grid in the vicinities of the cities of Kahului, Honolulu, and Hilo. These three cities have tide gage recordings of the 1964 tsunami, and Honolulu has a recording of the 1960 tsunami. The locations of the tide gages are shown in the figure.

MODEL VERIFICATION FOR TSUNAMIS

The finite element model was verified in the present study by comparing numerical model simulations of the 1960 Chilean and 1964 Alaskan tsunamis with tide-gage recordings of these tsunamis in the Hawaiian Islands. These two tsunamis are the only major tsunamis for which reliable information concerning source-generating characteristics exists, with much more information existing for the Alaskan source.

A deepwater recording of a tsunami far from the perturbing influences of a coastal area has never been made. Therefore, prototype wave records in deep water of the 1960 Chilean and 1964 Alaskan tsunamis are not available for use as input to verify the finite element model. However, a finite difference numerical model has been used in previous studies (Hwang, 1970 and Houston, 1974) to simulate the uplift deformation of the ocean water surface caused by the permanent vertical displacement of the ocean bottom during an earthquake and the subsequent propagation of the resulting tsunami across the deep ocean. The permanent deformation (permanent in the sense that the time scale associated with it is much longer than the period of the tsunami) is considered to be the important parameter governing far-field wave characteristics and not the transient movements within the time-history of the ground motion. These transient movements occur over periods of time of the order of seconds, whereas, tsunami

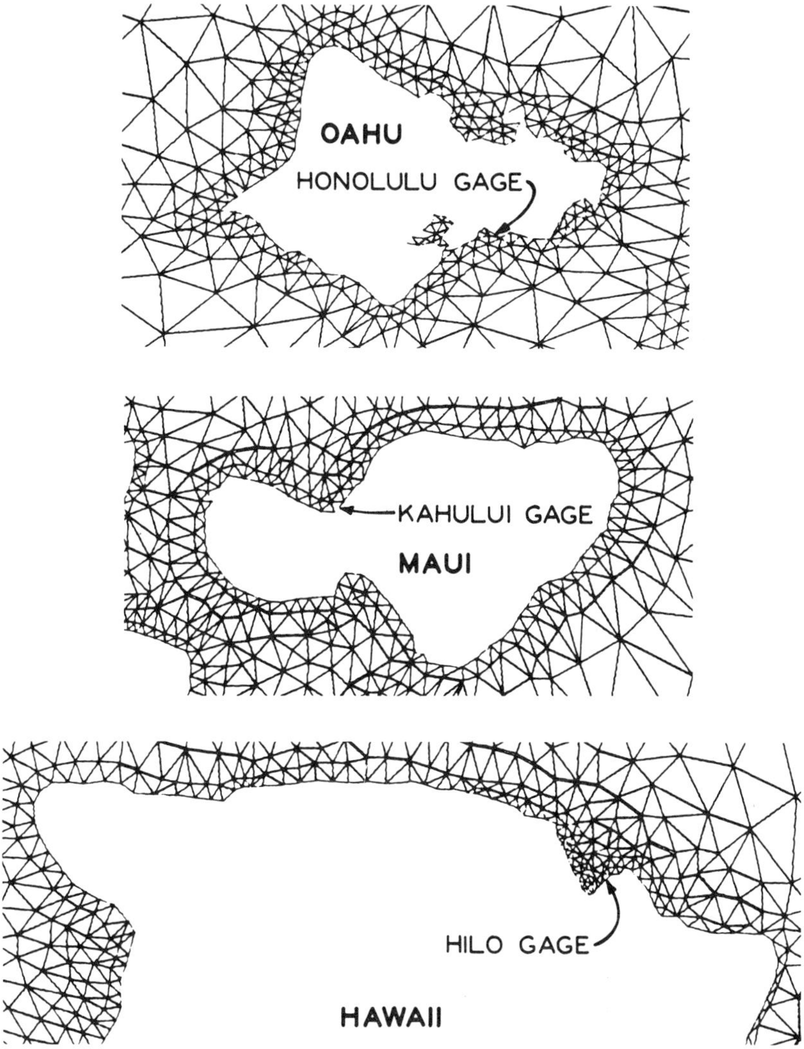

Figure 2. Gage locations.

wave periods are of the order of tens of minutes. Experimental investigations of tsunami generation by Hammack (1972) show that for spatially large ground displacements occurring over a short period of time, the transient ground movements do not influence far-field characteristics of resulting tsunamis. The permanent deformation of the ocean's bottom in the source region as a function of spatial location was taken from Plafker (1969) for the Alaskan source and Plafker and Savage (1970) and Hwang (1970) for the Chilean source. Grids with cells measuring 1/3° by 1/3° were used for the calculations.

Equation (8) involves a Fourier decomposition of a time series. This decomposition was accomplished for the time-histories of the 1960 Chilean and 1964 Alaskan tsunamis using a least squares harmonic fitting procedure. The time-history of the Alaskan tsunami in deep water was decomposed into 18 components with periods ranging from 14.5 minutes to the time length of the record. The variance of the residual (difference between the actual record and a recomposition of the 18 components) was approximately 0.2 percent of the variance of the record. Therefore, virtually all the energy in the wave record was contained in the 18 components. The time-history of the Chilean tsunami in deep water was decomposed into 11 components with periods ranging from 15.5 minutes to the time length of the record. The variance of the residual was less than 0.1 percent of the variance of the record. For both cases, the original time-history and a time-history constructed from a recomposition of the components were virtually indistinguishable.

Figures 3-5 present comparisons between tide gage recordings and numerical model calculations of the 1964 Alaskan tsunami at Kahului, Maui; Honolulu, Oahu; and Hilo, Hawaii, respectively. The largest waves recorded at each of these sites are shown. The waves arriving at later times are all much smaller than the waves shown. Whenever tide gage limits were encountered, the recordings were linearly extended. The tide gage locations are shown in Figure 2.

The comparisons shown in these three figures are in remarkable agreement, especially considering the fact that the ground displacement of the 1964 earthquake was not precisely known. Since the Hilo breakwater was not included in the numerical model grid, the numerical model calculations for Hilo are probably too large; however, this breakwater was undoubtedly highly permeable to the 1964 tsunami. The numerical model results appear to be too large in Hilo by some constant factor, since the tide gage recording and the numerical model calculations have the same form with the first wave crest and trough having approximately the same amplitude and being proportionately smaller than the second crest.

Figure 6 shows a comparison between the tide gage recording of the 1960 Chilean tsunami at Honolulu and the numerical model calculations. The Honolulu gage was the only tide gage in the Hawaiian Islands not destroyed by the 1960

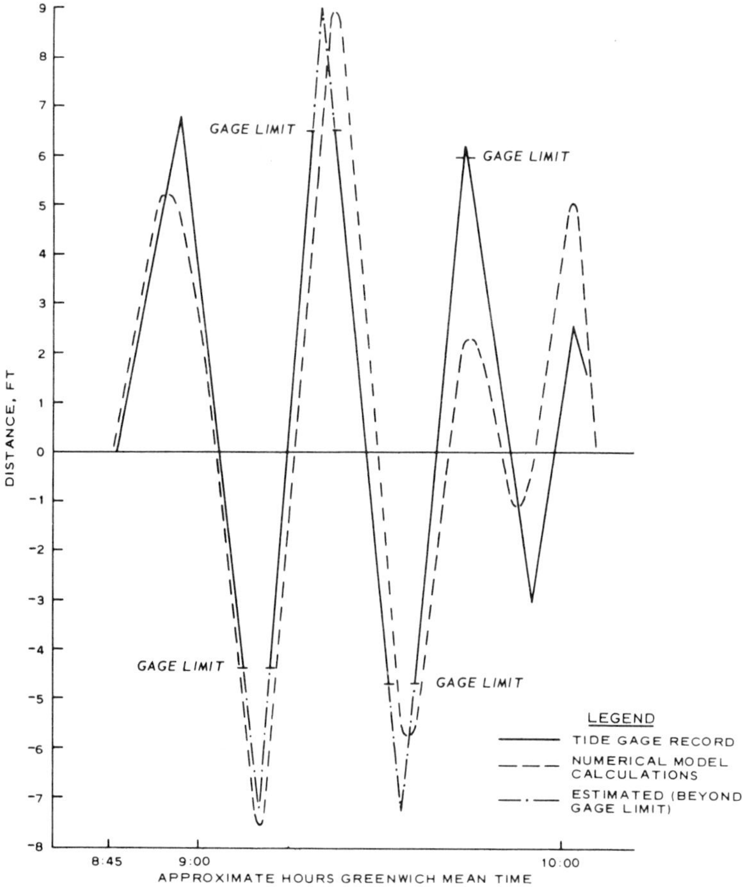

Figure 3. 1964 tsunami from Alaska recorded at Kahului, Maui.

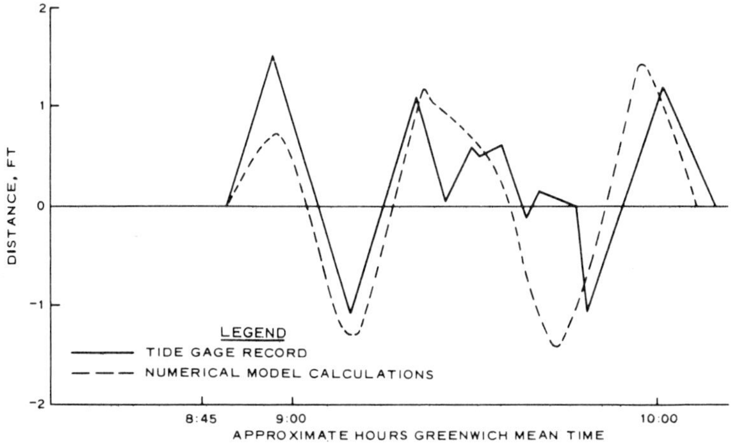

Figure 4. 1964 tsunami from Alaska recorded at Honolulu, Oahu.

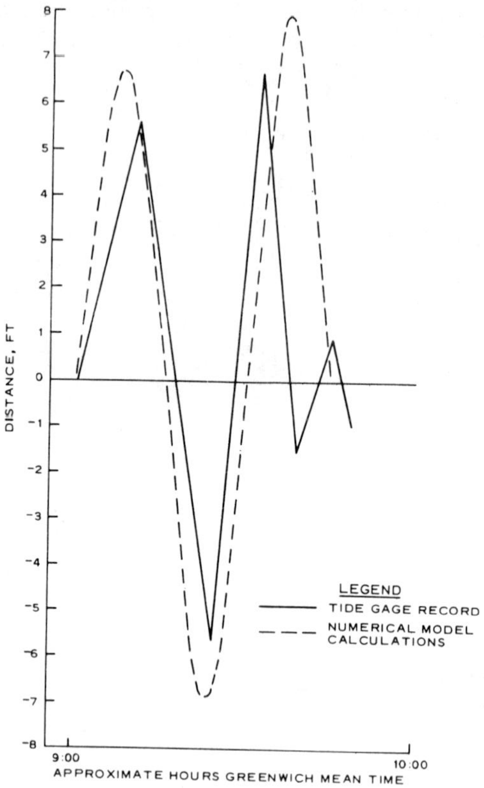

Figure 5. 1964 tsunami from Alaska recorded at Hilo, Hawaii

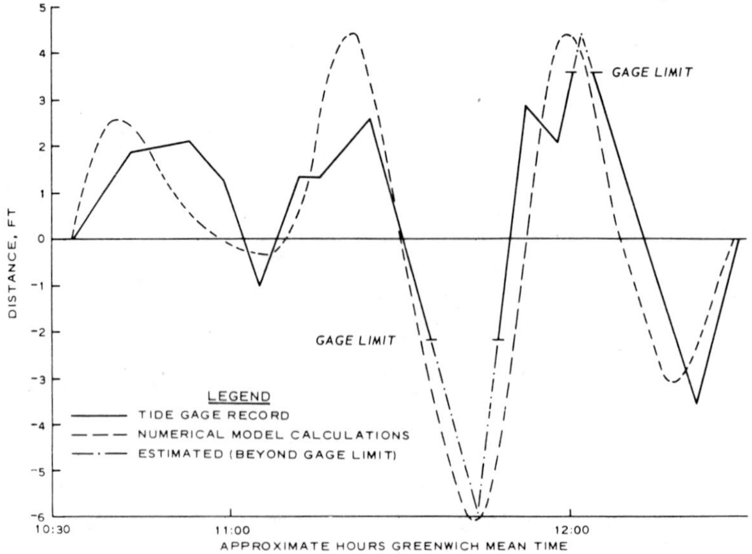

Figure 6. 1960 tsunami from Chile recorded at Honolulu, Hawaii.

tsunami. Again, the largest waves recorded are shown. The permanent ground motion of the 1960 earthquake is not known nearly as well as the ground motion for the 1964 earthquake. The agreement indicated in Figure 6 is, therefore, quite good considering the uncertainity of the uplift in the source region.

Tide gages, of course, do not record tsunamis perfectly since they are nonlinear devices with responses that depend upon both the period and amplitude of the disturbances they measure. However, simple calculations based upon the paper of Noye (1968) show that the distortion is small for the tsunamis shown in Figures 3-6.

HARBOR RESONANCE

Various numerical models have been used in the past to calculate resonant oscillations of small harbors or small sections of large harbors. It has not been possible to use these models to calculate the resonant response over a wide range of incident wave periods of a very large harbor of complicated shape, such as the Los Angeles and Long Beach Harbors complex (harbor occupies a region approximately 11 km by 11 km), because the computational time required by the models makes such computations economically infeasible. However, the computational requirements of the finite element model discussed earlier are an order of magnitude less than those required by previous numerical models. Therefore, it was possible to use the finite element model to calculate the response of all or large sections of the harbors complex to long wave excitation over a wide range of incident wave periods (1 to 10 min.).

Figure 7 shows one of the finite element grids used to study resonant oscillations in the LA-LB Harbors complex. Several grids were used because the calculation time of the model depends upon the fourth power of the element size. As the period of the incident wave increases, the grid can be made cruder while maintaining a desired resolution. A maximum triangle side length of one-eighth the local wavelength was selected for the lowest wave period tested in any particular grid. Grids covering halves of the harbors complex were used for incident waves with periods from 1 to 3 minutes and grids covering the entire harbors area for incident waves with periods from 3 to 10 minutes.

The finite element method allows very accurate representation of land-water boundaries in the very complicated, multiconnected LA-LB Harbors complex. Telescoping features of finite element grids allow proper resolution of rapid water surface or velocity gradients. Elements can also be concentrated in special regions of interest in the grid such as ship terminals.

Breakwaters in the Harbors complex are represented in various ways in the numerical grid depending upon their characteristics. A detached breakwater surrounding a proposed

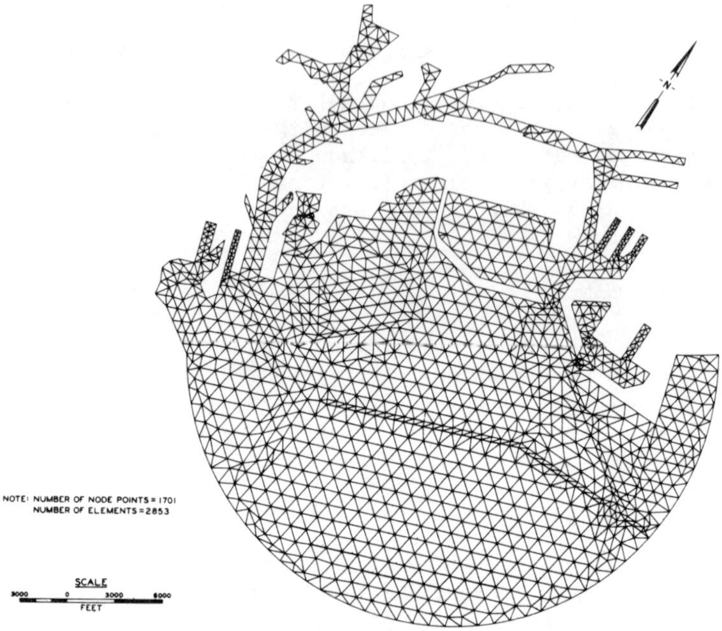

Figure 7. Finite element grid for harbor.

oil tanker terminal is represented as a solid barrier since the breakwater core extends above the still-water level and is impermeable. Part of the outer breakwater does not have a solid core and calculations using the formula of Keulegan (1973) for wave transmission through rock structures indicates that this breakwater should be highly permeable to very small amplitude long-period waves. Therefore, the breakwater is represented as being totally permeable. Although the breakwater appears impervious from the water surface as a result of the fitted concrete blocks extending up from the waterline, Wilson (1968) noted that this breakwater has long been considered an ineffective filter against long waves. Another part of the outer breakwater has a solid core to -26 ft mean lower low water (mllw) and is covered with a crown consisting of rock the same size or larger than the rock of the permeable section of the breakwater. This breakwater is represented by a depth change to -26 ft mllw with the core impermeable and the crown totally permeable. Small elements are placed along the breakwater in the numerical grid so that the depth transition can be accurately represented.

A wave period increment of 2 sec was used in this study for incident waves with periods from 1-3 min and an increment of 3 sec for longer period waves. Resonant peaks were further defined by using wave period increments as small as a fraction of a second. Frequency response plots of amplification factors and current velocities were made for numerous locations around the harbors. Contour plots of amplification factors and vector plots of velocity for the entire harbors area were made for wave periods corresponding to significant resonant peaks.

COMPARISONS WITH HYDRAULIC MODEL

Resonant oscillations in the LA-LB Harbors complex also were studied (Chatham, 1977) in a distorted hydraulic model that has a vertical scale of 1:100 and a horizontal scale of 1:400. The hydraulic model reproduces the entire harbor area and an offshore region out to approximately the 100-meter contour. Waves are generated by a 65-meter-long electrohydraulic wave generator composed of 14 individual sections that can be positioned to reproduce curved and variable-height wave fronts.

There are significant differences between test conditions used for the numerical model study and the hydraulic model wave test series. In the numerical model, the wave front is assumed to be straight and of constant height as the wave approaches the harbor. The depth is assumed constant seaward of the harbor area included in the finite-element grid. In the hydraulic model, depth variation seaward of the harbor is reproduced out to the -300 ft mllw contour to allow development of a convergence zone seaward of the harbor as found in wave refraction analyses. The wave front is curved and wave height varied along the wave front to simulate wave refraction occurring seaward of the model limits and to compensate for model distortion and the model depth limit of 300 ft prototype. The initial direction of wave propagation is the same in each of the models.

Wave-height amplification in the numerical model is undamped by frictional energy dissipation (boundary and internal), and the calculated response is expected to be larger in the numerical model than in the hydraulic model. The amplitude of resonant response is limited in the numerical model by radiation dissipation.

The wave-height amplification factor in the numerical study is defined at a point in the harbor as the wave height at the point divided by twice the incident wave height. This definition of amplification factor is traditional and results from the condition that the standing wave height for a straight coast with no harbor would be twice the incident wave height due to superposition of the incident and reflected waves.

Due to curvature of the wave front and wave-height variation along the wave front, wave-height amplification R in the hydraulic model is defined as

$$R = \frac{H_s}{H_o}$$

where H_s is the wave height at the gage location and H_o is the wave height that would have occurred along the initial straight wave front used in refraction analyses.

Direct correlation between numerical and hydraulic model data should not be expected because of frictional effects and the differences discussed above. However, the periods of resonant response and the relative form of the modal shape of wave height amplification at resonance should be similar.

Figure 8 is a comparison of numerical and hydraulic model wave-height amplification in the 60- to 280-sec period range for a location in slip 7 (see Figure 9) in the southeast basin. The periods of the major resonant peaks are similar with the largest response at a period just below four minutes. Figure 9 shows a comparison between the wave-height amplifications for this resonant peak predicted by the two models. The wave-height amplification patterns are quite similar with the primary oscillation involving a fundamental oscillation of slip 7. Figure 10 shows a similar comparison for an oscillation involving much of the southeast basin outside slip 7.

In general, it was found that there was good agreement between predictions by the numerical and hydraulic models of the periods of maximum harbor response and of modal configurations. The magnitude of wave-height amplification was not in good agreement, being greater for the numerical model. This difference may be attributable to the neglect of frictional dissipation in the numerical model. The exaggerated wave heights in the hydraulic model (necessary because of the practical difficulties of modeling waves that are only a few centimeters in height in the prototype) also may result in greater than actual dissipation due to exaggerated flow separation effects (Bowers, 1977).

The numerical model was found to provide an excellent relative indication of the results to be expected in hydraulic model tests. Thus, the numerical model can be used to assist in wave-gage placement in hydraulic models and selection of critical wave periods that should be tested, since numerical model tests can be performed much more rapidly and much cheaper than hydraulic model tests.

ACKNOWLEDGEMENT

The authors wish to acknowledge the Pacific Ocean Division of the U. S. Army Corps of Engineers for authorizing and funding the numerical studies of tsunami interaction with the Hawaiian Islands; the U. S. Army Engineer District, Los Angeles, for authorizing the hydraulic model study of Los Angeles and Long Beach Harbors; and the Cities of Los Angeles and Long Beach, California for authorizing and funding

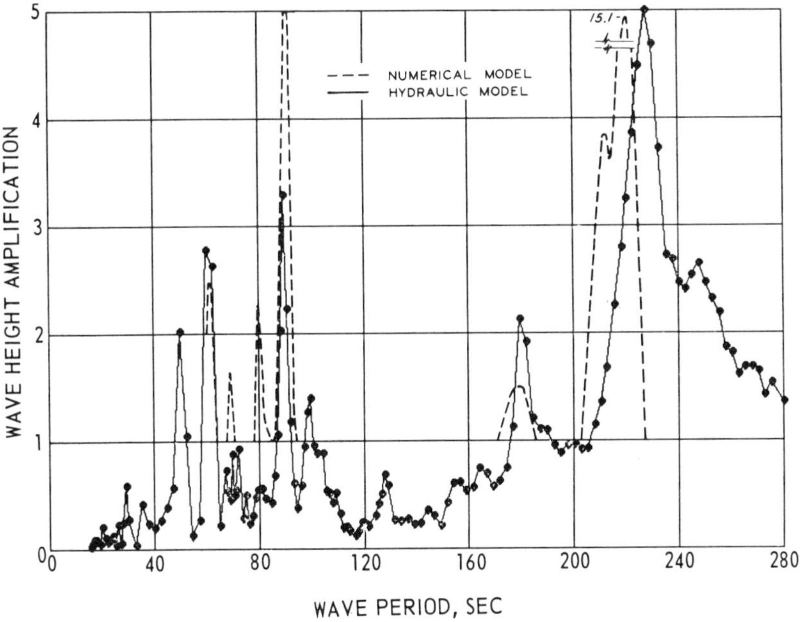

Figure 8. COMPARISON OF NUMERICAL AND HYDRAULIC MODEL WAVE-HEIGHT AMPLIFICATION IN THE 60- TO 280-SEC PERIOD RANGE FOR A GAGE LOCATION AT THE REAR OF SLIP 7 IN THE SOUTHEAST BASIN.

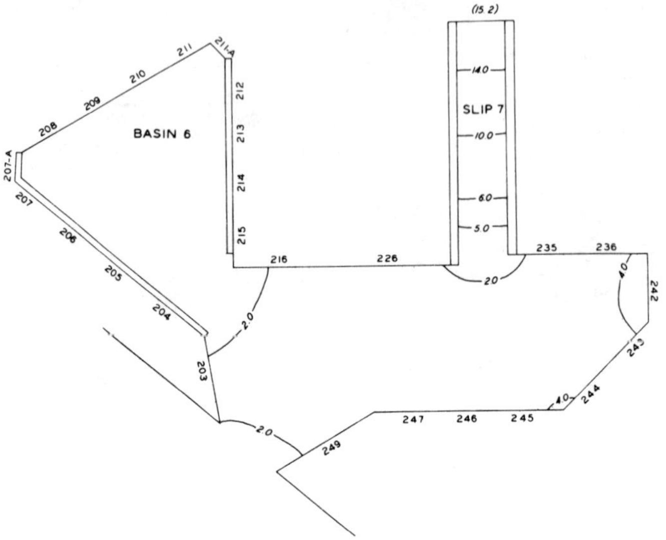

NUMERICAL MODEL DATA

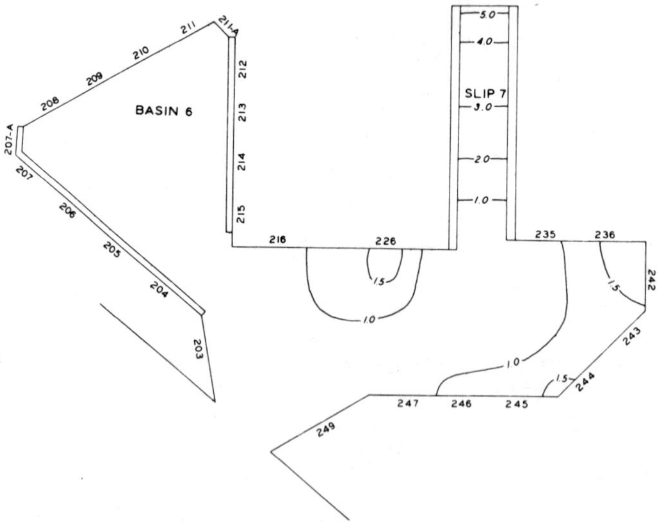

HYDRAULIC MODEL DATA

Figure 9. Comparison of numerical and hydraulic model predictions of wave-height amplification for an an incident wave with a period of 220 seconds.

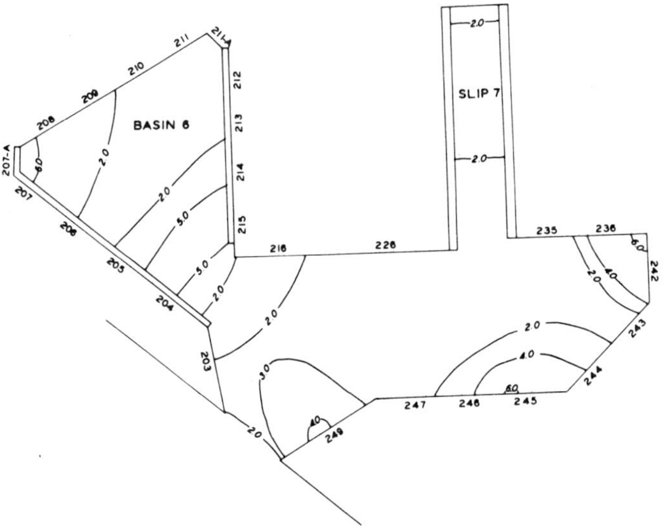

NUMERICAL MODEL DATA

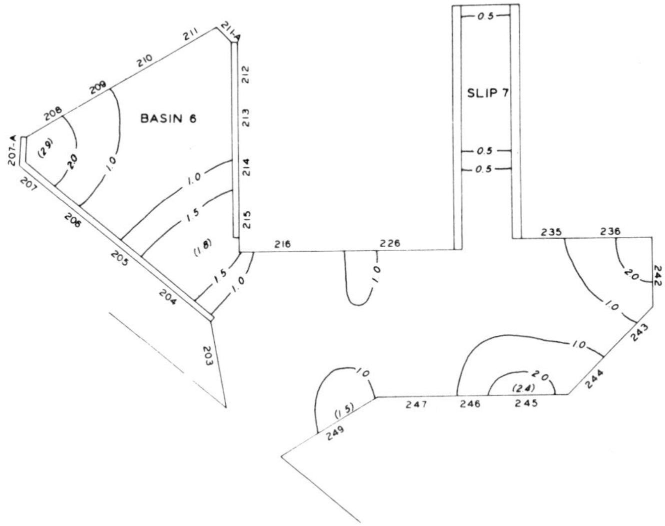

HYDRAULIC MODEL DATA

Figure 10. Comparison of numerical and hydraulic model predictions of wave-height amplification for an incident wave with a period of 80 seconds.

the numerical harbor oscillation studies reported herein. The Office, Chief of Engineers, is gratefully acknowledged for authorizing publication of this paper.

REFERENCES

Bernard, E. N. and A. C. Vastano (1977) Numerical Computation of Tsunami Response for Island Systems. J. Phys. Oceanogr., 7, 389-395.

Bowers, E. C. (1977) Harbour Resonance Due to Set-down Beneath Wave Groups. J. Fluid Mech., 79, pt. 1, 71-92.

Chatham, C. E. (1977) Los Angeles Harbor and Long Beach Harbor: Design of the Hydraulic Model. PORTS' 77 Symposium of the American Society of Civil Engineers, Vol. 1, 47-64.

Chen, H. S. and C. C. Mei (1974) Oscillations and Wave Forces in a Manmade Harbor in the Open Sea. Proc. 11th Symp. on Naval Hydrodynamics, 573-594.

Hammack, J. L. (1972) Tsunamis-A Model of their Generation and Propagation. Rep. KH-R-28, Calif. Inst. of Tech.

Houston, J. R. and A. W. Garcia (1974) Type 16 Flood Insurance Study: Tsunami Predictions for Pacific Coastal Communities. Tech Rep H-74-3, Waterways Experiment Station, Vicksburg, MS.

Hwang, L. S., D. Divoky, and A. Yuen (1970) Amchitka Tsunami Study. Rep. TC-177, Tetra Tech Inc.

Keulegan, G. H. (1973) Wave Transmission Through Rock Structures; Hydraulic Model Investigation. Res. Rep. H-73-1, Waterways Experiment Station, Vicksburg, MS.

Noye, B. J. (1968) The Frequency Response of a Tide-well- Third Australasian Conference on Hydraulics and Fluid Mechanics, Sydney, Australia, 65-70.

Plafker, G. (1969) Tectonics of the March 27, 1964 Alaska Earthquake. U. S. Geological Survey Professional Paper 543-I, 11-174.

Plafker, G. and J. C. Savage (1970) Mechanism of the Chilean Earthquake of May 21 and 22, 1960. Bull. Geo. Soc. Amer., 81 1001-1030.

Wilson, B. W., et. al. (1968) Wave and Surge-Action Study for Los Angeles and Long Beach Harbors, Vol. 2, Science Engineering Associates, San Marino, Calif.

COMPUTATION OF STATIONARY WATER WAVES DOWNSTREAM OF A TWO-DIMENSIONAL CONTRACTION

P.L. Betts,

University of Manchester Institute of Science and Technology,
Manchester M60 1QD, England.

INTRODUCTION

Binnie, Davies and Orkney (1955) investigated the possibility of leading water through an open topped contraction to a horizontal channel with parallel sides. In this way they hoped to provide Naval Architects with a ship testing flume which would provide the same benefits as are provided by wind tunnels in the field of aeronautics. Based on this work, a full size flume was built at the Ship Division of the National Physical Laboratory and is used for experiments which would be difficult to perform in the transient conditions of the usual towing tank facility. However, Binnie et al. found that when the stream Froude number in the working section exceeded about 0.5, but was less than unity, a stationary train of waves spontaneously appeared there. As the Froude number was increased, the amplitude and length of the waves increased, as did the distance of the first crest from the end of the contraction. The occurrence of these waves restricts the maximum speed at which ship models can be tested in such a channel and causes even greater limitations on the testing of Civil Engineering structures (such as vortex drops) where the required stream Froude number may be close to unity.

Binnie et al tested a range of control devices downstream but could detect no effect on the waves in the working section. They concluded that the waves were caused by the contraction through which the water had passed. Benjamin (1956) argued that the occurrence of the waves did not require the action of viscosity but was a phenomenon similar in principle to that of the wave resistance of an obstacle spanning an otherwise uniform free surface flow. Brady (1973) performed experiments with a simple two-dimensional contraction, consisting of a smooth change in bottom level set within the parallel wall section of a laboratory flume, and obtained results similar to those of Binnie et al. His attempts at numerical modelling, with finite difference techniques, were only partially successful as he was unable to model the flow over a prescribed contraction with waves downstream. His computations were performed in the complex potential plane, and this necessitated the use of a triple stack of iterations which would not converge with waves downstream of the contraction.

In the present paper, the Finite Element Method is applied to Brady's two-dimensional problem. A variational principle in the physical plane is presented, and the Finite Element formulation of this provides the stability necessary for convergent computations with waves on the flow. Comparisons are made with wave theory and also with Brady's experimental results.

VARIATIONAL PRINCIPLE AND FINITE ELEMENT FORMULATION

Betts (1976) has shown that the correct variational principle for a steady free-surface flow under the influence of gravity takes the form of an extremum of the functional

$$\chi = \tfrac{1}{2} \iint_{\Omega(h)} \left\{ \left(\frac{\partial\psi}{\partial x}\right)^2 + \left(\frac{\partial\psi}{\partial y}\right)^2 \right\} dx\, dy - \tfrac{1}{2} \int_{x_D}^{x_C} g(H-h)^2 dx \qquad (1)$$

when expressed in terms of the stream function ψ . In Equation 1, $\psi(x,y)$ is the stream function of the fluid flowing through the domain Ω , defined by the boundaries Γ_1, Γ_2, Γ_3 and Γ_4 in Figure 1. The end boundaries, Γ_2 and Γ_4, are chosen such that $\partial\psi/\partial n = 0$ and are not necessarily vertical or straight, whilst Γ_1 and Γ_3 are streamlines. Both $\psi(x,y)$ and $h(x)$ are allowed to vary subject to the restrictions $\delta\psi = 0$ on the fixed bottom surface Γ_1 ($\psi=0$) and on the variable surface Γ_3 where $y=h(x)$ ($\psi=Q$ say). The constant H is the elevation of total head (stagnation level) above y datum. Although the domain $\Omega(h)$ is dependent on the variable $h(x)$, variations in $\psi(x,y)$ and $h(x)$ are only interdependent in so far as is required by continuity of the first derivatives of variations in ψ at the variable boundary Γ_3. Satisfaction of this variational principle automatically requires Laplace's equation to be satisfied throughout the flow (irrotationality) and Bernoulli's equation to be satisfied with zero gauge pressure along the surface streamline. Betts also showed that numerical computations, based on a finite element formulation of the functional, converged to correct solutions regardless of whether the Froude number of the flow was greater or less than unity (cf. Southwell and Vaisey, 1946).

It may be noted that a negative sign appears before the surface line integral in Equation 1 whereas the sign is positive in the corresponding variational principle for the velocity potential ϕ (Luke, 1967). This difference is associated with the application of Green's identity, which requires a fixed domain. When allowance is made for variations of the domain ($\delta\Omega$ in Figure 1) an additional term in $\partial\psi/\partial n$ appears in the line integral obtained for Green's identity. This additional term leads to the negative sign in Equation 1. The corresponding term in Luke's principle, $\partial\phi/\partial n$, is zero on a steady flow free surface, and consequently the sign of his surface line integral is positive. Equation 1 can also be shown to be an applicable version of the more general functional J_2 stated by O'Carroll (1976) and moreover appears to be a direct statement of the principle used in an indirect fashion by Varoḡlu and Finn (1977), who also considered variations in the integrated flow rate Q.

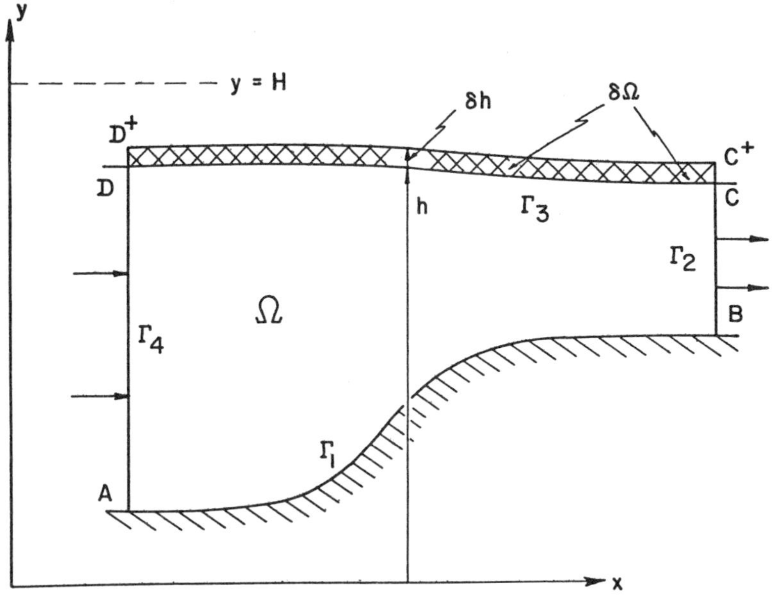

Figure 1 Flow domain with free surface

Finite element formulation

Since the highest derivatives in Equation 1 are first derivatives, the flow region can be divided into triangular elements over each of which a linear distribution of $\psi(x,y)$ is assumed. Since the variable surface Γ_3 (Figure 1) is then automatically represented by a connected series of straight lines, finite element approximations are available for all terms in the functional χ and the integrals can be replaced by appropriate summations.

When the finite element formulation is applied to Equation 1, internal nodal co-ordinates x_i, y_i can be chosen freely and are constants, whereas only x_i is constant for free surface nodes but the nodal value of ψ is also known and constant. Consequently, internal values of ψ_i and the values of y_i on the free surface are the unknowns subjected to variation.

This formulation satisfies the requirement of continuity of first derivatives at the variable **boundary** Γ_3.

When Equation 1 is differentiated to obtain the extremum, variations in the elevation of the free surface introduce non-linear terms into the summations corresponding to both integrals and an iterative method of solution is required. The chosen solution technique was of Newton-Raphson type and is similar in principle to that described by Ikegawa and Washizu (1973); differences in application have been described by Betts (1976) who also listed the essential equations. Only a brief outline is therefore given here.

An initial distribution is assumed for the unknown nodal parameters (ψ_i or y_i). The true nodal parameters are then considered as the sum of the initial value and a small correction term. When the variational procedure is applied to Equation 1, terms of the same order as the square of the correction terms are neglected in comparison with lower order terms, in the same way that the squares of variation terms are normally neglected. This leads to a complete set of linear equations for the unknown corrections, which can be solved by any of the standard methods. An improved distribution of nodal parameters is then obtained and the procedure repeated until the magnitude of the correction falls below a predetermined level.

It is naturally desirable that the initial distribution of ψ_i at internal nodes should be a reasonable approximation to the true values for flow beneath the initially assumed free surface. Since the non-linear terms only appear as a consequence of variations of free surface elevation, the correct internal streamfunction values can be obtained directly by suppressing variations in free surface level at the first iteration. Similarly, if in the course of correcting the free surface elevations, elements close to the free surface become distorted compared with adjacent internal elements, a revised distribution of internal nodal co-ordinates can be generated and appropriate values of ψ_i computed immediately.

APPLICATION TO THE WAVE PROBLEM

When water flows over an obstacle or change in bottom level and converges asymptotically to a uniform flow downstream, there is no particular difficulty in applying the finite element method which has been described, provided the end boundaries are placed sufficiently far from the obstacle for perturbations to parallel flow to be negligible. This was the situation considered by Betts (1976) who chose flow rates such that the Froude numbers were either below 0.5 everywhere or above 1.0. No waves then occur on the flow and the relationship between flow rate and total head is known. In the present case, with downstream Froude numbers between 0.5 and 1.0, the flow does not converge to a uniform stream downstream, but is known to have a train of stationary waves, of unknown phase and amplitude, superposed on it.

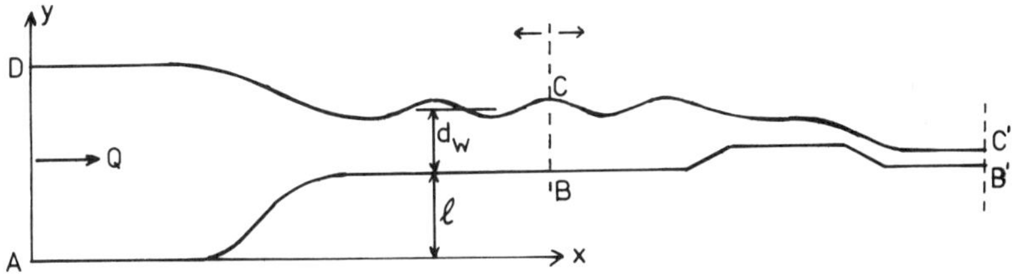

Figure 2 Alternative formulations of problem.

There are two possible approaches that could be used for the problem with stationary waves. These are illustrated in Figure 2. The first alternative (Domain AB'C'D) reflects the physical reality that if the flow downstream of the contraction is subundal (F_w<1) there must be an outlet control device of some form further downstream. Thus the whole flow could be modelled with the downstream end boundary in uniform superundal flow (F>1) well downstream of the outlet device. The disadvantages to this approach are that a large number of additional elements would be required to cover the added region near and downstream of the outlet device and that an additional outer iteration would be required to determine the unknown discharge coefficient of the device, which relates the flow rate to the total head (cf. Varoğlu and Finn, 1977). The second alternative was the one used and is illustrated by Domain ABCD in Figure 2, where the downstream boundary is located beneath a crest (or trough) of the wave train so that although not uniform the flow is purely horizontal ($\partial\psi/\partial n = 0$). In this way only the domain of interest needs to be included in the finite element formulation, which reduces the storage requirements considerably. Although the flow rate and total head can now be specified, an outer iteration is still required as the location of the downstream crest is not initially known.

The advantages of a variational principle with natural end boundary conditions become particularly apparent when the method of locating the downstream boundary is considered. As is shown by Lamb (1932), the problem is not fully specified until wave free conditions have been imposed upstream. Consequently, if the downstream boundary is located near, but not at, the correct crest position, a real flow is possible with a crest downstream but with an additional superposed wave train from end to end of the flow. This additional train of waves would vary in amplitude and wave length through the contraction, have a crest or trough at the upstream boundary ($\partial\psi/\partial n = 0$), and have a surface inclination downstream equal and opposite to the main stationary wave train at the downstream boundary, so that the sum of the two wave trains produces the required natural boundary condition there.

The way in which the correct position of the downstream

boundary can be located is now apparent. A position for the upstream boundary is chosen and remains constant. An initial position for the downstream boundary is selected and the finite element calculations are made. In general, if the downstream Froude number is not much more than 0.5, these iterations will converge for any downstream boundary position but there will be waves on the upstream surface. A new downstream boundary is then tried and the correct position is determined by the usual interpolation methods which reduce the upstream wave amplitude to zero. In practice it is convenient to use the difference between the surface elevation at the corner D (Figure 2) and the upstream asymptotic level as an initial criterion. However, the final criterion should be based on making the upstream surface monotonic as storage requirements at higher Froude numbers make it desirable to place AD at a position where the fluid velocities can be considered effectively horizontal, rather than further upstream where D may be considered at the asymptotic level (cf. Lamb 1932).

When this method was used to locate the correct position of the downstream boundary, it was found that comparatively large elements could be used upstream, as accurate modelling of the upstream wave train was not required in order to reduce its amplitude to zero. A slight difficulty can occur if the distance between the ends of the problem is such that an end-to-end wave train is possible with a crest or a trough at each end. The problem is then no longer fully specified. This situation arose in the course of the computations and is discussed later. Consideration is also give to the way in which a reasonable estimate of the position of the downstream crest can be obtained.

COMPUTATIONS AND NUMERICAL RESULTS

Computations were undertaken of the flow through a two-dimensional contraction of the same shape as that used by Brady (1973) in his experiments. This contraction consisted of a straight section inclined at 45° to the horizontal and joined to the horizontal beds upstream and downstream by circles of radius $\ell/3$ and $\ell/2$ respectively, where ℓ is the increase of bed level through the contraction. No attempt was made to allow for the effects of the thick boundary layers that were known to exist in Brady's experiments. Computations were performed in dimensionless form, with length scale ℓ and time scale such that the dimensionless gravitational acceleration equalled unity.

Betts (1976) has demonstrated that a simple horizontal line provides a sufficiently good initial approximation to the free surface for finite-element computations. Since Brady had shown that, in finite difference calculations, it is desirable for the initial approximation to have radii of curvature no less than the final free surface, this approximation was retained in the present work. In the results presented here, the level of the initial surface was taken as that of the nominal mean level downstream. Some computations were also performed with the initial surface at the upstream level, but these were sometimes found to produce marginally less consistent convergence. Three series of tests were conducted. For the first the upstream depth was constant at a dimensionless value of 2.0 (Series A) as the flow rate was increased, while in the other two the nominal downstream depths were 0.8 and 0.6 respectively (series 8A and 6A). The nodes were distributed horizontally in such a way that the spacing was proportional (1/8) to the difference between the bed level and the initial surface level (Figure 3). Nodes were then distributed uniformly along vertical lines between the bed and the initial free surface (9 nodes, 8 elements). Coarse variation of the downstream boundary position was obtained by adding or removing columns of elements there, and refined variation by altering the horizontal length of all elements in the downstream region by a constant amount, which was always less than $2\frac{1}{2}\%$ of the horizontal length of an element. This downstream region was considered to start at unit distance downstream of the nominal end of the contraction (i.e. where the extension to the 45° straight line intersects the downstream bed level). For a given flow rate, the total head H above the upstream bed level was calculated for uniform flow at the depth which had already been defined.

In the initial iteration, the free surface was fixed so that ψ values at internal nodes could be obtained. Large movements of the free surface occurred at the second iteration, due partly to the waves and partly to the difference in mean level that had to be established between upstream and downstream. Consequently internal nodes were redistributed vertically before the third iteration, for which the free surface was again fixed. Therefore the internal nodal co-ordinates were only altered if the surface elements became too distorted relative to the internal elements. The criterion for this was taken to be that the height of a surface element was either less than 0.7 or more than 1/0.7 of the height of an internal element. Typically for computations with the downstream boundary near its correct position, this additional redistribution was not required and a converged solution was obtained within a total of 7 iterations (maximum free surface movement reduced to less than 10^{-5}, Central Processor Execution Time on a CDC 7600 between 2.5 and 4 seconds depending on the distance to the downstream boundary).

Results

An example of the final free surface and internal nodal distribution is shown on Figure 3, where typical element shapes have been added to the computer display. The search for the correct downstream boundary is illustrated on Figure 4 at a greatly enhanced scale. It is noticeable that for this downstream Froude number, where the stationary waves are large, quite small differences in the downstream boundary position produce

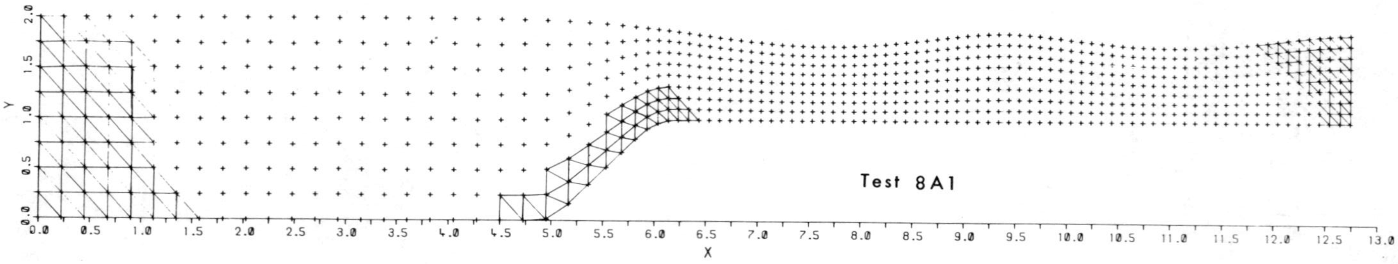

Figure 3 Final nodal distribution and sample elements.

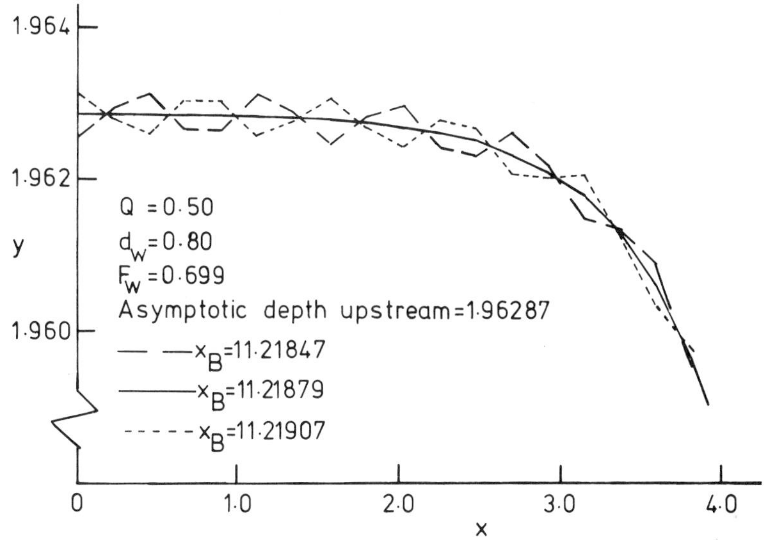

Figure 4 Upstream surface profiles during Test 8A6.

quite large undulations upstream. Although this provides a very sensitive test for the position of the downstream boundary, it also means that a reasonably good estimate of this position is required to obtain a converged solution, even with undulations upstream. This will be discussed shortly.

Figure 5 shows the free surface profiles, with exaggerated vertical scale, for all the tests in Series 8A (with a downstream depth of 0.8). As with the other two series, they exhibit the typical characteristics found in the experiments of Binnie et al (1955) and Brady (1973), namely that as the Froude number increases, the first crest of the stationary waves moves downstream as the amplitude and wave length increase. The location of the contraction is the same as that shown in Figure 3, and with one exception, the upstream boundary is a fixed distance from the contraction, so that the nodal distribution along the contraction bed is fixed. Test 8A2 is the exception and is an example of the situation discussed earlier, where the original upstream boundary was at a position such that an additional wave train throughout the length of the domain, and with extrema at each end, is a physical possibility. This situation is recognized by the fact that the amplitude of the upstream undulations reaches a minimum, rather than becoming zero, as the downstream boundary is altered. Normally this would be overcome by adding or subtracting a column of elements at the upstream end. However, in test 8A2 the natural half wave length of the upstream waves was within ½% of the horizontal nodal spacing there. The distance to the upstream boundary was therefore increased by half an element length, with a consequent slight change in the element representation of the contraction.

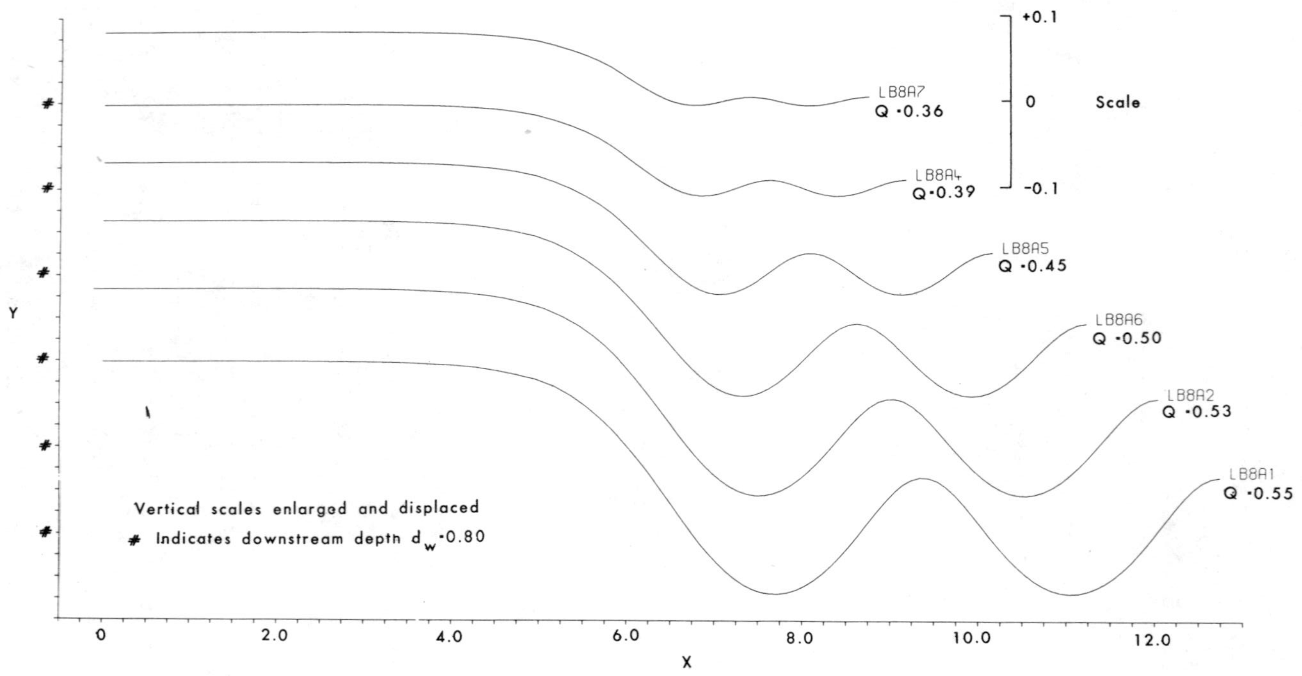

Figure 5 Free surface profiles (Series 8A).

In the results presented, the downstream boundary was placed at the second crest of the stationary waves. For all three series the distance between this and the downstream end of the contraction ($x_B - x_3$) has been plotted on Figure 6 against the theoretical length of the waves, λ_{th}. The wave length λ_{th} is that for infinitesimal waves on a stream of depth equal to the nominal uniform depth of the downstream flow and is given in dimensionless form by the equation

$$\frac{Q^2}{d_w^2} = \frac{\lambda_{th}}{2\pi} \tanh\left(\frac{2\pi d_w}{\lambda_{th}}\right) \tag{2}$$

where Q is the volumetric flow rate/unit width and d_w is the nominal depth downstream. It is immediately apparent that all the points lie very close to a common straight line. Although closer inspection reveals that the curve through the points increases in gradient slightly with λ_{th} and that values for a shallow depth series lie slightly below deeper depth values, the relationship is sufficiently accurate to provide a good estimate of the correct position for the downstream boundary under the second crest. The implication is that the downstream waves have constant phase (a crest) at a position close to the end of the contraction, so that once one or two surface profiles have been established, which is relatively easy at Froude numbers near 0.5, the position of the downstream boundary for higher Froude numbers can be estimated sufficiently well for convergent computations to be obtained. It should be noted in this context that the value of λ_{th} increases more rapidly than F^2.

Considerations of accuracy

As a first test of the accuracy of the final free surface, the total head was calculated at each free surface node. The horizontal surface velocity was obtained from a parabolic equation through the ψ values of the top three nodes of a vertical column. The adjacent surface nodes were used to compute the local surface gradient and hence the vertical velocity (cf. Betts, 1976). Values of total head calculated from these velocities and the elevations of the free surface never varied by more than 2% of the local element height from the prescribed total head. This is well within the error expected from the finite element method, especially when one realizes that the total head depends on the square of the velocity. It was also noticed that when the downstream boundary was incorrectly placed, the error at the upstream corner increased considerably while those downstream remained unaltered. This observation further increases confidence in the way the correct downstream boundary was located.

A further test was suggested by the fact that in test 8A4 the correct downstream boundary position lay at a position such that either the lengths of all elements in the downstream region could be increased by 2½% of their original value or an

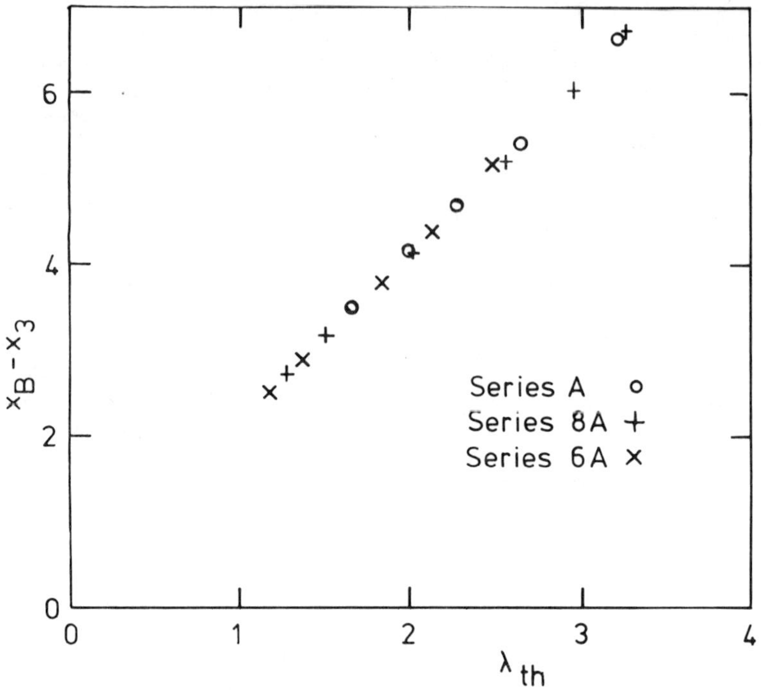

Figure 6 Downstream boundary position vs. theoretical wave length.

extra column of elements could be added and the lengths shortened by approximately the same amount. The difference in the position of the downstream boundary from the two calculations was only 4% of a basic element length, even though test 8A4 was at a low Froude number, 0.545, for which the crest to trough amplitude of the waves was only 0.019 (19% of mean element height).

The lengths of the computed waves downstream were also compared with theoretical values. For infinitesimal amplitude waves, the theoretical values are those given by Equation 2 above. The computed waves were slightly longer than the theoretical ones, which at first sight was surprising, as one would expect any finite amplitude effects to reduce the length of the waves for a given mean velocity. A more detailed comparison was therefore made for test A5, with an upstream depth of 2.00 and a flow rate Q of 0.55. For these conditions, the nominal depth downstream d_w, from a one-dimensional analysis, was 0.8036, with a Froude number based on this depth of 0.7635 and a theoretical wave length from Equation 2 of 3.207. However, the mean depth evaluated from the computed results over the final full and half wavelengths was 0.796 with a wave length of 3.32. On physical grounds one should expect the mean level downstream to be slightly less than the nominal value, as for a fixed total head, or energy level, the depth of the flow has to be reduced, in order to transfer energy from the mean flow into wave energy. When the computed mean depth d_m is used in Equation 2, one obtains a theoretical wave length of 3.30, which is very close to the computed value.

The work of De (1955) on 5th order Stokes waves can be used to quantify these comparisons further. In the computations the peak-to-trough wave amplitude H was 0.129, corresponding to $2\pi H/\lambda = 0.245$, and $2\pi\, d_m/\lambda$ was 1.507. De tabulates values of total head level and wave speed for $2\pi\, d_m/\lambda = 1.5$ over a range of dimensionless amplitudes. Interpolation of De's results leads to the conclusion that the total head level, above bottom, for waves of this amplitude should exceed the level required for uniform flow at the same mean depth by 0.30%. The corresponding calculation from the computed results shows that the specified total head level exceeded that for uniform flow, with Q = 0.55 and depth $d_m = 0.796$, by 0.28%. This difference (0.2% of an element height) is entirely negligible, which indicates that the finite element method has correctly modelled the mean depth downstream.

Comparison with De's wave speed results is complicated by the fact that his calculations were for a fixed wave length, whereas we require the effects of finite amplitude on the wave length at a fixed flow rate. However, when his results are converted to this form, they indicate an expected wave length of 3.26 for waves on a flow of mean depth 0.796. Thus the computed waves were 1.7% longer than those expected from a full finite amplitude wave analysis. Although this difference represents about half an element length, the comparison must be considered satisfactory with the element sizes used, since a difference in the depth upstream of the contraction of only ½%, for the same flow rate, would change the wavelength downstream by 9%. Moreover, the elements in the wave region were spaced approximately uniformily across vertical sections (8 elements/section), whereas the effects of surface waves decay exponentially with distance below the surface. Smaller elements near the surface would therefore have been needed for more accurate representation of the wave train.

Comparison with experiments

It has already been noted that the surface profiles obtained from computations (Figure 5) agree qualitatively with the experimental profiles of Brady (1973). A quantitative comparison is made on Figure 7, where the computed peak-to-trough wave amplitude divided by downstream depth is plotted against downstream Froude number (cf. Binnie et al 1955) and compared with Brady's experimental results for the same shape of contraction. It is then apparent that the computed results follow the same trend as the experimental ones but with a somewhat larger amplitude. This is to be expected since the analysis of Benjamin (1956) indicates that the occurrence of stationary waves in an inviscid phenomenon, so that viscosity might be expected to reduce their amplitude. Wave decay was indeed observed along the length of Brady's channel. In addition, the thick boundary layer along the bottom of his channel might be expected to produce an effective contraction shape with less severe corners and a consequent reduction in wave amplitude.

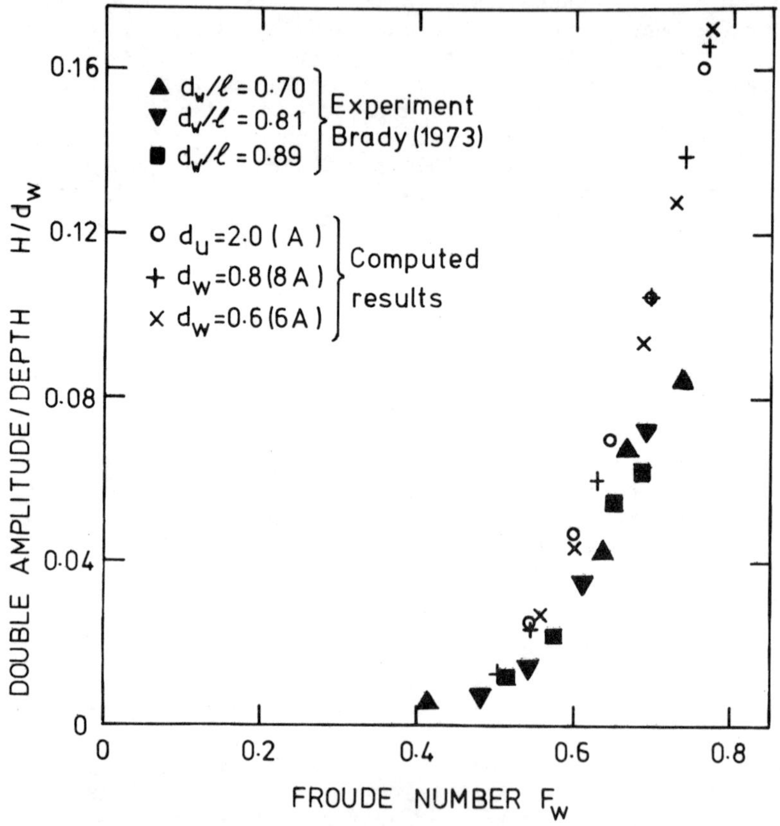

Figure 7 Comparison of wave amplitude with experiments.

Fuller comparisons with Brady's experimental results are unfortunately not possible for a number of reasons. His contraction was 0.203m high and was placed 1.5 m downstream of the start of a long horizontal channel. Consequently, the thick boundary layer through the contraction affected both the shape and the effective height of the contraction as well as the effective depths upstream and downstream. Moreover, Brady's experimental error, as indicated by comparing flow rates measured upstream and downstream of the contraction, was about 2%, and as was previously shown in the theoretical comparisons with test A5, quite small changes in upstream conditions will produce large differences in the lengths of the downstream waves (e.g. 2% on Q for fixed upstream depth leads to 13% change in wave length). A further effect of the thick boundary layers in Brady's experiments may be deduced from the work of Binnie and Cloughley (1970) who measured the length of the waves in their flume and found them to be shorter than the appropriate theoretical values. They later showed that this was at least partially due to the distortion of the bottom boundary layer by the waves (Binnie and Cloughley, 1971). The work of French (1977) shows that the same effect is produced by the boundary layers on the vertical side walls of the channel.

CONCLUDING REMARKS

It has been shown that the finite element method can be applied to an appropriate variational principle to model the occurrence of stationary waves downstream of a two-dimensional contraction. As was shown by Southwell and Vaisey (1946) such waves produce a particularly difficult numerical problem. Indeed, this is believed to be the first successful numerical model of water waves over a prescribed non-horizontal bottom. Previous finite difference computations of stationary waves, even over a horizontal bottom, have required a good initial estimate of the surface profile, whereas a simple horizontal line was used in the present work. It seems that the finite element representation of the free surface by a series of straight lines enhances the stability of the computation over the finite difference representation by a series of nodes.

Comparisons with theoretical data have shown that the accuracy of the computed results is well within that usually expected from the discretisation involved in the finite element method, even though the present work involves a variable surface boundary. Comparisons with experimental data were favourable, but indicated that more accurate experiments are needed for a detailed comparison. Any such experiments should be performed in a wide channel and should include measurement of the boundary layer along the bottom. This boundary layer could then be modelled in the computations as a displacement thickness.

An additional benefit of the method described in this paper should also be mentioned. Binnie and Sumer (1974) placed a hump on the bed of an otherwise monotonically converging contraction, so that the waves caused by the hump cancelled the main contraction waves at a specific value of the downstream Froude number. However, the amplitude of the waves in the working section was increased at other, off-design, Froude numbers. Binnie and Sumer determined the correct size and position of the hump by experimental trial and error. The experimental advantages of such a "wave-free" compound contraction are obvious.

Tugtekin (1974) discussed the theoretical requirements for the design of a two-dimensional "wave-free" contraction and showed that the Streamline Curvature Technique could be used for this. However, this method required considerable smoothing and could give no indication of the waves that would occur at off-design Froude numbers. In contrast, the present method can readily be extended to provide a more direct approach to the design problem. A simple basic contraction shape would be chosen, probably with a maximum gradient less than 45°, and the waves produced at the design Froude number would be calculated. The compound shape of the "wave-free" contraction would then be determined by systematic variation of the elevations of nodes in the contraction, and the off-design performance readily computed by the direct method of this paper.

REFERENCES

Benjamin, T.B. (1956) On the Flow in Channels when Rigid Obstacles are placed in the Stream. J.Fluid Mech., 1:227-248.

Betts, P.L. (1976) A Variational Principle in Terms of Stream Function for Free-Surface Flows and its Application to the Finite Element Method. University of Calgary, Mech.Eng.Rep. 84.

Binnie, A.M. and Cloughley, T.M.G. (1970) A Comparison of Ship-Model Tests in a Slotted-Wall Channel and in a Towing Tank. Trans.Roy.Inst.Nav.Archit., 112 1:101-111.

Binnie, A.M. and Cloughley, T.M.G. (1971) The Lengths of Stationary Waves on Flowing Water. J. Hydraulic Res. 9:35-41.

Binnie, A.M., Davies, P.O.A.L. and Orkney, J.C. (1955) Experiments on the Flow of Water from a Reservoir through an Open Horizontal Channel. I. The Production of a Uniform Stream, Proc.Roy.Soc.Lond. A230:225-236.

Binnie, A.M. and Sumer, B.M. (1974) A Method of Improving the Uniformity of the Stream in an Open Channel. J.Hydraulic Res., 12:299-313.

Brady, J.A. (1973) Flow Transitions in Open Channels. Ph.D. Thesis, The University of Manchester Institute of Science and Technology.

De, S.C. (1955) Contributions to the Theory of Stokes Waves. Proc.Camb.Phil.Soc. 51:713-736.

French, M.J. (1977) Hydrodynamic Basis of Wave-Energy Converters of Channel Form. J.Mech.Eng.Sci. 19:90-92.

Ikegawa, M. and Washizu, K. (1973) Finite Element Method Applied to Analysis of Flow over a Spillway Crest. Int.J. Num.Methods in Eng. 6:179-189.

Lamb, H. (1932) Hydrodynamics. 6th Ed., Art 245, Cambridge U.P.

Luke, J.C. (1967) A Variational Principle for a Fluid with a Free Surface. J.Fluid Mech. 27:395-397.

O'Carrol, M.J. (1976) Variational Principles for Two-Dimensional Open Channel Flows. Proc.2nd International Symposium on Finite Element Methods in Flow Problems, Santa Margherita Ligure, Italy, June.

Southwell, R.V. and Vaisey, G. (1946) Relaxation Methods Applied to Engineering Problems. XII Fluid Motions Characterized by 'Free Streamlines'. Phil.Trans.Roy.Soc.Lond.A240:117-160.

Tugtekin, S.K. (1974) The Computational Design of a Bottom Configuration to Provide Wave Free Open Channel Flow. M.Sc. Dissertation, The University of Manchester Institute of Science and Technology.

Varoğlu, E. and Finn, W.D.L. (1977) An Inverse Variable Domain Finite Element Technique for Numerical Analysis of Irrotational Flow Problems with a Free Surface. Symp. in Innovative Numerical Analysis, Paris.

S E S S I O N   3

FLUID MECHANICS

A FINITE ELEMENT MODEL OF TURBULENT FLOW IN PRIMARY SEDIMENTATION BASINS

D.R. Schamber, B.E. Larock

Civil Engineering Department, University of California, Davis, CA. 95616

INTRODUCTION

Primary sedimentation by gravity in large basins and tanks has been an integral part of most major water and sewage treatment plants for many decades. Current understanding of the process is incomplete, however, and is impeded primarily by a lack of detailed knowledge of the velocity field in the basin. The long-term goal of the present project is to develop a predictive numerical model of sedimentation basin performance, including both prediction of the velocity field and the efficiency of particle removal, and to compare the results with appropriate experimental data.

This paper presents a two-dimensional, steady state model for the turbulent and, in general, stratified flow that occurs in the basins. The equations in the model are first developed; the finite element discretization and outline of the solution technique follow. The paper concludes with a relatively simple computational example. Suitable experimental data for verification of a full-scale computation will be unavailable until later this year; consequently, an expensive large-scale example has not yet been attempted.

BACKGROUND

Thirty years ago Camp (1946) summarized the state of then-existing knowledge for the design, and the behavioral analysis, of sedimentation basins. His simplified theory, including the assumption of a uniform velocity distribution across each vertical section of the tank, is still in everyday use now. However, both observation and experiments (Wills and Davis, 1965; Clements and Khattab, 1968) confirm that the true flow pattern is usually recirculatory and not uniform. Recent temperature measurements taken at the Sacramento, CA, central treatment plant (by the first author) in June 1977 show in addition that the flow is sometimes thermally stratified with a maximum vertical temperature difference of about 3°C in a vertical section. Because influent sediment concentrations are typically 2 x 10^{-4} gm/cm^3, it is felt that stratification effects due to suspended particle concentration differences are negligible.

Sedimentation basins may be either rectangular or circular in plan. The length of rectangular basins ranges up to 90 m, and a width of 6 m is typical. Circular tanks may range in diameter from 10 m to 30 m or more. The mean depth of most basins is near 3 m. Characteristic fluid velocities in these basins may be 0.3 to 1.5 meters per minute. These data lead to characteristic Reynolds numbers (based on mean depth and a kinematic viscosity of 1.1 x 10^{-6} m^2/s) between 14,000 and 70,000; despite their quiescent outward appearance these flows are assuredly turbulent. Larsen (1976) confirms this conclusion with recently reported measurements of the time-averaged turbulent fluctuations in such basins.

The actual flow in a sedimentation basin is thus non-uniform, recirculatory, turbulent, and in some cases thermally stratified. It is no wonder that previous simplified theoretical treatments have failed to depict the true nature and behavior of these flows. It is also clear that an improved understanding, and prediction, of particle removal efficiencies depends primarily on an improved understanding of the true velocity field, for once the flow pattern is known, it is a relatively straightforward matter to compute sediment concentrations throughout the basin by solving the now-linear advection-diffusion equation (Ariathurai and Krone, 1976; Sarikaya, 1977).

In one sense even the present model is unrealistic, in that it neglects the inherently three-dimensional nature of flow near rectangular corners; the study of circular basins later this year should not suffer from this defect, however.

MATHEMATICAL MODEL

Mean Flow Equations

The flow is governed by the physical principles of conservation of mass, momentum and energy. For an incompressible fluid these equations are respectively (Bird et al, 1960)

$$\frac{\partial \tilde{u}_i}{\partial x_i} = 0 \qquad (1)$$

$$\frac{\partial \tilde{u}_i}{\partial t} + \tilde{u}_j \frac{\partial \tilde{u}_i}{\partial x_j} = -\frac{1}{\rho}\frac{\partial \tilde{p}}{\partial x_i} + g_i + \nu \frac{\partial^2 \tilde{u}_i}{\partial x_j \partial x_j} \qquad (2)$$

$$\frac{\partial\tilde{T}}{\partial t} + \tilde{u}_j \frac{\partial\tilde{T}}{\partial x_j} = \frac{k_T}{\rho\hat{C}_v} \frac{\partial^2\tilde{T}}{\partial x_j \partial x_j} \qquad (3)$$

in which x_i = coordinate in ith direction, t = time, $\tilde{u}_i$ = instantaneous velocity component, $\tilde{p}$ = instantaneous pressure, $\tilde{T}$ = instantaneous temperature, ρ = fluid density, g_i = component of gravity, ν = kinematic viscosity, k_T = thermal conductivity, and $\hat{C}_v$ = fluid heat capacity at constant volume. The Einstein summation convention is applied to repeated subscripts. Viscous dissipation terms are deleted from Equation 3, as they are significant only in the presence of large velocity gradients.

In principle procedures exist for solving Equations 1-3 directly. Proper resolution of the small-scale turbulent motion, which may be about one millimeter in size, would require an impractically fine computational mesh for the present problem, however. To study the mean properties of the flow on a reasonable grid size, each dependent variable, represented by $\tilde{B}$, is decomposed into a time-averaged mean B and an instantaneous fluctuation b about the mean; that is,

$$\tilde{B} = B + b \qquad (4)$$

If a bar denotes time averaging, then $\bar{\tilde{B}} = \bar{B} = B$ and $\bar{b} = 0$. Use of this decomposition in Equations 1-3, followed by time-averaging of the equations, leads to the following equations for steady, two-dimensional flow in a vertical x-z (x_1 = x, x_3 = z) plane:

$$F_p = \frac{\partial U}{\partial x} + \frac{\partial W}{\partial z} = 0 \qquad (5)$$

$$F_u = U\frac{\partial U}{\partial x} + W\frac{\partial U}{\partial z} + \frac{1}{\mathbb{F}^2}\frac{\partial P}{\partial x} + \frac{\partial\overline{uu}}{\partial x} + \frac{\partial\overline{uw}}{\partial z} = 0 \qquad (6)$$

$$F_w = U\frac{\partial W}{\partial x} + W\frac{\partial W}{\partial z} + \frac{1}{\mathbb{F}^2}\frac{\partial P}{\partial z} + \frac{\partial\overline{uw}}{\partial x} + \frac{\partial\overline{ww}}{\partial z} + \frac{1}{\mathbb{F}^2}(1-\beta^*T) = 0 \qquad (7)$$

$$F_T = U\frac{\partial T}{\partial x} + W\frac{\partial T}{\partial z} + \frac{\partial\overline{uT'}}{\partial x} + \frac{\partial\overline{wT'}}{\partial z} = 0 \qquad (8)$$

Here $\tilde{u}_1 = U + u$ and $\tilde{u}_3 = W + w$, but $\tilde{T} = T + T'$. Equations 5-8 are non-dimensionalized with a characteristic velocity U_o and length h. Both T and T' now represent temperature differences from a reference temperature T_o, and are non-dimensionalized with respect to a convenient reference temperature difference $\Delta T = T_1 - T_0$. Equation 7 incorporates the Boussinesq approximation, in which only the body force term is affected by the density changes caused by a change in mean temperature. The density changes have been related to mean temperature changes via a linear equation of state. In the original dimensional variables this equation is $\rho = \rho_0 + \rho_0 \beta^*(T-T_0)/\Delta T$, in which $\beta^* = \beta\Delta T$ and β = volumetric expansion coefficient. Finally, the squared Froude number $\mathbb{F}^2 = U_0^2/(gh)$. In a fully turbulent flow the viscous terms in Equations 6 and 7 and the heat flux terms due to molecular diffusion in Equation 8 are insignificant and thus are omitted.

Rigorous treatment of the free surface would require satisfaction of $dz_s/dx = W/U$ on the surface as well as P = constant at the free surface location z_s. In this paper, however, the free surface is assumed to be known a priori. By direct measurement the slope is known to be extremely small.

Turbulence Model

Time averaging has introduced additional unknown correlations ($\overline{uu}$, $\overline{uw}$, $\overline{ww}$, $\overline{uT'}$ and $\overline{wT'}$) into Equations 6-8 and created a closure problem. Historically the most common approach uses the scalar turbulent (eddy) viscosity concept in which the velocity correlations (called Reynolds stresses) are related to mean flow parameters (e.g. King et al., 1974; Young et al., 1976) by writing

$$- \overline{u_i u_j} = \nu_t \left(\frac{\partial U_i}{\partial x_j} + \frac{\partial U_j}{\partial x_i} \right) \qquad (9)$$

where ν_t = turbulent kinematic viscosity, which is known to be flow dependent and to vary from point to point in a flow domain. (This equation is not strictly correct unless the trace of the stress tensor is subtracted, but Reynolds (1970) comments that this difference is immaterial in a simple model.) The velocity-temperature correlations may be treated analogously (Young et al., 1976). Users of this approach must prescribe a priori the behavior of ν_t, and proper determination of this behavior for each flow requires extensive and difficult calibration via experimentation and/or intuition. However, it is often assumed constant to sidestep much of the calibration problem. An alternative modeling approach that retains more of the physics of turbulence appears desirable in this project.

Deardorff (1970) developed the sub-grid-scale approach to model plane turbulent channel flow. Rather than time-average the equations of motion, this method applies a space-averaging operator to the equations in such a way that time-dependent fluctuations with a scale smaller than the finite-difference grid are filtered out. Thus even a steady mean-flow problem is described by a transient three-dimensional set of equations. Long computing times are the primary drawback of this approach (Deardorff, 1973).

This work will employ a single-point closure model. In these models the local value of $\overline{u_i u_j}$ or $\overline{u_i T'}$ is related to mean and turbulent properties of the flow via differential transport equations. The equation describing the transport of $\overline{u_i u_j}$ is obtained by multiplying the equation of motion for $\tilde{u}_i$ by $\tilde{u}_j$, multiplying the equation of motion for $\tilde{u}_j$ by $\tilde{u}_i$, adding the products together and time-averaging the result to obtain for steady flow

$$U_k \frac{\partial \overline{u_i u_j}}{\partial x_k} = - \left[\overline{u_j u_k} \frac{\partial U_i}{\partial x_k} + \overline{u_i u_k} \frac{\partial U_j}{\partial x_k} \right] - \beta \left[\overline{u_i T' g_j} + \overline{u_j T' g_i} \right]$$

Convection of turbulence by mean flow (I) — production by shear (II) — production by buoyant forces (III)

$$- 2\nu \overline{\frac{\partial u_i}{\partial x_k} \frac{\partial u_j}{\partial x_k}} + \overline{\frac{p}{\rho} \left(\frac{\partial u_i}{\partial x_j} + \frac{\partial u_j}{\partial x_i} \right)} \tag{10}$$

viscous dissipation (IV) — pressure scrambling (V)

$$- \frac{\partial}{\partial x_k} \left[\overline{u_i u_j u_k} - \nu \frac{\partial}{\partial x_k} \left(\overline{u_i u_j} \right) + \delta_{ik} \overline{\frac{p}{\rho} u_j} + \delta_{jk} \overline{\frac{p}{\rho} u_i} \right]$$

diffusive transport (IV)

in which δ_{ij} = Kronecker delta. The physical interpretation of each quantity is listed below the term.

The equation for the transport of $\overline{u_i T'}$ is obtained by multiplying the equation of motion for $\tilde{u}_i$ by $\tilde{T}$, multiplying the energy equation by $\tilde{u}_i$, adding these two equations and time-averaging the result to obtain for steady flow

$$U_j \frac{\partial \overline{u_i T'}}{\partial x_j} = - \left[\overline{u_i u_j} \frac{\partial T}{\partial x_j} + \overline{u_j T'} \frac{\partial U_i}{\partial x_j} \right] - \beta \, \overline{T'^2} g_i$$

Convection of heat flux by mean flow (I) — production by shear (II) — production by buoyant forces (III)

$$- (\lambda + \nu) \overline{\frac{\partial u_i}{\partial x_j} \frac{\partial T'}{\partial x_j}} + \overline{\frac{p}{\rho} \frac{\partial T'}{\partial x_i}} \qquad (11)$$

dissipation (IV) — pressure scrambling (V)

$$- \frac{\partial}{\partial x_j} \left[\overline{u_i T' u_j} + \delta_{ij} \frac{\overline{pT'}}{\rho} - \nu \overline{T' \frac{\partial u_i}{\partial x_j}} - \lambda \overline{u_i \frac{\partial T'}{\partial x_j}} \right]$$

diffusive transport (VI)

in which λ = thermal diffusivity = $k_T/(\rho \hat{C}_v)$. Again the physical interpretation is given below each term. Excepting terms I, II and III in Equation 10 and terms I and II in Equation 11, all the processes which influence $\overline{u_i T'}$ and $\overline{u_i u_j}$ contain unknown correlations of fluctuating quantities; these factors must all be expressed in terms of mean flow quantities and double correlations of $\overline{u_i T'}$ or $\overline{u_i u_j}$. The closure approximations of Launder and his co-workers are now adopted.

One of the more fundamental closure schemes uses the scalar viscosity concept and is called the k-ε model (Launder and Spalding, 1974). In this model ν_t in Equation 9 is written as

$$\nu_t = c_\mu \frac{k^2}{\varepsilon} \qquad (12)$$

in which c_μ = constant = 0.09, $k = \overline{u_i u_i}/2$ is the turbulent kinetic energy, and

$$\frac{2}{3} \delta_{ij} \varepsilon = 2\nu \overline{\frac{\partial u_i}{\partial x_k} \frac{\partial u_j}{\partial x_k}} \qquad (13)$$

is the dissipation rate of turbulence energy. (Models presented herein assume the viscous dissipation is isotropic and thus are truly valid only for high turbulent Reynolds number flows.) For non-stratified flows this closure uses two differential equations for k and ε together with mean flow Equations 5-7. A two-dimensional non-stratified flow involves five unknowns per computation point. This model has demonstrated success in modeling various flows, including free shear flows, flows in closed conduits, and recirculating flows (Launder and Spalding, 1974).

Launder et al (1975) also propose a more complete closure. In this model a differential equation is developed for each of the Reynolds stress components $\overline{u_i u_j}$, and they are solved simultaneously with the mean flow equations, giving a total of eight unknowns per computational node for a non-stratified flow in two dimensions. (The extension to include temperature effects is given by Launder (1975).)

The simultaneous solution of eight differential equations is a formidable task for all but simple problems. To reduce computational cost and still retain a level of closure consistent with the difficulty of the sedimentation problem, we choose to implement the algebraic stress model of Gibson and Launder (1976). In order of complexity the model is between the k-ε model and the complete $\overline{u_i u_j}$ differential equation model. The example presented herein will show that use of this model is promising, although it still must be tested for a full-scale basin. In the long run the simpler k-ε model may be adequate, but currently the algebraic model seems more appropriate because the normal stresses are expected to exert a significant influence on the flow. In this model k and ε are determined from their respective differential transport equations. For two-dimensional steady flow the non-dimensionalized equations are

$$F_k = U\frac{\partial k}{\partial x} + W\frac{\partial k}{\partial z} - \frac{c_\mu}{\sigma_k}\left[\frac{\partial}{\partial x}\left(\frac{k^2}{\varepsilon}\frac{\partial k}{\partial x}\right) + \frac{\partial}{\partial z}\left(\frac{k^2}{\varepsilon}\frac{\partial k}{\partial z}\right)\right] - P_r + \varepsilon = 0 \quad (14)$$

$$F_\varepsilon = U\frac{\partial \varepsilon}{\partial x} + W\frac{\partial \varepsilon}{\partial z} - \frac{c_\mu}{\sigma_\varepsilon}\left[\frac{\partial}{\partial x}\left(\frac{k^2}{\varepsilon}\frac{\partial \varepsilon}{\partial x}\right) + \frac{\partial}{\partial z}\left(\frac{k^2}{\varepsilon}\frac{\partial \varepsilon}{\partial z}\right)\right] - c_{\varepsilon_1}\frac{\varepsilon}{k}P_r + c_{\varepsilon_2}\frac{\varepsilon^2}{k} = 0 \quad (15)$$

in which the production term P_r is given by

$$P_r = -\overline{uu}\frac{\partial U}{\partial x} - \overline{uw}\left(\frac{\partial U}{\partial z} + \frac{\partial W}{\partial x}\right) - \overline{ww}\frac{\partial W}{\partial z} + \frac{\beta^*}{\mathbf{F}^2}\overline{wT'} \quad (16)$$

The constants are $c_\mu = 0.09$, $\sigma_k = 1.0$, $\sigma_\varepsilon = 1.3$, $c_{\varepsilon_1} = 1.45$ and $c_{\varepsilon_2} = 1.90$. These values were selected after extensive examination of data for turbulent free shear flows; however, Launder and Spaulding (1974) indicate that use of these values should also lead to satisfactory predictions for flows near walls. Equation 14 is half the sum of the normal stress equations for $\overline{u_1u_1}$, $\overline{u_2u_2}$ and $\overline{u_3u_3}$ given by Equation 10. The pressure scrambling term in Equation 10 (which redistributes energy among the stress components) is absent in Equation 14 due to continuity considerations; the P_r and ε terms in Equation 14 serve as a source and sink for turbulent energy. The diffusive transport term in Equation 10 is modeled by the square-bracketed quantity in Equation 14. Equation 15, the dissipation equation, is the weakest part of the model. Details of its origin are given hy Hanjalic and Launder (1972).

The Reynolds stress and heat flux terms in Equations 6-8 remain to be modeled. Gibson and Launder (1976) have reduced Equations 10 and 11 to a series of algebraic (in the stress and heat flux terms) equations given for two-dimensional steady flow by

$$F_{\overline{uu}} = (\overline{uu} - \tfrac{2}{3}k)(\hat{c}_1\varepsilon + P_r) + \hat{c}_2 k[2(\overline{uu}\frac{\partial U}{\partial x} + \overline{uw}\frac{\partial U}{\partial z}) + \tfrac{2}{3}P_r] = 0 \quad (17)$$

$$F_{\overline{ww}} = (\overline{ww} - \tfrac{2}{3}k)(\hat{c}_1\varepsilon + P_r) + \hat{c}_2 k[2(\overline{uw}\frac{\partial W}{\partial x} + \overline{ww}\frac{\partial W}{\partial z}) - \frac{2\beta^*}{\mathbb{F}^2}\overline{wT'} + \tfrac{2}{3}P_r] = 0 \quad (18)$$

$$F_{\overline{uw}} = \overline{uw}(\hat{c}_1\varepsilon + P_r) + \hat{c}_2 k[\overline{uu}\frac{\partial W}{\partial x} + \overline{ww}\frac{\partial U}{\partial z} - \frac{\beta^*}{\mathbb{F}^2}\overline{uT'}] = 0 \quad (19)$$

$$F_{\overline{uT'}} = \overline{uT'}[(c_{1T} - \tfrac{1}{2})\varepsilon + \tfrac{1}{2}P_r] + k[\overline{uu}\frac{\partial T}{\partial x} + \overline{uw}\frac{\partial T}{\partial z}] \quad (20)$$

$$+ (1-c_{2T})k[\overline{uT'}\frac{\partial U}{\partial x} + \overline{wT'}\frac{\partial U}{\partial z}] = 0$$

$$F_{\overline{wT'}} = \overline{wT'}[(c_{1T} - \tfrac{1}{2})\varepsilon + \tfrac{1}{2}P_r] + k[\overline{uw}\frac{\partial T}{\partial x} + \overline{ww}\frac{\partial T}{\partial z}] \quad (21)$$

$$+ (1-c_{2T})k[\overline{uT'}\frac{\partial W}{\partial x} + \overline{wT'}\frac{\partial W}{\partial z} + \frac{\beta^*}{\mathbb{F}^2} c_T' \frac{k}{\varepsilon}(\overline{uT'}\frac{\partial T}{\partial x} + \overline{wT'}\frac{\partial T}{\partial z})] = 0$$

in which $\hat{c}_1 = c_1 - 1$ and $\hat{c}_2 = 1 - c_2$. The constants $c_1 = 2.2$ and $c_2 = 0.55$ are determined by reference to data on normal stresses in a nearly homogeneous shear flow and on the sudden distortion of isotropic turbulence. Coefficients $c_{1T} = 3.2$ and $c_{2T} = 0.50$ were chosen by reference to experimental data for a nearly homogeneous shear flow in which the temperature increased linearly with height under essentially non-buoyant conditions. Finally, coefficient $c_T' = 1.6$ was selected by reference to data on the decay of temperature fluctuations behind a grid. An equation set similar to Equations 17-21 has been used successfully by Hossain and Rodi (1976) to predict turbulence intensities in a heated surface jet and a vertical buoyant jet.

The complete equation set for two-dimensional steady flow consists of Equations 5-8, 14, 15, and 17-21 which must be solved simultaneously for properly specified boundary conditions. The total of 11 unknowns consists of four mean flow variables U, W, P and T, and seven turbulence variables k, ε, $\overline{uu}$, $\overline{ww}$, $\overline{uw}$, $\overline{uT'}$ and $\overline{wT'}$. A twelfth variable $\overline{vv}$ is readily computed from the definition of the turbulence kinetic energy once k, $\overline{uu}$ and $\overline{ww}$ are known.

The arguments used to develop the approximation to the pressure scrambling term in Equations 17-19 preclude their use in the vicinity of a rigid boundary. The presence of a rigid wall causes the component of stress normal to the wall to be damped while the streamwise component of stress is augmented. Launder et al (1975) present a more complex expression for the pressure scrambling term which includes wall proximity effects. In this work it was felt that initially the use of the more complicated formulation was unwarranted; in any event the velocity near a rigid boundary is matched to the universal law of the wall.

FINITE ELEMENT FORMULATION

The present turbulence model has been applied to a variety of turbulent flows by Launder and co-workers using a form of the Pantakar-Spalding (1970) finite difference algorithm. Surprisingly, in the past very little has been reported on the solution of turbulent flows using advanced turbulence models (other than specified eddy viscosity) by the finite element method. Hutton (1976) described a Galerkin formulation of essentially the k-ε model but reported no numerical results. Taylor et al (1977) present a Galerkin solution for fully-developed pipe flow using a k-ℓ model, i.e., a transport equation for k is solved with a transport equation for the turbulence length scale ℓ.

Galerkin approach

To effect a finite element solution for this problem, an integral representation of the equations must be found. Since no exact variational formulation for the Navier-Stokes equations or extensions thereof exists (Finlayson, 1972), the problem has been recast here via the Galerkin form of the method of weighted residuals. Experience with laminar flows indicates that the basis functions used to interpolate the pressure field should be one degree lower than those used for the velocity field (Hood and Taylor, 1974). For the present equation set C^0 continuous basis functions are sufficient to assure compatibility and completeness. Figure 1a depicts the eight node general quadrilateral element used to partition the flow domain. Linear basis functions M_i spanning the element corner nodes are used to interpolate the pressure field; quadratic basis functions N_i are used for all other variables.

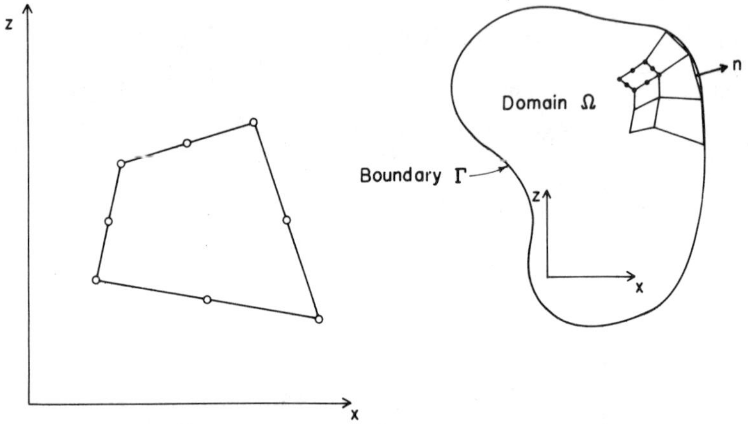

(a) (b)

Figure 1. (a) Eight node element, (b) Flow domain.

(See e.g. Huebner (1975) for the basis functions.) The trade-offs between use of more elements and a linear representation of the turbulence variables and the present scheme can only be examined meaningfully after greater experience is acquired.

Application of the Galerkin method to Equations 5-8, 14, 15, and 17-21 over a domain Ω bounded by Γ results in

$$f_p = \int_\Omega M_i \hat{F}_p \, d\Omega \tag{22}$$

$$f_u = \int_\Omega \hat{N}_i F_u \, d\Omega, \quad f_w = \int_\Omega N_i \hat{F}_w \, d\Omega, \quad f_T = \int_\Omega N_i \hat{F}_T \, d\Omega \tag{23}$$

$$f_k = \int_\Omega \hat{F}_{k,\Omega} \, d\Omega + \int_\Gamma \hat{F}_{k,\Gamma} \, d\Gamma \tag{24}$$

$$f_\varepsilon = \int_\Omega \hat{F}_{\varepsilon,\Omega} \, d\Omega + \int_\Gamma \hat{F}_{\varepsilon,\Gamma} \, d\Gamma \tag{25}$$

$$f_{\overline{uu}} = \int_\Omega N_i \hat{F}_{\overline{uu}} \, d\Omega, \quad f_{\overline{ww}} = \int_\Omega N_i \hat{F}_{\overline{ww}} \, d\Omega, \quad f_{\overline{uw}} = \int_\Omega N_i \hat{F}_{\overline{uw}} \, d\Omega \tag{26}$$

$$f_{\overline{uT'}} = \int_\Omega N_i \hat{F}_{\overline{uT'}} \, d\Omega, \qquad f_{\overline{wT'}} = \int_\Omega N_i \hat{F}_{\overline{wT'}} \, d\Omega \tag{27}$$

The term $\hat{F}$ in each integrand is the residual error in the point equation that results from substitution of the approximate finite element representation of each variable into the original equation. In Equations 24 and 25 the second-order terms have been integrated by parts using Green's theorem in the plane. The boundary residuals from each element (due to element boundary discontinuities in $\partial k/\partial x$, $\partial k/\partial z$, $\partial \varepsilon/\partial x$ and $\partial \varepsilon/\partial z$) have been added to the integrated k and ε equations in such a way that boundary integrals along adjacent element interfaces cancel (Finlayson, 1966; Ames, 1972). The final expressions contain only boundary integrals along Γ which can be used to introduce natural boundary conditions on k and ε. The other terms in Equations 24 and 25 are

$$F_{k,\Omega} = N_i \left[U\frac{\partial k}{\partial x} + W\frac{\partial k}{\partial z} - P_r + \varepsilon \right] + \frac{c_\mu}{\sigma_k} \frac{k^2}{\varepsilon} \left[\frac{\partial N_i}{\partial x} \frac{\partial k}{\partial x} + \frac{\partial N_i}{\partial z} \frac{\partial k}{\partial z} \right] \tag{28}$$

$$F_{k,\Gamma} = -\frac{c_\mu}{\sigma_k} N_i \frac{k^2}{\varepsilon} \left[\frac{\partial k}{\partial x} \ell_x + \frac{\partial k}{\partial z} \ell_z \right] \tag{29}$$

$$F_{\varepsilon,\Omega} = N_i \left[U\frac{\partial \varepsilon}{\partial x} + W\frac{\partial \varepsilon}{\partial z} - c_{\varepsilon_1} \frac{\varepsilon}{k} P_r + c_{\varepsilon_2} \frac{\varepsilon^2}{k} \right]$$

$$+ \frac{c_\mu}{\sigma_\varepsilon} \frac{k^2}{\varepsilon} \left[\frac{\partial N_i}{\partial x} \frac{\partial \varepsilon}{\partial x} + \frac{\partial N_i}{\partial z} \frac{\partial \varepsilon}{\partial z} \right] \tag{30}$$

$$F_{\varepsilon,\Gamma} = -\frac{c_\mu}{\sigma_\varepsilon} N_i \frac{k^2}{\varepsilon} \left[\frac{\partial \varepsilon}{\partial x} \ell_x + \frac{\partial \varepsilon}{\partial z} \ell_z \right] \tag{31}$$

in which ℓ_x and ℓ_z are the direction cosines of the outer unit normal n shown in Figure 1b. It is interesting that no second derivatives of the stresses or velocities appear in this model, in contrast to the result obtained by use of the scalar eddy viscosity, Equation 9.

Newton-Raphson solution

The problem defined by Equations 22-27 is thoroughly nonlinear and coupled. Presently the writers use a Newton-Raphson solution procedure which updates the entire equation set after each iterative computational cycle. It is not surprising that this approach is rather expensive in terms of computer storage when one solves an 11 degree-of-freedom problem for each corner node. For this reason it may be productive to consider more thoroughly schemes whereby the mean flow and turbulence variables are alternately updated within a Newton-Raphson context; the maximum number of unknowns per node would then be seven.

Methods for solving finite element systems of equations are now relatively standardized (e.g. King et al., 1973; Taylor and Hood, 1973). The coefficients and residuals in the Newton-Raphson approach are formed element by element. The so-called element stiffness matrix and load vector for the nth iteration takes the form

$$\mathbf{A}_e^n \, \delta\mathbf{X}_e^n = - \mathbf{f}_e^n \tag{32}$$

The portion of the element coefficient matrix $\mathbf{A}_e$ associated with a single element node j is

$$\mathbf{A}_{ej}^n = \begin{bmatrix} \dfrac{\partial f_p}{\partial P_j} & \dfrac{\partial f_p}{\partial U_j} & \dfrac{\partial f_p}{\partial W_j} & \cdots & \dfrac{\partial f_p}{\partial \overline{wT'_j}} \\ \dfrac{\partial f_u}{\partial P_j} & \dfrac{\partial f_u}{\partial U_j} & \dfrac{\partial f_u}{\partial W_j} & \cdots & \dfrac{\partial f_u}{\partial \overline{wT'_j}} \\ \vdots & & & \ddots & \vdots \\ \dfrac{\partial f_{\overline{wT'}}}{\partial P_j} & \dfrac{\partial f_{\overline{wT'}}}{\partial U_j} & \dfrac{\partial f_{\overline{wT'}}}{\partial W_j} & \cdots & \dfrac{\partial f_{\overline{wT'}}}{\partial \overline{wT'_j}} \end{bmatrix} \tag{33}$$

The corresponding portions of the correction vector $\delta\mathbf{X}_{ej}^n$ and residual vector $\mathbf{f}_{ej}^n$ are

$$\delta\mathbf{X}_{ej}^n = \left[\delta P_j^n, \ \delta U_j^n, \ \delta W_j^n, \ \ldots, \ \delta \overline{wT'_j}^n\right]^T \tag{34}$$

and

$$\mathbf{f}_{ej}^n = \left[f_{pj}^n, \ f_{uj}^n, \ f_{wj}^n, \ \ldots, \ f_{\overline{wT'_j}}^n \right]^T \tag{35}$$

Listing all terms in the element coefficient matrix is a bulky prospect; however, the partial derivatives of f_u and $f_{\overline{ww}}$, which are typical, are given in Appendix I. Each partial derivative is formed for each of eight element nodes (Exception: Equation 22 is formed only for corner nodes). Each function is also evaluated for each of the eight weight functions N_i (Exception: Equation 22 is evaluated for four weight functions M_i). For a two-dimensional stratified flow the complete coefficient matrix is 84 x 84 at the element level. The integrals in $\mathbf{A}_e^n$ and $\mathbf{f}_e^n$ are evaluated by 3 x 3 Gaussian integration.

The system matrix and residual vector are formed by appropriate addition of the element contributions. The system

$$\mathbf{A}^n \, \delta\mathbf{X}^n = - \, \mathbf{f}^n \tag{36}$$

is solved for the nodal corrections $\delta\mathbf{X}$. The new estimate for $\mathbf{X}$ is then

$$\mathbf{X}^{n+1} = \mathbf{X}^n + \delta\mathbf{X}^n \tag{37}$$

Iterations cease when the corrections become sensibly small.

It is assumed that the domain boundary Γ may be divided into two parts, a kinematic part on which velocities are specified and a mechanical part on which stresses are specified. It is also possible to treat a mixed boundary condition which is a linear combination of the two. At each node along Γ where velocities, stresses or both are prescribed, the corresponding coefficient equation and residual are deleted from the global matrix and vector. As indicated earlier, derivative boundary conditions on k and ε enter the equation set via the line integrals in Equations 24 and 25.

Near a rigid wall the local turbulence Reynolds number [$\equiv k^2/(\nu\varepsilon)$, where ν = kinematic viscosity] is so small that viscous effects dominate turbulence effects, even though the present model assumes a fully turbulent flow. Hence boundary conditions for this model are properly applied at the edge of the turbulent flow regime, a distance δ from the wall. Neglecting convection, the mean momentum equation yields a formula for wall shear stress τ_w or friction velocity U_τ, since $\tau_w = \rho U_\tau^2$. Launder et al (1975) finds $k \simeq 4.2U_\tau^2$ best represents the turbulence kinetic energy in the near-wall region. The mean velocity V parallel to the wall is then given by the law of the wall (e.g. a smooth wall)

$$V/U_\tau = \frac{1}{\kappa} \ell n(yU_\tau/\nu) + C \tag{38}$$

where $\kappa = 0.334$, $C = 5.5$ and $y = \delta$. Equating the production and dissipation rates then fixes ε.

EXAMPLE

Often the strength of a numerical model is also its weakness. If it is designed to produce a wealth of information about a process which is difficult to measure experimentally, how does one determine whether it works properly and the extent of its limitations? Model verification can be difficult. Although in this instance model validation is still incomplete, an instructive but relatively simple non-stratified flow example is presented.

Fully developed plane channel flow

Figure 2 depicts a section of a fully developed plane channel flow between smooth boundaries; the finite element discretization and the boundary conditions are also shown. Symmetry

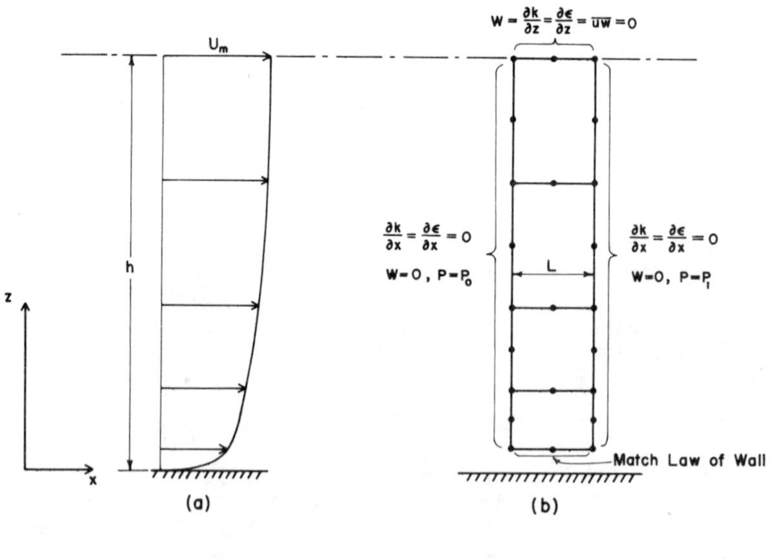

Figure 2. Plane channel flow. (a) Schematic mean velocity profile, (b) Finite element discretization.

allows one to consider only half the cross-section. Specification of the streamwise pressure gradient uniquely defines a solution, i.e. the channel Reynolds number $R = U_m h/\nu$ is a function only of $(P_1-P_0)/L$ for smooth channel walls. Boundary conditions for the stress components are in general <u>not</u> required because the equations describing their transport are algebraic. Specifying one of the stress components is equivalent in some instances to specifying the normal derivative of a velocity component. For example, the specification of $\overline{uw}$ on the channel centerline is a symmetry condition corresponding to $\partial U/\partial z = 0$, as can be seen from Equation 19. The one-dimensional nature of this problem simplifies the equations immediately by allowing one to set all x-derivatives to zero.

Figures 3 and 4 present results for flow at a channel Reynolds number of 12,500. Comparison with Laufer's (1951) experimental data is good, considering the neglect of the wall

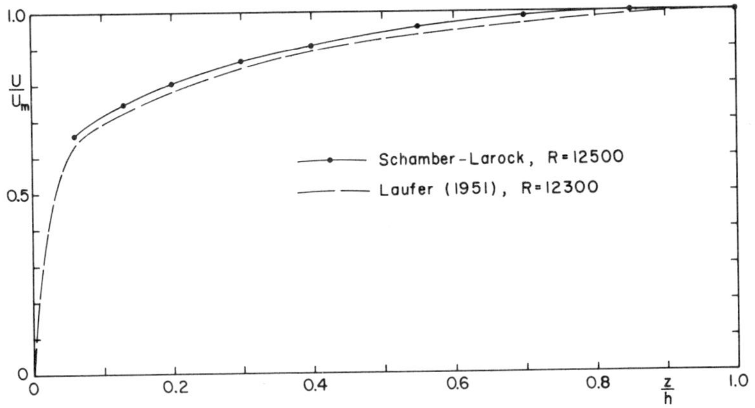

Figure 3. Mean velocity profile, theory and experiment.

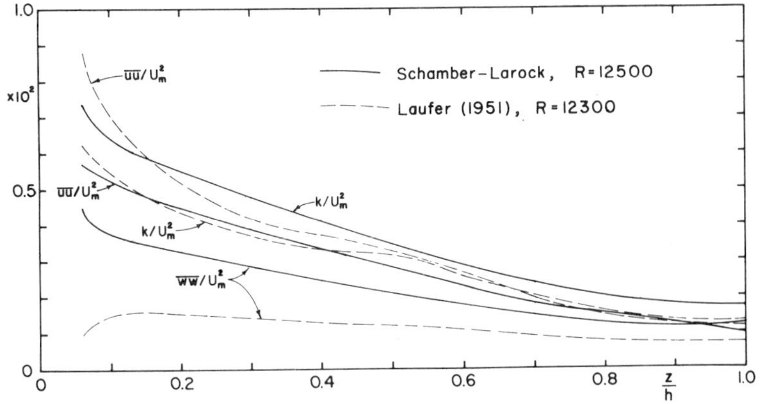

Figure 4. Turbulence parameters, theory and experiment.

proximity effect on the pressure scrambling terms. In this example the value of C_2 has been adjusted somewhat to $C_2 = 0.76$. All other constants remain as previously defined. For this turbulence model, the $\overline{vv}$ and $\overline{ww}$ correlations are equal even though experimental measurements show a slight separation between these normal stress components. Convergence to a final solution was sensitive to the initial estimate. For fully developed plane channel flow 16 iterations were required to achieve correction components such that $|\delta \mathbf{X}_i^{(16)}|/|\mathbf{X}_i^{(16)}| < 10^{-4}$. The total number of unknowns in this problem was 126.

SUMMARY

A finite element model for turbulent, stratified two-dimensional flow is formulated. It employs Launder's k-ε turbulence closure model with nonlinear algebraic relations for the Reynolds stresses and turbulent heat flux correlations. The computer program uses the Newton-Raphson technique to solve the resulting equation set. Several significant tradeoffs and alternative courses of action including choice of order of approximation vs. element size and solution strategy, deserve further study. However, preliminary verification tests for the model hold forth the promise of successfully modeling flow in prototype sedimentation basins in the near future.

ACKNOWLEDGEMENTS

This paper is based on work supported by the National Science Foundation under Grant No. ENG7618846. The aid of I. King and W. Norton in initial development of the present computer program is appreciated.

REFERENCES

Ames, W.F. (1972) Nonlinear Partial Differential Equations in Engineering, Vol. 2, Academic Press, p. 149.

Ariathurai, R. and R.B. Krone (1976) Finite Element Model For Cohesive Sediment Transport. J. Hyd. Div., ASCE, 102, HY3, March, pp. 323-338.

Bird, R.B., Stewart, W.E. and E.N. Lightfoot (1960) Transport Phenomena, John Wiley.

Camp, T.R. (1946) Sedimentation and the Design of Settling Tanks. Trans. ASCE, 111, pp. 895-936.

Clements, M.S. and A.F.M. Khattab (1968) Research into Time Ratio in Radial Flow Sedimentation Tanks. Instn. Civ. Engrs., 40, Aug., pp. 471-494.

Deardorff, J.W. (1970) A Numerical Study of Three-Dimensional Turbulent Channel Flow at Large Reynolds Numbers. J. Fluid Mech., 41, pp. 453-480.

Deardorff, J.W. (1973) The Use of Subgrid Transport Equations in a Three-Dimensional Model of Atmospheric Turbulence. J. Fluids Engineering, ASME, Sept., pp. 429-438.

Finlayson, B.A. (1972) Existence of Variational Principles for the Navier-Stokes Equation. Physics of Fluids, 15, no. 6, June, pp. 963-967.

Finlayson, B.A. and L.E. Scriven (1966) The Method of Weighted Residuals - A Review. Applied Mechanics Reviews, 19, no. 9, Sept., pp. 735-748.

Gibson, M.M. and B.E. Launder (1976) On the Calculation of Horizontal, Turbulent, Free Shear Flows Under Gravitational Influence. J. Heat Transfer, ASME, Feb., pp. 81-87.

Hanjalic, K. and B.E. Launder (1972) A Reynolds Stress Model of Turbulence and its Application to Thin Shear Flows. J. Fluid Mech., 52, pp. 609-638.

Hood, P. and C. Taylor (1974) Navier - Stokes Equations Using Mixed Interpolation, in Finite Element Methods in Flow Problems, Oden, T. et al., ed., UAH Press, Alabama, pp. 121-132.

Hossain, M.S. and W. Rodi (1976) Influence of Buoyancy on the Turbulence Intensities in Horizontal and Vertical Jets. Proceedings 1976 ICHMT Seminar on Turbulent Buoyant Convection, Dubrovnik, Yugoslavia.

Huebner, K.H. (1975) The Finite Element Method for Engineers, John Wiley, pp. 167-171.

Hutton, A.G. (1976) Finite Element Analysis of Turbulent, Incompressible Flow, Bounded by Smooth Walls. Second International Symposium on Finite Element Methods in Flow Problems. Santa Margherita Ligure, Italy, June 14-18, pp. 527-538.

King, I.P., Norton, W.R, and K.R. Iceman (1974) A Finite Element Model for Two-Dimensional Flow, in Finite Element Methods in Flow Problems, Oden, T. et al, ed., UAH Press, Alabama, pp. 133-137.

King, I.P., Norton, W.R. and G.T. Orlob (1973) A Finite Element Solution for Two-Dimensional Density Stratified Flow. Water Resources Engineers, Inc., Final Report for the U.S. Dept. of Interior, Office of Water Resources Research.

Larsen, Peter (1976) Research on Settling Basin Hydraulics. Tenth Anniversary Papers on Research in Progress, Dept. of Water Res. Engng, Lund Inst. of Tech., Bull. Serie A No. 55, Lund, Sweden, July, pp. 137-149.

Laufer, J. (1951) Investigation of Turbulent Flow in a Two-Dimensional Channel. NACA Report no. 1053, pp. 1247-1266.

Launder, B.E. (1975) On the Effects of a Gravitational Field on the Turbulent Transport of Heat and Momentum. J. Fluid Mech., 67, pp. 569-581.

Launder, B.E., Reece, G.J. and W. Rodi (1975) Progress in the Development of a Reynolds-stress Turbulence Closure. J. Fluid Mech., 68, pp. 537-566.

Launder, B.E. and D.B. Spalding (1974) The Numerical Computation of Turbulent Flows. Comp. Meth. Appl. Mech. and Engng., 3, pp. 269-289.

Patankar, S.V. and D.B. Spalding (1970) Heat and Mass Transfer in Boundary Layers, 2nd Ed. Intertext.

Reynolds, W.C. (1970) Computation of Turbulent Flows--State-of-the-Art, 1970. Rep. MD-27, Thermosciences Div., Dept. Mech. Engng., Stanford University, Stanford, CA. (Reprinted Feb. 1972), p. 8.

Sarikaya, H.Z. (1977) Numerical Model for Discrete Settling. J. Hyd. Div., ASCE, 103, HY8, August, pp. 865-876.

Taylor, C. and P. Hood (1973) A Numerical Solution of the Navier-Stokes Equations Using the Finite Element Technique. Computers and Fluids, 1, pp. 73-100.

Taylor, C., Hughes, T.G. and K. Morgan (1977) A Predictive Model for Turbulent Flow Utilizing The Eddy Viscosity Hypothesis and the Finite Element Method. Application of Computer Methods in Engineering, Vol. 2, University of Southern California, August, pp. 971-982.

Wills, R.F. and C. Davis (1962) Flow Patterns in a Rectangular Sewage Sedimentation Tank. Int. Conf. on Water Pollution Research, London, pp. 335-385.

Young, D., Liggett, J.A. and R.H. Gallagher (1976) Steady Stratified Circulation in a Cavity. J. Engng. Mech. Div., ASCE, 102, February, pp. 1-17.

APPENDIX I

Partial derivatives of f_u and $f_{\overline{ww}}$ are given here.

$$\frac{\partial f_u}{\partial P_j} = \mathbf{F}^{-2}\int_\Omega N_i \frac{\partial M_j}{\partial x}\, d\Omega, \qquad \frac{\partial f_u}{\partial W_j} = \int_\Omega N_i N_j \frac{\partial U}{\partial z}\, d\Omega$$

$$\frac{\partial f_u}{\partial U_j} = \int_\Omega N_i \left[N_j \frac{\partial U}{\partial x} + U\frac{\partial N_j}{\partial x} + W\frac{\partial N_j}{\partial z}\right] d\Omega$$

$$\frac{\partial f_u}{\partial T_j} = \frac{\partial f_u}{\partial k_j} = \frac{\partial f_u}{\partial \varepsilon_j} = \frac{\partial f_u}{\partial \overline{ww}_j} = \frac{\partial f_u}{\partial \overline{uT'_j}} = \frac{\partial f_u}{\partial \overline{wT'_j}} = 0$$

$$\frac{\partial f_u}{\partial \overline{uu}_j} = \int_\Omega N_i \frac{\partial N_j}{\partial x}\, d\Omega, \qquad \frac{\partial f_u}{\partial \overline{uw}_j} = \int_\Omega N_i \frac{\partial N_j}{\partial z}\, d\Omega$$

$$\frac{\partial f_{\overline{ww}}}{\partial P_j} = \frac{\partial f_{\overline{ww}}}{\partial T_j} = \frac{\partial f_{\overline{ww}}}{\partial \overline{uT'_j}} = 0$$

$$\frac{\partial f_{\overline{ww}}}{\partial U_j} = -\int_\Omega N_i \left(\overline{ww} - \frac{2}{3} c_2 k\right)\left(\overline{uu}\frac{\partial N_j}{\partial x} + \overline{uw}\frac{\partial N_j}{\partial z}\right) d\Omega$$

$$\frac{\partial f_{\overline{ww}}}{\partial W_j} = -\int_\Omega N_i \left(\overline{ww} - \frac{2}{3} k - \frac{4}{3}\hat{c}_2 k\right)\left(\overline{uw}\frac{\partial N_j}{\partial x} + \overline{ww}\frac{\partial N_j}{\partial z}\right) d\Omega$$

$$\frac{\partial f_{\overline{ww}}}{\partial k_j} = \int_\Omega N_i N_j \left\{ -\frac{2}{3}(\hat{c}_1 \varepsilon + P_r) + \hat{c}_2 \left[2\left(\overline{uw}\frac{\partial W}{\partial x} + \overline{ww}\frac{\partial W}{\partial z}\right) - \frac{2\beta^*}{\mathbb{F}^2}\overline{wT'} + \frac{2}{3} P_r \right] \right\} d\Omega$$

$$\frac{\partial f_{\overline{ww}}}{\partial \varepsilon_j} = \int_\Omega N_i N_j\, \hat{c}_1 \left(\overline{ww} - \frac{2}{3}k\right) d\Omega, \qquad \frac{\partial f_{\overline{ww}}}{\partial \overline{uu}_j} = -\int_\Omega N_i N_j \frac{\partial U}{\partial x}\left(\overline{ww} - \frac{2}{3} c_2 k\right)$$

$$\frac{\partial f_{\overline{ww}}}{\partial \overline{ww}_j} = \int_\Omega N_i N_j \left[(\hat{c}_1 \varepsilon + P_r) + \frac{\partial W}{\partial z}\left(\frac{4}{3}\hat{c}_2 k - \overline{ww} + \frac{2}{3} k\right) \right] d\Omega$$

$$\frac{\partial f_{\overline{ww}}}{\partial \overline{uw}_j} = -\int_\Omega N_i N_j \left[\left(\overline{ww} - c_2 \frac{2}{3} k\right)\left(\frac{\partial U}{\partial z} + \frac{\partial W}{\partial x}\right) - 2\hat{c}_2 k \frac{\partial W}{\partial x} \right] d\Omega$$

$$\frac{\partial f_{\overline{ww}}}{\partial \overline{wT'_j}} = \int_\Omega N_i N_j \frac{\beta^*}{\mathbb{F}^2}\left[\overline{ww} - \frac{2}{3} k - \frac{4}{3}\hat{c}_2 k \right] d\Omega$$

SIMULATION OF STRATIFIED TURBULENT FLOWS IN CLOSED WATER BODIES USING THE FINITE ELEMENT METHOD

A. N. Findikakis, J. B. Franzini, R. L. Street
Stanford University, Stanford, Ca. 94305

INTRODUCTION

Several finite element (FE) models for simulating flow and water quality in reservoirs, lakes and estuaries have been developed over the last few years. The main motivations for developing FE models have been the flexibility of the FE method in dealing with solution domains of practically any shape and the possibility of focusing the analysis on some areas of interest by increasing the density of the FE net. Some investigators (King et al. 1973, Baker et al. 1975) have developed FE solutions of the full momentum and heat balance equations, while others (Smith et al. 1973, Leimkuhler et al. 1975, Lam et al. 1975) have simply solved the convection-diffusion equation. Although all these modelers developed models for simulating turbulent flows, they did not elaborate on the description of the effect of turbulence on the general flow pattern. The existing FE hydrodynamic models use eddy viscosity and eddy diffusivity coefficients to describe the effect of turbulence. These coefficients are constant over either the entire solution domain, or large parts of it, and are selected quite arbitrarily, apparently, in such a way as to obtain the desirable simulation results. This is a severe limitation on the general applicability of these models. Of course, in some instances the solution of the hydrodynamic equations is dominated by the convection terms and proper modeling of turbulent transport is not a critical factor for the success of a flow simulation. However, this is not always the case, especially in confined water bodies, where convective velocities are small compared to those found, for example, in estuaries, and the general circulation pattern consists of large, slowly rotating eddies. The purpose of this paper is to present a FE model for turbulent flow simulation requiring a significantly smaller amount of empiricism in the selection of the constants involved than the models mentioned above, and therefore, of more general applicability.

Significant progress in turbulence modeling has been made over the last decade by meteorologists and other turbulence modelers. An important contribution in this field has been the development of the large eddy simulation (LES) method, originated in meteorological applications (Smagorinsky 1963, Deardoff 1972, 1973), and refined by a group of investigators in the Department of Mechanical Engineering at Stanford University (Leonard, 1974, Kwak, Reynolds and Ferziger 1975). The basic idea of LES is that, since it is not practically possible to simulate all turbulent motions from the largest scales down to the smallest which are governed by viscous dissipation, we simulate motions of scales larger than some cutoff value and model the effect of turbulent motions of smaller scales. This approach has been used successfully to simulate some laboratory flows, mostly with open boundaries (Kwak et al. 1975, Shaanan et al. 1976). The same approach has been used for simulating turbulence in the planetary boundary layer (Deardoff, 1972, 1973, Sommeria 1976) and in some channel flows (Deardoff 1970, Schumman 1975). All these models have used finite difference methods for the numerical solution of the large scale (LS) flow equations. The present work explores the use of the FE method in LES applied to closed stratified water bodies.

LARGE EDDY SIMULATION

It is important to make clear what exactly is simulated in a numerical simulation of a turbulent flow. To answer this question we first analyze each field variable F into a LS component $\overline{F}$ and a sub-grid-scale (SGS) component f, i.e.,

$$F = \overline{F} + f \tag{1}$$

The object of a turbulent flow simulation is to predict the LS field. The term 'sub-grid' arose because in the early large eddy simulations motions of scales larger than the grid size were simulated, while the effect of scales smaller than the grid (sub-grid-scales) was modeled. It was realized later that a more formal definition of the LS field was needed. Leonard (1974) defined the LS component $\overline{F}$, as the convolution of F with a filter function G(x). Thus,

$$\overline{F}(\underline{x}) = \int_V G(\underline{x} - \underline{x}')\, F(\underline{x}')\, d\underline{x}' \tag{2}$$

where the integration takes place over the entire flow regime. The filtered value $\overline{F}$ defined by Equation (2) is not identical to the conventional mean value and it is not subject to all the averaging rules applicable to conventional means. Important differences arise when we compare the filtered value of non-linear terms with their conventional mean value. A commonly used filter function is the 'top hat' or 'box' filter, defined by

$$G(\underline{x}-\underline{x}') = \begin{cases} \dfrac{1}{\Delta_1 \Delta_2 \Delta_3} & \text{for } |x_i - x_i'| < \dfrac{\Delta_i}{2} \\[2ex] 0 & \text{for } |x_i - x_i'| < \dfrac{\Delta_i}{2} \end{cases} \tag{3}$$

Use of the 'box' filter is equivalent to spatial averaging over the volume $\Delta_1 \Delta_2 \Delta_3$. This averaging concept (with $\Delta_1 = \Delta_2 = \Delta_3$) has been used in most of the early LES models (Lilly 1967, Deardoff 1970, 1973). Kwak et al. (1975) pointed out that the filter size need not be equal to the size of the grid. In fact they obtained better agreement with experimental results by using a filter equal to twice the grid size. Here we define the filter at the nodes of the elements as equal to the maximum rectangular parallilepiped that fits in the space of the elements surrounding each node. We also assume that the variation of the filter size is continuous throughout the solution domain. We approximate the filter size within each element in terms of its value at the nodes, assuming the form of its variation.

Equations for the LS flow in a thermally stratified fluid can be derived by filtering the continuity, the Navier-Stokes and the heat balance equations, after making the Boussinesq approximation for the density. The result is

$$\frac{\partial \overline{U}_i}{\partial x_i} = 0 \tag{4}$$

$$\frac{\partial \overline{U}_i}{\partial t} + \frac{\partial \overline{U_i U_j}}{\partial x_j} = \frac{1}{\rho_o}\left(\overline{F}_i - \frac{\partial \overline{P}}{\partial x_i}\right) + \beta g_i \overline{\Theta} + \frac{\partial}{\partial x_j}\left(\nu \frac{\partial \overline{U}_i}{\partial x_j}\right) \tag{5}$$

$$\frac{\partial \overline{\Theta}}{\partial t} + \frac{\partial \overline{U_i \Theta}}{\partial x_i} = \overline{Q}_H + \frac{\partial}{\partial x_i}\left(k \frac{\partial \overline{\Theta}}{\partial x_i}\right) \tag{6}$$

where $\overline{U}_i$ is LS velocity, $\overline{P}$ is LS pressure measured above hydrostatic pressure, $\overline{\Theta}$ is LS temperature measured above a reference temperature, $\overline{F}_i$ is Coriolis force, g_i is acceleration of gravity, β is coefficient of thermal expansion, ρ_o is reference density, ν is kinematic viscosity, k is thermal diffusivity, and $\overline{Q}_H$ is net heat inflow from different sources.

The filtered value of the nonlinear term in Equation (5) can be represented, by introduction of Equation (1), as

$$\overline{U_i U_j} = \overline{\overline{U}_i \overline{U}_j} + \overline{u_i \overline{U}_j} + \overline{\overline{U}_i u_j} + \overline{u_i u_j} \tag{7}$$

In contrast to what is true for conventional mean values, $\overline{\overline{U}_i \overline{U}_j} \neq \overline{U}_i \overline{U}_j$, $\overline{\overline{U}_i u_j} \neq 0$ and $\overline{u_i \overline{U}_j} \neq 0$, because the LS variables are not constant in space, but are running averages over the filtering volume. An approximation to $\overline{\overline{U}_i \overline{U}_j}$ can be obtained by observing that the filtered value of a smooth function F(x) can be approximated by carrying out a Taylor series expansion about the center of the filter volume x_o (Clark et al., 1977, p. 36). Using the 'box' filter defined by Equation (3) we have

$$\overline{F}(\underline{x}_o) = F(\underline{x}_o) + \frac{\Delta_k^2}{24}\left(\frac{\partial^2 F(\underline{x})}{\partial x_k \partial x_k}\right)_{\underline{x}_o} + O(\Delta_1^4, \Delta_2^4, \Delta_3^4) \tag{8}$$

Equation (7) for $F = \overline{U}_i \overline{U}_j$ gives

$$\overline{\overline{U}_i \overline{U}_j} \approx \overline{U}_i \overline{U}_j + \frac{\Delta_k^2}{24} \frac{\partial^2 \overline{U}_i \overline{U}_j}{\partial x_k \partial x_k} \tag{9}$$

This approximation, for the case of an isotropic filter ($\Delta_1 = \Delta_2 = \Delta_3$), was first proposed by Leonard (1974). The second term in Equation (9) has been referred to in the literature as the Leonard term (Reynolds, 1976).

Following Clark, Ferziger and Reynolds (1977) we can model the cross terms involved in Equation (7) by

$$\overline{\overline{U}_i u_j} \approx -\frac{\Delta_k^2}{24} \overline{U}_i \frac{\partial^2 U_j}{\partial x_k \partial x_k} \tag{10}$$

As pointed out by Clark et al. (1977) this approximation is not as good as the Leonard approximation for $\overline{\bar{U}_i\bar{U}_j}$, but it is still better than neglecting the cross terms completely. Combining Equations (7), (9) and (10) we have

$$\overline{U_iU_j} \approx \bar{U}_i\bar{U}_j + L_{ij} + \overline{u_iu_j} \tag{11}$$

where

$$L_{ij} = \frac{\Delta_k^2}{12} \frac{\partial\bar{U}_i}{\partial x_k} \frac{\partial\bar{U}_j}{\partial x_k} \tag{12}$$

Similarly we can approximate the filtered value of the nonlinear terms in the heat balance equation by

$$\overline{U_i\Theta} \approx \bar{U}_i\bar{\Theta} + L_{i\theta} + \overline{u_i\theta} \tag{13}$$

where

$$L_{i\theta} = \frac{\Delta_k^2}{12} \frac{\partial\bar{U}_i}{\partial x_k} \frac{\partial\bar{\Theta}}{\partial x_k} \tag{14}$$

Thus Equations (5) and (6) become

$$\frac{\partial\bar{U}_i}{\partial t} + \bar{U}_j \frac{\partial\bar{U}_i}{\partial x_j} = \frac{1}{\rho_o}\left(\bar{F}_i - \frac{\partial\bar{P}}{\partial x_i}\right) + \beta g_i\bar{\Theta}$$

$$+ \frac{\partial}{\partial x_j}\left(\nu \frac{\partial\bar{U}_i}{\partial x_j} - L_{ij} - \overline{u_iu_j}\right) \tag{15}$$

$$\frac{\partial\bar{\Theta}}{\partial t} + \bar{U}_i \frac{\partial\bar{\Theta}}{\partial x_i} = \bar{Q}_H + \frac{\partial}{\partial x_i}\left(k \frac{\partial\bar{\Theta}}{\partial x_i} - L_{i\theta} - \overline{u_i\theta}\right) \tag{16}$$

where L_{ij} and $L_{i\theta}$ are given by Equations (13) and (14). The terms $\overline{u_iu_j}$ and $\overline{u_i\theta}$ appearing in Equations (15) and (16) are the result of purely SGS interactions and must be modeled.

The simplest available model for $\overline{u_i\theta}$ and $\overline{u_iu_j}$ is the eddy viscosity-eddy diffusivity model

$$\overline{u_iu_j} - \frac{2}{3}\delta_{ij}\overline{E} = \overline{\tau}_{ij} = -\epsilon\overline{S}_{ij} \tag{17}$$

$$\overline{u_i\theta} = -\epsilon_\theta \frac{\partial\overline{\Theta}}{\partial x_i} \tag{18}$$

where $\overline{S}_{ij} = \frac{\partial \overline{U}_i}{\partial x_j} + \frac{\partial \overline{U}_j}{\partial x_i}$, $\overline{E} = \frac{1}{2}\overline{u_iu_j}$ is the SGS turbulent kinetic energy, and ϵ and ϵ_θ are the eddy coefficients. The first attempt to relate the eddy coefficients to the characteristics of the LS flow field and the size of the grid used for the numerical solution of the LS flow equations was made by Smagorinsky (1963), who proposed the following expression for the eddy viscosity:

$$\epsilon = (c\Lambda)^2\left(\frac{1}{2}\overline{S}_{ij}\overline{S}_{ij}\right)^{1/2} \tag{19}$$

where c is a constant and Λ is a length scale which is characteristic of the grid. Smagorinsky, as well as many others, let Λ equal the cube root of the product of the typical finite difference cell dimensions. Kwak et al. (1975), Clark et al. (1977) set Λ equal to the width of the filter which was twice the grid spacing. In these two simulations an isotropic filter was used. It is still open to question if, and if so how, one should modify Smagorinsky's model in the case of using a highly anisotropic filter, which is unavoidable in simulations of many geophysical flows of interest. The constant c has been given values in the range of 0.10 to 0.20. The eddy diffusivity coefficient can be expressed in terms of the eddy viscosity coefficient by

$$\epsilon_\theta = \frac{\epsilon}{a_T} \tag{20}$$

where a_T is a turbulent Prandtl number. The basic weakness of the eddy viscosity-eddy diffusivity model, described by Equations (18), (19) and (20), is that it does not include the effect of stratification on the SGS Reynolds stresses and heat fluxes. This may be a severe drawback in simulating stably stratified flows, where turbulent transport of momentum and heat is inhibited by stratification. To account for the effect of stratification on the SGS quantities, one has to go beyond Smagorinsky's expression for the eddy viscosity and develop a more sophisticated model. Perhaps, the simplest, theoretically sound, approach is to consider the differential equations for the SGS Reynolds stresses and turbulent heat fluxes, make certain simplifying assumptions to reduce them to algebraic equations, and then derive a set of explicit expressions for τ_{ij} and $\overline{u_i\theta}$ in terms of the LS velocity and temperature field (Sommeria 1976, Findikakis and Street, 1978).

In the present work we follow Sommeria to develop a simple algebraic model for the SGS correlations. This model is based on the differential equations for the SGS Reynolds stresses, turbulent heat fluxes, turbulent kinetic energy, and $\overline{\theta^2}$. These equations are simplified by adoption of the following assumptions:

a) Turbulence in the sub-grid space is in stationary equilibrium, i.e., the total time rate of change and net advection of the SGS turbulent quantities is zero.

b) Viscous dissipation is isotropic. In addition the filter size is small enough to lie in an inertial subrange. In this case it is possible to use dimensional arguments to derive expressions for the dissipation of the turbulent kinetic energy and $\overline{\theta^2}$.

c) Diffusive flux of all SGS quantities is negligible.

d) Production of SGS turbulence is isotropic.

e) Stratification affects turbulent momentum transport only indirectly through its effect on the magnitude of the turbulent kinetic energy. Thus, the buoyancy terms are neglected in the equations for the Reynolds stresses, but they are retained in the equations for the turbulent heat fluxes.

g) The pressure fluctuations can be analyzed into three components, the first arising from purely SGS interactions, the second due to the interaction between SGS turbulence and LS flow, and the third generated by buoyancy forces due to temperature fluctuations. Consequently the pressure-velocity gradient and pressure-temperature gradient correlations can be analyzed into three components corresponding to the three parts of the pressure fluctuation. The first part of the pressure-velocity gradient correlation is assumed proportional to the anisotropy in the Reynolds stress (Rotta, 1951), the second part is modeled after Mellor and Herring (1973) and Deardoff (1973), and the third is neglected in agreement with assumption (e). The first part of the pressure-temperature gradient correlation is assumed proportional to $\overline{u_i\theta}$, the second is neglected and the third is assumed proportional to the rate of production of anisotropy by the buoyancy forces. These assumptions make it possible to reduce the differential equations for the SGS quantities of interest to a system of algebraic equations. These are

$$\bar{\tau}_{ij} \text{ eq.: } -c_M \frac{\bar{E}^{1/2}}{\Lambda} \bar{\tau}_{ij} + (c_1 - \tfrac{2}{3}) \bar{E}\bar{S}_{ij} = 0 \tag{21}$$

$$\overline{u_i\theta} \text{ eq.: } -c_S \frac{\bar{E}^{1/2}}{\Lambda} \overline{u_i\theta} - \frac{2}{3} \bar{E} \frac{\partial \bar{\Theta}}{\partial x_i} + (1 - c_2)\beta g_i \bar{\theta}^2 = 0 \tag{22}$$

$$\bar{E} \text{ eq.: } -c_E \frac{\bar{E}^{1/2}}{\Lambda} \bar{E} - \bar{\tau}_{ij} \frac{\partial \bar{U}_i}{\partial x_j} + \beta g_i \overline{u_i\theta} = 0 \tag{23}$$

$$\bar{\theta}^2 \text{ eq.: } -c_\theta \frac{\bar{E}^{1/2}}{\Lambda} \bar{\theta}^2 - 2\, \overline{u_i\theta} \frac{\partial \bar{\Theta}}{\partial x_i} = 0 \tag{24}$$

where c_M, c_S, c_θ, c_E, c_1, and c_2 are constants and Λ a length scale associated with the dissipation of SGS turbulence. Here, Λ is set equal to $(\Delta_1 \Delta_2 \Delta_3)^{1/3}$. The solution of system (21-24) can be set in the form of Equations (17) and (18) with

$$\epsilon = f^{1/2} (c\,\Lambda)^2 \left(\frac{1}{2} \bar{S}_{ij} \bar{S}_{ij}\right)^{1/2} \tag{25}$$

$$\epsilon_{\theta_i} = \frac{\epsilon}{a_T} \quad \text{for} \quad i = 1, 2 \tag{26}$$

$$\epsilon_{\theta_i} = \frac{\epsilon}{a_T} \varphi \quad \text{for} \quad i = 3 \tag{27}$$

where

$$c = \left(\frac{\frac{2}{3} - c_1}{c_M} \right)^{3/4} \frac{1}{c_E^{1/4}}$$

$$a_T = \left(1 - \frac{3}{2} c_1\right) \frac{c_M}{c_S} \tag{28}$$

and f and φ are functions which reflect the effect of stratification. They are given in terms of the local Richardson numbers

$$Ri = \beta g \frac{\dfrac{\partial \bar{\Theta}}{\partial x_3}}{\frac{1}{2}\bar{S}_{ij}\bar{S}_{ij}} \qquad Ri* = \beta g \frac{\left(\dfrac{\partial \bar{\Theta}}{\partial x_k}\dfrac{\partial \bar{\Theta}}{\partial x_k}\right)^{1/2}}{\frac{1}{2}\bar{S}_{ij}\bar{S}_{ij}} \tag{29}$$

by

$$f = \frac{1}{2}\left\{(1-(c_3+c_4)Ri + \left[[1-(c_3+c_4)Ri]^2 + 4c_4 Ri \right.\right.$$
$$\left.\left. + 4c_3c_4(Ri*^2 - Ri^2)\right]^{1/2}\right\} \tag{30}$$

$$\varphi = \frac{1 - c_4(Ri*^2 - Ri^2)/Ri\, f)}{1 + c_4\, Ri/f} \tag{31}$$

where

$$c_3 = \frac{\frac{2}{3}}{\left(\frac{2}{3} - c_1\right)}\frac{c_M}{c_S} \qquad c_4 = \frac{2(1-c_2)}{\left(\frac{2}{3} - c_1\right)}\frac{c_M}{c_S}\frac{c_E}{c_\theta} \tag{32}$$

For flow in a thermally uniform fluid $f = 1$ and Equation (25) is reduced to the Smagorinsky model. A basic weakness of our model is that, because of assumptions (d) and (e), it does not predict the total suppression of turbulence in shear flows for high Richardson numbers. It also neglects the terms, similar to the Leonard terms, that arise in the derivation of the τ_{ij} and $\overline{u_i\theta}$ equations by filtering nonlinear terms. Despite these shortcomings, this model has been used successfully by Sommeria (1976) in atmospheric flow simulations. A more complete algebraic model for SGS turbulence is currently being tested at Stanford University (Findikakis and Street 1978).

DEVELOPMENT OF FINITE ELEMENT EQUATIONS

Several FE solutions of the Navier-Stokes equations have been presented over the last decade. Olson (1976) has reviewed and compared some of these solutions.

Hood and Taylor (1974) have pointed out that for error consistency in solving the Navier-Stokes equations, the maximum error associated with the residual of each variable must be the same. Carrying out a Taylor series expansion of the approximation functions they showed that, to achieve the same accuracy for all variables, the approximation function for the velocity, which has second order derivatives in the Navier-Stokes equations, must be one order higher than the approximation function for the pressure, which has first order derivatives only. The same criterion, applied to the heat balance equation, leads to an approximation function for the temperature of the same order as the approximation function for the velocity. The simplest choice satisfying this requirement is a linear approximation function for the pressure and quadratic functions for the velocity and the temperature. Thus, the velocity, the temperature and the pressure can be approximated within each element by:

$$\overline{U}_i = N_k \overline{U}_{i_k} \qquad \overline{\Theta} = N_k \overline{\Theta}_k \qquad \overline{P} = \overline{M}_\ell \overline{P}_\ell \tag{33}$$

where $\overline{U}_{i_k}$ is the value of the ith component of the LS velocity at node k, $\overline{\Theta}_k$ is the LS temperature at node k, $\overline{P}_\ell$ is the LS pressure at node 1, N_k is a quadratic approximation function used for the velocity and the temperature, and M_ℓ is a linear approximation function for the pressure.

The most commonly used methods to derive equations for the unknown nodal variables are the variational approach and the method of weighted residuals. The first is based on the existence of a functional, which one seeks to extremize. Since a preliminary investigation indicated that there is no apparent appropriate functional for Equations (4), (15) and (16) we followed the method of weighted residuals. The basic idea of this method is to find the values of the unknown nodal variables which minimize the residual (error) in the governing equations over the entire solution domain. This condition is expressed by

$$\int_V H^1 \frac{\partial \overline{U}_i}{\partial x_i} \, dV = 0 \tag{34}$$

$$\int_V H^2 \left\{ \frac{\partial \overline{U}_i}{\partial t} + \overline{U}_j \frac{\partial \overline{U}_i}{\partial x_j} + \frac{\partial}{\partial x_j} \left[L_{ij} + \overline{u_i u_j} - \nu \frac{\partial \overline{U}_i}{\partial x_j} \right] + \frac{1}{\rho_o} \left(\frac{\partial \overline{P}}{\partial x_i} - F_i \right) - \beta g_i \overline{\Theta} \right\} dV = 0 \tag{35}$$

$$\int_V H^3 \left\{ \frac{\partial \bar{\Theta}}{\partial t} + \bar{U}_i \frac{\partial \bar{\Theta}}{\partial x_i} + \frac{\partial}{\partial x_i} \left[L_{i\theta} + \overline{u_i \theta} - k \frac{\partial \bar{\Theta}}{\partial x_i} \right] - Q_H \right\} dV = 0 \qquad (36)$$

where H^1, H^2 and H^3 are weighting functions. A widely used approach is to select the weighting functions equal to the approximation functions used for the unknown variables (Galerkin method). This choice, however, is not simple in the case of having more than one variable, approximated by different functions. Hood and Taylor (1974) have suggested that, in order to assign a consistent accuracy to all the equations, the residual arising from each equation must be weighted according to the maximum error occurring in the equation. This criterion applied to Equations (34), (35) and (36) leads to

$$H^1 = M \qquad H^2 = H^3 = N \qquad (37)$$

Using Equation (37) and applying the Gauss theorem on Equations (35) and (36) to remove second order derivatives of velocity and temperature and first order pressure derivatives, we obtain

$$\int_V M_\ell \frac{\partial \bar{U}_i}{\partial x_i} \, dV = 0 \qquad (38)$$

$$\int_V N_k \left[\frac{\partial \bar{U}_i}{\partial t} + \bar{U}_j \frac{\partial \bar{U}_i}{\partial x_j} - \frac{1}{\rho_o} \bar{F}_i - \beta g_i \bar{\Theta} \right] - \frac{\partial N_k}{\partial x_j} \left[L_{ij} + \overline{u_i u_j} - \nu \frac{\partial \bar{U}_i}{\partial x_j} \right]$$

$$- \frac{\partial N_k}{\partial x_i} \frac{1}{\rho_o} \bar{P} \, dv + \int_A N_k \left[L_{ij} + \overline{u_i u_j} - \nu \frac{\partial \bar{U}_i}{\partial x_j} \right] \eta_j + N_k \frac{1}{\rho_o} \bar{P} \eta_i \, dA = (\qquad (39)$$

$$\int_V N_k \left[\frac{\partial \bar{\Theta}}{\partial t} + \bar{U}_i \frac{\partial \bar{\Theta}}{\partial x_i} - Q_H \right] - \frac{\partial N_k}{\partial x_i} \left[L_{i\theta} + \overline{u_i \theta} - k \frac{\partial \bar{\Theta}}{\partial x_i} \right] dv$$

$$+ \int_A N_k \left[L_{i\theta} + \overline{u_i \theta} - k \frac{\partial \Theta}{\partial x_i} \right] \eta_i \, dA = 0 \qquad (40)$$

where η_i is the direction cosine of the surface dA in the ith direction.

BOUNDARY CONDITIONS

In low Reynolds number flows the boundary elements can be selected small enough to lie within the laminar sublayer. Thus, the velocity at the boundary can be assumed equal to zero. At high Reynolds numbers, however, it is not practically possible to have the required very fine discretization of the solution domain to describe properly the variation of the velocity from the laminar sublayer all the way through to the turbulent zone. In this case, the boundary of the solution domain must be set at a distance from the physical boundary. We assume that the component of the velocity, normal to the numerical boundary is zero and that the shear stress at the numerical boundary can be approximated in terms of the LS tangential velocity by

$$L_{ns_i} + \overline{u_n u_{s_i}} - v \frac{\partial U_{s_i}}{\partial x_n} = c_D \left(\bar{U}^2_{s_1} + \bar{U}^2_{s_2} \right)^{1/2} \bar{U}_{s_i} \qquad (41)$$

where n is the direction normal to the boundary and s_1 and s_2 are the directions of two perpendicular axes on the boundary, and c_D is an empirical coefficient. More vigorous boundary conditions can be applied to flows with certain well defined mean flow patterns. For example in one dimensional channel or pipe flow we have

$$\langle L_{ns} + \overline{u_n u_s} - \frac{\partial \bar{U}_s}{\partial x_n} \rangle = - \left(\frac{f}{8}\right)^{1/2} |\langle V_s \rangle| \langle V_s \rangle \qquad (42)$$

where $\langle\ \rangle$ indicates averaging in the s direction, $\langle V_s \rangle$ is the maximum velocity of the s-direction averaged velocity profile and f is a friction factor.

If we are interested in simulating the LS eddy structure in large closed water bodies, we can neglect the details of the flow field structure at the air-water interface and assume that a rigid lid is imposed on the free water surface. The lid is lowered or raised as a whole when we have net outflow or inflow, respectively. The LS velocity component normal to the lid is set equal to zero when there is no change in the total volume of the water body, or equal to a specified value which is estimated from the rate of total volume change. The shear stress at the lid is assumed equal to the wind stress, i.e.,

$$L_{ns_i} + \overline{u_n u_{s_i}} - \frac{\partial \bar{U}_{s_i}}{\partial x_n} = \tau_w \cos(s_w, s_i) \qquad (43)$$

where s_w is the direction of the wind.
A more rigorous boundary condition can be obtained if the wind stress is constant along the lid; then,

$$\langle L_{ns_i} + \overline{u_n u_{s_i}} - \nu \frac{\partial \bar{U}_{s_i}}{\partial x_n} \rangle = \tau_w \cos(s_w, s_i) \tag{44}$$

where ⟨ ⟩ indicates averaging in the direction of the wind.
Boundary conditions for the heat balance equation are either specified temperature at the boundary or specified heat flux through the boundary. The latter condition is expressed by

$$L_{i\theta} + \overline{u_i \theta} - k \frac{\partial \bar{\Theta}}{\partial x_i} = q_i \tag{45}$$

where q_i is the heat flux through the boundary in the ith direction.

NUMERICAL SOLUTION OF THE FE EQUATIONS

Equations (38), (39), (40) are solved using a semi-implicit scheme allowing large time steps. However, for any realistic simulation the time step must be limited to a fraction of the rotation period of the faster moving LS eddies.

Equations (6) and (7) can be written as

$$\frac{\partial \bar{U}_i}{\partial t} = G_i(\bar{U}_j, \bar{P}, \bar{\Theta}) \tag{46}$$

$$\frac{\partial \bar{\Theta}}{\partial t} = H(\bar{U}_j, \bar{\Theta}) \tag{47}$$

where G_i and H include all the terms of Equations (6) and (7) except the time derivatives terms. Using simple finite differencing for the time derivatives, we approximate Equations (46), and (47) by:

$$\frac{\partial \bar{U}_i}{\partial t} \approx \frac{\bar{U}_i^{n+1} - \bar{U}_i^n}{\Delta t} = G_i\left(\alpha_1 \bar{U}_j^n + \alpha_2 \bar{U}_j^{n+1}, \alpha_1 \bar{P}^n + \alpha_2 \bar{P}^{n+1}, \alpha_1 \bar{\Theta}^n + \alpha_2 \bar{\Theta}^{n+1}\right) \tag{48}$$

$$\frac{\partial \bar{\Theta}}{\partial t} \approx \frac{\bar{\Theta}^{n+1} - \bar{\Theta}^n}{\Delta t} = H\left(\alpha_1 \bar{\Theta}^n + \alpha_2 \bar{\Theta}^{n+1}, \alpha_1 \bar{U}_j^n + \alpha_2 \bar{U}_j^{n+1}\right) \tag{49}$$

where $\bar{U}_i^n$ is the value of the ith component of the velocity at the beginning of the time step Δt, $\bar{U}_i^{n+1}$ is its value at the end of the time step Δt, and α_1 and α_2 are parameters characterizing the implicitness of the scheme ($\alpha_1 + \alpha_2 = 1$). The solution scheme is fully implicit for $\alpha_2 = 1$, and explicit for $\alpha_2 = 0$. Applying the approximations (48), (49) on Equations (39), (40) and writing the continuity equation for $\alpha_1 \overline{U}^n + \alpha_2 \overline{U}^{n+1}$, we obtain:

$$\int_V \alpha_2 M_\ell \frac{\partial \overline{U}_i^{n+1}}{\partial x_i} dV = - \int_V \alpha_1 M_\ell \frac{\partial \overline{U}_i^{n}}{\partial x_i} dV \tag{50}$$

$$\begin{aligned}
&\int_V N_k \Bigg[\frac{\overline{U}_i^{n+1}}{\Delta t} + \alpha_1 \alpha_2 \left(\overline{U}_j^n \frac{\partial \overline{U}_i^{n+1}}{\partial x_j} + \overline{U}_j^{n+1} \frac{\partial \overline{U}_i^{n}}{\partial x_j} \right) + \alpha_2^2 \overline{U}_j^{n+1} \frac{\partial \overline{U}_i^{n+1}}{\partial x_j} \\
&- \alpha_2 \beta g_i \overline{\Theta}^{n+1} \Bigg] + \alpha_2 \frac{\partial N_k}{\partial x_j} \left[(\epsilon + \nu) \frac{\partial \overline{U}_i^{n+1}}{\partial x_j} + \epsilon \frac{\partial \overline{U}_j^{n+1}}{\partial x_i} \right] \\
&- \frac{\partial N_k}{\partial x_j} L_{ij} - \frac{\alpha_2}{\rho_o} \frac{\partial N_k}{\partial x_i} \overline{P}^{n+1} dV + \int_A \frac{\alpha_2}{\rho_o} N_k \overline{P}^{n+1} \eta_i \, dA \\
&= \int_V N_k \left[\frac{\overline{U}_i^n}{\Delta t} - \alpha_1^2 \overline{U}_j^n \frac{\partial \overline{U}_i^n}{\partial x_j} + \frac{1}{\rho_o} (\alpha_1 \overline{F}_i^n + \alpha_2 \overline{F}_i^{n+1}) + \alpha_1 \beta g_i \overline{\Theta}^n \right] \\
&- \alpha_1 \frac{\partial N_k}{\partial x_j} \left[(\epsilon + \nu) \frac{\partial \overline{U}_i^n}{\partial x_j} + \epsilon \frac{\partial \overline{U}_j^n}{\partial x_i} \right] + \frac{\alpha_1}{\rho_o} \frac{\partial N_k}{\partial x_i} \overline{P}^n dV \\
&- \int_A \frac{\alpha_1}{\rho_o} N_k \overline{P}^n \eta_i \, dA
\end{aligned} \tag{51}$$

$$\begin{aligned}
&\int_V N_k \left[\frac{\Theta^{n+1}}{\Delta t} + \alpha_1 \alpha_2 \left(\overline{U}_i^n \frac{\partial \overline{\Theta}^{n+1}}{\partial x_i} + \overline{U}_i^{n+1} \frac{\partial \overline{\Theta}^{n}}{\partial x_i} \right) + \alpha_2^2 \overline{U}_i^{n+1} \frac{\partial \overline{\Theta}^{n+1}}{\partial x_i} \right] \\
&+ \alpha_2 \frac{\partial N_k}{\partial x_i} (\epsilon_{\theta_i} + k) \frac{\partial \overline{\Theta}^{n+1}}{\partial x_i} - \frac{\partial N_k}{\partial x_i} L_{i\theta} \, dV = \int_V N_k \Bigg[\frac{\overline{\Theta}^n}{\Delta t} \\
&- \alpha_1^2 \overline{U}_i^n \frac{\partial \overline{\Theta}^n}{\partial x_i} + \alpha_1 Q_H^n + \alpha_2 Q_H^{n+1} \Bigg] - \alpha_1 (\epsilon_{\theta_i} + k) \frac{\partial N_k}{\partial x_j} \frac{\partial \overline{\Theta}^n}{\partial x_i} dV
\end{aligned} \tag{52}$$

Equation (51) has been written for the case of fixed velocities at the boundaries. If we assume a limited velocity at the boundary, the contribution of the corresponding area integrals over the boundaries must be added to Equation (51). Similarly, Equation (52) has been written for the case where no heat is gained or lost through the boundaries.

Because of the dependence of the eddy coefficient on the velocity field and the presence of the Leonard terms, the nonlinearities in Equation (51) are of a higher order than in the Navier-Stokes and heat balance equations for laminar flow. Equations (50)-(52) are solved using an iterative procedure. The nonlinear terms are linearized at each iteration by approximating $\overline{U}_j^{n+1}$ in the convective terms and ϵ in the SGS Reynolds stress term by their respective values at the previous iteration. The Leonard terms are expressed explicitly in terms of the values of the velocity gradients at the previous iteration. A special problem arises in the case that we use C^0 elements. In this case there is no inter-element continuity in the variation of the eddy coefficients and the Leonard terms because these terms are functions of the velocity and temperature gradients which are not continuous. This may have rather severe implications for the stability of the numerical solution. It was found, for example, that in the simulation of stably stratified flows, $L_{i\theta}$ caused artificial accumulation or dissipation of heat (depending on the sign of the velocity gradient, at the boundary between elements because of the lack of inter-element continuity of the velocity gradient. This resulted, through the interaction of the velocity and the temperature field, in a growing instability. This instability is removed if the velocity and temperature gradients are continuous. Since it is not feasible to use C^1 elements because of the large number of the gradients involved, a method to smooth the velocity and temperature gradients is employed. In the two-dimensional applications described in the next Section, we used the smoothing technique proposed by Hinton and Campbell (1974). This method is based on the observation that the 2 x 2 Gaussian integrating points are the optimal sampling points for the gradient of the approximated variable. Locally smoothed values of the gradient at the corner nodes are obtained by bilinear extrapolation of the values of the gradient at the 2 x 2 Gauss points. Then the values of the gradient obtained for each node from different elements are averaged to obtain unique values of the gradient at all nodes. The smoothed values of the velocity and temperature gradients are used to estimate the eddy coefficients and the Leonard terms.

The linearized system obtained from Equations (50) and (51) is solved first. The velocity field obtained from the solution of this system is used to formulate the FE heat balance equations. The heat balance equations form a linear system which is solved to obtain the new values of the temperature field. To accelerate convergence of the iterative solution, we introduce an underrelaxation scheme. After we solve Equations (50) and (51), we adjust the values of all the unknowns according to the following formula:

$$^{q+1}X = {}^{q}X + \omega\,({}^{q+1}\tilde{X} - {}^{q}X) \tag{53}$$

where X is the vector of all the unknowns U and P, the superscrips indicate iterations [$^{q+1}\tilde{X}$ is the solution of Equations (50) and (51) at the $(q+1)^{th}$ iteration] and ω is an underrelaxation parameter. The optimum convergence rate was obtained for $\omega = 0.5$. It was found that the adjustment of the unknowns by Equation (53) accelerated the solution of the nonlinear equations significantly. The same underrelaxation scheme was applied on the solution of Equation (53).

An attempt to solve Equations (50) and (51) using the Newton-Raphson method was unsuccessful because the solution did not converge unless the initial choice of the unknowns was close to the final solution.

APPLICATION

As a first step towards developing a FE model for large eddy simulations, we have developed a two-dimensional model. 2-D simulations may describe actual flows realistically under two conditions, viz., a) when we deal with flows having predominantly 2-D LS structure (This may be the case when the flow is driven by a continuous and strong force acting in one plane.) and b) when we include in the formulation an additional eddy viscosity to account for the three dimensional structure of turbulence, i.e., Equation (19) should be modified to

$$\epsilon = (c\Lambda)^2 \left(\frac{1}{2}\bar{S}_{ij}\bar{S}_{ij}\right)^{1/2} + K \tag{54}$$

K of course, has to be selected empirically and it may not be a constant throughout the flow regime, since the structure of turbulence is not the same everywhere. Here, we assume that most of the energy cascaded down to the SGS level is spent to generate turbulent motions in our solution plane and that $K = 0$.

Eight-node, quadratic, isoparametric elements are used for the velocity and the temperature, and four-node, linear, subparametric elements for the pressure. The corner nodes of the two elements coincide. The filter size is defined at the corner nodes and we assume that its variation within each

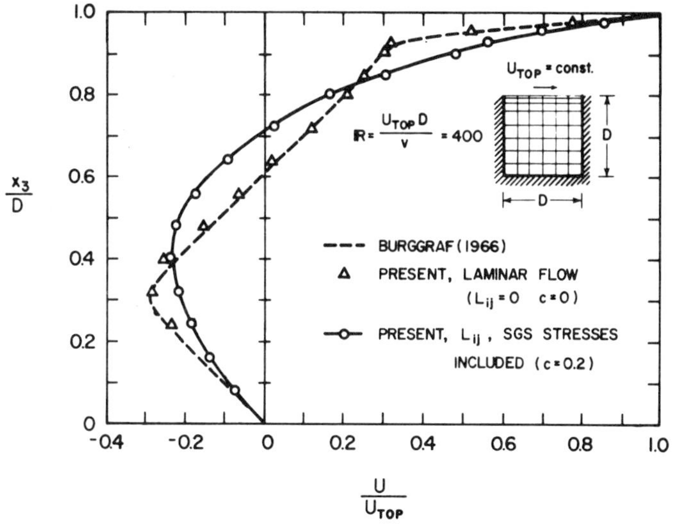

Fig. 1 Midplane velocity profile in a thermally uniform square cavity. Flow driven by the upper boundary moving at constant speed.

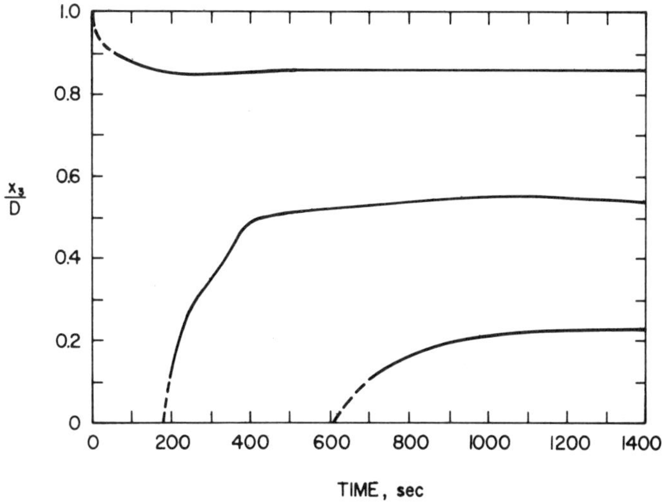

Fig. 2 Level of center of eddies in stably stratified square cavity vs. time. Flow driven by constant shear stress at the top.

element is linear. For the numerical solution in time $\alpha_2 = 1$ is assumed.

The first application of our model was to the flow in a thermally uniform square cavity, driven by moving the upper boundary at constant speed. This flow has been the subject of many studies (Burgraff 1966, Pan and Acrivos 1967, Greenspan 1969, Bozeman and Dalton 1973). All the numerical solutions of this problem have treated the flow as laminar. We checked our model by checking it against the laminar flow solution obtained by several other investigators. As shown in Fig. 2 our solution is almost identical to the finite difference solution by Burgraff (1966) and the FE solution by Gartling and Becker (1974). However, when we included the SGS Reynolds stresses, as modeled by Smagorinsky's expression (19), we obtained a significantly different velocity profile. Turbulent momentum transport results in a smoother velocity profile. Also, the center of the core eddy is at a higher level than in the laminar flow solution. Figure 2 shows the midplane velocity profile obtained for $\mathbb{R} = \frac{U_{top}}{\nu} = 400$, where U_{top} is the velocity of the top wall, and D the width of the cavity.

Even more dramatic differences in the predicted velocity field by the present and conventional models were found in wind driven flows in thermally stratified cavities. Young, Ligget and Gallagher (1976a, b) have presented a FE solution of this problem. They modeled the effect of turbulence by replacing the kinematic viscosity by an eddy viscosity, different in the two directions, but constant throughout the entire flow field. For a wind stress of 1.0 dynes/cm^2 acting on the top of a 10 by 10 m cavity, with an initial temperature gradient 0.4 C/m, the flow pattern developed after an initial transient period consisted of two large eddies. We verified this solution by adjusting our model to their eddy viscosity formulation. However, when we applied the present model, i.e., modeling the SGS Reynolds stresses by equations (25)-(27) and including the Leonard terms, we obtained a dramatically different velocity field. The LS flow pattern consisted of three eddies, one on top of the other. This difference in the flow pattern is due entirely to the different way of modeling turbulent transfer transport mechanisms. The constant eddy viscosity formulation allows relatively high momentum transport in the lower part of the cavity, making thus the lower eddy strong enough to maintain itself. In contrast, the eddy coefficients predicted by Equations (25)-(27) are much smaller in the lower part of the cavity than in the upper part because velocity gradients become significantly smaller as we move towards the bottom of the cavity. This results in less momentum transfer to the second

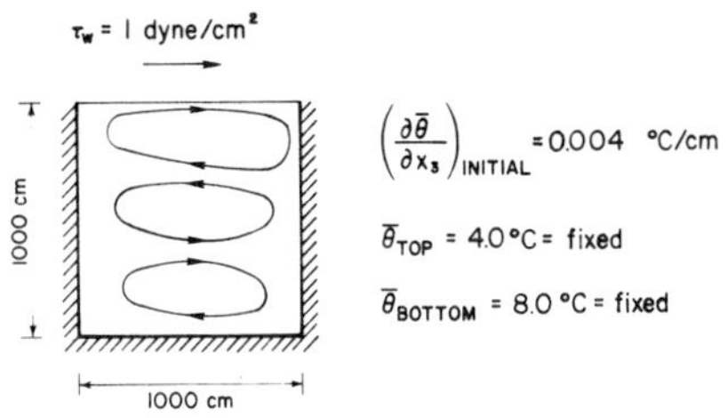

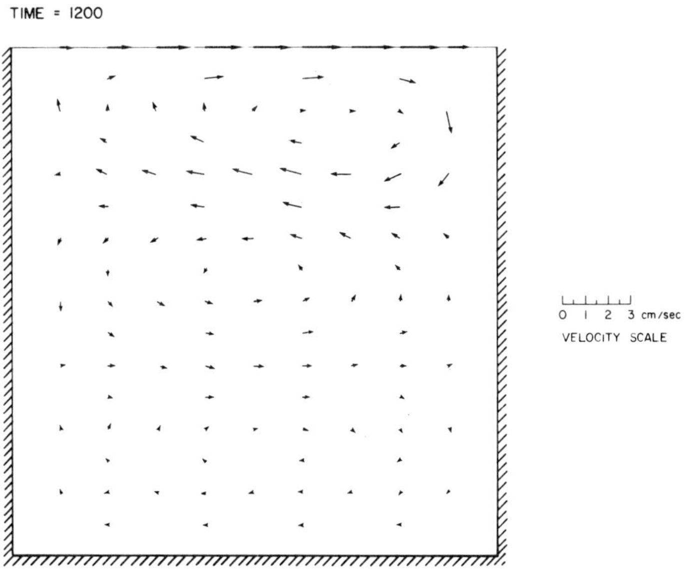

(a)

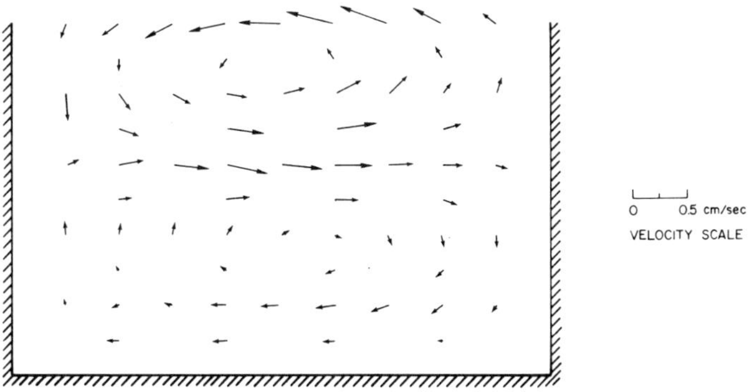

(b)

Fig. 3 Velocity field in stably stratified cavity at time 1200 sec: (a) entire cavity, (b) lower part of the cavity (larger velocity scale).

eddy than with the constant eddy viscosity model. Thus, the second eddy is weaker and after it is formed, it starts shrinking in the vertical direction, giving space for the formation of another eddy underneath. Figure 2 shows the variation of the level of the center of the three eddies with time. The velocity field at time 1000 sec is shown in Figures 3(a) and 3(b). Figure 3(b) shows the velocity field in the lower part of the cavity using a larger velocity scale. In the selection of the values of the constants we followed Sommeria (1976) and used $c_1 = 2/5$, $c_2 = 1/3$, $c_M/c_E = 5.71$, $c_S/c_E = 5.71$, $c_\theta/c_E = 1.71$, but we adjusted c_E to 0.503; instead of 0.7 that was used by Sommeria, in order to have $c = .2$ for neutral conditions (Eq. (28)). Apparently the total number of simulated eddies depends on our assumption for the relationship between the rate of turbulent momentum transport and the local structure of the velocity and temperature field. Unfortunately, there are no experimental data available for flow in stratified cavities. An experimental program for this problem is under way at Stanford University.

ACKNOWLEDGEMENTS

This work was supported in part by the Water Resources, Urban and Environmental Engineering Program, Engineering Division, National Science Foundation through Grant ENG77-13880.

REFERENCES

Baker, A. J., 1975: Predictions in Environmental Hydrodynamics using the Finite Element Method - Part I: Theory. AIAA J., 13, 1.

Bozeman, J. D., C. Dalton, 1973: Numerical Study of Viscous Flow in a Cavity. J. of Comp. Physics, 12, pp. 348-363.

Burggraf, O. R., 1966: Analytical and Numerical Studies of the Structure of Steady Separated Flows. J. of Fluid Mech., 24, 1, pp. 113-151.

Clark, R. A., J. H. Ferziger, W. C. Reynolds, 1977: Evaluation of Sub-Grid-Scale Turbulence Models Using a Fully Simulated Turbulent Flow. Stanford Univ., Dept. Mech. Eng. Report TF-9.

Deardoff, J. W., 1970: A Numerical Study of Three-Dimensional Turbulent Channel Flow at Large Reynolds Numbers. J. Fluid Mech. 41, pp. 453-480.

Deardoff, J. W., 1972: Numerical Investigation of Neutral and Unstable Planetary Boundary Layers. J. Atm. Sci., 29, pp. 91-115.

Deardoff, J. W., 1973: The Use of Subgrid Scale Transport Equations in a Three-Dimensional Model of Atmospheric Turbulence. J. Fluids Eng., Trans. ASME, pp. 429-438.

Findikakis, A. N., R. L. Street, 1978: An Algebraic Model for Sub-Grid-Scale Turbulence in Stratified Flows, Submitted for publication.

Gartling, D. K., E. B. Becker, 1976: Finite Element Analysis of Viscous Incompressible Fluid Flow, Part 2: Applications. Comp. Methods in Appl. Mech. and Eng., 8, pp. 127-138.

Greenspan, D., 1969: Numerical Studies of Prototype Cavity Flow Problems. Hinton E., J. S. Campbell, 1974: Local and Global Smoothing of Discontinuous Finite Element Functions Using a Least Squares Method. Int. J. Num. Meth. in Eng., 8, pp. 461-480.

Hood, P., C. Taylor, 1974: Navier-Stokes Equations Using Mixed Interpolation. Proc. Int. Symposium on FEM in Flow Problems, Swansea.

King, I. P., W. R. Norton, G. Orlob, 1973: A Finite Element Solution of Two-Dimensional Stratified Flows, Water Resources Engineers Inc., Walnut Creek, Ca.

Kwak, D., W. C. Reynolds, J. V. Ferziger, 1975: Three Dimensional Time-Dependent Computation of Turbulent Flow. Stanford Univ. Dept. Mech. Eng. Report TF-5.

Leihmkuhler, W., J. Connor, J. Wang et al., 1975: Two-Dimensional Finite Element Dispersion Model. ASCE Waterways, Harbors and Coastal Eng. Div., Conf on Modeling, San Francisco.

Lam, D. C. L., 1976: Comparison of Finite Element and Finite Difference Methods for Nearshore Advection-Diffusion Transport Models. Int. Conf. on Finite Elem. in Water Res., Princeton Univ.

Leonard, A., 1974: Energy Cascade in Large-Eddy Simulations of Turbulent Flows. Adv. Geophys., 18A, pp. 237-248.

Mellor, G. L., H. J. Herring, 1973: A Survey of the Mean Turbulent Field Closure Models. AIAA J., 11, 5, pp. 600-609.

Olson, M. D., 1976: Comparison of Various Finite Element Solution Methods for the Navier-Stokes Equations. Proc. Symp. on FEM in Water Resources, Princeton Univ.

Pan, F., A. Acrivos, 1967: Steady Flows in Rectangular Cavities. J. of Fluid Mech., 28, 4, pp. 643-655.

Reynolds, W. C., 1976: Computation of Turbulent Flows. Annual Rev. of Fluid Mech., 8, pp. 183-208.

Rotta, J. C., 1951: Statistische Theorie nichthomogener Turbu-lenz, Zeitschrift für Physik, 129, pp. 547-572.

Shaanan, S., J. H. Ferziger, W. C. Reynolds, 1975: Numeric simulation of homogeneous turbulence with rotation. Stanford Univ. Dept. Mech. Eng. Report TF-6.

Schumann, U., 1975: Subgrid Scale Model for Finite Difference Simulations of Turbulent Flows in Plane Channels and Annuli. of Comp. Physics, 18, pp. 376-404.

Smagorinsky, J., 1963: General Circulation Experiments with t Primitive Equations: I - The Basic Experiment. Mon. Wea. Re 91, pp. 99-164.

Smith, I. M., R. V. Faraday, B. A. O'Conner, 1975: Rayleigh Ritz and Galerkin Finite Elements for Diffusion-Convection Prob-lems. Water Res. Research, 9, 3, pp. 593-606.

Sommeria, G., 1976: Three-Dimensional Simulation of Turbule Processes in an Undisturbed Trade Wind Boundary Layer. J. Atm. Sci., 33, pp. 216-241.

Young, D.-L., J. A. Ligget, R. H. Gallagher, 1976a: Steady Stratified Circulation in a Cavity. ASCE J. Eng. Mech. Div., 102, EM1, pp. 1-17.

Young, D.-L., J. A. Ligget, R. H. Gallagher, 1976b: Unsteady Stratified Circulation in a Cavity. ASCE J. Eng. Mech. Div., 102, EM6, pp. 1009-1023.

SOLUTION OF THE TIME-DEPENDENT NAVIER-STOKES EQUATIONS VIA F.E.M.

P. M. Gresho, R. L. Lee, T. W. Stullich

Lawrence Livermore Laboratory, University of California, Livermore, CA 94550

and R. L. Sani

University of Colorado, Boulder, CO 80302

INTRODUCTION

As a further step toward generating a time-dependent, three-dimensional numerical model of the atmospheric boundary layer, we have developed a finite element computer code for solving the incompressible, two-dimensional, time-dependent Navier-Stokes equations. This code employs the primitive variables (u,v,p) and includes three types of elements, all defined on isoparametric quadrilaterals: (1) 4-node, bilinear velocity with piecewise constant pressure, (2) 8-node serendipity for velocity with 4-node bilinear pressure (we have essentially stopped using this element however, for reasons discussed in Huyakorn et al. (1978)), (3) 9-node biquadratic velocity with 4-node bilinear pressure.

Our goal was to generate accurate, yet efficient solutions to the (semi-discretized) time-dependent problem with an algorithm which would be stable for any grid spacing and any Reynolds number. Based on our successful earlier experience with a simpler FEM code, which solved the (linear) advection-diffusion equation (Gresho et al., 1978), wherein we first tested our new time integration ideas, we have extended and generalized these techniques to the nonlinear Navier-Stokes equations. When this earlier code was used to solve the transient heat equation, we were able to generate an accurate simulation, from small time all the way to steady state with a reasonable number (e.g. 50) of time steps (they start very small, but grow approximately exponentially). The technique employed, which also takes advantage of the inherent implicitness (via the mass matrix) of FEM systems, is completely stable, second-order accurate, and includes an automatic time step selection algorithm which is based solely on accuracy requirements. This method is in contrast to an earlier, explicit method, advocated by Baker (1974) in the stream function - vorticity approach, which (effectively) requires mass lumping and is stability-limited.

The following sections will describe the algorithm in detail, but an overview might be useful. By an appropriate combination of two common integration techniques, the (implicit) trapezoid rule and an (explicit) Adams-Bashforth formula, we are able to vary the integration time step based on an estimate of the local (single step) time truncation error. The non-linear equations engendered by the implicit step are solved via a modified Newton-Raphson iterative method (chord method), and the associated linear equations are solved using a direct method (Gaussian elimination via the frontal technique).

We believe that this technique will reliably generate stable and accurate simulations. It is expected to be especially efficient for flows which approach a steady state, since the time step will then increase quite rapidly (with essentially no loss in accuracy and without the oscillatory behavior associated with the conventional use of the trapezoid rule). In fact, this technique or a similar one, may turn out to be the most effective way to calculate large Reynolds number flows when only the steady solution is of interest. (This is to be contrasted with schemes which solve the steady state equations via iterative methods and by incrementally increasing the Reynolds number. These techniques are not always reliable since convergence is usually not assured.) The technique will also be useful for fully time-dependent flows, such as vortex shedding (the "correct", and variable time step would be selected based on the instantaneous "physics" of the flow) or flows with time-varying boundary conditions (e.g. the diurnal cycle in the atmospheric boundary layer).

In the remainder of this paper, we explain our algorithm in some detail and present some numerical results.

NAVIER-STOKES EQUATIONS AND FEM SPATIAL DISCRETIZATION

The two-dimensional equations of motion and continuity for a constant property incompressible Newtonian fluid are the Navier-Stokes equations, written here in stress-divergence form,

$$\rho\left(\frac{\partial \underline{u}}{\partial t} + \underline{u} \cdot \nabla \underline{u}\right) = \nabla \cdot \underline{\underline{\tau}}$$

$$\nabla \cdot \underline{u} = 0 \qquad (1)$$

where

$$\underline{u} = (u,v) \quad \text{and} \quad \tau_{ij} = -P\delta_{ij} + \mu\left(\frac{\partial u_i}{\partial x_j} + \frac{\partial u_j}{\partial x_i}\right)$$

is the symmetric stress tensor. The density (ρ) and viscosity (μ) are physical properties of the fluid. Equation (1) can be used, given appropriate initial and boundary conditions, to obtain the two velocity components (u and v in the x- and y-directions, respectively) and the pressure (P).

Note that in an incompressible fluid the pressure is an intrinsic and independent variable of the motion (Aris, 1962) and is not related to any thermodynamic equation of state; it is an implicit variable which instantaneously "adjusts itself" in such a way that the incompressibility constraint (continuity equation) remains satisfied. This is one characteristic of an incompressible flow that invariably makes the problem difficult to solve; another is the nonlinear advection terms, $\underline{u} \cdot \nabla\underline{u}$.

The finite element discretization of these equations is performed via the Galerkin method, wherein the velocity and pressure are approximated by

$$u = \sum_{j=1}^{N} u_j(t)\ \phi_j(x,y) \tag{2}$$

with a similar equation for v, and

$$P = \sum_{j=1}^{M} P_j(t)\ \psi_j(x,y)$$

where there are N velocity nodes and M pressure "nodes" in the discretized domain. The basis functions for velocity approximation, $\phi_i(x,y)$, are piecewise polynomials which are one degree higher than those for pressure approximation, $\psi_i(x,y)$, for reasons enumerated previously (e.g. Olson, 1977). The resulting Galerkin equations, written in a compact matrix form, are

$$\begin{aligned} M\dot{u} + [K + N(u)]u + CP &= f \\ C^T u &= 0 \end{aligned} \tag{3}$$

where u is the global vector (length 2N) of u_i and v_i, P is a global vector (length M) of the P_i, and f is a 2N global vector which incorporates the appropriate boundary conditions in velocities or surface tractions. M is the (2N x 2N) mass or inertia matrix, K is the (2N x 2N) viscous matrix, N(u) is the (2N x 2N) nonlinear advection matrix, C is the (2N x M) pressure gradient matrix and its transpose, C^T, is the (M x 2N) divergence matrix. Details of the formulation and matrix definitions have been published too often to require repetition here (see, for example, Gray and Pinder, 1977); we prefer to utilize our limited space to present something new.

Equation (3), which describes a nonlinear system of ordinary differential equations (ODE's) with algebraic constraints, can also be represented in the following partitioned matrix form, as a full ODE system,

$$\begin{bmatrix} M & 0 \\ 0 & 0 \end{bmatrix} \begin{Bmatrix} \dot{u} \\ \dot{p} \end{Bmatrix} + \begin{bmatrix} K + N(u) & C \\ C^T & 0 \end{bmatrix} \begin{Bmatrix} u \\ p \end{Bmatrix} = \begin{Bmatrix} f \\ 0 \end{Bmatrix} \tag{4}$$

which is called a time-singular system, since the coefficient matrix of $(\dot{u},\dot{p})^T$ is singular (the continuity equations, $C^Tu = 0$, still appear only as algebraic constraints which, indirectly, determine the pressure).

Because of the inherent implicitness of the pressure, it is *not* possible to solve Eq. (3) or (4) by any purely explicit (time-marching) time integration scheme; this is especially clear in the time-singular form, Eq. (4). This fact, combined with the mass matrix which couples the nodal accelerations, suggests that implicit time integration techniques are appropriate. There are of course other reasons to consider implicit methods, most of them focusing on their greater stability - which can often be used to advantage when an explicit method would impose unrealistically (or unaffordably) small time steps. Unfortunately, this greater stability exacts a significant computational price; viz. implicit techniques applied to nonlinear ODE's generate nonlinear algebraic systems whose solution must be obtained once per time step.

Faced with the prospect of solving large nonlinear algebraic systems, and the fact that, as a minimum, the pressure must be solved implicitly, one could (and should) consider compromise techniques. The compromise, from a fully implicit method, generally results in exchanging stability advantages for computational simplicity. Some obvious compromises to consider when solving Eq. (3) are:

1. Treat viscous and advection terms explicitly. This leads to a *linear* system of equations (for u_{n+1} and P_{n+1}) with a *symmetric* (but indefinite) coefficient matrix. Depending on both Reynolds number and nodal spacing, the time step will be limited by either advective or diffusive stability limits.
2. Treat only the advection matrix explicitly. Again, a linear system with a symmetric, indefinite matrix will result. This scheme would be better than the first if it avoids an (otherwise) artificially small diffusion time step (as could occur in our PBL modeling or in any *boundary* layer flow analysis).
3. As in the first scheme except: (a) lump (diagonalize, in some simple, ad hoc manner) the mass matrix, (b) solve for u_{n+1} in terms of P_{n+1} (and a given right-hand-side vector), (c) insert the result into the continuity constraint, $C^Tu_{n+1} = 0$ to obtain, effectively,

the appropriate Poisson equation for the pressure, $(C^TC)P_{n+1} = g$. A variant of this solution algorithm is also used in the Marker and Cell (MAC) method which is widely used in finite difference simulations. Either method of spatial discretization could be subject to a grid Reynolds number instability, however, which is usually suppressed via an upwind difference technique on the advection terms, which introduces additional (pseudo) viscosity. The principal advantage of this approach is that the resulting linear system involves a symmetric and *positive definite* matrix, which opens the door to other solution methods (we have assumed that only direct solution techniques could be confidently applied to the indefinite matrix systems); e.g. a block successive over relaxation (SOR) technique. Such a scheme could be useful for 3-D simulations or for very large 2-D simulations when the sheer cost of direct methods makes them nearly unaffordable.

4. Invoke a penalty method to eliminate both the pressure and the continuity equation (Malkus and Hughes, 1978). In this formulation, the penalty term must be treated implicitly. If the advection terms are treated explicitly, this approach generates a linear system with a symmetric and positive-definite matrix and again, iterative solution methods can at least be considered.

While we have considered, and are still considering, these and other compromise techniques, in this paper we concentrate on a no-compromise, implicit integration scheme. This will serve as a point of reference for possible future compromises, as well as being useful in its own right.

THE TIME INTEGRATION SCHEME

As mentioned earlier, we employ a completely stable (A-stable, to be precise), second-order accurate implicit time integration scheme - the trapezoid rule. This scheme, of course, is far from new in FEM programs. What *is* new is our approach to time step selection: we do not use a constant Δt, nor even an occasionally user-modified Δt. Since stability is no longer a concern, we can focus on a more desirable criterion for step size selection - systematic and automatic control of temporal accuracy. To do this, we rely on the well-developed theory of ODE's, as expounded in Shampine and Gordon (1975) for example, and implemented in general software packages, such as EPISODE (Byrne and Hindmarsh, 1977). Rather than considering *general* algorithms (variable step, variable order) however, we decided to extract only a small, but relevant portion of this theory; hopefully a near optimum selection for systems which are only moderately stiff and for which numerical damping is not desired.

If one were to invoke the penalty method as an alternative technique for solving the Navier-Stokes equations, for example, it might be more appropriate to select an ODE method *with* damping, so that the spurious transient effects associated with the penalty term (the pressure gradient term, in effect) can be quickly and efficiently damped into the "noise level". Such a scheme could be developed, for example, using the implicit Euler method (with an explicit Euler predictor formula) which, although only first order accurate, would still permit stable time steps of variable size. Another alternative would be to use the stiffly stable, higher order backward differentiation formulas as discussed, for example, by Byrne and Hindmarsh (1977).

A summary of the ODE theory employed in the following sections is presented in Appendix 1.

The Adams-Bashforth (AB) Predictor Step

This second-order accurate scheme, being explicit, is applied only to the velocities in Eq. (3):

$$u^P_{n+1} = u_n + \frac{\Delta t_n}{2}\left[\left(2 + \frac{\Delta t_n}{\Delta t_{n-1}}\right)\dot{u}_n - \frac{\Delta t_n}{\Delta t_{n-1}}\dot{u}_{n-1}\right] \quad . \qquad (5)$$

Note that two history vectors, $\dot{u}_n$ and $\dot{u}_{n-1}$ must be computed and stored. The manner of computing them, which does not require inverting the mass matrix in Eq. (3), will be presented shortly. The vector u^P_{n+1} has two uses: (1) it forms a first guess (predictor) for the implicit step (corrector) to follow and (2) it provides a portion of the information required to estimate the local time truncation error. Note that this explicit predictor step is quite cheap in comparison to the implicit step to be described next; hence error estimation and the associated time step control are, in this sense, almost free of extra cost. Also note that Eq. (5) cannot be applied until the second time step (n=1) and then $\dot{u}_o$ must be available; again, we return to this point shortly.

Trapezoid Rule (TR) Corrector Step and the Solution for Pressure

The TR, being an implicit method, can be applied to the full system in Eq. (3): this gives

$$\begin{aligned} \frac{2}{\Delta t_n} M(u_{n+1} - u_n) + [K + N(u_n)]\, u_n + [K + N(u_{n+1})]\, u_{n+1} \\ + C(P_n + P_{n+1}) = 2f \\ C^T u_{n+1} = 0 \end{aligned} \qquad (6)$$

In Eq. (6) we have taken the boundary conditions to be independent of time and we have assumed that the continuity equation was satisfied at the previous step - $C^T u_n = 0$. For reasons discussed in Appendix 2, we insist that the initial velocity field, u_o, satisfy $C^T u_o = 0$, from which it follows by induction that $C^T u_n = 0$. Rearrangement of Eq. (6) gives

$$\begin{bmatrix} \frac{2}{\Delta t_n} M + K + N(u_{n+1}) & C \\ C^T & 0 \end{bmatrix} \begin{Bmatrix} u_{n+1} \\ P_{n+1} \end{Bmatrix}$$

$$= \begin{bmatrix} \frac{2}{\Delta t_n} M - K - N(u_n) & -C \\ C^T & 0 \end{bmatrix} \begin{Bmatrix} u_n \\ P_n \end{Bmatrix} + \begin{Bmatrix} 2f \\ 0 \end{Bmatrix} \qquad (7)$$

which we recognize and interpret as a nonlinear system $A(X)X = b$. We defer the details of solving Eq. (7) to the next section, as it is a digression from our discussion of the time integration scheme. Once u_{n+1} and P_{n+1} are available (and they are the final, "reported" solution) from Eq. (7), we can form the acceleration vectors required for the next AB predictor step via the "inversion" of the TR; viz

$$\dot{u}_{n+1} = \frac{2}{\Delta t_n} (u_{n+1} - u_n) - \dot{u}_n \qquad (8)$$

where $\dot{u}_n$ is available from the previous application of Eq. (8). This of course ultimately leads to the need to compute $\dot{u}_o$, which is in fact done, as described in Appendix 2. (It may be successfully argued that we do indeed invert the mass matrix in order to obtain our acceleration history vectors; this is true, but only once per problem, and with a payoff which we believe makes it worthwhile.)

Local Time Truncation Error and Time Step Selection

As discussed in Appendix 1, once we have u^P_{n+1} and u_{n+1}, we can compute the local time truncation error, its norm, and consider a time step variation. We use the following RMS relative error norm,

$$||d_T||^2 \equiv \frac{1}{N_u + N_v} \left[\frac{1}{U^2_{max}} \sum_{i=1}^{N_u} (du_i)^2 + \frac{1}{V^2_{max}} \sum_{i=1}^{N_v} (dv_i)^2 \right], \qquad (9)$$

where N_u, N_v are the number of variable u-nodes and v-nodes, respectively and U_{max}, V_{max} are estimates of the maximum x- and y-velocity components, respectively. Also du_i and dv_i are the components of the local truncation error vector; e.g.

$$du_i \equiv \frac{u_i^{n+1} - u_i^{(P)n+1}}{3\left(1 + \frac{\Delta t_{n-1}}{\Delta t_n}\right)} \quad ; \qquad i = 1,2,\ldots,N_u$$

Upon comparing $||d_T||$ with the input value, ε, we can adjust the time step, up or down, according to

$$\Delta t_{new} = \Delta t_{old} \left(\varepsilon / \|d_T\|\right)^{1/3} \tag{10}$$

Since the predictor formula cannot be applied at the first time step, as mentioned earlier, error estimates begin with the completion of the second time step (cf. Appendix 2).

The remaining details of the strategy employed for time step changes include some additional economic considerations and are thus deferred until we discuss our solution method for obtaining u_{n+1} and P_{n+1}, from Eq. (7).

SOLUTION OF THE NONLINEAR SYSTEM

To solve Eq. (7), we employ a Newton-Raphson iterative method, in a modified form called the chord approximation, which we now describe. The Newton method for solving $A(X)X = b$ leads to the following iterative sequence:

$$J_k \, \delta X^{m+1} = b - A(X^m)X^m \tag{11}$$

where $\delta X^{m+1} = X^{m+1} - X^m$ is the change in X between iterations and $J_k \equiv \partial/\partial X^k \, [A(X^k)X^k]$ is the Jacobian matrix. If the Jacobian were to be fully updated for each iteration, we would use $k = m$ and be performing "full" Newton iterations. Although this is the only way to realize the (asymptotic) quadratic convergence rate associated with Newton's method, it is obviously expensive since J_m must be formed and factored (via an LU decomposition) at each iteration within a single time step. This is probably not a cost-effective technique and we therefore plan to use an out-of-date Jacobian and accept the slower convergence rate (asymptotically linear) associated with such a chord method. This is justifiable since (a) we have a good starting guess from an AB predictor formula (if Δt isn't too large), (b) we can afford more iterations since they now involve only "back-substitutions" which are much cheaper than a full factorization plus back substitutions, and (c) we obtain the same result in the converged limit (as $m \to \infty$, $\delta X^{m+1} \to 0$ and we have the solution to $A(X)X = b$ for *any* convergent method). Hence we will employ an outdated Jacobian, hereafter designated by J. In fact, not only do we avoid full Jacobian updates within a time step, we will even permit J to lag several time steps behind (note that, from Eq. (7), J is a function of Δt) and only update it when the chord iteration sequence is converging too slowly or not at all. The final form of our chord system, from Eqs. (7) and (11) is

$$\left[\begin{array}{c|c} \frac{2}{\Delta t} M + K + N(u) + N'(u) & C \\ \hline C^T & 0 \end{array}\right] \left\{\begin{array}{c} \delta u_{n+1}^{m+1} \\ P_{n+1}^{m+1} \end{array}\right\}$$

$$= \left\{\begin{array}{c} 2f + \left[\frac{2}{\Delta t_n} M - K - N(u_n)\right] u_n - CP_n \\ \hline 0 \end{array}\right\} - \left\{\begin{array}{c} \left[\frac{2}{\Delta t_n} M + K + N(u_{n+1}^m)\right] u_{n+1}^m \\ \hline 0 \end{array}\right\} \quad . \qquad (12)$$

The first vector on the RHS is "b" and is evaluated only once per time step, while the second vector corresponds to $A(X^m)X^m$ and must be reevaluated at every iteration. Other noteworthy features of the chord system above are: (1) both Δt and u are left unsubscripted in the outdated Jacobian; they were computed at some earlier time in general so that J remains constant for "several" time steps (each of which entails "several" iterations) and is stored in factored form, (2) the matrix $N'(u)$ in J represents the new terms arising from differentiation of $N(u)u$ (which is quadratic in the velocities) - whereas $N(u)$ corresponds to the operator $\underline{u} \cdot \nabla$ in Eq. (1), $N'(u)$ corresponds to terms like $\partial u/\partial x$, etc. We compute our nonlinear Jacobian matrix terms analytically, from the velocity basis functions, element-by-element, and store all resulting triply subscripted arrays on disk (we are not yet sure whether this approach is more efficient than recomputing them whenever needed), (3) since the pressure variable appears only linearly, it is more efficient to take advantage of this fact - thus P_{n+1}^{m+1} appears in the solution vector rather than δP_{n+1}^{m+1} and there is no CP_{n+1}^m term on the RHS, (4) since $C^T u_o = 0$, it can be shown that $C^T u_{n+1}^m = 0$ for all $m,n \geq 0$; hence we've dropped the $C^T u_{n+1}^m$ term from the bottom portion of the $A(X^m)X^m$ vector, finally (5) the value of u_{n+1}^o in $A(X^m)X^m$ is u_{n+1}^P; no predictor value is required for pressure. We solve the linear algebraic systems using a disk-based unsymmetric frontal solver (Hood, 1976) without pivoting.

We monitor the convergence rate of the chord system and accept the solution as "converged" when the norm (again, relative, RMS) of the total velocity "iterates" is $\leq$ 10% of the time truncation error. The philosophy here is to "solve" Eq. (7), via Eq. (12), only accurately enough so that the Newton convergence error doesn't contaminate the estimation of the local time truncation error. Notice that the pressure plays no role in either of the convergence tests. At this time, all that can be said is that we hope and believe that a sufficiently accurate velocity solution will ensure an acceptably accurate pressure field, although we are quite aware of the notorious "sensitivity of the pressure" in numerical solutions of the Navier-Stokes equations. Hence, we will monitor convergence rates of the pressure as we test the algorithm on more problems and, if necessary, will modify our Newton convergence strategy.

We can now return to a brief discussion of our overall strategy for time step changes. First, noting that the RHS of Eq. (12) always needs to be recomputed when commencing a new time step, we see that it costs us nothing to increase Δt whenever our accuracy test, Eq. (10), indicates that an increase is permissible; hence we increase Δt whenever possible, according to Eq. (10) but we do so *only* on the RHS of Eq. (12). The same general strategy applies to time step reductions except that in this case, the current Δt has failed the local accuracy criterion and the entire time step should be repeated; we repeat the step only if the reduction is "significant" enough (say 20% or more), otherwise we "push on": furthermore, if a Δt change is "too large" or "too small" (e.g. $\sim$ 50%), we will force a Jacobian update so that our chord approximation doesn't fall too far behind.

Perhaps even more relevant than these update criteria, however, are those employed in the solution of Eq. (12): (1) if the solution doesn't converge within a prescribed number of iterations (e.g. 3 or 4), but is converging, we update the Jacobian in the middle of a time step and continue the iterations, (2) if the solution is diverging after several iterations, we start the step over with an updated Jacobian, (3) if the Jacobian is updated via (1) or (2) above and the solution still is converging too slowly or is diverging, we stop the calculation to study the cause. It is to be emphasized that we thus far have limited experience in exercising these new algorithms and the details of our strategies at this point are tentative and will surely change as we learn more about their behavior on a range of practical problems.

NUMERICAL RESULTS

For our test problem we selected a simple, but non-trivial example: the flow development, starting from rest, in a channel with a sudden expansion. The ultimate Reynolds number, while low ($\sim$ 60), is large enough to cause the development of a significant recirculation region in the expansion corner. For this case, $\rho = \mu = 1$ and the channel is six units long by one (or two) units high. The flow is initiated by imposing normal stress (and zero tangential velocity) boundary conditions at the inlet and outlet boundaries ($f_n = 715$ at the inlet and zero at the outlet), where $\pm f_n = - P + 2\mu \partial u/\partial x$ is the normal stress. The value of f_n at the inlet was obtained from an earlier run with our steady state code at Re = 60, for which the maximum velocity is 60. This (natural) boundary condition corresponds, approximately, to imposing a step change in ΔP across the channel since, for $v = 0$ at x = constant, $\partial v/\partial y = 0 \Rightarrow \partial u/\partial x = 0 \Rightarrow f_n = \pm P$. The domain is discretized with 20 equal-sized (1.0 by 0.5) quadratic (9-node) elements. There is a total of 101 velocity nodes and 31 pressure nodes which leads to a system with 233 equations (which includes all boundary nodes).

The nature of the flow development is revealed in Fig. 1, which shows the time history of velocities and pressure at certain selected locations. It is seen that the flow initially fills the channel and that there is no (resolvable) separation. Before long, however, as the velocity increases and the inertia terms become important, the flow can no longer negotiate the corner and separation occurs (presumably somewhere in the neighborhood of t = .2-.3); of course with this coarse grid we could easily miss earlier separation. From this time forth, the recirculation region grows steadily and the entire flow field monotonically approaches a steady state.

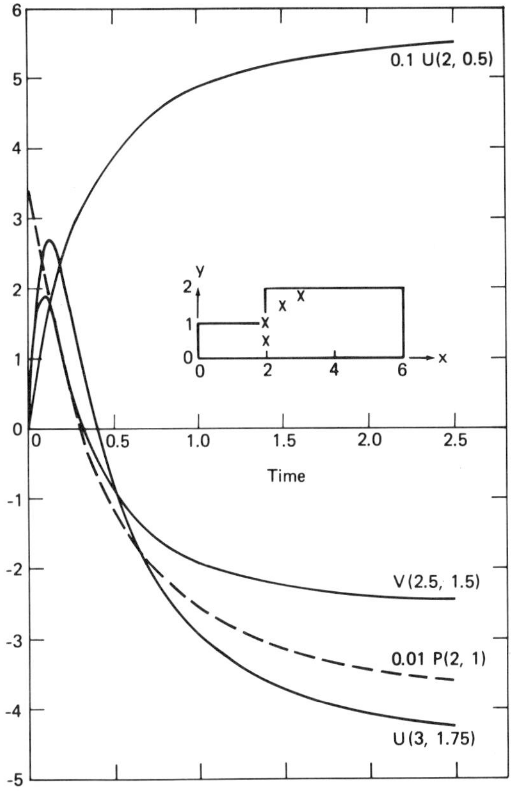

FIGURE 1. Selected Time Histories

Figures 2 and 3 show streamlines and pressure contours at small time (t = .02), when the maximum velocity is only ∿ 3.4. The streamlines fill the entire expansion region and the pressure is little different from that at t = 0, which is essentially a monotonic profile with a steeper gradient in the narrow part of the channel.

These features are repeated at t = 0.4 in Figs. 4 and 5 at which time the maximum velocity is ∿ 36; here, the eddy formation, although very weak, has clearly begun. The pressure

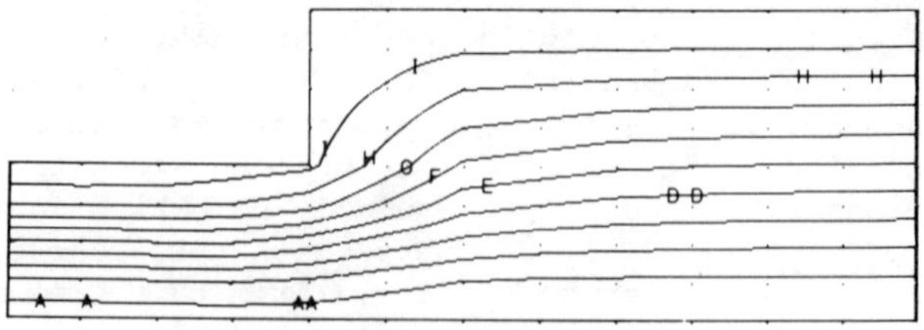

FIGURE 2. Streamlines (ψ) at small time (t = 0.02); the contour interval is Δψ = 0.278.

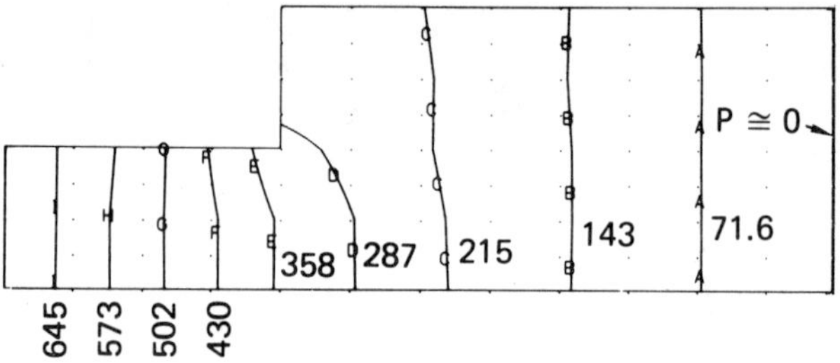

FIGURE 3. Pressure contours at small time (t = 0.02); the contour interval is ΔP = 71.6.

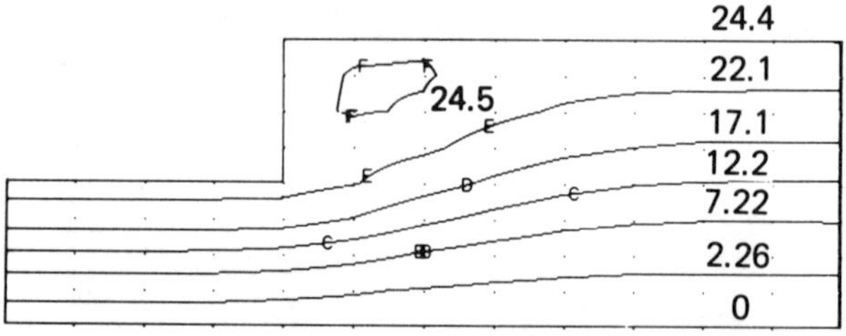

FIGURE 4. Streamlines at intermediate time (t = 0.4).

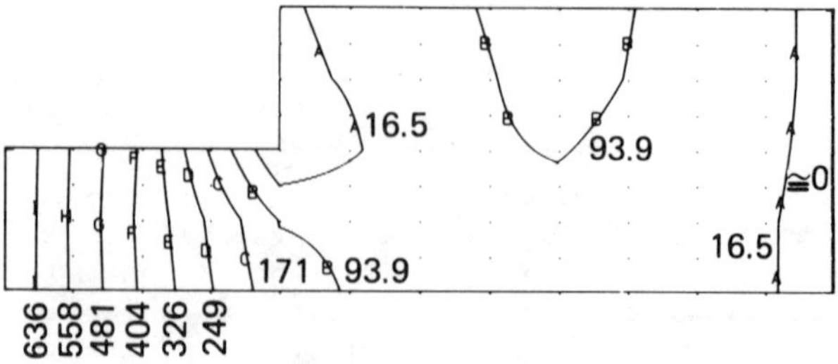

FIGURE 5. Pressure contours at intermediate time (t = 0.4).

distribution has changed dramatically, reflecting the flow redistribution.

Finally, at t = .8, at which time the maximum velocity is ∿ 47, the results are shown as a vector plot and pressure contours in Figs. 6 and 7. At this time the qualitative form of

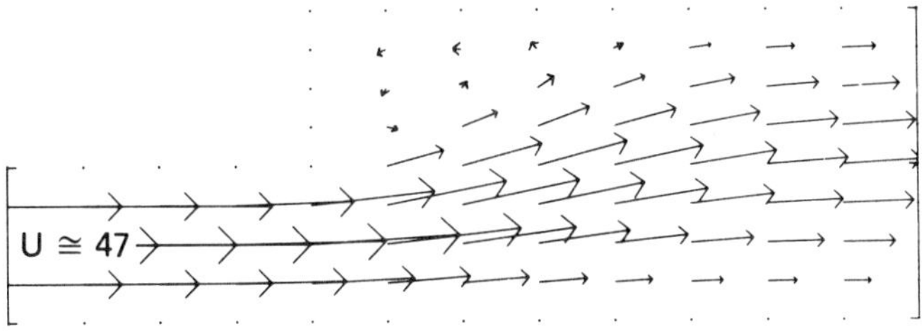

FIGURE 6. Velocity vectors at t = 0.8.

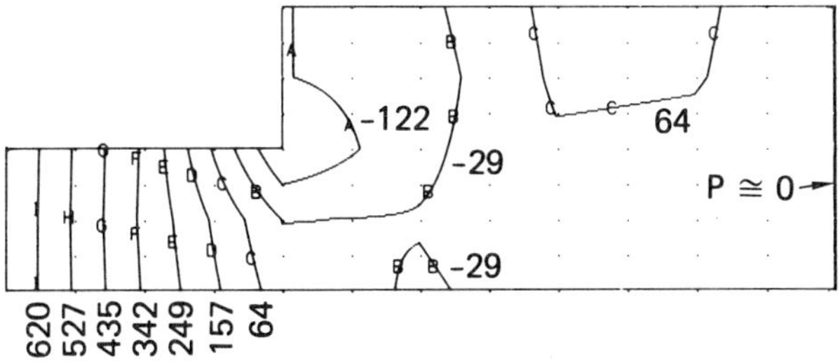

FIGURE 7. Pressure contours at t = 0.8.

of the flow is essentially final - only the magnitudes will change. The recirculation region, revealed by the vector plot, will grow somewhat in strength, but will not change shape significantly. The pressure distribution has again changed significantly, tending toward that which exists at steady state; viz., there is a large pressure drop in the inlet portion of the channel and a pressure rise after the expansion.

Unfortunately, our algorithm for the chord method was not quite debugged at the time of this writeup. Thus, the results presented here are those for "full Newton" iterations; i.e., the Jacobian was updated at every iteration. For this case, convergence was, as expected, quite fast and dependable - all time steps required two iterations.

As mentioned earlier, we compute and monitor the pressure convergence rate. For this example, the relative norm of the change in pressure from the first to the second iteration was about the same size as that for the velocities, suggesting that our hypothesis is thusfar true that convergence in velocity implies convergence in pressure.

This calculation was also repeated using the lumped mass approximation. The results were little different (< 1-3%) and may suggest that mass lumping is not too detrimental for low Reynolds number flows. Flows dominated by advection, however, may be another matter. As we have demonstrated previously (Gresho et al., 1978), mass lumping can be highly deleterious to the accuracy (at least for certain elements) when solving the advection-diffusion equation. We plan to test the adequacy of mass lumping, with both linear and quadratic elements, on a more advection-dominated flow in the near future (e.g. vortex shedding).

We now turn to a discussion of the performance of the variable time step algorithm for the transient flow results presented above. The calculations were performed for three values of the relative error control parameter: $\varepsilon = 10^{-1}$, 10^{-3}, and 10^{-5} (the last one was used for the figures presented above). Figure 8 shows the manner in which step size varied

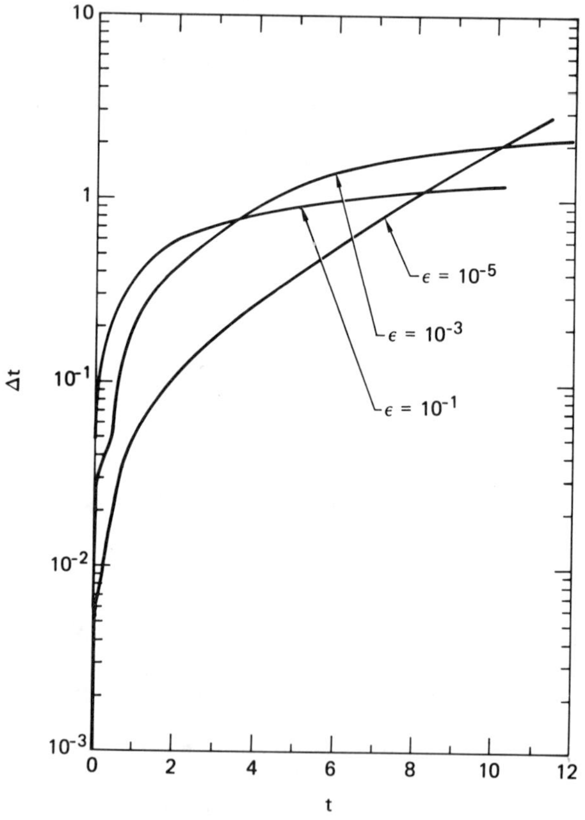

FIGURE 8. Time step variations.

with time for these three cases; all three values of ε cause initially rapid step size increases. Only with the smallest time steps ($\varepsilon = 10^{-5}$), however, does the time step variation even approach a simple exponential growth (corresponding to the decay of the smallest, dominant eigenvalue - at least for linear problems). For larger ε, where the asymptotic theory is less well satisfied, the time step variation is quite different. It is also interesting, but not yet fully understood, that the slopes of these curves, for $\varepsilon = 10^{-1}$ and 10^{-3}, continually decrease with time and in a manner which leads to the observed crossovers. It must be recognized, however, that the theory is being "pushed" quite hard for large ε and at large times, when the time step could become large enough to invalidate the local truncation error estimates. Nevertheless, the algorithm seems to behave in a useful and conservative manner. Table 1 presents a summary of the time step selection results. If the steady

TABLE 1. Summary of time steps selected.

t	Number of steps, from t = 0, to reach time t		
	$\varepsilon = 10^{-1}$	$\varepsilon = 10^{-3}$	$\varepsilon = 10^{-5}$
0	(10^{-2})	(3×10^{-3})	(6×10^{-4})
	(Initial step size)		
0.1	3	7	25
1	7	20	75
4	12	27	100
10	18	32	111

state is "chosen" to correspond to t = 4 (as Fig. 1 might suggest), then the entire transient can be computed in only 12 time steps if one is willing to use $\varepsilon = .10$ (we discuss accuracy shortly). Also noteworthy is that it is relatively inexpensive to go to a much larger "final time", especially if ε is small.

If we accept the results for $\varepsilon = 10^{-5}$ as "exact", we can compare relative accuracy quite easily. In general, we observed the following: the $\varepsilon = 10^{-3}$ results were correct to two or three significant figures and the $\varepsilon = 10^{-1}$ results were accurate to one or two figures. A more quantitative comparison is presented in Table 2, where we have compared a normalized absolute difference between the $\varepsilon = 10^{-5}$ solution and the other two solutions for two of the nodal values shown earlier (in Fig. 1) at several (quadratically interpolated) output times. We

TABLE 2. Approximate global time integration error

t	U(3,1.75)		U(2,0.5)	
	$\varepsilon = 10^{-1}$	$\varepsilon = 10^{-3}$	$\varepsilon = 10^{-1}$	$\varepsilon = 10^{-3}$
0.02	2.9×10^{-3}	2.2×10^{-3}	2.7×10^{-3}	1.9×10^{-4}
0.07	7.2×10^{-2}	9.0×10^{-4}	5.2×10^{-4}	1.4×10^{-4}
0.20	1.9×10^{-2}	2.0×10^{-3}	6.8×10^{-3}	6.2×10^{-4}
0.80	4.8×10^{-3}	2.2×10^{-3}	7.9×10^{-3}	9.4×10^{-4}
2.0	4.8×10^{-3}	2.0×10^{-3}	2.6×10^{-3}	1.1×10^{-3}
4.0	1.3×10^{-2}	8.2×10^{-4}	2.4×10^{-3}	4.8×10^{-4}
10.0	1.7×10^{-2}	8.0×10^{-6}	3.7×10^{-3}	2.0×10^{-5}

normalized the U(3,1.75) result by 5 and the other by 50. From this we infer that the $\varepsilon = 10^{-3}$ solution is surely accurate enough for most purposes and that even the $\varepsilon = 10^{-1}$ solution is probably acceptable, especially if considered against spatial accuracy. The $\varepsilon = 10^{-1}$ results are still somewhat in error as the flow approaches steady state (e.g. t = 10) since the TR oscillations are not fully damped for these large time steps.

For this simple problem, with a small number of elements, it appears that an acceptably accurate transient simulation, from $t = 0$ to $t \to \infty$ can be accomplished with very few time steps. For a larger number of elements, however, with the associated better spatial resolution, the calculation will not be so easy. This is because the algorithm tends to "balance" spatial and temporal accuracy in that more elements generally introduce larger eigenvalues into the matrix (the system becomes "stiffer" in the ODE sense) which will lead to the requirement of using smaller time steps at small time (for a fixed ε) - this assumes of course that the eigenvectors associated with the large eigenvalues contribute significantly to the solution, which they generally do at small time. For this extra price, however, one would obtain commensurate increased overall accuracy in both time and space.

SUMMARY AND CONCLUSIONS

We have presented and (partially) demonstrated a new time integration algorithm for solving the time-dependent Navier-Stokes equations. While the presentation is based on a finite element spatial discretization, it is by no means so limited and could be just as easily applied to a finite difference discretization. Although we are still in the development stage, we believe that the concept is fundamentally sound and that this method of treating time-dependent flows, and perhaps even steady flows, has much to offer with respect to accuracy, stability, and computational efficiency.

ACKNOWLEDGMENT

The design and implementation of the time integration algorithm presented here could not have been accomplished without the willing and very capable assistance of Dr. Alan Hindmarsh of the Lawrence Livermore Laboratory. Thanks are due him also in many related areas associated with the linear and nonlinear algebra generated by the FEM.

This work was performed under the auspices of the U. S. Department of Energy by the Lawrence Livermore Laboratory under contract no. W-7405-Eng-48.

REFERENCES

Aris, R. (1962) *Vectors, Tensors, and the Basic Equations of Fluid Mechanics*, Prentice-Hall, Englewood Cliffs, N.J., p. 129.

Baker, A. J. (1974) A Highly Stable Explicit Integration Technique for Computational Continuum Mechanics, in *Numerical Methods in Fluid Dynamics*, Brebbia and Connor (ed.), Pentech Press, London, pp. 99-121.

Byrne, G. and A. Hindmarsh (1977) A Comparison of Two ODE Codes: GEAR and EPISODE. *Comp. and Chem. Eng.*, *1*, pp. 133-147.

Gray, W. and G. Pinder (ed.) (1977) *Finite Elements in Water Resources*, Pentech Press, London.

Gresho, P., R. Lee, and R. Sani (1978) Advection Dominated Flows, with Emphasis on the Consequences of Mass Lumping, in *Finite Elements in Fluids, Vol. 3*, John Wiley and Sons (in press).

Hood, P. (1976) Frontal Solution Program for Unsymmetric Matrices. *Int. J. Num. Meth. Engng., 10*, pp. 379-399.

Huyakorn, P., C. Taylor, R. Lee and P. Gresho (1978) A Comparison of Various Mixed Interpolation Finite Elements in the Velocity-Pressure Formulation of the Navier-Stokes Equations. *Computers and Fluids* (in press).

Malkus, D. and T. Hughes (1978) Mixed Finite Element Methods - Reduced and Selective Integration: A Unification of Concepts. *Comp. Meth. Appl. Mech. and Engng.* (to appear).

Olson, M. (1977) Comparison of Various Finite Element Solution Methods for the Navier-Stokes Equations, in the Gray and Pinder reference; p. 4.185.

Shampine, L. F. and M. K. Gordon (1975) *Computer Solution of Ordinary Differential Equations: The Initial Value Problem*, W. H. Freeman and Co., San Francisco, CA.

APPENDIX 1. DEVELOPMENT OF TIME STEP SELECTION ALGORITHM

Consider the ODE $\dot{y} = f(y,t)$, assume the exact solution is available at t_n, that h_{n-1} is the step size used in going from t_{n-1} to t_n, and that a step size h_n is to be used to advance to t_{n+1}. The explicit Adams-Bashforth formula is given by

$$y^p_{n+1} = y_n + \frac{h_n}{2}\left[\left(2 + \frac{h_n}{h_{n-1}}\right)\dot{y}_n - \frac{h_n}{h_{n-1}}\dot{y}_{n-1}\right] ,$$

where the superscript p indicates a predictor result (one that will ultimately be discarded). Similarly, the trapezoid rule is given by

$$y_{n+1} = y_n + \frac{h_n}{2}\left[\dot{y}_n + \dot{y}_{n+1}\right] ,$$

where, of course $\dot{y}_{n+1}$ is not yet available and is the feature which makes this scheme implicit. Each of the above schemes is second-order accurate, as shown via the following Taylor series expansions: here $y(t_{n+1})$ is the (unknown) exact solution at t_{n+1}, and $y_n = y(t_n)$ is presumed known. The local time truncation error for the AB scheme is

$$y^p_{n+1} - y(t_{n+1}) = -\frac{1}{12}\left(2 + 3\frac{h_{n-1}}{h_n}\right) h_n^3 \dddot{y}_n + O(h^4)$$

where the third derivative at t_n, $\dddot{y}_n$ is also unknown. Similarly, the local truncation error for the TR is

$$y_{n+1} - y(t_{n+1}) \equiv d_{n+1} = \frac{1}{12} h_n^3 \dddot{y}_n + O(h^4) \quad .$$

Since both algorithms are employed to advance a time step, y^p_{n+1} and y_{n+1} are available and the above two equations can be solved for $y(t_{n+1})$ and $\dddot{y}_n$ (to $O(h^4)$ and $O(h)$ respectively). The quantity we really need, however, is the local truncation error, d_{n+1} of the result (TR) which we plan to use. This turns out to be

$$d_{n+1} = \frac{y_{n+1} - y^p_{n+1}}{3(1 + h_{n-1}/h_n)} + O(h^4)$$

and is used to select the time step in the following way: After taking a step of size h_n, we can compute d_{n+1} and *consider* repeating the step at a different step size, say h'_n, and predict its local error, d'_{n+1} from

$$|d'_{n+1}| = |d_{n+1}| \; (h'_n/h_n)^3$$

(an appropriate vector norm is required in general). Now we wish to select our step size, h'_n, in order to maintain a specified accuracy. We employ a relative error control criterion, as follows: $|d'_{n+1}| \leq \varepsilon y_{max}$, where y_{max} is the expected maximum value of $y(t)$ and is constant, and ε is the (maximum) relative error permitted in a single step (e.g. an ε of .001 says that at no time do we wish to exceed a local error of .1% of the maximum value expected anywhere). The selection of ε (and to a lesser extent, of y_{max}) is of course the remaining task and is left up to the "user"! At this point there is no substitute for intuition and numerical experience with a certain class of problems. We have used 10^{-1} to 10^{-8} for ε for various problems, and generally find that 10^{-2}-10^{-4} will generate sufficiently accurate solutions (of course a larger ε could, in theory, be used to generate a steady-state solution, using larger steps, in less time). Unfortunately there is little or no theory to advise us on the real issue - global truncation error, $(y_{n+1} - y(t_{n+1})$, when starting from $t = 0$. Again, numerical experimentation is required to establish reasonable estimates and guidelines.

Finally then, we adjust our time step, increasing it when possible and decreasing it when necessary in order to maintain the desired relative error, from

$$h'_n = h_n \; (\varepsilon y_{max}/|d_{n+1}|)^{1/3} \quad .$$

Note also that one could obtain a third order accurate result by simply subtracting the local truncation error from the TR value. This procedure, while perhaps useful at output times, would reduce the stability of the scheme (it is then no longer a TR scheme).

APPENDIX 2. INITIALIZATION AND THE FIRST TWO TIME STEPS

In order to employ the AB predictor algorithm during the second time step, we need the initial acceleration vector, $\dot{u}_o$. We obtain this from Eq. (3), applied at t = 0, as follows (we use the fact that if $C^T u_o = 0$, then $C^T \dot{u}_o = 0$):

$$\begin{bmatrix} M & C \\ C^T & 0 \end{bmatrix} \begin{Bmatrix} \dot{u}_o \\ p_o \end{Bmatrix} = \begin{Bmatrix} f - [K + N(u_o)]\, u_o \\ 0 \end{Bmatrix} ,$$

which is a linear system in $\dot{u}_o$ and p_o. This, we believe, is the only consistent way to initialize the calculation and it provides, as a useful bonus, the initial *compatible* pressure field (compatible with the initial solenoidal, via $C^T u_o = 0$, velocity field). Having $\dot{u}_o$ we then take two time steps using the normal TR (both steps at the initial user-input value of Δt_o) and the second step ($u_1 \rightarrow u_2^p$) using also the AB scheme. The predictor value for u_1 is simply obtained via $u_1^p = u_o + \Delta t_o \dot{u}_o$ (i.e., explicit Euler). After the second step, we perform our standard truncation error estimate to see if the initial time step is "good enough". If so, we increase Δt and proceed; if not, we return to t = 0 with a new time step, from Eq. (10), and repeat the first two steps ($\dot{u}_o$ is saved, of course).

A final comment should be made here regarding the requirement that the initial velocity satisfy the continuity equation (the *approximate* one, $C^T u_o = 0$) since this constraint has been referred to repeatedly. If $C^T u_o \neq 0$, it can be shown that the ODE system, Eq. (3), is ill-posed in that any implicit scheme (recall that explicit schemes are inapplicable) will generate pressure and velocity fields (at the first time step) which blow up, like $O(C^T u_o/\Delta t)$ as $\Delta t \rightarrow 0$. Hence we insist that our initial conditions satisfy the discretized continuity equation, $C^T u_o = 0$.

OPTIMAL CONTROL OF THE BOUNDARY LAYER EQUATIONS BY THE OUTER VELOCITY USING FINITE ELEMENT METHOD

G.M. ASSASSA[1], C.M. BRAUNER[2], B. GAY[3]

[1] Laboratoire de Mécanique des Fluides, Ecole Centrale de Lyon 69130 ECULLY - France
[2] Laboratoire de Mathématiques, Ecole Centrale de Lyon, 69130 ECULLY - France
[3] Centre de Mathématiques, Institut National des Sciences Appliquées, 69621 VILLEURBANNE - France

INTRODUCTION

The application of finite element technique in the calculation of viscous fluid flows is a recent approach and is a one of considerable interest in the domain of numerical analysis. Because there is no variational formulation of the Navier-Stokes equations (Finlayson, 1972), numerous investigations based on the method of weighted residuals or on pseudo variational methods have been developed. A formal formulation is proposed by Baker 1973, whereas Lynn 1974 used a finite element method based on the criterium of least squares.

Certain classical problems in the field of mathematical physics can be better approached by employing new numerical methods. In particular, the methods of optimal control (Lions, 1968) as applied to non linear problems are now available for application in the fields of physics of plasma, oceanography and fluid mechanics (Glowinski-Pironneau, 1974). In this paper, a fluid mechanics problem is dealt with which concerns the development of a boundary layer flow. The governing equations to be used are those of Prandtl after writting them in the Von Mises form (Schlichting, 1968). The development of the boundary layer can then be controled through either the wall suction velocity or the longitudinal pressure gradient as exerted by the outer flow. This pressure gradient is directly related to the outer velocity gradient at the edge of the boundary layer through Bernoulli equation. In a previous work, the authors (1977) presented the control by the wall suction velocity whereas in this paper the control problem through the outer velocity gradient will be considered.

In the field of fluid mechanics and in particular of aeronautics and turbomachinary, one of the major problems is the specification of the outer velocity gradient in a manner to avoid certain undesired phenomenon such as the separation of the boundary layer. Thus if we have certain initial conditions and certain desired properties at a given fixed station (control station), the outer velocity is controled in such a way to yield at the best these required properties. As far as the numerical technique is considered the governing equations are numerically handled using a finite element approach based on the general formulation of the weighted residuals method. In the following paragraphs, the finite element technique as applied to the boundary layer equations will be presented. Next, the definitions, the formulation of the optimal control problem as well as related subjects will be given. Examples of numerical results are shown also to display the control procedure.

BOUNDARY LAYER EQUATIONS

The governing equations (equations of motion and continuity) of steady, incompressible, two dimensional laminar boundary layer flow can be written as follows :

$$u \frac{\partial u}{\partial x} + v \frac{\partial u}{\partial y} = - \frac{1}{\rho} \frac{\partial p}{\partial x} + \nu \frac{\partial^2 u}{\partial y^2} \qquad (1)$$

$$\frac{\partial u}{\partial x} + \frac{\partial v}{\partial y} = 0 \qquad (2)$$

where u and v are the velocity components respectively in the stream wise direction x and its normal y ; p, ρ and ν denote respectively the pressure, the density and the kinematic viscosity. Equation (1) is a parabolic partial differential equation whose integration necessitates the specification of boundary and initial conditions. The boundary conditions are given by :

$$\begin{aligned} y = o &: u(x,o) = o \quad ; \quad v(x,o) = v_p(x) \\ y = \infty &: u(x,\infty) = U_e(x) \end{aligned} \qquad (3)$$

with $v_p(x)$ representing the normal suction or blowing at the wall and $U_e(x)$ is the local velocity outside the boundary layer (see figure 1). The initial condition is :

$$x = o : u(o,y) = U_o(y) \qquad (4)$$

At the edge of the boundary layer Bernoulli equation yields :

$$\frac{1}{\rho}\frac{dp}{dx} + U_e \frac{dU_e}{dx} = o \qquad (5)$$

By introducing a modified Von Mises variables defined by :

$$u(x,y) = \frac{\partial \psi}{\partial y} \quad ; \quad v(x,y) - v_p(x) = \frac{\partial \psi}{\partial x} \qquad (6)$$

and the function : $f(x,\psi) = u^2/2$

the continuity equation (2) is automatically satisfied where as the equation of motion (1) yields :

$$\frac{\partial f}{\partial x} + v_p(x)\frac{\partial f}{\partial \psi} = \frac{1}{2}\frac{d}{dx}U_e^2 + \nu\sqrt{2f}\,\frac{\partial^2 f}{\partial \psi^2} \qquad (7)$$

The boundary and initial conditions are thus given by :

$$\psi = o \quad : \quad f = o$$

$$\psi = M\ (x) \quad : \quad f = U_e^2/2$$

where, for the purpose of numerical calculation, the boundary condition at infinity is replaced by the one taken at the edge of the boundary layer $\psi_e(x)$ which is noted as M(x). It must be noted that the non linear parabolic equation (7) degenerates at the wall as mentioned by Gay and Assassa 1977.

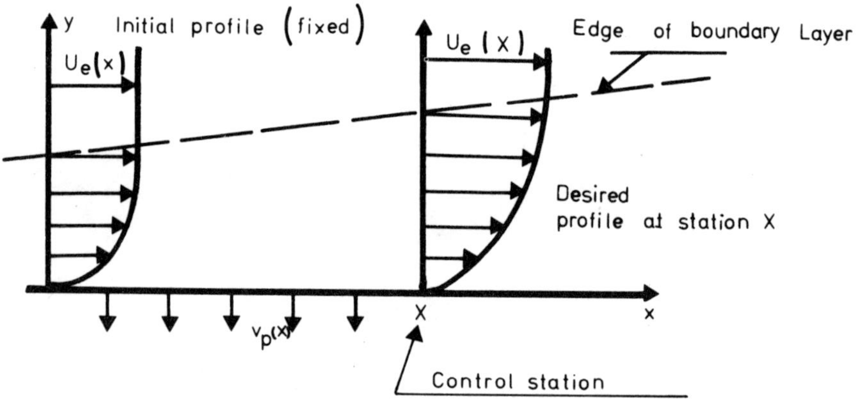

Figure 1 : Boundary layer development and control station.

FINITE ELEMENT FORMULATION

The finite element approach adopted in this work is based on the method of Galerkin as was applied to the linear problem of Stefan by Jamet and Bonnerot 1975. Thus if we define certain shape functions $\Phi(x,\psi)$ in such a way that they vanish at the wall (ψ = o) and at the edge of the boundary layer ($\psi = M(x)$), équation (7) can be put, after multiplication by the function $\Phi(x,\psi)$ and integration on the whole element $[(x_1,x_2),(o,M(x))]$, in the following form :

$$- \int_{x_1}^{x_2} \int_0^{M(x)} f \frac{\partial \Phi}{\partial x} d\psi dx + \int_0^{M(x_2)} \psi(x_2,\psi) f(x_2,\psi) d\psi$$

$$- \int_0^{M(x_1)} \Phi(x_1,\psi) f(x_1,\psi) d\psi \quad - \int_{x_1}^{x_2} \int_0^{M(x)} f v_p(x) \frac{\partial \Phi}{\partial \psi} d\psi dx$$

$$- \frac{1}{2} \int_{x_1}^{x_2} \int_0^{M(x)} \Phi \frac{d}{dx} U_e^2 \quad d\psi dx = - \int_{x_1}^{x_2} \int_0^{M(x)} \frac{\partial f}{\partial \Psi} \frac{\partial \Phi}{\partial \Psi} \sqrt{2f} \, d\Psi dx$$

$$- \int_{x_1}^{x_2} \int_0^{M} \Phi \frac{(\partial f/\partial \Psi)^2}{\sqrt{2f}} d\Psi dx \qquad (8)$$

The different integrals of the equation (8) are numerically evaluated after employing the transformation shown in figure 2.

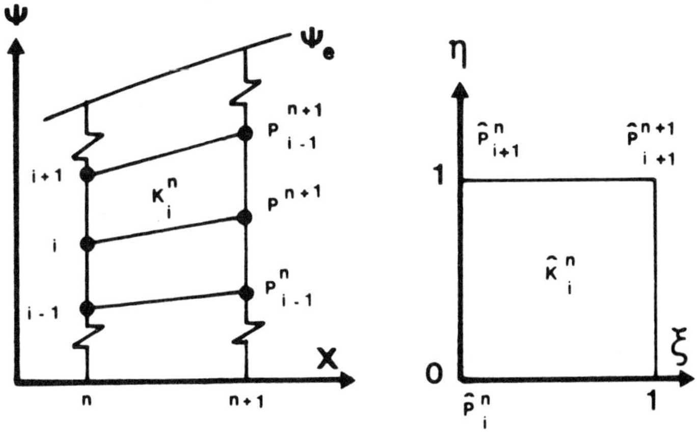

Figure 2 : Transformation of the element K_i^n into a square element $\hat{K}_i^n$ of unit length.

An element K_i^n in the plane (x,ψ) is transformed into a square of unit length in the plane (ξ,η) using the transformation :

$$x = (1-\xi)x^n + \xi x^{n+1}$$

$$\Psi = (1-\eta)\ [(1-\xi)\Psi_i^n + \xi\ \Psi_i^{n+1}] + \eta[(1-\xi)\Psi_{i+1}^n + \xi\Psi_{i+1}^{n+1}] \quad (9)$$

The Jacobian of this transformation is given by :

$$d\Psi dx = J_i^n(\xi,\eta)\,d\xi d\eta \ = k(\Psi_{i+1}^{n+\xi} - \Psi_i^{n+\xi})\,d\xi d\eta$$

where k is the step in the x direction $k = x^{n+1} - x^n$ and

$$\Psi_i^{n+\xi} = (1-\xi)\Psi_i^n + \xi\Psi_i^{n+1}$$

In the plan (ξ,η) a function $F(\xi,\eta)$ defined in K_i^n will be approximated by the polynomial form :

$$F(\xi,\eta) = a+b\xi+c\eta+d\xi\eta \quad (10)$$

It follows that F is determined by its four values at the corners of the square K_i^n : $[(0 \leqslant \xi \leqslant 1)\ ,\ (0 \leqslant \eta \leqslant 1)]$ as :

$$F(\xi,\eta) = (1-\eta)\ (1-\xi)F(0,0) + (1-\eta)\xi F(0,1)$$
$$+ \ \eta(1-\xi)F(1,0) + \xi\eta F(1,1) \quad (11)$$

As far as the integrals (8) in the plane (ξ,η) are concerned, they are numerically evaluated for each element K_i^n by the formula :

$$\int_{K_i^n} F(x,\Psi)\ d\Psi dx = \int_{\hat{K}_i^n} F(\xi,\eta)J_i^n\ d\xi d\eta$$

$$\simeq \frac{1}{4}\sum_{\substack{\xi=0,1\\ \eta=0,1}} F(\xi,\eta)J_i^n\ (\xi,\eta) \quad (12)$$

where $\hat{K}_i^n$ is the element in the plane (ξ,η) that corresponds to the element K_i^n in the plane (x,ψ). The shape functions $\Phi(\xi,\eta)$ used in the calculation are presented in figure 3 for the element $\hat{K}_i^n$.

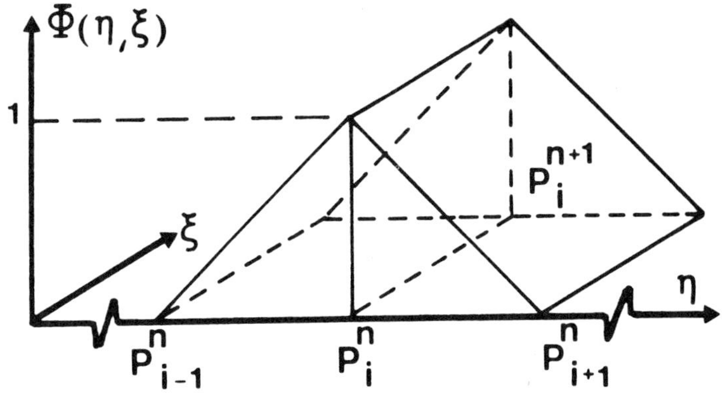

Figure 3 : Shape function at the point P_i^n

Thus, the initial equation (8) can be put in the linear tridiagonal matrix form :

$$A_i\ f_{i-1}^{n+1} + B_i\ f_i^{n+1} + C_i\ f_{i+1}^{n+1} = D_i \tag{13}$$

with i = 2, 3,..., (I-1) where I is the number of nodes taken accross the boundary layer. The coefficients of equation (13) are given by :

$$A_i = \frac{1}{4}\,(\psi_{i-1}^{n+1} - \psi_i^n) - \frac{k}{4}\,v_o^{n+1} - \frac{\nu k}{4}\,\frac{s_{i-1}^n + s_i^n}{\psi_i^{n+1} - \psi_{i-1}^{n+1}} - \frac{\nu k}{4s_i^n}\,\frac{f_i^n - f_{i-1}^n}{\psi_i^n - \psi_{i-1}^n} \tag{14.a}$$

$$B_i = \frac{\nu k}{4}\left(\frac{s_i^n + s_{i+1}^n}{\psi_{i+1}^{n+1} - \psi_i^{n+1}} + \frac{s_{i-1}^n + s_i^n}{\psi_i^{n+1} - \psi_{i-1}^{n+1}}\right) + \frac{\nu k}{4s_i^n}\left(\frac{f_i^n - f_{i-1}^n}{\psi_i^n - \psi_{i-1}^n} - \frac{f_{i+1}^n - f_i^n}{\psi_{i+1}^n - \psi_i^n}\right)$$

$$+ \frac{1}{2}(\Psi_{i+1}^{n+1} - \Psi_{i-1}^{n+1}) \tag{14.b}$$

$$C_i = \frac{-1}{4}(\Psi_{i+1}^{n+1} - \Psi_{i+1}^{n}) + \frac{k}{4} v_p^{n+1} - \frac{\nu k}{4} \frac{s_{i-1}^{n} + s_i^{n}}{\Psi_i^{n+1} - \Psi_{i-1}^{n+1}}$$

$$- \frac{\nu k}{4 s_i^n} \frac{f_i^n - f_{i-1}^n}{\Psi_i^n - \Psi_{i-1}^n} \tag{14.c}$$

$$- D_i = \frac{-1}{4} f_{i+1}^n (\Psi_{i+1}^{n+1} - \Psi_{i+1}^{n}) + \frac{1}{4} f_{i-1}^n (\Psi_{i-1}^{n+1} - \Psi_{i-1}^{n})$$

$$+ \frac{k}{4} v_p^n (f_{i+1}^n - f_{i-1}^n) + \frac{\nu k}{4} [(s_{i-1}^n + s_i^n) \frac{f_i^n - f_{i-1}^n}{\Psi_i^n - \Psi_{i-1}^n}$$

$$- (s_i^n + s_{i+1}^n) \frac{f_{i+1}^n - f_i^n}{\Psi_{i+1}^n - \Psi_i^n}] + \frac{\nu k}{4 s_i^n} [\frac{(f_{i+1}^n - f_i^n)^2}{\Psi_{i+1}^n - \Psi_i^n} + \frac{(f_i^n - f_{i-1}^n)^2}{\Psi_i^n - \Psi_{i-1}^n}]$$

$$- \frac{1}{2} f_i^n (\Psi_{i+1}^n - \Psi_{i-1}^n) + \frac{g^n - g^{n+1}}{8} [(\Psi_{i+1}^{n+1} - \Psi_{i-1}^{n+1})$$

$$+ (\Psi_{i+1}^n - \Psi_{i-1}^n)] \tag{14.d}$$

with $g^n = (U_e^2/2)^n$ and $s_i^n = (\sqrt{2f})_i^n$. The last coefficient is modified to take into consideration the outer boundary condition.

OPTIMAL CONTROL

The control problem as applied to the development of a boundary layer flow can be stated in the following terms. Consider figure 1 where the initial velocity profile is fixed at the station x = 0. At the station x = X we look for a "desired" velocity profile $U_d(x,y)$. As stated in the introduction, the control can be applied either through the distribution of the wall suction (or blowing) velocity or through the distribution of the outer velocity. In the two cases, a corresponding optimal distribution is seeked to approach at the best the desired velocity profile U_d. It must be noted, however, that such a profile is not always realisable.

Optimal control of the Von Mises equation

In this work, the chosen control is that of the outer velocity. However, for the purpose of generality, the equations will contain terms corresponding to wall suction (or blowing). Reconsider the equation of Von Mises which can be written along with the boundary and initial conditions in the form :

$$\frac{\partial u}{\partial x} - \nu \frac{\partial}{\partial \psi} \left(u \frac{\partial u}{\partial \psi}\right) + v_p \frac{\partial u}{\partial \psi} - \frac{1}{u}\frac{d}{dx}(U_e^2/2) = 0 \qquad (15)$$

$$\psi = o : u(x,o) = o \quad ; \quad v(x,o) = v_p(x) \qquad (16a)$$

$$\psi = \infty : u(x,\infty) = U_e(x) \qquad (16b)$$

$$x = o : u(o,\psi) = U_o(\psi) \qquad (16c)$$

the boundary condition at $\psi = \infty$ will, again for the purpose of numerical calculation, be replaced by the one $\psi = M$, where M is taken outside the edge of the boundary layer (for exemple : M = M(X)). Thus $u(x,M) = U_e(x)$. We shall assume also that there is a continuous matching between the initial condition and the boundary condition there, that is :

$$U_o(\psi = M) = U_e(x = o) \qquad (17)$$

Let the position $X > o$ be fixed and let the desired velocity profile $U_d(X,\psi)$ be given. One can introduce the cost function $J(U_e)$, which characterizes the difference between the calculated and the desired velocity profiles at the control station X, as follows :

$$J(U_e) = \int_0^M \{u((U_e,X),\psi) - U_d(X,\psi)\}^2 d\psi \qquad (18)$$

The control function $U_e(x)$ which yields the minimum value of the cost function $J(U_e)$ will be called the optimal control.

As can be seen from equations (15) and (16b), the control function appears simultaneously in the equation (15), by the term $1/u.\ d/dx(U_e^2/2)$, and in the boundary condition by the term $u(x,M) = U_e(x)$. This leads to a more difficult problem than that when the control function was the suction velocity at the wall. It could be possible to eliminate this difficulty by imposing regularity conditions at infinity of the type $\partial u/\partial\psi = o$ and $\partial^2 u/\partial\psi^2 = o$ as ψ goes to infinity.

However, numerical experiments show that these conditions are not of vital importance and that, in fact, it is preferable to deal with boundary conditions of the type Dirichlet ($u(x,M) = U_e(x)$) than that of the type Neumann ($\partial u/\partial\psi(x,M) = o$); the reason being that with the latter conditions the results are very sensible to the choice of M and that even the solution can be physically inacceptable. That is why we shall retain equation (15), which is called the state equation, with the following boundary and initial conditions :

$$\begin{aligned} u(x,o) &= o & \qquad v(x,o) &= v_p(x) \\ u(x,M) &= U_e(x) & & \\ u(o,\psi) &= U_o(\psi) & & \end{aligned} \qquad (19)$$

<u>The gradient $J'(U_e)$ of the cost function and the adjoint equation</u>

To get the minimum of the cost function $J(U_e)$ defined by equation (18), one has to calculate its gradient $J'(U_e)$. Thus we can write :

$$\int_0^X J'(U_e)\ P(x)dx = 2\int_0^M (u(X,U_e) - U_d)(u'(U_e).\ P)(X)d\psi \qquad (20)$$

where $P(x)$ is an element of $L^2(o,X)$.

The state equation (15) can be put into the following condensed form :

$$G(u,U_e) = o \qquad (21)$$

Using the theoremof implicit functions, one can differentiate with respect to U_e the form (21) to get :

$$\frac{\partial G}{\partial u} \cdot \frac{\partial u}{\partial U_e} + \frac{\partial G}{\partial U_e} = o \qquad (22)$$

or, for all $P(x)$ of $L^2(o,X)$, we have also

$$\frac{\partial G}{\partial u}(u'(U_e).P) + \frac{\partial G}{\partial U_e} \cdot P = o \qquad (23)$$

To simplify the notations, $u'(U_e)P$ shall be denoted by R. Thus the state equation (15) gives, after differentiation with respect to U_e as stated by (23), the form :

$$\frac{\partial R}{\partial x} - \frac{\partial^2}{\partial \psi^2}(uR) + v_p \frac{\partial R}{\partial \psi} + \frac{1}{2}\frac{R}{u^2}\frac{d}{dx}(U_e) - \frac{1}{u}\frac{d}{dx}(U_e P) = o \qquad (24)$$

with the conditions :

$$R(x,\psi = o) = o$$

$$R(x,\psi = M) = P$$
$$R(x = o,\psi) = o$$

The evaluation of the gradient $J'(U_e)$ can be simply carried out by introducing an adjoint state $q = q(U_e)$ defined as the solution of the equation :

$$-\frac{\partial q}{\partial x} - u\frac{\partial^2 q}{\partial \psi^2} - v_p \frac{\partial q}{\partial \psi} + \frac{1}{2}\frac{d}{dx}(U_e^2) \frac{q}{u^2} = o \qquad (25)$$

with the conditions

$$q(x,\psi = o) = q\ (x,\psi = M) = o$$
$$q(X,\psi) = u\ (X,\psi) - U_d \qquad (26)$$

The introduction of the adjoint equation is done in such a manner that its integration yields equation (24) (without the last term) from which the gradient $J'(U_e)$ can be calculated once the adjoint state q is determined.

Integration of equation (25) after multiplication by R on the domain $\{(o,X),(o,M)\}$ yields :

$$\int_0^M\int_0^X \left(-\frac{\partial q}{\partial x} - u\frac{\partial^2 q}{\partial \psi^2} - v_p\frac{\partial q}{\partial \psi} + \frac{1}{2}\frac{q}{u^2}\frac{d}{dx}U_e^2\right) R\ dxd\psi = o \qquad (27)$$

Carrying out integration by parts we get, after taking into account the boundary conditions of q and R, the following terms :

$$\int_0^M\int_0^X -\frac{\partial q}{\partial x} R\ dxd\psi = \int_0^M\int_0^X q\frac{\partial R}{\partial x} dxd\psi - \int_0^M R(X)\{u(X)-U_d\}d\psi$$

and

$$-\int_0^X\int_0^M UR\frac{\partial^2 q}{\partial \psi^2} dx\ d\psi = -\int_0^X \left\{uR\frac{\partial q}{\partial \psi}\right\}_0^M dx$$

$$+\int_0^X \int_0^M \frac{\partial q}{\partial \psi} \frac{\partial}{\partial \psi} (uR) \, dx d\psi$$

$$= -\int_0^X U_e(x) \, P(x) \frac{\partial q}{\partial \psi} (M) - \int_0^X \int_0^M q \frac{\partial^2}{\partial \psi^2} (uR) \, dx d\psi$$

and

$$-\int_0^X \int_0^M v_p \, R \frac{\partial q}{\partial \psi} \, dx d\psi = -\int_0^X \{v_p \, Rq\}_0^M$$

$$+\int_0^X \int_0^M v_p \, q \frac{\partial R}{\partial \psi} \, dx \, d\psi$$

It is supposed that the behaviour of u near $\psi = o$ is sufficiently regular so that the terms neglected in the above equations are negligible. In particular, the term $qR \, \partial u/\partial \psi$ at $\psi = o$ is assumed to be zero (Misiti, 1978). With the aid of the above integrals, equation (27) can be written as :

$$\int_0^M \int_0^X \left(\frac{\partial R}{\partial x} - \frac{\partial^2}{\partial \psi^2} (uR) - v_p \frac{\partial R}{\partial \psi} + \frac{1}{2} \frac{R}{u^2} \frac{d \, U_e^2}{dx}\right) q \, dx \, d\psi$$

$$= \int_0^M (u(X) - U_d) \, R(X) \, d\psi + \int_0^X U_e P \frac{\partial q}{\partial \psi} (M) \, dx \qquad (28)$$

Comparing equations (20) and (24) to (28), we get :

$$\int_0^M \int_0^X \frac{q}{u} \frac{d}{dx} (U_e P) \, dx \, d\psi = \frac{1}{2} \int_0^X J'(U_e) \, P \, dx$$

$$+\int_0^X U_e P \frac{\partial q}{\partial \psi}(M)\, dx \tag{29}$$

which yields after integration by parts :

$$\int_0^M \int_0^X \frac{q}{u} \frac{d}{dx}(U_e P)\, dx\, d\psi = - \int_0^M \int_0^X \frac{d}{dx}\left(\frac{q}{u}\right) U_e P\, dx\, d\psi + I \tag{30}$$

with

$$I = \int_0^M \left\{\frac{q}{u} U_e P\right\}_0^X d\psi = \int_0^M \left\{\frac{q(X)}{u(X)} U_e(X)P(X) - \frac{q(o)}{U_o(\psi)} U_e(o)P(o)\right\} d\psi$$

At the control station $x = X$, we impose the matching condition

$$U_e(X) = U_d(M)$$

which is equivalent to : $P(X) = o$ (which means that the outer velocity at the control station is fixed (as the one for the initial station) and not to be controled). This, in turns, gives $I = 0$.

Going back to equation (29), the gradient $J'(U_e)$ is seen to be given by :

$$\frac{1}{2} J'\{U_e\}(x) = - U_e(x) \int_0^M \frac{d}{dx}\left(\frac{q}{u}\right) d\psi - U_e(x) \frac{\partial q}{\partial \psi}(M) \tag{31}$$

Summing up, the solution of the state equation (15) with the boundary and initial conditions (19) is first found. The gradient of the cost function J is then evaluated from equation (31), where the adjoint state $\sigma(x,\psi)$ is found as the solution of the adjoint state equation (25) with its boundary and initial conditions (26).

NUMERICAL ALGORITHM

Knowing the initial conditions at $x = o$ and the desired velocity profil U_d at the control station $x = X$, the problem to be solved is to find out the outer velocity distribution $U_e(x)$ which verifies at the best the imposed conditions. To get the solution, we proceed by iteration as stated below. Let an outer initial velocity distribution denoted by $U_e^k(x)$ be assumed, where k is the level of iteration (k = 1 for the initial approximation). With $U_e^k(x)$ a first iteration can be carried out during which the state equation and the adjoint state equation are numerically solved, using the finite element method presented above, thus yielding the cost function $J(U_e^k)$ and its gradient $J'(U_e^k)$. The problem now is to determine the optimal control that will operate on $U_e^k(x)$ so that $U_e^{k+1}(x)$ can be calculated. However, it must be noted that the uniqueness of the optimal control is not guaranteed. That is why the optimal control will be evaluated by the method of steepest descent such that $J(U_e^{k+1}) < J(U_e^k)$. Since the direction of steepest descent is determined by the gradient $J'(U_e)$, then we can write.

$$U_e^{k+1} = U_e^k - \lambda J'(U_e^k) \qquad (32)$$

the choice of the optimal λ is a difficult problem which can be solved in the simple following way. The local (during the iteraction k + 1) optimal λ can be chosen such that it corresponds to the minimum cost function J. Thus, if we consider the cost function during the iteration k + 1 as a function of λ and denote it by $\hat{J}(\lambda)$, we can write :

$$\hat{J}(\lambda) = J\,(U_e^k - \lambda J'(U_e^k)) \qquad (33)$$

Assuming that $\hat{J}(\lambda)$ is locally convex (having a unique extremum) in the neighbourhood of $\lambda = o$, and that it can approximated by a cubic, we can solve the state equation that corresponds to U_e^{k+1} for four values of λ, say $\lambda_1 = o$, λ_2, λ_3, λ_4. This yields four values of $\hat{J}$: $\hat{J}_1$, $\hat{J}_2$, $\hat{J}_3$ and J_4. The choice of λ is determined by the condition that the minimum of $\hat{J}$ lies in the interval (o, λ_4), that is : $o < \lambda_2 < \lambda_3 < \lambda_4$ and $\hat{J}(\lambda_3) > \hat{J}(\lambda_2)$ or $\hat{J}(\lambda_3) > \hat{J}(\lambda_1)$. Thus the cubic, is defined and the local optimal λ is determined once the minimum $\hat{J}$ is found.

NUMERICAL RESULTS

Numerical experiments were carried out using fortran IV language program on a C.D.C. 7600 computer. To apply the control technique, one must fix the desired velocity profile U_d at the control station $x = X$ and the initial conditions ; thus the equations will yield the outer velocity distribution $U_e(x)$. To test the validity of the approach, the desired velocity profile is fixed by calculating it for a given initial conditions and a given outer velocity distribution. These given conditions are prescribed from the similarity solution (of the Falkner and Skan familiar equation) which is characterized by the value of the pressure gradient parameter β;($U_e \sim x^m$; $m = \beta/(2-\beta)$). Solving the control equations gives us the outer velocity distribution $U_e(x)$ to be compared to the prescribed one of certian value of β. Here, we present only results corresponding to two tests of different values of β. In these examples the initial station is considered at x=1 cm, whereas the control station is fixed at X = 10 cm.

<u>Favorable pressure gradient flow (β = 0,8)</u>

This case is shown in figure 4 and 5. The initial outer velocity distribution ($U_e^1(x)$) was taken constant (β=o). These two figures show the evolution of $U_e(x)$ during iterations (from 2 to 300). Figure 4 corresponds to the solution obtained with constant value of the descent parameter λ which is calculated during the second iteration and is kept constant as long as a $J(U_e)$ decreases (otherwise, a new value of the constant is determined). Figure 5 shows the results obtained with a seeked optimal λ during each iteration which is calculated as explained in the text. For the 300 iterations considered in the two cases ; we got in the first case : $J(300)/J(1) \simeq 10^{-4}$, whereas in the second case $J(300)/J(1) \simeq 10^{-8}$.

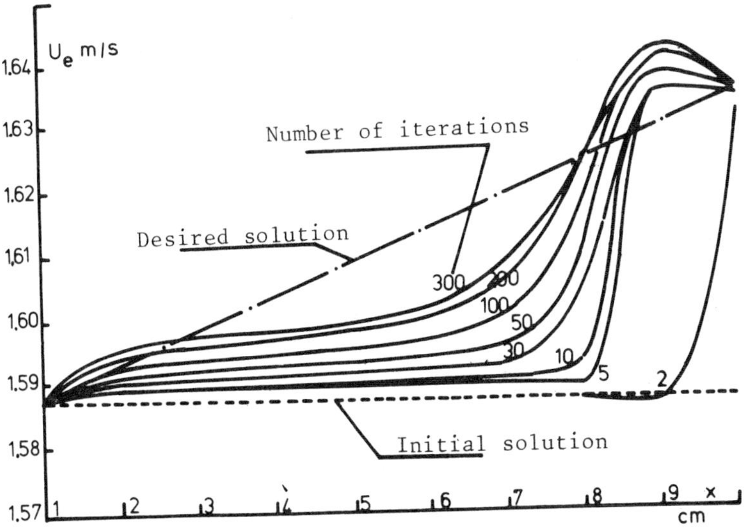

Figure 4 : Evolution of the control function $U_e(x)$ during 300 iterations using constant λ (The desired U_e corresponds to the desired U_d).

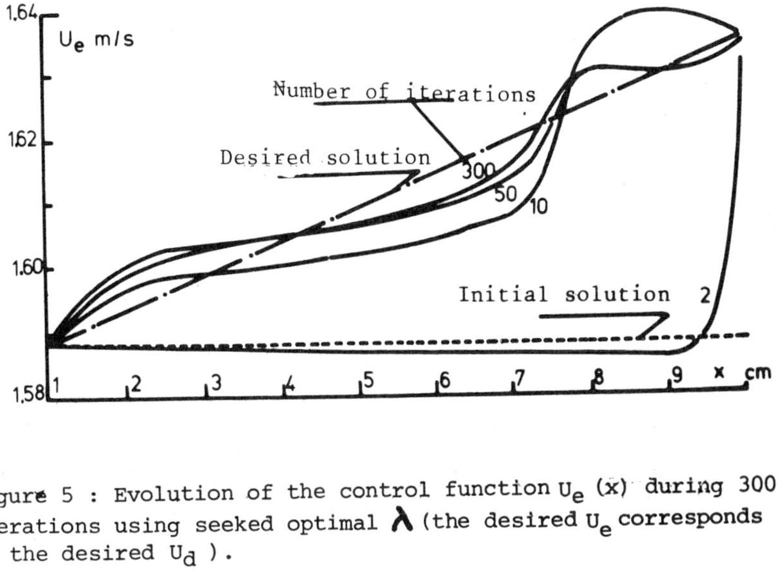

Figure 5 : Evolution of the control function U_e (x) during 300 iterations using seeked optimal λ (the desired U_e corresponds to the desired U_d).

Defavorable pressure gradient (β = - 0.16)

The results of this case are displayed on figure 6 . The initial outer velocity distribution ($U_e^1(x)$) corresponds to a value of β = -0.19 (the desired U_d is obtained from β = -0.16). One hundred iterations are carried out using a seeked optimal λ during each iteration. For this case we got : J(100)/J(1) = 0.5 10^{-6}.

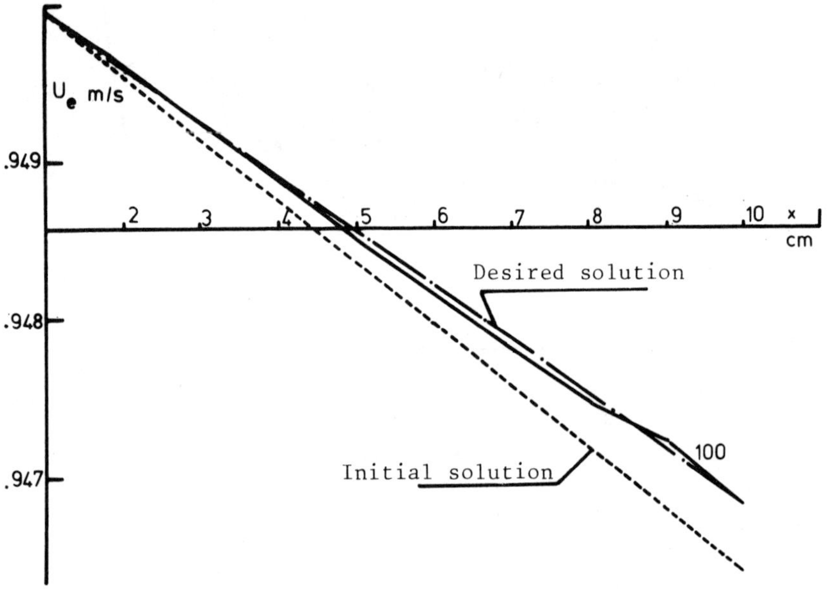

Figure 6 : The control function $U_e(x)$ after 100 iterations using seeked optimal λ (the desired U_e corresponds to the desired U_d).

CONCLUSION

In a previous work, the control of the development of the boundary layer was carried out through the wall suction (or blowing) velocity. The results obtained were encouraging so that we tried in the present work to deal with the control problem through the outer velocity distribution which appears, in this case, in the equations to be solved and in the boundary conditions. The numerical technique, used to solve the governing equations, is the finite element one. Numerical tests presented show that the control approach through the outer velocity distribution is also encouraging. Examination of the results shows that improvements of the form of the distribution $U_e(x)$, to get a more smooth one, are possible is we introduce additional constraints. Another modification which can decrease the necessary number of iterations is the improvement of the algorithm of descent. These possibilities will be investigated in a futur work.

REFERENCES

ASSASSA G.M., BRAUNER C.M. and GAY B., (1977) A new approach to the boundary layer equations : Optimal control of the von Mises equation by wall suction and numerical solution using finite difference and finite element techniques. 6th Australasian Hydraulics and Fluid Mechanics Conference, The Institute of Engineers, Adelaide, South Australia, 5-9 Dec.

ASSASSA G.M., and GAY B. (1977) Calcul d'une couche limite laminaire avec aspiration ou soufflage parietal en presénce d'un gradient de vitesse extérieure par une méthode d'éléments finis. Journal de Mécanique Appliquée vol. 1, n° 4 pp.357-376.

BAKER A.J. (1973) Finite element solution algorithm for viscous incompressible fluid dynamics. Int. J. Num. Meth. Engng. Vol. 6, p. 89-101.

BONNEROT R. and JAMET P. (1975) A second order finite element method for the one dimensional Stefan problem. Int. J. for Num. Math. in Eng. Vol. 8 n°4, pp. 811-820.

BRAUNER C.M. and GAY B. (1977) Contrôle optimal d'équations paraboliques non linéaires dégénérées. International Symposium on Innovative Numerical Analysis in Applied Engineering Science. Versailles France.

FINLAYSON B.A. (1972) The method of weighted residuals and variational principles. Academic press.

GAY B. and ASSASSA G.M. (1977) Finite element solution of boundary layer equations. International conference on applied Numerical modelling, Southampton, England.

GLOWINSKI R. and PIRONNEAU O. (1975) Sur le calcul des carenes optimales en écoulement laminaire visqueux. Rapport de recherche IRIA n° 137.

LIONS J.L. (1968) Contrôle optimal des systèmes gouvernés par des équations aux dérivées partielles. DUNOD.

LYNN P.P. (1974) Least squares finite element analysis of laminar boundary layer flows. Int. J. for Meth. in Ing., Vol.8 n° 4, pp. 865-876.

MISITI R. (1978) Ph. D. Thesis (in preparation) Ecole Centrale de Lyon. France.

SCHLICHTING H. (1968) Boundary layer theory. 6e Ed. Mc Graw-Hill.

STRESSES IN OSCILLATORY CONVERGING OR DIVERGING FLOW BY FINITE ELEMENT SIMULATION

M. Durin, J. Ganoulis

School of Technology, Aristotle University, Thessaloniki-Greece

INTRODUCTION

This work is stimulated by the problems occuring in biomechanics and especially the problem of pulsatile flow in blood vessels. Our study concerns principally the evaluation of the influence of a non rectilinear oscillatory flow on the stresses applied on the walls. Shear stresses are considered as a very important parameter in biomechanics since they are linked to the intimal injury of the arterial walls (Fry, 1968).

It is well known that physiological channels form networks with complex geometry, having some stenoses, curvatures, ramifications and variations of their cross section. The understanding of the flow into these systems can be easier using a schematisation of the geometry. In the past, attention was confined in experimental or numerical study of the flow near an abrupt or regular constriction, which simulate a stenosis (Golia and Evans, 1973; Cheng R.T.S., 1972; Cheng L.C., Clark and Robertson, 1972). In our work, the acceleration or deceleration effects of a pulsatile converging or diverging flow are studied by numerical method.

When finite differences are used, serious limitations can arise in order to take in account the irregular topography of the walls. For this reason, the finite element technique is adopted and the numerical algorithm build is proved to be very precise and efficient. It is the aim of this paper to describe the methodology used and to present the obtained results.

MATHEMATICAL FORMULATION

It was decided to study the problem in a two dimensional geometry, because the numerical solution becomes more simple, and the principal characteristics of the phenomenon are conserved. After eliminating the pressure, the problem is formulated in terms of Ψ stream function and ζ vorticity. Introducing the dimensionless variables :

$$x = \frac{x'}{L},\ y = \frac{y'}{L},\ u = \frac{u'}{U},\ v = \frac{v'}{U},\ t = \omega t',\ \Psi = \frac{\Psi'}{UL},\ \zeta = \frac{\zeta'}{\omega}$$

where (x', y') are cartesian coordinates, (u', v') velocity components, t' the time, ω the angular frequency of boundary conditions, L and U reference length and velocity, the equations of the problem can be written as :

$$\alpha^2 \frac{\partial \zeta}{\partial t} + Re \{ u \left(\frac{\partial \zeta}{\partial x}\right) + v \left(\frac{\partial \zeta}{\partial y}\right) \} = \nabla^2 \zeta \qquad (1)$$

$$\zeta = - \frac{Re}{\alpha^2} \nabla^2 \Psi \qquad (2)$$

$Re = UL/\nu$ is the mean Reynolds number and $\alpha = L(\omega/\nu)^{1/2}$ the unsteadiness parameter of the flow.

Now, the equations (1) and (2) are written in variational form, using the Galerkin method. Calling Ω the integration domain fixed by the boundary Γ, the auxiliary functions Ψ^* and ζ^* are considered, which are arbitrarly valued everywhere in Ω and take the value 0 on the boundary Γ. Multiplying the two members of equation (1) by ζ^*, integrating over Ω and using the Green theorem, the continuity equation and the fact that $\zeta^* = 0$ on Γ, the following equation is obtained :

$$\int_\Omega \{ \alpha^2 \zeta^* \frac{\partial \zeta}{\partial t} + Re \left[\zeta^* \frac{\partial}{\partial x} (u\zeta) + \zeta^* \frac{\partial}{\partial y} (v\zeta) \right] + \left[\frac{\partial \zeta^*}{\partial x} \cdot \frac{\partial \zeta}{\partial x} + \frac{\partial \zeta^*}{\partial y} \cdot \frac{\partial \zeta}{\partial y} \right] \} d\Omega = 0 \qquad (3)$$

Using the same procedure, after multiplication of both members of equation (2) by Ψ^*, the following equation is derived :

$$\int_\Omega \Psi^* \zeta d\Omega = \int_\Omega \frac{Re}{\alpha^2} \left(\frac{\partial \psi^*}{\partial x} \cdot \frac{\partial \psi}{\partial x} + \frac{\partial \Psi^*}{\partial y} \cdot \frac{\partial \Psi}{\partial y} \right) d\Omega \qquad (4)$$

Boundary conditions express the non slip velocity conditions on the solid walls, the prescribed velocity distribution in the entry and the continuation of the flow in the exit. During a typical computation, a parabolic Poiseuille velocity distribution is imposed in the entry. The non stationary mean velocity $U(t)$ is computed according to the following expression :

$$Rei = Re\,(1 + \gamma \sin\omega t) \tag{5}$$

where $Rei = UL/\nu$ is the instantaneous Reynolds number, $Re = U_o L/\nu$ the mean Reynolds number, γ the amplitude factor and ω the angular frequency. Equations (3) and (4) along with the boundary conditions are now solved using the finite element method.

FINITE ELEMENT METHODOLOGY

The flow domain Ω, defined by the closed boundary Γ is divided into a finite number of triangular elements and the unknown functions ζ and Ψ appearing in equations (3) and (4) are approximated using linear interpolation. The terms $u\zeta$ and $v\zeta$ in equation (3) are computed over the three nodes of every element and they are linearly interpolated. Noting φ_m the variable on the element m, the following expressions can be written :

$$\varphi_m = \{L_m\}\{\varphi\} = \{L_1, L_2, L_3\} \begin{Bmatrix}\varphi_1\\ \varphi_2\\ \varphi_3\end{Bmatrix}$$

$$\varphi_{m,x} = \{L_{m,x}\}\{\varphi\} = \{\frac{\partial L_1}{\partial x}, \frac{\partial L_2}{\partial x}, \frac{\partial L_3}{\partial x}\} \begin{Bmatrix}\varphi_1\\ \varphi_2\\ \varphi_3\end{Bmatrix} \tag{6}$$

$$\varphi_{m,y} = \{L_{m,y}\}\{\varphi\} = \{\frac{\partial L_1}{\partial y}, \frac{\partial L_2}{\partial y}, \frac{\partial L_3}{\partial y}\} \begin{Bmatrix}\varphi_1\\ \varphi_2\\ \varphi_3\end{Bmatrix}$$

where L_1, L_2, L_3 are the surface coordinates relative to the triangular element and φ_1, φ_2, φ_3 the nodal values of φ. Now φ can be replaced by one of the functions Ψ, ζ, Ψ^*, ζ^*. Introducing the expressions (6) of these functions into the equations (3) and (4), after assembling the elements into the global matrix, the following system of linear equations is obtained:

$$\alpha^2 \{A\}\{\frac{\partial \zeta}{\partial t}\} = -\{B\}\{\zeta\} - Re\left[\{C\}\{u\zeta\} + \{D\}\{v\zeta\}\right] \tag{7}$$

$$\{B\}\{\Psi\} = \frac{Re}{\alpha^2}\{A\}\{\zeta\} \tag{8}$$

where :

$$\{A\} = \sum_m \int_{\Omega_m} \{L_m\}^T \{L_m\}\, d\Omega_m$$

$$\{B\} = \sum_m \int_{\Omega_m} \left[\{L_{m,x}\}^T \{L_{m,x}\} + \{L_{m,y}\}^T \{L_{m,y}\}\right] d\Omega_m \tag{9}$$

$$\{C\} = \sum_m \int_{\Omega_m} \{L_m\}^T \{L_{m,x}\} d\Omega_m$$

$$\{D\} = \sum_m \int_{\Omega_m} \{L_m\}^T \{L_{m,y}\} d\Omega_m$$

Now the computation is organised according to the following explicit algorithm. Knowing the initial flow conditions and using the boundary conditions of stream function Ψ, first the system equations (8) is inversed by the Gauss method. Vorticity distribution along the boundaries is computed using the values of Ψ and the derivative $\partial\zeta/\partial t$ is obtained according to the equation (7). Then, new values of ζ at time $t + \Delta t$ are computed according to the simple expression :

$$\zeta_{t+\Delta t} = \zeta_t + \frac{\partial \zeta}{\partial t}\Big|_t \cdot \Delta t$$

Boundary conditions

Figure 1 shows the flow domain and the boundaries used in order to study the converging or diverging flow. Along the entry and exit boundaries (1) and (2) Poiseuille conditions are imposed

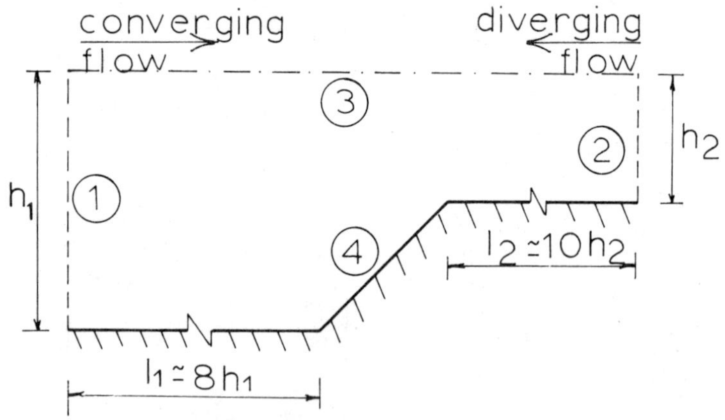

Figure 1 Definition of the flow domain.

for the velocity distribution.

For example, along the entry (1) the following expressions are used :

$$u\ (y,t) = \frac{3}{2}\ \frac{U(t)}{h_1^2}\ y\ (2h_1 - y)$$

$$v\ (y,t) = 0$$

$$\Psi\ (y,t) = \frac{3}{2}\frac{U(t)}{h_1^2}\ y^2\ (h_1 - \frac{y}{3}) \qquad (9)$$

$$\zeta\ (y,t) = 3\ \frac{U(t)}{h_1^2}\ (h_1 - y)$$

where U(t) is computed according to the equation (5).
Along the symetry axis (3) the stream function is constant and the vorticity is equal to zero. At every point M of the solid wall (4) the stream function is taken equal to zero and the vorticity is computed using a first order development along the direction perpendicular to the wall. If M_1 is a point in a distance Δn from the wall we have :

$$\zeta\ (M) = \frac{2}{\Delta n^2}\ \{\Psi(M_o) - \Psi(M_1)\}$$

All the above boundary conditions are of Dirichlet type and they are directly incorporated into the matrix.

PRELIMINARY VERIFICATION OF THE NUMERICAL PRECISION

Before starting the computations in a non-rectilinear flow, the numerical algorithm is tested during pulsatile flow between two parallel walls. Analytical expressions of the velocity, vorticity and stream functions can be obtained in this case. If $y = \pm h$ are the equations of the walls, at every time t, velocity distribution u(y,t) is given by the Navier-Stokes equation:

$$\frac{\partial u}{\partial t} = -\frac{1}{\rho}\cdot\frac{\partial p}{\partial x} + \nu\ \frac{\partial^2 u}{\partial y^2} \qquad (10)$$

with

$$u\ (h,\ t) = u\ (-h,\ t) = 0 \qquad (11)$$

Consider the unsteady flow which is developed by the imposed pressure gradient and perform the decomposition :

$$\frac{\partial p}{\partial x}\ (t) = \frac{\partial p_o}{\partial x} + \frac{\partial p_1}{\partial x}\ (t) \qquad (12)$$

where $\partial p_o/\partial x$ is the steady pressure gradient and $\partial p_1(t)/\partial x$ is the oscillatory pressure gradient of the form :

$$\frac{\partial p_1}{\partial x}\ (t) = A\cdot e^{-i\omega t}$$

Using the principle of superposition, the following velocity distribution satisfies the system equations (10) and (11) :

$$u(y, t) = u_1(y, t) + u_o(y) =$$

$$= \frac{iA}{\omega} e^{-i\omega t} \left[1 - \frac{\cos ky}{\cos kh} \right] + \frac{3}{2} \frac{U_o}{h^2} (h^2 - y^2) \qquad (13)$$

where

$$k^2 = \frac{i\omega}{\nu} \qquad (14)$$

The mean velocity of the flow is :

$$U(t) = \frac{iA}{\omega} e^{-i\omega t} \left[1 - \frac{1}{kh} \frac{\sinh kh}{\cosh kh} \right] + U_o \qquad (15)$$

Using (13) it's easy to obtain the vorticity and stream function distribution in the following form :

$$\zeta\,(y,t) = - \frac{\partial u}{\partial y} = - \frac{iA}{\omega} e^{-i\omega t} \left[\frac{k \sin ky}{\cos kh} \right] - \frac{3U_o}{h^2} y \qquad (16)$$

$$\Psi\,(y,t) = \int u\,dy = \frac{iA}{\omega} e^{-i\omega t} \left[y - \frac{1}{k} \frac{\sin ky}{\cos kh} \right] + \frac{3}{2} \frac{U_o}{h^2} \cdot (h^2 y - \frac{y^3}{3}) + C \qquad (17)$$

where C is a constant which can be derived in order to impose $\Psi = 0$ on $y = -h$.

Examination of the analytical solutions (13), (16) and (17), shows that the unsteady terms depend on the value of the parameter

$$\alpha = (\omega/\nu)^{1/2} \cdot h$$

Two asymptotic cases can be considered :

(i) $\alpha \rightarrow 0$

The unsteady perturbation can be deduced as the response in a steady excitation

(ii) $\alpha \rightarrow \infty$

The flow resistance takes a very high value and the mass transport is confined into a very fine layer near the walls.

Figure 2 shows numerical and analytical velocity profils which are computed using the finite element algorithm and according to the (13) formula. The mean Reynolds number and the unsteadiness parameter defined on the channel width are Re = 100 and α = 79.3 respectively. The amplitude factor γ entering in equation (5) is taken γ = 1.

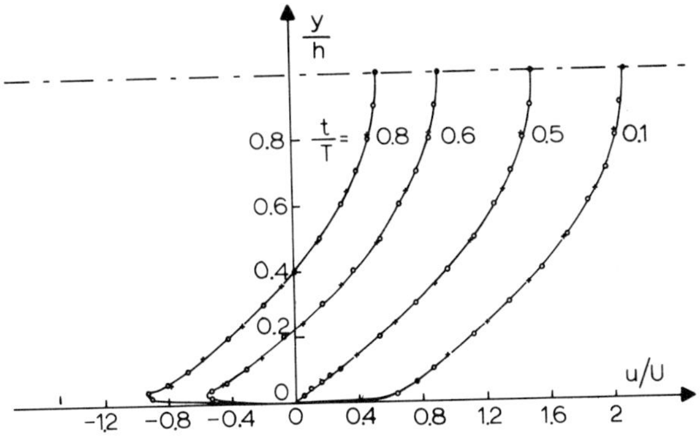

Figure 2. Velocity profils for pulsatile flow between two parallel walls. Re = 100, α = 79.3, γ = 1. + Finite Element solution, -o-Analytical solution.

A good agreement can be concluded for this case, while the precision of the numerical computation is a function of several parameters like the Reynolds number and the parameter α, which caracterizes the unsteady importance of the pulsatile flow. The influence of α on the vorticity or shear stress distribution along the wall is shown in figure 3.

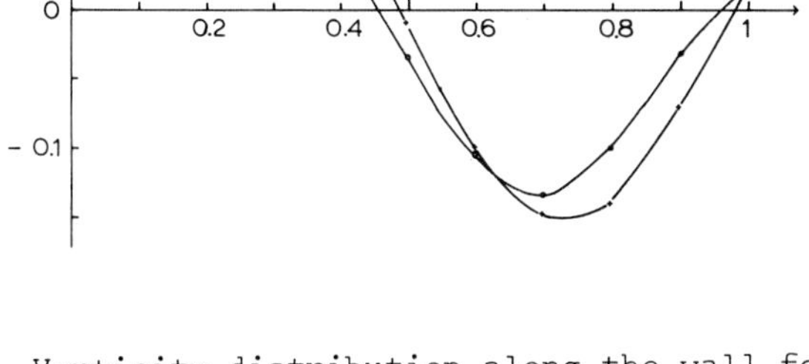

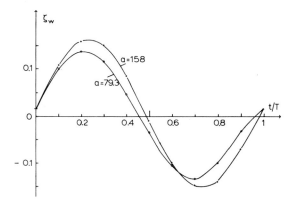

Figure 3. Vorticity distribution along the wall for pulsatile flow between parallel plates.

NUMERICAL RESULTS

The finite element mesh used to compute the pulsatile converging flow, is automatically generated on the computer. In order to obtain a more accurate description of the flow within the boundary layer, an exponential mesh spacing is used in the direction perpendicular to the walls. Figure 4 shows a typical finite element mesh near the converging region.

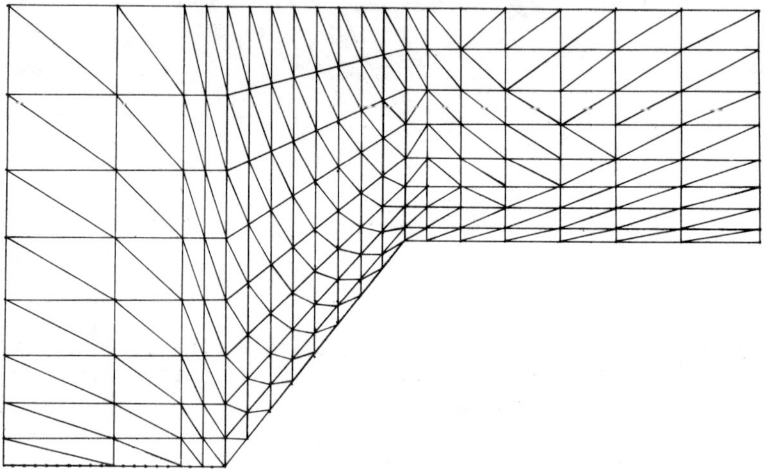

Figure 4. The Finite Element mesh

For every Re and α chosen the method of solution consists of the following steps :

(a) stationary solution is obtained for the mean Reynolds number of the flow
(b) pulsatile flow is computed during two periods and only the results obtained during the second period are taken into consideration.

During the computation the periodic character of the flow is verified with a good precision. Although the program developed gives information about velocity and pressure distribution throughout the oscillation cycle, only the streamlines and the stress distributions will be presented.

Figure 5 shows the streamline configuration during pulsatile converging flow.

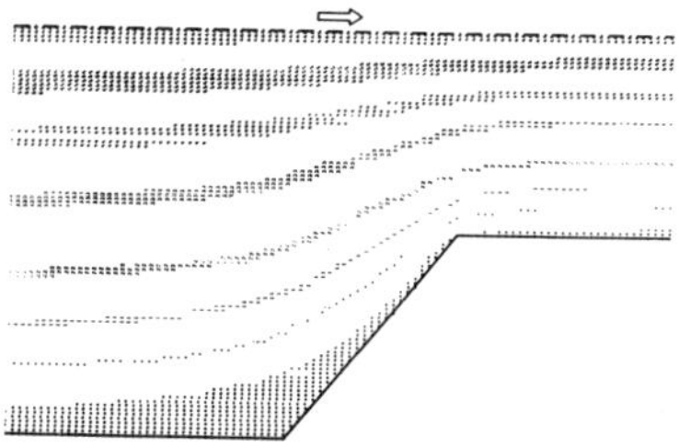

t/T = 0, Rei = 100

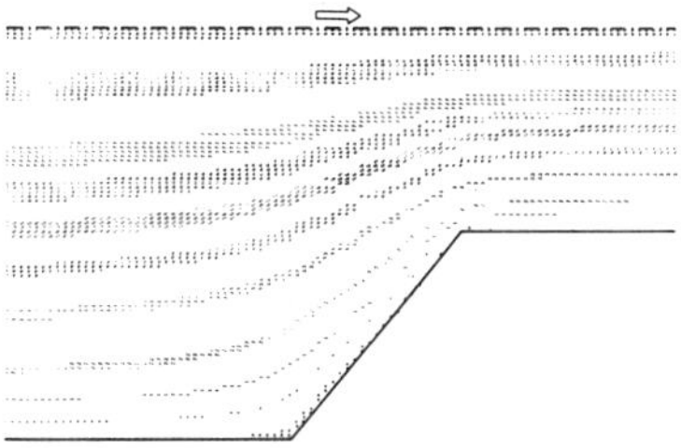

t/T = 0.3, Rei = 195

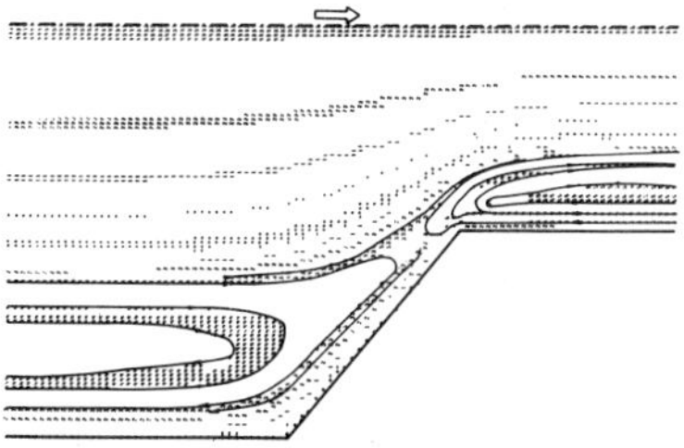

t/T = 0.6, Rei = 41.2

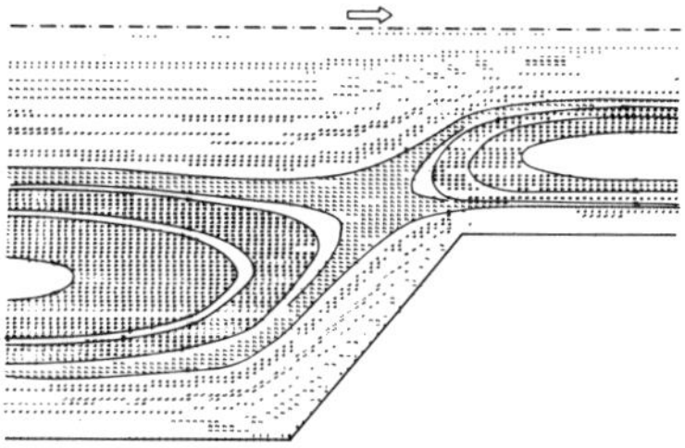

t/T = 0.7, Rei = 4.89

Figure 5. Streamlines in converging pulsatile flow.

The streamlines are drawn automatically on the line printer of UNIVAC 1100 computer of Aristotle University of Thessaloniki using a special subroutine. The geometry of the flow is defined by the angle of the converging wall equal to 45° and the ratio between the constricted and the enlarged channel width, which is equal 1 : 2. The mean Reynolds number and the parameter α are defined on the enlarged channel width and take the values 100 and 79.3 respectively. Instantaneous Reynolds numbers Rei are indicated at every time t/T in figure 5. We can remark that during the acceleration stage, the streamlines are pushed against the wall and no separation occurs. When the movement is decelerated, two large eddies appear and the flow direction is reversed near the wall. Shear stress distribution along the wall appearing in figure 6, shows the influence of the converging flow on the magnitude of stresses. The high stress region is situated near the corner of the constricted section at the exit of the converging flow.

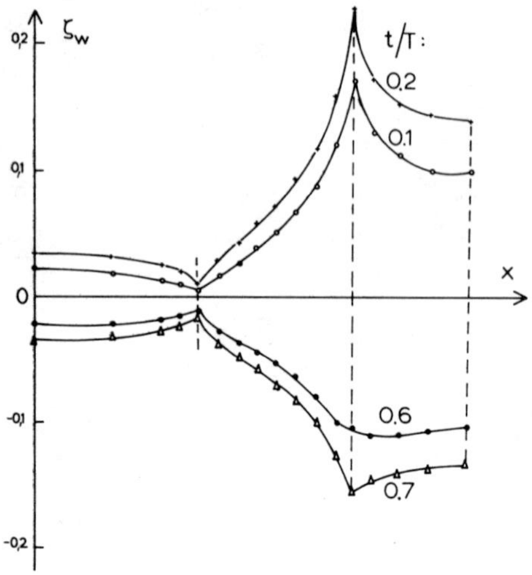

Figure 6. Shear stress distribution along the wall in converging pulsatile flow. Re = 100, α = 58.

Figures 7 and 8 indicate the streamline development and the shear stress distribution along the wall for pulsatile diverging flow.

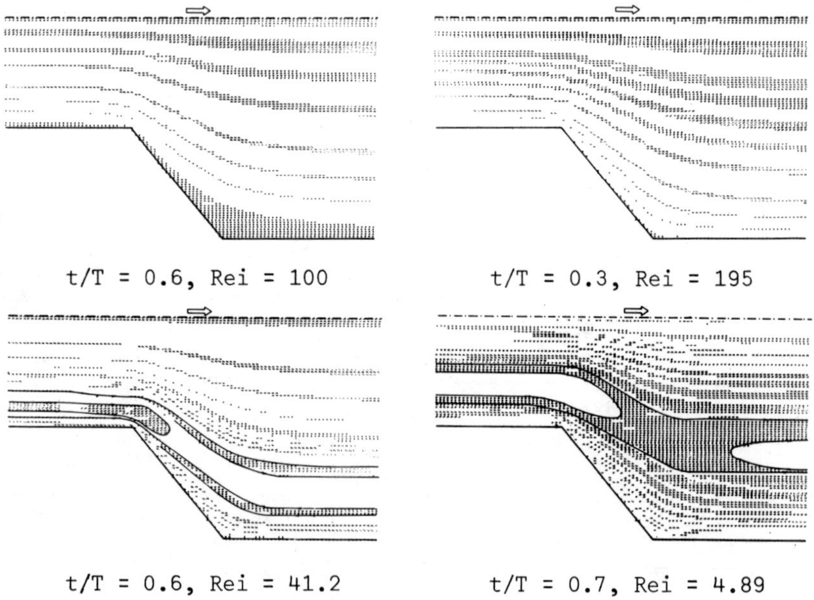

Figure 7. Streamlines in diverging pulsatile flow. Re = 100, α = 79.3

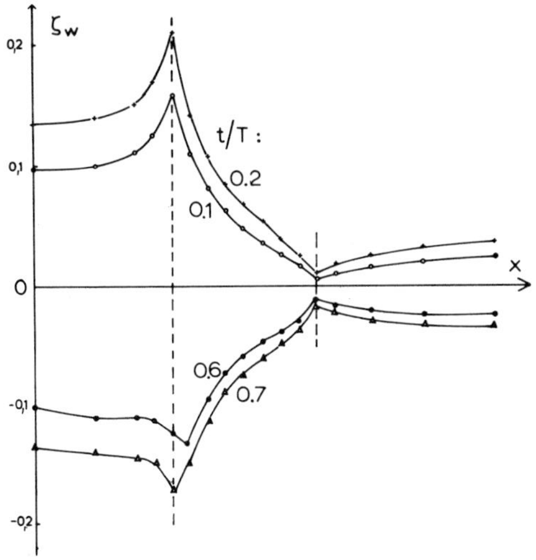

Figure 8. Shear stress distribution along the wall in diverging pulsatile flow. Re = 100, α = 58.

CONCLUSIONS

The pulsatile converging or diverging flow is accurately simulated by the finite element algorithm presented in this work. The obtained results show that the acceleration or deceleration effects of the non rectilinear flow influence in a significant manner, the magnitude of the shear stresses applied on the wall. Improvements are now making on the finite element algorithm, in order to extend the range of computation at lower values of parameter α and greater Reynolds numbers.

REFERENCES

1. Cheng, R.T.S. (1972) Numerical Solution of the Navier-Stokes Equations by the Finite Element. Phys. Fluids, 15, 12: 2098 - 2105.
2. Cheng, L.C., Clark, M.E., Robertson, T.M. (1972) Numerical Calculations of Oscillating Flow in the Vicinity of Square Wall Obstacles in Plane Conduits, J. Biomech., 5.
3. Fry, D.L. (1968) Acute Vascular Endothelial Changes Associated with Increased Blood Velocity Gradients, Circulation Research, 22 : 167 - 197.
4. Golia, C., Evans, N.A. (1973) Flow Separation Through Annular Constrictions in Tubes, Experimental Mechanics : 157-162.

THE NUMERICAL TREATMENT OF FREE SURFACE FLOWS BY FINITE ELEMENTS

H.J. Diersch, H. Martin

Technical University of Dresden, Dresden, G.D.R.

INTRODUCTION

Jets flowing from slots, nozzles or valves; flows under sluice gates; flows over steps or sills in the open channel; spillway flows are contraction flows involving free surface where the effects of inertia predominate over the influences of fluid viscosity. With a close approximation, they allow a treatment as irrotational flows, which was assumed also for these problems in the solutions obtained so far, i.e., the assumption of the validity of the potential theory.

Various analytical solution methods (e.g., Strelkoff 1964; Fangmeier and Strelkoff 1968) can take into account only horizontal and vertical boundaries, whereas the finite-difference relaxation techniques (e.g., Southwell and Vaisey 1946; Rouse and Abul-Fetouh 1950; McNown et al. 1955) are subject to problems of accuracy for curvilinear field geometries. The possibility of analysing curved boundary contours by means of the numerical conformal mapping (e.g., Watters and Street 1964; Moayeri 1973; Larock 1969,1970) is limited in practice due to the necessity of establishing the mapping relations required.

Chan(1971) developed a finite-element model for solving two-dimensional and axisymmetric free surface flows, such as jets flowing from orifices, nozzles or valves, the effect of gravity having been taken into account or not. Among the flow problems treated, discharge was assumed to be known. Simple trial-and-error calculations were used by McCorquodale and Li (1971) in the finite-element analysis of flows under plane sluice gate structures for determining the free surface profil. In addition, the outflowing jet was approximated by an ell
Brebbia and Spanos (1973) used irregular triangular n
works for the modelling of a spillway flow. The iter
method used was based upon the calculation of ground

water flows involving a free surface, but without any reference to the discharge and network controls. The problem of the flow over a spillway was treated more in detail by Ikegawa and Washizu (1973) with the aid of the finite-element method. The initially unknown discharge could be determined by studying the behaviour of the free surface during the iteration process. On the basis of the iteration algorithm suggested by Chan with the potential function as the primary unknown, later Larock (1975) used curved isoparametric elements for the analysis of two-dimensional flows over gated spillways to obtain a more precise modelling of the boundary geometry, whereas Sarpkaya and Hiriart (1975) used straight-sided quadratic triangular elements for the analysis of deflectors and thrust reversers with two free surfaces. Then, isoparametric quadratic elements were used by Larock and Taylor (1976) for high-speed three-dimensional free surface flows by the example of the gravity-influenced flow from a circular orifice at Froude numbers, F_o, larger than 1.2 . Chan's iteration algorithm is extended by Diersch et al. (1977) to a great number of complex two-dimensional and axisymmetric flows involving a free surface with an initially unknown discharge. The finite-element model presented by Isaacs (1977) for the solution of flows under sluice gates differs from the other models by the use of curved cubic triangular elements, and the adjustment of the free surface differs from them by the choice of the stream function as the primary unknown.

In the present paper, the numerical and physical conditions and correlations in the finite-element analysis of high-speed (supercritical) and low-speed (subcritical) free surface flows will be discussed, iteration algorithms for securing the numerical convergence for flows within ranges of small F_o numbers and for determining the initially unknown discharge value will be formulated and the possibility of the convergence acceleration will be investigated. The features of the finite-element method proposed are shown by illustrative examples for flows in open channels with a varying bottom configuration and for spillway flows.

PROBLEM FORMULATION AND FINITE-ELEMENT APPROACH

Stationary two-dimensional and axisymmetric irrotational flows of a homogeneous, inviscid and incompressible fluid are considered. With the velocity potential function ϕ as the primary unknown, the solution is reduced to the Laplacian equation under Cauchy boundary conditions

$$\nabla^2\phi = 0 \qquad (1a)$$

with

$$\nabla^2 = \frac{\partial^2}{\partial x^2} + \frac{\partial^2}{\partial y^2} \qquad (1b)$$

for two-dimensional flow and

$$\nabla^2 = \frac{\partial^2}{\partial x^2} + \frac{1}{r}\frac{\partial}{\partial r} + \frac{\partial^2}{\partial r^2} \qquad (1c)$$

for axisymmetric flow.

If Φ is known, the velocity components can be calculated in the horizontal, $u = \partial\Phi/\partial x$, or axial direction, $v_x = \partial\Phi/\partial x$, and in the vertical, $v = \partial\Phi/\partial y$, or radial direction, $v_r = \partial\Phi/\partial r$. The pressure p in the flow field is calculated from the Bernoulli equation

$$\frac{1}{2} q^2 + p/\varrho + g z = g H \quad , \qquad (2)$$

where q = fluid speed, i.e., $q^2 = u^2 + v^2$ or $q^2 = v_x^2 + v_r^2$, ϱ = constant fluid density, g = gravitational acceleration, z = vertical Cartesian coordinate, measured above a constant reference level, and H = total energy head. The finite-element discretization of the differential equation (1) is performed with the aid of simple straight-sided quadratic quadrilateral elements (Figure 1) composed of four triangular elements.

As a result of the finite-element approximation, whether it is a Ritz formulation or a Galerkin one, a linear, symmetric and banded system of equations is obtained for solving the nodal values Φ_i of the velocity potential function. The development of these finite-element working equations is well known and documented, e.g., by Chan (1971); Diersch et al. (1977). At the flow boundaries, node-dependent Φ_i values and/or their outer normal derivatives $(\partial\Phi/\partial n)_i$ can be specified in a simple way. The condition for a solid impermeable boundary with $(\partial\Phi/\partial n)_i = 0$ is automatically satisfied as a natural boundary condition. The local values of the velocity components and, together with them, pressure p, can be immediately determined as derivative fields.

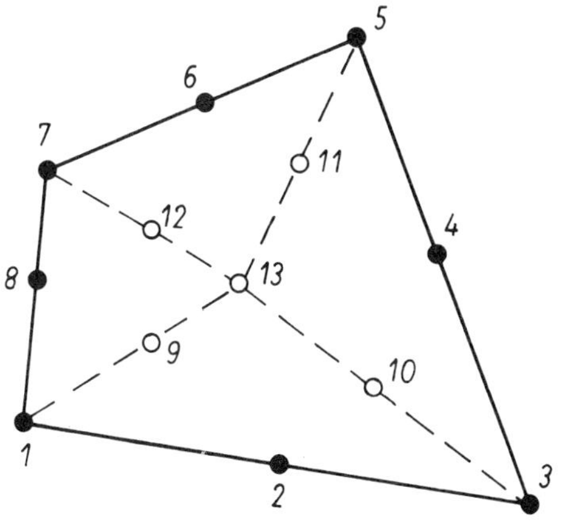

Figure 1 Eight-node quadrilateral element.

Whereas Φ varies quadratically in the finite element and a continuous transition takes place from one element to the other, the velocities derived show a linear behaviour between adjacent nodes in the element and, in addition, the transition behaviour is discontinuous. Since high velocity gradients have to be recorded for the present flow problems, a linear velocity distribution in the finite element constitutes a minimum demand (constant velocities and thus a linear distribution of Φ in the element proved insufficient). It was demonstrated that the velocity distribution over the flow field can be regarded to be continuous with a good approximation, when a sufficient number of finite elements is used. The following examples show that it is not imperative to apply generalized variational formulations, as suggested by Meissner (1973), with which the velocity components appear as primary unknowns besides the potential function.

ITERATION SCHEME

Free-surface flows of fluids lead to non-linear boundary value problems, since the surface profile is initially unknown. On the free surface, the following two boundary conditions have to be met: (1) Normal velocity component $q_n = \partial \Phi / \partial n = 0$; and (2) p = const or $p_u = 0$, i.e., the pressure corresponds to the atmospheric pressure.

For the practical solution, one starts from a assumed free surface, which is being iteratively corrected until a prescribed maximum error criterion is satisfied. From this, for the iterative finite-element solutions in the mutual satisfying of the two boundary conditions there result two fundamental ways of action:

Method I

(i) Assumption of the initial free-surface location;

(ii) Introduction of the normal derivative $q_n = 0$ (natural boundary condition) for all free-surface nodes;

(iii) Solving of the system of equations for Φ_i ;

(iv) Calculation of the velocity components u , v or v_x , v_r , respectively, and from them the determination of the resultant velocity q ;

(v) Checking of the condition $p_u = 0$ in each node i of the free surface according to equation (2) by means of

$$z_i = H - q_i^2/2g \quad . \tag{3}$$

Since, as a rule, the z_i^{n+1} coordinates of the n+1 iteration do not agree with the z_i^n values of the preceding iteration, the measure of the coordinate adjustment of the free surface is obtained as

$$\epsilon_i^{n+1} = z_i^{n+1} - z_i^n$$

This iteration method corresponds to the way of solution as applied to the calculation of flows in a porous medium (Taylor and Brown 1967; Brebbia and Spanos 1973), there, however, with the simplification that the velocity head can be neglected. This is an explanation for the fact that this simple iteration algorithm is successful there. For the potential flows investigated here, however, the application of this method causes increasing errors in the form of oscillations (Figure 2), which cannot even be suppressed by the smoothing of the surface profile (e.g., by a curve-fitting scheme).

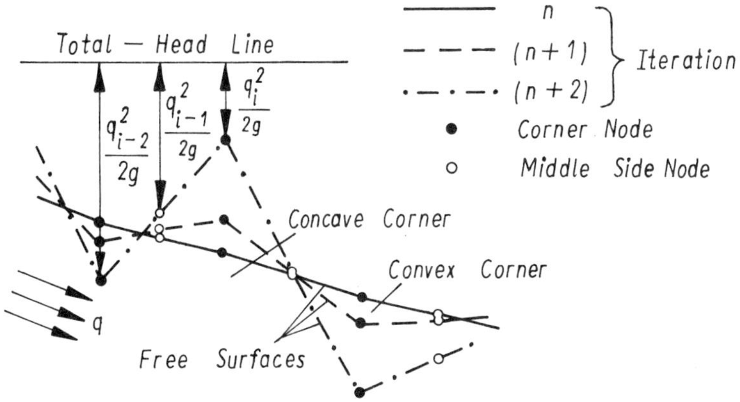

Figure 2 Increasing errors in the adjustment of the surface coordinates according to method I .

Method II

(i) Assumption of the initial free-surface location;

(ii) Calculation of the adequate resultant velocities according to the Bernoulli equation, successively starting from a reference value

$$q_{i+1} = \left[q_i^2 - 2g \, (z_{i+1} - z_i) \right]^{1/2} \qquad (4)$$

(iii) Determination of the pertinent ϕ values along the free surface according to $q = \partial\phi / \partial s$, where s = tangential direction, successively starting from a reference level:

$$\Phi_i - \Phi_{i+1} = \int_{S_{i+1}}^{S_i} q\left[z(x)\right] ds \qquad (5)$$

with

$$ds = \left[1 + \left(\frac{dz}{dx}\right)^2\right]^{1/2} dx \qquad (6)$$

Here, a linear velocity distribution is assumed between two adjacent nodes of the free surface

$$q = \frac{(q_{i+1} - q_i)\, s}{\Delta s} + q_i \quad , \qquad (7)$$

where Δs designates the linear polygonal segment between the nodes i and $i+1$. Consequently, we obtain

$$\Phi_i = \Phi_{i+1} - \frac{(q_i + q_{i+1})}{2} \Delta s \; ; \qquad (8)$$

(iv) Introduction of the Φ_i values according to equation (8) into the system of equations. This corresponds to the fulfilment of the condition $p =$ const on the free surface;

(v) Solving of the system of equations for all Φ_i ;

(vi) Calculation of the velocity components;

(vii) The free surfaces or parts of them are approximated by polynomial curves $z = \beta(x)$ of the order σ. Then the slope

$$S = \frac{dz}{dx} = \gamma(x) \qquad (9a)$$

of the order $\sigma - 1$. If the local slopes

$$S_k = v_k/u_k \quad \text{or} \quad S_k = v_{rk}/v_{xk} \, , \qquad (9b)$$

respectively, are known, the functions $\gamma(x)$ or $\beta(x)$ can be determined, the number of nodes of the free surface being $k \geq \sigma$. For a quadrilateral element, $k = \sigma = 3$ with

$$S(x) = 3\,A\,x^2 + 2\,B\,x + C \qquad (10)$$

is chosen. The unknown coefficients A, B and C result from the three unknown local slopes S_k. Finally, for the coordinate adjustment one obtains

$$\epsilon_i^{n+1} = \omega\,(z_i^{n+1} - z_i^{n}) \qquad (11a)$$

where $$z_i^{n+1} = z_{i-2}^{n+1} + \Delta z_i^n \qquad (11b)$$

with $$\Delta z_i^n = (S_{i-2}^n + 4\, S_{i-1}^n + S_i^n)\, \frac{\Delta x_i}{6} \qquad (11c)$$

and ω = 1.0, ... , 2.0 overrelaxation factor.

The three surface nodes i-2, i-1, i belong to a quadrilateral element. In some cases it may become necessary to choose an even higher accuracy. Therefore, Hiriart and Sarpkaya (1974) used a fitting polynomial curve for $z = \beta(x)$ through each set of nine successive nodal points (k=9) of the free surface. The form chosen here with k=3 corresponds to that used by Chan (1971). According to equation (11b), the z_i^{n+1} coordinates (i.e., y_i^{n+1} or r_i^{n+1}) are calculated successively from a fixed initial value (e.g., the separation point). By this, the condition $\partial\phi/\partial n$ is approximately satisfied on the free surface. If ϵ_i^{n+1} exceeds a prescribed error

$$\max \left| \epsilon_i^{n+1} \right| \geq 1 \times 10^{-\varkappa} ,$$

then a new cycle has to be started.

The iteration techniques belonging to method II prevail for the problems present (Chan 1971; Sarpkaya and Hiriart 1975; Larock and Taylor 1976; Isaacs 1977). They distinguish themselves by their good stability and their good convergence behaviour. Thus, the method becomes relatively widely applicable. As described hereinafter, however, there are limits of the application of the iterative finite-element solution for free-surface flows with low flow velocities (subcritical flow).

TEST PROBLEM

The convergence behaviour of method II is investigated by the use of a simple two-dimensional channel flow, as it is shown in Figure 3 . Since the bottom is horizontal, also the free surface has to become horizontal according to the physical laws. We assume an "estimated" location of the free surface, which results in a deflection from the horizontal surface by $\pm\,\delta$. Six different initial deflections $\varepsilon_o = \pm\,\delta / h$ are investigated in dependence on the Froude number in the approach channel $F_o = q_o/\sqrt{gh}$, where h is the channel depth. The relative deflections $\varepsilon / \varepsilon_o$ are represented in dependence on the number of iterations I in Figure 3 in the diagrams a) to f) (simple relaxation ω = 1.0).

For supercritical flows ($F_o > 1$) convergent solu-

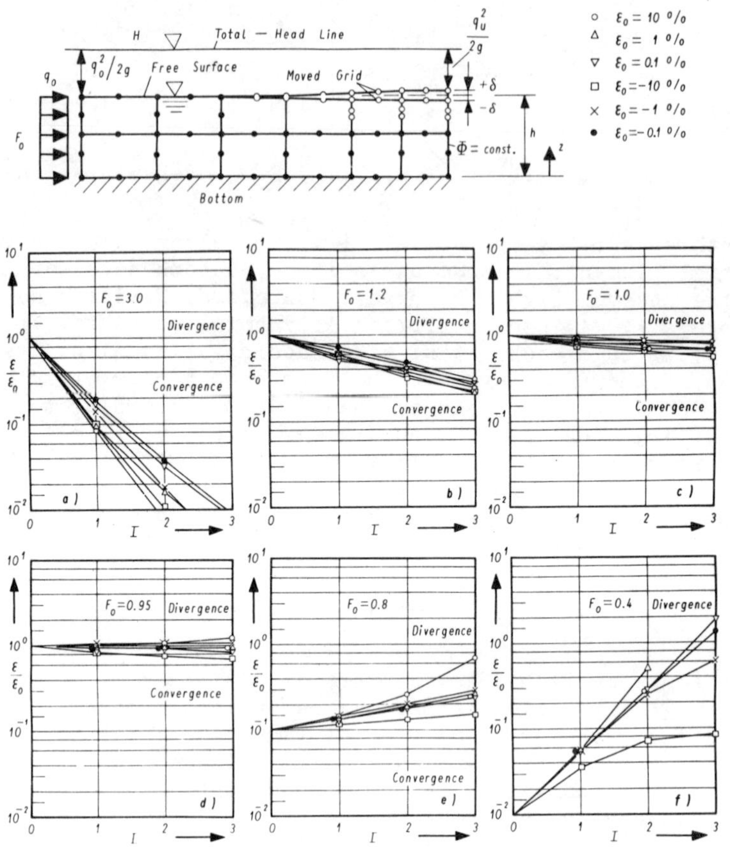

Figure 3 Development of the relative deflection $\varepsilon/\varepsilon_0$ as a function of the initial deflection $\pm\varepsilon_0$ and the Froude number, F_0, for a simple channel flow (relaxation ω = 1.0).

tions are obtained, without exception (Figures 3a, b), i.e., $d(\varepsilon/\varepsilon_0)/dI \leq 1$. The rate of convergence increases with increasing F_0 numbers. For subcritical flows ($F_0 < 1$) (Figures 3e, f), in general no convergent solutions can be obtained, i.e., $d(\varepsilon/\varepsilon_0)/dI > 1$. The rate of divergence increases with decreasing F_0 numbers. It is about constant, whether ε_0 is larger or smaller. For all $+\varepsilon_0$, a "bulging" of the free surface is caused. Due to it, the energy level can be intersected after few iterations. If $-\varepsilon_0$ is assumed, the free surface rapidly lowers and moves towards a theoretically possible water depth d_{c2}, as it will be described later. For the case ε_0 = - 10%, the rate of divergence (as against the horizontal) is lower because the solution is already in the neighbourhood of d_{c2} (i.e., convergence as against d_{c2}). For $F_o = 1$ (Figure 3c) convergent solutions were obtained, too, and also for $F_o = 0.95$ (Figure 3d). Here, the rates of convergence are already unsatisfactory.

It is remarkable that solutions are possible also for $F_o < 1$, but under the conditions $\varepsilon_o \ll 1$ with F_o not being too small, i.e., an as good initial estimation as possible is required. This is obvious also due to the fact that the solution is always stable for all $F_o < 1$, if $\varepsilon_o = 0$. The following reasons are responsible for the non-convergence at small F_o numbers:

(a) Influences of mathematical solutions which are physically not possible:

According to Figure 3, when the continuity and energy equation is applied one will obtain

$$\frac{q_o^2}{2g} + h = d + \frac{h^2 q_o^2}{2 g d^2} \quad \text{with the water depth } d = h \pm \delta \qquad (12)$$

After the introduction of the dimensionless quantities F_o and $d_c = d/h$ there follows

$$d_c^3 - \left(\frac{F_o^2}{2} + 1 \right) d_c^2 + \frac{F_o^2}{2} = 0 \,. \qquad (13)$$

The following solutions are obtained for equation (13):

$$d_{c1} = 1 \qquad (14a)$$

$$d_{c2,3} = \frac{F_o}{4} \left(F_o \pm \sqrt{F_o^2 + 8} \right) . \qquad (14b)$$

From them only the water depth d_{c1} is possible physically. As already shown above, under certain conditions, however, the solution d_{c2} is aimed at in the subcritical range of flow. Here, in the range of the surface deflection there are obtained Froude numbers $F_u = q_u/\sqrt{g\, d_{c2}\, h}$ with $F_u = F_o/d_{c2}^{3/2}$ which become greater than one for $F_o < 1$. Therefore, it is mathematically easily possible always to switch over from subcritical to supercritical flow at $F_o < 1$, which, however, occurred only at $-\varepsilon_o$ in the example, and never at $+\varepsilon_o$, whereas the solution never switch over to d_{c2} at $F_o \geq 1$, i.e., from supercritical to subcritical flow. This behaviour can be physically interpretated. It is well known that the transition from subcritical to supercritical flow is continuous (in the sense of a smooth surface, which actually originates in the solution). On the other hand, physically a transition from supercritical to subcritical flow is possible only discontinuously by "losses" of hydraulic energy, which should lead to a jump of the free surface (a hydraulic jump). Such a discontinuity is impossible with the mathematical model.

(b) Influences due to the dominance of errors of the assumed initial free-surface location upon the kinematic boundary conditions:
This dominance can be made clear from equation (4). At low flow velocities q_i the 2nd term $2g(z_{i+1} - z_i)$ predominates over q_i^2, whereas at high velocities it is small in comparison with q_i^2. Consequently, estimation errors in the subcritical range of flow exert a great influence on the kinematic boundary conditions. From this point of view, also the statements and results by Watters and Street (1964) concerning flows over sills in open channels are to be understood. They obtained solutions for $F_u < 1$ only when the assumed free surface almost coincided with the true surface. Otherwise, always divergence resulted.

(c) Influences due to the approximation error in the calculation of velocities from ϕ_i:
As a result of the discretization τ, for the velocity components approximation errors $\eta(\tau)$ or $\xi(\tau)$ are obtained, respectively,

$$\begin{aligned} \tilde{u}(\tau) &= u + \eta(\tau) \\ \tilde{v}(\tau) &= v + \xi(\tau), \end{aligned} \tag{15}$$

in general $(\eta/u, \xi/v) \ll 1$, where u, v designate the exact components and $\tilde{u}$, $\tilde{v}$ are the approximated components. Then the slope of the free surface is given by

$$1/S = \frac{u + \eta(\tau)}{v + \xi(\tau)} . \tag{16}$$

At increasingly lower flow velocities, it can be observed for certain τ that the approximation errors gain in influence on the quotient of equation (16), since they do not get smaller with decreasing (u, v).
Thus

$$1/S = f\left(\frac{u}{v}\right) \longrightarrow 1/S \approx f\left(\frac{\eta}{\xi}\right)$$

As a whole, the conclusion can be drawn that the iteration according to method II for $F_o \geq 1$ is applicable without any greater restrictions. For $F_o < 1$, convergence to the exact solution is present only when the estimation errors are very small, when F_o is not too small and when a good degree of approximation in τ is present.

STEP-UP TECHNIQUE, AN ADAPTATION METHOD FOR SPILLWAY FLOWS

At spillways a transition from subcritical to supercritical flow occurs. At the same time, discharge is initially unknown. In contrast to the possibility of the automatic determination of discharge for flows under sluice gates in the iteration process, Diersch et al.(1977); Isaacs (1977), (as we known, there always $F_o \geq 1$, since the opening ratio is C_c a/h ≤ 1, where C_c = coefficient of contraction ≤ 1, and a = gate opening), the discharge for spillway flows can be determined only from the study of behaviour of the free surface (Cassidy 1965; Ikegawa and Washizu 1973; Diersch et al. 1977). The flow diagram for the sequence of iterations of the proposed finite-element step-up technique for spillway flows is shown in Figure 4. The main content of this

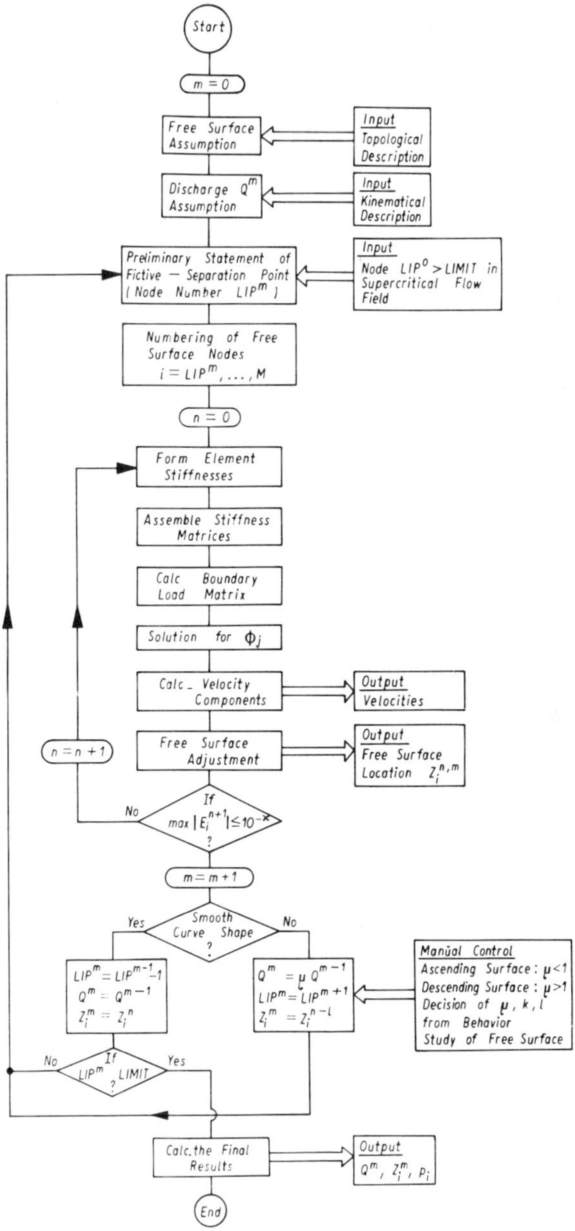

Figure 4 Flow diagram of the step-up adaptation for spillway problems

technique consists in the following: Starting from the supercritical range, stable solutions are generated with a given and afterwards improved discharge Q , and the flowing range can be calculated by the shifting of a fictive separation point LIP upstream. Here, the surface profile calculated for an LIP node when the error criterion in z_i is satisfied is used as an initial solution for the next upstream surface node, i.e., LIP-1 . To satisfy the error criterion in z_i , n iterations become necessary for each LIP node. If so far the free surface is sufficiently smooth for a specific Q, we go to the next surface node; if the surface lowers, Q has to be increased; if the surface rises, Q has to be reduced. In the last two cases, the LIP node has to be shifted again slightly downstream, if necessary. At the same time, an equivalent surface line, which was computed before, has to be taken as the starting solution. With each change of LIP a new series of iterations m begins. This is done until an LIP node = LIMIT will be got to. Virtually, this is the node from which upstream the surface is sufficiently horizontal. The investigations have shown that a variation of Q by ±0.5% from the convergent discharge effects a pronounced distortion of the smoothness of the free surface.

In contrast to problems without any discharge variation, with this problem the number of iterations required is larger by the factor m , i.e., $I = m \times n$. If the free surfaces are to be determined within a precision of four decimals, then $n = 6, \ldots, 13$ and $m = 10, \ldots, 18$ should be expected.

A FEW EXAMPLES

Flow from a circular orifice

In order to check the accuracy of the solutions by the use of straight-sided quadrilateral elements and the iteration technique proposed, the flow from a circular orifice was computed, the effect of gravity having been neglected. The free surface coordinates obtained from a discretization of the flow field with 125 quadrilateral elements and an opening area a / pipe area A - ratio of 1/100 are compared with two other solutions in Table 1 .

Whereas Hunt (1968) solved the problem by means of conformal mapping and singularity distribution, Jeppson (1970) used an inverse problem formulation in connection with finite-difference techniques. The surface coordinates given in Table 1 according to the present finite-element method for x = distance from the separation point (LIP) were determined partly by the straight-line interpolation from the computed nodal values. Table 1 shows the accuracy of the solutions obtained. Ten iterations were necessary for the solution (coordinates within a precision of four decimals). The sequence of iterations is shown in Figure 5 for various overrelaxation factors. As we see, no convergence acceleration can be found with the aid of overrelaxation for such problems. In each case ten iterations were necessary.

Table 1 Surface coordinates for $a/A \rightarrow 0$ and $g = 0$ (r_o = orifice radius).

	r/r_o		
x/r_o	Hunt	Jeppson	Present Solution
0.00	1.000	1.000	1.000
0.0125	0.971	0.962	0.981
0.0400	0.938	0.933	0.941
0.0813	0.905	0.902	0.904
0.1400	0.873	0.873	0.874
0.1975	0.850	0.851	0.850
0.2725	0.829	0.831	0.832
0.3575	0.813	0.813	0.814
0.4588	0.799	0.798	0.798
0.5700	0.788	0.786	0.787
0.6950	0.780	0.776	0.780
0.8188	0.774	0.769	0.774
0.9438	0.770	0.765	0.770
1.0713	0.767	0.763	0.768
1.2000	0.765	0.762	0.766
2.7000	-	-	0.7629

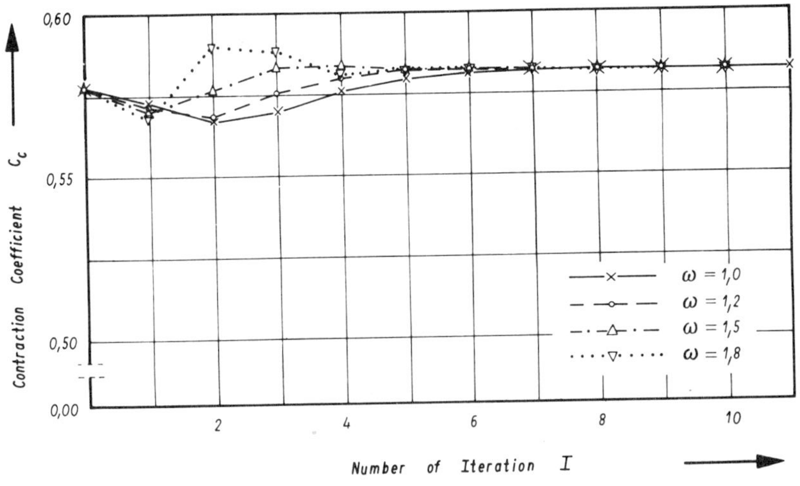

Figure 5 Overrelaxation convergence study for circular orifice with $a/A = 1/100$, $g = 0$.

Flow in the open channel with varying bottom configuration

Figure 6 shows the finite-element idealization for the flow over a smooth sill. The number of quadrilateral elements is 124 . The free surfaces computed for four supercritical

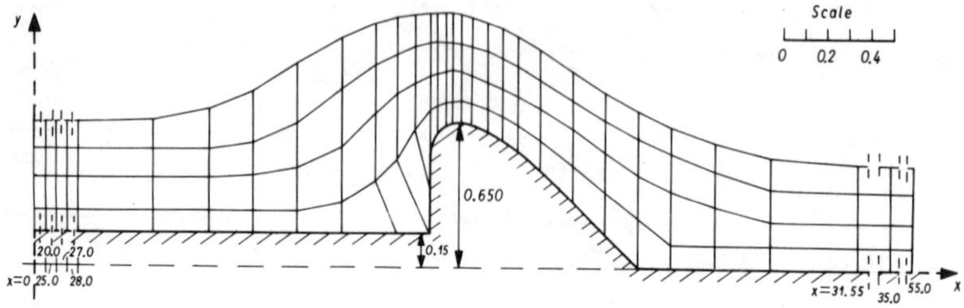

Figure 6 Finite-element idealization with an assumed free surface for flow over a smooth sill.

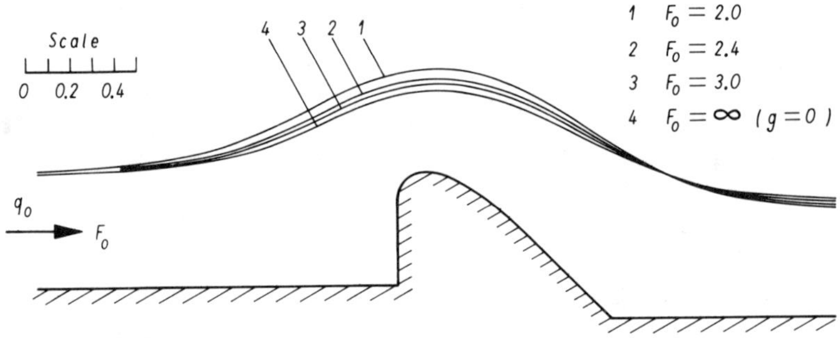

Figure 7 Computed free surfaces for flow over a smooth sill with various Froude numbers ($g=9.81 ms^{-2}$)

Froude numbers are shown in Figure 7. No difficulties arose in the iterative calculation. The number of iterations required ranged between 8 and 12 . A flexible network control had to be applied, however, by the treatment of internal field network-lines as streamlines. In this way, excessive stretching and squashing (distortions) of the finite elements can be avoided.

The convergence behaviour with regard to the tailwater depth d_c during the flow over a positive step with the height $\Omega = 0.26\ h$ for $F_0 = 1.7$ is shown in Figure 8 . In spite of the intentionally unfavourable assumption of the free surface, the iteration method is rapidly converging. Pronounced improvements of convergence are obtained due to the application of overrelaxation. Whereas in the case of the simple relaxation ($\omega = 1.0$) the solution showed a poor convergence in the neighbourhood of the exact

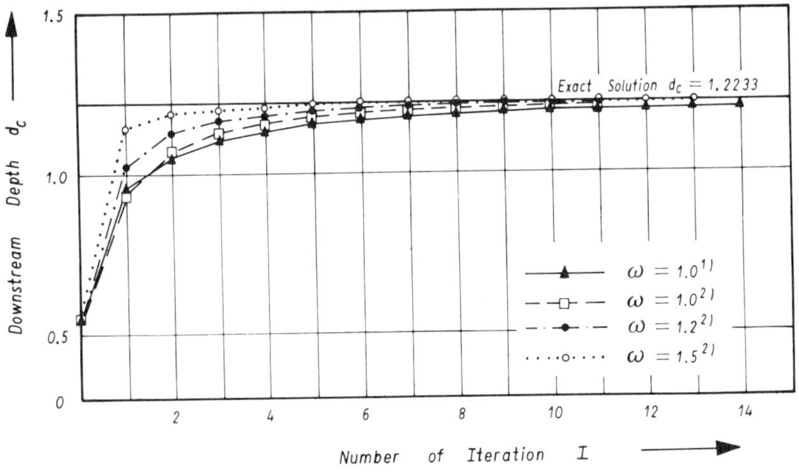

Figure 8 Convergence behaviour of the tailwater depth d_c for the flow over a step with $\Omega = 0.26$ h , $F_o = 1.7$, $g = 9.81 \text{ ms}^{-2}$

1) Calculation of q_u from the continuity condition
2) Calculation of q_u from the energy condition.

tailwater depth and did not show any variation of the surface coordinates within a precision of four decimals on 14 iterations, the computed free surface is in agreement with the exact tailwater depth already on six iterations in the case of $\omega = 1.5$.

Investigations of flow over steps at subcritical Froude numbers have shown that convergent solutions cannot be obtained because of the conditions of subcritical flow present throughout. An adaptation in the mode of the step-up technique was not successful. Whereas within the range of $0.5 < F_o < 0.9$ due to the condition

$$\Omega_{max}/h = 1 + \frac{F_o^2}{2} - \frac{3\,F_o^{2/3}}{2}$$

only slight heights of steps Ω can be flown over, obviously the relatively poor initial estimations of the free surface were responsible for the non-convergence. Within the range of small F_o numbers ($F_o < 0.5$) the solutions were considerably divergent, in spite of the surface which was almost horizontal there. Obviously this is due to the influences of the errors of discretization. It can be assumed that such flows cannot be calculated by the application of conventional finite elements. A similar assumption was expressed already by Larock and Taylor (1976).

Spillway flow

In Figure 9 the finite-element idealization of a spillway

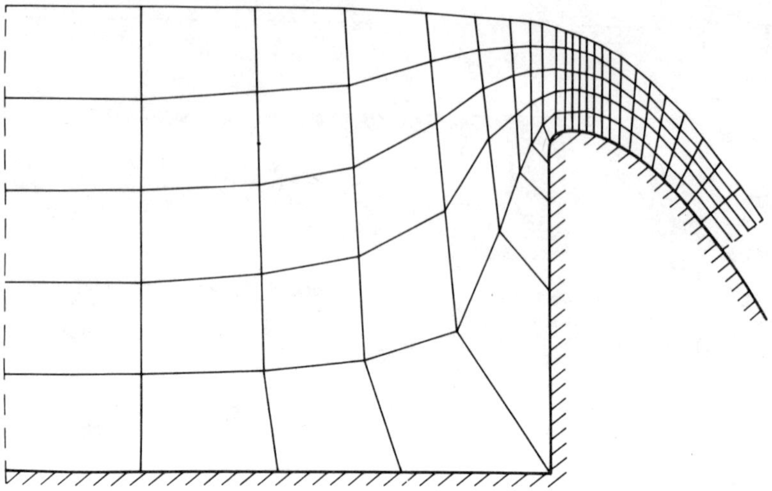

Figure 9 Finite-element idealization of a spillway flow with h_o/h_{oE}=1.5 .

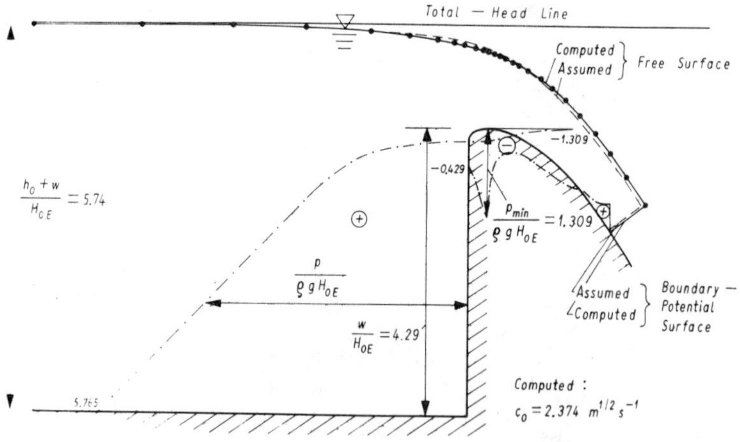

Figure 10 Computed free surface and pressure distribution of a spillway flow with h_o/h_{oE}=1.5 .

flow is represented for a spillway head h_o / design spillway head h_{oE} - ratio of 1.5 (125 quadrilateral elements). The spillway contour used corresponds to that according to Creager. The pressure distribution along the spillway crest computed and the free surface are shown for this example in Figure 10. The solution was obtained according to the step-up technique. Altogether 51 iterations (m × n) were required. Corresponding to the spillway equation, $Q=c_o B H_o^{3/2}$, where c_o = discharge coefficient, B = 1. spillway width, $H_o = h_o + q_o^2/2g$ spillway energy head, a unique c_o value of 2.394 $m^{1/2} s^{-1}$ could be determined. This value agrees well with experimental values; e.g., Schirmer (1976) found c_o = 2.32 $m^{1/2} s^{-1}$. The minimum pressure computed on the spillway contour with $p_{min}/\rho g H_{oE}$ = - 1.309 , however, is below the experimental value of - 0.883 .

CONCLUSIONS

The iterative finite-element model investigated can be successfully applied to various free surface flows, especially to flows from orifices and open channel flows with $F_o > 1$. Under supercritical conditions, no important restrictions exist by the geometry of the flow field.

Open channel flows with subcritical Froude numbers cannot be computed by means of the iterative finite-element model because of the subcritical flow condition presents throughout. It was estimated that for $F_o < 0.5$ mainly discretization errors are responsible for non-convergence.

For spillway flows, the range with $F_o < 1$ can be overcome by means of the step-up technique, since here the free surface can be approximated as almost horizontal and only the normal condition $\partial \Phi / \partial n = 0$ has to be met (prescription of the LIMIT node). The correctness of this assumption can be checked from the pressure distribution then determined.

Perhaps, an improvement of the finite-element analysis of subcritical flows over steps can be expected from hybrid formulations (coupling with analytical techniques). Improvements are obtained also by a transformation of the problem on the complex potential plane, but by this the essential advantages of the finite-element method are lost.

REFERENCES

Brebbia, C.A.; Spanos, K.A. (1973) Application of the Finite Element Method to Steady Irrotational Flow, Rev. Roum. Sci. Techn.-Mech. Appl., Bucarest, 18, 3:463-489.

Cassidy, J.J. (1965) Irrotational Flow over Spillways of Finite Height. J. of Engng. Mech. Div., ASCE, 91, 6:155-173

Chan, S.T.K. (1971) Finite Element Analysis of Irrotational Flows of an Ideal Fluid. Ph. D. Diss., Univ. Calif., Davis.

Diersch, H.J.; Schirmer, A.; Busch, K.F. (1977) Analysis of Flows with Initially Unknown Discharge. J. of Hydr. Div., ASCE, 103, 3:213-232.

Fangmeier, D.D.; Strelkoff, T.S. (1968) Solution for Gravity Flow Under a Sluice Gate, J. of Engng. Mech. Div., ASCE,

94, 1:153-176.

Hiriart,G.; Sarpkaya,T. (1974) Jet Impingement on Axisymmetric Curved Deflectors. Int. Symp. on Finite Elements in Flow Problems, Swansea, U.K., Proceed. 153-157.

Hunt,B.W. (1968) Numerical Solution of an Integral Equation for Flow from a Circular Orifice, J. of Fluid Mechanics, Part 2, 31, 361-377.

Ikegawa,M.; Washizu,K. (1973) Finite Element Method Applied to Analysis of Flow over a Spillway Crest. Int. J. Num. Meth. Engng., 6, 179-189.

Isaacs,L.T. (1977) Numerical Solution for Flow Under Sluice Gates, J. of Hydr. Div., ASCE, 103, 5:473-481.

Jeppson,R.W. (1970) Inverse Formulation and Finite Difference Solution for Flow from a Circular Orifice, J. of Fluid Mechanics, Part 1, 40, 215-223.

Larock,B.E. (1969) Jets from Two-Dimensional Symmetric nozzles of Arbitrary Shape, J. of Fluid Mech., Part 3,37,479-489.

Larock,B.E. (1970) A Theory for Free Outflow beneath Radial Gates, J. of Fluid Mechanics, 41, 4:851-864.

Larock,B.E. (1975) Flow over Gated Spillway Crests. Dev. in Mech., 8, Proceed. 14th Midwestern Mech, Conf., 437-451.

Larock,B.E.; Taylor,C. (1976) Computing Three-Dimensional Free Surface Flows. Int. J. Num. Meth. Engng., 10,1143-1152.

McCorquodale,J.A.; Li,C.Y. (1971) Finite Element Analysis of Sluice Gate Flow. Trans. Engng. Inst. Can., 14, No. C-2.

McNown,J.S.; Hsu,E.Y.; Yih,C.S. (1955) Applications of the Relaxation Technique in Fluid Mechanics. Trans. ASCE, 120, 650-686.

Meissner,U. (1973) A Mixed Finite Element Model for Use in Potential Flow Problems. Int. J. Num. Meth. Engng., 6, 467-473.

Moayeri,M.S. (1973) Flow in Open Channels with Smooth Curved Boundaries, J. of Hydr. Div., ASCE, 99, 12:2217-32.

Rouse,H.; Abul-Fetouh,A. (1950) Characteristics of Irrotational Flow through Axially Symmetric Orifices. J.of Appl. Mechanics, 17, 4:421-426.

Sarpkaya,T.; Hiriart,G. (1975) Finite Element Analysis of Jet Impingement on Axisymmetric Curved Deflectors. Finite Elements in Fluids, John Wiley & Sons, 1, 265-279.

Schirmer,A. (1976) Wirkungsweise und Leistungsgrenzen rundkroniger Überfälle an Talsperren bei Überlastung. Diss. Techn. Univ. of Dresden, Dresden.

Southwell,R.; Vaisey,G. (1946) Relaxation Methods Applied to Engineering Problems. XII. Fluid Motions Characterized by Free Streamlines. Phil. Trans. Roy. Soc., Lon., Ser. A, 240, 117-161.

Strelkoff,T.S. (1964) Solution of Highly Curvilinear Gravity Flows. J. of Engng. Mech. Div., ASCE, 90, 3:195-221.

Taylor,R.L.; Brown,C.B. (1967) Darcy Flow Solutions with a Free Surface. J. of Hydr. Div., ASCE, 93, 2:25-33.

Watters,G.Z.; Street,R.L. (1964) Two-Dimensional Flow Over Sills in Open Channels. J. of Hydr. Div., ASCE, 90, 4:107-140.

S E S S I O N 4

NUMERICAL TECHNIQUES

A FINITE ELEMENT METHOD FOR THE DIFFUSION-CONVECTION EQUATION

E. Varoḡlu, W.D.L. Finn

Faculty of Applied Science, The University of British Columbia, Vancouver, B.C., V6T 1W5, Canada

INTRODUCTION

The prediction of dispersal of pollutants in air, bodies of water or in the ground by seepage is one of the most difficult problems of environmental control. Dispersion occurs by both convection and diffusion. The distribution of pollutant concentration in the host medium at any time is governed by the diffusion-convection equation and appropriate boundary and initial conditions.

The numerical solution of the diffusion-convection equation has attracted considerable attention in the water resources research literature because of its wide applicability to pollutant transport problems in lakes, in near shore zones and in surface and subsurface hydrology.

Several authors (Price et al, 1966, Chaudhari, 1971, Keller, 1971, Book et al, 1975, and Martin, 1975) derived and employed finite difference schemes in space and time for the numerical solution of the diffusion-convection equation. On the other hand, a combination of Galerkin type finite elements in space and finite differences in time were employed by Price et al, 1968, Adey and Brebbia, 1973, Dailey and Harleman, 1973, and Smith et al, 1973. A Galerkin finite element technique, in space and time, is derived in Gray and Pinder, 1974, and Bonnerot and Jamet, 1974, for the diffusion equation and in Bruch and Zyvlovski, 1975, for the diffusion-convection equation.

A review and comparison of the numerical methods for the solution of the diffusion-convection equation can be found in Lam, 1977, Ehlig, 1977, van Genuchten, 1977, Mercer and Faust, 1977, and Smith, 1977. These papers examined and contrasted several numerical methods. It is shown that most of the numerical methods give acceptable numerical results for diffusion dominated flow problems but when convection is a strong component of dispersion the existing numerical solutions have an intrinsically unstable character which exhibits oscillation, overshoot, undershoot, artificial diffusion, negative concentrations and clipping errors even in one-dimensional cases. Reduction of the time step or reduction of the space increment is not sufficient to eliminate oscillations in the conventional finite element solution of some immiscible flow problems (Mercer and Faust, 1977). Neither higher order finite elements (van Genuchten, 1977), nor higher order integration schemes in time (Smith, 1977) remedy the oscillations and overshoots in convection dominated flow problems.

In Lam, 1977, it is shown that the central differencing scheme and box scheme are oscillatory; upstream differencing introduces large artificial diffusion and when flux correction is used to control oscillations in upstream differencing schemes then sharp concentration fronts are smeared and clipping errors and artificial diffusion is created even in one-dimensional convection dominated flows. Therefore, it is desirable to develop a numerical method which is capable of solving numerically the diffusion-convection equation for the whole spectrum of dispersion from diffusion only - through mixed dispersion -to pure convection.

In view of the lack of success in eliminating the difficulties with using existing finite difference and conventional finite element methods for the solution of convection dominated flow problems, we attempted to develop a new solution technique which, in the limit, would degenerate into efficient solutions for either the convection problem or the diffusion problem. We have succeeded in developing a method which combines the method of characteristics which is applicable to the hyperbolic equations which govern convection with the finite element method which works well for the parabolic equations which govern diffusion. This can be achieved in the framework of an unconventional Galerkin finite element method in space and in time. In the case of pure diffusion, the method reduces to the finite element method given in Gray and Pinder, 1974, for the diffusion equation. Our method differs from the finite element method given in Bruch and Zyvlovski, 1975, because of the incorporation of the method of characteristics intrinsically into the finite element method. The derivation of the method will be given here for the one-dimensional diffusion-convection equation.

STATEMENT OF THE PROBLEM

Consider the one-dimensional diffusion-convection equation

$$\frac{\partial C}{\partial t} = -u\frac{\partial C}{\partial X} + D\frac{\partial^2 C}{\partial X^2}, \quad 0<X<L, \quad t>0 \tag{1}$$

with the initial condition

$$C(X,0) = c^{o}(X), \quad 0<X<L \tag{2}$$

and the boundary conditions

$$C(0,t) = g(t), \quad t>0 \tag{3}$$

$$C(L,t) = 0, \quad t>0 \tag{4}$$

Here, t denotes the time and X is the spatial co-ordinate, C(X,t) is the pollutant concentration, u and D are the velocity of the flow and the eddy diffusivity coefficient, respectively. In the case of pure convection D=0 and, in this case, it should be noted that only one boundary condition is required to have a well posed problem.

The derivation of the method of solution will be given to solve the initial boundary value problem defined by Equations 1-4. This simple initial boundary value problem facilitates greatly the presentation of the fundamental ideas of the proposed method. Moreover, it has been extensively used to compare the numerical results of the finite difference and the finite element methods (Lam, 1977, van Genuchten, 1977, Smith, 1977, Ehlig, 1977).

TRANSFORMATION OF THE PROBLEM

The initial-boundary value problem for the diffusion-convection equation (Equations 1-4) can be transformed by taking

$$x = X - ut \tag{5}$$

and

$$C(X,t) = c(X-ut,t) \tag{6}$$

into the initial-boundary value problem for the diffusion equation in the x-t plane

$$\frac{\partial c}{\partial t} = D\frac{\partial^2 c}{\partial x^2}, \quad -ut<x<L-ut, \quad t>0 \tag{7}$$

$$c(x,0) = c^{o}(x), \quad 0<x<L \tag{8}$$

$$c(-ut,t) = g(t), \quad t>0 \tag{9}$$

$$c(L-ut,t) = 0, \quad t>0 \tag{10}$$

The convective term has disappeared from the governing equation which is now reduced to the simple diffusion form. However, the effect of the convective term is still included by the altered geometry of the region in which the originally vertical boundaries are now sloped with a slope equal to -1/u. Once the solution of Equations 7-10, c(x,t), is found, then C(X,t) is readily obtained by employing Equations 5 and 6. The original problem in the X-t plane and the transformed problem in the x-t plane are illustrated in Figure 1(a,b).

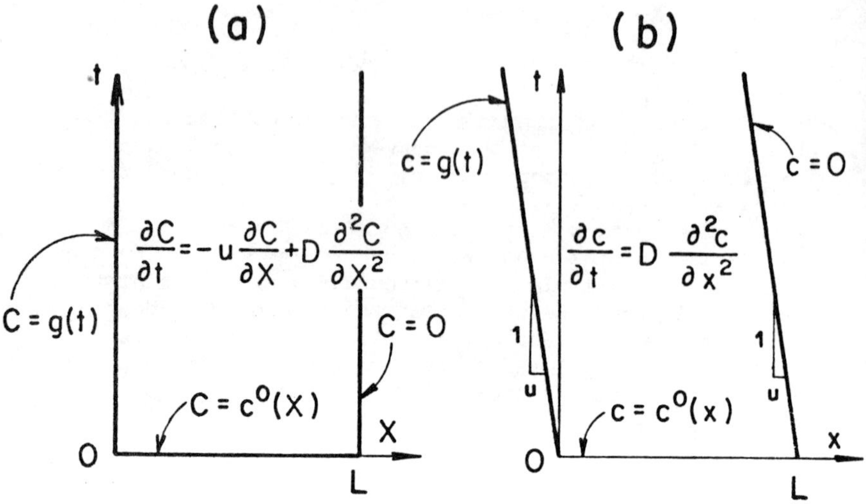

FIGURE 1 The initial boundary value problem.
a) in the X-t plane;
b) in the x-t plane.

FINITE ELEMENTS IN SPACE AND TIME

To solve Equations 7-10, the Galerkin method of finite elements is used. Let $\hat{c}(x,t)$ be an approximation to the solution $c(x,t)$ of Equation 7 which also satisfies the initial condition Equation 8 and the boundary conditions Equations 9-10. The vanishing of the weighted residual of Equation 8 with a continuous weighting function $\phi(x,t)$ defined in $-ut<x<L-ut$, $t>0$ can be expressed as

$$\int_{t=\tau_1}^{\tau_2} \int_{x=-ut}^{L-ut} \left(\frac{\partial \hat{c}}{\partial t} - D\frac{\partial^2 \hat{c}}{\partial x^2}\right) \phi(x,t)\, dx\, dt = 0 \qquad (11)$$

for all $0 \leq \tau_1 < \tau_2$.

Letting $\phi(x,t)$ vanish on the boundary $x = L-ut$ and employing integration by parts, Equation 11 becomes

$$\int_{t=\tau_1}^{\tau_2} \int_{x=-ut}^{L-ut} \left(-\hat{c}\frac{\partial \phi}{\partial t} + D\frac{\partial \hat{c}}{\partial x}\,\frac{\partial \phi}{\partial x}\right) dx\, dt$$

$$+ \int_{x=-u\tau_2}^{L-u\tau_2} \hat{c}(x,\tau_2)\ \phi(x,\tau_2)dx - \int_{x=-u\tau_1}^{L-u\tau_1} \hat{c}(x,\tau_1)\ \phi(x,\tau_1)dx$$

$$- \int_{x=-u\tau_2}^{-u\tau_1} \hat{c}(x,-\frac{x}{u})\,\phi(x,-\frac{x}{u})dx + D\int_{t=\tau_1}^{\tau_2} \phi(-ut,t)\frac{\partial \hat{c}}{\partial x}(-ut,t)dt$$

$$= \quad 0 \qquad (12)$$

The problem will be solved step by step in time. At each time step the initial conditions will be determined by the solution obtained at the previous time step. Let $\tau_1 = t^n$ and $\tau_2 = t^{n+1}$ denote the beginning and the end of a typical time step. The region $t^n<t<t^{n+1}$, $-ut<x<L-ut$ will be discretized by the spatial temporal elements K_0^n, $K_1^n,\ldots,$ K_{I-1}^n as illustrated in Figure 2.

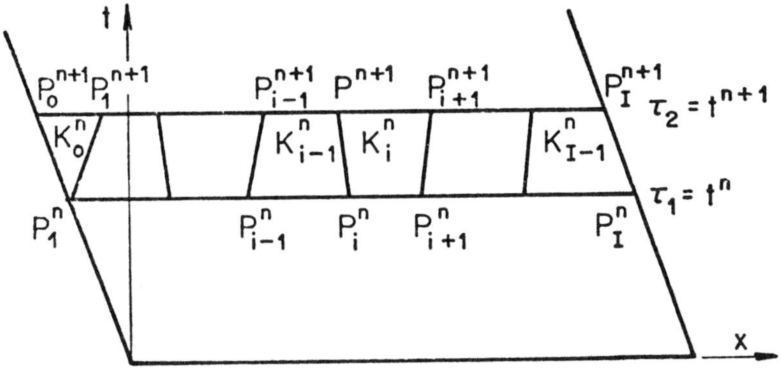

FIGURE 2 Finite elements in space and time at a typical time step.

At a typical time step, we consider two types of elements, triangular and trapezoidal. A triangular element is required next to the left-hand boundary in order to ensure that a constant number of elements can be used throughout all the time steps without any of the elements becoming unduly stretched. The triangular element, K_0^n, has two nodes, P_1^n and P_0^{n+1} on the boundary $x = -ut$. The finite element procedure for the trapezoidal elements has been developed by Bonnerot and Jamet, 1974, and is summarized here. The trapezoidal finite elements are transformed from the x-t global co-ordinate system to a η-ξ local co-ordinate system such that the transformed element in the η-ξ system is a unit square ($0\leq\eta\leq1$, $0\leq\xi\leq1$) as shown in Figure 3.

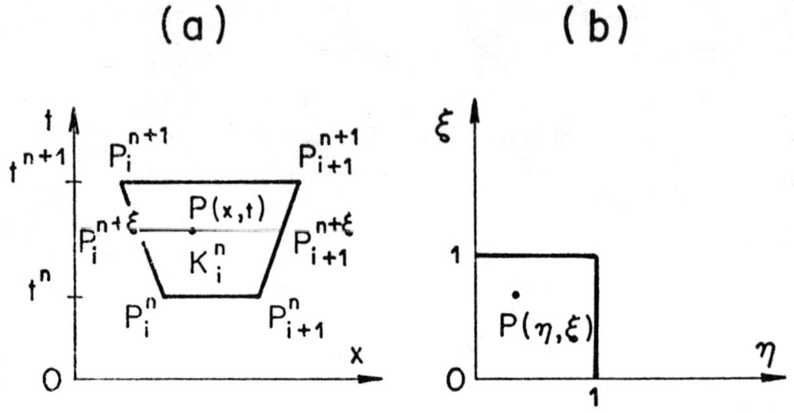

FIGURE 3 A typical trapezoidal element.
a) Global co-ordinates;
b) Local co-ordinates.

The co-ordinate transformation which maps a unit square to the typical trapezoidal element, K_i^n, in the x-t plane can be expressed as

$$t = t^n + \xi(t^{n+1} - t^n) \tag{13}$$

$$x = (1-\eta)\,[(1-\xi)x_i^n + \xi x_i^{n+1}] + \eta[(1-\xi)x_{i+1}^n + \xi x_{i+1}^{n+1}] \tag{14}$$

or equivalently

$$\xi = (t - t^n) / (t^{n+1} - t^n) \tag{15}$$

$$\eta = (x - x_i^{n+\xi}) / (x_{i+1}^{n+\xi} - x_i^{n+\xi}) \tag{16}$$

where

$$x_i^{n+\xi} = (1 - \xi)\, x_i^n + \xi x_i^{n+1} \tag{17}$$

and x_j^k denotes the x-coordinate of the point P_j^k in the x-t plane. In the local co-ordinates, the approximate solution at a typical trapezoidal element K_i^n will be taken as a polynomial of the form

$$\hat{c}(\eta,\xi) \; = \; (1-\eta)(1-\xi)c_i^n + (1-\eta)\xi c_i^{n+1} + \eta(1-\xi)c_{i+1}^n + \eta\xi c_{i+1}^{n+1} \qquad (18)$$

in which c_j^k is the value of $\hat{c}$ at the node P_j^k. It is clear that $\hat{c}(x,t)$ is linear along each side of the trapezoidal K_i^n. Since the shape functions used in Equation 18 are also used in the co-ordinate transformations (Equations 13 and 14), the finite elements are isoparametric.

Similarly, the concentration in the triangular element K_o^n is approximated by the polynomial

$$\hat{c}(\eta',\xi) \; = \; (1-\eta')c_o^{n+1} + (1-\xi')c_i^n + (-1+\eta'+\xi)c_1^{n+1} \qquad (19)$$

in the local co-ordinates $(\eta'-\xi)$ where η' is defined as

$$\eta' \; = \; 1 + (x - x_1^{n+\xi}) \;/\; (x_1^{n+1} - x_o^{n+1}) \qquad (20)$$

with $x_1^{n+\xi}$ and ξ defined by Equations 17 and 15, respectively.

Let the space V denote the space of all the continuous functions defined on the finite elements K_o^n, $K_1^n, \ldots, K_{I-1}^n$ by Equations 19 (on K_o^n) and 18 (on $K_1^n, \ldots, K_{I-1}^n$). A function ϕ in V is uniquely determined by its values at all the nodes P_i^n $(i=1,2,\ldots,I)$ and P_i^{n+1} $(i=0,1,\ldots,I)$ and function ϕ is linear along the sides of the finite elements.

Let the approximate solution $\hat{c}$ to be evaluated at time step n $(t^n \leq t \leq t^{n+1})$ satisfy the boundary conditions

$$c_1^n \; = \; g(t^n), \quad c_o^{n+1} = g(t^{n+1}), \quad c_I^n = c_I^{n+1} = 0 \qquad (21)$$

Since the nodal values c_i^n $(i=2,3,\ldots,I-1)$ are evaluated at the previous time step and therefore are known, there is one unknown for each i $(c_i^{n+1}, i=1,2,\ldots,I-1)$ to be evaluated at each time step. We define the weighting function $\phi(x,t)$ for each i $(i=1,2,\ldots,I-1)$ as the function of space V such that

$$\phi^{(i)}(P_j^n) = \phi^{(i)}(P_j^{n+1}) = \begin{cases} 1, & j=i \\ 0, & j\neq i \end{cases} \quad \begin{array}{l} i=1,2,\ldots,I-1 \\ j=0,1,\ldots,I \end{array} \quad n \geq 0 \qquad (22)$$

To define the discrete analog of the integral identity (Equation 12), we replace $\phi(x,t)$ by $\phi^{(i)}$ $(i=1,2,\ldots,I-1)$ and evaluate the contributions to each integral from the finite elements by a numerical quadrature. This standard procedure is explained in Zienkiewicz, 1971. The non-vanishing contributions of the triangular finite element K_o^n to each integral in Equation 12 and the evaluation of the boundary integrals (taking $\tau_1 = t^n$ and $\tau_2 = t^{n+1}$) are given as

$$\iint_{K_o^n} \hat{c}\frac{\partial\phi^{(1)}}{dt}\, dx\, dt = -\frac{1}{6}(x_1^{n+1}-x_1^n)(c_1^n+c_o^{n+1}+c_2^{n+1}) \tag{23}$$

$$\iint_{K_o^n} \frac{\partial\hat{c}}{\partial x}\frac{\partial\phi^{(1)}}{\partial x}\, dx\, dt = \frac{1}{2}\, k\, \frac{c_1^{n+1}-c_o^{n+1}}{x_1^{n+1}-x_o^{n+1}} \tag{24}$$

$$\int_{x_o^{n+1}}^{x_1^{n+1}} \hat{c}(x,t^{n+1})\,\phi^{(1)}(x,t^{n+1})\,dx = \frac{1}{2}\,(x_1^{n+1}-x_o^{n+1})\,c_1^{n+1} \tag{25}$$

$$\int_{x_o^{n+1}}^{x_1^{n}} \hat{c}(x,-\frac{x}{u})\,\phi^{(1)}(x,-\frac{x}{u})\,dx = \frac{1}{2}\,(x_1^n-x_o^{n+1})\,c_1^n \tag{26}$$

$$\int_{t^n}^{t^{n+1}} \phi^{(1)}(-ut,t)\,\frac{\partial\hat{c}}{\partial x}(-ut,t)\,dt = \frac{1}{2}\,k\,\frac{c_1^{n+1}-c_o^{n+1}}{x_1^{n+1}-x_o^{n+1}} \tag{27}$$

where

$$k = t^{n+1} - t^n \tag{28}$$

The contributions from the trapezoidal elements to integrals in Equation 12 are evaluated in Bonnerot and Jamet, 1974, and will not be repeated here.

When the contributions from the trapezoidal and triangular elements to the integrals in Equation 12 (ϕ replaced by $\phi^{(i)}$, $i=1,2,\ldots,I-1$) are written explicitly, we obtain the discrete equations

$$\frac{1}{6}\,(x_1^{n+1}-x_1^n)(c_1^n+c_o^{n+1}+c_1^{n+1}) + \frac{1}{4}\Big[(c_1^n+c_1^{n+1})(x_1^{n+1}-x_1^n) + (c_2^n+c_2^{n+1})(x_2^{n+1}-x_2^n)\Big] - \frac{1}{2}\,(x_2^n-x_o^{n+1})\,c_1^n + \frac{1}{2}\,(x_2^{n+1}-x_o^{n+1})\,c_1^{n+1} + \frac{1}{2}\,kD\left[2\,\frac{c_1^{n+1}-c_o^{n+1}}{x_1^{n+1}-x_o^{n+1}} - \frac{c_2^n-c_1^n}{x_2^n-x_1^n} - \frac{c_2^{n+1}-c_1^{n+1}}{x_2^{n+1}-x_1^{n+1}}\right] = 0 \quad \text{(for } i=1\text{)} \tag{29}$$

$$\frac{1}{4}\left[(x_{i+1}^{n+1}-x_{i+1}^{n})(c_{i+1}^{n}+c_{i+1}^{n+1}) - (x_{i-1}^{n+1}-x_{i-1}^{n})(c_{i-1}^{n}+c_{i-1}^{n+1})\right]$$

$$-\frac{1}{2}\left[(x_{i+1}^{n+1}-x_{i-1}^{n+1})c_i^{n+1} - (x_{i+1}^{n}-x_{i-1}^{n})c_i^{n}\right] + \frac{1}{2}kD\left[\frac{c_{i+1}^{n}-c_i^{n}}{x_{i+1}^{n}-x_i^{n}}\right.$$

$$\left. - \frac{c_i^{n}-c_{i-1}^{n}}{x_i^{n}-x_{i-1}^{n}} + \frac{c_{i+1}^{n+1}-c_i^{n+1}}{x_{i+1}^{n+1}-x_i^{n+1}} - \frac{c_i^{n+1}-c_{i-1}^{n+1}}{x_i^{n+1}-x_{i-1}^{n+1}}\right] = 0, \quad i=2,3,\ldots,I-1 \tag{30}$$

in the unknowns c_i^{n+1}, $(i=1,2,\ldots,I-1)$.

METHOD OF CHARACTERISTICS IN FINITE ELEMENTS

The difficulties arise in existing numerical methods when the convection is the dominant factor in dispersion. Therefore, we study the limiting case D=0 which, by substitution into Equation 7, yields

$$\frac{\partial c}{\partial t} = 0, \quad -ut<x, \quad t>0 \tag{31}$$

This corresponds to pure convection in the X-t plane and the boundary condition at X=L must be dropped. The transformed exact solution of the pure convection problem is

$$c(x,t) = \begin{cases} c^{o}(x) & 0<x \\ g(-x/u) & -ut<x<0 \end{cases}, \quad t>0 \tag{32}$$

which expresses that the initial concentration, $c^{o}(x)$ and the concentration $g(-x/u)$ fed in at $t=x/u$ along the boundary move parallel to the t-axis without changing in the x-t plane.

Now, we want to ensure that the finite element method (Equations 29-30) will give the exact solution

$$c_i^{n+1} \rightarrow c_i^{n}, \quad i=1,2,\ldots,I-1 \tag{33}$$

for $D \rightarrow 0$. This will be accomplished if the element sides joining the points P_i^n to P_i^{n+1}, $i=1,2,\ldots,I-1$ are vertical. With this orientation of element sides at a typical time step n, we have

$$x_i^{n+1} = x_i^{n}, \quad i=1,2,\ldots,I-1 \tag{34}$$

which substituted into Equations 29 and 30 yield Equation (33) for $D \rightarrow 0$. Equations 29 and 30, in view of Equation 34, reduce to

$$\frac{1}{2}\,kD\left[\frac{c_2^n-c_1^n}{x_2^n-x_1^n}+\frac{c_2^{n+1}-c_1^{n+1}}{x_2^n-x_1^n}-2\,\frac{c_1^{n+1}-c_o^{n+1}}{x_1^n-x_o^{n+1}}\right]$$

$$-\frac{1}{2}\,(x_2^n-x_o^{n+1})\,(c_1^{n+1}-c_1^n) = 0\,,\quad \text{(for } i=1\text{)} \qquad (35)$$

$$\frac{1}{2}\,kD\left[\frac{c_{i+1}^n-c_i^n}{x_{i+1}^n-x_i^n}+\frac{c_{i+1}^{n+1}-c_i^{n+1}}{x_{i+1}^n-x_i^n}-\frac{c_i^n-c_{i-1}^n}{x_i^n-x_{i-1}^n}-\frac{c_i^{n+1}-c_{i-1}^{n+1}}{x_i^n-x_{i-1}^n}\right]$$

$$-\frac{1}{2}\,(x_{i+1}^n-x_{i-1}^n)\,(c_i^{n+1}-c_i^n) = 0\,,\quad i=2,3,\ldots,I-2 \qquad (36)$$

$$\frac{1}{2}\,kD\left[\frac{c_{I-1}^n}{x_I^n-x_{I-1}^n}+\frac{c_{I-1}^n-c_{I-2}^n}{x_{I-1}^n-x_{I-2}^n}+\frac{c_{I-1}^{n+1}}{x_I^{n+1}-x_{I-1}^{n+1}}+\frac{c_{I-1}^{n+1}-c_{I-2}^{n+1}}{x_{I-1}^{n+1}-x_{I-2}^{n+1}}\right]$$

$$+\frac{1}{2}\left[(x_I^{n+1}-x_{I-2}^{n+1})\,c_{I-1}^{n+1}-(x_I^n-x_{I-2}^n)\,c_{I-1}^n\right]= 0,\quad \text{(for } i=I-1\text{)} \qquad (37)$$

These are the finite element equations for the general diffusion-convection problem (Equations 7-10) obtained by constraining the orientation of the sides of the finite elements ($P_i^n\,P_i^{n+1}$, $i=1,2,\ldots,I-1$) to be vertical in the x-t plane. These vertical lines along which concentrations propagate without change in the limiting case $D\to 0$ correspond to the characteristic lines of slope $1/u$ of the original hyperbolic equation (Equation 1 for $D\to 0$) in the X-t plane. Thus, the method of characteristics is incorporated into the finite elements so that the transformed solution of Equations 35 and 36 (for D=0) yield the propagation of concentration without change along the characteristics of the resulting hyperbolic equation when D=0 is substituted into Equation 1. It is worth noting that, for equally spaced nodes, Equation 36 reduces to the Crank-Nicolson scheme.

In the case of the diffusion alone (u=0), the boundaries in the x-t plane are parallel to the t-axis, therefore, $P_1^{n+1} = P_o^{n+1}$ and the triangular area element K_o^n vanishes and the unknowns of the problem C_i^{n+1}, $i=2,3,\ldots,I-1$ can be evaluated from the set of equations (Equations 36 and 37) which is basically equivalent to the set of equations to be obtained in conventional finite element methods (Gray and Pinder, 1974).

EXAMPLES

We give numerical results for the standard test problems, of the plume and the rectangular wave using the finite element method incorporating characteristics [FEMIC] developed above.

The plume problem is defined as

$$c^o(X) = 0, \qquad 0<X<L$$

$$C(0,t) = 1, \quad C(L,t) = 0, \qquad t>0 \tag{38}$$

and the rectangular wave problem

$$c^o(X) = \begin{cases} 1, & 0.095<X<0.205 \\ 0, & 0<X<0.095, \quad 0.205<X<L \end{cases} \tag{39}$$

uses the same wave form employed in Lam, 1977. The data used in these test problems is summarized in Table 1. The space and time discretizations given in Table 1 are chosen such that a direct comparison of the results from FEMIC with published results from several other numerical methods is possible.

TABLE 1 Data for Test Problems

PROBLEM TYPE	Problem No.	D m^2/sec	u m/sec	L m	Number of Ele.	Time Step sec
PLUME	1	0.01	0.05	2.5	52	0.5
	2	1	10^4	0.64	35	5×10^{-7}
RECTANGULAR WAVE	3	1	10^3	1	100	6×10^{-6}
	4	0	1	1	100	6×10^{-3}

Test problem 1 is a diffusion dominated flow case and another finite element solution of this problem can be found in Adey and Brebbia, 1973. The numerical results from FEMIC and their comparison to the exact solution (Ehlig, 1977) are illustrated in Figure 4. The solution obtained from FEMIC is very accurate which also is the case for several other numerical schemes.

Test problem 2 is the highly convection dominated flow case. The numerical solutions for test problem 2 obtained from several numerical schemes are illustrated in Figure 5 which has been reproduced from Lam, 1977. The results shown in Figure 5 illustrate the numerical difficulties encountered in the convection dominated flow problems. None of the numerical solution schemes compared in Lam, 1977 are suitable for test problem 2 except the upstream differencing scheme plus flux-corrected transport (UDS+FCT). The numerical results for test problem 2 obtained from FEMIC are illustrated in Figure 6 and compared to exact solution. These results are a better approximation to the exact solution than those from any of the other methods including UDS+FCT. From Figure 6, it

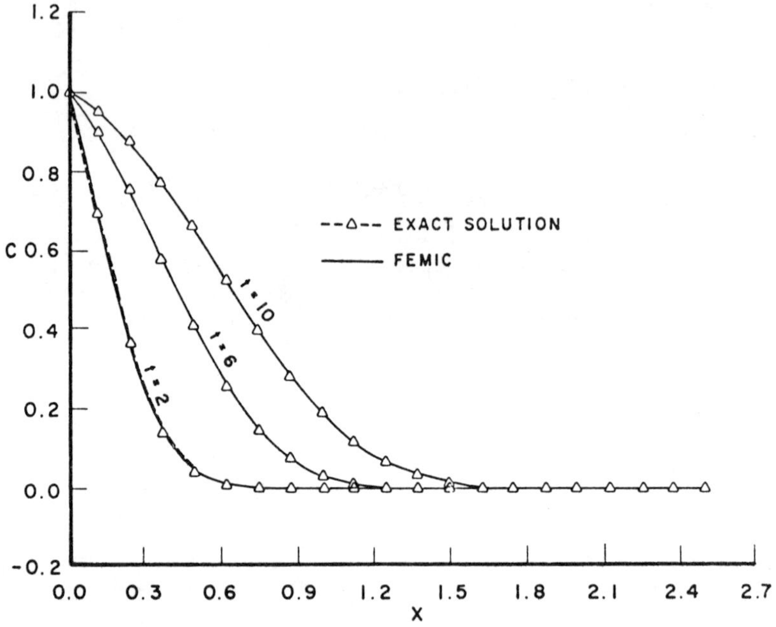

FIGURE 4 Test problem 1

is clear that the solution from FEMIC does not exhibit any undesirable aspects of the other solution methods such as oscillations, undershoot and overshoot.

Test problem 3 corresponds to a diffusion dominated flow with an initial rectangular wave. The conventional finite element solution of this problem is very satisfactory although the upstream differencing scheme plus flux-corrected transport exhibit some clipping error. The numerical solution of test problem 3 from FEMIC is illustrated in Figure 7. This solution is very accurate and does not exhibit any clipping error.

Test problem 4 is a pure convection case in which the initial rectangular wave concentration should propagate without change. The numerical results from conventional finite elements (FEM) and from the upstream differencing plus flux-corrected transport (UDS+FCT) are illustrated in Figure 8 after Lam, 1977. The numerical results from FEMIC are illustrated in Figure 9. This solution exhibits the translation of the initial wave form without change at t = 0.6 sec and is exact.

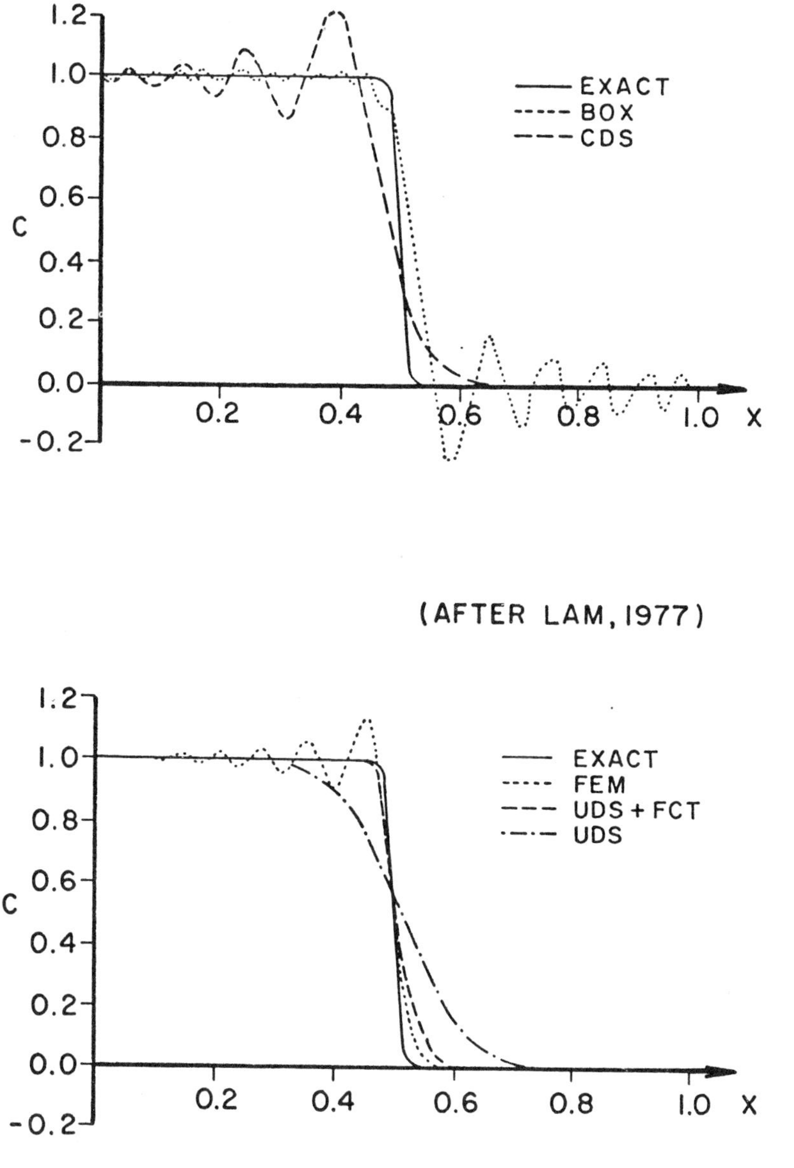

FIGURE 5 1-D Plume results, test problem 2, $t=5X10^{-5}$ $u=10^{4}$, $D=1$, $h=1/50$, $\Delta t=5X10^{-7}$
BOX - box scheme; CDS - central differencing scheme; FEM - finite element (Galerkin) methods with linear basis function; UDS+FCT - upstream differencing scheme plus flux-corrected transport; UDS - upstream differencing.

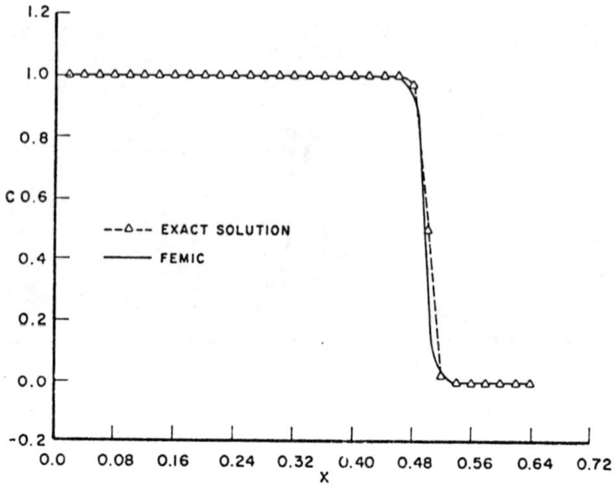

FIGURE 6 Test problem 2, $t=5X10^{-5}$ sec

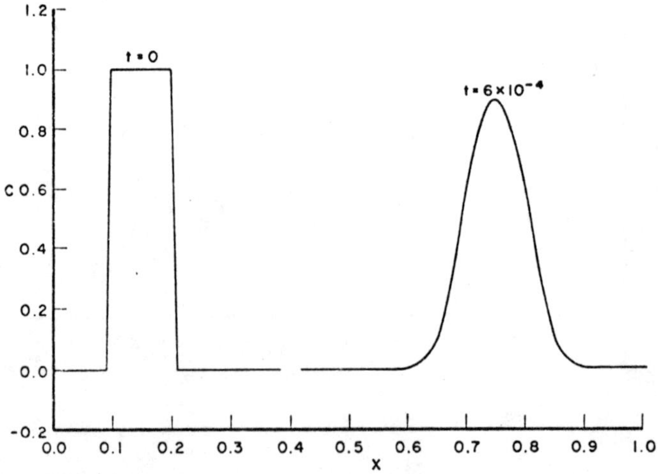

FIGURE 7 Test problem 3, rectangular wave at t=0 and solution from FEMIC at $t=6X10^{-4}$ sec

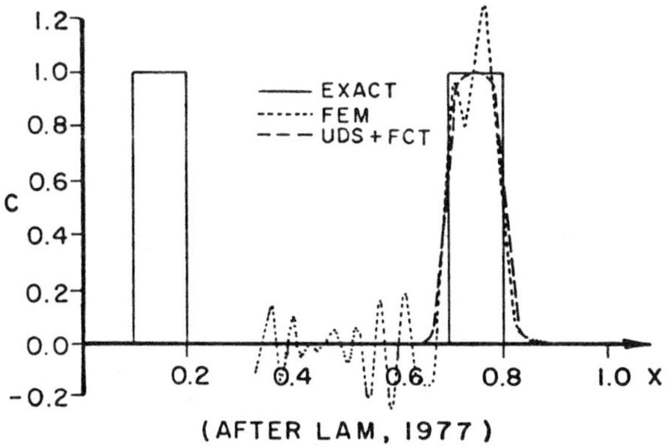

FIGURE 8 Rectangular wave, test problem 4, u=1, D=0, h=1/100, $\Delta t=6X10^{-3}$, t=0 and t=0.6 sec.

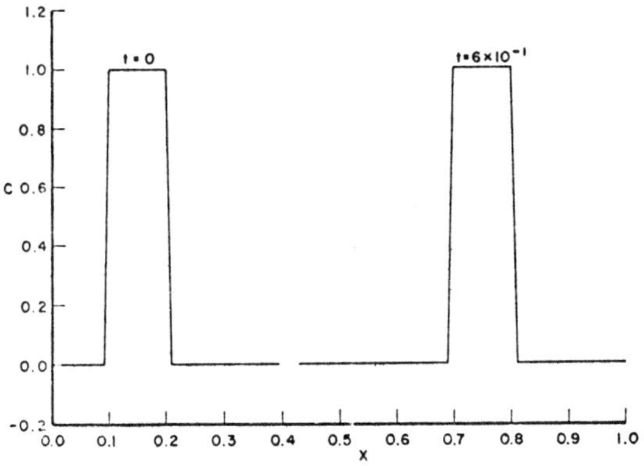

FIGURE 9 Test problem 4, rectangular wave at t=0 and solution from FEMIC at t=0.6 sec.

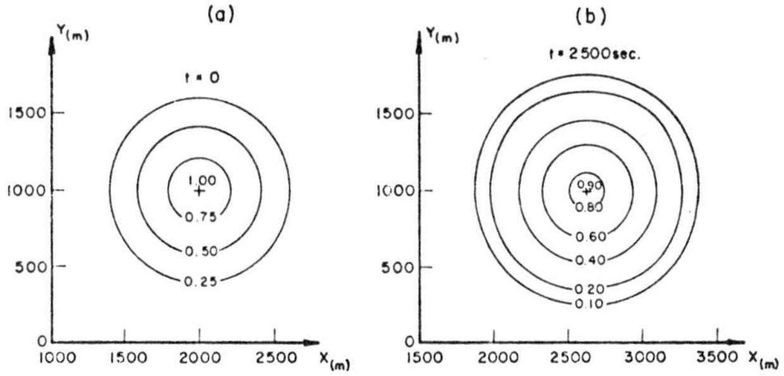

FIGURE 10 2-Dimensional example

a) Cone-shaped concentration distribution at t=0;
b) Concentration distribution at t=2500 sec.

DISCUSSION AND GENERALIZATIONS

A new finite element method (FEMIC) for the solution of the diffusion-convection equation is introduced. The method employs spatial-temporal elements and the method of characteristics is incorporated by orienting the sides of the elements joining the nodes at subsequent time levels in a particular direction.

The derivation of the method is given as it was first developed by the authors. Further studies have since showed that the transformation from the X-t plane to the x-t plane is not essential. However, it has the advantage of making clear what conditions (Equation 34) the finite element mesh must satisfy so that the concentration is propagated without change in the limiting case of pure convection.

The initial boundary value problem of diffusion-convection described by Equations 1-4 can be solved directly in the X-t plane employing spatial-temporal elements with sides (joining the nodes at subsequent time levels) oriented along the characteristic lines which are determined by the parameters of the problem. For the case of constant convection velocity, u, the set of equations obtained in the X-t plane are the inverse transform of Equations 35-37 obtained for the x-t plane. Direct formulation of the equations for the system in the X-t plane has the advantage of applicability to the case of a variable velocity of convection (Varoglu and Finn, 1978a).

The novel idea of orienting sides (joining the nodes at subsequent time levels) of the spatial-temporal elements along the characteristic lines of the hyperbolic equation resulting for D=0 is general enough to allow the development of a finite element method incorporating method of characteristics in two and three spatial dimensions. The development of the method for the diffusion-convection equation in two spatial dimensions is given in Varoglu and Finn, 1978b in which the automatic mesh generation at each time step and the other novel features of the computer program are explained in detail. Here, we reproduce the numerical results for a test problem in two spatial dimensions. A cone-shaped concentration distribution is used as the initial distribution as shown in Figure 10a. For eddy diffusivities, $D_x = D_y = 2\ m^2/sec$ and velocities, $u_x = 0.25$ m/s, $u_y = 0$, the conventional finite element results (Smith *et al*, 1973) and the numerical results by finite difference methods (Lam, 1977) of this test problem are available. The spatial discretization, $\Delta x = \Delta y = 200$ and the temporal discretization, $\Delta t = 500$ sec are employed. The concentration distribution from 2-D FEMIC is illustrated in Figure 10b. The results from FEMIC do not produce any oscillations or negative concentrations which is unavoidable in the conventional finite element results (Lam, 1977). The axial symmetry of the concentration distribution at t = 2500 sec with respect to top of the cone is far better than any of the other numerical solutions given in Lam, 1977.

It is shown by examples that the method developed gives very satisfactory numerical results for the whole spectrum of dispersion. The orientation of the sides of the spatial-temporal elements in a particular direction does not introduce any restrictions on the element sizes. The associated computer program has an algorithm to generate automatically a spatially variable mesh at each time step.

ACKNOWLEDGEMENTS

The authors wish to thank Dr. D.C.L. Lam, Applied Research Division, Canada Centre for Inland Waters, Burlington, Ontario for giving permission to reproduce some of his results for comparison purposes. Also, thanks are due to Mr. W. Deacon, Computer Science Department, University of British Columbia for his helpful contribution to the automatic plotting of some of the figures in this paper and to Miss D. Cheung for the careful typing of the manuscript.

REFERENCES

Adey, R.A. and Brebbia, C.A. (1973) Finite Element Solution for Effluent Dispersion. Numerical Methods in Fluid Dynamics, (C. Brebbia and J.J. Connor, eds.), Pentech Press, London, 325-354.

Bonnerot, R. and Jamet, P. (1974) A Second Order Finite Element Method for the One-Dimensional Stefan Problem. Int. J. Num. Meth. in Engineering, 8, 811-820.

Book, D.L., Boris, J.P. and Hain, K. (1975) Flux-Corrected Transport II: Generalization of the Method. J. Comp. Phys., 18, 248-283.

Price, H.S., Varga, R.S. and Warren, J.E. (1966) Application of Oscillation Matrices to Diffusion-Convection Equation. J. Maths. and Physics, 45, 301-311.

Price, H.S., Cavendish, J.C. and Varga, R.S. (1968) Numerical Methods of Higher Order Accuracy for Diffusion-Convection Equations. Soc. Petrol Eng. J., 243(3), 293-303.

Bruch, C.J. and Zyvoloski, G. (1975) A Finite Element Solution to a General Two-Dimensional Nonsymmetric Parabolic Partial Differential Equation. Computers and Fluids, 3, 217-224.

Chaudhari, N.M. (1971) An Improved Numerical Technique for Solving Multidimensional Miscible Displacement Equations. Soc. Petrol. Eng. AIME J., 11(3), 277-284.

Dailey, J.E. and Harleman, D.R.F. (1973) A Numerical Model of Transient Water Quality in a One-Dimensional Estuary Based on the Finite Element Method. Numerical Methods in Fluid Dynamics (C. Brebbia and J.J. Connor, eds.), Pentech Press, London, 412-439.

Ehlig, C. (1977) Comparison of Numerical Methods for Solution of the Diffusion-Convection Equation in One and Two Dimensions. Finite Elements in Water Resources, (W.G. Gray, G.F. Pinder and C.A. Brebbia, eds.), Pentech Press, London, 1.91-1.102.

Gray, W.G. and Pinder G.F. (1974) Galerkin Approximation of the Time Derivation in the Finite Element Analysis of Groundwater Flow. Water Resources Res. 10(4), 821-828.

Keller, H.B. (1971) A new Difference Scheme for Parabolic Problems, (B. Hubbard, ed.), Academic Press, N.Y., 327-350.

Lam, D.C.L. (1977) Comparison of Finite-Element and Finite-Difference Methods for Nearshore Advection-Diffusion Transport Models. Finite Elements in Water Resources, (W.G. Gray, G.F. Pinder and C.A. Brebbia, eds.), Pentech Press, London, 1.115-1.129.

Martin, B. (1975) Numerical Representations which Model Properties of the Solution to the Diffusion Equation. J. Comp. Phys., 17-4, 358-383.

Mercer, J.W. and Faust, C.R. (1977) The Application of Finite-Element Techniques to Immiscible Flow in Porous Medium. Finite Elements in Water Resources, (W.G. Gray, G.F. Pinder and C.A. Brebbia, eds.), Pentech Press, London, 1.21-1.57.

Smith, I.M., Faraday, R.V. and O'Connor, B.A. (1973) Rayleigh-Ritz and Galerkin Finite Elements for Diffusion-Convection Problems. Water Resources Res., 9(3), 593-606.

Smith, I.M. (1977) Integration in Time of Diffusion and Diffusion-Convection Equations. Finite Elements in Water Resources, (W.G. Gray, G.F. Pinder and C.A. Brebbia, eds.), Pentech Press, London, 1.3-1.20.

van Genuchten, M.Th. (1977) On the Accuracy and Efficiency of Several Numerical Schemes for Solving the Convective-Dispersive Equation. Finite Elements in Water Resources, (W.G. Gray, G.F. Pinder and C.A. Brebbia, eds.), Pentech Press, London, 1.71-1.90.

Varoglu, E. and Finn, W.D.L. (1978a) Finite Elements Incorporating Characteristics for 1-D Diffusion-Convection Equation (submitted to publication).

Varoḡlu, E. and Finn, W.D.L. (1978b) Finite Element Solution Incorporating Characteristics for Diffusion-Convection Equation in 2 Spatial Dimensions, (submitted to publication).

Zienkiewicz, O.C. (1971) The Finite Element Method in Engineering Science, McGraw-Hill, London.

SOLUTION OF THE CONVECTION - DIFFUSION EQUATION USING A MOVING COORDINATE SYSTEM

O. K. Jensen, B. A. Finlayson

University of Washington, Seattle, WA 98195

INTRODUCTION

The numerical solution of convection-diffusion problems is notoriously difficult when convection dominates because the equation then assumes a hyperbolic character. The convection-diffusion equation is

$$\frac{\partial c}{\partial t} + \lambda \frac{\partial c}{\partial x} = \frac{\partial^2 c}{\partial x^2} \tag{1}$$

The steady state equation is

$$\lambda \frac{\partial c}{\partial x} = \frac{\partial^2 c}{\partial x^2} \tag{2}$$

with the boundary conditions

$$c = 1 \quad \text{at } x = 0$$
$$c = 0 \quad \text{at } x = 1$$

Clearly at high values of λ the transient equation is hyperbolic in character.

Price, et al. (1966) were the first to recognize that the difficulties are mainly due to the spatial discretization. They proved that a finite difference solution with a central difference approximation will not oscillate provided

$$\lambda \Delta x \leqslant 2. \tag{3}$$

Christie, et al. (1976) analyzed the steady state equations (2) in the manner described below. They solved the difference equations and showed that a Galerkin finite element method with linear basis functions will not oscillate if Eq. (3) is satisfied. For quadratic trial functions, they showed that the solution would oscillate unless

$$\lambda \Delta x \leqslant 4 \tag{4}$$

Other methods have been used for examining the oscillations. Stone, et al. (1963) analyzed the transient equation with a Fourier-series. Gray and Pinder (1976/77), Gresho, et al. (1976), and Runchal (1977) made similar comparisons of finite difference formulations with weighted residual methods. The general conclusion is that weighted residual methods are better. Lantz (1971) and Chaudhari (1971) examined the truncation error of the finite difference method applied to the transient equations. They found that backward schemes introduced artificial dispersion and thereby smoothed the front.

Lantz (1971) combined the spatial and temporal truncation errors in Eq. (1) and Van Genuchten (1976) and Laumbach (1975) have both made equivalent combinations (using higher order temporal integrations) to improve the accuracy of the calculations. As yet these manipulations have not been extended to non-linear problems, although Lantz (1971) achieves some success in special non-linear problems.

To summarize, we know how to eliminate oscillations in the solution due to the temporal integration, and for some methods we know the spatial increment needed to eliminate oscillations due to the convective term. Unfortunately, the spatial increment (Δx) needed is frequently much too small and we are forced to degrade the quality of the solution by introducing artificial dispersion in the numerical scheme. We first present the exact solution for the transient problem (1), which gives the eigenvalues for this problem. An extension of the investigation by Christie, et al. (1976) presents the oscillation limits for the steady-state equation and a variety of methods. The eigenvalues of the resulting matrix are calculated to relate the steady-state analysis to the transient analysis. Finally, the new method is explained and illustrated.

Exact solution

The exact solution to Equation (1) can be found by separation of variables after transforming the spatial part to self-adjoint form.

$$c(x,t)= \frac{e^{\lambda}-e^{\lambda x}}{e^{\lambda}-1} - 4e^{\lambda x/2-\lambda^2 t/4}\sum_{n=1}^{\infty} \frac{n\pi}{\lambda^2/4+(n\pi)^2}\, e^{-n^2\pi^2 t} \sin n\pi x \qquad (5)$$

Although not useful for computational purposes, it gives the eigenvalues for this problem.

$$\mu_i = -\lambda^2/4 - n^2\pi^2 \ , \ n = 1, 2, --- \qquad (6)$$

For computational purposes a solution with Laplace transform is generally used.

Difference Equations and Eigenvalue Analysis

By the numerical solution of the steady-state convection diffusion equation (2) together with an eigenvalue analysis of the equation (1), the importance of mesh spacing can be considered.

The steady state equation (2) when solved by numerical schemes gives a system of equations

$$\underline{\underline{M}} \cdot \underline{c} = \underline{b}$$

As an example, the method of orthogonal collocation on finite elements (Carey and Finlayson, 1975) using Lagrange basis function and one interior collocation point gives

$$\left(\frac{4}{\Delta x} + \lambda\right) c_{i-1} - \frac{8}{\Delta x} c_{1-1/2} + \left(\frac{4}{\Delta x} - \lambda\right) c_i = 0 \text{ (Diff. eqn.)}$$

$$c_{i-1} - 4c_{i-1/2} - 8c_i - 4c_{i+1/2} + c_{i+1} = 0 \text{ (Continuity)}$$

$$\left(\frac{4}{\Delta x} + \lambda\right) c_i - \frac{8}{\Delta x} c_{i+1/2} + \left(\frac{4}{\Delta x} - \lambda\right) c_{i+1} = 0 \text{ (Diff. eqn.)}$$

This arbitrary set of equations (nodes i-1, i, i+1) can be combined to give

$$\left(1 + \frac{\lambda \Delta x}{2}\right) c_{i-1} - 2c_i + \left(1 - \frac{\lambda \Delta x}{2}\right) c_{i+1} = 0 \qquad (7)$$

The theoretical solution is

$$c_i = R + S\,[\phi(\lambda \Delta x)]^i$$
$$\phi(\lambda \Delta x) = (1 + \lambda \Delta x/2) \,/\, (1 - \lambda \Delta x/2) \qquad (8)$$

When the function $\phi(\lambda \Delta x)$ changes signs in Eq. (8) [c not monotone] the solution oscillates. Thus the criterion for no oscillations is

$$\phi(\lambda \Delta x) \geqslant 0 \text{ or } \lambda \Delta x \leqslant 2 \qquad (9)$$

The same type of analysis has been made for various Galerkin/finite difference schemes by Christie, et al. (1976). The results obtained are summarized in Table 1. New calculations are presented in Table 1 for OCFE Lagrange and Hermite basis functions.

The eigenvalues for the transient equation (1) have been found analytically Eq. (6), but it is of interest to examine eigenvalues of a numerical approximation. Using OCFE to discretize (1) gives a set of coupled ordinary difference equations.

$$\frac{d\underline{c}}{dt} = \underline{\underline{M}} \cdot \underline{c} \qquad (10)$$

For large Δx the eigenvalues are complex, but for small enough Δx (depending on the parameters λ) the eigenvalues are real and corresponds to the analytical results. For each method there was a distinct limit between getting real or complex eigenvalues. Price, et al. (1966) found this distinct limit for real or complex eigenvalues theoretically for the finite difference method. We have calculated the limit for other methods as well (see Table 1) and find it corresponds quite

Table 1. Oscillation Limits for Different Methods

Scheme	Criteria Diff. Eqn.	Criteria Eigenvalue	Order of Method (SS)	Reference
Finite Difference				
- Upwind Diff.	None		O(h)	Price, et al. (1966)
- Forward	$\lambda\Delta x<1$		O(h)	
- Central	$\lambda\Delta x<2$	<2	$O(h^2)$	
Galerkin-FD				
- Var. Upwind linear base functions	$\lambda\Delta x<\frac{2}{1-\alpha}$		O(h) $\alpha=0$ $O(h^2)$	Chien (1977) Christie, et al. (1976) Huyakorn (1976)
Galerkin				
- linear, C^o	$\lambda\Delta x<2$		$O(h^2)$	Christie, et al. (1976)
- Quadratic, C^o	$\lambda\Delta x<4$		$O(h^3)$	
OCFE-Lagrange				
$NCOL^{INT} = 1$	$\lambda\Delta x<2$	$\leqslant 2.12\pm.04$	$O(h^3)$	This report
$NCOL^{INT} = 2$	$\lambda\Delta x<4.39$	$\leqslant 4.01\pm.16$	$O(h^4)$	
$NCOL^{INT} = 3$	$\lambda\Delta x<4.64$	$\leqslant 4.77\pm.23$	$O(h^5)$	
OCFE-Hermite				
1st (cubic)	$\lambda\Delta x<4.39$		$O(h^4)$	This report
2nd (quartic)	$\lambda\Delta x<4.64$		$O(h^5)$	

well to the result expected from the steady-state equations. Consequently, we henceforth use the result from the analysis of steady-state equations as criteria for eliminating oscillations in both steady state and transient calculations.

SOLUTION METHODS

A great variety of solution methods have been applied to the convection-diffusion equation. The finite difference formulation initiated with Peaceman and Rachford (1962) were followed by schemes of increasing complexibility. Stone and Brian (1963) introduced higher order spatial schemes while Laumbach (1975) used high order in both space and time. Unfortunately, the complex schemes cannot always be adopted to more difficult non-linear problems in two dimensions. The Galerkin finite element method was introduced by Price, *et al*. (1968) and provided better results, especially for large λ, 10^3-10^6.

All conventional schemes can solve the convection-diffusion equation, but they all must use a large number of elements, nodes, and timesteps when convection dominates. Therefore, additional techniques have been tried.

A variable interpolation technique introduced by Price, *et al*. (1968) improved the efficiency remarkably but it is not easily adopted to more difficult problems. Garder, *et al*. (1964) solved the convection-diffusion equation by method of characteristics and employing a marker and cell technique to track the concentration front. The accuracy of the calculation is limited to about 2%. Here we employ a moving coordinate system but then we solve the diffusion equation. The basic idea is that if the front has the same shape over a long period of time, the description in a moving coordinate system should be fairly stationary. If this is possible, more points can be located at the front and elsewhere only a few elements are needed. Extension of this method for non-linear problems can be made by introducing a time dependent velocity of the coordinate system. The velocity must then be calculated in order to follow the front. For two dimensional problems the coordinate system can be moved as to locate the front in the region of sharp change of the solution.

Moving Coordinate System

We introduce a moving coordinate system to equation (1) with new coordinates ξ, η

$$\xi = x - \lambda t, \quad \eta = t. \tag{11}$$

Here λ is taken as a constant.

Since $\dfrac{\partial c}{\partial t} = \dfrac{\partial c}{\partial \xi}\dfrac{\partial \xi}{\partial t} + \dfrac{\partial c}{\partial \eta}\dfrac{\partial \eta}{\partial x} = -\lambda\dfrac{\partial c}{\partial \xi} + \dfrac{\partial c}{\partial \eta}$

$$\frac{\partial c}{\partial x} = \frac{\partial c}{\partial \xi}$$

we obtain

$$\frac{\partial c}{\partial \eta} = \frac{\partial^2 c}{\partial \xi^2} \tag{12}$$

The boundary conditions are transformed to

$$\begin{array}{lll} c = 0 & \xi > 0 & \eta = 0 \\ c = 1 & \xi = -\lambda\eta & \\ & & \eta > 0 \\ c = 0 & \xi = 1 - \lambda\eta & \end{array} \tag{13}$$

Now the spatial domain ξ is [$-\infty$, 1] and front will always be near $\xi = 0$ for $\eta < 1/\lambda$.

For almost all purposes computation to $\eta = 1/\lambda$ is sufficient as the solution is then near steady state.

We can apply any numerical method to solve Eq. (12, 13). As we are only interested in $\eta < 1/\lambda$ the discretization only covers [-1, 1]. We choose to apply OCFE, giving a set of ordinary differential equations and algebraic equations and use a Crank-Nicholson time discretization.

All the equations can then be written as

$$\underline{\underline{C}}\ \underline{c}^{n+1} = \underline{\underline{D}}\ \underline{c}^{n} \tag{14}$$

A similar set of equations like (14) could be developed using finite-difference or Galerkin finite-element methods.

The boundary conditions are applied by modifying the matrices $\underline{\underline{C}}$, $\underline{\underline{D}}$ and $\underline{c}^n$ at each timestep. For points to the left of the location of the boundary, $i<I$, $\xi_i < -\lambda\eta$, we replace $\underline{\underline{C}}$ and $\underline{\underline{D}}$ by the identity matrix $\underline{\underline{I}}$ and $\underline{c}^n$ by 1. Since the location of the boundary changes in time, the LU decomposition must be performed every time the boundary locations pass another collocation point.

Results

The advantages of using a MCS for problems with dominating convection is illustrated with comparisons using the same spatial discretization (here collocation) with both a MCS (Moving Coordinate System) and a fixed coordinate system. The convection-diffusion equation is solved with $\lambda = 877.9$ and 87790.

In Fig. 1 is shown a comparison of a conventional scheme versus MCS-scheme. It is apparent that only a relative few number of elements are needed to give a non-oscillatory "fair" solution.

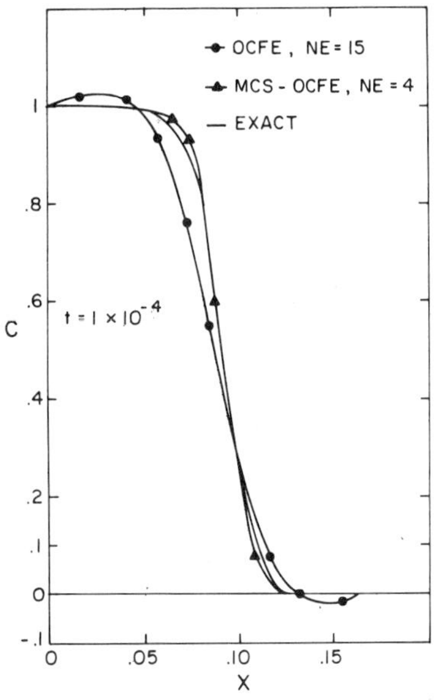

Fig. 1. Comparison of standard OCFE versus moving coordinate system.
λ=877.9, NCOL = 2, $\Delta t=2.5 \times 10^{-6}$.

In contrast to the method of characteristics by Gardér, et al., the error of MCS decreases with increasing number of elements and with smaller timestep. The method of characteristics has a minimum error of 2%. When compared to the method of variable interpolation Price, et al. (1968) the MCS is better. Figure 2 shows how the error decreases with time for the same number of elements, and the MCS is very much more accurate at small times (when the front is steepest).

The elements are not all the same size in MCS-OCFE. If small elements are used near the front the solution will be good at small times (when all changes occur near the front). At later times, however, the changes also occur near the front. Thus the best discretization may depend on time, and this is shown in Figure 3.

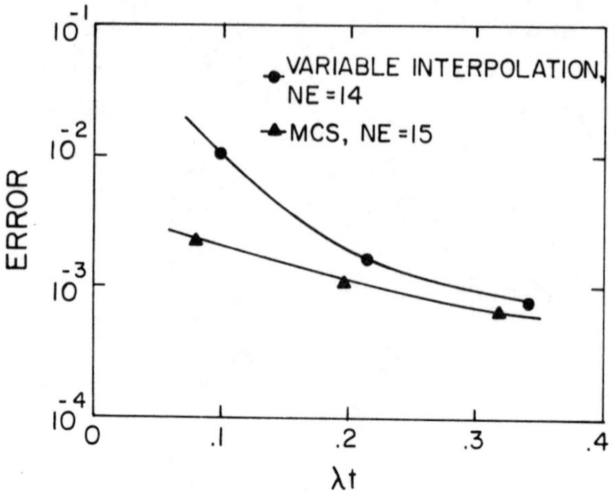

Fig. 2. Comparison of moving coordinate system versus variable interpolation (Price, et al. (1968).) λ=877.9

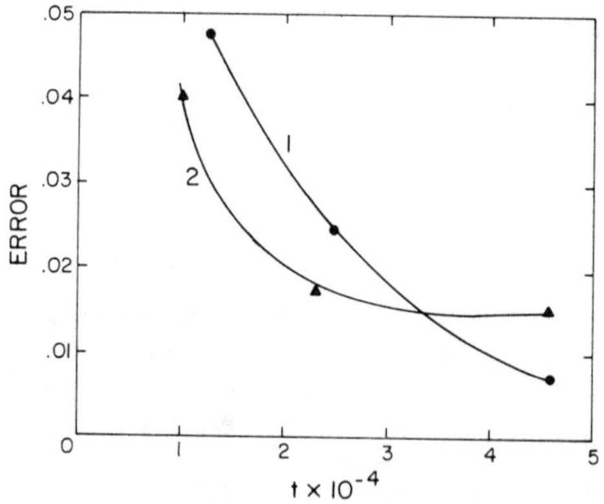

Fig. 3. Error as function of time. Effect of mesh location. λ=877.9, NE=5, Δt=2.5 x 10^{-6}.
Curve 1 element nodes at ξ=-1., -.2, -.05, .05, .2, 1
2 element nodes at ξ=-1., -.12, -.04, .04, .12,

In Table 2 are shown results, errors, computation time, numbers of elements and Δt. It is clear that MCS is superior to the conventional schemes. By having the same number of elements (entries 5 and 8 for 15 elements) the error is decreased by a factor 20-40, but the computation time is only two times longer. For the same computation time (entries 1 and 6 for ~3 secs.) the conventional scheme is unreliable (10%) while MCS gives fairly accurate results (3%). Comparing at 16 secs. (entries 5 and 8) the error is decreased by a factor of 4-12 and MCS gives no oscillations while the conventional scheme does.

For $\lambda = 87790$ the MCS is even better. Table 2 shows results for both MCS and conventional OCFE. Comparing entries 5 and 2 the standard OCFE gives excessive oscillation, but MCS gives only 1% error and used 20 times less computer time.

CONCLUSIONS

The steady state convection-diffusion equation is used to provide criteria for the elimination of oscillations. Small element sizes (Δx) are necessary and criteria are given for finite difference, Galerkin and collocation methods.

Solving the convection diffusion equation on a moving coordinate system rather than a fixed coordinate system can give better solutions at less cost. Often factors of 10 or higher in improved accuracy or reduced cost are possible.

TABLE 2
Comparison of MCS with Conventional Scheme (Both using OCFE)

Entry	Scheme	NCOL	NE	for λ=877.9 Δtx10^6	CPU*	Error at $t=1x10^{-4}$	Error at $t=5x10^{-4}$	Maximum Oscillations at $t=5x10^{-4}$
1	MCS	2	4	2.5	3.3	.03	.04	–
2		2	6	2.5	5.1	.01	.005	–
3		2	6	10	1.0	.04	.01	–
4		2	8	2.5	6.7	.008	.006	–
5		2	15	1.25	15.1	.004	.001	–
6	OCFE	2	5	2	3.2	~.1	~.1	~.1
7		2	15	2.5	8.1	.07	.04	.04
8		3	20	2.5	17.0	.03	.004	.0022
				for λ=87790 Δtx10^8			Error at $t=5x10^{-8}$	
1	MCS	2	12	5	4.5	.5	.02	.005
2	–	2	14	2.5	8.1	.5	.01	.004
3	–	2	16	1.25	17.0	.5	.004	–
4	OCFE	3	50	1	70.5	0.0	.003	.004
5	–	3	50	.5	152.0	.5	~.1	~.1
6	–	5	50	1	140.0	0.0	.003	–

*CPU on CDC 6400

References

Carey, G. F. and B. A. Finlayson, "Orthogonal collocation on finite elements," Chem. Eng. Sci., 30, 587-596 (1975).

Chaudhari, N. M., "An improved numerical technique for solving multi-dimensional miscible displacement," Soc. Pet. Eng. J. (Sept. 1971) 277-284.

Chien, T. C., "A general finite difference formulation for Navie-Stoke's equation," Computers & Fluid 5, p. 15 (1977).

Christie, I., D. F. Griffiths, A. R. Mitchell, O. C. Zienkiewics, "Finite element methods for second order differential equations with significant first derivatives," Int. J. Num. Meth. Engng. 10, 1389-1396 (1976).

Garder, A. O., D. W. Peaceman, A. L. Pozzi, "Numerical calculation of multi-dimensional miscible displacement by method of characteristics," Soc. Pet. Eng. J. (Mar. 1964) 26-36.

Gresho, P. M., R. L. Lee, R. L. Sani, "Advection-dominating flows, with emphasis on the consequences of mass lumping" from Proceeding Finite Element Methods in Flow Problems, Italy, June 14-18, 1976.

Huyakorn, P. S., "Upwind finite element scheme for improved solution of convection-diffusion equation," Wat. Res. Prog. #76-WR-Z.

Lantz, R. B., "Quantitative evaluation of numerical diffusion," Soc. Pet. Eng. J. (Sept. 1971) 315-320.

Laumbach, D. D., "A high accuracy finite difference technique for treating the convection-diffusion equation," Soc. Pet. Eng. J., 15, 517 (1975).

Peaceman, D., H. H. Rachford, "Numerical calculation of multi-dimensional miscible displacement," Soc. Pet. Eng. J. (Dec. 1962) 327-339.

Pinder, G. F. and W. G. Gray, "Finite element simulation in surface and subsurface hydrology," Academic Press, New York, 1977.

Price, H. S., J. C. Cavendish, R. S. Varga, "Numerical methods of higher order accuracy for diffusion-convection equation," Soc. Pet. Eng. J., 8, 293 (1968).

Price, H. S., R. S. Varga, J. E. Warren, "Application of oscillation matrices to diffusion-convection equations," J. Math. & Physics, 45, 301-311, 1966.

Runchal, A. K., "Comparative criteria for finite-difference formulations for problems of fluid flow," Int. J. Num. Meth. Eng. 11, 11, 1667-1681 (1977).

Stone, H. L., P. L. T. Brian, "Numerical solution of convective transport problems," A.I.Ch.E. J., 9, 5, 681-688 1963).

Van Genuchten, M. T., "On the accuracy and efficiency of several numerical schemes for solving convective-dispersive equation," Int. Conf. on Finite Elements in Water Resources, July 1976.

Young, L. C., Private communications.

VARIABLE DOMAIN FINITE ELEMENT ANALYSIS OF UNSTEADY COMPRESSIBLE FLUID FLOW PROBLEMS

G. Van Goethem

J.R.C. EURATOM, Ispra, Italy and
Université Catholique de Louvain, Belgium

ABSTRACT

A variable domain finite element method is presented for Solving transient, compressible, non-viscous fluid flow problems involving moving boundaries (free surfaces, fluid-fluid and fluid-solid interfaces). Novel characteristics of the method are the use of a variational principle defined on an arbitrarily moving domain of integration, and the subsequent development of a transition element that allows Lagrangian and Eulerian meshes to be compatible along their (slip) interface. The method is based on a mixed Eulerian-Lagrangian description of the hydrodynamic conservation laws.

Quadrilateral isoparametric elements with bilinear velocity field and uniform pressure and density are considered in axisymmetric geometries. The Galerkin procedure is used to formulate a variational statement associated with the momentum equation and its boundary conditions in connection with an accelerating frame of reference. For numerical stability reasons it was found convenient to divide the calculational scheme into two steps at each time interval. First an estimation is computed for the values of the different hydrodynamic quantities by using a simple time extrapolation. These values are then introduced into the conservation equations which are integrated in time by means of an explicit finite difference method in connection with a lumped mass matrix.

For the computation of discontinuous solutions (shocks), linear and quadratic pseudoviscosities are employed. Good agreement with analytical results is obtained for the one - and two- dimensional fluid impact problems described here.

INTRODUCTION

Presented here is a numerical method to analyse the unsteady states of axisymmetric fluid motions for those complicated cases where part of the configuration is bounded by continuously changing surfaces. Such problems often arise in continuum mechanics when several media are involved, the unknown free boundary being the interface between the media. This configuration, as is well known, constitutes at the same time both a reflection of the solution of the dynamic equations, and also implicitly the prerequired information needed to generate the solution. Simple examples of flows involving moving interfaces are free jet, gravity flow over a dam and fluid flow in distensible tubes.

In the framework of the nuclear safety programme carried out at JRC-Ispra, a finite element code is under development with the aim of investigating some fluid-fluid and fluid-solid interface problems related to the hypothetical core disruptive accident. For simplicity we limit attention to what may be regarded as an idealized interface, namely one that can be defined and located as the surface of zero thickness separating two immiscible materials. In addition the study of free surfaces is simplified by assuming no free surface energy.

This type of initial-/ boundary-value problems exhibits two main difficulties even if we ignore the effect of temperature on material properties. Firstly the material response is in general non-linear, and, secondly the problem involves mixed boundary conditions. The problem of locating the unknown moving interface and computing the velocity field has been solved analytically only in very particular cases (Birkhoff, Zarantonello, 1957). Hence a numerical method has to be used for this class of problems.

If dissipative forces such as viscosity and heat conduction are neglected, the flow of compressible fluids is governed by partial differential equations which are hyperbolic in type. The basic variables are chosen to be the density, ρ , the internal specific energy, ρe, and the fluid velocity, $\bar{v}$ .Other thermodynamic quantities like the internal pressure, p, and the sound speed, C, are determined through the equation of state :

$$p = f(\rho, e) \qquad (1)$$

CHOICE OF A NUMERICAL METHOD

The nature of the boundary conditions largely determines which numerical course we must follow. Because of the desirability of an accurate solution near the moving boundary, we need to have a fine discretization of the field along this irregular surface. This strongly suggests a finite element approach rather than a finite difference one using an uniform mesh. Other important advantages of the finite element method are its ability to accept complex geometries and mixed material characteristics. In addition this method provides the user with an extreme flexibility in chosing the basis functions. Therefore in very complicated domains or for problems with complicated interfaces, this procedure might even be the only feasible one, mainly when an isoparametric transformation is used which allows the boundary conditions to be satisfied exactly on curved sides.

As no true extremum principle exists for the general hydrodynamic equations, Galerkin's method of weighted residuals is applied to formulate a variational statement associated with the momentum equation and its boundary conditions. This leads as usually (Brebbia, 1976) to the well known Principle of Virtual Power of which we present here an extension to a variable domain of integration.

Finally one has to be clear about whether Lagrangian (material) whether Eulerian (spatial) coordinates are being used. Lagrangian techniques allow to keep track of moving material interfaces but they are restricted to flow problems with relatively small mesh distortions. When the physical situation involves slip surfaces or other severe distortions of the original mesh the Eulerian approach is preferred. The space containing the material then is divided into finite elements through which the material points flow, while in a Lagrangian formulation they are pieces of material always composed of the same material. Since the fluid flows under consideration are subjected to high transients and bounded by moving interfaces we need both types of coordinates, and, in order to make them compatible one with each other we introduce a so-called transition element based on a mixed Eulerian-Lagrangian description.

TREATMENT OF THE MOVING INTERFACES

The material interface problems that we are dealing with, are thus intrinsically non-linear initial-/boundary-value problems in which part of the boundary, the moving interface, is unknown and must be determined as part of the solution. On a moving material interface two conditions are specified, namely one on the normal velocity component and one on the internal stress components while in the other cases only one condition is needed, e.g. the velocity or the surface force. Since the moving boundary is a material interface, there can be no flow across it, and as far as the stress condition is concerned, the guiding principle is that of stress continuity.

The finite element method is applied in order to obtain a tentative location of the moving interface and the distribution of the dependent variables is determined for the resulting fixed domain. The trial fixed domain solution will not usually satisfy both boundary conditions simultaneously and, therefore, the trial free surface is adjusted and a solution obtained for the revised fixed domain. This process is repeated until the required adjustment in the free boundary is considered negligible. Thus the solution of the non-linear free surface problem is performed by a series of solutions to simplified problems with prescribed boundaries.

Each of the interacting media is divided into quadrilateral finite elements defined by four nodes and straight line boundaries. The Eulerian coordinate system (time t, space location $\overline{x}$) is adopted to describe flow at nodes inside the fluid domain. The corresponding finite elements were shown to work very satisfactorily (Van Goethem, 1976). The Lagrangian coordinates (t, initial material location $\overline{a}$), on the other hand, are useful to follow the nodes of moving boundaries as free surfaces, fluid-fluid or fluid - structure interfaces. The corresponding structural finite elements are taken from an already existing code (Donea, 1976).

In order to carry out the compatibility of an Eulerian and a Lagrangian finite element mesh along their interface, we introduced a layer of transition elements. A transition element is defined in a mixed Eulerian - Lagrangian mesh as a fluid element that in general possesses 2 fixed nodes, located inside the fluid domain, and 2 moving nodes at the velocity $\overline{W}$ to be determined on the moving boundary. Along a fluid - solid interface only one transition layer is used and the velocity $\overline{W}$ is equal to the structural velocity $\overline{V}_s$. Along a fluid - fluid interface, on the other hand, two transition layers are needed, namely one on each side, and there only the normal components of $\overline{W}$ and of the various fluid velocities $\overline{V}_f^{\,i}$, $i = 1, 2$, have to be identified.

When the interfaces are moving over large distances, things become more tractable if one uses two overlapping finite element meshes. Here one net follows the boundary and is overlapped by an ordinary regular fixed mesh covering the entire space of possible configurations. When an interface moves outside or nearly overlaps the spatial region of the regular mesh occupied by the transition elements one just has to redefine the transition zone which once again will include the interface. When two moving interfaces come close together they are separated by one special transition layer containing elements with four moving nodes.

This method thus requires to formulate the hydrodynamic equations in connection with a moving frame of reference at the velocity $\overline{W}$. Let us notice that such a frame turns out to be the Eulerian one when $\overline{W} = \overline{0}$ and the Lagrangian one when $\overline{W} = \overline{V}$. The Galerkin procedure then is used to formulate in an accelerating frame a variational statement associated with the local form of the momentum equations and its boundary conditions. The numerical solution is performed by means of linear isoparametric quadrilateral elements. This finite element method thus leads to a continuous velocity field formulation while the density and the specific internal energy are assumed uniform over each element. These two last quantities are updated at each time step by means of the conservation laws of mass and total energy; their values thus must be determined only once per element.

KINEMATICAL PRELIMINARIES

Fluid flow is a physical notion which is represented mathematically by a continuous transformation of three-dimensional Euclidean space into itself. In order to describe this transformation analytically let us introduce the coordinate systems (a_1, a_2, a_3) and (x_1, x_2, x_3) referring to the undeformed configuration $\Omega^3(0)$ and the deformed one $\Omega(t)$, respectively, the parameter t being the time. It is possible to select a different coordinate system $\overline{x}$ at each time t, or equivalently, to view the motion in terms of a coordinate system in motion with respect to the common inertial frame. To do so is in our class of problems very convenient as we already demonstrated.

Let a typical material point $\overline{a}$ be carried into the position $\overline{x}$ at time t. The flow then may be represented by the bijective transformation:

$$\Omega(0) \longrightarrow \Omega(t), \ \overline{x} = \overline{F}(\overline{a}, t) \text{ with } a_i = F_i(a_1, a_2, a_3, 0) \quad (2)$$

Of frequent use is the absolute scalar J whose numerical value is the Jacobian $|\overline{x}/\overline{a}|$:

$$J = \frac{D(x_1,x_2,x_3)}{D(a_1,a_2,a_3)} = \frac{D(F_1,F_2,F_3)}{D(a_1,a_2,a_3)} \qquad (3)$$

From the assumption that Eq.(2) possesses a differentiable inverse it follows that :

$$0 < J < \infty \qquad (4)$$

Through this jacobian, there exists a simple relation between the infinitesimal volumes dvol, depending on $(\overline{F},t)$ in the $\Omega(t)$ configuration, and dvol°, depending on $(\overline{a},t)$ in the $\Omega(0)$ configuration :

$$\text{dvol} = J\,\text{dvol}^\circ \qquad (5)$$

Let us introduce the frame velocity vector $\overline{W}(\overline{a}, t)$ defined in the common frame as :

$$W_i = \frac{\partial F_i}{\partial t} = \frac{\partial x_i}{\partial t}, \quad i = 1, 2, 3 \qquad (6)$$

As particular cases this definition involves the fixed spatial (or Eulerian) frame $\overline{x}$ and the moving material (or Lagrangian) frame $\overline{a}$ when F_i is chosen so that $W_i = 0$ and $W_i = V_i$, respectively. In order to avoid any confusion the coordinates of the general moving frame ($W_i \neq V_i \neq 0$) are called mixed Eulerian-Lagrangian and denoted (F_1,F_2,F_3).

Let $G = G(\overline{F},t)$ be a given scalar (or vector) function of the mixed variables $\overline{F}(\overline{a}, t)$ and t, and let us denote :

$$G^*(\overline{a}, t) = G(\overline{F}(\overline{a}, t), t) \qquad (7)$$

We will now derive some kinematical relations between the mixed Eulerian-Lagrangian representation of G and the purely Lagrangian one depending on the four independent variables (a_1,a_2,a_3,t). Let us therefore calculate the time and space derivatives of the composite function $G^*(\overline{a}, t)$ using the chain rule of differentiation :

$$\frac{\partial G^*(\overline{a},t)}{\partial t} = \frac{\partial G}{\partial t}(\overline{F}(\overline{a},t),t) + \sum_{j=1}^{3} W_j(\overline{a},t)\frac{\partial G}{\partial x_j}(\overline{F}(\overline{a},t),t) \qquad (8)$$

where the time derivatives have to be understood with the a_i fixed,

$$\frac{\partial G^*(\overline{a},t)}{\partial a_i} = \sum_{j=1}^{3} \frac{\partial G}{\partial x_j}(\overline{F}(\overline{a},t),t)\frac{\partial F_j}{\partial a_i}(\overline{a},t), \quad 1 < i < 3 \qquad (9)$$

Solving system (9) with respect to $\partial G/\partial x_j$ one gets :

$$\frac{\partial G}{\partial x_1} = \frac{1}{J} \frac{D\,(G^*,x_2,x_3)}{D(a_1,a_2,a_3)}, \text{and so on for} \frac{\partial G}{\partial x_2}, \frac{\partial G}{\partial x_3} \quad (10)$$

Last formula (10) gives an important relationship between the partial derivatives of the function G with respect to the x_j and the a_j, respectively. On the other hand, it follows from the derivation rule of a determinant:

$$\frac{1}{J}\frac{\partial J}{\partial t} = \frac{1}{J}\left\{\frac{D(W_1,x_2,x_3)}{D(a_1,a_2,a_3)} + \frac{D(x_1,W_2,x_3)}{D(a_1,a_2,a_3)} + \right.$$

$$\left. + \frac{D(x_1,x_2,W_3)}{D(a_1,a_2,a_3)}\right\} = \text{div}\,\overline{W} \quad (11)$$

which is an extension of the well known cubic dilatation velocity formula.

Substituting Eq.(11) into Eq.(8) and decomposing the scalar product $\{\overline{W}\cdot\overline{\text{grad}}\;G\}$ one obtains an important relation including the local time derivative $\frac{\partial G}{\partial t}$, the position $\overline{x} = \overline{F}(\overline{a},t)$ being fixed (formula independent of any meaning attached to G),

$$J\,\frac{\partial G}{\partial t}\,(\overline{F},t) = \frac{\partial\,(J\;G(\overline{a}\;,t))}{\partial t} - J\,\text{div}\;G\;(\overline{F},t)\;\overline{W}\} \quad (12)$$

Other kinematical relations have been derived in the same way (Van Goethem, 1978) for the volume integral of an intensive propriety g defined as :

$$I\;(t) = \int_{V(t)} g\;(\overline{F}\;(\overline{a},t),\;t)\;dvol \quad (13)$$

where V(t) represents the volume of a variable integration domain bounded by the surface s(t) of unit normal vector $\overline{n}$. In the particular case of the frame being attached to the interface s(t), one has obtained a formula equivalent to Leibnitz theorem:

$$\frac{d\;I\;(t)}{d\;t} = \int_{V(t)} \frac{\partial g(\overline{F},t)}{\partial t}\;dvol + \int_{s(t)} g(\overline{F},t)\overline{W}(\overline{a},t)\overline{n}\;ds \quad (14)$$

to be compared with Reynold's transport theorem concerning the material derivative (denoted D/Dt) of the function I (t) now defined in a material volume.

The subtraction of Eq. (14) from Reynold's well known equation gives a useful relation :

$$\frac{DI(t)}{Dt} = \frac{dI(t)}{d\ t} + \int_{s(t)} g(\overline{v} - \overline{w})\ \overline{n}\ ds \qquad (15)$$

Eq.(15) allows a formulation in terms of a control volume moving with the continuum, i.e. an accelerating control volume.

HYDRODYNAMIC EQUATIONS IN A MOVING FRAME

Let us first examine an Eulerian mesh of finite elements and recall the corresponding mass and internal energy equations (in integral form) :

Mass conservation. If $M(t)= \int_{V(t)} \rho\, dvol$, one has :

$$\frac{DM}{Dt} = 0 \qquad (16)$$

Internal energy equation. If $E(t)= \int_{V(t)} \rho e\ dvol$, one has:

$$\frac{DE}{Dt} + \int_{V(t)} p \frac{\delta v_j}{\delta x_j}\ dvol = 0 \qquad (17)$$

where the repetition of an index means a summation.

Since we are concerned with a time dependent domain of integration we need equations developed for an accelerating control volume. Let g in formula (15) represent the density and the specific internal energy, successively. The substitution of Eqs. (16) and (17) into (15) then gives successively the time derivative of the total mass and internal energy contained in a finite element V(t) delimited by a moving boundary at the velocity $\overline{W}$:

$$\frac{dM}{dt} = -\int_{s(t)} \rho\ (\overline{v} - \overline{w})\ \ \overline{n}\ ds \qquad (18)$$

$$\frac{dE}{dt} = -\int_{s(t)} \rho e\ (\overline{v} - \overline{w})\ \overline{n}\ ds - \int_{V(t)} p\ \frac{\delta v_j}{\delta x_j}\ dvol \qquad (19)$$

For the equation of motion we will use the differential form and its corresponding boundary conditions in order to allow the application of Galerkin's method of weighted residuals. Under the assumption of perfect fluid (friction-less flow) the stress tensor components σ_{jk} obey the following constitutive equation :

$$\sigma_{jk} = - \delta_{jk}\, p \quad \text{with} \quad \delta_{jk} \text{ Kronecker's delta} \tag{20}$$

so that the Eulerian representation of the governing equation and its boundary condition are written as,

$$\frac{\delta}{\delta t}(\varrho v_k) = - \frac{\delta}{\delta x_j}(\varrho v_k v_j) - \frac{\delta p}{\delta x_k} + \varrho g_k \tag{21}$$

$$T_k = - n_j \quad \delta_{jk} \quad p \qquad \text{on} \quad s(t) \tag{22}$$

where $\overline{g}$ and $\overline{T}$ are the gravity acceleration and the external contact force, respectively.

It should be emphasized that Eq. (21) is valid only when referred to axes moving without acceleration, since the usual form of Newton's law holds under these conditions. For the case of an accelerating control volume, some terms have to be added in order to take into account the consequent mass variation. Let G in formula (12) represent the k-th component of momentum. The substitution of Eq.(21) into (12) then gives the local form of the momentum equation in a moving frame at the velocity $\overline{w}$,

$$\frac{\delta}{\delta t}(\varrho v_k J) + J\frac{\delta}{\delta x_j}\left\{\varrho v_k (v_j - w_j)\right\} + J\frac{\delta p}{\delta x_k} - \varrho J g_k = 0 \tag{23}$$

Performing similar computations one can derive the local forms of the mass and energy equations in a moving frame as well (Van Goethem, 1978). The mixed Eulerian-Lagrangian description presented here involves both Eulerian and Lagrangian descriptions as particular cases. For obtaining the Eulerian representation (21) of the momentum equation, one just has to cancel $\overline{W}$ in (23), observing that the jacobian J will drop since $\overline{W} = \overline{0}$ yields $\partial J/\partial t = 0$ through formula (11). In an analogous way, for obtaining the Lagrangian representation in (23), one just has to cancel the transport term since $\overline{w} = \overline{v}$.

When the equation of state (1) and the moving frame relation (6) are given, the integral equations (18),(19) and the differential one (23), together with the necessary initial values and boundary conditions, yield a complete set of equations which determine in principle the evolution of the hydrodynamic system.

VARIATIONAL STATEMENT AND FINITE ELEMENT MODEL FOR THE MOMENTUM EQUATIONS

Let us write for an arbitrary finite element the variational statement associated with the governing equations of motion (23) and the pressure boundary condition (22) in a Galerkin's form by considering arbitrary admissible variations of the velocity field.

Decomposing the time derivative of Eq. (23) and substituting the formula (11) into it, we first obtain:

$$\frac{\delta}{\delta t}(\varrho v_k J) = J\left\{\varrho\frac{\delta v_k}{\delta t} + \left(\frac{\delta\varrho}{\delta t} + \varrho\frac{\delta w_j}{\delta x_j}\right)v_k\right\} \quad (24)$$

Let us introduce this latter result in (23) and apply Galerkin's method on the initial volume V_o of an arbitrary finite element, i.e. in the underformed reference configuration $\Omega(o)$. One then obtains the basic variational formulation :

$$\int_{V_o}\left\{\varrho\frac{\delta v_k}{\delta t} + \left(\frac{\delta\varrho}{\delta t} + \varrho\,\mathrm{div}\;\overline{w}\right)v_k + \frac{\delta}{\delta x_j}\left\{\varrho v_k (v_j - w_j)\right\} + \frac{\delta}{\delta x_j}(\delta_{jk}\,p) - \varrho g_k\right\}\Delta v_k\, J\, d\mathrm{vol}^{\circ} = \int_S (T_k - S_k)\,\Delta v_k\; ds \quad (25)$$

where $\Delta\overline{v}$ is the virtual velocity and $\overline{S}$ in the surface traction vector ($S_k = n_j\,\sigma_{jk}$).

The lefthandside of Eq. (25) contains a rather cumbersome integrand that can be simplified as follows.

a) We can get out of the jacobian J by adopting the mixed Eulerian-Lagrangian description using Eq. (5)

b) The presence of the pressure gradient term a priori doesn't allow to approximate the internal pressure p by means of a piecewise constant function. This difficulty is solved by introducing the boundary condition (22) into Eq. (25) and applying integration by parts.

The final result turns out to be an extension of the Principle of Virtual Power, now defined on a variable domain of integration $V(t)$:

$$\int_{V(t)} \varrho \frac{\delta v_k}{\delta t} \Delta v_k \, dvol + \int_{V(t)} \left\{ \frac{\delta \varrho}{\delta t} + \varrho \, div \, \overline{W} \right\} v_k \Delta v_k \, dvol + \int_{s_v(t)} \varrho v_k (v_j - w_j) \Delta v_k \, n_j \, ds - \int_{V(t)} \left\{ \varrho v_k (v_j - w_j) \frac{\delta \Delta v_k}{\delta x_j} \right\} dvol - \int_{V(t)} p \frac{\delta \Delta v_k}{\delta x_k} dvol = \int_{s_t(t)} T_k \Delta v_k \, ds + \int_{V(t)} \varrho g_k \Delta v_k \, dvol. \quad (26)$$

Statement (26) applies at any time for any arbitrary finite element in a mixed Eulerian -Lagrangian description. Its right hand side is the external virtual power, i.e. the virtual work rate of the prescribed external force $\overline{T}$ and the acceleration $\overline{g}$. The left hand side is the internal virtual power which results from four different types of contributions: the local time variation of the momentum, the effect of the control volume dilatation, the momentum transport and the action of the internal pressure.

Let us notice that the variational formulation (26) includes automatically at the element level the boundary conditions associated with any problem of continuum mechanics, i.e. the external velocity (or displacement) and force boundary conditions, expressed in the s_v - and s_t- surface integrals, respectively. This is moreover an important characteristic of the finite element method. Let us recall in addition that these conditions usually appear separately on complementary parts of the boundary while, in our class of mixed boundary value problems, they apply together to the same part of the boundary.

SPACE AND TIME INTEGRATION SCHEMES

The numerical solution of (26) is accomplished by discretizing the flow domain into quadrilateral finite elements with four nodes and straight line boundaries. Within these elements the velocity field is approximated by means of the well known isoparametric shape functions N_i (p,q), defined in terms of normalised coordinates (p,q) which have values $\pm$ 1 on the faces of the quadrilateral. In this way, the continuity of the velocity components is ensured through all the mesh.

In order to establish the global equilibrium equations for a continuum subdivided into different elements, one has to assemble the element matrices and to apply the external boundary conditions. This is done by applying the principle of virtual power over the entire configuration with the virtual power due to the discontinuities in pressures between elements neglected. In other words the surface forces $\bar{T}$ inside the fluid domain are supposed to appear there in equal and opposite pairs (action and reaction) so that they cancel everywhere except on the external boundary. The assembly process finally consists thus in summing Eq. (26) for all the elements, noting that the T_k - term will drop everywhere except where an element surface coincides with an external surface. Obviously the same reasoning applies on the s_v- surface integral containing the momentum transport terms. After cancellation of the global nodal virtual velocity vector, one finally obtains the characteristic finite element matrix system which approximates the momentum equation.

For direct integration procedures, the type of mass matrix that should be used depends on the method of temporal integration(explicit or implicit). The explicit integration techniques are particularly suited for non-linear problems, for unlike implicit methods, no significant additional costs are introduced by non-linearities. Moreover, the time step required to follow adequately high transients, like those considered here, is often of the order of the stability limit so that the numerical stability condition, inherent in every explicit method, is not at all prejudicial. According to the interesting conclusions of (Krieg,73), in explicit methods it is desirable from the viewpoints of convergence as well as accuracy to use a lumped diagonal mass matrix rather than a consistent non-diagonal one.

The fully non-linear case, involving quadratic matrices due to the momentum transport terms of Eq. (26) would require the use of tedious numerical algorithms. Solution of the problem is considerably simplified at the outset by substituting a previously known or estimated velocity in the different transport terms of the hydrodynamic equations.

For realizing this the calculational scheme is

divided into two steps at each time cycle by using a modified trapezoidal rule. First intermediate values are calculated for the velocity, the density and the specific internal energy by means of a simple time extrapolation. Second, the three governing equations (18), (19) and (26) are solved by assuming that the three hydrodynamic quantities (ρ , ρe, $\rho \bar{v}$), evaluated at the first intermediate step, are transported through the nodal mesh at the estimated intermediate velocity. The results of this second step then serve to reinitialize each element in readiness for the computation of the next time cycle, and so on.

In dealing with transport effects, it is necessary to determine at each interface which are the donor and the acceptor elements: therefore one needs to know the sign of the flux integrals. In addition, for numerical stability reasons, the value taken for the hydrodynamic quantity g transported through each interface must be a weighted average between those of the donor and the acceptor,

$$g^{final} = \frac{1}{2}\left\{(1+\alpha)g^{donor} + (1-\alpha)g^{acceptor}\right\} \qquad (27)$$

where the coefficient α is proportional to the velocity flux through the interface (Roache, 1972) .

It is well known that most of the numerical schemes give rise to oscillations when used for the computation of solutions which are discontinuous or rapidly changing. To remedy this difficulty it is classical to introduce a dissipative effect (pseudoviscosity) in the numerical scheme. The pseudoviscous term that was found most suitable is that of (Wilkins, 1964), containing both a linear and a quadratic component which are defined only at a compression or at a shock region.

INITIAL AND BOUNDARY CONDITIONS

A computer program has been developed to construct and solve the set of governing equations (6), (18), (19) and (26). In order to start the solution process, initial values are given as input data to the position of the moving interfaces, and to the hydrodynamic quantities ρ , ρe and $\rho\bar{v}$.

In addition boundary conditions are derived in such a way as to describe the various surfaces or walls that may occur . These are of several types :

1) Applied force boundary. In this case the s_t-

integral in Eq. (26) is non zero since both components of $\overline{T}$ are imposed.

2) Prescribed input (or output) boundary. Here the s_v-integral in Eq. (26) is non zero since values are prescribed to both components of $\overline{v}$ and $\overline{w}$.

3) Rigid wall. At a structural rigid boundary fluid may slip freely along the wall ($\overline{v} \cdot \overline{n} = \overline{w} \cdot \overline{n} = 0$ along s) or it may stick to it ($\overline{v} = \overline{w} = \overline{v}_{solid}$ on s_v). In both cases the s_v-integral drops and the s_t-integral yields the reaction force of the wall.

4) moving material interface (adjacent to a vacuum, to an applied constant pressure, to an other fluid or to a deformable structure). Here the boundary conditions are of the mixed type, as already mentioned: no mass flow and stress continuity.

The no mass flow condition is equivalent to determining the motion of the interface by the necessary and sufficient condition for a surface to be material (Lagrange's criterion). In other words the local speed of propagation of the surface, υ , which is the normal speed of the surface with respect to the particles instantaneously situate on it, must be zero (Truesdell, 1960) :

$$\upsilon = w_n - v_n = 0 \qquad (28)$$

Moreover this relation (28) simplifies considerably the basic variational statement (26) since it allows to suppress the tedious s_v - integral (taken on a slip surface).

As far as the stress continuity condition is concerned, the external surface pressure, given by the s_t - integral of Eq. (26), has to be equalized with that of the neighbouring continuum. This equalization process results, as already mentioned, from a series of tentative solutions which all must converge to the simultaneous satisfaction of both the mixed boundary conditions.

NUMERICAL RESULTS

1- Shock tube problem.

A cylindrical rigid tube, divided into two parts, is filled with a perfect gas, the initial pressure in the two parts being different. At zero time, the diaphragm separating the two parts is removed. This first problem which doesn't involve any moving material interface, is a good test of dispersion and noiseintroduced by the numerical discretisation at discontinuities, that is, at the shock moving into the lower pressure part and at the head and tail of the rarefaction moving into the higher pressure part of the tube.

Figures 1 and 2 show an excellent agreement between the Eulerian solution and the theoretical predictions of the shock profile at two different times before reflection. In addition, the total energy (kinetic + internal) and the total mass were shown to remain perfectly constant during the numerical calculation and no radial dependence was found in the two-dimensional Eulerian mesh. The presented method has operated with a time step Δt of $10 \mu s$ which, by comparison with other codes (Apricot project - ERDA 1974), was found to be very close to the stability limit of $20 \mu s$ suggested by the Courant criterion. It was also concluded that an Eulerian technique needs much less artificial viscosity than its Lagrangian counterpart: this is due to the additional smoothing introduced in the solution by the mass convection algorithm (not space centered scheme).

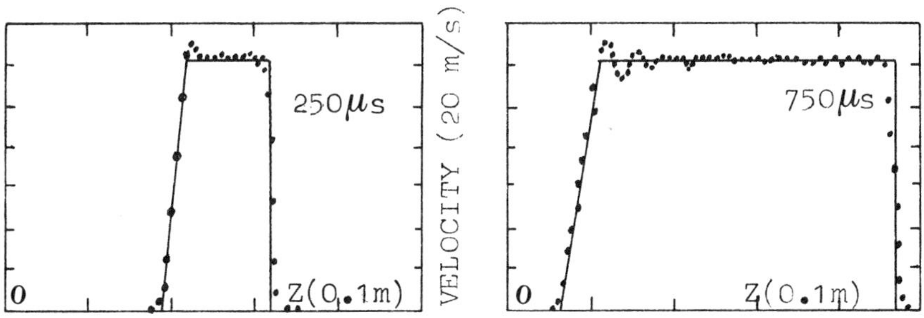

FIGURE 1. VELOCITY PROFILES IN THE SHOCK TUBE

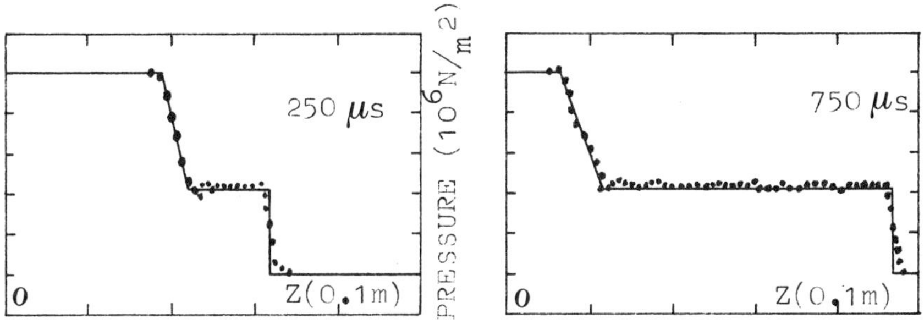

FIGURE 2 . PRESSURE PROFILES IN THE SHOCK TUBE

2- Roof impact of a one-dimensional waterhammer.

In the framework of fast reactor safety studies, it is crucial to simulate roof impact phenomena. They are generated by well-defined energy releases in partially water-filled containment vessels. Of particular interest are the high compression waves produced, immediately after impact, on both sides of the impact surface. Theoretically, the pressure behind these compression waves is the largest pressure that exists throughout the entire flow process. Since the waterslug is of finite dimensions, rarefaction waves are also generated which result in the detachment of the free surface from the roof at the impact surface.

Since it is usually difficult to calculate the fine detail of an impact process, especially close to to the symmetry axis, we first tested the variable domain finite element method on a simple problem with known analytical solution: the axisymmetric normal impact problem. The conditions after impact can be determined from the conservation laws, better known as the Rankine-Hugoniot relations (Truesdell, 1960):

Conservation of momentum; $P = \varrho(C + v)v$ (29)

where P is the impact pressure, C is the wave speed and v is the material velocity.

FIGURE 3.
INITIAL MESH
AND BOUNDARY
CONDITIONS

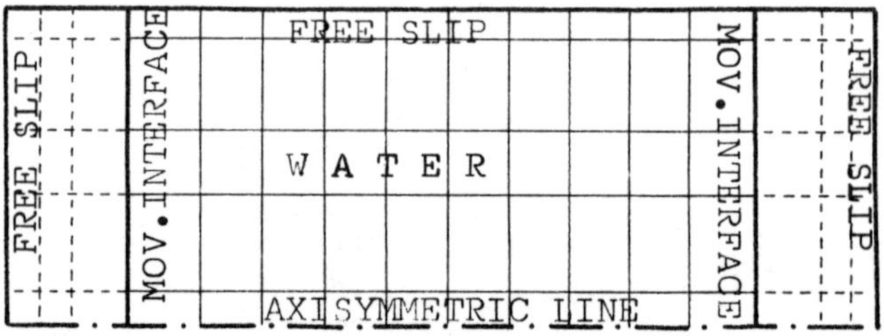

In our calculational example, described in Figure 3, two void regions are assumed above and below the cylindrical waterslug. Giving an initial velocity $v = 1.5 \times 10^2$ m/s and assuming a normal sound speed of $C = 1.5 \times 10^3$ m/s, we obtain through formula (29) an impact pressure $P = 2.475 \times 10^8$ N/m^2 which is correctly reproduced by the numerical solution plotted in Figure 4. In addition Figure 5 gives the balance of energy in function of time: shortly after impact all the kinetic energy of the traveling water-hammer is of course transformed into internal energy. This last figure shows also that the value of the time integration step Δt tends to decrease when the total internal energy increases, and vice versa.

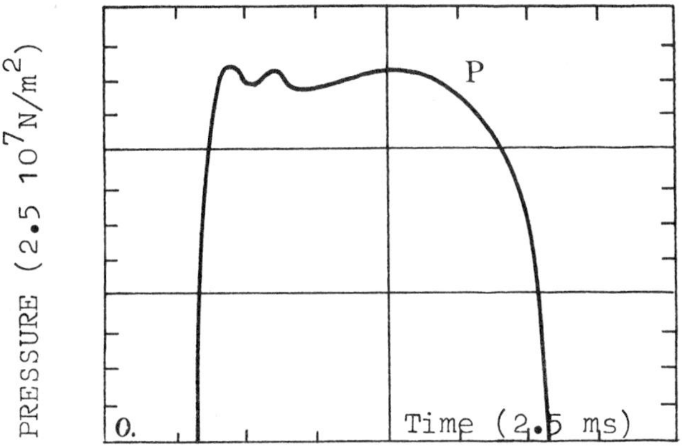

FIGURE 4. NORMAL IMPACT PRESSURE VERSUS TIME

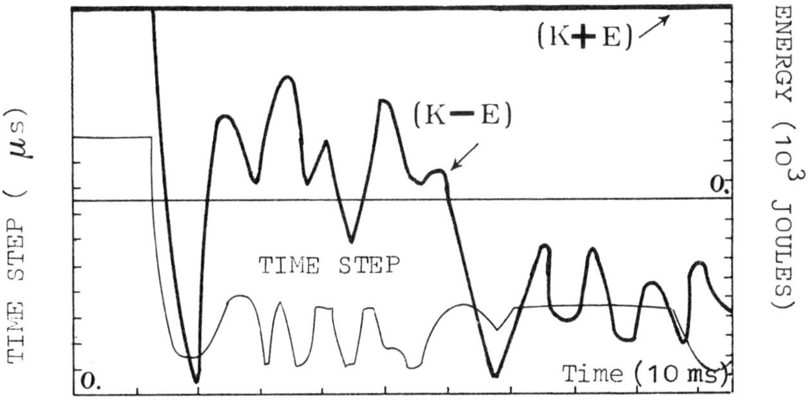

FIGURE 5. ENERGY DISTRIBUTION AND Δt VERSUS TIME.

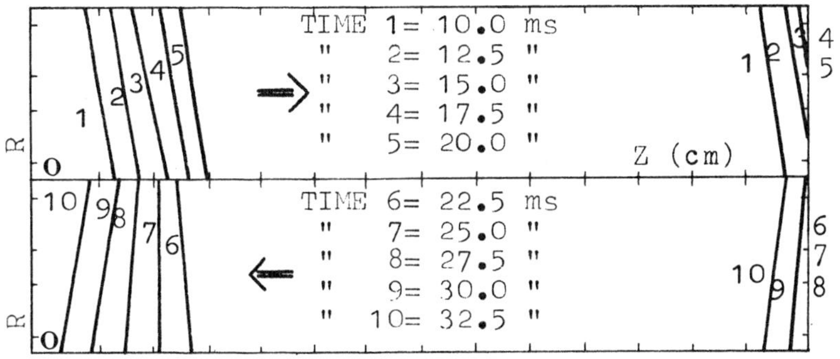

FIGURE 6. PROPAGATION OF AN OBLIQUE FREE BOUNDARY

3- Two-dimensional fluid slug impact.

In the oblique impact process the free surface nodes arrive at the roof at discrete times when a continuous phenomenon is required, and a similar problem in reverse occurs on detachment of the free surface from the roof. As a consequence, compression waves resulting from the first contact with the roof are reflected from the wall before impact is completed, leading to a complex interplay of oblically propagating waves. In addition the nodes reaching the roof at impact time move radially and the flow pattern becomes very complicated. Nevertheless the presented method succeedsin analysing accurately this impact process by taking the frame velocity $\overline{w}$ equal to the structural velocity $\overline{v}_s$ of the inner roof surface: in this way the meshes of the fluid and of the structure remain thoroughly compatible one with each other.

Figure 6 shows a time sequence of the different profiles assumedbyan initially oblique free surface before, during and after impact: the see-sawingphenomenon which can be described analytically by the method of characteristics, is particularly well reproduced by the numerical model. Conservation of total system energy is used once again as a measure of accuracy and stability. This is shown on Figure 7, together with the time evolution of the total mass M;as in the previous Figure 5,K and E denote the total kinetic and internal energies, respectively.

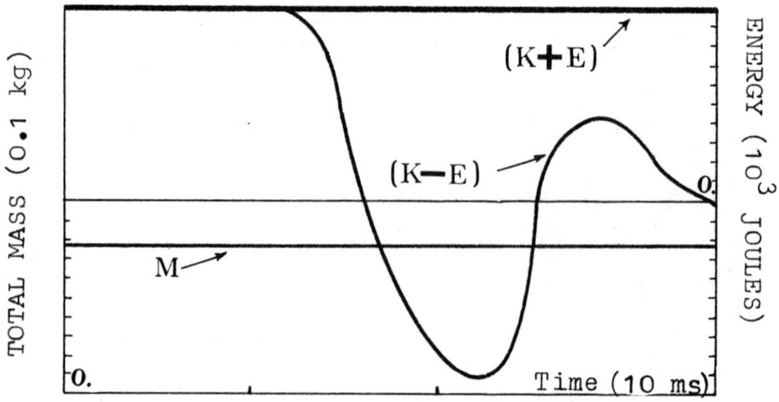

FIGURE 7. MASS AND ENERGY BALANCES VERSUS TIME (OBLIQUE IMPACT PROBLEM)

CONCLUSIONS

In the present paper we have applied Galerkin's method of weighted residuals to the mixed Eulerian-Lagrangian representation of the momentum equation and to its boundary conditions. The resulting variational statement then served as a basis for a variable domain finite element formulation of the dynamic equations. The mass and internal energy distributions were calculated through the usual flux equations which were integrated over each finite element. The substitution of previously estimated values into the transport terms and the use of an explicit time integration scheme make the method a flexible and effective tool for analysing large deformation flows.

Since the mesh may move arbitrarily, the treatment of slip interfaces, in particular, is considerably simplified. For this purpose we developed a transition element which allows Eulerian and/or Lagrangian meshes to remain compatible along their common slip surface. The method presented retains the benefits of both Eulerian and Lagrangian descriptions while avoiding their drawbacks, and it has proven to be very stable and accurate for solving one- and two-dimensional fluid impact problems.

ACKNOWLEDGMENTS

Many of the ideas of the present work are incorporated in a Ph. D. thesis at the University of Louvain written under the supervision of Professor M.CROCHET. The author is also grateful to Drs.J. DONEA and P.FASOLI of JRC-ISPRA for many interesting discussions. Thanks are finally due to the Commission of the European Communities and to the S.A. BELGONUCLEAIRE which provided financial assistance during the course of this investigation.

REFERENCES

1-Apricot programme (1974). Analysis of Primary Containment Transients (APRICOT) is an ERDA sponsored project which compares a variety of numerical solutions of shock wave problems in the domain of nuclear reactor safety.

2-Birkhoff,G., and Zarantonello,E.H.(1957) Jets, wakes and cavities. Academic Press, New-York.

3-Connor, J.J., and Brebbia,C.A.(1976) Finite Element Techniques For Fluid Flow. Newness-Butterworths, London.

4-Donea, J.,Giuliani, S.and Halleux, J.P.(1976) "Prediction of the Nonlinear Dynamic Response of Structural Components Using Finite Elements", Nuclear Engng. and Design, 37, 95-114

5-Krieg, R.D. and Key, S.W.(1973)"Transient Shell Response by Numerical Time Integration", Int. J. Num. Methods in Engng. 17, 273-286

6-Roache P.J.(1972) Computational Fluid Dynamics. Hermosa Publ. Albuquerque.

7-Truesdell C.(1960)"The Classical Field Theories", in Encyclopedia of Physics, Vol. III/1. Springer Verlag. Berlin.

8-Van Goethem G.(1976)"An Eulerian Finite Element Method For Unsteady Compressible Flow Problems" in Proc. of the Int. Meet. on Fast Reactor Safety and Related Physics, Chicago (conf. 761001) Vol.III, 1261-1268.

9-Van Goethem G. (1978) Variable Domain Finite Element Analysis Of Transient, Compressible, Non-Viscous Fluid-Flow Problems Involving Moving Boundaries. Ph. D. thesis to be published by the Université Catholique de Louvain, Belgium.

10-Wilkins M. (1964) "Calculation of Elastic-Plastic Flow", in Methods in Computational Physics, Vol.III Academic Press, New-York.

A TIME-SPLIT FINITE ELEMENT ALGORITHM FOR ENVIRONMENTAL RELEASE PREDICTION

A. J. Baker, M. O. Soliman and D. W. Pepper

University of Tennessee, Knoxville, Tennessee and
Department of Energy, Savannah River Laboratories,
Aiken, South Carolina, USA

INTRODUCTION

The accurate and economical prediction of environmental transport of accidental and/or designed releases is becoming a requirement for site certification. The finite element solution algorithm has emerged as a viable candidate for establishment of a numerical solution of the governing differential equations. This paper presents a time-split variant of the finite element algorithm that retains the inherent accuracy features while effecting a consequential reduction in both computer storage and execution CPU requirements. Accuracy and convergence rate with discretization refinement are established in appropriate norms, and numerical predictions are presented for advection of single and multiple distributions in several divergence-free rotational and irrotational velocity fields.

PROBLEM STATEMENT

This paper is addressed to solution of the advection-diffusion equation governing transport of a scalar field variable, e.g., nuclear, thermal, and/or pollutant release. In Cartesian coordinates, the partial differential equation is

$$L(q) = \frac{\partial q}{\partial t} + \vec{U} \cdot \nabla q - \nabla \cdot k \nabla q + f = 0 \tag{1}$$

where $q = q(\vec{x},t)$ is the species fraction, $\vec{U}$ is the velocity vector, k is the effective diffusion coefficient and f is any forcing function. The boundary conditions are contained within

$$\ell(q) = a^1 q + k \nabla q \cdot \hat{n} + a^3 = 0 \tag{2}$$

where the a^i are specified constants. An initial condition is required as

$$q(\vec{x},0) = q_o(\vec{x}) \tag{3}$$

Primary emphasis in this analysis is on determination of the accuracy, convergence properties and economy of solution of the equations (1)-(3) in the limit of inviscid flow, i.e., k = 0, for a range of velocity fields of practical impact in release analysis. For this purpose, five two-dimensional velocity fields were selected as:

constant: $$\vec{U}_1 = U_\infty(\hat{i} + \hat{j}) \tag{4}$$

solid body rotation: $$\vec{U}_2 = r\Omega\hat{\theta} \tag{5}$$

irrotational flow about a cylinder:

$$\vec{U}_3 = U_\infty \nabla \times y\left(1 - \frac{\alpha^2}{r^2}\right)\hat{k} \tag{6}$$

irrotational flow with circulation:

$$\vec{U}_4 = U_\infty \nabla \times \left[y\left(1 - \frac{\alpha^2}{r^2}\right) + \frac{\Gamma}{2\pi}\ln\left(\frac{r}{\alpha}\right)\right]\hat{k} \tag{7}$$

rotational flow over terrain:

$$\vec{U}_5 = U_\infty\left[\hat{i} - \frac{a|\gamma|}{2\pi L}\cos\frac{2\pi x}{a}\exp(-|\gamma|)\hat{i} + \sin\frac{2\pi x}{a}\exp(-|\gamma|)\hat{j}\right] \tag{8}$$

In equations (4)-(8), Ω is the constant angular velocity, the domain spans $0 \leq x \leq a$, $0 \leq y \leq b$, Γ is the circulation, α is cylinder radius, $L \equiv b/40$, and $\gamma \equiv (y - b/2)/L$.

SOLUTION ALGORITHM

A finite element solution algorithm is employed for equations (1)-(3). On subdomains $\Omega_e \equiv R_e^2 \times t$ of the solution domain $\Omega \equiv R^2 \times t \in [x,y] \times [t,0]$, assume $q(\vec{x},t)$ represented as

$$q_e(\vec{x},t) \equiv \left\{N_k(x)\right\}^T\left\{Q(t)\right\}_e \tag{9}$$

Attention is restricted to N_k a linear or bi-linear function on R^1 and R^2, respectively. The Galerkin-Weighted Residuals formalism is employed to cast solution of equations (1)-(3) as

$$S_e\left[\int_{\Omega_e}\{N_k\}L(q_e)d\tau + \lambda\int_{\partial\Omega_e}\{N_k\}\ell(q_e)d\sigma\right] \equiv \{0\} \tag{10}$$

Using a Green-Gauss theorem, the matrix equivalent ordinary differential equation becomes

$$S_e\left[[C]_e\{Q\}'_e + \left([U]_e + [K]_e\right)\{Q\}_e + \{f\}_e\right] \equiv \{0\} \tag{11}$$

where S_e is the finite element assembly operator, and $[C]_e$, $[U]_e$ and $[K]_e$ are the element matrix equivalents of the initial-value, velocity and diffusion terms in equations (1)-(2).

The implicit trapezoidal integration algorithm is employed for equation (11) as

$$\{Q\}_{j+1} = \{Q\}_j + \frac{h}{2}\left[\{Q\}'_{j+1} + \{Q\}'_j\right] \tag{12}$$

where h is the integration time step and j is the time index. Substitution of equation (11) into (12) yields the matrix solution algorithm

$$[J]\{Q\}_{j+1} = \{F\}_j \tag{13}$$

where [J] is the global Jacobian

$$[J] \equiv S_e\left[[C]_e + \frac{h}{2}\left([U]_e + [K]_e\right)\right] \tag{14}$$

and

$$\{F\} \equiv S_e\left[[C]_e\{Q\}_e - \frac{h}{2}\left([U]_e + [K]_e\right)\{Q\}_e + \{f\}_e\right] \tag{15}$$

For the conventional application of the finite element solution algorithm, on two-dimensional space for example,

$$[C]_e \equiv \int_{R_e} \{N_k(\vec{x})\}\{N_k(\vec{x})\}^T d\tau$$
$$\equiv A_e[B200]_e \tag{16}$$

where for $k = 1$, [B200] is a 4 x 4 matrix and A_e is the area of the quadrilateral finite element. For the time-split algorithm, [J] and {F} are factored as, for example,

$$[J] \Rightarrow [J_1] \otimes [J_2] \tag{17}$$

where $\otimes$ signifies the tensor matrix product. In this instance, the initial-value operator matrix equivalent is

$$[C]_e \equiv [C_1]_e \otimes [C_2]_e \tag{18}$$

and

$$[C_i]_e \equiv \int_{R_e^1} \{N_k(x_i)\}\{N_k(x_i)\}^T dx_i$$
$$\equiv \Delta_e [A200I]_e \tag{19}$$

In equation (19), i(I) is not a tensor summation index, and for $k = 1$, each [A200I] is a 2×2 matrix with measure (length) Δ_e. All other matrices comprising the algorithm are similarly expressed, and the concept of time-splitting is readily extended to three-dimensional space.

NUMERICAL RESULTS

The first requirement is to attest the character of accuracy and convergence with discretization refinement of the developed time-split algorithm. Gresho, et al. (1976) were the first to document the superiority of the banded initial-value matrix structure produced by the finite element algorithm, compared to the more familiar (finite difference) diagonalized form. This remains valid in the time-split form as well. The first case corresponds to pure advection ($k \equiv 0$) of a concentration cone with initial distribution

$$q_0(x,y,0) \equiv 100\left[\sin^2\left(\frac{2\pi r}{\lambda}\right)\right] \tag{20}$$

and where $0 \leq r \leq \lambda$ is the radial coordinate with origin (x_0,y_0) and λ is spanned by M finite elements. Figure 1.a illustrates the initial condition in equation (20) for $M = 8$ and $(x_0,y_0) = (7,7)$ on a uniform square mesh of span $0 \leq a,b \leq 80{,}000$ m. This corresponds to a rather coarse discretization, as is common in environmental analyses, and places significant demands upon the algorithm. Figure 1.b shows the fidelity with which the conventional multi-dimensional bi-linear finite element algorithm predicts advection for $U_\infty = 2$ m/s and after 150 time steps with h = 125 s. Numerical diffusion is absent (the peak level is retained unaltered) and the trailing wakes, indicative of dispersion error, possess ±3% relative extrema. Figure 1.c illustrates the final solution obtained by the conventional algorithm, but with the initial-value matrix $[C]_e$ diagonalized, equation (16). This operation is common for use with explicit integration and/or implicit finite difference algorithms, and is observed to essentially obliterate solution fidelity on this grid. (Of course, all algorithms discussed are consistent and converge in a predictable manner under discretization refinement. This is where the economy aspect becomes of vital importance.)

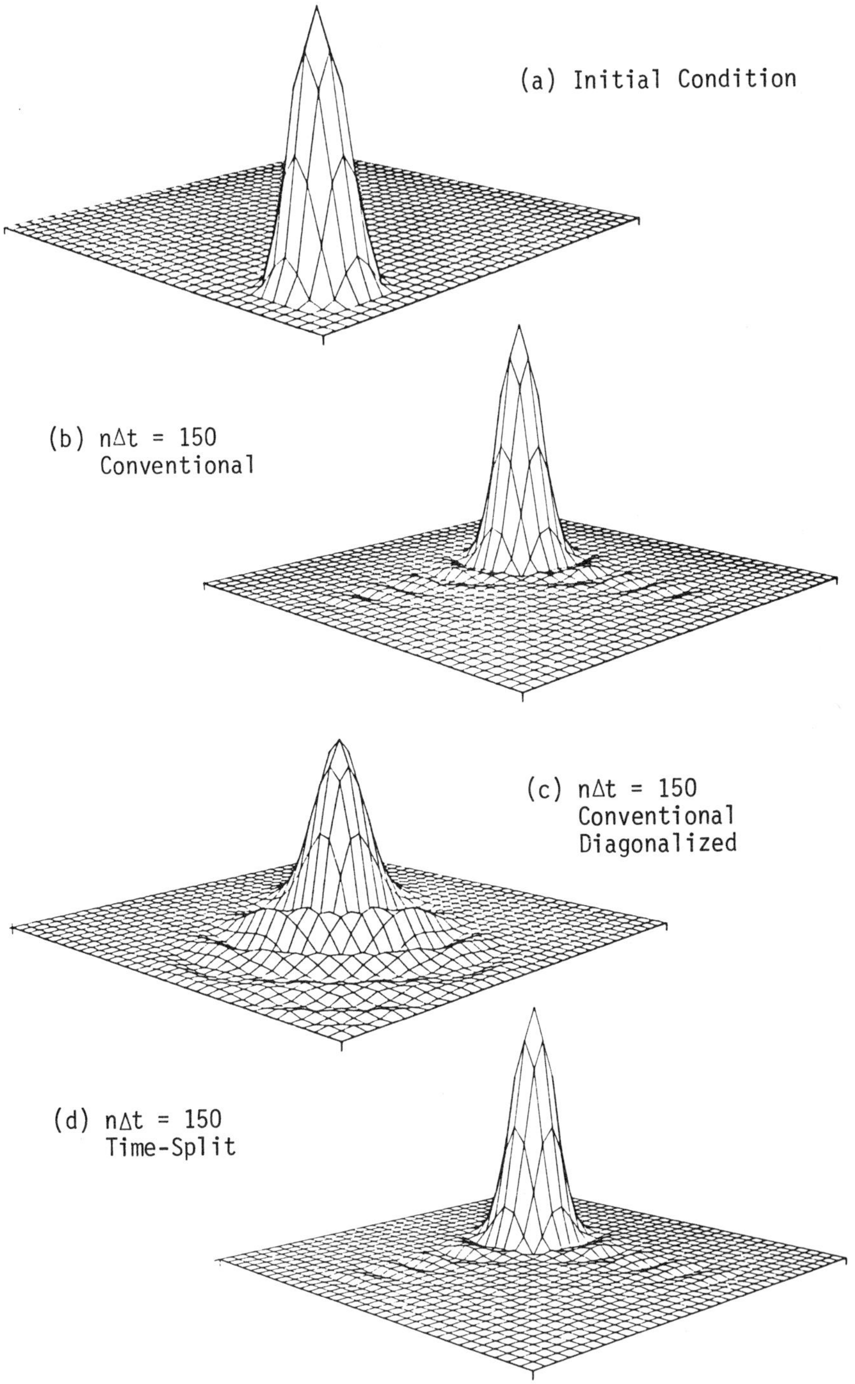

Figure 1. Advection of Environmental Release in Constant Velocity Field $\vec{U}_1$, C = 0.1

Figure 1.d shows the final solution obtained with the time-split finite element algorithm. It is virtually identical to the conventional results, and was obtained at approximately one-fourth expenditure of computer CPU and one-fifth the computer core requirement. This difference is essentially a function of matrix bandwidth of [J], hence becomes progressively favorable as the mesh is increased.

The computed convergence properties with discretization refinement of the split algorithm are illustrated in Fig. 2, for both the full and diagonalized (Crank-Nicholson) $[C]_e$ matrix structure. The three error norms are

$$\text{Energy} \equiv \frac{1}{2} \sum_e \int_{R_e^n} \nabla q_e \cdot \nabla q_e d\tau \tag{21}$$

$$\text{Sum} \equiv \sum_e \int_{R_e^n} q_e d\tau \tag{22}$$

$$\text{Sum}^2 \equiv \frac{1}{2} \sum_e \int_{R_e^n} q_e^2 d\tau \tag{23}$$

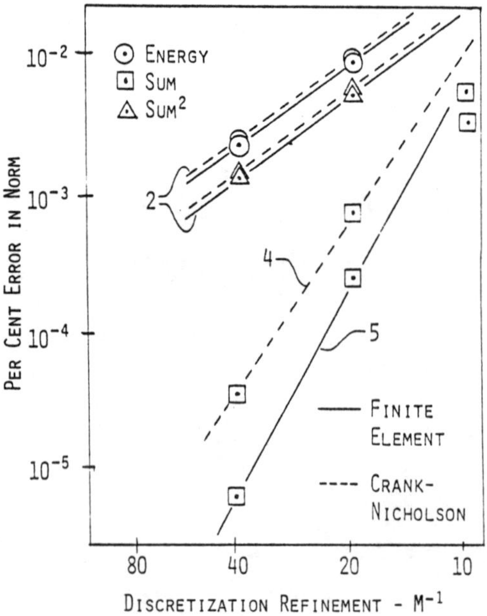

Figure 2. Accuracy and Convergence of Split Implicit Finite Element Solution Algorithm

Computed convergence in energy and sum squared is quadratic and insensitive to $[C]_e$ structure. Convergence in the sum norm is fifth order, and this norm is useful in quantizing the differences illustrated in Fig. 1.c,d. The split algorithm converges quadratically with Courant number C.

$$C \equiv U_\infty h/\Delta_e \tag{24}$$

The results illustrated in Figure 1 were obtained for C = 0.1, and discretization refinement may require a corresponding reduction in integration time step to control dispersion, hence increase the cost of computer execution. The superior convergence property illustrated in Figure 2 helps ameliorate this dilemma.

Dispersion error is the bain of environmental release prediction in a diffusion-free analysis, and the results obtained for the solid-body rotation flow field $\vec{U}_2$ provide quantization. The solution parameters (q_0,a,b,U_∞,h,M) remain identical, the diagonalized algorithm is relegated to history, and Figure 3 illustrates the solution obtained at the quarter, three-quarter and full 360° rotation of the release field. The initial-distribution would appear identical to Fig. 1.a, moved to the 9 o'clock position, and the exact solution would be advection with no alteration or distortion. The split finite-element algorithm is again free of numerical diffusion (the peak level remains intact). The ripple structure in the ground-plane is dispersion error, and while modest in comparison to other solution algorithms, cf., Long and Pepper (1976), has become borderline on acceptability. Filters can be constructed to annihilate short period waves, e.g., Raymond and Garder (1976), and Fig. 3.d illustrates the substantial improvement accrued at the three-quarter turn of the split-filtered algorithm. The peak value has been reduced by about 2%, due to the filtering, and a somewhat modified immediate trailing wake remains identifiable.

An important assessment pertains to preservation of symmetries and skew-symmetries by the split algorithm. The irrotational velocity fields $\vec{U}_3$ and $\vec{U}_4$ serve the purpose, and the former corresponds to flow about a cylinder without circulation. The cylinder diameter is $4\Delta_e$ and centered at the centroid of the grid. Two release concentrations were symmetrically disposed about the stagnation streamline, and all other solution parameters remain identical (q_0,a,b,U_∞,h,M). The computed split-filtered solution for $\vec{U}_3$ is shown in Fig. 4 for select time steps. Figure 4.a shows the solution just after start-up. Figure 4.b corresponds to the extremum deflection, Fig. 4.c shows near coalescence near the downstream stagnation point, and Fig. 4.d illustrates the essentially exact symmetry obtained after the complete transit. The far-downstream peak values are within 2% of the initial level, and dispersion error

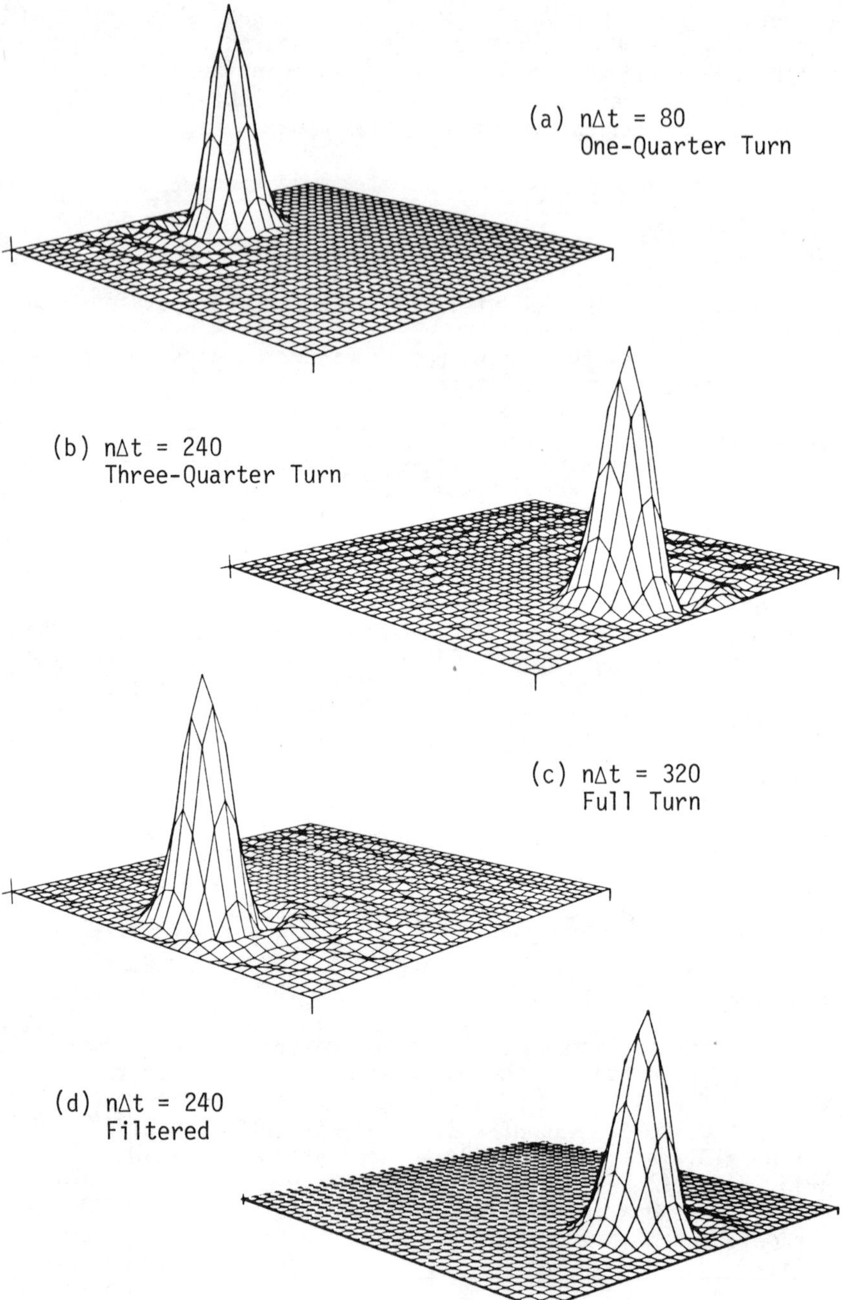

Figure 3. Advection of Environmental Release in Solid-Body Rotation Velocity Field $\vec{U}_2$, C = 0.1

extrema are approximately ±4%. (The largest ones actually lie inside the cylinder and should have been suppressed on print-out.) Figure 5 illustrates results obtained correspondingly, by the split-filtered algorithm, for the irrotational velocity field $\vec{U}_4$ with circulation. Very large gradients are illustrated supported with acceptable dispersion error and loss of peak value. Again, the large dispersion spikes in the figures actually lie interior to the cylinder and should be suppressed as meaningless.

The final predictions were generated with velocity field $\vec{U}_5$, which is rotational, and was constructed to simulate a wind-shear downdraft and resultant reflection off the ground-plane. Such a condition occurs, for example, when smoke from a stack is observed to bend towards the ground in a strong cross-wind. The velocity field $\vec{U}_5$ is highly rotational and produces strong accelerations about the vertical mid-plane. The computed results are shown in Figure 6, and all solution parameters are again identical. In Fig. 6.b, the release enters the downdraft, in Fig. 6.c it has rotated near the ground plane, and Fig. 6.d illustrates the considerably flattened distribution convected within the resultant updraft. Evidence of disperson error is minimal, and very large gradients in the concentration are supported. No other quantitative observations can be readily made.

CONCLUSIONS

A time-split implicit finite element solution algorithm has been evaluated for environmental release transport prediction. It displays considerable economy, in terms of computer CPU and storage, in comparison to the conventional finite element algorithm, with negligible loss in accuracy. Accuracy and convergence properties of the split algorithm are quantized, and numerous numerical solutions for a variety of velocity fields have illustrated its potential usefulness in release transport prediction.

ACKNOWLEDGMENTS

Initial research on the split algorithm was supported by NASA Langley Research Center under Grant NSG-1391 to University of Tennessee. Additional contractual support was provided through Computational Mechanics Consultants, Inc.

REFERENCES

Gresho, P. M., Lee, R. L., Sani, R. L. (1976) Advection Dominated Flows with Emphasis on the Consequences of Mass Lumping. Proceedings, 2nd Int. Symp. F.E. Mtd. in Flow, Italy.

Long, P. E., Pepper, D. W. (1976) A Comparison of Six Numerical Schemes for Calculating Advection of Atmospheric Pollution. Proceedings, 3rd Symp. Atmos. Turbulence, Diffusion and Air Quality, USA.

Raymond, W. H., Garder, A. (1976) Selective Damping in a Galerkin Method for Solving Wave Problems with Variable Grids. Monthly Weather Review, V. 104, pp. 1583-1590.

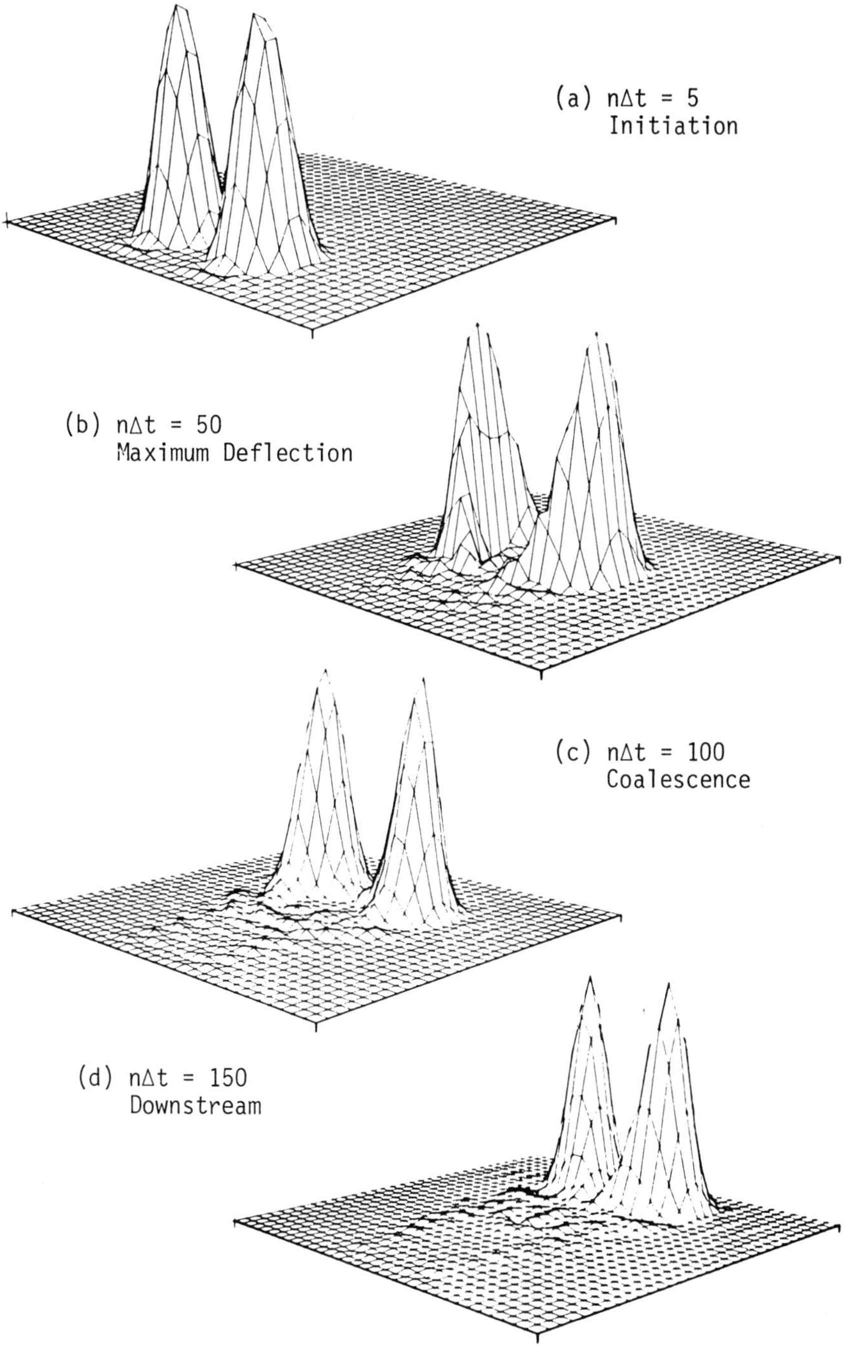

Figure 4. Advection of Environmental Release Pair in Irrotational Velocity Field $\vec{U}_3$, C = 0.1

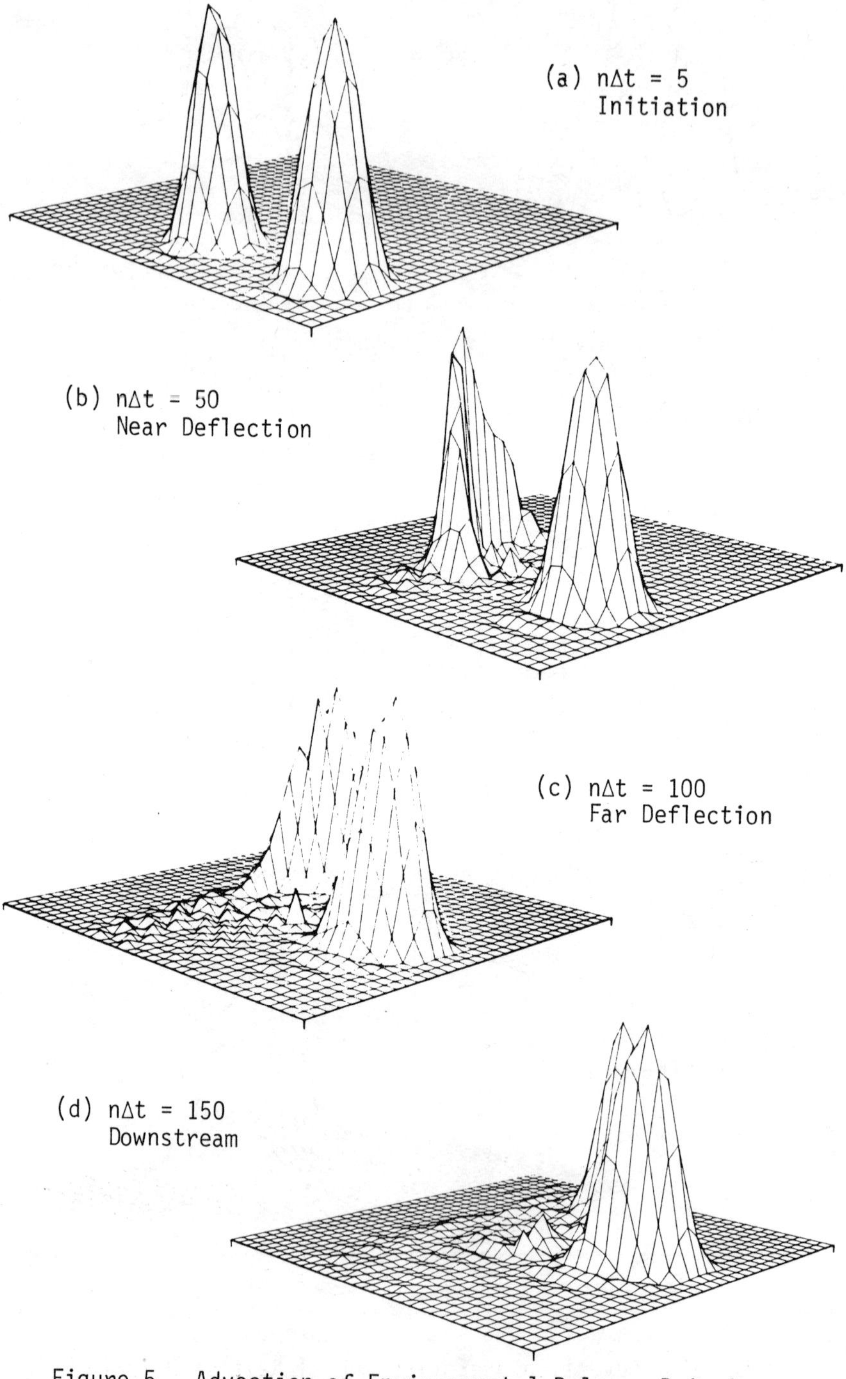

Figure 5. Advection of Environmental Release Pair in Irrotational Velocity Field $\vec{U}_4$ with Circulation, C = 0.1

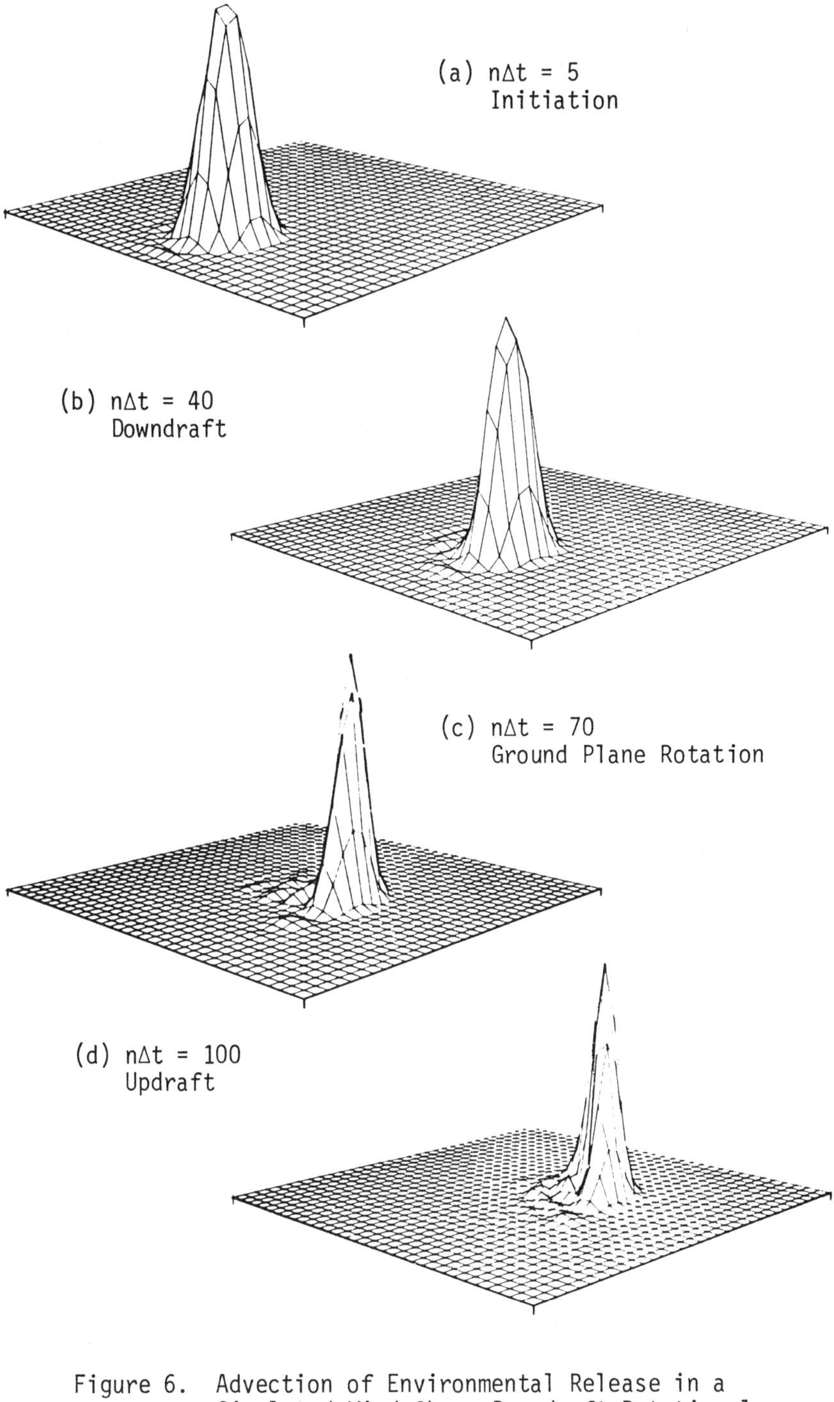

Figure 6. Advection of Environmental Release in a Simulated Wind-Shear Downdraft Rotational Velocity Field $\vec{U}_5$, C = 0.1

APPLICATIONS OF BOUNDARY ELEMENTS IN FLUID FLOW

by C.A. Brebbia and L.C. Wrobel

Southampton University, UK

1. INTRODUCTION

Analysts have recently become aware of the possibilities of boundary solutions and their advantages when compared with domain-type methods for many practical applications. At the same time they have developed ways of combining both techniques to optimize the computational effort. This has been possible because of the better understanding of the common principles underlining all approximate methods.

Engineering applications of boundary solutions can be attributed to the works of Jaswon, Symm and Ponter [1] [2] [3] on integral equations methods applied to potential problems. Later on the same techniques were successfully applied by several researchers to elasticity problems. A review of this work can be found in reference [4]. At the same time the method was generalized to what is now called the "direct" solution, which in contrast with the original or "indirect" formulation, works with the physical variables instead of sources or dipoles.

At Southampton University the technique was developed to allow for arbitrary variations of the unknowns over the surface of the body. The body was divided into elements and over each of them interpolation functions were assumed. This generalization naturally leads to using the name boundary elements to describe the new technique. This work, started at the beginning of the 1970's is reviewed in detail in [5]. A simple exposition of the boundary elements for potential problems can be seen in reference [6].

The method as described in references [5] and [6] is based on the weighted residuals technique. In this way, it is easy to relate boundary elements to more conventional methods such as finite elements, which can also be described as a weighted residual technique. The same approach is used in this paper and generalized for an arbitrary self-adjoint operator (the limitation of self-adjointness is mainly due to the difficulty of finding appropriate fundamental solutions for non-adjoint problems).

In the present paper the application of the method for the solution of fluid flow problems is discussed. Four different examples are shown to illustrate the potentialities of the method. They are a perfect fluid flow problem; a free surface problem; Poiseuille's type flow in a pipe of elliptical cross-section and the viscous incompressible flow at low Re numbers. These examples show the main advantages of the method i.e., small number of unknowns required, accuracy of the solution and simplicity of the input data.

2. FUNDAMENTALS OF THE METHOD

Consider a differential equation represented by an operator $\mathcal{L}$, such that,

$$\mathcal{L}(u_o) = b \qquad \text{in } \Omega \tag{1}$$

with boundary conditions,

$$\begin{array}{lll} \text{i) Essential conditions} & S(u_o) = \bar{u} & \text{on } \Gamma_1 \\ \text{ii) Natural conditions} & G(u_o) = \bar{q} & \text{on } \Gamma_2 \end{array} \tag{2}$$

We will now try to find an approximate solution u for u_o and minimize the error by weighting the differential equation (1) by a new function v. Both u and v are combinations of sets of functions multiplied by a series of parameters. The parameters corresponding to v are arbitrary [5]. The weighting gives,

$$< \mathcal{L}(u) - b \ , \ v >_\Omega = 0 \tag{3}$$

The brackets indicate a scalar product in the Ω domain. Integrating (3) by parts as many times as needed we find,

$$<\mathcal{L}(u) - b, v >_\Omega = < u, \mathcal{L}^*(v) >_\Omega - <b,v>_\Omega + <G(u), S^*(v)>_\Gamma - < S(u), G^*(v)>_\Gamma \tag{4}$$

If $\mathcal{L}(\ ) = \mathcal{L}^*(\ )$ the operator is called self-adjoint and we also have that $G(\ ) = G^*(\ )$ and $S(\ ) = S^*(\ )$. Equation (4) can then be written for self adjoint problems as follows,

$$< u, \mathcal{L}(v) >_\Omega - < b,v >_\Omega + < G(u),S(v) >_\Gamma - < S(u),G(v) >_\Gamma = 0 \tag{5}$$

Notice that we can take the boundary conditions (2) into consideration, which gives,

$$< u, \mathcal{L}(v) >_{\Omega} + < \bar{q}, S(v) >_{\Gamma_2} + < G(u), S(v) >_{\Gamma_1} =$$

$$= < b, v >_{\Omega} + < \bar{u}, G(v) >_{\Gamma_1} + < S(u), G(v) >_{\Gamma_2} \tag{6}$$

In order to render the problem a boundary solution we need to find the fundamental solution such that,

$$\mathcal{L}(v) + \delta_i = 0 \tag{7}$$

where δ_i is the dirac delta function. Hence the first term in equation (6) becomes

$$< u, \mathcal{L}(v) >_{\Omega} = - u_i \tag{8}$$

u_i indicates the value of the unknown function at a point 'i'.
Note that equation (6) becomes,

$$u_i + < b, v >_{\Omega} + < \bar{u}, G(v) >_{\Gamma_1} + < S(u), G(v) >_{\Gamma_2} =$$

$$= < \bar{q}, S(v) >_{\Gamma_2} + < G(u), S(v) >_{\Gamma_1} \tag{9}$$

To illustrate the application of the above relationship, let us consider the case of $\mathcal{L}(\ )$ being a Laplace operator [6], i.e.,

$$\mathcal{L}(u) - b = \frac{\partial^2 u}{\partial x^2} + \frac{\partial^2 u}{\partial y^2} - b \tag{10}$$

The boundary conditions are,

$$G(u) = \frac{\partial u}{\partial n} = q, \quad S(u) = u \tag{11}$$

Hence expression (9) produces,

$$u_i + \iint_{\Omega} b\, v\, d\Omega + \int_{\Gamma_1} \bar{u} \frac{\partial v}{\partial n} d\Gamma + \int_{\Gamma_2} u \frac{\partial v}{\partial n} d\Gamma = \int_{\Gamma_2} \bar{q} v\, d\Gamma + \int_{\Gamma_1} q v\, d\Gamma \tag{12}$$

v is the fundamental solution of the Laplace operator, i.e.

$$\frac{\partial^2 v}{\partial x^2} + \frac{\partial^2 v}{\partial y^2} + \delta_i = 0 \qquad (13)$$

which gives

$$v = \frac{1}{2\pi} \ln\left(\frac{1}{r}\right) \qquad (14)$$

where r is the distance from the point of application of the unit potential to the point under consideration.

Equations (9) and (12) are valid for an internal 'i' point. In general we have,

$$c_i u_i + \langle b, v \rangle_\Omega + \langle \bar{u}, G(v) \rangle_{\Gamma_1} + \langle S(u), G(v) \rangle_{\Gamma_2} = \\ = \langle \bar{q}, S(v) \rangle_{\Gamma_2} + \langle G(u), S(v) \rangle_{\Gamma_1} \qquad (15)$$

or

$$c_i u_i + \iint_\Omega b\ v\ d\Omega + \int_{\Gamma_1} \bar{u} \frac{\partial v}{\partial n} d\Gamma + \int_{\Gamma_2} u \frac{\partial v}{\partial n} d\Gamma = \\ = \int_{\Gamma_2} \bar{q}\ v\ d\Gamma + \int_{\Gamma_1} q\ v\ d\Gamma \qquad (16)$$

The value of c_i is

$c_i = 1$ for i inside the Ω domain.

$c_i = 0$ for i outside the Ω domain

$c_i = \frac{1}{2}$ for i on the (smooth) boundary Γ.

Boundary Elements

Equations (15) or (16) can now be applied on the boundary of the domain under consideration. The boundary can be divided into n elements. The points where the unknown values are considered are called 'nodes' and are similar to those of finite elements. The main difference is that now elements and nodes

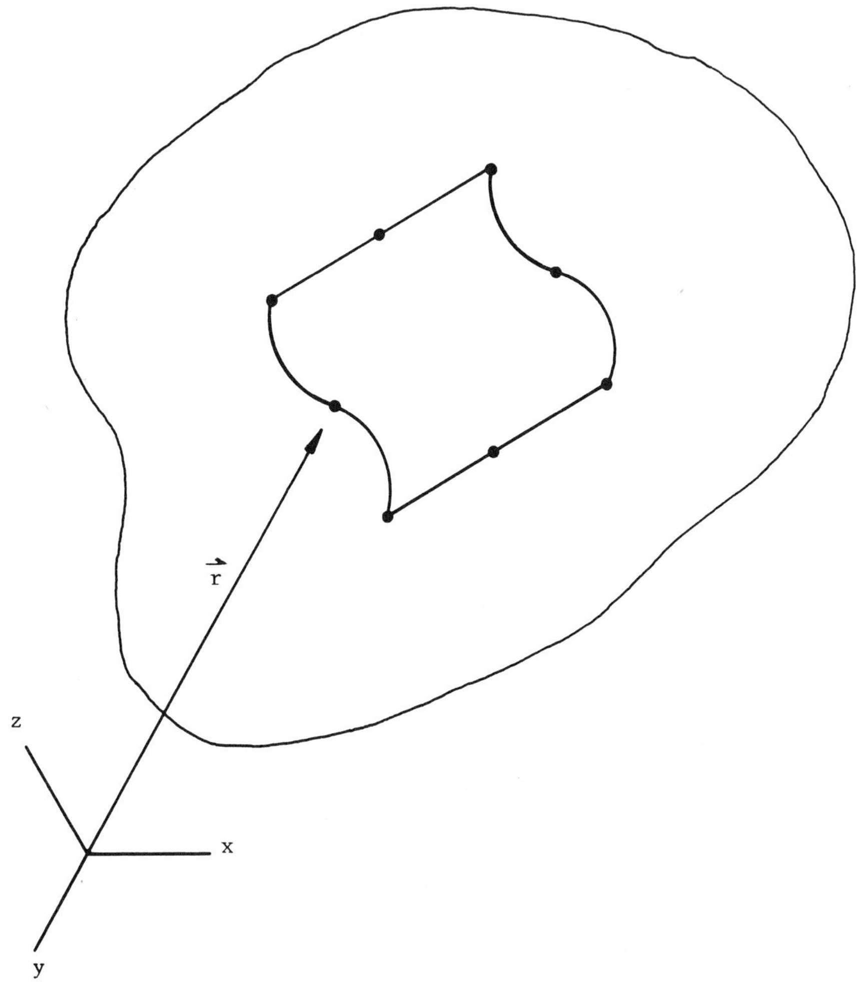

Figure 1 Boundary Element

are defined only on the surface of the domain (figure 1). Consider the potential case for simplicity. The functions u and q over a boundary element are defined as,

$$u = \underset{\sim}{\phi}^T \underset{\sim}{u}^n$$
$$q = \underset{\sim}{\psi}^T \underset{\sim}{q}^n \tag{17}$$

and equation (16) is discretized as follows,

$$c_i u_i + \sum^{m} (bv)_m + \sum^{n} \left(u \frac{\partial v}{\partial n}\right)_n = \sum^{n} (q\ v)_n \tag{18}$$

In this work both interpolation functions $\underset{\sim}{\phi}$ and $\underset{\sim}{\psi}$ are assumed to be linear.
Note that m internal elements or cells need to be defined to compute the integrals in Ω. These elements do not introduce any further unknown, hence the problem is still a boundary problem.

We can substitute the u and q values given by (17) into (18) and carry out the integrations (usually numerically). This gives for each node the following equations, after assembling,

$$c_i u_i + \left[\hat{H}_{i1} \quad \hat{H}_{i2} \cdots \hat{H}_{in}\right] \begin{Bmatrix} u_1 \\ u_2 \\ \vdots \\ u_n \end{Bmatrix} = \left[G_{i1} \quad G_{i2} \cdots G_{in}\right] \begin{Bmatrix} q_1 \\ q_2 \\ \vdots \\ q_n \end{Bmatrix} + B_i \tag{19}$$

The u_i and q_i values are the nodal unknowns.

This equation can be written as,

$$c_i u_i + \sum_{j=1}^{n} \hat{H}_{ij} u_j = \sum_{j=1}^{n} G_{ij} q_j + B_i \tag{20}$$

By writing the two left hand side terms together we have,

$$\sum_{j=1}^{n} H_{ij} u_j = \sum_{j=1}^{n} G_{ij} q_j + B_i \tag{21}$$

where

$H_{ij} = \hat{H}_{ij}$ for $i \neq j$ and $H_{ij} = \hat{H}_{ij} + c_i$ for $i = j$.

The whole set can be written in matrix form as follows,

$$B_\theta = B_o\left(1 - \frac{\theta}{\pi}\right)$$

$$R_\theta^{inner} = \left(R_c - \frac{B_o}{2}\right) + \frac{B_o}{2} \cdot \frac{\theta}{\pi}$$

$$R_\theta^{outer} = \left(R_c + \frac{B_o}{2}\right) - \frac{B_o}{2} \cdot \frac{\theta}{\pi}$$

R_c = 1.50 m

B_o = 0.90 m

Q = 31.3 ℓ/sec

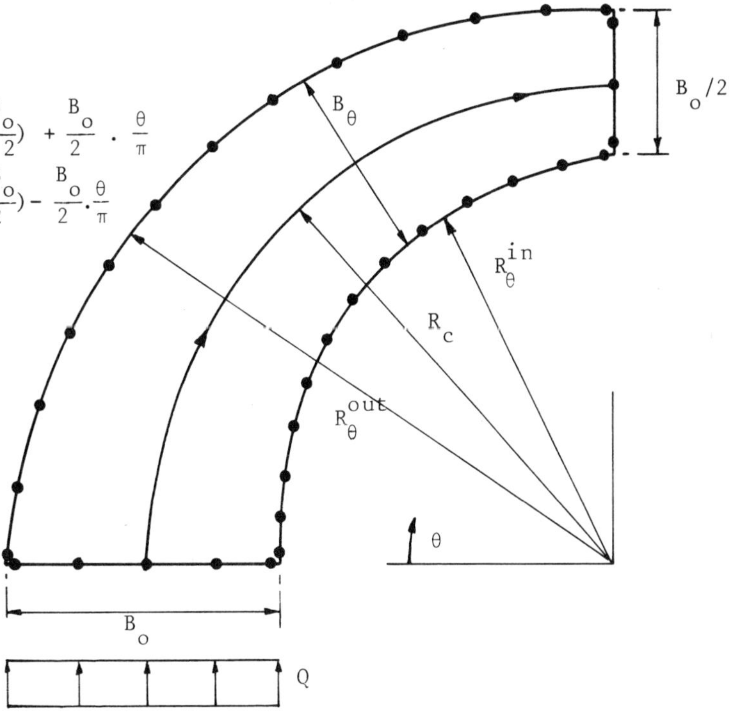

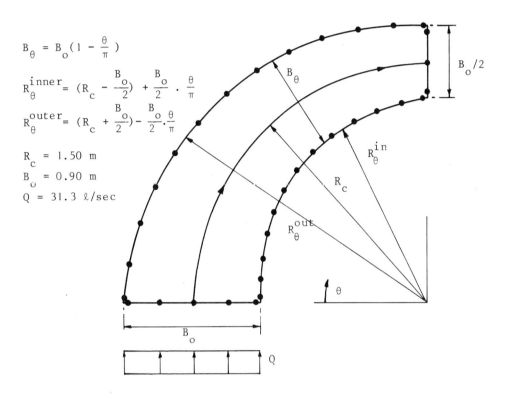

Figure 2 Geometry of channel and boundary elements division

$$\underset{\sim}{H}\,\underset{\sim}{U} = \underset{\sim}{G}\,\underset{\sim}{Q} + \underset{\sim}{B} \qquad (22)$$

Note that unless the surface is smooth at the point 'i' the value $c_i = \frac{1}{2}$ on Γ is not valid. However we can always calculate the diagonal terms of $\underset{\sim}{H}$ by the fact that when a uniform potential is applied over the whole boundary, the normal derivatives must be zero. Hence equation (22) in the absence of body forces becomes,

$$\underset{\sim}{H}\,\underset{\sim}{U} = \underset{\sim}{0} \qquad (23)$$

Thus the sum of all the elements of H in any row ought to be zero, and the value of the coefficient on the diagonal can be easily calculated once the off-diagonal coefficients are all known, i.e.

$$H_{ii} = - \sum_{\substack{j=1 \\ (i \neq j)}}^{n} H_{ij} \qquad (24)$$

3. APPLICATIONS

In order to illustrate how boundary elements can be applied to study some fluid flow problems, the following four different problems will be discussed:

i) Perfect flow in an experimental curved channel;
ii) Free surface flow over a spillway;
iii) Poiseuille's type flow in a pipe of elliptical cross-section;
iv) Viscous incompressible flow at low Re numbers.

These problems although different can all be attempted using boundary elements.

i) Flow in a curved channel

The channel showed in figure 2 was analysed experimentally at the Department of Civil Engineering, University of Southampton. The dimensions of the channel are shown in the figure. The depth varies slightly over the region (from 10.12 cm to 12.53 cm).

In order to perform the solution with a two-dimensional boundary element program, an averaged depth was considered. The flow was assumed to be potential, i.e. incompressible and inviscid. Hence the problem can be represented by a Laplace's equation in the stream function ψ. The rigid walls are considered to be streamlines.

Results for the longitudinal velocity distribution in some cross-sections are compared with experimental velocities in figure 3 and table 1, showing good agreement, although the

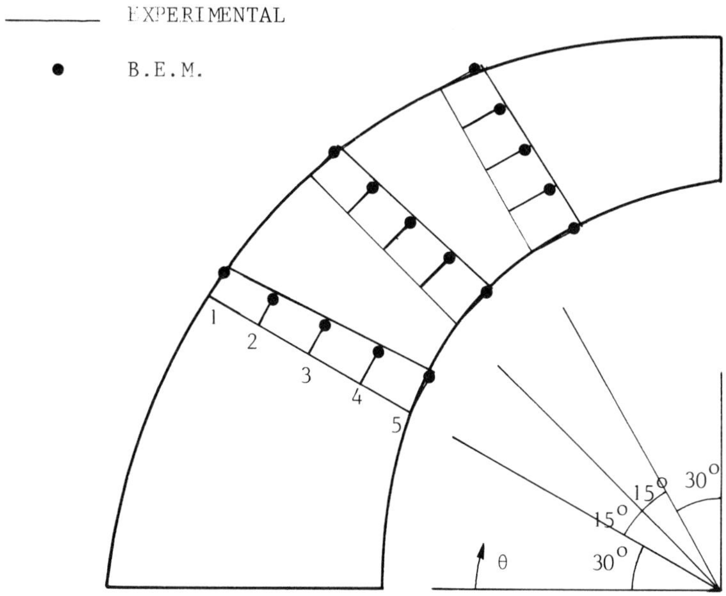

Figure 3 Velocity distributions at some cross-sections

experimental velocities are depth averaged.

The method can be easily extended to study the three-dimensional potential flow, in which case only surface elements are required.

ii) <u>Flow over a spillway</u>
Figure 4 shows the flow region under study for a typical spillway problem. The nappe profile corresponds to a standard spillway as defined in reference [7].

Of great interest and importance in analysing this problem is the determination of the top flow line and the pressure distribution over the crest. The condition to find the free surface is that at any point on it the potential head u has to equal the elevation head. The potential is defined as:

$$u = y + \frac{p}{\rho g} \tag{25}$$

where y is the elevation above a fixed reference plane, ρ is the density of the fluid and p is the pressure above atmospheric. Notice that since the pressure on the free surface is usually atmospheric the second term on the right hand side of equation (25) vanishes. It can be shown that the behaviour of u is governed by the Laplace's equation $\nabla^2 u = 0$.

The flow across the sections S' and S" (figure 4) was assumed to be uniform (i.e. constant potential head). There is no flux through the rigid walls and the free surface so $q = 0$ on these boundaries. The problem can be solved in an iterative way. Initially one has to guess a top flow line and solve the problem for u. After each iteration the calculated potential at every nodal point on the free surface is compared against its elevation; if the difference between these two values is greater than a maximum acceptable error, this difference is algebraically added on the elevation of the nodal point.

When solving the problem with finite elements [8] the top nodal points are shifted upwards or downwards and all interior nodal points are moved proportionally. As a result all the elements below the free surface are slightly "stretched" or "squashed" so that the element network still fits the new flow region. When using boundary elements only the nodes on the free surface will be shifted as all other boundaries are fixed. The elements on the G and H matrices - equation (22) - corresponding to the influence of fixed boundary nodes on other fixed boundary nodes will remain constant during the analysis. Hence they can be computed once and stored, saving computer time. The potentials at internal points are calculated using equation (20) after the correct position of the free surface has been found. With finite elements instead the internal potentials

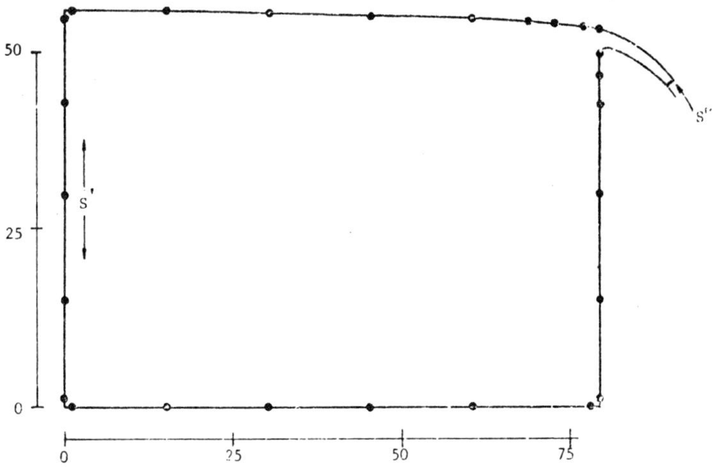

Figure 4   Boundary elements division

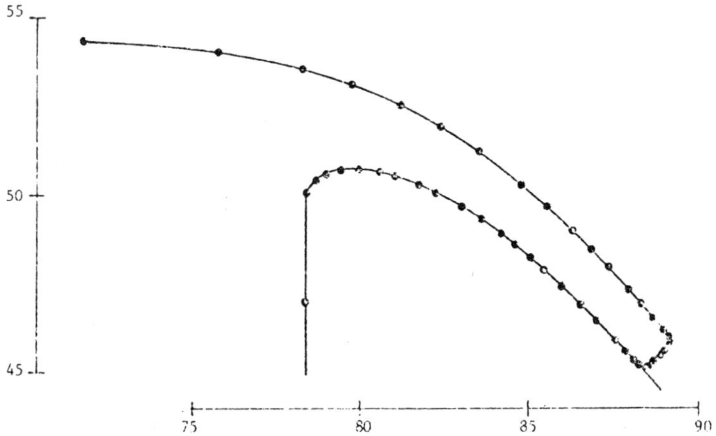

Figure 5   Boundary elements over the crest

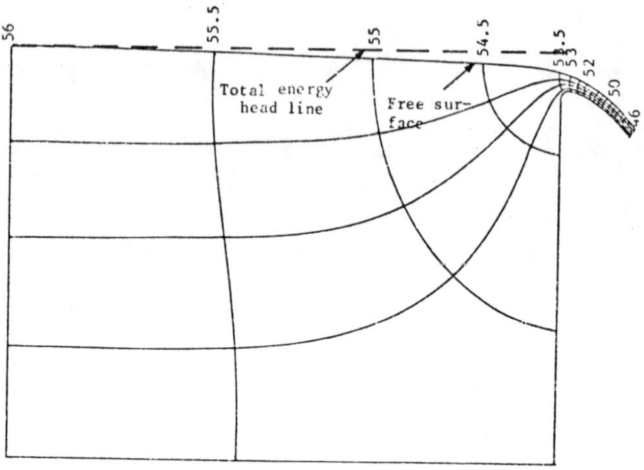

Figure 6 Flow over a spillway

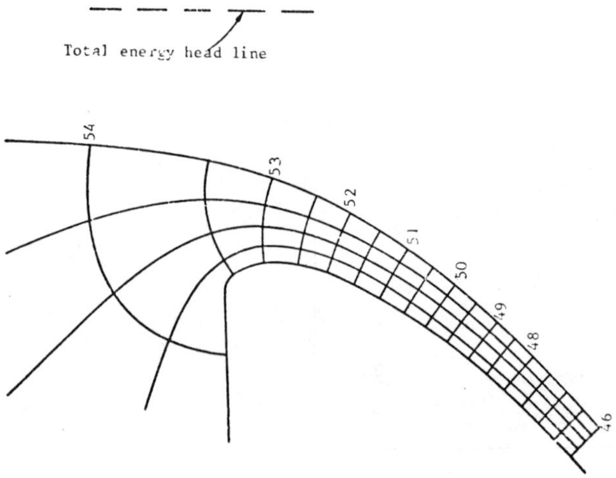

Figure 7 Detail of problem illustrated in Fig. 6

need to be computed during all iterations.

The discretization employed for the solution with boundary elements is shown in figures 4 and 5 where a larger number of elements were used over the crest to take into account the considerable difference between the hydraulic gradient in the reservoir and over the crest. The solution converges fast when starting with a reasonable approximation for the free surface profile. After the 3rd iteration the difference between the computed potential head and the elevation of each nodal point along the free surface was less than 0.08% of the elevation and the computer run was terminated (figures 6 and 7).

iii) <u>Flow in a pipe of elliptical cross-section</u>
The equation of motion of a uniform incompressible viscous fluid in steady unidirectional flow can be written as [9]:

$$-\frac{\partial p}{\partial z} + \mu\left(\frac{\partial^2 w}{\partial x^2} + \frac{\partial^2 w}{\partial y^2}\right) = 0 \tag{26}$$

where μ is the viscosity of the fluid, $\partial p/\partial z = -G$ is a constant pressure gradient and w is the velocity component on z direction. This equation can be seen as a Poisson-type equation:

$$\nabla^2 w = -\frac{G}{\mu} \tag{27}$$

For a pipe of elliptical cross-section the velocity distribution is of the form:

$$w = \frac{G}{2\mu} \cdot \frac{1}{(a^{-2} + b^{-2})}\left(1 - \frac{x^2}{a^2} - \frac{y^2}{b^2}\right) \tag{28}$$

where a and b are the semi axes of the ellipse.

The value of G/μ was taken to be equal to 2 and the semi axes a = 2 and b = 1. Hence the problem to be solved is:

$$\nabla^2 w = -2 \tag{29}$$

with the following analytical solution:

$$w = \frac{1}{5}(4 - x^2 - 4y^2) \tag{30}$$

In order to find the boundary element solution of the Poisson-type equation (29) the whole domain needs to be divided into cells in addition to discretizing the boundary with the usual elements. These cells, which must not be confused with finite elements as their inclusion does not introduce any internal unknowns, are needed to compute numerically the domain integral. The discretization adopted for the boundary element solution is shown in figure 8. The boundary condition is $w = 0$ over all the boundary.

In table 2, the boundary element results are compared against finite element results [10] and the analytical solution. The finite and boundary elements results for the velocity are in agreement with the analytical solution. The tangential stresses σ_{xz} and σ_{yz} however are proportional to the velocity derivatives and for them the finite element solution is not as accurate as the boundary elements one (tables 3 and 4).

iv) Viscous incompressible flow at low Re numbers

The governing equation for two-dimensional transient incompressible viscous flow (Navier-Stokes equation) can be written in terms of stream function ψ and vorticity ω as [11], [12]:

$$\frac{\partial \omega}{\partial t} + \frac{\partial \psi}{\partial y}\frac{\partial \omega}{\partial x} - \frac{\partial \psi}{\partial x}\frac{\partial \omega}{\partial y} = \nu \nabla^2 \omega \qquad (31)$$

where ν is the kinematic viscosity of the fluid. The relationship between vorticity and stream function is given by:

$$\omega = - \nabla^2 \psi \qquad (32)$$

and holds at any time t. Hence it can be differentiated with respect to time, i.e. [13]:

$$\frac{\partial \omega}{\partial t} = - \frac{\partial}{\partial t}(\nabla^2 \psi) \qquad (33)$$

Substituting equation (33) in (31) one obtains:

$$\frac{\partial}{\partial t}(\nabla^2 \psi) = \frac{\partial \psi}{\partial y} \cdot \frac{\partial \omega}{\partial x} - \frac{\partial \psi}{\partial x} \cdot \frac{\partial \omega}{\partial y} - \nu \nabla^2 \omega \qquad (34)$$

Using a first order approximation for the time derivative one can write:

$$\Delta(\nabla^2 \psi) = \Delta t \left(\frac{\partial \psi}{\partial y} \cdot \frac{\partial \omega}{\partial x} - \frac{\partial \psi}{\partial x} \cdot \frac{\partial \omega}{\partial y} - \nu \nabla^2 \omega \right)_t \qquad (35)$$

where Δt is the time increment.

The time cycle for time integration will therefore consist of solving the above equation (35) for the increment in stream function $\Delta\psi$ in a time Δt and then use equation (32) to solve for the vorticity at time $t + \Delta t$.

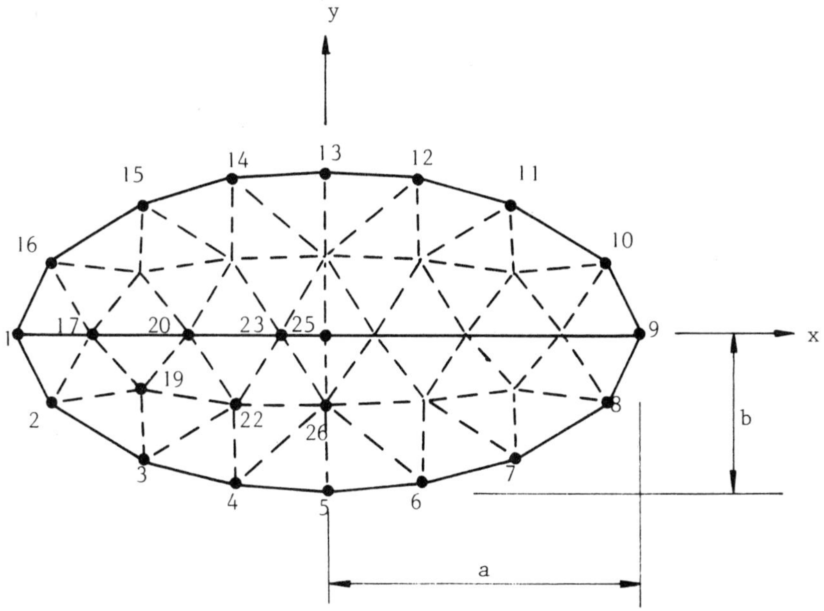

Figure 8 Boundary elements and cells for the elliptical section

Equation (35) can be interpreted as a non-linear Poisson-type equation, i.e.:

$$\nabla^2 \psi_{t+\Delta t} = p(\psi, \omega, t) \qquad (36)$$

In order to perform the solution with the boundary element method one begins with the potential problem ($\nabla^2 \psi = 0$) and then finds a vorticity distribution using equation (32). Next one integrates on time (36) to compute $\psi_{t+\Delta t}$ with the p term evaluated by integrating over the domain in the same way as before, i.e. by dividing the whole domain in cells and computing the integral numerically.

In this way during each iteration one only has to solve a system of equations such as (22), the vorticities being computed directly by equation (32). Notice that the matrices $\underset{\sim}{H}$ and $\underset{\sim}{G}$ remain constant during all the analysis. There is no need to specify boundary conditions for vorticities as the only unknowns in the system are stream functions.

4. CONCLUSIONS

The present paper discusses how boundary elements can be applied to solve fluid flow problems. The theory presented and the examples solved show that:

i) The boundary element method needs a smaller number of unknowns than the domain-type techniques (i.e. finite elements, finite difference, etc.) for similar solution accuracy. The advantages are more remarkable in three-dimensional applications where the number of unknowns for finite elements for instance may become prohibitive.

ii) The technique allows for a better representation of regions of high flux gradients due to the characteristics of the fundamental solution. Polynomial-type expressions such as those used in finite elements or simple finite difference expansions can instead be grossly inaccurate in those regions.

iii) The data preparation for boundary elements is much simpler than that required for domain techniques, as only the boundary points need to be prescribed. Even in those cases where cells are defined inside the domain (i.e. case with body forces type terms) their description is simple and does not involve concepts such as node numbering or connectivity.

iv) The method can be applied for non-linear problems such as those due to geometrical changes (free surface) or to time dependent non-linearities (Navier-Stokes flow) but more work is still required to solve a series of non-linear problems for which the known types of fundamental solutions are not adequate. It is hoped that in the next few years researchers will concentrate in solving these and other problems connected with the method.

5. ACKNOWLEDGEMENTS

The second author is indebted to C.N.Pq., Conselho Nacional de Desenvolvimento Cientifico e Tecnologico, Brasil, for the financial support provided during his studies at Southampton University.

REFERENCES

1. Jaswon, M.A., "Integral Equation Methods in Potential Theory: I", Proc. Royal Soc., A, 275, 1963.

2. Symm, G.T., "Integral Equation Methods in Potential Theory: II", Proc. Royal Soc., A, 275, 1963.

3. Jaswon, M.A. and Ponter, A.R., "An Integral Equation Solution of the Torsion Problem", Proc. Royal Soc., A,273, 1963.

4. Cruse, T.A., "Application of the Boundary Integral Equation Solution Method in Solid Mechanics", Var. Methods in Eng., edited by C.A. Brebbia and H. Tottenham, Southampton University Press, 1973.

5. Brebbia, C.A., "The Boundary Element Method for Engineers", Pentech Press, London, 1978.

6. Brebbia, C.A. and Dominguez, J., "Boundary Element Methods for Potential Problems", Applied Math. Modelling, vol.1, No.7, Dec. 1977.

7. Cassidy, J.J., "Irrotational Flow over Spillways of Finite Height", Proc. ASCE J. of Eng. Mech. Div., vol.91, Dec. 1965.

8. Brebbia, C.A. and Spanous, K.A., "Application of the Finite Element Method to Steady Irrotational Flow", Rev. Roum. Sci. Tech. - Méc. Appl., Tome 18, No.3, Bucarest, 1973.

9. Batchelor, G.K., "An Introduction to Fluid Dynamics", Cambridge University Press, 1967.

10. Brebbia, C.A. and Ferrante, A.J., "Computational Methods for the Solution of Engineering Problems", Pentech Press, 1978.

11. Smith, S.L. and Brebbia, C.A., "Improved Stability Techniques for the Solution of Navier-Stokes Equations", Appl. Math. Modelling, vol.1, No.5, June 1977.

12. Connor, J.J. and Brebbia, C.A., "Finite Element Techniques for Fluid Flow", Newnes-Butterworths, 1976.

13. Tong, P., "On the Solution of the Navier-Stokes Equations in Two-Dimensional and Axial-Symmetric Problems", Int. Symp. on Finite Element Meth. in Flow Problems, Swansea, 1974.

	30°		45°		60°	
POINT	EXP.	B.E.M.	EXP.	B.E.M.	EXP.	B.E.M.
1	0.29	0.28	0.35	0.31	0.43	0.36
2	0.33	0.31	0.38	0.35	0.47	0.41
3	0.37	0.35	0.42	0.39	0.49	0.44
4	0.41	0.39	0.46	0.43	0.52	0.47
5	0.46	0.45	0.51	0.50	0.57	0.54

Table 1 - Longitudinal velocities (m/sec) for curved channel

NODE	EXACT	B.E.M.	F.E.M.
17	0.350	0.334	0.341
19	0.414	0.401	0.392
20	0.638	0.626	0.627
22	0.566	0.557	0.561
23	0.782	0.772	0.789
26	0.638	0.629	0.665
25	0.800	0.791	0.793

Table 2 - Solution for the velocity w

NODE	EXACT	B.E.M.	F.E.M.
1	0.800	0.759	0.683
17	0.600	0.602	0.587
19	0.480	0.490	0.475
20	0.360	0.370	0.364
22	0.240	0.255	0.279
23	0.120	0.119	0.152
25	0.000	0.000	0.000

Table 3 - Solution for the derivative $\partial w/\partial x$

NODE	EXACT	B.E.M.	F.E.M.
5	1.600	1.611	1.209
19	0.720	0.720	0.750
22	0.720	0.715	0.699
26	0.560	0.562	0.550
25	0.000	0.000	0.000

Table 4 - Solution for the derivative $\partial w/\partial y$

INTEGRAL EQUATION SOLUTIONS TO NON-LINEAR FREE SURFACE FLOWS

P. L-F. Liu

Cornell University, Ithaca, N.Y. 14853

INTRODUCTION

The Boundary Integral Equation Method (BIEM) using the fundamental singularity is developed herein as an efficient numerical technique for solving nonlinear wave problems. The flow problems are formulated within the framework of potential theory. The initial boundary value problem is transformed into an integral equation by applying Green's formula. The fundamental solution to the Laplace equation, i.e. ln r for two-dimensional problems and 1/R for three-dimensional problems, is used as the kernel in the integral equation. The advantages of choosing the fundamental solution as the kernel are many-fold: (1) the Green function for nonlinear wave problems and the problems with complex geometry is not available in closed form, (2) the computations of the new kernel function involving ln r (or 1/R) and its normal derivative are simple and straight-forward, and (3) the phenomenon of irregular frequencies is unlikely to occur because of the new kernel.

The advantage of the BIEM is that only the variables at the boundaries are used in the computation. Hence it reduces the effective dimension of the problem by one; two-dimensional problems require only a line integration, and three-dimensional problems require a surface integration.

In this paper, the BIEM is used to solve two-dimensional nonlinear waves generated by either atmospheric pressures or ground movements. Excellent agreement has been found between numerical results by the BIEM and the available experimental data.

MATHEMATICAL FORMULATION OF THE FREE SURFACE GRAVITY WAVE PROBLEMS

The initial and boundary value problem describing free surface gravity waves is summarized in this section. Consider an inviscid, incompressible fluid domain R bounded by a free surface S_f and by a solid boundary S_b. As shown in Figure 1, Cartesian coordinates (x, z) embedded on the undisturbed free surface, z = 0, are employed. The gravity waves may be generated by (i) the sea floor movements, (ii) the atmospheric pressure on the free surface. On the other hand, the incident waves may be prescribed and the diffraction and refraction of water waves by natural boundaries and man-made structures can be studied. If the flow is assumed to be irrotational, the velocity $\underline{u}$ = (u, w) can be expressed as $\underline{u} = \nabla\phi$, where ϕ is the velocity potential.

The governing equation and boundary conditions are

$$\nabla^2\phi = \frac{\partial^2\phi}{\partial x^2} + \frac{\partial^2\phi}{\partial z^2} = 0, \text{ in the fluid R} \tag{1}$$

$$\frac{\partial\zeta}{\partial t} = |\nabla F_s| \frac{\partial\phi}{\partial n_s}, \tag{2}$$

$$\frac{\partial\phi}{\partial t} + \frac{1}{2}|\nabla\phi|^2 + g\zeta = -\frac{P_a}{\rho}, \tag{3}$$

(2) and (3) on the free surface $F_s = z-\zeta\ (x, t) = 0$

$$\frac{\partial h}{\partial t} + |\nabla F_b| \frac{\partial\phi}{\partial n_b} = 0, \quad \text{on the solid boundary } F_b = z+h\ (x, t) = 0 \tag{4}$$

where $\underline{n}_b$ and $\underline{n}_s$ are the unit normal vectors along the solid boundary and the free surface, respectively. If the domain R is unbounded at infinity, the radiation condition must be imposed, i.e. the scattered waves or the radiated waves must be outgoing waves.

For a prescribed ground movement or a prescribed atmospheric pressure on the free surface, the resulting free surface displacement, ζ, and the wave velocity field can be calculated from Equations (1)-(4) in principle. It is, however, an inherently difficult task due to the non-linear terms in the free surface boundary conditions and the fact that the position of the free surface is unknown prior to the solution of the problem. Remedies traditionally include applying the boundary conditions at an approximate surface location and then iterating, successively solving the complete problem and relocating the approximate surface. The inefficiency of such a numerical technique is obvious.

The boundary integral equation method is used to write equations for the location of discrete points on the free

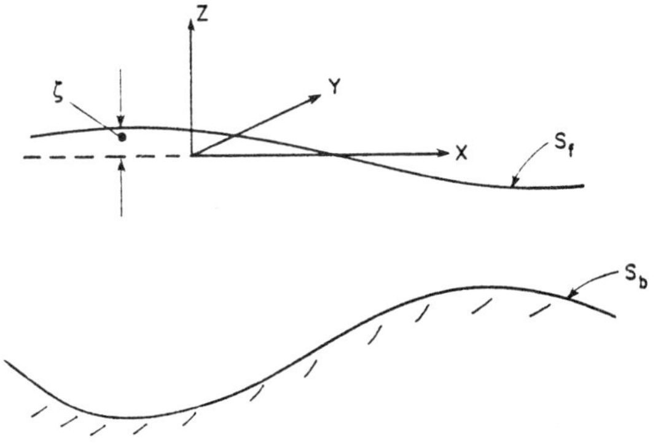

Fig. 1. Schematic drawing of the flow motion and symbol definitions

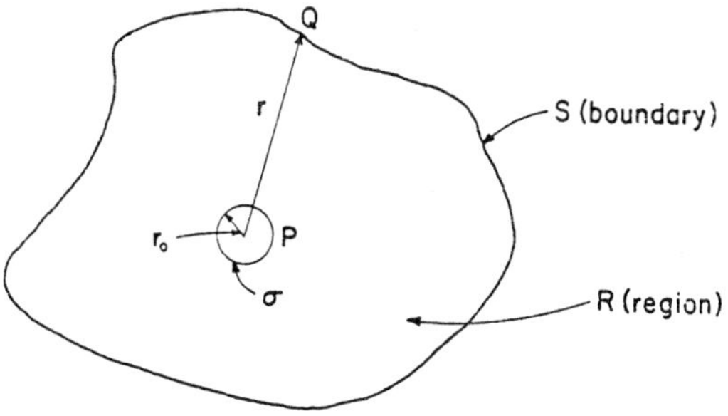

Fig. 2. The solution region.

surface. These equations depend only on the boundary data and thus the free surface can be located without solving the complete problem. Also, the resulting algebraic equations completely define the location of the free surface and no iteration is necessary for transient problems.

FORMULATION OF THE BOUNDARY INTEGRAL EQUATION METHOD

The BIEM is not a new technique. It is closely related to methods using Green's function (Wehausen, 1971; Berkhoff, 1973) and has recently gained considerable popularity in the solution of water wave problems with linearized free surface boundary conditions (e.g. Bai and Yeung, 1974; Yeung, 1977). More recently, the BIEM has been employed to solve nonlinear free surface groundwater problems (e.g. Liggett, 1977; Liu and Liggett, 1977; Liggett and Liu, 1977; Liu and Liggett, 1978).

The mathematical basis of the BIEM stems from Green's identity (see, e.g. Kellogg, 1929). Referring to Figure 2, p is a point inside the fluid region K which is bounded by the boundary S. For two-dimensional problems, the application of Green's second identity to the functions ϕ and ln r, which is the fundamental solution to Laplace equation, in the fluid region results in

$$2\pi\phi(p) = -\oint \frac{\partial\phi}{\partial n} \ln r \, dS + \oint \phi\frac{\partial}{\partial n} (\ln r) \, dS \qquad (5)$$

in which $r = \overline{PQ}$ and Q is a point on the boundary, $\underline{n}$ is the unit outward normal from R.

Letting the point p approach the boundary S, one obtains the following integral equation for ϕ and $\partial\phi/\partial n$ along the boundary S:

$$\alpha\phi(p) = -\oint \frac{\partial\phi}{\partial n} \ln r \, dS + \oint \phi\frac{\partial}{\partial n} (\ln r) \, dS \qquad (6)$$

where $\alpha(>0)$, in general, takes a value of π if p is on a line with continuous derivatives but otherwise is the value of the solid angle inside the fluid in the neighborhood of the point p (Liggett, 1977). The solution of Equation (6) yields the unknown value of either ϕ or $\partial\phi/\partial n$ along the boundary. The values of velocity potential can then be readily calculated from Equation (7).

TREATMENT OF NONLINEAR FREE SURFACE BOUNDARY CONDITIONS

To find the relation between ϕ and $\partial\phi/\partial n$ along the free

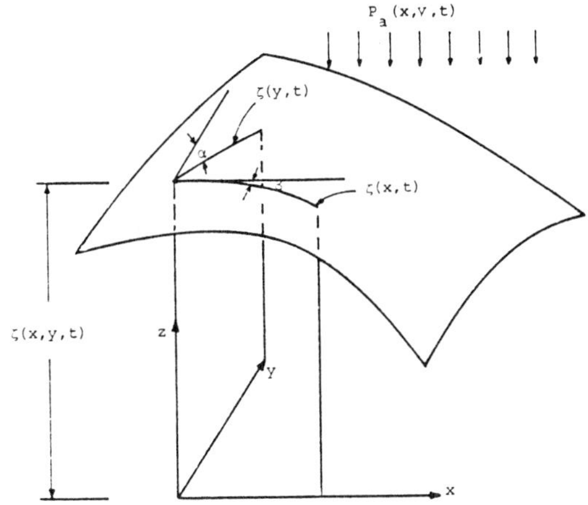

Fig. 3. Sketch of the free surface geometry and the coordinates system.

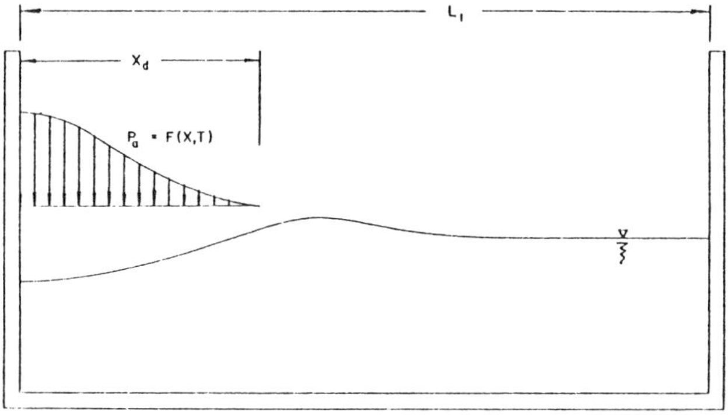

Fig. 4. Setup of the pressure pulse problem.

surface, special care must be taken. Referring to Figure 3, one can define

$$\frac{\partial \zeta}{\partial x} = - \tan \beta \tag{7}$$

where β is the angle the free surface makes with the x-axes. Substitution of Equation (7) into Equation (2) results in

$$\frac{\partial \zeta}{\partial t} = \frac{\partial \phi}{\partial n_s} / \cos\beta \tag{8}$$

Now, one may rewrite equations (3) and (8) in finite difference forms as follows

$$\phi^{k+1} = \phi^k - \Delta t \{ [(\frac{\partial \phi}{\partial x})^2 + (\frac{\partial \phi}{\partial z})^2]^k + \theta_1 [g\zeta + P_a/\rho]^{k+1} + (1 - \theta_1) [g\zeta + P_a/\rho]^k \} \tag{9}$$

and

$$\zeta^{k+1} = \zeta^k + \Delta t [\theta_2 (\frac{\partial \phi}{\partial n_s})^{k+1} + (1-\theta_2) (\frac{\partial \phi}{\partial n_s})^k]/\cos \beta^k \tag{10}$$

in which k is the number of the time step. θ_1 and θ_2 are weighting factors which choose between an explicit scheme ($\theta_{1,2} = 0$) and implicit scheme ($1>\theta_{1,2}>0$). Note that the angle β and $|\nabla\phi|^2$ are computed at the time step k even though the equations are written for the time level k + 1. Although this problem can be avoided by iteration, a small time step Δt may provide sufficient accuracy. Substitution of Equation (10) into Equation (9) yields the relationship between ϕ and ∂ϕ/∂n along the free surface.

$$\phi^{k+1} = g (\Delta t)^2 \theta_1 \theta_2 (\frac{\partial \phi}{\partial n_s})^{k+1} / \cos \beta^k + \phi^k - \Delta t \{ [(\frac{\partial \phi}{\partial x})^2 + (\frac{\partial \phi}{\partial z})^2]^k + g\zeta^k + \Delta t (1-\theta_2) \theta_1 g (\frac{\partial \phi}{\partial n_s})^k / \cos \beta^k + \theta_1 (P_a/\rho)^{k+1} + (1-\theta_1)(P_a/\rho)^k \} \tag{11}$$

EXAMPLES

Water Waves Generated By Simple Harmonic Surface Pressure

As an example, simple harmonic surface pressure was used to generate a train of oscillatory waves in a channel of uniform depth (Figure 4). The water is entirely at rest at t = 0 at which time an atmospheric pressure distribution

$$\frac{P_a(x,t)}{\rho} = \begin{cases} gh\,P_o \sin\left(\frac{2\pi t}{T_p}\right)\left[\frac{1+\cos(\pi x/x_d)}{2}\right], & 0<x<x_d \\ 0 \end{cases} \quad (12)$$

is applied to the free surface. Here ρghP_o is the amplitude of the pressure, T_p is its period and x_d denotes the region of the free surface subject to the pressure. For the numerical computations, a channel length $L_1/h = 30.0$. $P_o = 0.1$, $T_p = 7.6$ and $x_d/h = 4.0$ were used in Equation (12). Referring to Figure 4, the boundary of the problem is divided into 68 segments and therefore only 68 points are used in the computation. Linear variations for the potential ϕ and its normal derivatives are assumed between a pair of nodal points, P_j and P_{j+1}. The integral equation, Equation (6), can be reduced to a system of algebraic equations (Liggett, 1977)

$$[H]\{\phi\} = [G]\{\frac{\partial\phi}{\partial n}\} \quad (13)$$

where matrices [H] and [G] are numerical coefficients matrices.

Employing the free surface boundary condition, Equation (11), and requiring the condition of zero normal flux along the solid boundaries, $\partial\phi/\partial n = 0$, one can solve equation (13) for $\partial\phi/\partial n$ along the free surface and for ϕ along the solid walls at time step (k+1). The free surface displacement, ζ , can be determined from Equation (10).

The example of Figure 4 was computed assuming that flow starts at rest. $\Delta t = 0.5$ was used in the computations and the free surface profiles for successive times are plotted in Figure 5. The free surface fluctuations at different stations are displaced in Figure 6. We should note here that the same problem was studied by Chen et al. (1970) using the method of markers and cells. In their method the computation domain consisted of 80 x 24 = 1920 cells and the time step $\Delta t < 0.05$ had to be used. In addition they had to move the markers at each time step in order to determine the location of the free surface. The efficiency of the BIEM (68 points and $\Delta t = 0.5$) for this problem is obvious.

Tsunamis Generated by Sea Floor Movement

For the problems concerning the tsunamis generated by the vertical ground movements, the atmospheric pressure P_a is assumed to be zero in Equations (9) and (11). Along the moving sea floor, the boundary condition becomes

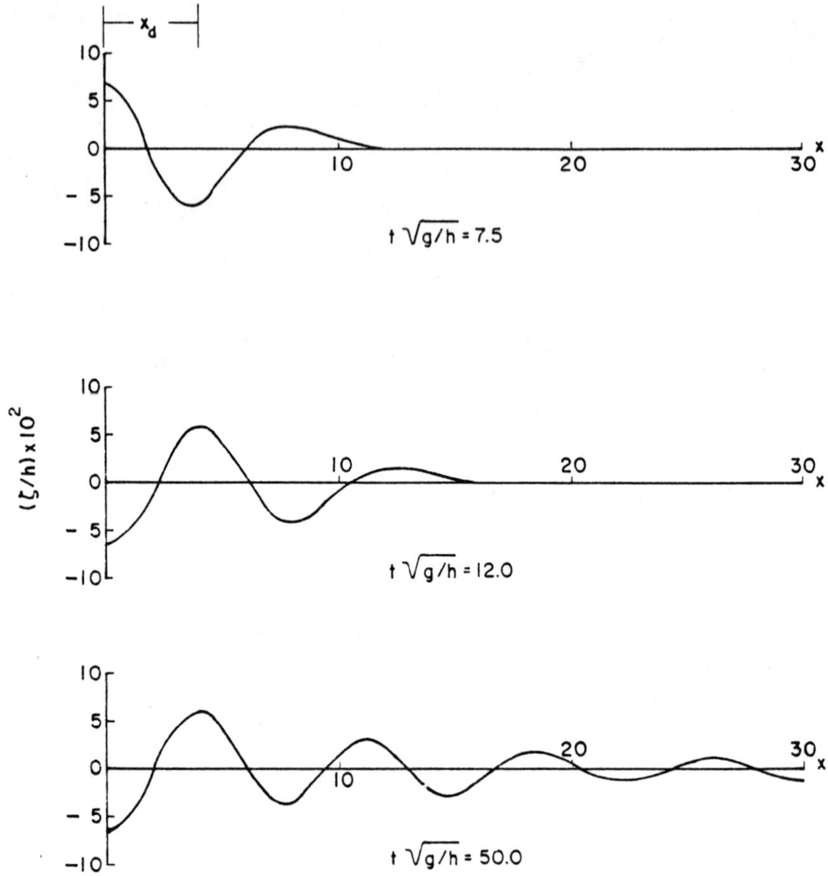

Fig. 5. Free surface profiles at different time.

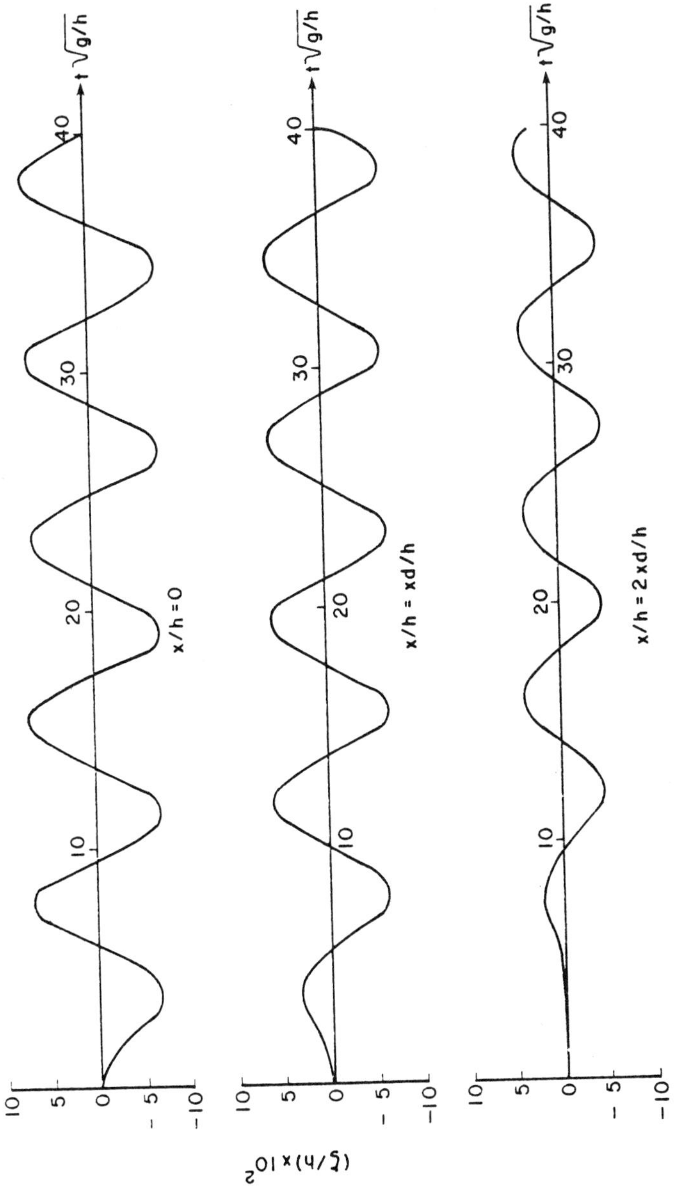

Fig. 6. Free surface fluctuation at different stations.

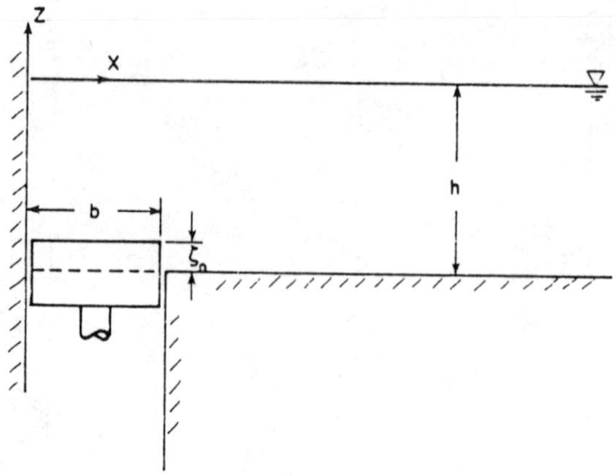

Fig. 7. A sketch of the wave generator.

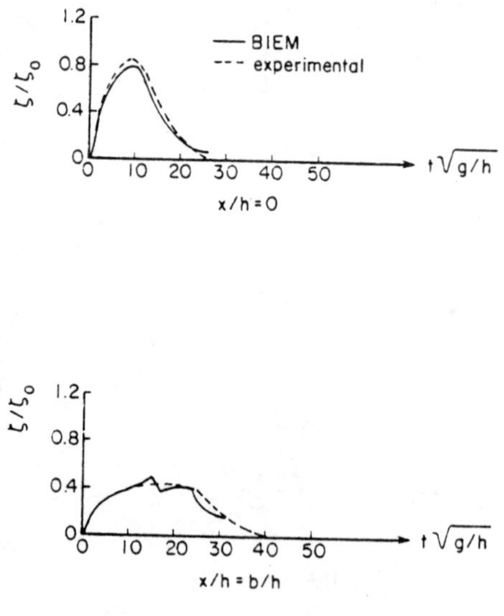

Fig. 8. Free surface displacement at different stations.

$$\frac{\partial \phi}{\partial n_b} = - \frac{\partial h}{\partial t} / \sqrt{1 + (\frac{\partial h}{\partial x})^2} \; , \; z = - h(x, t) \tag{14}$$

For a prescribed ground movement, i.e. h(x, t), the right hand side of the above equation is known.

For demonstration purposes, a simple example is presented here such that the numerical results obtained from the BIEM can be compared with the experimental data of a hydraulic model (Hammack, 1972). The geometry of the model is sketched in Figure 7. The fluid was at rest initially and the sea floor displacement was described mathematically by

$$h(x, t) = h_o - \zeta_o (1 - e^{-\alpha t}) \; H(b^2 - x^2), \; t>0 \tag{15}$$

where h_o is the constant water depth and $H(b^2-x^2)$ is the heavyside step function defined by

$$H(b^2-x^2) = \begin{cases} 1 & , \; b^2 - x^2 > 0 \\ 0 & , \; b^2 - x^2 < 0 \end{cases} \tag{16}$$

The variations of water surface with time are shown in Figure 8. The experimental data (Hammack, 1972) are also plotted in the same figure. Excellent agreements are observed.

CONCLUDING REMARKS

The Boundary Integral Equation Method (BIEM) is demonstrated as an efficient numerical technique for solving nonlinear water wave problems. Although the illustrations herein are for two-dimensional problems, the method is also efficient for solving three-dimensional problems, which requires surface integration instead of line integration.

REFERENCES

Bai, K. J., and R. W. Yeung (1974) "Numerical Solutions to Free Surface Flow Problems", 10th Naval Hydrodynamics Symposium, M.I.T. Cambridge, Mass.

Berkhoff, J. C. W. (1973) "Computation of Combined Refraction-Diffraction", Proc. 13th Coastal Engrg. Conf. Vol. 1, pp. 471-490.

Chan, R. K.-C., Street, R. L. and J. E. Fromm (1970) "A Summary of Studies of Long Waves by the SUMMAS Method", Proc. of the Symposium on Long Waves, U. of Delaware, Newark, DE.

Hammack, J. L. (1972) "Tsunamis - A Model of Their Generation and Propagation" Rep. No. KH-R-28, W. M. Keck Laboratory of Hydraulics and Water Resources, California Institute of Technology.

Kellogg, O. D. (1929) "Foundations of Potential Theory" Verlag Von Julius Springer, Berlin.

Liggett, J. A. (1977) "Location of Free Surface in Porous Media" J. of Hydraulics Division, ASCE, 103, 4.

Liggett, J. A. and P. L-F. Liu (1977) "Unsteady Free Surface Flow Through a Zoned Dam Using Boundary Integral" Symp. on Application of Computer Meth. in Engrg. L. A. California.

Liu, P. L-F. and J. A. Liggett (1977) "Boundary Integral Solutions to Groundwater Problems" Int. Conf. on Appl. Numer. Modeling, Southampton, U.K.

Liu, P. L-F. and J. A. Liggett (1978) "An Efficient Numerical Method of Two Dimensional Steady Groundwater Problems" Water Resrouces Research (in press).

Wehausen, J. V. (1971) "The Motion of Floating Bodies" Annual Review of Fluid Mechanics, Vol. 3, pp. 237-268.

Yeung, R. W. (1975) "Hybrid Integral-Equation Method for Time-Harmonic Free-Surface Flows" 1st Int. Conf. on Numer. Ship Hydrodynamics, Maryland, pp. 581-607.

FINITE ELEMENT FORMULATIONS OF FLOWS WITH SINGULARITIES

J. E. Akin

University of Tennessee, Knoxville, Tenn. 37916, U.S.A.
Brunel University, Uxbridge, Middlesex, UB8 3PH, England.

INTRODUCTION

Fluid flow problems often involve boundary conditions or geometry that introduces point or line singularities into the solution domain. These singularities have a pollution effect, on the overall accuracy, that can be overcome by giving proper consideration to the treatment of the singularity. This paper reviews the effects of the singularities, outlines various approaches for coping with them and shows typical numerical results.

FLOW SINGULARITIES

Most flow singularities are associated with the presence of re-entrant corners or edges in the flow domain. Insight into the general flow case can be obtained by examining certain special cases of engineering interest. For the sake of simplicity only two-dimensional problems with point singularities will be illustrated. The corresponding three-dimensional problems with line singularities can be clearly identified.

Darcy and Potential Flows

Of course, the simplest flow problems are modelled by Laplace's equation. One of the first flow problems to be solved by finite elements was that of Darcy flow seepage through a porous media. These problems often involve a sheet pile wall or dam foundation intruding into the flow domain as illustrated in Figure 1. As shown later, both of these introduce point singularities in the velocity field. These generally correspond to regions of maximum velocity and should be modelled as accurately as possible.

Potential flow problems are still of practical interest in their own right and also as intermediate approximations in more complicated problems. Both of these simple flow problems can be represented by Laplace's equation in the flow domain, Ω.

That is

$$\frac{\partial^2 H}{\partial x^2} + \frac{\partial^2 H}{\partial y^2} = 0 \quad , \qquad (x,y) \in \Omega \ . \tag{1}$$

When the boundary, $\partial\Omega$, contains a re-entrant corner having an internal angle of β the solution near the corner point is known to be of the form

$$H(r,\theta) = \sum_{i=0}^{\infty} a_i \, r^{in} \sin (in\theta) \tag{2}$$

where $n = \pi/\beta$ and r and θ are polar coordinates at the corner point. Thus the function H (hydraulic head, or velocity potential) remains finite at the corner but its derivative (the velocity) is singular and has an order of

$$\frac{\partial H}{\partial r} = 0(r^{n-1}) . \tag{3}$$

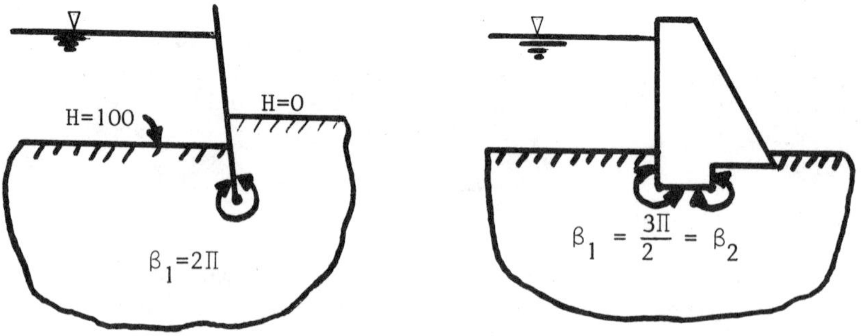

Figure 1. Typical Darcy Flow Geometries Introducing Singularities in The Velocity Field.

Referring to Figure 1 one notes that for the sheet pile and dam $\beta = 2\pi$ and $\beta = 3\pi/2$, respectively. Thus the corresponding flow singularities are $0(r^{-\frac{1}{2}})$ and $0(r^{-\frac{1}{3}})$, respectively. In the first case the standard error analysis for the problem is no longer useful. For example, Oden and Reddy, 1976, give an error estimate for a linear elliptic boundary value problem of order m:

$$||e|| = ||H-H_h||_{m,\Omega} \leq C_k h^{\mu} ||H||_{p,\Omega} \tag{4}$$

where $\mu = \min\,(k+1-m,\ p-m)$ (5)

and where h is the element size, k is the order of the interpolating polynomial, $m = 2$ is the order of the operator and p is the order of the highest derivative that is square integrable. Usually $p \geq 2m$ so that the error is governed by the polynomial order. That is usually $\mu = (k-1)$. Thus for a quadratic element $\| e \| = O(h^1)$. However, due to the singularity in this example $p = 1$ so the estimates suggests that $\| e \| = O(h^{-1})$. Thus the singularity is clearly polluting the accuracy of the solution. The above estimate indicates that increasing the order of the polynomial, k, will not in general improve the results when a singularity is present. This type of behaviour has been observed by Daley, 1973.

By considering the loss of regularity it is possible to obtain a sharper error estimate. This has been done by Schatz and Wahlbin, 1978. For Laplace's equation in a polygonal domain such as in Figure 2 they show one obtains different estimates in the interior, Ω_0, and near the corner regions, say Ω_j.

$$\Omega = \bigcup_{j=0}^{M} \Omega_j$$

$$0 < \beta_1 \leq \beta_2 \leq \ldots \leq \beta_M < 2\Pi$$

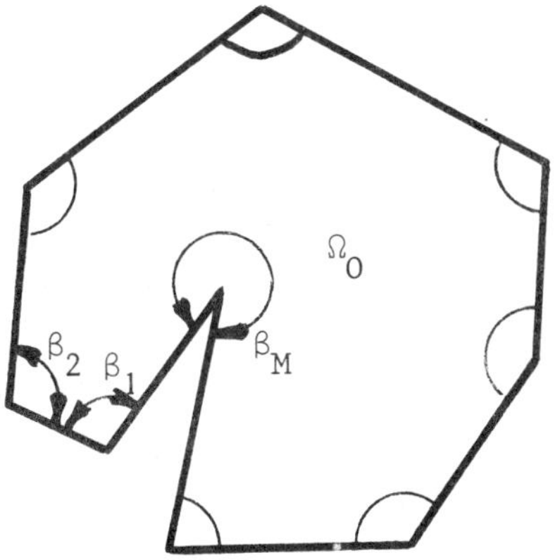

Figure 2. A Polygonal Domain with Singularities.

For a maximum interior angle of β_M these estimates give

$$\| e \|_j = \| H-H_h \|_{L_\infty(\Omega_j)} \leq C_j h^{\mu_j} \tag{6}$$

where

$$\begin{aligned} \mu_0 &= \min(k-1, 2\pi/\beta_M) \\ \mu_j &= \min(k-1, \pi/\beta_j, 2\pi/\beta_M) \ , \quad j > 0 \end{aligned} \tag{7}$$

Thus in the above example when the elements are quadratic and $\beta_M \doteq 2\pi$ one has an interior estimate of

$$\| e \|_0 = O(h) \tag{8}$$

while near the maximum singularity one has

$$\| e \|_M = O(h^{\frac{1}{2}}) \quad . \tag{9}$$

Near the other corners the estimates lie between these limits.
The size of the pollution region, Ω_M, is still unclear. It has been shown by Barnhill and Whiteman, 1973, and others that the full O(h) estimates can be recovered if one gives special treatment to the known singularities. This will be considered in a later section.

Creeping Flow

Another limiting case of interest in flow problems is that of creeping viscous flows. In that case one introduces a stream function, ψ, and the governing equation becomes the biharmonic equation:

$$-\Delta^2\psi = g \quad . \tag{10}$$

Then the velocity components are given by the derivations of ψ. In regions of re-entrant corners one again encounters derivative singularities, but in this case they appear in the second derivatives and not the first. Thus one wants to find procedures that can treat a known singularity in a specified derivative.

Navier-Stokes Flow

These well known flow relations can either be formulated in terms of primative variables or a stream function-vorticity model. In the latter case one obtains a set of nonlinear Poisson equations. Thus the same type of boundary singularities as in the potential flow case must be considered. Problems with such singularities have been discussed by Hutton, 1975. Studying the high Reynolds number flow through a step expansion he noted that "... the corner region played a major role in inhibiting the convergence...". He also suggested that "... removal of the corner singularity should increase the performance of the vorticity-stream function to an acceptable level".

Of course there are several other flow problems that with boundary singularities. For example shallow water circulation problems and tidal flow studies often involve geometries with sharp man-made, or natural, re-entrant corners. Space does not permit a detailed study of these cases. However, the above special cases can often be used to determine the type of singularity present.

METHODS FOR TREATING SINGULARITIES

To recover the standard convergence rate one must give special consideration to the treatment of the singularity. A number of procedures have been developed to accomplish this goal. These include grid refinement, function augmentation, distorted elements, special interpolation functions, etc.

A comparison of Equations 8 and 9 suggests that a systematic grid refinement procedure near the singularity may be fruitful. A disadvantage of this procedure is that the number of extra degrees of freedom can rapidly increase. One can also over-refine the mesh to the detriment of the accuracy away from the singularity. This has been observed by Whiteman, 1975.

If the exact form of the local solution is known it can be included in the finite element analysis. Such an approach was used by Hutton, 1974, in the analysis of flow near a reentrant corner. It has also be employed by Fix, Gulati and Wokoff, 1973, and by Barnhill and Whiteman, 1973. This procedure is not easily automated and can significantly increase the bandwidth of the equation. Thus other procedures seem more practical.

As an introduction to these procedures a brief review of element interpolation procedures will be given to define the notation and concepts. Generally the elements are defined in a local space, p-q, a global system, x-y. This is shown in Figure 3. In the usual isoparametric procedure the quantity

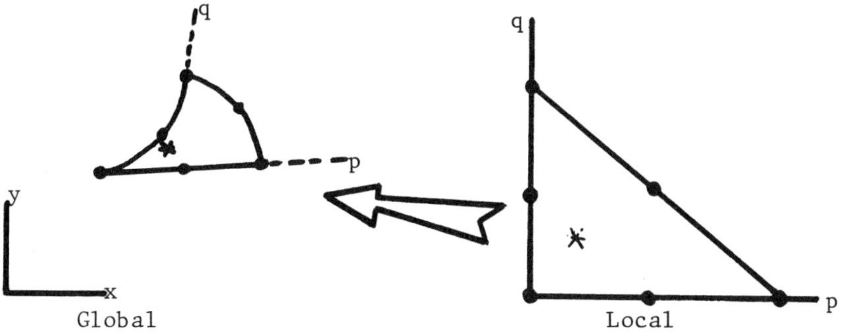

Figure 3. Local to Global Transformation.

of interest, say u, is interpolated from its nodal values. That is

$$u(p,q) = \sum_i N_i(p,q)U_i = [N]\{U\} \qquad (11)$$

and similarly the global coordinates are defined by their nodal values:

$$t(p,q) = \sum_j M_j(p,q)\ T_j = [M]\{T\}\ , \quad t = x,y\ . \qquad (12)$$

For an isoparametric element one has $M_i = N_i$.

Clearly one can evaluate the local derivatives of u, x and y. Denote the local derivatives as

$$\{\partial_\ell(\ )\}^T = \left[\frac{\partial(\ )}{\partial p} \quad \frac{\partial(\ )}{\partial q}\right] . \tag{13}$$

Then

$$\{\partial_\ell u\} = [\partial_\ell N]\{U\} \quad . \tag{14}$$

However, the singularity and governing integral statement involves the global derivatives. These two sets of derivatives are related by the chain rule of calculus. For example, recall that $\partial u/\partial p = (\partial u/\partial x)(\partial x/\partial p) + (\partial u/\partial y)(\partial y/\partial p)$, etc. Expressing these identities in matrix form gives

$$\{\partial_\ell u\} = [J]\{\partial_g u\} \tag{15}$$

where $\{\partial_g u\}^r = [\partial u/\partial x \quad \partial u/\partial y]$ denotes the desired global derivatives and [J] is the Jacobian of the coordinate transformation. That is,

$$[J] = [\{\partial_\ell x\} \vdots \{\partial_\ell y\}] = [\partial_\ell M][X \vdots Y] \tag{16}$$

and is a known quantity. By inverting Equation 15 one obtains the desired results, namely

$$\{\partial_g u\} = [J]^{-1}\{\partial_\ell u\}$$

or

$$\{\partial_g u\} = [J]^{-1}[\partial_\ell N]\{U\} \quad . \tag{17}$$

Therefore, the global derivatives are defined in terms of the product of two known quantities. If one wishes to introduce a known global derivative singularity then Equation 17 shows that one must work with the local derivatives, $\{\partial_\ell u\}$, or the coordinate transformation. Generally one is singular and the other is regular. These two approaches will now be illustrated.

Special Interpolation

Introducing the singularity through the local derivative requires special element interpolation functions. A number of special elements have been developed for the form $u = 0(r^{\frac{1}{2}})$ which occurs frequently in stress analysis (Blackburn, 1973). However, most such elements can not be generalized to other singularities. The first conforming element having a general form of $u = 0(r^a)$ was given by Akin, 1976. This models at least the leading term singularity in a first or second derivative term. This procedure is valid for, and compatible with, an element that satisfies the usual condition that $\sum_i N_i = 1$. This easily programmed procedure is outlined here.

If the singularity occurs at node k one defines a function

$$w(p,q) = 1 - N_k(p,q) \tag{18}$$

which is zero at node k and equal to unity at all other nodes.

To obtain the power singularity one can divide by a function $R = W^a$. If N_i denotes the standard functions, then the modified singular functions, H_i, are defined as

$$H_i = [1 - 1/R]\delta_{ki} + N_i/R \quad . \tag{19}$$

These also satisfy $\sum_i H_i = 1$ and near node k give a solution of the form

$$\phi(r,\theta) = u_k + c(\theta)r^{(1-a)} \tag{20}$$

so that the first derivative is of order

$$\frac{\partial\phi}{\partial r} = O(r^{-a}) \quad . \tag{21}$$

Thus if $a > 0$ the first derivative is singular and if $a < 0$ the higher derivatives can be singular. Of course, if $a = 0$ the standard element is obtained. The local derivatives are easily obtained from the chain rule (Akin, 1976).

a) Akin Modified Quadratic

$$N_1 = 1 + (2p^2 - 3p)/R$$
$$N_2 = (2p^2 - p)/R$$
$$N_3 = (4p - 4p^2)/R$$
$$R = (3p - 2p^2)^a$$

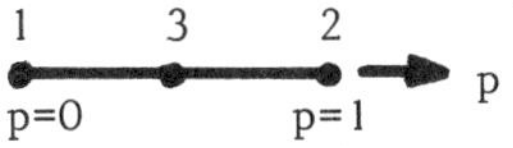

b) Tracey Element and Akin Modified Linear Element

$$N_1 = 1 - p^a$$
$$N_2 = p^a$$

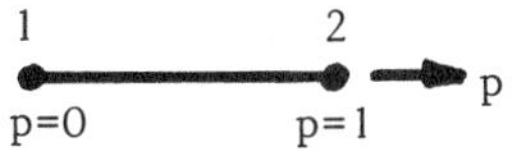

c) Stern Element

$$N_1 = 1 + [(c^a-1)p + (1-c)p^a]K$$
$$N_2 = [-c^a p + cp^a]K$$
$$N_3 = [p-p^a]K$$
$$K = 1/(c-c^a) \ , \quad c = 1/2 \quad , \ a \neq 1$$

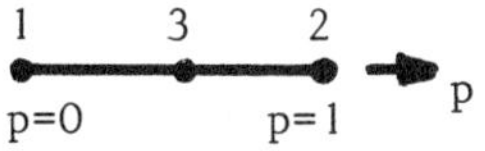

Figure 4 Typical 1-D Examples for Power Law Singularities

A three node singularity triangle has been presented by Tracey and Cook, 1977. It contains terms of $O(r^a)$ and is compatible with the standard linear triangle. Stern, 1978, has generalized this concept to a (5+K) node triangular element. The additional K nodes are introduced on the non-singular side so as to be compatible with surrounding standard isoparametric elements. It contains terms of both $O(r)$ and $O(r^a)$.

The one-dimensional forms of these three approaches are illustrated in Figure 4. In one-dimension, the Tracey element corresponds to the linear element modification of Akin. The procedures of Tracey and Stern can be generalized to other elements. Usually triangular elements are better for singularity applications since they can maintain straight line mappings. However, quadrilateral elements can also be useful. A new singularity quadrilateral element is given in Figure 5. If one sets a = 2 it reduces to the standard eight node isoparametric element. With all of the above special functions one retains standard geometric interpolation.

$$N_1 = 1 - (p+q)(1+\frac{K}{2}) + pq(1+K) + \frac{K}{2}(p^a+q^a-pq^a-qp^a)$$
$$N_2 = p(1-\frac{K}{2}) + pq(\frac{K}{2}-3) + \frac{K}{2}(1-q)p^a + 2p^2q$$
$$N_3 = pq(2p+2q-3)$$
$$N_4 = q(1-\frac{K}{2}) + pq(\frac{K}{2}-3) + \frac{K}{2}(1-p)q^a + 2pq^2$$
$$N_5 = K(1-q)(p-p^a)$$
$$N_6 = 4pq(1-p)$$
$$N_7 = 4pq(1-q)$$
$$N_8 = K(1-p)(q-q^a)$$

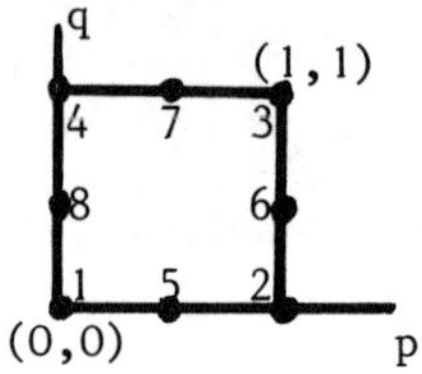

$K = 1/(C - C^a)$
$C = 1/2$
$a = 2$ for standard element.

Figure 5 A quadrilateral for $u = b_1 + b_2r + b_3r^a$.

Jacobian Manipulation

Returning to Equation 16 one notes that there are also two methods to manipulate [J] and thus $[J]^{-1}$. These are associated with distorted elements and strained coordinates. Henshell, 1975, was among the first to note that some desired singularities can be generated through the use of distorted elements. Generally the order of the singularity is governed by the order of the geometric interpolation functions, [M]. One can control the location of the singular point by moving nodes in the global coordinates. The singular point usually lies outside the element and one must force it to occur on an element boundary or node.

To illustrate this concept consider a one-dimensional quadratic element shown of length h. The coordinate transformation of a standard element with an interior node at ah is

$$x(p) = h(4a-1)p + 2h(1-2a)p^2 \tag{22}$$

and the Jacobian is

$$J = \frac{\partial x}{\partial p} = h(4a-1) + 4h(1-2a)p \quad . \tag{23}$$

Thus if one wants a singularity to occur at p = 0, i.e. node 1, one must set $a = \frac{1}{4}$. That means that the global coordinate of node 3 must be specified so that $x_3 = h/4$. Then one has $x = hp^2$ so that $p = (x/h)^{\frac{1}{2}}$ and the inverse Jacobian becomes

$$J^{-1} = \frac{\partial p}{\partial x} = \tfrac{1}{2}(hx)^{-\frac{1}{2}} \tag{24}$$

which has a singularity of order $0(x^{-\frac{1}{2}})$ at x = 0. The derivative singularity can also be seen by noting that the quadratic interpolation gives

$$\begin{aligned} u(p) &= \alpha_1 + \alpha_2 p + \alpha_3 p^2 \\ &= \alpha_1 + \beta_2 x^{\frac{1}{2}} + \beta_3 x \quad . \end{aligned} \qquad (25)$$

As the above example illustrates, the algebra involved in this type of procedure can be rather involved in higher dimensional elements.

Equation 16 shows that a nonlinear transformation, [M], could also generate some types of singularity. This will be referred to as the method of strained coordinates (Akin, 1977). For simplicity consider a line element with two nodes. Using standard interpolation for u gives

$$u(p) = [N]\{U\} = c_1 + c_2 p \quad . \qquad (26)$$

Next use a nonlinear (strained) coordinate mapping of the form

$$x(p) = [M]\{X\} = x_1 + (x_2 - x_1)p^a \quad . \qquad (27)$$

if $x_1 = 0$ and $x_2 = h$ then $x = hp^a$. Solving for p(x) and substituting into Equation 26 gives

$$u = c_1 + c_2(x/h)^{1/a} \qquad (28)$$

so that selecting a = 3/2 would give a singularity of $u' = 0(x^{-1/3})$.

TYPICAL NUMERICAL RESULTS

Since most practical flow problems involve singularities ranging from $r^{-2/3}$ to $r^{-1/2}$ these two common limits will be considered. Assume a region with a re-entrant slit that involves potential flow or Darcy flow. In such a case one will be solving Laplace's equation for H with a $r^{-1/2}$ singularity. Various solutions are available for comparison, but most involve simple geometries. Consider a rectangle with quarter symmetry as shown in Figure 6.

Figure 6. A Rectangular domain with an $r^{-1/2}$ singularity.

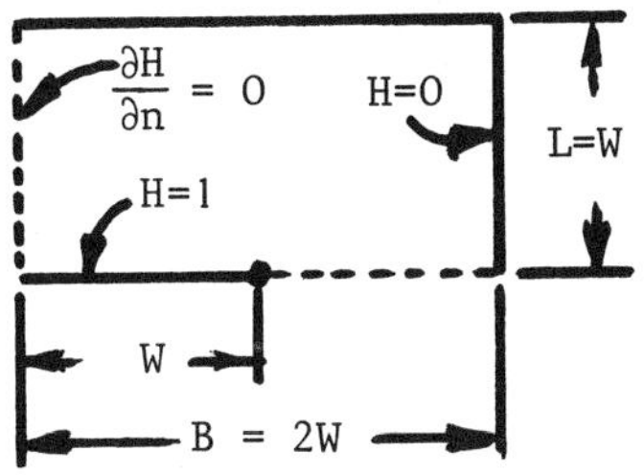

This provides a useful comparison for three reasons. First, it has a significant singularity. Secondly, the energy

$$E = \int_{\Omega} \left[(H_{,x})^2 + (H_{,y})^2 \right] da$$

has a known value and can be used as a error norm. Finally, a finite element solution of this problem which shows the effects of continued mesh refinement has been given by Daley, 1973.

Daly considered a region proportioned such that B = 2L = 2W which B and L denote the width and height of the outer boundary and W is the width of the centre boundary. In that case, the exact value of the energy is E = 1.46922. This value provides a useful norm for comparing the accuracy of various solutions. In his analysis Daly established a mesh and then obtained separate solutions using linear, quadratic, and cubic triangles which had common corner node locations. His uniform grid solutions will be compared with the Akin procedure which was based on bilinear quadrilaterals and quadratic triangles. Results showing the percent error in the energy are presented in Table I. The right hand column and the first two lines represent the present solution and the other values are from Daly.

Note that the non-singular higher order element solutions given by Daly converge rather slowly for uniform meshes. This is due to the predominance of the singularity error. To show the significance of the singularity error the problem was first run with quadratic triangles having a corner mesh grid of 3×5 points. When no special attention is given to the singularity the results are over four percent in error. Yet, when four of the sixteen elements are modified to account for their being adjacent to the singular point, the error is reduced to less than one percent and is better than any of the values obtained by mesh refinement alone.

For the first crude mesh the quadratic elementds modified to reflect the presence of the singularity occupy a fourth of the total solution domain. The ability of the quadratic and higher order elements to accurately model both the angular and radial behaviour of the singularity makes them quite important for problems with known singular points. Thus, it is recommended that at least quadratic elements be used when possible and that they be modified to account for the presence of the singularity.

Of course, the modification procedure of Akin can be applied to lower order elements but the improvements are less dramatic. For the linear triangle and bilinear quadrilateral the angular variation along edges not containing the singular point, are approximated linearly.

To illustrate the accuracy of the simpler element the above problem was solved using the bilinear quadrilateral and the same number of total degrees of freedom. As shown in

Table I
Percent Error in Energy

Corner Nodes	No. Sing. Elements	Type of Element LT	QT	CT	BQ
3 × 5	0	--	4.02	--	--
3 × 5	4	--	0.76	--	--
5 × 9	0	7.00	4.01	--	3.73
5 × 9	2	--	--	--	3.03
7 × 13	0	4.57	2.66	2.01	--
8 × 15	0	--	--	--	2.15
8 × 15	2	--	--	--	1.75
9 × 17	0	3.39	1.29	--	--

Uniform Grid
B = 2L = 2W
LT = Linear Triangle
QT = Quadratic Triangle
CT = Cubic Triangle
BQ = Bilinear Quadrilateral

Table I the error was 3.73% when the singularity was not included. However, this decreased to 3.03% when two elements adjacent to the singularity were modified. There were 32 elements in that mesh, so the singularity was directly considered in only 6% of the total area. The results clearly show the benefit of accounting for known singularities, especially when the elements are quadratic or higher.

To illustrate a milder $r^{-1/3}$ singularity, and compare local grid refinement with singularity modifications consider and L-shaped region considered by Whiteman, 1975, and Akin, 1976. Both used bilinear quadrilaterals in a finite element solution of the Laplace equation. The standard finite element results were compared against singularity modifications, using Akin's procedure, and local grid refinement near the corner. The latter employed five node bilinear transition quadrilaterals developed by Whiteman. Full details are given in the references. Here, only the solutions near the singularity will be considered. They are compared in Figure 6 with an 'analytic' solution obtained by numerical conformed mapping by Papamichael and Sideridis, 1978.

As expected, both singularity modifications and local grid refinement significant improvements in the accuracy of the finite element solutions. In this case, the modification procedure is more cost effective. In general it would probably be best to have limited local grid refinement, say two or three levels, combined with a singularity modification.

CONCLUSIONS

It has been shown that singularities in the flow fields are common in problems of engineering interest. Several finite element methods for accounting for the singularities have been reviewed. These procedures have been shown to be easily im-

P = Point Number
Q = Bilinear Quadrilaterals, 79 DOF
RQ = Q with five local refinements, 144 DOF
SQ = Q with three modified Q, 79 DOF
NCM = Numerical Conformal Mapping

P	Q	SQ	RQ	NCM
26	8725	8820	8824	8843
37	8409	8565	8553	8586
48	8496	8467	8448	8487
27	8440	8528	8535	8553
38	7953	8095	8112	8154
49	7928	7948	7923	7961
28	8135	8200	8199	8210
39	7456	7557	7553	7565
50	6635	6667	6667	6667
56	5118	4887	4894	4870
62	3674	3601	3603	3580
29	7864	7901	7895	7898
40	7043	7070	7077	7066
51	6048	6026	6039	6020
57	4891	4835	4811	4881
63	3637	3571	3569	3550
30	7658	7673	7677	7672
41	6768	6772	6787	6772
52	5780	5766	5776	5757
58	4693	4663	4659	4642
64	3541	3507	3499	3486

P	
Q	SQ
RQ	NCM

h = 1/10

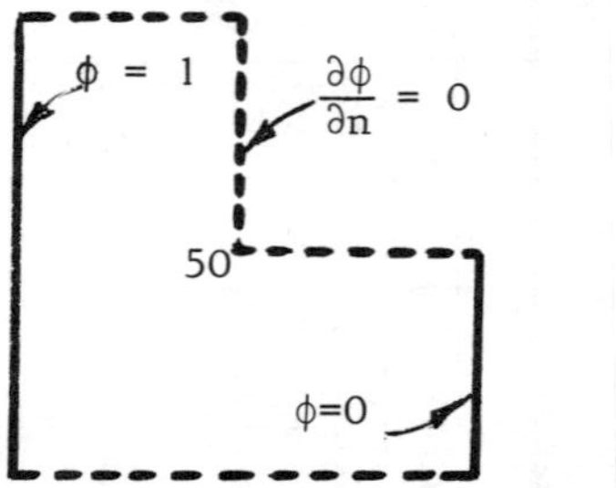

Figure 7. Comparison of Standard Elements, Local Refinement, and Singularity Modifications Near A Singularity of $O(r^{-1/3})$.

plemented and at the same time yield improved accuracy.

Similar problems exist in three-dimensional problems. However, the order of the point and line singularities may not be known exactly and may even vary with position. All of the above procedures are applicable for three-dimensional problems. Some typical three-dimensional singularity elements have been presented (Akin, 1978). Due to the increased cost of the three-dimensional solutions singularity modifications, or Jacobian manipulations will probably be more cost effective than local grid refinement.

ACKNOWLEDGEMENTS

Portions of this work were supported by the UK Science Research Council grant number GR/A45784 and the US National Science Foundation grant number SMI77-17378. The use of facilities at the Institute of Computational Mathematics, Brunel University, and the Texas Institute for Computational Mechanics, University of Texas at Austin, are also gratefully acknowledged.

REFERENCES

Akin, J.E. (1976) Generation of Elements with Singularities. I.J.Numer.Meth.Eng., 10, 1249-1259.
Akin, J.E. (1976) Finite Element Analysis of Fields with Boundary Singularities. Proc.Int.Conf.Num.Meth. in Elect. and Magnetic Field Prob., 61-72.
Akin, J.E. (1977) Physical Bases for the Design of Special Finite Element Interpolation Functions. Proc.Soc.Eng.Sci. Annual Mtg.
Akin, J.E. (1978) Elements for the Analysis of Line Singularities. Proc. 3rd Int.Conf.Math. of Finite Elements and Applics.
Barnhill, R.E. and Whiteman, J.R. (1973) Error Analysis of Finite Element Methods with Triangles for Elliptic Boundary Value Problems. Mathematics of Finite Elements and Applications, J.R. Whiteman (Ed), Academic Press, 83-112.
Blackburn, W.S. (1973) Calculation of Stress Intensity Factors at Crack Tips using Special Finite Elements. Mathematics of Finite Elements and Applications, J.R. Whiteman (Ed), Academic Press, 327-336.
Daly, P. (1973) Singlarities in Transmission Lines. Mathematics of Finite Elements and Applications, J.R. Whiteman (Ed), Academic Press, 337-350.
Fix, G.J., Gulati, S. and Wakoff, G.I. (1973) On the Use of Singular Functions with Finite Element Approximations. J.Comp. Physics, 13, 209-228.
Henshell, R.D. and Shaw, K.G. (1975) Crack Tip Elements are Unnecessary. I.J.Num.Meth.Eng., 9, 495-509.
Hutton, A.G. (1974) On Flow Near Singular Points of a Wall Boundary. Finite Element Methods in Flow Problems, Oden, J.T., et al(Eds), University of Alabama Press, 67-83.

Hutton, A.G. ((1975) A General Finite Element Model for Vorticity and Stream Function Applied to a Laminar Separated Flow. Central Electric. Generating Board, UK, Report RD/B/N3050.
Oden, J.T. and Reddy, J.N. (1976) An Introduction to the Mathematical Theory of Finite Elements, John Wiley & Sons.
Papamichael, N. and Sideridis, A. (1978) The Use of Conformal Transformations for the Numerical Solution of Elliptic Boundary Value Problems with Boundary Singularities. Technical Report TR/74, Mathematics Department, Brunel University.
Schatz, A.H. and Wahlbin, L.B. (1978) Maximum Norm Estimates in the Finite Element Method on Plane Polygonal Domains. Math. of Comp., to appear.
Stern, M. (1978) Families of Consistent Conforming Elements with Singular Derivative Fields. I.J.Num.Meth.Eng., to appear.
Whiteman, J.R. (1975) Numerical Solution of Steady State Diffusion Problems Containing Singularities. Finite Elements in Fluids, Oden, J.T. et al (Eds), 2, 101-119

COMPARISON OF FINITE ELEMENT AND FINITE DIFFERENCE METHODS IN THERMAL DISCHARGE INVESTIGATIONS

L. D. Spraggs

McGill University, Montreal, Quebec, Canada H3A 2K6

INTRODUCTION

Ultimately, the objective of scientific research into numerical methods must be to provide workers in the field with reliable credible tools which can be used to analyse complex problems. In the field of water resources, there is the further need to be able to assess potential environmental impacts before irreversible modifications are made to existing water systems. In addition, the analysis methodology must be credible, economical and achievable in a reasonable period of time. In this study, the use of numerical simulation models for analysing the impact of a proposed thermal-electric power plant on the proposed cooling reservoir are investigated.

The underlying objective of this present study is to provide an estimate of the thermal regime and subsequent evaporation for a cool-fired power plant with a potential for producing 1200 MW. However, the secondary objective was to determine whether existing finite element models or existing finite difference models could be used more effectively in meeting the stringent time frame imposed on completion of the study. It is this second objective which will be considered here. It is not the purpose of this present paper to make an assessment of the environmental implications of the proposed power plant. Rather, the proposed power plant system is to be the media to make a comparative study of the significance of existing, available, numerical models.

An initial investigation of the reservoir, shown schematically in Figure 1, indicated a potential for two possible hydrodynamic regimes. First, it was possible that the reservoir would stratify,resulting in a structure of layer homogeneity. Second, it was possible that, because of the relatively high persistent winds and the large thermal effluent, the reservoir

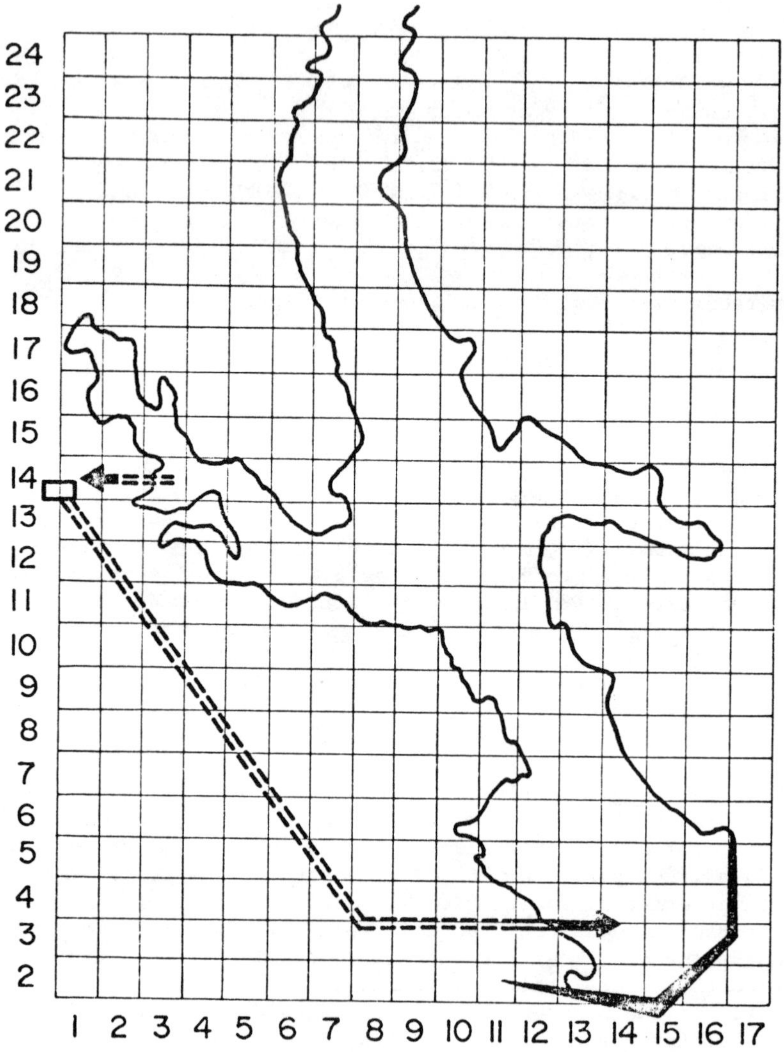

Figure 1. Poplar River Reservoir

would at any location in the reservoir be homogeneous throughout its depth. Therefore, two types of integrated models would be required to determine the expected hydrodynamic regime for the reservoir.

Before continuing with the model exposition, it is necessary to review peripheral information with relevance to this study. The power plant at the reservoir is to begin operation in the early summer of 1979 at which time it is expected that the reservoir will finish filling for the first time. Hydrologic flows in the river are small and average approximately 12,500 acre-feet for the period of record from 1931 through 1976. Most of the runoff into the reservoir occurs in a relatively short period of time in the spring, thus resulting in a closed system for the remainder of the year. Water quality in the reservoir is generally good, However, due to the small inflows and due to plant treatment there is a potential for an accumulation of TDS in general and Boron in particular.

Based on the above, two conclusions can be drawn about the methodology required for the study. First, as there is no possibility of obtaining comparative data, every effort will have to be made to generate collaborating information. Second, because of the small external inflow and because of the small reservoir, errors in the temperature prediction will have a pronounced effect on the volume of water in the reservoir because of the increased evaporation. Also, as the surface of the reservoir begins to drop, the potential for short-circuiting in the reservoir will increase, resulting in a further elevation in temperature and an increase in evaporation.

The period of record for the current study is 1931 through 1976. Simulation of the hydrodynamic regime for this extended period is economically unjustifiable and philosophically questionable. It was felt that simulating the steady state hydrodynamics of the reservoir under two unit (600 MW) plant operation with no hydrologic input would be representative of the questionable summer period. Consequently, the above philosophy was adopted for the two types of models previously mentioned and discussed subsequently.

MATHEMATICAL MODELS

The mathematical models considered for use in the study are presented in this section. As the basic concepts and model developments are fully documented elsewhere, the interested reader is directed to the references for a more complete discussion. Here the concern is with the use of the models and the information generated rather than on the theoretical basis of the models.

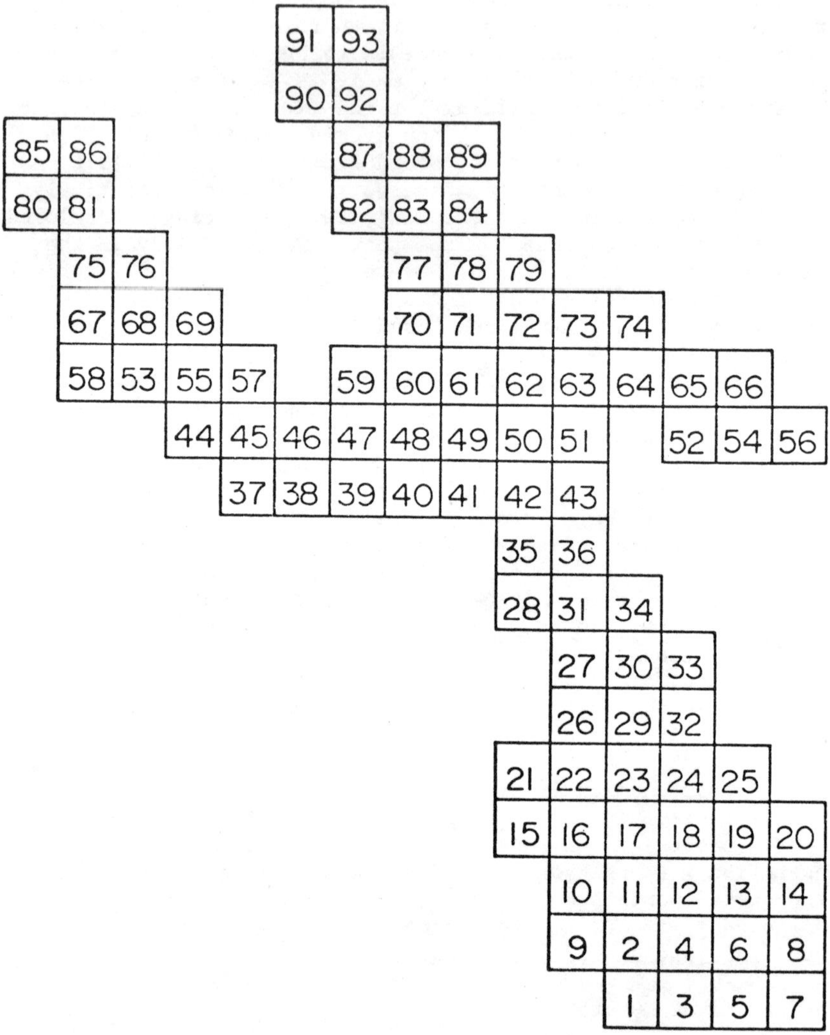

Figure 2. Discrete Mesh For Models

Two-Dimensional Vertically Integrated Finite Difference Model (HT2DV)

For the analysis of vertically mixed bodies of water, Edinger et al, 1972, have developed a two-dimensional vertically integrated model which solves for the velocity field and thermal regime simultaneously. The modified model (HT2DV) was taken as being representative of the type of finite difference model that could be used for an analysis of this kind.

The HT2DV model is variable in both space and time and it can be used to simulate the temperature and evaporation on a daily basis. Unfortunately, the cost of simulating for an extended period of time is large and in general precludes the use of the model for lengthy simulations. Therefore, the HT2DV model was used to simulate the hydrodynamics of the reservoir and to predict the excess temperature field for the case when the hydrologic flow into the reservoir was negligible.

The excess temperature in the reservoir was found using the HT2DV model and then the predicted temperature was found by incrementing simulated monthly temperatures. In this particular study, the reservoir was initialized to 20°C and the heat flux at the surface was determined to maintain this reference temperature. When the simulation was made with the power plant operational the excess temperature could be found as the difference between the reference temperature and the simulated temperature.

Two-Dimensional Finite Element Model

The objective of rapidly ascertaining the hydrodynamics and thermal regime of the reservoir using an existing finite element program was considered. However, it became clear at an early stage that if such a program did exist it was not readily available. Consequently, an attempt was made to use existing simple theory combined with a simplified heat transport model to solve the problem. If the approach was reasonable then subsequent work would involve refinement and development of a good analytical tool.

As a first attempt at solving for the hydrodynamics of the reservoir it was decided to use the isoparametric quadrilateral element STIF55 from the ANSYS library (Desalvor and Swanson, 1973). A model of the reservoir corresponding to the exact shape used in the finite difference simulation is shown in Figure 2. Assuming that the concept of potential flow is valid in this case then an analogy can be drawn between the heat flow which is actually being solved and the velocity potential which is what we desire. Notice in Figure 2 that Nodes 9 and 58 are inserted to simulate the source and sink respectively. While obvious, the analogy is given below for comparative purposes:

Heat Flow	Fluid Flow	
$\frac{\partial^2 T}{\partial x^2} + \frac{\partial^2 T}{\partial y^2} = 0$	$\frac{\partial^2 \phi}{\partial x^2} + \frac{\partial^2 \phi}{\partial y^2} = 0$	(1)

The temperature of the exterior nodes of elements 9 and 58 are chosen such that the analogous fluid flow into and out of the reservoir will correspond to the plant requirements. Consequently, the fluid flow potential field will be solved when the corresponding temperature distribution is obtained.

Because we are interested in extending the model to include time-dependent situations the identical mathematical model was selected and solved using Galerkin's method.
At this juncture the transition to triangular elements was made and the inlet and outlet were refined to reduce the possibility of numerical problems. Due to the time constraint the model was limited to the simple steady state case discussed above with a provision to simulate the excess temperature regime in the reservoir.

The expected thermal regime in the reservoir can be computed using a very simple model proposed by Edinger, 1971. Using a surface heat exchange coefficient (K) which accounts for all the various heat transfer mechanism at the air-water interface, the rate of change of reservoir temperature with respect to area can be defined as

$$\frac{\partial T}{\partial A} = \frac{-K}{Qp}(T - Tn) \qquad (2)$$

in which Qp is the plant pumping rate and Tn is the natural lake temperature. Rearranging Equation 2, integrating and dividing by the specific weight and heat of the water leads to

$$\Theta_i = C_1 e^{-KAi/Qp} \qquad (3)$$

where C_1 is a constant to be determined, Θ_i is the average excess temperature in the zonal area under consideration (a_i) and Ai is the accumulated area along the main path up to and including zone i, i.e.,

$$A_i = \sum_{j=1}^{i} a_j \qquad (4)$$

Evaluation of the constant of integration C_1 is a natural consequence of the operating design of the proposed plant.

This simplified model for the excess temperature would be used in conjunction with the finite element model and average monthly temperatures to quickly evaluate various operating techniques. The zones of considerations can be delineated by the equipotential lines in the simulation of the velocity field. While this method of analysis will result in some peripheral analysis now, a computer program could easily be devised to automatically calculate the position of the equipotential lines and the area between those lines. For this study our objective was the practical implementation of existing techniques to provide credible economical estimates of the particular problem at hand. Thus, extensive developmental work was precluded by the imposed practical and economic constraints.

Summary

In summary, this paper presents a discussion of the applicability of existing finite element and finite difference models for simulating the forced evaporation from a cooling reservoir. It is not an objective here to draw conclusions about the environmental implications of the proposed plant. Rather, the objective is to determine if there are readily available numerical models that can be used to solve the problem.

RESULTS

The results of the finite difference simulation for the velocity distribution are shown in Figure 3. These results are basically what would be expected for a steady state condition. It is of interest that a considerably larger area is active for the two unit case than for the one unit case.

The corresponding simulation for the two dimensional finite element model using the same grid as the finite difference simulation is shown in Figure 4. Also shown on this figure are the equipotential lines delineating the zones of cooling in the reservoir. The area between pairs of these lines is assumed to be of homogeneous heat content. This figure does not show the mesh refinement used to simulate the velocity at the inlet and outlet.

From these two simulations it was possible to calculate the expected excess temperature distribution for a steady state condition. Figure 5 indicates the predicted excess temperatures for the two unit case, calculated using the finite difference technique.

Figure 6 is the analogous case calculated using the finite element formulation and the simple heat transport model discussed earlier. The average monthly temperature calculated using the finite element model is shown in Table 1 with the corresponding evaporation.

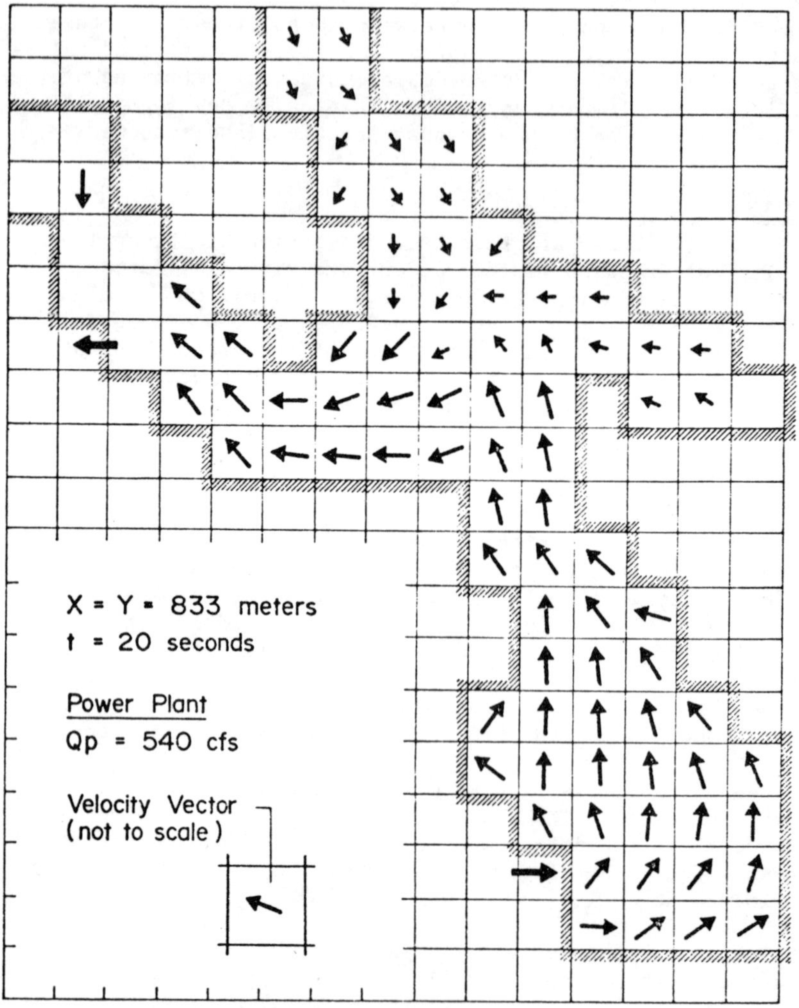

Figure 3. Finite Differences - Velocity Field

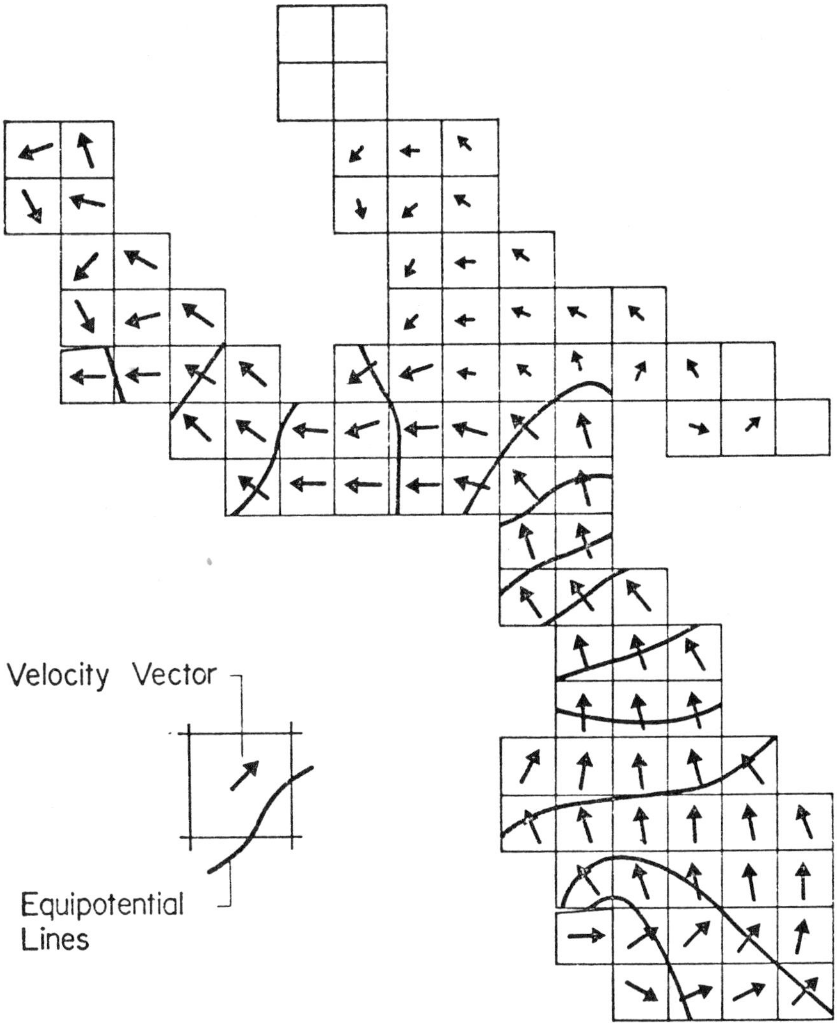

Figure 4. Finite Elements - Velocity Field

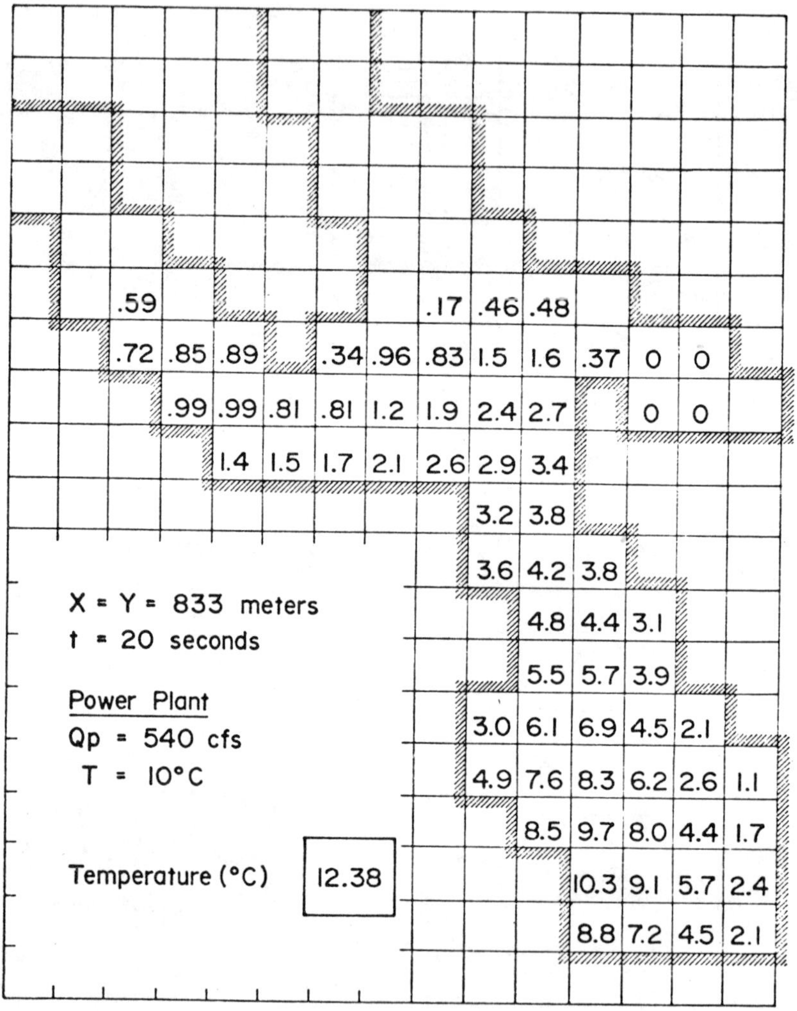

Figure 5. Finite Difference - Excess Temperature

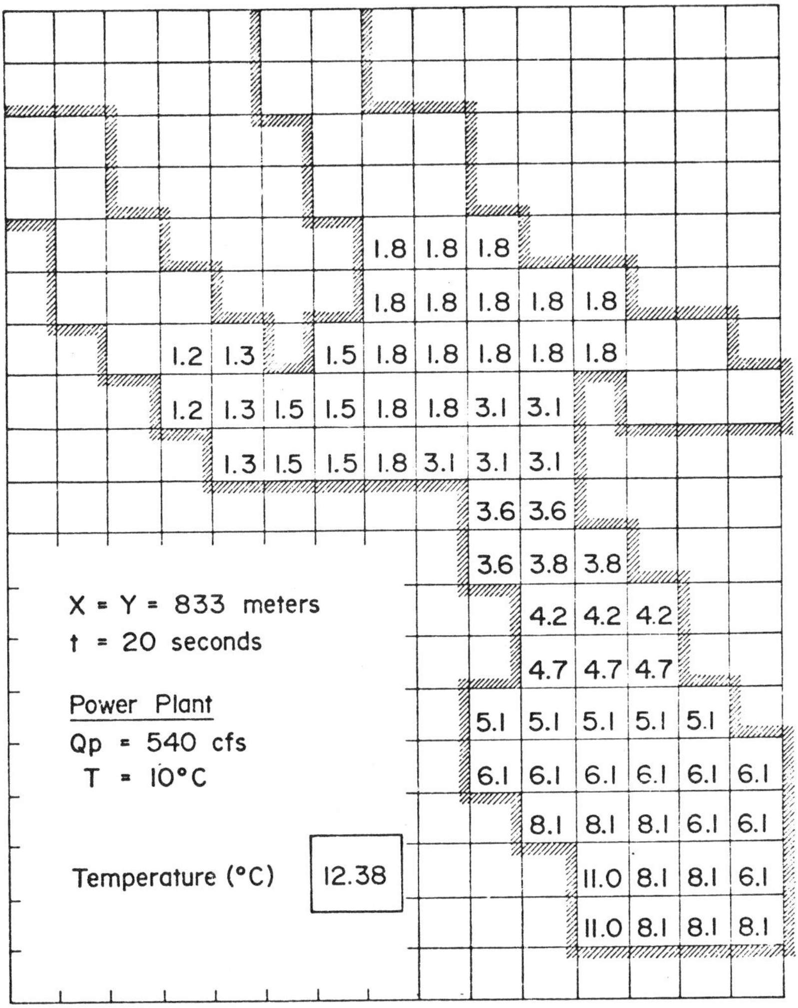

Figure 6. Finite Elements - Excess Temperature

Month	Zone 1		Zone 3		Zone 5		Zone 7		Zone 9		Zone 11		Zone 13	
	Ts	He	Ts	He	Ts	He	Ts	He	Ts	He	Ts	He	Ts	He
Jan	-	-	-	-	-	-	-	-	-	-	-	-	-	-
Feb	-	-	-	-	-	-	-	-	-	-	-	-	-	-
Mar	-	-	-	-	-	-	-	-	-	-	-	-	-	-
Apr	15.95	29.8	10.07	76.7	8.68	21.9	7.85	19.8	7.06	29.8	5.55	23.3	5.2	13.8
May	21.95	45.4	16.07	117.5	14.68	33.5	13.85	30.4	13.06	45.8	11.55	35.8	11.2	21.20
June	27.6	54.1	21.67	14.2	20.28	40.7	19.45	36.9	18.66	55.8	17.15	43.8	16.8	25.9
July	31.95	66.7	26.07	181.9	24.68	52.7	23.85	48.27	23.06	73.63	21.55	59.0	21.2	35.2
Aug	28.05	51.0	22.17	131.3	20.78	37.3	19.95	33.7	19.16	50.6	17.66	39.0	17.3	23.0
Sept	23.65	46.9	17.77	121.3	16.38	34.6	15.55	31.3	14.76	47.2	13.25	36.8	12.90	21.8
Oct	17.55	32.0	11.67	79.9	10.28	22.5	9.45	20.2	8.66	30.1	7.2	22.9	6.80	13.5
Nov	12.95	26.6	7.07	74.1	5.68	21.7	4.85	20.1	4.06	30.8	2.55	25.2	2.2	15.1
Dec	-	-	-	-	-	-	-	-	-	-	-	-	-	-

Table 1. Finite Element Prediction

Finally, in Table 2 a comparison is made between the predicted evaporation for the two different models. Notice in particular that the evaporation for the four winter months has been calculated elsewhere from a heat budget approach. While these four months may contain considerable error, they will not influence the comparison between the two procedures.

	m/mo Natural	Ac/ft Natural	Finite Element	Finite Difference
Jan	0	0	131	131
Feb	0	0	131	131
Mar	0	0	118	118
Apr	.039	207.3	213	116
May	.086	457.1	324	300
June	.151	802.6	386	333
July	.206	1094.9	459	378
Aug	.178	946.1	374	352
Sept	.135	717.5	336	312
Oct	.083	441.1	238	269
Nov	.070	372.1	174	161
Dec	0	0	131	131
Total	0.95	5039	3015	2732

Table 2. Evaporation Predictions

CONCLUSION

A comparison of the results indicates that the very simple finite element model used for this study produces information which is as good as the more sophisticated finite difference model. The velocity fields produced by the two models are nearly identical even though two totally different approaches were used. The predicted excess temperatures for the reservoir are in general very close. There is some difference in the temperatures calculated for the regions away from the inlet but this is probably due to the fact that the finite difference model has not achieved a steady state condition. Indeed, comparison with a simpler finite difference model which calculated the transient case for a year shows remarkable agreement with the calculations obtained from the finite element case.

While a very simple finite element model has been used for an actual practical application, two points are notable. First, the state-of-the-art of finite element modelling is far ahead of the availability of these models for practical application. Second, given a simple finite element model, it can be used effectively and economically to evaluate the effect of large thermal flows on a relatively small impoundment. Finite difference models do exist for analysing these problems but in general they are expensive to use and they cannot easily accommodate subtleties in the boundary configuration.

Already, work is progressing towards enhancing the model used herein, to make it more useful as a managerial tool. In the first place, a more sophisticated heat equation is being included with the ability to include the average monthly meteorological conditions. In addition, the model is being changed to operate on a pseudo time-dependent basis. That is, the model will operate on a yearly basis but it will calculate the average monthly steady state conditions for the imposed hydrological, meteorological and physical conditions (plant operating characteristics). It is expected that the revised model will allow the assessment of environmental implications of proposed operating schedules on a practical, economic basis.

This study has had an objective of implementing an existing finite element model for a practical real problem. However, it is clear that the technological advancements in the finite element method are far ahead of practical, attainable models. Consequently, instead of being able to use the existing technology one must use quite simple models unless sufficient time and money are available to implement and test the more sophisticated models which have been published.

REFERENCES

DeSalvo, G. J. and J. A. Swanson, "ANSYS: Information Systems Manual", Computer Science Corp. Manual No. E00173-01, Jan. 1973.

Edinger, J. E., Buchak, E. M., Kaplan, E., and G. Socratous, "Generic Emergency Cooling Pond Analysis", University of Pennsylvania, School of Engineering and Applied Science, Civil Engineering, Rept. No. COO-2224-1, Oct. 1972.

Edinger, J. E., "Shape Factors for Cooling Lakes", ASCE, J. Power Division, PO-4, Dec. 1971, pp. 861-867.

NUMERICAL SMOOTHING TECHNIQUES APPLIED TO SOME FINITE ELEMENT SOLUTIONS OF THE NAVIER-STOKES EQUATIONS

R. L. Lee, P. M. Gresho

Lawrence Livermore Laboratory, University of California, Livermore, CA 94550

and R. L. Sani

University of Colorado, Boulder, CO 80302

SUMMARY

Two categories of smoothing techniques which generate continuous approximations (i.e., nodal values) of vorticity and pressure from finite element solutions of the Navier-Stokes equation are considered. The simplest schemes, developed for quadrilateral elements, are those based on combinations of linear extrapolation and/or averaging algorithms which bring element-wise Gauss point evaluations out to node points. More complicated schemes based on a global smoothing technique which employ the mass matrix (consistent or lumped) are also presented.

An initial assessment of the accuracy of several schemes is obtained by comparing with an analytical function. Next, qualitative comparisons are made from numerical solutions of the steady state driven cavity problem. Finally, applications of smoothing techniques to discontinuous pressure fields are demonstrated.

INTRODUCTION

In finite element problems, it is often necessary to display fields which are described by discontinuous functions - resulting in nonunique values at node points. For hydrodynamic problems, several examples in which such situations arise are: (i) Those finite element formulations of the Navier-Stokes equations which employ piecewise continuous (C^0) basis functions for velocity generate vorticity and stress fields (from derivatives of velocities) that exhibit discontinuities across element boundaries; (ii) The lowest order consistent mixed-interpolation in the primitive variable (u,v,p) formulation (Huyakorn, et al., 1978) requires velocities to vary linearly while pressures are represented as piecewise constants ("defined" at the element centroid rather than at nodes) within elements; (iii) Any of the family of elements advocated by Malkus and Hughes (1978) wherein the pressure is recovered at the appropriate Gauss points by a post-processing procedure applied to the "penalty" term. For each of the described cases, rational and consistent procedures are required in order to generate unique values of the discontinuous variable at node points; i.e., a continuous nodal representation of an initially discontinuous function is desired.

Practioners of finite element methods have frequently developed their own simple schemes for extracting first derivative information from node point values of the primary dependent variable calculated. Typical examples are stresses from displacements, heat flux from temperature, and vorticity from velocity fields. Apparently, to date, no serious effort has been made to document such algorithms and to assess their accuracy in a systematic manner, perhaps, in part, owing to their often ad hoc nature. For structural problems, it has been established (Barlow, 1976) that stresses computed at the appropriate Gauss points have higher accuracy than at any other points. Thus, stresses are normally reported with confidence only at those points. However, a mesh of non-nodal points, such as Gauss points, is generally inconsistent with the requirements of most contour-plotting packages. For this reason, only those schemes which generate values at node points for quadrilateral elements are considered in this paper.

The application of least squares techniques for the smoothing of discontinuous stress fields via the finite element basis functions was considered by Hinton (1968) and Oden and Brauchli (1971). In a subsequent paper (Hinton and Campbell, 1974) global and local smoothing schemes for stress calculations based on a least squares formulation are discussed and compared. It was concluded that local stress smoothing in conjunction with nodal averaging is preferred, the alternative (global smoothing) being much more expensive with no significant gain in overall accuracy. Theoretical justification of local smoothing was also suggested showing that, when "reduced integration" is applied, the scheme is simply a bilinear extrapolation of the more accurate 2 x 2 (in two dimensions) Gauss point stresses.

Little attention has heretofore been devoted to the development of schemes useful for smoothing fields other than stress, such as heat flux or vorticity. It is generally implied that the techniques for stress smoothing would be applicable to these other cases. However, it has yet to be established whether Gauss point values have special significance for predicting quantities unrelated to stresses. Nevertheless, these are the natural sampling points in the finite element methodology. Therefore, several simple smoothing schemes which employ Gauss point extrapolations and/or averaging are considered in this paper. It is to be noted that although only two-dimensional problems are presented in this study, the schemes may all be extended to three dimensions in a straightforward manner. Similarly, whereas the examples here are taken from fluid mechanics applications, the techniques obviously apply to other areas of application.

VORTICITY SMOOTHING

Solutions of the Navier-Stokes equations with the primitive equation formulation are given as velocities and pressures at node points (or Gauss points if the pressure field is discontinuous). In many instances, one is interested in the post-processing of these solutions to obtain the associated vorticity field. The following procedures, developed in the spirit of the finite element methodology, generate such a vorticity field.

The two-dimensional velocity approximations over an element are given by:

$$u(x,y) = \sum_{i=1}^{n} \phi_i(x,y)\, u_i$$

$$v(x,y) = \sum_{i=1}^{n} \phi_i(x,y)\, v_i \tag{1}$$

where u_i, v_i are the nodal values of the two velocity components, ϕ_i is the set of basis functions with n being the total number of nodes for an element. The vorticity field can be calculated from Eq. (1) as follows:

$$\omega \equiv \frac{\partial v}{\partial x} - \frac{\partial u}{\partial y} = \sum_{i=1}^{n} \left(\frac{\partial \phi_i}{\partial x} v_i - \frac{\partial \phi_i}{\partial y} u_i \right) \tag{2}$$

When piecewise continuous (C°) basis functions are used for ϕ_i, ω is discontinuous at element boundaries and some sort of smoothing is required if unique values at node points are to be obtained. Several schemes for smoothing of these discontinuous functions are now described.

Global Smoothing

A global approximation of the smoothed vorticity field may be represented as:

$$\omega^*(x,y) = \sum_{i=1}^{N} \Phi_i(x,y)\, \omega_i^* \tag{3}$$

where Φ_i is a set of global basis functions and ω_i^* are the (unknown) N nodal values. Although Φ_i may be of a different order than the basis functions associated with the velocities, in this study they are chosen to be the same. For a given unsmoothed vorticity field $\omega(x,y)$, the smooth function which gives a best fit in the least squares sense over a domain Ω is obtained from a minimization of the functional:

$$J = \iint_{\Omega} (\omega^* - \omega)^2 \, dA \quad . \tag{4}$$

This procedure, which requires that

$$\frac{\partial J}{\partial \omega_i^*} = 0 \qquad \text{for} \qquad i = 1,2,\ldots,N$$

results in a linear system of equations given by

$$\sum_{j=1}^{N} M_{ij}\omega_j^* = f_i \qquad ; \qquad i = 1,2,\ldots,N \tag{5}$$

where M_{ij} (the mass matrix) and f_i are global quantities formed by assembling element level results,

$$M_{ij}^e = \iint_{\Omega_e} \phi_i \phi_j \, dA \tag{6}$$

and

$$f_i^e = \iint_{\Omega_e} \omega\phi_i \, dA \tag{7}$$

Note that this result is equivalent to applying a Galerkin weighting procedure, using (3), to an arbitrary given function, $\omega(x,y)$. The ϕ_i's of Eq. (6) are now the element basis functions (for velocity) and ω is given by Eq. (2). In order to obtain ω^*, the solution of the linear system of Eqs. (5) must be obtained. If, however, a lumped mass matrix [obtainable at element level from either row-sum or diagonal scaling (Gresho et al., 1978)] is used, the solution of Eq. (5) becomes almost trivial. It may be shown that either mass lumping scheme, applied to rectangular elements, leads to a form of area-weighted averaging for the nodal values.

Local Smoothing

In contrast to global smoothing, local smoothing techniques attempt to smooth functions using individual element information only. Some form of nodal averaging is finally required, however, since unique values at node points are desired. In accordance with several algorithms suggested for smoothing stresses, the local techniques used in this paper are based on a linear extrapolation and/or averaging of Gauss point values. Described below are several methods of generating nodal values for each individual 4-node (bilinear) or 8- or 9-node (quadratic) element. In each case, simple (nodal) averaging based on the number of elements joined at a given node is employed to obtain the final result at that node.

Smoothing Schemes for Bilinear Elements

(a) The vorticity in each element [from Eq. (2)] is computed at the 2 x 2 Gauss points and the result averaged. The averaged value (originally assumed to apply at the centroid of the element) is then assigned to each of the four corner nodes of the element.

(b) The centroidal value obtained from scheme (a) is linearly extrapolated through each Gauss point value to the corresponding corner node.

Smoothing Schemes for Quadratic Elements (8- or 9-node)

(c) Vorticity sampled at the 2 x 2 Gauss points are assigned to each nearest corner node.

(d) As described for scheme (b) except 3 x 3 Gauss points are used (which includes the centroidal value) and the results are available at 9 nodes rather than 4.

(e) A bilinear extrapolation through the 2 x 2 Gauss point values (Hinton et al., 1975) is used to generate corner values.

Schemes (a) and (c) are recommended by Hallquist and Goudreau for stress smoothing (1977). Schemes (b) and (d) were suggested by Callabresi (1978) who noted that a similar technique is employed by the MARC Finite Analysis Program (1976) with the exception that a more sophisticated weighted nodal average (based on the angles at the node) is used. The smoothing schemes, listed in decreasing order of computational complexity, are summarized in Table 1. This ordering is also used in all subsequent figures.

NUMERICAL RESULTS

Two examples are now considered to demonstrate the accuracy of these schemes applied to 4- and 9-node elements. A hypothetical flow field described by an analytical stream function is first presented. With this, an exact vorticity field (computed analytically) is compared against numerically smoothed vorticities generated from exact nodal velocity values. Next, a finite element solution of the Navier-Stokes equation for the driven cavity problem is used to compute the vorticity field. The results are then compared with those from a finite differ-

TABLE 1. Summary of various smoothing schemes.

Scheme	Form of Smoothing	Element Application	Number of Node Values Calculated	Algorithm
1	Global	Linear or Quadratic	4 or 9	Least squares smoothing with consistent mass matrix.
2	Global	Linear or Quadratic	4 or 9	"Area-weighted" smoothing with lumped mass matrix.
3	Local	Quadratic	4	Bilinear extrapolation through 2 x 2 Gauss point values [Scheme (e)].
4	Local	Linear or Quadratic	4 or 9	Linear extrapolation from centroid through 2 x 2 (linear) or 3 x 3 (quadratic) Gauss points [Schemes (b) and (d)].
5	Local	Quadratic	4	2 x 2 Gauss point values applied to nearest corner node [Scheme (c)].
6	Local	Linear	4	Centroidal value applied to corner nodes [Scheme (a)].

Note: For the 8-node quadratic element, Schemes 1, 2 and 4 could be used to generate 8 nodal values, if desired (in the obvious ways). 9-node results may be more compatible with plotting packages, however.

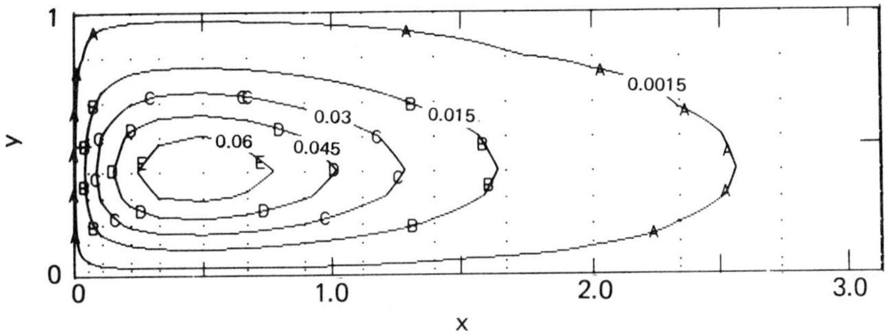

FIGURE 1. Streamlines for the analytic problem.

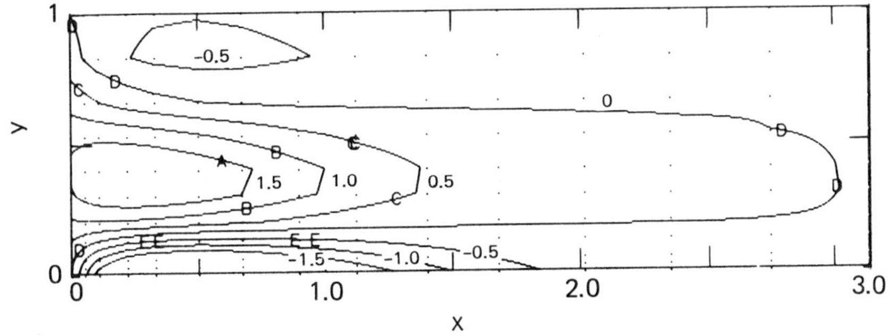

FIGURE 2. Exact vorticity field for the analytic problem.

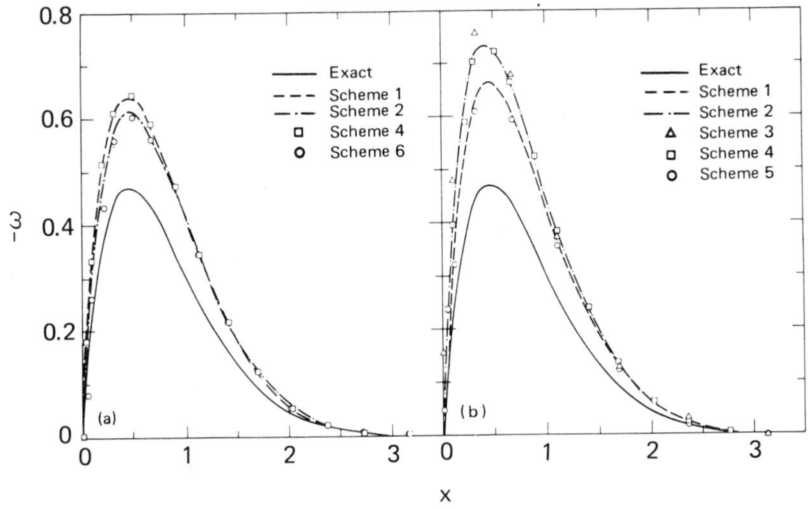

FIGURE 3. Vorticity profiles at top wall (y = 1.0) for various schemes; (a) 4-node fine mesh; (b) 9-node fine mesh.

ence calculation in which vorticities are obtained directly with the stream function-vorticity formulation.

Example 1. Analytic Function

Consider a flow field described by a stream function $\psi(x,y)$, where

$$\psi(x,y) = e^{-2(x+y)} \sin x \sin^2 (\pi y) \qquad \begin{matrix} 0 \le x \le \pi \\ 0 \le y \le 1 \end{matrix} \quad . \tag{8}$$

The exact velocity (u,v) and vorticity fields are calculated from

$$\frac{\partial\psi}{\partial y} , - \frac{\partial\psi}{\partial x}$$

and $- \nabla^2\psi$ respectively.

Contour plots of the ψ and ω fields are presented in Figs. 1 and 2. For this problem, two rectangular meshes ("coarse" and "fine") with an expanding grid containing finer zones close to the origin (where the largest gradients occur) are employed. Each mesh contains 4- or 9-node elements with identical nodal locations. The fine mesh has 11 x 15 mesh points whereas every other mesh point is used for the coarse mesh (the elements are twice as large). Exact nodal velocities (obtained from differentiating Eq. (8)) are used for the vorticity smoothing schemes. The results are presented in Figs. 3 through 6.

The agreement between exact and smoothed vorticities is seen to be poorest at the top wall (y = 1.0) where the elements are largest and at y = 0.4 where the spatial variation is the most rapid. At other stations, the schemes are generally close enough to each other to make comparisons more difficult. Nodal root-mean square (RMS) errors for the various schemes are displayed in Table 2.

As expected, the coarse mesh results are significantly less accurate than the fine mesh results. Moreover, schemes 1, 2 and 4 are usually more accurate than the other schemes. It may appear paradoxical at first, that global smoothing via the consistent mass matrix (scheme 1), with its associated high cost, would fail to generate consistently the most accurate set of nodal values. Theory stipulates, however, that only the *integral* of the square of the errors is minimized, therefore, the nodal errors need not be smallest. As we are primarily interested in *nodal* results, we didn't bother computing global (integrated) RMS errors. Although only rectangular meshes are presented here, the results obtained from general distorted isoparametric element meshes were similar. In addition, results from either the row-sum or diagonal lumping schemes for the mass matrix (being different only for nonrectangular meshes) were virtually identical.

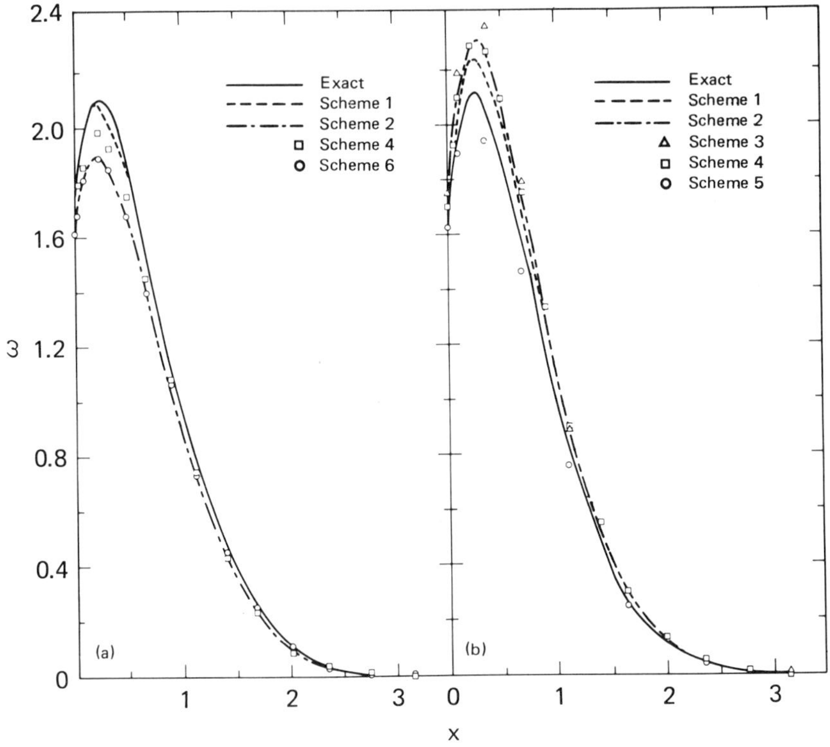

FIGURE 4. Vorticity profiles at y = 0.4 for various schemes; (a) 4-node fine mesh; (b) 9-node fine mesh.

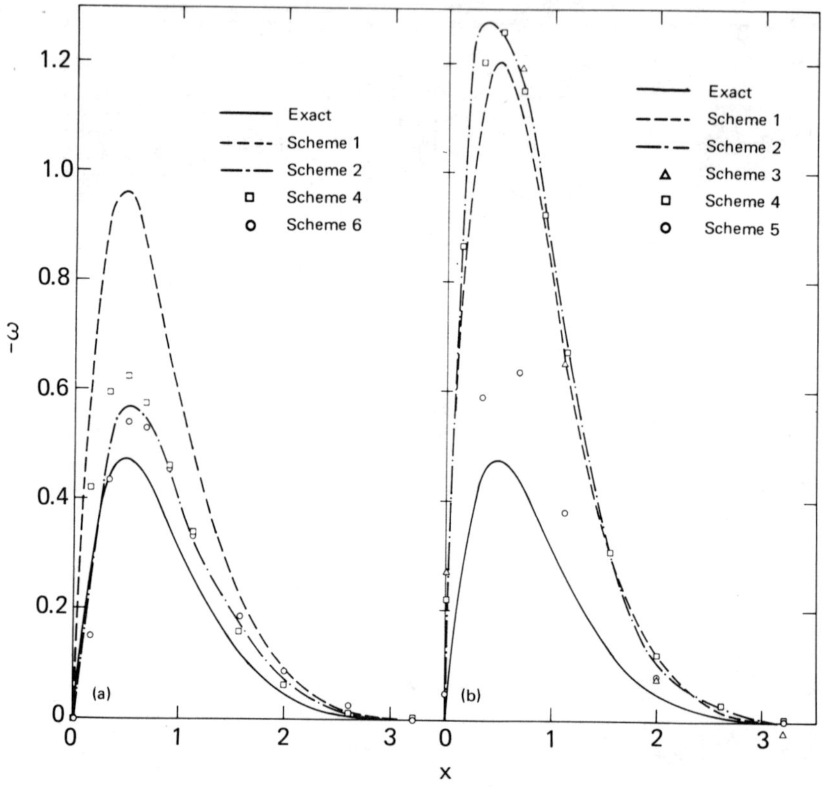

FIGURE 5. Vorticity profiles at top wall (y = 1.0) for various schemes; (a) 4-node coarse mesh; (b) 9-node coarse mesh.

TABLE 2. Nodal RMS errors for various schemes.

Mesh	Scheme					
	1	2	3	4	5	6
4-node fine	.081	.128	-	.086	-	.120
9-node fine	.069	.091	.125	.083	.113	-
4-node coarse	.213	.195	-	.296	-	.351
9-node coarse	.197	.226	.299	.221	.364	-

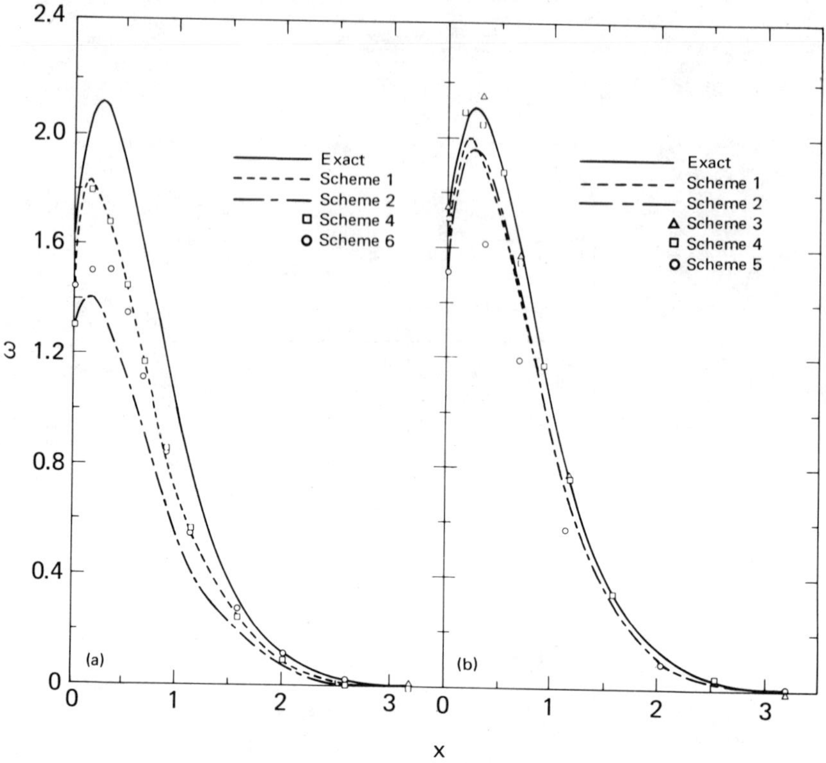

FIGURE 6. Vorticity profiles at y = 0.4 for various schemes; (a) 4-node coarse mesh; (b) 9-node coarse mesh.

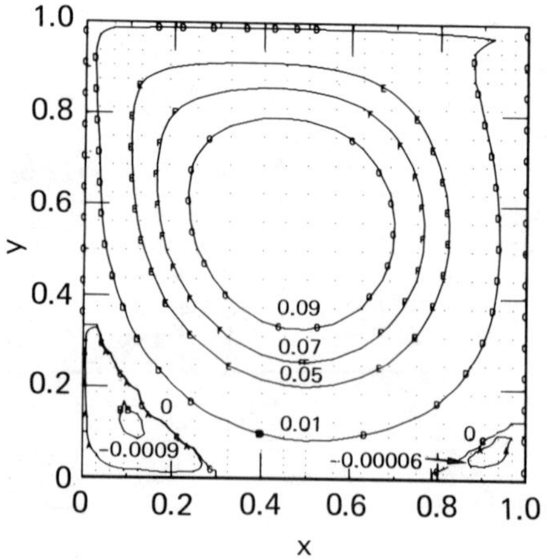

FIGURE 7. Streamlines for the steady state driven cavity problem at Re = 1000.

Example 2. A Driven Cavity Problem

The smoothing schemes are next applied to a solution of the Navier-Stokes equations. Nodal velocities of the steady state driven cavity problem at a Reynolds number of 1000 are obtained from a finite element formulation (Lee and Gresho, 1978) giving streamlines as shown in Fig. 7. A graded rectangular mesh containing 10 x 11 Lagrange (9-node) elements is used for the calculation. In this case, the smoothed vorticities are compared with a finite difference (stream function-vorticity) solution (de Vahl Davis and Mallinson, 1976) calculated on a 31 x 31 nonuniform mesh (Fig. 8a). Since vorticity contour plots of schemes 1, 2 and 4 are virtually undistinguishable, only scheme 1 is displayed (Fig. 8b). For the same reason, of the plots for schemes 3 and 5, only scheme 3 is shown (Fig. 8c).

Generally the smoothed vorticities are in good qualitative agreement with the finite difference solution. However, for this given mesh, the coarseness of node spacing near the center of the region presented problems for the contour package resulting in rugged contour lines. This is particularly evident for schemes 3 and 5 in which vorticities are calculated only at the four corners of each element. A vertical cross-section of the vorticity profiles through the primary vortex center ($x = 0.45$) displayed in Fig. 8d shows distinctly the differences between the solutions. As in the analytic problem, the various schemes agree closely with each other; however, the finite difference solution appears to be slightly smoother, as expected, since these values were obtained directly.

DISCONTINUOUS PRESSURE SMOOTHING

Discontinuous basis functions are frequently used for pressure interpolation in finite element formulations. For example, mixed interpolation applied to the 4-node bilinear element requires piecewise constant pressure over the element. A similar situation arises if linear elements are used in conjunction with the penalty method as proposed by Hughes et al. (1976). It is natural to define, in such cases, pressures at the centroid of each element. Discontinuous pressures are also associated with higher order elements in the penalty formulation; the 9-node biquadratic element for velocities incorporates bilinear pressures using the 2 x 2 Gauss points for the pressure "nodes". An interesting problem has been known to plague the discontinuous pressure solutions in these elements. Calculations (Huyakorn et al., 1978) show, for example, that the 4-node element exhibits "checkerboard" pressure modes for certain prescriptions of boundary conditions. In order to generate acceptable pressure solutions, these checkerboard modes must be filtered through application of smoothing algorithms. Of course, even if these modes are absent, a smoothing technique would serve to bring Gauss point values to the (more convenient) node points.

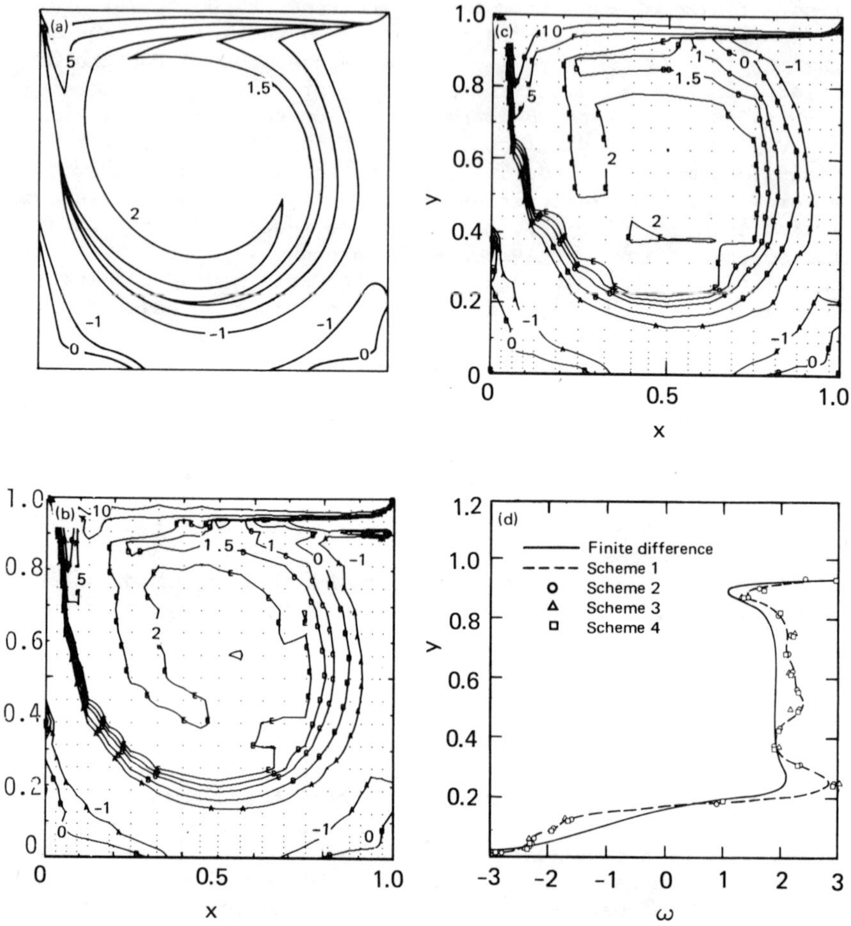

FIGURE 8. Comparison of vorticities from various schemes for the driven cavity problem at Re = 1000; (a) finite difference solution of de Vahl Davis; (b) global smoothing, Scheme 1; (c) local smoothing, Scheme 3; (d) vorticity profiles through the center of the primary vortex.

A least squares smoothing of piecewise constant (or otherwise discontinuous) pressures leads to a global set of equations for the N nodal pressures p^* given by:

$$\sum_{j=1}^{N} M_{ij}\, p_j^* = f_i \qquad i = 1,2,\ldots,N \quad . \tag{9}$$

As before, the corresponding element level matrices are:

$$M_{ij}^e = \int_{\Omega_e} \phi_i \phi_j \, dA \tag{10}$$

$$f_i^e = \int_{\Omega_e} p\phi_i \, dA \tag{11}$$

with

$$p = \sum_{i=1}^{m} \psi_i p_i \tag{12}$$

where ϕ_i is a set of continuous (velocity) basis functions. Here, ψ_i is a set of discontinuous basis functions with the summation taken over the "m" Gauss point pressures for each element. As before, a lumped mass matrix may be used in Eq. (10) leading to, for rectangular elements, an area-weighted averaging scheme.

The lumped mass smoothing algorithm, when applied to domains which contain corner nodes along a boundary, does not adjust (average) the corner node values. Good corner values may be obtained, after the initial smoothing process, by employing a local weighted least squares bilinear fit to the eight neighboring nodes and linearly extrapolating to the corner node in question. Since it is reasonable to weight the closest node values more heavily, each of the eight nodal equations is multiplied by a factor of $1/r$ where r is the distance from the corner node in question. If more substantial weighting is desired, a factor of $1/r^2$ is suggested. This "corner" procedure requires formation and solution of a 4 x 4 (overdetermined least squares) linear system for a set of 4 coefficients from which *only* the corner node value is computed and used. More details regarding discontinuous pressures and checkerboard smoothing will be presented in a future paper. Although our present program contains options for either the $1/r$ or $1/r^2$ weighting, no serious attempt has yet been made to determine which option is "better"; both weighting schemes have given good results on the few problems examined thus far.

A solution of the driven cavity problem for a mesh containing 10 x 11 linear elements at a Reynolds number of 400 is presented in Fig. 9a. The boundary conditions for this problem result in a checkerboard pressure node of magnitude $\sim \pm 90$

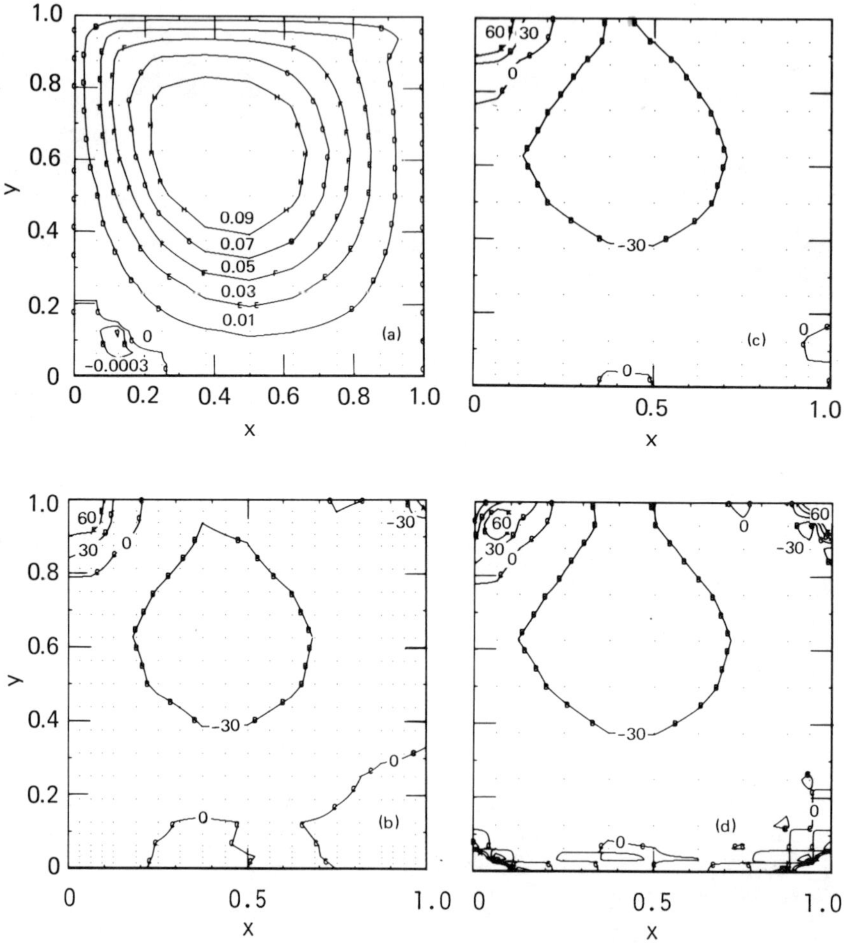

FIGURE 9. Pressure smoothing for the driven cavity problem at Re = 400. (a) streamlines; (b) pressure solution, 9-node fine mesh; (c) smoothed pressure, lumped mass matrix; (d) smoothed pressure, consistent mass matrix.

alternating from element to element. The smoothed pressures (using 1/r weighting), shown in Figs. 9c and 9d are compared with the pressure solution from a 9-node (10 x 11 element) fine mesh calculation (Fig. 9b). Of course, no smoothing is required for the 9-node problem since continuous pressure interpolations are used and no checkerboard pattern is generated.

Of the two schemes, it is apparent that lumped mass smoothing results in a much better pressure field. It is known that a least squares fit (via the consistent mass matrix) to a discontinuous function with alternating signs typically exhibits an excessive amount of noise (oscillations); again, this result, at the nodes, is a direct consequence of obtaining a global approximation with a minimum *integrated* square error. On the other hand, a straight averaging scheme, associated with lumped mass smoothing, attempts to filter out checkerboard oscillations thus generating a more smooth solution. Either scheme appears to generate acceptable pressures for problems in which no checkerboard mode is present.

Although the simpler local smoothing schemes, as described previously, have not yet been tested for the post-processing of pressures, it is conceivable that those schemes may be useful here too and they will be considered in the future. However several initial calculations appear to indicate that the suggested lumped mass smoothing, with its associated area-weighting, is usually more reliable than simple averaging schemes.

CONCLUSIONS

Comparison of several derivative smoothing schemes of varying computational complexity reveal that no single scheme is significantly more accurate than the others. Scheme 1 gives generally the best, but probably not the most cost-effective vorticity results. Therefore, it is recommended that this scheme be employed only on small problems. For large problems in which the costs associated with storage and subsequent solution of a linear system are significant (or prohibitive), either scheme 2 or 4 is recommended. Of the two, scheme 4, being cheaper to use, is preferred.

Discontinuous pressure solutions for linear or quadratic elements may be globally smoothed with either the consistent or lumped mass matrix. If the pressure field contains a checkerboard mode, the lumped mass matrix is recommended, in which case corner nodes require some additional special treatment.

ACKNOWLEDGMENTS

This work was performed under the auspices the the U. S. Department of Energy by the Lawrence Livermore Laboratory under contract no. W-7405-Eng-48.

REFERENCES

Barlow, J. (1976) Optimal Stress Locations in Finite Element Models. *Int. J. Num. Meth. Engng., 10*, 243-251.

Callabresi, M. (1978) Private Communication, Sandia Laboratory, Livermore, CA.

de Vahl Davis, G. and G. D. Mallinson (1976) An Evaluation of Upwind and Central Difference Approximations by a Study of Recirculation Flow. *Comp. and Fluids, 4*, 29-43.

Gresho, P. M., R. L. Lee and R. L. Sani (1978) Advection-Dominated Flows with Emphasis on the Consequences of Mass Lumping. *Finite Elements in Fluids, Vol. 3.* J. Wiley & Sons (in press).

Hallquist, J. O. and G. L. Goudreau (1977) SAPP - A Post-Processor for Two-Dimensional Finite Element Codes, Lawrence Livermore Laboratory Report UCRL-52318.

Hinton, E. (1968) Least Squares Analysis Using Finite Elements. M.Sc. Thesis, Civil Engineering Department, University College of Swansea.

Hinton, E. and J. S. Campbell (1974) Local and Global Smoothing of Discontinuous Finite Element Functions Using a Least Squares Method. *Int. J. Num. Meth. Engng., 8*, 461-480.

Hinton, E., F. C. Scott and R. E. Ricketts (1975) Local Least Squares Stress Smoothing for Parabolic Isoparametric Elements, *Int. J. Num. Meth. Engng., 9*, 235-238.

Hughes, T. J. R., R. L. Taylor and J. F. Levy (1976) A Finite Element Method for Incompressible Flows, Preprints of the Second International Symposium of Finite Elements in Flow Problems, S. Margherita Liqure, Italy.

Huyakorn, P. A., C. Taylor, R. L. Lee and P. M. Gresho (1978) A Comparison of Various Mixed Interpolation Finite Elements in the Velocity - Pressure Formulation of the Navier-Stokes Equations. *Computers and Fluids* (in press).

Lee, R. L. and P. M. Gresho (1978) The Development of a Three-Dimensional Model of the Atmospheric Boundary Layer Using the Finite Element Method, Lawrence Livermore Laboratory Report UCRL-52366.

MARC-CDC General Purpose Finite Element Analysis Program - User Information Manual, Vol. 1 (1976), MARC Analysis Corp.

Malkus, D. S. and T. J. R. Hughes (1978) Mixed Finite Element Methods - Reduced and Selective Integration Techniques: A Unification of Concepts. *Comp. Meth. Appl. Mech. and Engng.* (to appear).

Oden, J. T. and H. J. Brauchli (1971) On the Calculation of Consistent Stress Distributions in Finite Element Applications. *Int. J. Num. Meth. Engng., 3*71-425.

A FINITE ELEMENT STUDY OF LARGE AMPLITUDE WATER WAVES

P.L.Betts and B.L. Hall

University of Manchester Institute of Science and Technology, Manchester M60 1QD, England.

A train of finite amplitude waves is considered in a frame of reference moving with the wave crest. The finite element method is applied to a variational principle in terms of stream function which is valid for waves of any length. Computations have been performed for wave lengths between the ranges of validity of Stokes theory and cnoidal theory, and comparisons are made with existing analytic series solutions.

INTRODUCTION

In the non-linear theory of oscillatory waves of permanent form and finite amplitude in water of finite depth, it is usual to classify the waves in terms of their relative wave length (Wiegel, 1964; Le Méhauté, 1976). For short waves, where the wave length is less than about five times the mean depth, it has been usual to apply expansions of the type first considered by Stokes (1847). However for long waves, with a relative length greater than about 10, the Stokes expansions do not converge and it is more appropriate to apply cnoidal theory to long wave approximations of the equations of motion (Benjamin and Lighthill, 1954). Very long waves may be treated as solitary waves.

In the intermediate region of wave-length between Stokes waves and cnoidal waves, the choice of valid theory depends on the magnitude of $(a/\lambda)(\lambda/h)^3$, where a is half the crest-to-trough amplitude, λ the wave length and h the mean depth. However as was shown graphically by Wiegel (1964), even at quite moderate finite amplitudes, the expansions of the two theories do not produce the same results (i.e. wave speed) at any wave length in the intermediate region.

A number of attempts have been made to treat the water wave problem by numerical methods. One of the most successful was that of Chan, Street and Strelkoff (1969), who used finite difference methods in the complex potential plane to compute the characteristics of solitary waves. This method requires an initial good approximation of the surface profile for convergence, and at low amplitudes they used the Boussinesq (1871) profile whereas at higher amplitudes, up to near the amplitude for wave breaking, they used scaled-up initial profiles obtained from their previous computations at slightly lower amplitudes. For waves of finite wave length, the most successful method has been the semi-analytic approach of Dean (1965, 1974), who expanded the stream function and free-surface elevation as Fourier series in the horizontal direction and truncated the series after the eleventh harmonic. The Fourier coefficients were then determined by minimising the error in the dynamic free-surface boundary condition by least squares over a number of points evenly spaced horizontally between the wave crest and trough. As with all methods based on Fourier series, the accuracy of the method reduces as the wave amplitude increases, since higher and higher harmonics are required to describe the rapid changes in curvature near the wave crest, whilst the intermediate harmonics become less important. Other numerical methods are mentioned by Le Méhauté (1976).

Longuet-Higgins and Cokelet (1976) have recently described a time-varying singularity method, which they used to compute the overturning of waves in water of infinite depth. However, the application of this method to water of finite depth has not yet been successful (Fenton and Mills, 1977).

The success of the finite element method in computing the stationary waves which spontaneously appear on an otherwise uniform flow downstream of a contraction (Betts, 1978), suggested that the method might also be used to compute finite amplitude wave characteristics in the region of wave lengths between those of Stokes and cnoidal theory. The advantages of the finite element method in this context are that the elements can be arbitrarily disposed, so that regions of rapid change can be covered by smaller elements than other regions, and that the method does not appear to require a very good initial surface profile as a first estimate (cf. Betts, 1978). In this way one is freed from the rectangular mesh used in finite difference methods (Chan, Street and Strelkoff, 1969) and from the need in series solutions for higher and higher Fourier harmonics, which are only required for the description of a relatively short region of the flow (Dean 1965, 1974).

During the course of this work, Cokelet (1977) published the results of his studies of steep gravity waves. He assumed a solution in terms of a series similar to that of Stokes (1847), but based on a perturbation parameter related to the fluid speed at a crest. In this way, he was able to derive recursion relationships for the expansion coefficients and sum the series to high order with the aid of Padé approximants. He was then able to obtain convergent series for wave lengths up to values well into the cnoidal range, although the accuracy declined for waves near the maximum amplitude at the longest lengths. He also found that the highest waves were not the fastest waves. This result has been confirmed for a solitary wave by an integral equation method independent of the use of Padé approximants (Byatt-Smith and Longuet-Higgins, 1976), but independent confirmation for waves of finite length by a direct approach, such as is offered by the finite element method, is highly desirable. Moreover, the finite element method offers the possibility of extension to the important field of waves on currents of variable depth (Provis and Radok, 1977), the linear theory of which was given by Lamb (1932).

In this paper the waves are considered in a frame of reference fixed relative to a crest, and the finite element method is applied to a previously derived variational principle. Two wave lengths are considered and computations have been extended up to about 90% of the maximum wave amplitude. Comparisons are made with the results of De (1955) for fifth-order Stokes waves and also with some of the results of Cokelet (1977). For large amplitude waves the present results are more accurate than those of De, even for relatively short waves where the Stokes expansion would be considered valid, and consideration is given to the way the disposition of the finite elements could be improved to increase the accuracy significantly without increasing the storage requirements or computation time. Only a relatively crude initial estimate of the wave profiles was found to be required and a simple sinusoid was used in all the present work.

WAVE THEORY

The work of Benjamin and Lighthill (1954) is essential to the understanding of finite amplitude gravity waves of permanent form, when they are considered in a frame of reference which is stationary relative to a wave crest. They showed that any inviscid steady two-dimensional flow over a flat horizontal bottom can be uniquely specified by three parameters, in their approach Q, the volumetric flow rate per unit span, H, the total head level and S the momentum flow rate per unit span corrected for pressure forces. Moreover, Q, H and S must lie between certain limits beyond which no stationary waves or steady flows are possible. Within these limits, the amplitude, 2a, length, λ, and mean depth, h, are dependent variables.

In the present work, symmetry allows us to consider only half of a wave, as illustrated in Figure 1. To apply the finite element method within a fixed horizontal length, it is convenient in this work to consider the total head H, the flow rate Q and the wave length λ as the independent variables. In this way H and λ can be fixed and as Q is varied, within the necessary limits, the mean depth h and peak-to-trough amplitude 2a will change. According to Benjamin and Lighthill (1954), for a given value of H and λ there is only one value of Q (and hence 2a) for which waves are possible on a flow of given h.

The "still-water depth", d_s, is also illustrated on Figure 1, since Cokelet (1977) and many other workers have displayed their results in terms of this dependent variable, because of

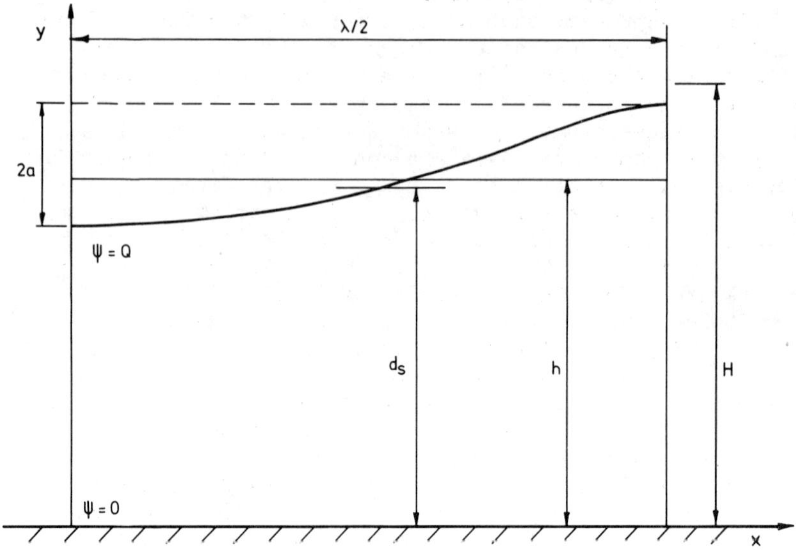

Figure 1 Defining sketch

its convenience for Stokes type expansions. The "still-water depth" is defined as the flow rate Q divided by the average stream velocity, taken over a wave length, at any vertical elevation less than the lowest free surface elevation. Since this average velocity (known as Stokes's "first definition" of wave velocity) ignores the low velocities immediately under the wave crest, it is higher than the wave speed defined by the mean velocity Q/h, and hence $d_s < h$. Both definitions of depth, and hence wave speed, appear in the literature and care must therefore be taken in making comparisons. De (1955) displays his results in terms of a fixed mean depth, but gives sufficient information for d_s to be calculated; whereas the results of Cokelet (1977) are for a fixed d_s, with the mean depth increasing with wave amplitude.

FINITE ELEMENT FORMULATION AND PROBLEM SPECIFICATION

The appropriate variational principle for the study of stationary waves, of the form illustrated by the steady flow situation shown in Figure 1, is that derived by Betts (1976). This principle, which is an extremum of the functional

$$\chi = \frac{1}{2} \iint_{\Omega(y_s)} \left\{ \left(\frac{\partial\psi}{\partial x}\right)^2 + \left(\frac{\partial\psi}{\partial y}\right)^2 \right\} dx\, dy - \frac{1}{2} \int_{x_1}^{x_2} g(H-y_s)^2 dx, \qquad (1)$$

has been presented in the related paper presented at this conference (Betts, 1978), to which reference can be made for further details. In the present context, the functional is subjected to variations such that the stream function $\psi = 0$ along the fixed horizontal bottom, and $\psi = Q$ on the variable free surface streamline. The parameters subjected to variation are the streamfunction values within the flow domain and the elevation y_s of the free surface streamline. These variations are independent, except in so far as is required by continuity of the first derivative of ψ at the free surface. The correct conditions $\partial\psi/\partial x = 0$ on the vertical end boundaries ($x_1 = 0$, $x_2 = \lambda/2$) occur as a natural consequence of the variational principle.

When the flow domain is divided into linear triangular finite elements, the variation of nodal parameters produces non-linear terms in the matrix equations. Consequently an iterative technique is required for their solution, and a Newton-Raphson method was chosen. This is also described in the related paper by Betts (1978) and the final equations were given by Betts (1976). An initial estimate for the free surface is required and fixed nodal positions are chosen within the flow. These positions are generated automatically by the program. In the first iteration the values of ψ at internal nodes are calculated for flow beneath a solid upper boundary. In subsequent iterations simultaneous corrections to ψ at the fixed internal nodes and y_s at the variable surface nodes (where $\psi = Q$) are calculated from the variation of Equation 1. The computation ceases when the maximum surface movement is less than a pre-determined value, in the present work 10^{-5}. In most of the present work, the nodes were distributed evenly along vertical lines equi-spaced in the horizontal direction and these internal nodal positions were regenerated at the third iteration, and also whenever the surface element became too distorted relative to the internal elements. The element was considered distorted if the length of the vertical side of the surface element was outside the range 0.7 to 1/0.7 times the vertical length of an internal element. This regeneration was only required for the larger amplitude waves. A typical final distribution of nodes and elements is shown in Figure 2.

COMPUTATIONS AND RESULTS

The computations were programmed in a dimensionless form such that the nominal mean depth h (Figure 1) and the gravitational acceleration g were equal to unity. Two wave lengths were considered such that $2\pi h/\lambda = mh = 2.0$ and 0.75. A wave-number mh of 2.0 is well within the range where Stokes expansions should be valid, whereas $mh = 0.75$ corresponds to a wave length in the intermediate region between Stokes waves and cnoidal waves. For each wave-number, the mean velocity for waves of infinitesimal amplitude was first computed, and from this the flow rate Q_o and total head level H_o for waves of zero amplitude

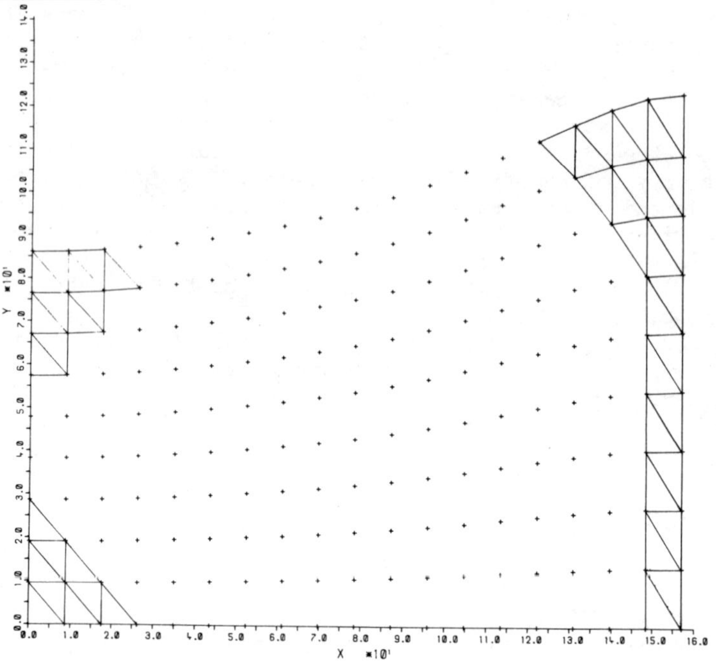

Figure 2 Example of final nodal distribution and sample elements. 190 nodes, mh = 2.0, $(H-H_o)/H_o = 0.029$.

were immediately obtained.

The total head level H was chosen in terms of $(H-H_o)/H_o$ and, with an appropriate initial estimate to the expected free surface, the computations were performed for increasing values of $(Q-Q_o)/Q_o$. Within certain limits of $(Q-Q_o)/Q_o$ a stationary wave train is obtained, as was expected from the work of Benjamin and Lighthill (1954). The mean depth of this wave train, and consequently the wave amplitude, varies with $(Q-Q_o)/Q_o$, and it is then a simple matter to determine the correct value for h = 1 by an interpolation procedure (cf. Figure 3). The results of De (1955) were used to provide a reasonable estimate for the limits on H and Q. Moreover, it was found that if Q was significantly higher than the correct value, the computations produced a uniform flow. This is consistent with the theoretical considerations since, with λ no longer a parameter, there will always be one depth of flow which corresponds to uniform flow for any given H and Q. It was also found that the range of Q for which true waves were obtained increased with $(H-H_o)/H_o$ (i.e. with wave amplitude). Again this is to be expected from the theoretical considerations, and in practice it made large amplitude waves easier to compute than small amplitudes ones. For all the results presented in this

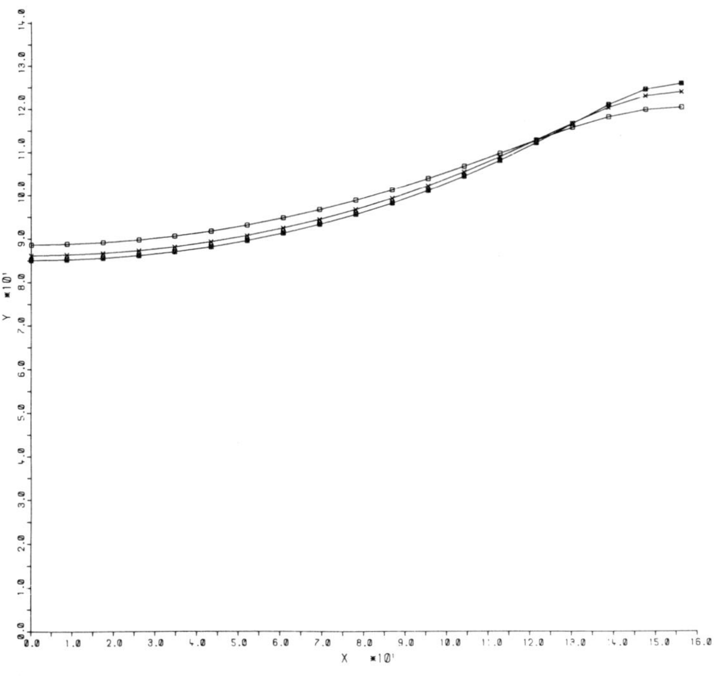

Figure 3 Depth and amplitude variations caused by changes in flow rate. Nominal mh = 2.0, 190 nodes, $(H-H_o)/H_o$ = 0.029. □, $(Q-Q_o)/Q_o$ = 0.0377; X, 0.0393; ⊠, 0.0489.

paper the initial free surface profile was taken as half of a negative cosine wave, with an amplitude near that expected for the final wave and a mean depth of unity (see Figure 4). In some of the earlier tests a simple horizontal line at the trough level was used, but it was then found necessary to fix the surface level of the node at the trough. Although convergent computations were obtained, the range of possible values of $(Q-Q_o)/Q$ was reduced to that provided by discretization error. This estimate was therefore abandoned in favour of the cosine profile.

Basic computations were performed with a total of 190 nodes in columns of 10 nodes (9 element sides) at equal horizontal intervals (see Figure 2). The resulting surface profiles for mh = 2.0 and 0.75 are shown on Figures 5 and 6. In each case the largest amplitude corresponds to about 90% of the maximum possible, and as the amplitude increases, the crest becomes more peaked as the trough becomes flatter. Consequently, the difference between the initial cosine profile and the final wave increases. This produced no difficulties until the highest wave for mh = 0.75, where the first free surface movement took the elevation of the crest above

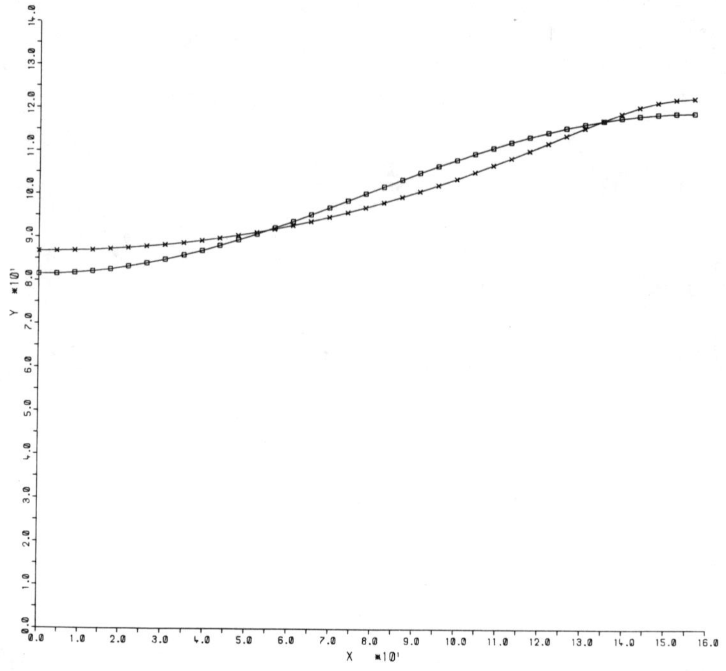

Figure 4 Initial and final free surface profiles. 703 nodes, mh = 2.0, $(H-H_o)/H_o$ = 0.028. —□— , Initial profile; —×— , final profile.

the total head level; as might be expected the computation would not then converge. The application of a damping factor if 1.2 (under-relaxation) reduces the initial movement sufficiently to prevent this, and the converged result shown was then obtained. Damping of the later iterations only increased the computation time, and a more efficient way of computing large amplitude waves would be to use an initial profile obtained by scaling up the results from a previous computation at a slightly small amplitude (cf. Chan, Street and Strelkoff, 1969).

The results for all the computations with mh = 2.0 and 0.75 are displayed on Figures 7 and 8 in terms of $(H-H_o)/H_o$ and $(Q-Q_o)/Q_o$ as functions of the peak-to-trough amplitude 2a. Strictly $(H-H_o)/H_o$ should be taken as the independent variable in the present work, but this form of display is more convenient for comparison with other workers' results. Also shown on these graphs are the results obtained by De (1955) using 5th order Stokes theory.

It is immediately obvious from these graphs that for small wave amplitudes our results lie below those of De, and that this is particularly true for the shorter wave (mh = 2.0).

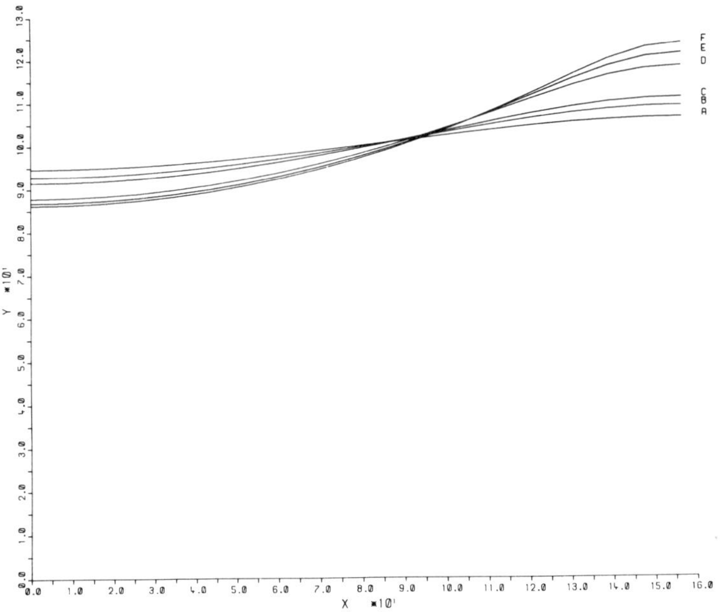

Figure 5 Free surface profiles, mh = 2.0, 190 nodes.
A, $(H-H_o)/H_o$= 0.00077; B, 0.0035; C, 0.00595; D, 0.018; E, 0.024; F, 0.029.

Although the difference is only about 0.4% of the basic wave speed, it can be attributed to the way the elements were distributed vertically. The perturbation velocity in a stationary wave train decays as a combined exponential with distance below the free surface for infinitesimal waves (Lamb, 1932); the vertical rate of change of the perturbation velocity is a maximum at the free surface and for short waves, for very long waves it beoomes negligible. Consequently, improved results are to be expected with elements graded in size exponentially, with the smallest elements nearest the surface. The difference in size between the small elements near the surface and the large ones near the bottom will greatest for the shortest waves. Preliminary work with elements graded in this way indicates an increase in accuracy comparable with doubling the number of elements in the vertical columns. The reduction of $(Q-Q_o)/Q_o$, caused by the use of nodes uniformly distributed in vertical columns, is a natural consequence of making the perturbation velocity more uniform at each vertical column, so that there is a tendency towards solving for the lower speed of the shallow water wave approximation.

It can also be seen from Figures 7 and 8 that at large wave amplitudes the computed values of $(H-H_o)/H_o$ and $(Q-Q_o)/Q_o$ lie above those given by De. This is not surprising when mh = 0.75 (Figure 8) since even 5th order Stokes theory is unlikely

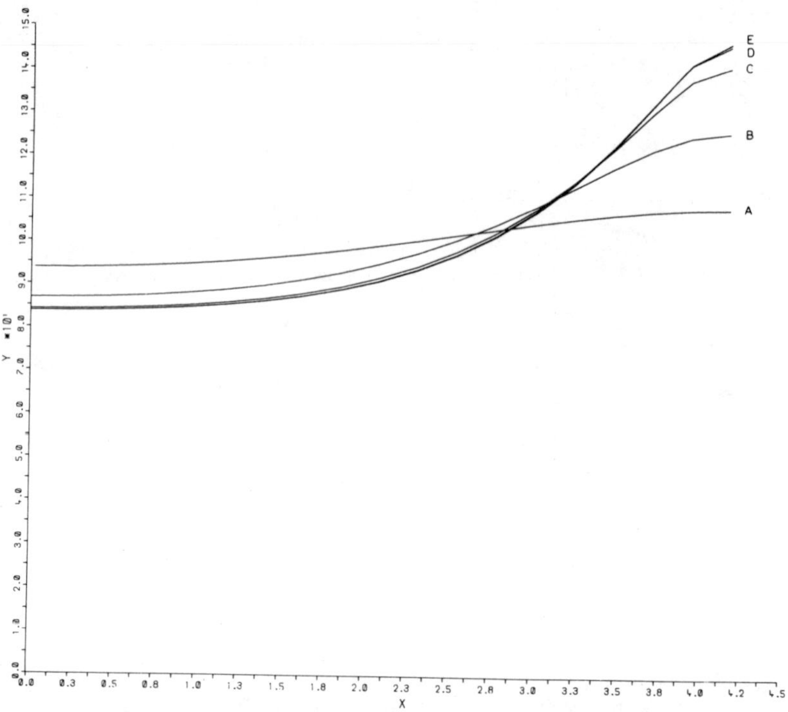

Figure 6 Free surface profiles, with exaggerated vertical scale, mh = 0.75, 190 nodes. A, $(H-H_o)/H_o$ = 0.003; B, 0.030; C, 0.060; D, 0.070; E, 0.071.

to be very accurate for such a long wave at such large amplitudes. However with mh = 2.0, Stokes expansion would normally be considered valid. Two computations were therefore performed with the flow divided into smaller elements, four times as many elements, 703 nodes. As can be seen on Figure 7, at a small amplitude this produced an approximately four fold improvement in the agreement with De's results. However at the large amplitude, the values of $(H-H_o)/H_o$ and $(Q-Q_o)/Q_o$ exceeded De's results to an even greater extent. Unfortunately, it is not possible to make a direct comparison with Cokelet's (1977) more accurate 28 term Padé approximant results at the same wavenumber. However, De's results can be interpolated in such a way as to be compared with Cokelet's at mh = 2.36. Since the waves are short, the comparisons are made in terms of dimensionless wave amplitude $4\pi a/\lambda$ = 2ma. For values of 2ma of 0.70 and 0.80, Cokelet's results exceed those of De by 15% and 20% respectively for $(H-H_o)/H_o$ and 24% and 36% for $(Q-Q_o)/Q_o$. Our results, with 703 nodes, mh = 2.0 and 2ma = 0.703 (ma = 0.352), exceeded De's by 13% for $(H-H_o)/H_o$ and 24% for $(Q-Q_o)/Q_o$.

This indirect agreement with Cokelet's results suggests that, for large amplitude waves, the simple finite element

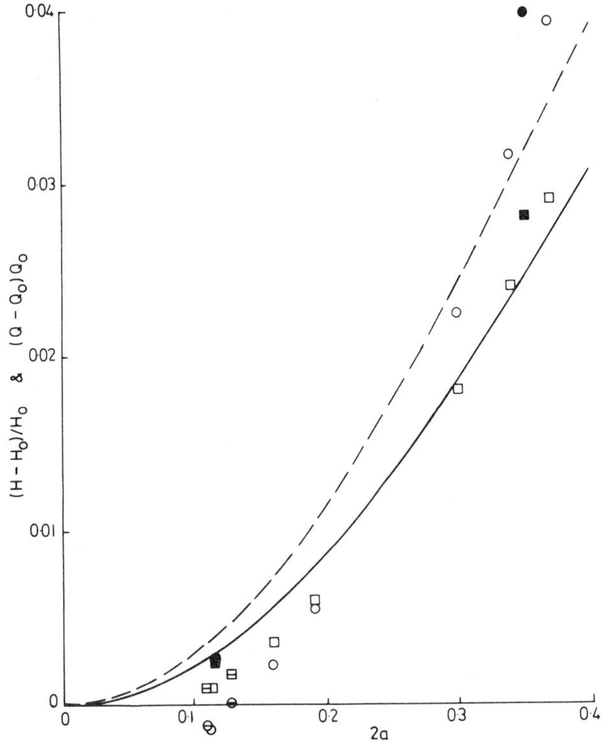

Figure 7 Total head and flow rate as functions of wave amplitude, mh = 2.0. $(H-H_o)/H_o$; □ , 190 nodes; ⊟ 210 nodes; ■ , 703 nodes; ———— , De (1955): $(Q-Q_o)/Q_o$; O, 190 nodes, θ, 210 nodes; ●, 703 nodes; — —, De (1955).

method is more accurate than the complicated Stokes series expansion used by De. The reason for this appears to be related to the curvature of the free surface near the crest, which becomes more and more severe as the wave amplitude increases. Consequently, an adequate description of this surface profile is essential at large amplitudes. 703 nodes corresponds to 36 element sides along half a wave length, which should be adequate to give information about a Fourier series up to about the 14th harmonic. In contrast, terms about the fifth harmonic are neglected in De's series. The question of accuracy of description will be considered further in the next section.

For our longer wave length results (mh = 0.75, Figure 8), the convergence of a Stokes series becomes doubtful and agree-

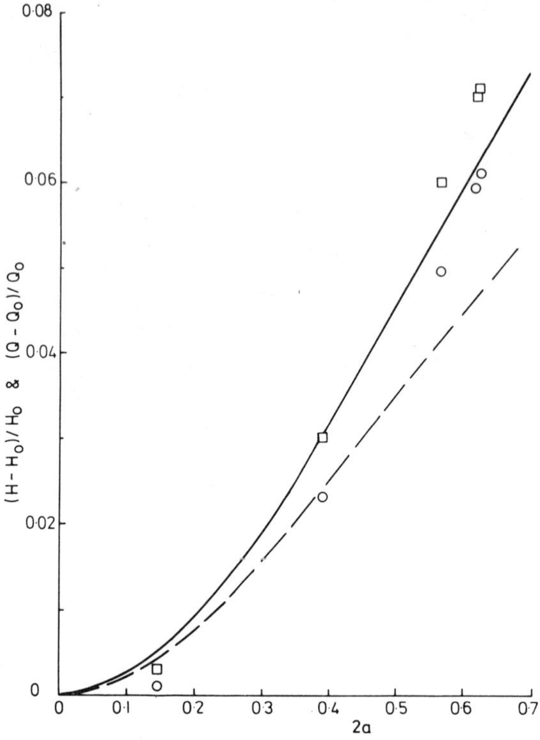

Figure 8 Total head and flow rate as functions of wave amplitude, mh = 0.75. Legend as on Figure 7.

ment with De's results was not to be expected. Rather than make further indirect comparisons with those of Cokelet, additional wave profiles were computed for mh = 0.717, $(H-H_o)/H_o$ = 0.065, in order to make a direct comparison. The surface profiles, obtained with 190 nodes and 703 nodes respectively, are shown on Figure 9. Some interpolation from Cokelet's tabulated results is still required, owing to his use of the "still-water depth", d_s, but for these values of mh and $(H-H_o)/H_o$ he obtained values of 2a and $(Q-Q_o)/Q_o$ of 0.5719 and 0.0581 respectively. Our computations with 198 nodes produced values of 0.5879 and 0.0549, whilst those with 703 nodes gave 0.5755 and 0.0571 respectively. This latter agreement (0.6% for 2a, 1.7% for $(Q-Q_o)/Q_o$ or 0.1% for Q) is within the accuracy expected from the present size and disposition of the finite elements. The way in which this accuracy can be improved without increasing the number of elements is considered in the next section.

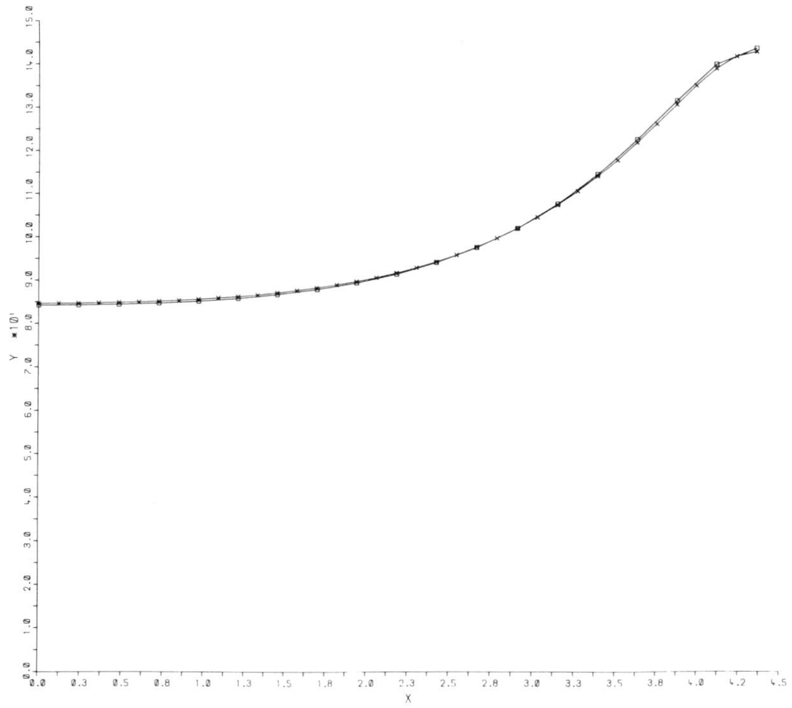

Figure 9 Free surface profiles with exaggerated vertical scales, mh = 0.717, $(H-H_o)/H_o$ = 0.065; □, 190 nodes; X,703 nodes.

ACCURACY AND ITS IMPROVEMENT

The basic difficulty with all computations of free surface flows is that the extremes of velocity occur at the free surface, where the location is a priori unknown and obtained through a non-linear boundary condition (cf. Larock and Taylor, 1976). Consequently, it is errors at the free surface that need to be assessed. This assessment is conveniently made in terms of the difference between locally computed values of the total head on the free surface and the prescribed value of H used in Equation 1. Figure 10 shows the horizontal distribution of this discrepancy, ΔH_s, for the surface profiles of Figure 9. The local values of total head on the free surface streamline were obtained from the nodal values of ψ in the following way (cf. Betts, 1976, 1978). Values of horizontal velocity, $\partial\psi/\partial y$, were obtained by differentiation from a quadratic for ψ through the top three nodes of a vertical column, whilst surface gradients were obtained from a similar curve for y_s through the surface node of interest and those on either side. The locally computed value of total head is then taken

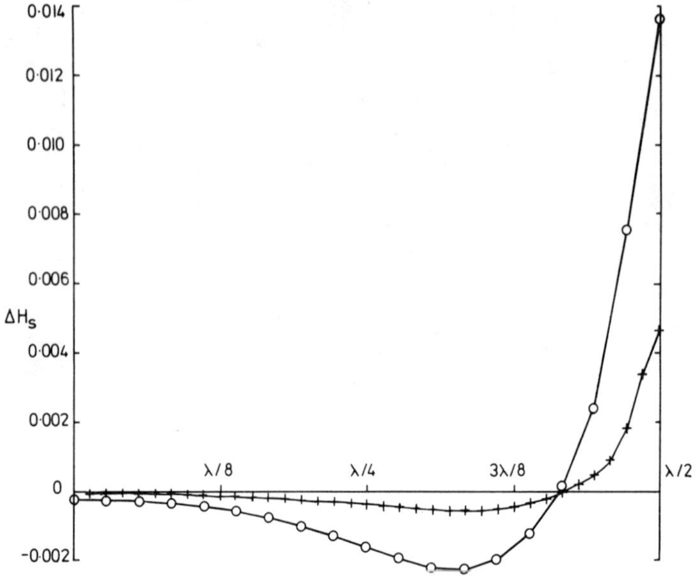

Figure 10 Distribution of estimated error in total head along free surface, mh = 0.717, $(H-H_o)/H_o = 0.065$: 0, 190 nodes; + , 703 nodes.

as the sum of the surface level, y_s, and the velocity head (velocity squared/2). The fact that the error, ΔH_s, is a length scale dependent on the square of differentiations, $\partial\psi/\partial y$ and $\partial y_s/\partial x$, is at the heart of the difficulties encountered in computing free surface flows.

The way in which the greatest free surface errors are concentrated round the crest of the wave is clearly shown in Figure 10, and the maximum error at the crest of 0.0046 with 703 nodes compares well with the 0.0036 difference between this amplitude and that of Cokelet. For lower amplitude waves the surface total head error is more uniformly distributed. It is therefore apparent from Figures 9 and 10 that, for large amplitude waves, the accuracy of a solution with uniform horizontal spacing of the elements along the free surface is limited by the description of the surface near the crest, while at the same time more elements are used than are needed over a wide range around the trough. Similar problems were encountered by Dean (1974) with his Fourier series for stream function up to the eleventh harmonic. However, with finite elements the problem can be overcome easily by increasing the horizontal spacing of the nodes around the trough and decreasing it near the crest, so that the total number of nodes does not need to be increased.

The desirability of a quasi-exponential vertical distribution of the nodes, with the smallest elements adjacent to the free surface,has already been mentioned. From the results obtained in the present work, it is estimated that the use of elements graded in size both vertically and horizontally, in the manner described, would provide an accuracy equal to or better than that which would be obtained by quadrupling the number of elements with a quasi-uniform size distribution. (2701 nodes, 5184 elements). The insertion of additional elements beneath the wave crest would further increase the accuracy, but this would be at the expense of either increasing the matrix bandwidth or of loosing the benefits of simplicity provided by the vertical columns of elements. However, sufficient accuracy is possible, even with only 703 nodes, to provide an independent test of Cokelet's conclusion, based on the use of Padé approximants, that the highest wave is not the fastest one (differences of order 4% of $(Q-Q_o)/Q_o$). Such a test would be of particular interest at the longer wave lengths where the effect is most pronounced but Cokelet's results are least accurate. In making these tests at the highest amplitudes, it would be advisable to use a better initial estimate for the free surface profile than the cosine wave used in this work, since for any value of $(H-H_o)/H_o$ there are likely to be two possible wave amplitudes at the same depth. The technique of Chan, Street and Strelkoff (1969), based on the profile computed at a slightly smaller amplitude, should be adequate for this purpose and is in any case desirable as a way of reducing the number of iterations required.

Although no computations have been performed with flow over an undular bottom streamline, it is readily apparent that the finite element method can easily be extended to such a problem. Any bottom profile that can be represented by a Fourier cosine series could be considered; the length of the domain would then be half this fundamental wave length (cf. Lamb, 1932). However, the increase in accuracy from grading the sizes of the elements would then be limited by the necessity for an adequate description of the bottom profile and of velocities adjacent to it.

CONCLUSIONS

The finite element method has been used to study the behaviour of a large amplitude train of stationary waves on an otherwise uniform stream. The method is based on a non-linear variational principle which is valid without approximation for all depths of water. Computations have been performed at three wave lengths between the ranges of validity of short wave Stokes theory and long wave cnoidal theory. Owing to the non-linearity of the variational principle, an iterative procedure is required for the finite element solution, and an initial estimate of the free surface profile is therefore needed. It was found that the form of this initial estimate was not critical, and a simple cosine wave was used in all the computations. Moreover, the larger amplitude waves were easier to compute than the smaller ones, because the physics of the formulation ensured that computations would converge to a solution, at a different mean depth, over a wider range of flow rates.

It has been shown that for large amplitude waves, the finite element results are more accurate than those of 5th order Stokes theory (De, 1955) even at the shortest wave lengths). Favourable comparisons have also been made with the tabulated results of Cokelet (1977), which are based on a 28 term (N/N) Padé approximant summation of a series solution and are the most accurate results available to date. A consideration of the accuracy of the finite element computations lead to the conclusion that this could be further improved by a simple redistribution of the elements. This would provide sufficient accuracy, with no increase in computer storage, for the method to provide a valuable independent test of Cokelet's results at the highest amplitudes, which indicated that the fastest wave has an amplitude less than the maximum.

Consideration has also been given to the application of the finite element method to the problem of stationary waves over an undular bottom. Such computations can provide valuable comparisons with the usual analytic approximation methods used to study waves on water of variable depth (Provis and Radok, 1977). All computations were performed on the CDC 7600 of the University of Manchester Regional Computing Centre.

REFERENCES

Benjamin, T.B. and Lighthill, M.J. (1954) On Cnoidal Waves and Bores. Proc.Roy.Soc.Lond.A224: 448-460.

Betts, P.L. (1976) A Variational Principle in Terms of Stream Function for Free-Surface Flows and its Application to the Finite Element Method. University of Calgary, Mech.Eng.Rep. 84.

Betts, P.L. (1978). Computation of Stationary Water Waves Downstream of a Two-dimensional Contraction. Proc. 2nd International Conference on Finite Elements in Water Resources, London, July.

Boussinesq, J. (1871) Theorie de l'intumescence liquide appelée onde solitaire ou de translation se propagant dans un canal rectangulaire. Institut de France, Académie des Sciences, Comptes Rendus, June 19:755.

Byatt-Smith, J.G.B. and Longuet-Higgins, M.S. (1976) On the Speed and Profile of Steep Solitary Waves. Proc.Roy.Soc.Lond. A350: 175-189.

Chan, R.K.C., Street, R.L. and Strelkoff, T. (1969) Computer Studies of Finite Amplitude Waves. Civil Engineering Tech. Rep.104,Stanford Univ., California.

Cokelet, E.D. (1977) Steep Gravity Waves of Arbitrary Uniform Depth. Phil. Trans.Roy.Soc.Lond. A286: 183-230.

De, S.C. (1955) Contributions to the Theory of Stokes Waves. Proc.Camb.Phil.Soc. 51: 713-736.

Dean, R.G. (1965) Stream Function Representation of Non-Linear Ocean Waves. J. Geophys. Res. 70: 4561-4572.

Dean, R.G. (1974) Evolution and Development of Water Wave Theories for Engineering Application. Vols. I and II. Special Report No.1, U.S. Army Coastal Engin.Res.Center, Fort Belvoir. Va.

Fenton, J.D. and Mills, D.A. (1977) Shoaling Waves: Numerical Solution of Exact Equation. Lecture Notes in Physics 64, Australian Academy of Science, Canberra/Springer-Verlag: 94-101.

Lamb, H. (1932) Hydrodynamics. 6th Ed., Cambridge U.P.

Larock, B.E. and Taylor C. (1976). Computing Three-Dimensional Free Surface Flows. Int.J.Num.Methods in Engg. 10: 1143-1152.

Le Méhauté, B. (1976). An Introduction to Hydrodynamics and Water Waves. Springer-Verlag, New York.

Longuet-Higgins, M.S. and Cokelet, E.D. (1976) The Deformation of Steep Surface Waves on Water, I. A Numerical Method of Computation. Proc.Roy.Soc.Lond. A350: 1-26.

Provis, D.G. and Radok, R. (1977).Eds. Waves on Water of Variable Depth.Lecture Notes in Physics 64,Australian Academy of Science, Canberra/Springer-Verlag, Berlin/Heidelberg/New York.

Stokes, G.G. (1847) On the Theory of Oscillatory Waves. Trans. Cam.Phil.Soc. 8:441-455.

Wiegel, R.L. (1964) Oceanographical Engineering. Prentice-Hall, London.

A DIGITAL SIMULATION APPROACH FOR A TRACER CASE IN HYDROLOGICAL SYSTEMS (Multi-compartmental Mathematical Model)

Y. Yurtsever and B.R. Payne

Electric Power Resources Survey and Development Administration, Ankara, Turkey and International Atomic Energy Agency Vienna, Austria.

ABSTRACT

Quantitative evaluation of the hydrodynamic parameters of a given hydrological system, by the use of tracers, requires development of a mathematical model to describe the tracer input-output relations for the system.

In the case of environmental tracers, and particularly for the use of environmental tritium in natural hydrological systems, the commonly employed functional relationship between tracer input and output is based on rather over-simplified assumptions describing the extreme cases of tracer behaviour in hydrological systems. In such linear lumped parameter models the response of the system is based on a single parameter in order to describe the hydrodynamic behaviour (mixing and dispersion of the tracer) of the system.

The approach proposed in the paper consists of the use of multi-compartmental models where the hydrodynamic behaviour of the system for a tracer case is simulated by the use of recursive equations based on mass transport principles, under conditions of non-steady state flow.

Available tritium data from the Glomma river (Norway) are used to illustrate the application of the proposed approach.

I. INTRODUCTION

The use of environmental isotopes in the study of hydrological systems offers a potential tool to extract information on various parameters associated with the dynamics of the system. In this regard, the environmental isotopes ^{18}O, ^{2}H and ^{3}H are generally the most favourable tracers to be employed for hydrological systems, because they are the isotopic species of the two constituents of the water molecule and, thus, their behaviour within a hydrologic system is very similar to that of the flow through the system.

They are naturally occurring (in the case of tritium it is also released into the environment by man-made activities) and are introduced into the system by precipitation. Consequently, the known spatial and time variations of the isotopic composition of precipitation provide valuable tracer-input information even for large-scale hydrological systems. A number of studies using environmental isotopes in regional-scale hydrological investigations have been reported in the literature [1,2,3,4,5,6].

A quantitative evaluation of the hydrodynamic parameters of a given hydrological system by the use of these tracers requires that a mathematical model be developed to describe the tracer input-output relations for the system. The most commonly employed mathematical models for this purpose, such as "Completely Mixed Reservoirs" or "Piston Flow", are linear lumped parameter models and they are based on rather over-simplified assumptions describing the extreme cases of tracer behaviour in hydrological systems. The system response function linking the tracer input with the tracer output in such models, includes a single parameter to describe the hydrodynamic behaviour of the system.

The use of multi-compartmental models where the composition response of the system is simulated by the use of recursive equations based on mass transport principles offers considerable advantages in arriving at quantitative interpretations of environmental isotopes, particularly tritium, in hydrological systems.

The approach discussed in the paper is to describe the system behaviour by an interconnected system of linear storage elements, whereby the various degrees of mixing in the system for non-steady state flow conditions can be simulated. Application of the proposed approach to available tritium data from Glomma river (Norway) is given as an example.

II. BASIC THEORETICAL CONSIDERATIONS

1. General

The general approach using environmental tritium in quantitative evaluations of hydrological systems, has so far been mainly of the "analysis" type. In most of the cases the tritium input concentration to the system (tritium content of precipitation) is known, and the tritium output concentration is recorded by measurements performed on samples collected from the outflow of the system. For a linear lumped-parameter model the tritium output from the system is described by the following convolution integral:

$$C_o(t) = \int_o^t C_i(t-\tau).F(\tau).e^{-\lambda t}.dt \qquad (1)$$

where $C_o(t)$ = tritium output concentration

$C_i(t)$ = tritium input concentration

$F(\tau)$ = system transfer function

$e^{-\lambda t}$ = radioactive decay correction factor

In most of the cased the problem is treated as a "prediction", where a system transfer function $F(\tau)$, which is essentially the "transit time distribution function", of the system is assumed. Tritium output curves for different trial values assigned to the parameters of the assumed transfer function are computed and compared with the actual observed output. The quantitative information obtained in this way obviously depends solely on the type of transfer function selected, i.e. the most commonly employed models such as "piston-flow" or "completely-mixed-reservoir" have their transfer functions described by a single parameter. The so-called "dispersive models" generally have their transfer functions selected so as to describe the dispersion of tracer within the system. Usually some form of probability distribution function is used for this purpose, the selection of which is rather arbitrary.

In the treatment of the problem as an "identification" type, as discussed by Erikson [5], the known tritium input to the system and the observed output from the system are used to construct the "transfer function" of the system for a tracer case so that information on the transit time distribution and mean transit time can be obtained.

In all of the above mentioned formulations of the problem, major limitations arise from the assumption that the system is in a steady-state flow and also from the rather oversimplified representation of the tracer behaviour with the generally employed simple forms of "transfer functions". Furthermore, in the "identification" type of approach, quite long-term observations on the tritium output are essential for most hydrological systems. Although the existing models can be considered close approximations for some simple systems, more realistic and closer representation of tracer behaviour would be most desirable for conditions of non-steady state flow.

The "synthesis" type of approach to the problem, where a working model is selected to represent the hydraulic response of the system and the tracer input is superimposed on the assigned flow pattern to evaluate the tracer output from the system, seems to offer considerable advantages in these respects.

2. Interconnected storate elements as a tool to describe the mixing process of a tracer in a hydrological system

The use of interconnected storage elements as conceptual description of the mixing process in hydrologic systems has already been discussed in an earlier paper by (Przewlocki-Yurtsever) 1974 [7]. A similar approach has already been proposed for evaluation of mixing in industrial process dynamics [8]. The approach is to use a cascade of storage elements as shown in Fig. 1, as a general conceptual model describing the mixing process of the tracer where each of the storage elements in the cascade is assumed to be uniformly mixed. The computation of tracer output from such a system is made from tracer-mass-balance considerations for each of the storage elements at selected discrete time intervals. The basic assumption in the above approach is that the incoming volume of flow during a given time interval, with its known tracer concentration, mixes uniformly with the volume of water within the storage, and an equal volume of water is discharged out of the storage. The approach enables the time variations both in input (recharge) into the system and in its tracer concentration to be taken into account. Furthermore, the storage elements can be interconnected and input (recharge) to the storage elements can be assigned, in any desired manner, so that the model could also represent actual known physical features of the hydrologic system under consideration. Computation of tracer output from such an adopted model can easily be handled with a recursive equation developed for this purpose. The discussion of such a simulation model and its use for hydrological purposes is also given by Simpson [9].

FIGURE - 1

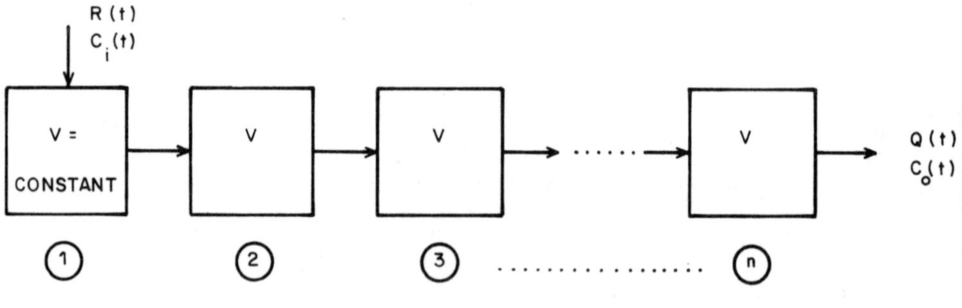

In such a treatment, input (recharge) to the system is assumed to be equal to the discharge from the system during the same time increment. Consequently the discharge-time curve for such a system is exactly the same as the recharge-time curve. During a period of no recharge, there is no discharge from the system. An approach which could more closely represent the hydraulic response of storage volumes associated with natural hydrologic systems would be a desirable refinement and would provide a more appropriate description of the system behaviour.

3. Interconnected Linear Storage Elements as a Tool to Describe a Hydrologic System

The functional relationship between the volume and outflow from a storage unit associated with a natural hydrological system is generally of the form:

$$V = K.Q^n \tag{2}$$

where K and n are constants. The most commonly employed and simple form of such a relation is the one where n = 1, so that:

$$V = K.Q \tag{3}$$

A storage element having the above property is the so-called "linear storage" element in which its volume is always directly proportional to the outflow. The behaviour of most of the storage units associated with natural hydrological systems can be closely approximated by either a single linear storage element or a combination of them. In the case of groundwater systems, the generally observed exponential recession characteristics would imply that the associated groundwater storage units are of such a linear type, since the recession equation from a linear storage element can be shown to be:

$$Q = Q_o\ e^{-t/K} \tag{4}$$

Similarly, a cascade of linear storage elements is generally a conceptual model for the hydraulic response of direct runoff from a catchment basin (Nash) [10]. Thus, a combination of such linear storage elements can be conveniently adopted to simulate the actual hydraulic response of natural hydrological systems.

The hydraulic response of such a linear storage element as shown in Fig. 2 is evaluated by Dooge [11], based on the volume routing of a uniform input (recharge) rate on a discrete time basis, having equal time intervals "T". The resulting equation describing the contribution to a given outflow Q_n during the n'th time interval from the past recharge volumes R_n, R_{n-1}, R_{n-2}, ... R_1 is [11]:

$$\begin{aligned} Q_n = {} & R_n\left[1 - \frac{K}{T}(1 - e^{-T/K})\right] \\ & + R_{n-1}\frac{K}{T}(e^{T/K} - 1)^2\ e^{-2T/K} \\ & + R_{n-2}\frac{K}{T}(e^{T/K} - 1)^2\ e^{-3T/K} \\ & + R_{n-3}\frac{K}{T}(e^{T/K} - 1)^2\ e^{-4T/K} \\ & + \text{etc.} \end{aligned} \tag{5}$$

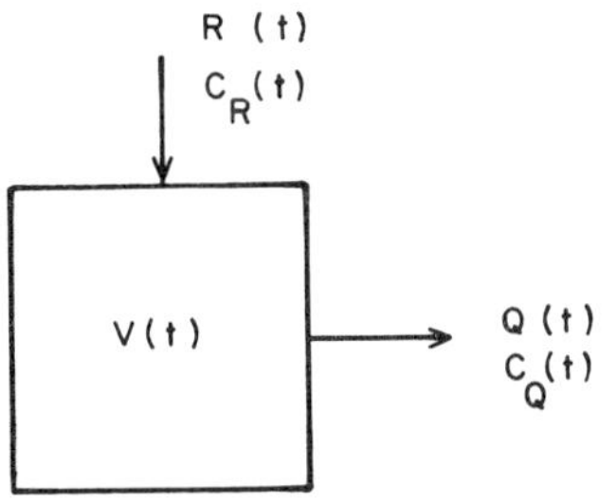

FIGURE - 2

The terms of the above equation, in fact, describe the volume fractions of water contributed from past recharge volumes to the outflow volume at the n'th time interval, which depends on the coefficient K and the selected time interval T. Further mathematical manipulations on the above equation results in the following simple routing equation [11]:

$$Q_n = C_o R_n + C_1 R_{n-1} + C_2 Q_{n-1} \tag{6}$$

where Q_n = outflow volume during n'th interval

R_n = recharge volume during n'th interval

R_{n-1} = recharge volume during previous interval (n-1)

Q_{n-1} = discharge volume during previous interval (n-1)

and C_o, C_1 and C_2 are the coefficients given by:

$$C_o = 1 - \frac{K}{T}(1 - e^{-T/K})$$

$$C_1 = \frac{K}{T}(1 - e^{-T/K}) - e^{-T/K} \tag{7}$$

$$C_2 = e^{-T/K}$$

Thus, starting from a selected initial hydraulic state, subsequent discharge volumes from a linear storage element for a given recharge pattern can be computed on a discrete time basis by the use of the above simple recursive equation. The coefficient "K" has a dimension of time, and it is generally termed the "average storage delay time" or "mean turnover time". Thus, for the case of non-steady flow, where the volume of the storage element also is time variable, it can be taken to be:

$$K = \frac{\bar{V}}{\bar{Q}} \tag{8}$$

where $\bar{V}$ and $\bar{Q}$ denote average values of volume and discharge throughout the whole time span considered. It should be noted that the sum of coefficients C_o, C_1 and C_2 in Eq. 6 is equal to unity:

$$C_o + C_1 + C_2 = 1 \tag{9}$$

For a steady state flow through the system, for example, ($Q_{n-1} = R_{n-1} = R_n$), Eq. 6 will yield:

$$Q_n = R_n = Q_{n-1} = R_{n-1}$$

for all values of n.

If uniform mixing is assumed for the tracer case in such a storage element, then the material balance for a tracer during the n'th time interval can be written as:

$$V_n \times C_{v_n} - V_{n-1} \times C_{v_{n-1}} = R_n \times C_{R_n} - Q_n \times C_{Q_n} \qquad (10)$$

where "C" represents the concentration of the tracer in the volume and flow terms respectively. Since, with the assumed uniform mixing condition of the storage element $C_{v_n} = C_{Q_n}$, the tracer concentration of the outflow can be accordingly solved as:

$$C_{Q_n} = C_{v_n} = \frac{V_{n-1} \times C_{v_{n-1}} + R_n \times C_{R_n}}{V_n + Q_n} \qquad (11)$$

The above equation is based on the assumption that mixing of recharge with the volume in the storage occurs prior to discharge in a given time interval.

Thus, for a given linear storage element, knowing the recharge input with its corresponding tracer concentration, the computation of the discharge from the system and tracer output concentration can be estimated from the following four recursive equations for selected equal time intervals:

(1) $Q_n = C_o R_n + C_1 R_{n-1} + C_2 Q_{n-1}$

(2) Change in storage $\Delta v_n = R_n - Q_n$

(3) $V_n = V_{n-1} + \Delta V_n$ (12)

(4) $C_{V_n} = C_{Q_n} = \dfrac{(V_{n-1} \times C_{v_{n-1}} + R_n \times C_{R_n})}{V_n + Q_n} e^{-\lambda T}$

The last term in Eq. (12.4) is the decay correction for a radioactive tracer for the selected time interval "T". For a hydrological system described as consisting of various interconnected linear storage elements, the above recursive equations are applied to each of the storage elements in the system. The above computational procedure is easily handled with a computer. For a given tracer input and steady state flow conditions the tracer output computed from the above equations will be identical to the approach by Przewlocki-Yurtsever [7], and the "linearity" assumption for storage elements in the system does not play any role. However, the advantage of the above recursive equations, is in the simulation of a non-steady state flow condition, where the hydraulic response of the storage elements is also taken into account, and the linearity assumption of each storage element is an approximation of the actual hydraulic response of natural storage units. To illustrate the behaviour of such a linear storage element, the tracer output concentration curves computed for a hypothetical case of varying recharge patterns are shown in Fig. 3.

The storage element with an assumed value of $\bar{K} = 10$ is initially assumed to be labelled with unit concentration of a non-radioactive tracer which is then washed out by different recharge patterns (having zero tracer concentration), each having the same long-term mean value $\bar{R} = 1$. For the steady state case, the computed tracer output concentration is very close to the curve given by the analytical solution:

$$\frac{C}{C_o} = e^{-t/k} \tag{13}$$

for such a uniformly mixed reservoir. However, for a variable recharge pattern, the observed tracer output concentration curves differ considerably depending on the variance of the input recharge. The discharge from such a storage element will also have an average assigned delay time as compared to recharge input, as shown in Fig. 3.

III. EXAMPLE OF APPLICATION - "A STUDY OF THE SUBSURFACE FLOW CHARACTERISTICS OF THE GLOMMA RIVER BASIN WITH THE AID OF ENVIRONMENTAL TRITIUM"

1. Introduction

The tritium data from the Glomma River in Norway is used with the aim of arriving at a quantitative interpretation of the subsurface (groundwater) flow characteristics of the catchment. Tritium (^3H) is a radioactive isotope of Hydrogen with a half-life of 12.26 years. Its occurrence in precipitation and consequently in terrestrial waters is both due to its natural production by interaction of cosmic radiation with atmospheric components and to mainly detonation of thermonuclear devices since 1952. Its concentration in terrestrial waters is generally expressed in Tritium Units (T.U.), where 1 T.U. = 1 ^{3}H atom per 10^{18} ^{1}H atoms.

The Glomma River was one of several rivers selected for tritium monitoring by the IAEA, within the scope of the UNESCO IHD programme, and available tritium data from this river cover a time span of about 7 years, beginning in May 1965. The data are for samples collected at Sarpsborg stream - gauging station. The tritium analyses were made in the Isotope Laboratory of IAEA.

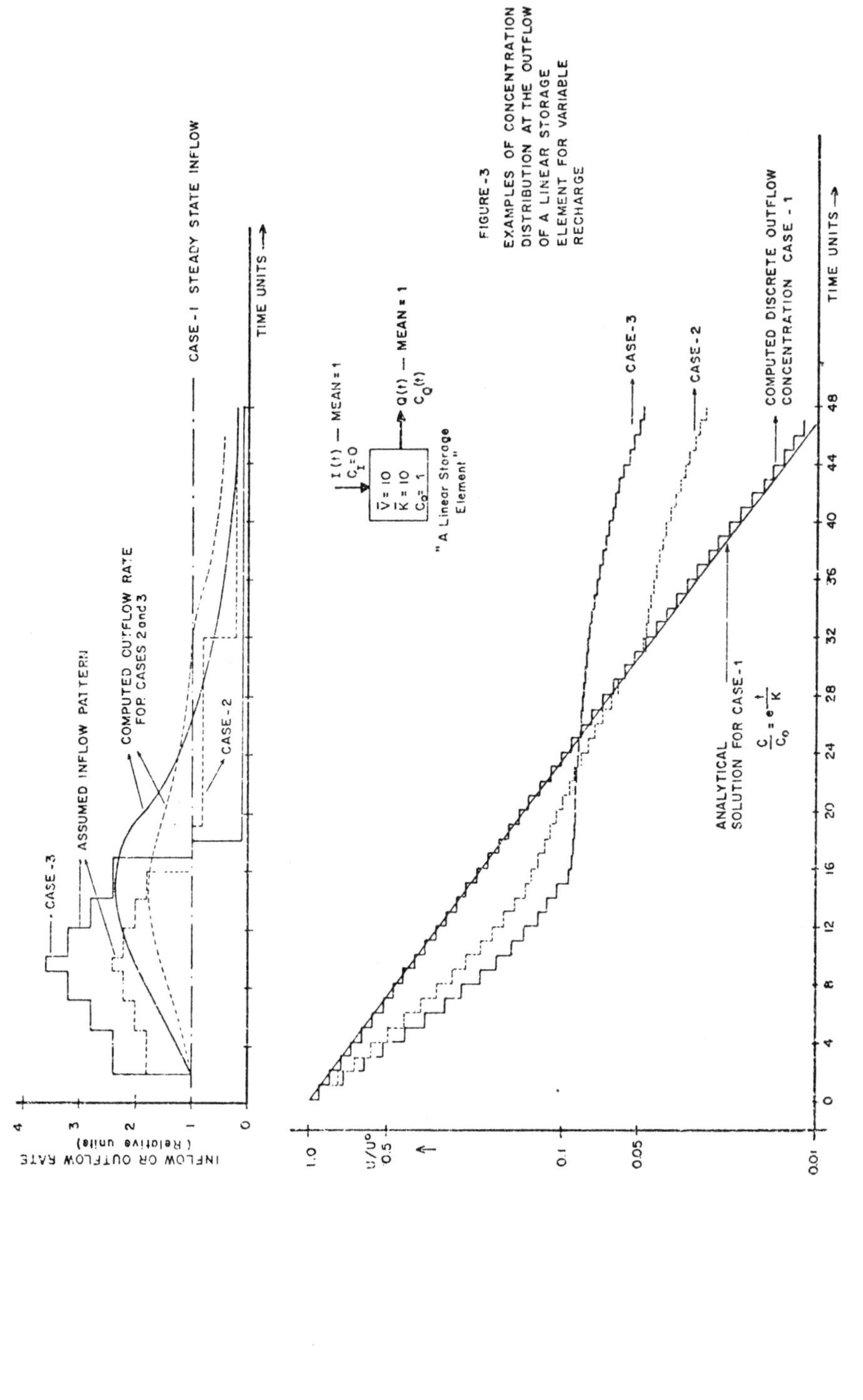

FIGURE - 3
EXAMPLES OF CONCENTRATION DISTRIBUTION AT THE OUTFLOW OF A LINEAR STORAGE ELEMENT FOR VARIABLE RECHARGE

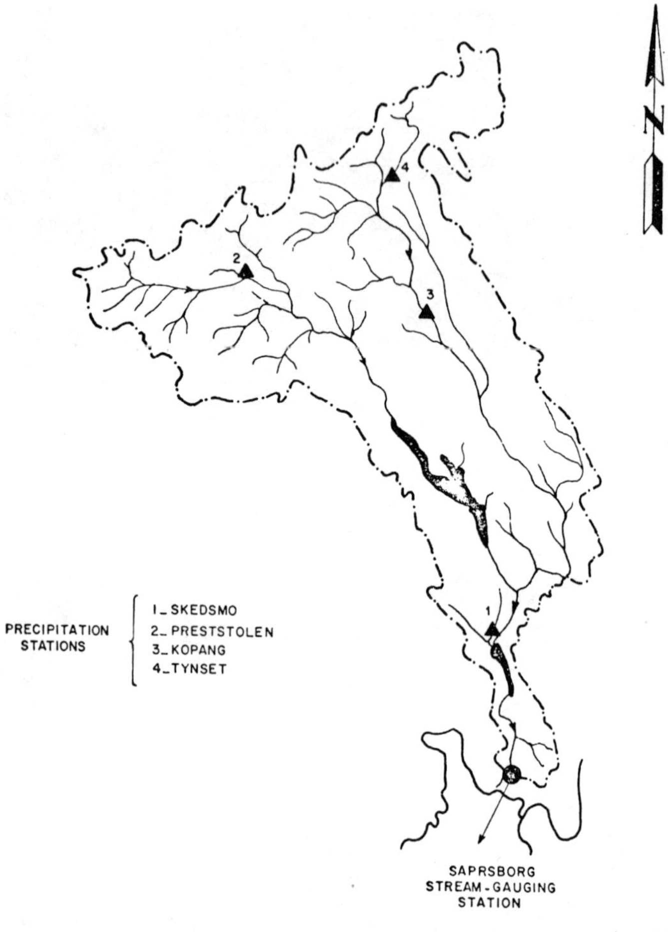

FIGURE-4 SKETCH OF GLOMMA RIVER BASIN, LOCATION OF SAMPLING SITE AND PRECIPITATION STATIONS

The Glomma River has a drainage area of 41,284 km . The average annual rainfall over the catchment area is estimated to be about 560 mm, and the annual flow of the river as the average depth of runoff is about 198 mm for the period 1953-1971. The location of the catchment area, of the sampling site and, of the four precipitation stations used in the evaluation, are shown in Fig. 4.

2. Available Tritium Results and Basic Hydrological Data

The available data on the tritium content of the river as measured at the sampling site of Sarpsborg gauging station and the corresponding discharges of the river at the time of sampling are given in Table 1. A graph showing the time variations of tritium content of the river is given in Fig.5, where the mean monthly discharge values of the river and the estimated monthly precipitation amounts over the catchment area are also shown. Monthly precipitation over the catchment area is estimated by taking the simple mean of the monthly data from four selected precipitation stations (Preststulen, Tynset, Skedsmo and Kopang) within the catchment, each located at different altitudes. Data on tritium content of the precipitation is available from the station "Trysil" in the catchment area, and in view of the known relatively small spatial variation of the tritium content of the rainfall, the data can be assumed to represent the tritium content of the rainfall over the whole catchment. Missing tritium data for this station are estimated from linear correlations with the WMO/IAEA Network stations Lista and Ottawa. Prior to 1953, the tritium content of the rainfall in the latitudes of the study area was probably about 8 to 10 T.U. It can be neglected for the type of evaluation discussed in this paper, because the high level of tritium input since 1952 completely masks these early values.

The tritium content of monthly rainfall for the study area is also shown in Fig. 5 (a step function) for the period where river tritium measurements are available.

It is interesting to note that after 1966, the tritium content of the river water is generally higher than the tritium content of the precipitation, which evidently indicates that part of the discharge from the catchment is a carry-over from earlier years when the tritium content of the rainfall was significantly higher. It is an expected feature of such a rather large catchment basin where various storage volumes associated with the basin would impose a certain delay between the input and the output. The tritium content of the river at low-flow periods, during which a relatively small amount of precipitation occurs, and the major fraction of the flow in the river is supplied by the sub-surface flow of groundwater, shows a smooth recession over the entire observation period, starting from about 600-650 T.U. at the beginning of 1966. Relatively larger monthly variations during the rainy periods obviously reflect the effect of a relatively fast

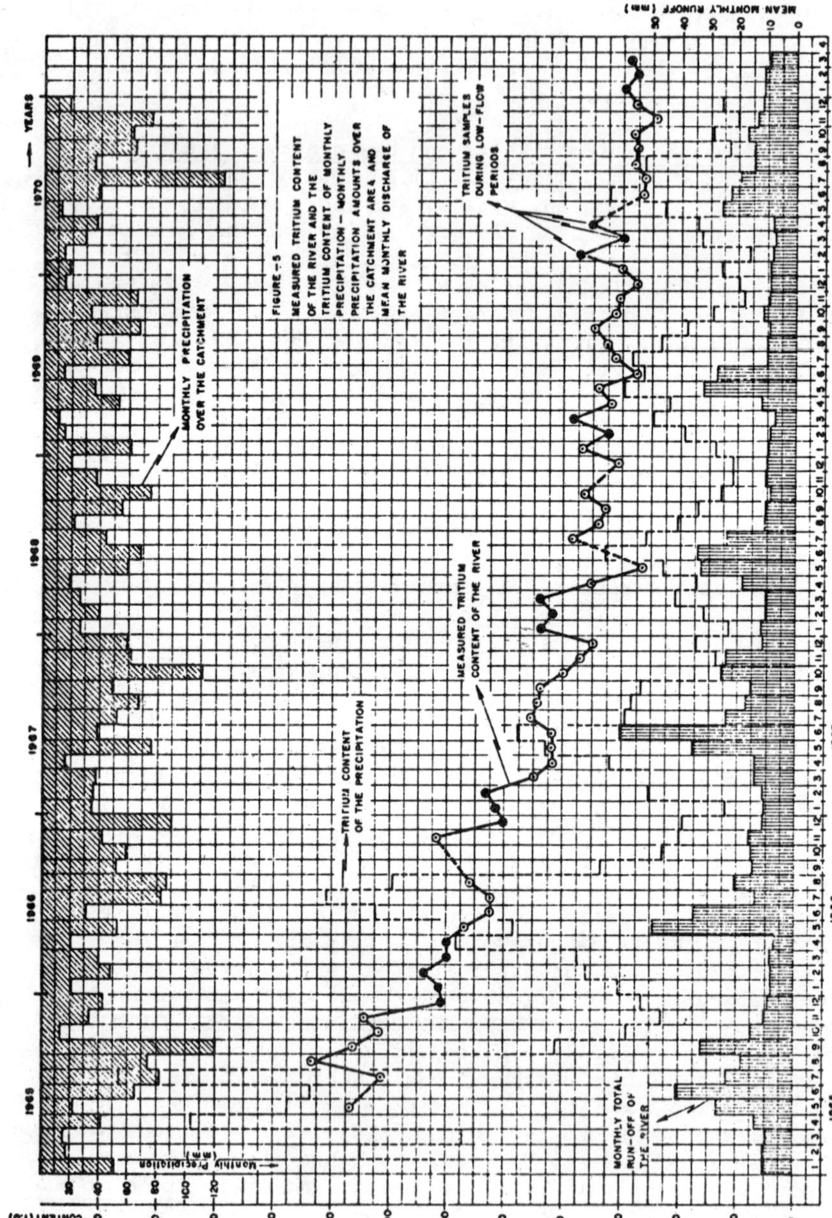

FIGURE-5 MEASURED TRITIUM CONTENT OF THE RIVER AND THE TRITIUM CONTENT OF MONTHLY PRECIPITATION – MONTHLY PRECIPITATION AMOUNTS OVER THE CATCHMENT AREA AND MEAN MONTHLY DISCHARGE OF THE RIVER

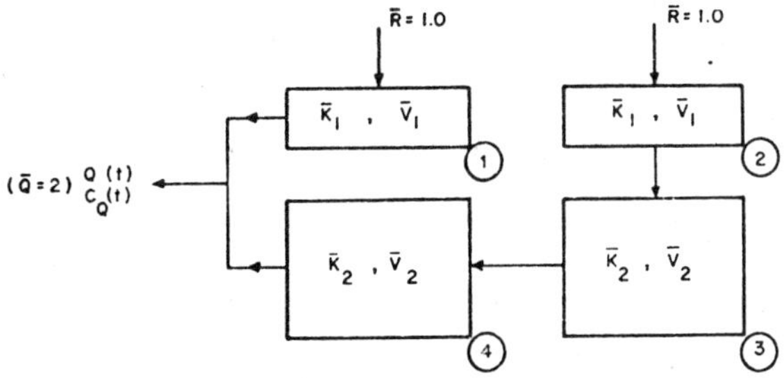

FIGURE - 6

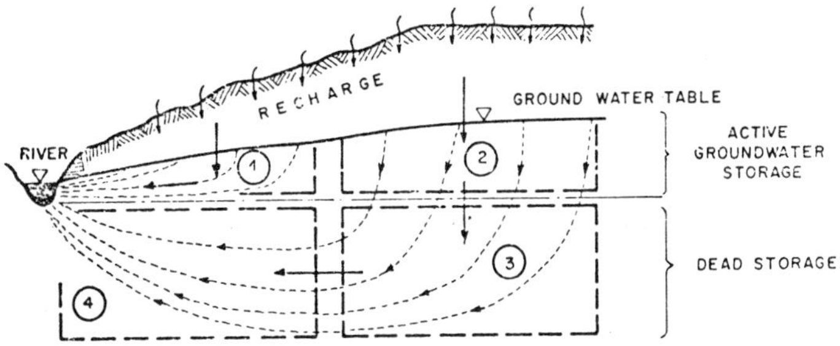

FIGURE - 7

surface runoff contribution to the total flow of the river. In the evaluation, certain dynamic parameters of the catchment pertinent to the groundwater flow component are evaluated by the aid of available tritium data during the base-flow period of the river, where the major contribution to river flow is from groundwater.

3. Evaluation of the Sub-Surface Flow (Groundwater) Component and Associated Sub-Surface Storage in the Basin

The general approach discussed in the first part of the paper is used in the evaluation where a model composed of inter-connected linear storage elements is used to represent the overall behaviour of the groundwater storage volumes in the catchment, and the known tritium input is superimposed on the assigned recharge and flow patterns so that tritium output curves are computed from the recursive equations given earlier. The working model selected for this purpose consists of four linear storage elements as shown in Fig. 6.

The selection for the layout of the working model is based on the general features of the common flow pattern of sub-surface drainage into a river channel as shown schematically in Fig. 7. The volume elements Nos.1 and 2 represent the "active storage" in the catchment (groundwater storage above the level of the river bed) and the volume elements Nos. 3 and 3 represent the "dead storage" (storage below the level of the river bed.) The flow pattern assigned to the selected working model reflects the fact that the recharge occurring in the areas adjacent to the river channel generally circulates through an active storage volume which is located at relatively shallow depths and has relatively shorter pathways to follow. However, the recharge occurring further away from the stream channels, would need to circulate through a deeper storage (volume elements Nos. 3 and 4 in Fig. 7) prior to reaching the stream channel. Consequently, the sub-surface flow into the river is treated as being the result of these two main components having two different pathways.

The computations are carried out on yearly time increments beginning in 1952. The mean tritium content of precipitation (of the recharge) for each year is computed from the monthly data as a weighted mean:

$$(\bar{C})\text{annual} = \frac{\sum_{i=1}^{12} P_i \times C_i}{\sum_{i=1}^{12} P_i} \qquad (14)$$

where: $(\bar{C})$ annual = weighted (for amount) mean tritium content for a year

P_i = monthly precipitation amounts

C_i = tritium content of the corresponding monthly precipitation.

TABLE 1

AVAILABLE TRITIUM RESULTS FROM GLOMMA RIVER, NORWAY AND DISCHARGE OF THE RIVER AT THE TIME OF SAMPLING

Sampling Date	Discharge at the time of sampling (m^3/sec)	Tritium content (T.U.)	Sampling Date	Discharge at the time of sampling (m^3/sec)	Tritium (T.U.)
3. 5.1965	900	733±9	1. 4.1968	675	355±10.9
31. 5.1965	1375	800±17	1. 5.1968	1400	268±10.1
1. 7.1965	1675	716±15	3. 7.1968	1375	387±12.6
1. 8.1965	1050	835±18	9. 8.1968	475	341±9.1
2. 9.1965	850	763±24	6. 9.1968	384	330±14.0
1.10.1965	1075	720±25	8.10.1968	343	366±9.6
1.11.1965	425	746±26	2.12.1968	425	307±8.4
1.12.1965	403	690±22	6. 1.1969	418	369±12.3
31.12.1965	342	610±18	1. 2.1969	389	324±13.3
1.1966	-	615±19	3. 3.1969	328	384±10.7
1. 2.1966	343	641±22	2. 4.1969	272	320±11.7
1. 3.1966	353	599±20	1. 5.1969	946	342±11.3
2. 4.1966	313	599±20	2. 6.1969	1650	278±12.8
4. 5.1966	935	571±23	1. 7.1969	525	314±8.2
2. 6.1966	2220	526±21	6. 8.1969	450	328±14.9
2. 7.1966	675	525±21	3. 9.1969	400	350±9.3
2. 8.1966	500	561±22	1.10.1969	450	316±8.6
1.11.1966	500	618±19	4.11.1969	400	309±9.0
1.12.1966	450	503±15	1.12.1969	410	278±9.0
3. 1.1967	475	517±15	13. 1.1970	370	305±17.3
2.1967	470	535±18	1. 2.1970	370	374±10.8
3.1967	445	449±16	9. 3.1970	330	301±16.0
4.1967	625		1. 4.1970	285	356±20.1
5.1967	650	419+17*	1. 6.1970	1400	268±13.0
6.1967	2900		3. 7.1970	800	266±16.1
1. 7.1967	1325	457±17.5	3. 8.1970	825	283±14.8
1. 8.1967	675	447±17.3	1. 9.1970	450	279±16.2
1. 9.1967	500	440±17.1	5.10.1970	600	284±16.5
1.10.1967	600	400±16.9	2.11.1970	500	245±18.6
1.11.1967	1500	372±14.7	1.12.1970	475	281±9.1
1.12.1967	475	350±16	4. 1.1971	480	299±8.9
2. 1.1968	475	440±13.4	1. 2.1971	490	277±8.4
1. 2.1968	475	421±16.8	2. 3.1971	440	289±8.4
1. 3.1968	395	442±17.3	7. 4.1971	380	247±6.9

* Composite sample

TABLE 1 - Cont'd

Sampling Date	Discharge at the time of sampling (m^3/sec)	Tritium content (T.U.)
1. 5.1971	490	247 ± 8.5
1. 5.1971	1600	230 ± 8.0
2. 7.1971	1350	270 ⊥ 8.2
2. 8.1971	1100	280 ± 7.6
1.10.1971	375	254 ± 7.9
1.11.	375	233 ± 8.2
6.12.1971	390	255 ± 7.9
3. 1.1972	390	263 ± 6.3
2. 2.1972	390	285 ± 8.7
1. 3.1972	335	275 ± 9.4

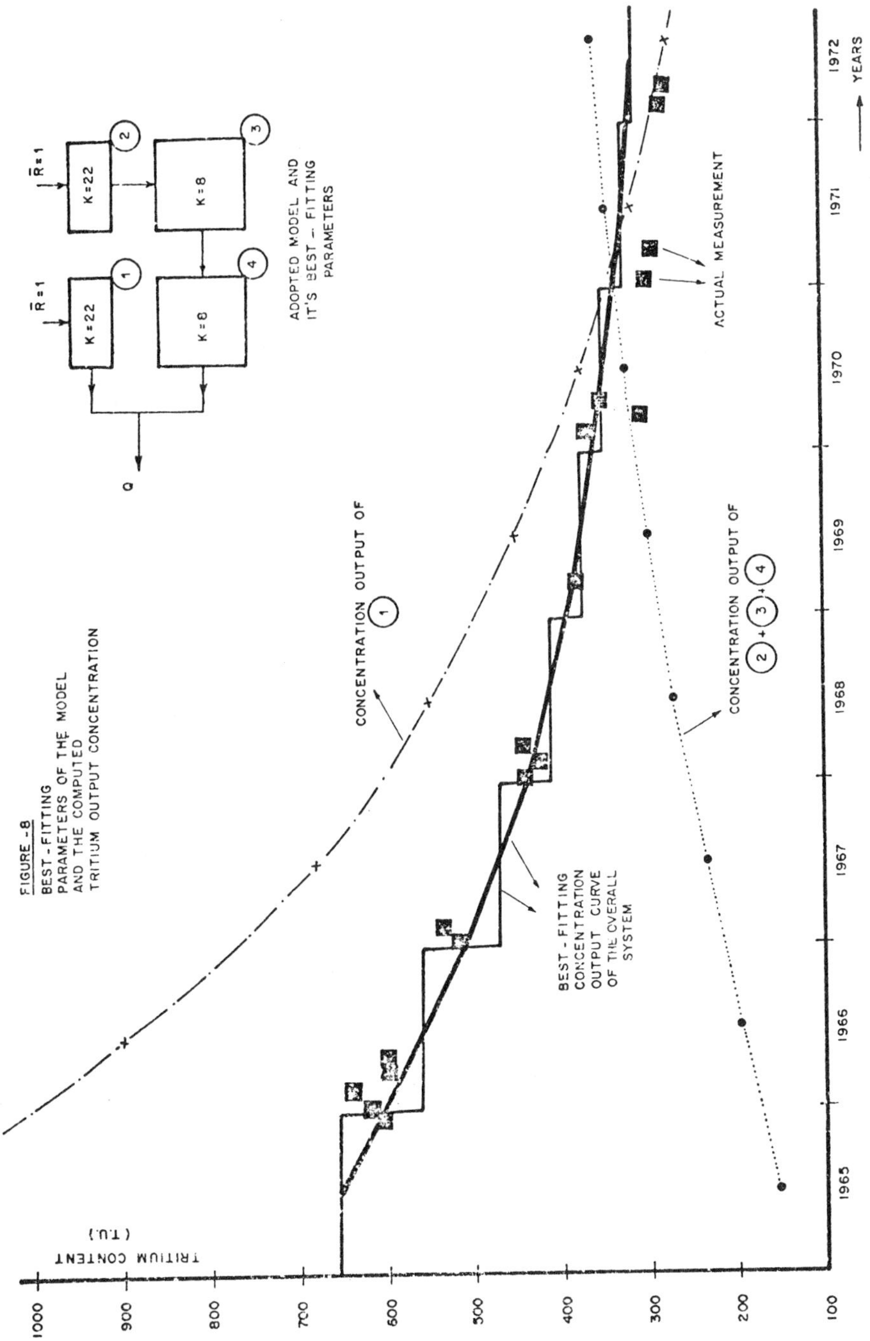

FIGURE - 8
BEST - FITTING
PARAMETERS OF THE MODEL
AND THE COMPUTED
TRITIUM OUTPUT CONCENTRATION

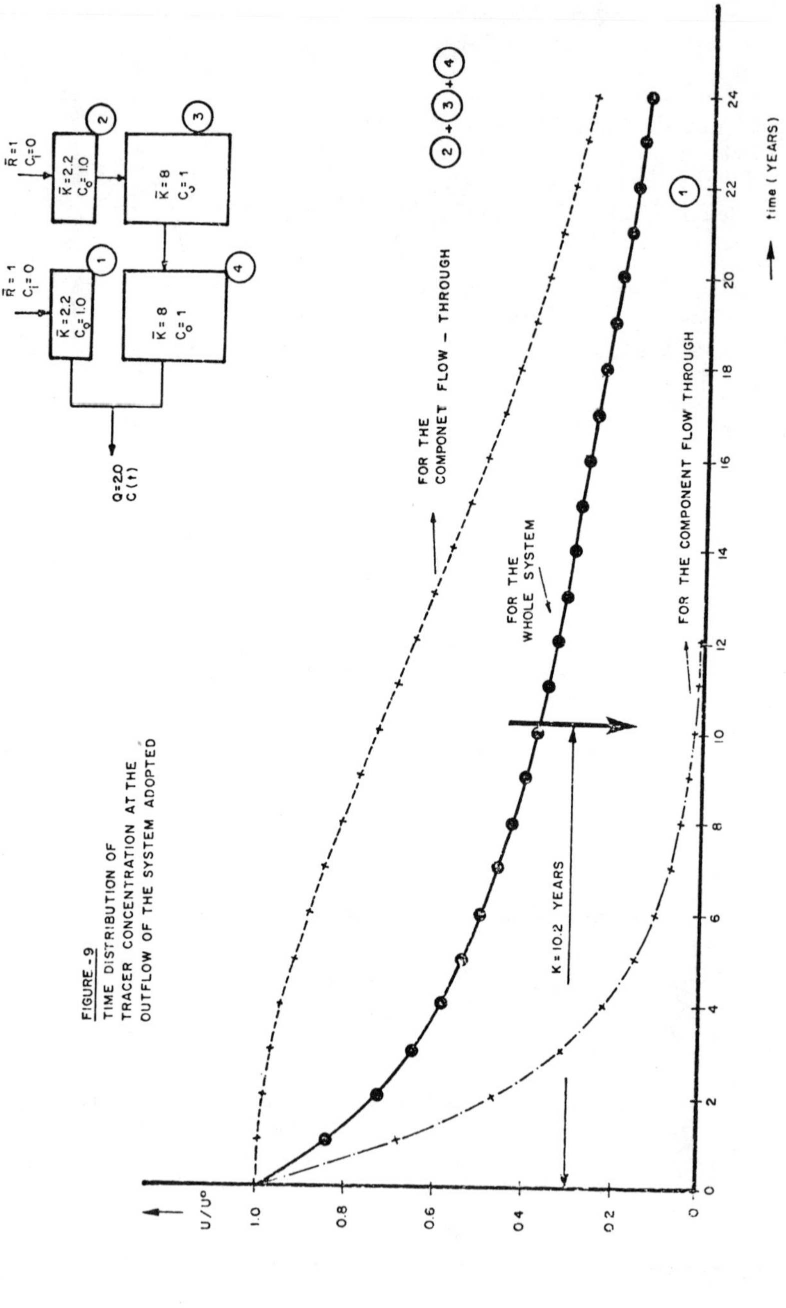

FIGURE - 9
TIME DISTRIBUTION OF
TRACER CONCENTRATION AT THE
OUTFLOW OF THE SYSTEM ADOPTED

These values, together with the annual precipitation amounts, are given in Table 2, and shown on Fig. 8 (step-curve).

A family of tritium output curves for the adopted model are computed by changing the values of storage and the ratio of recharge to volume elements 1 and 2, until a good fit is obtained between the computed and observed output. The values assigned to storage volume and yearly recharge are all in relative units. The relative recharge values assigned for each year represent the actual observed yearly variations of the precipitation over the catchment area, as can be seen from Table 2, and, thus, it is assumed that recharge in the catchment is directly proportional to the annual precipitation.

The parameters of the best fitting model and the computed tritium output curve are shown in Fig. 8, together with the actual measured tritium values of the river during low flow periods. Thus the overall effect of the sub-surface storage in the catchment basin can be well simulated by four linear storage elements. It should be mentioned that with a model consisting of only one linear storage element, an acceptable fit to the observed tritium output cannot be obtained, regardless of the value assigned to the volume of such a single storage element. From the parameters of the best fitting model (see Fig. 8), the mean turnover time of the overall sub-surface storage in the basis is computed to be 10.2 years.

The minimum observed discharge of the river during low flow amounts to about 3.4×10^9 m^3/year , and if it is assumed to be totally supplied by the sub-surface flow from the catchment, the associated volume of the total sub-surface storage in the catchment would be:

$$10.2 \times 3.4 \times 10^{P} = 34.7 \times 10^9 \ m^3$$

The working model adopted for the sub-surface flow and its estimated parameter is used to construct the "transit time distribution" of the system. The transit time distribution curve is computed by assuming the system is initially labelled by a non-radioactive tracer and then washed out by steady flow (tracer free). The concentration distribution at the outlet computed for such a case in fact represents the shape of the transit time distribution function. The curve computed for the best fitting model shown in Fig. 9, has an average transit time of 10.2 years. The transit time distribution function can be obtained by normalizing the concentration time distribution curve given in Fig. 9.

CONCLUSIONS

The digital simulation approach for evaluation of tracer data offers a potential tool for making quantitative estimates of turn-over time and transit-time distribution of a system.

The interconnected linear storage elements can be congeniently used as an operational model for a hydrological system.

The tracer input-output relations for such a model can be easily handled by a set of recursive equations, which take into account the hydraulic response of the system and as well enable simulation of mixing processes of a tracer in the system for a non-steady state flow.

Information obtained in this way on groundwater storage in a catchment area and its transit time distribution has practical hydrological implications with respect to the estimation of the potential amount of groundwater in the catchment, its turnover rate, and of the minimum expected groundwater flow into the river during extended dry periods, which is also important from the stand point of pollution.

ACKNOWLEDGEMENTS

The tritium samples of the Glomma River were collected by the staff of the Sarpsborg Kommune, Byingenioren, Norway, and the background hydrological data for the river basin were provided by them.

The contribution of the staff members of the Section of Isotope Hydrology, International Atomic Energy Agency for the invaluable remarks and comments is greatly appreciated.

REFERENCES

1. Begeman, F., and Libby, W.F. "Continental Water balance - Groundwater inventory and storage time, surface ocean mixing rates and world-wide water circulation patterns from cosmic-ray and bomb tritium", Geochim. et Cosmochim. Acta, Vol. 21, 1957.

2. Erikson, E. "The possible use of tritium for estimating groundwater storage", Tellus, 10, 1958.

3. Erikson, E. "Natural reservoir and their characteristics" Geofisica International, I (2), Mexico, 1961.

4. Brown, R.M. "Hydrology of tritium in the Ottawa Valley", Geochim. et Cosmochim. Acta, 21, 1961.

5. Erikson, E. "Atmospheric tritium as a tool for the study of certain hydrologic aspects of river basins", Tellus, XV, 1963.

6. Erikson, E. "Large-scale utilization of tritium in hydrologic studies", Geophysical Monograph Series No. 11, A.G.D., 1967.

7. Przewlocki, K. and Yurtsever, Y. "Some conceptual mathematical models and digital simulation approach in the use of tracers in hydrological systems", Proceedings of a Symposium on Isotope Techniques in Groundwater Hydrology, Vol.2, 1974.

TABLE 2

MEAN ANNUAL TRITIUM INPUT CONCENTRATIONS FOR THE CATCHMENT AREA AND ANNUAL RAINFALL AMOUNTS

Years	Annual Rainfall (mm)	Annual Fainfall in relative units	Annual Mean Tritium Input Concentration (T.U.)
1952	559	1.000	10
1953	631	1.128	20
1954	618	1.105	191
1955	355	0.635	40
1956	491	0.878	188
1957	634	1.134	118
1958	504	0.901	539
1959	522	0.933	386
1960	614	1.098	151
1961	616	1.101	137
1962	544	0.973	705
1963	612	1.094	3350
1964	611	1.092	1614
1965	578	1.033	669
1966	626	1.119	452
1967	638	1.141	245
1968	491	0.878	194
1969	467	0.835	195
1970	525	0.937	213
1971	560*	1.000	200
1972	560*	1.000	195*
MEAN	559.8	1.000	

* Assumed

8. Himmelblau, D.M., and Bischoff, K.B. "Process Analysis and Simulation-Deterministic Systems", Wiley, New York, 1968.

9. Simpson, E.G. "Finite State Mixing-Cell Models", Bilateral USA-Yugoslavian Seminar in Karst Hydrology and Water Resources, Dubrovnik, 1975.

10. Nash, J.E. "The form of instantaneous unit hydrograph", I.A.S.H., Pub. 45, Vol. 3, 1957.

11. Dooge, J.C.I. "The routing of groundwater recharge through typical elements of linear storage", Int. Ass. Sci. Hydrology, Helsinki, Publication No.52, 1960.